Lecture Notes in Electrical Engineering

Volume 615

The book series *Lecture Notes in Electrical Engineering* (LNEE) publishes the latest developments in Electrical Engineering - quickly, informally and in high quality. While original research reported in proceedings and monographs has traditionally formed the core of LNEE, we also encourage authors to submit books devoted to supporting student education and professional training in the various fields and applications areas of electrical engineering. The series cover classical and emerging topics concerning:

- Communication Engineering, Information Theory and Networks
- Electronics Engineering and Microelectronics
- Signal, Image and Speech Processing
- Wireless and Mobile Communication
- Circuits and Systems
- Energy Systems, Power Electronics and Electrical Machines
- Electro-optical Engineering
- Instrumentation Engineering
- Avionics Engineering
- Control Systems
- Internet-of-Things and Cybersecurity
- Biomedical Devices, MEMS and NEMS

For general information about this book series, comments or suggestions, please contact leontina.dicecco@springer.com.

To submit a proposal or request further information, please contact the Publishing Editor in your country:

China
Jasmine Dou, Associate Editor (jasmine.dou@springer.com)

India
Swati Meherishi, Executive Editor (swati.meherishi@springer.com)
Aninda Bose, Senior Editor (aninda.bose@springer.com)

Japan
Takeyuki Yonezawa, Editorial Director (takeyuki.yonezawa@springer.com)

South Korea
Smith (Ahram) Chae, Editor (smith.chae@springer.com)

Southeast Asia
Ramesh Nath Premnath, Editor (ramesh.premnath@springer.com)

USA, Canada:
Michael Luby, Senior Editor (michael.luby@springer.com)

All other Countries:
Leontina Di Cecco, Senior Editor (leontina.dicecco@springer.com)
Christoph Baumann, Executive Editor (christoph.baumann@springer.com)

**** Indexing: The books of this series are submitted to ISI Proceedings, EI-Compendex, SCOPUS, MetaPress, Web of Science and Springerlink ****

More information about this series at http://www.springer.com/series/7818

Walter Zamboni • Giovanni Petrone
Editors

ELECTRIMACS 2019

Selected Papers - Volume 1

Springer

Editors
Walter Zamboni
DIEM
Università degli Studi di Salerno
Fisciano, Italy

Giovanni Petrone
DIEM
Università degli Studi di Salerno
Fisciano, Italy

ISSN 1876-1100 ISSN 1876-1119 (electronic)
Lecture Notes in Electrical Engineering
ISBN 978-3-030-37160-9 ISBN 978-3-030-37161-6 (eBook)
https://doi.org/10.1007/978-3-030-37161-6

This Springer imprint is published by the registered company Springer Nature Switzerland AG.
The registered company address is: Gewerbestrasse 11, 6330 Cham, Switzerland

Preface

ELECTRIMACS is the short and well-known name of the international conference of the IMACS TC1 Committee. The conference is focussed on the theory and application of modelling, simulation, analysis, design, optimisation, identification and diagnostics in electrical power engineering. The conference is a meeting point for researchers to share ideas and advances in the broad fields of electric machines and electromagnetic devices, power electronics, transportation systems, smart grids, electric and hybrid vehicles, renewable energy systems, energy storage, batteries, supercapacitors and fuel cells.

ELECTRIMACS 2019 was held in Salerno, Italy, from 21st to 23rd May 2019. Three tutorial sessions, three plenary sessions with thought leaders from academia and research centres, four technical tracks, two forums and nine special sessions were included in the conference programme. The conference hosted 133 oral presentations of papers, selected among 169 submissions received. The review process involved at least three reviewers per paper.

The main institutional sponsor of the conference is the *Università degli studi di Salerno—Dipartimento di ingegneria dell'Informazione ed Elettrica e Matematica applicata* (*DIEM*). The conference received also the technical co-sponsorship from two important scientific societies: *IMACS* and *IEEE Industrial Electronics Society* (*IES*), and a financial co-sponsorship from *Institut Français—Italia and the Ambassade de France en Italie* in the framework of *Programma CASSINI*. Many industries and private companies sponsored the event or took part in the industrial exhibit.

This book collects a selection of 56 papers presented at ELECTRIMACS 2019 Salerno. These papers are particularly focussed on electrical engineering simulation aspects and innovative applications.

The collection is organized in six thematic parts. The first three parts: *Modelling and simulation*, *Control and power management* and *Design, optimisation, identification and diagnosis* are dedicated to conference papers mainly selected among technical tracks papers. The last three parts: *Advances in photovoltaic applications, Advances in power electronics for energy storage and hardware emulation* and *Advances in batteries and supercapacitor technologies, graphene-*

based devices, and safety issues in electrical engineering contain papers focussed on new technologies for renewable energy, storage systems and electronic devices, mainly selected from the special session papers.

Organising Committee

General Chairs

Giovanni Petrone, Università degli studi di Salerno, *Italy*
Walter Zamboni, Università degli studi di Salerno, *Italy*

Technical Program Chairs

Giovanni Spagnuolo, Università degli studi di Salerno, *Italy*
Éric Monmasson, Université de Cergy-Pontoise, *France*
Benoît Robyns, HEI Lille, *France*
Ramon Blasco-Gimenez, Universitat Politécnica de Valéncia, *Spain*

Track Chairs

Federico Baronti, Università di Pisa, *Italy*
Efstratios Batzelis, Imperial College, London, *United Kingdom*
Pavle Boškoski, Jožef Stefan Institute, Ljubljana, *Slovenia*
Mario Cacciato, Università degli studi di Catania, *Italy*
Maria Carmela Di Piazza, ISSIA-CNR Palermo, *Italy*
Seiichiro Katsura, Keio University, *Japan*
Eftichios Koutroulis, Technical university of Crete, *Greece*
Marie-Cecile Pera, Université de Franche-Comté, *France*
Carlos Andres Ramos Paja, Universidad National de Colombia, *Colombia*
Bruno Sareni, ENSEEIHT, *France*
João Pedro Trovão, University of Sherbrooke, *Canada*
Dmitri Vinnikov, Tallinn University of Technology *Estonia*

Special Session Chair

Ilhem Slama-Belkodja, *Université de Tunis El Manar*, Tunisia

Local Organising Committee

Gina Scorziello
Raffaele Raimo
Andrea Contrada
Antonio Guarino
Rudy Alexis Guejia Burbano
Vittorio Mattei
Luigi Mattia
Brian Ospina Agudelo
Carmine Russomando

Scientific Committee

Yacine Amara, *France*
Federico Baronti, *Italy*
Efstratios Batzelis, *United Kingdom*
Ramon Blasco-Gimenez, *Spain*
Pavle Boskoski, *Slovenia*
Alain Bouscayrol, *France*
Mario Cacciato, *Italy*
Carlo Cecati, *Italy*
Chandan Chakraborty, *India*
Gerard Champenois, *France*
Bruno Dehez, *Belgium*
Louis-A. Dessaint, *Canada*
Maria Carmela Di Piazza, *Italy*
Christian Dufour, *Canada*
Maurice Fadel, *France*
Gabriel Garcerá, *Spain*
Leopoldo García Franquelo, *Spain*
Guillaume Gateau, *France*
Luis Gomes, *Portugal*
Gabriele Grandi, *Italy*
Francesco Grasso, *Italy*
Hamid Gualous, *France*
Lennart Harnefors, *Sweden*
Sergio Junco, *Argentina*
Đani Juričić, *Slovenia*
Hadi Y. Kanaan, *Lebanon*
Seiichiro Katsura, *Japan*
Samir Kouro, *Chile*
Eftichios Koutroulis, *Greece*
Marco Liserre, *Germany*
Luc Loron, *France*
Óscar Lucía, *Spain*
Massimiliano Luna, *Italy*
Chengbin Ma, *China*
Jean Mahseredjian, *Canada*
Mariusz Malinowski, *Poland*
Patrizio Manganiello, *Belgium*
Sébastien Mariethoz, *France*
Fabrizio Marignetti, *Italy*
Bogdan Marinescu, *France*
Pascal Maussion, *France*
Tuomas Messo, *Finland*
Rosario Miceli, *Italy*

Eric Monmasson, *France*
Teresa Orlowska-Kowalska, *Poland*
Nicolas Patin, *France*
Ruben Peña, *Chile*
Marie-Cecile Pera, *France*
Giovanni Petrone, *Italy*
Serge Pierfederici, *France*
Mircea Radulescu, *Romania*
Carlos Andres Ramos-Paja, *Colombia*
Nicolas Retière, *France*
Xavier Roboam, *France*
Benoit Robyns, *France*
Francesco Roca, *Italy*
Georges Salloum, *Lebanon*
Bruno Sareni, *France*
Manuela Sechilariu, *France*
Ilhem Slama-Belkhodja, *Tunisia*
Giovanni Spagnuolo, *Italy*
João Pedro F. Trovão, *Canada*
Maria I. Valla, *Argentina*
Alex Van den Bossche, *Belgium*
Philippe Viarouge, *Canada*
Dmitri Vinnikov, *Estonia*
Walter Zamboni, *Italy*

Guest Editors
Fisciano, Italy

Giovanni Petrone
Walter Zamboni

Contents

About the Editors

Walter Zamboni received the PhD degree in Electrical Engineering from the Università degli studi di Napoli Federico II, Italy, in 2004. From 2016 he is Associate Professor of Electrical Engineering at the Università degli studi di Salerno, Italy, where he served as assistant professor from 2008 to 2016. He is a member of IEEE and IMACS. At present, his main scientific interests include battery modelling and SoC–SoH estimation, identification and diagnostics of batteries and fuel cells, and energy and power management in systems with energy storage. He co-authored more than 80 papers published in international journals and conference proceedings, and one patent.

Giovanni Petrone is Associate Professor of Circuit Theory at the University of Salerno, Italy. His research areas include analysis and design of power electronics and controls for photovoltaic, fuel cell and wind systems, tolerance analysis of electronic circuits, non-linear control techniques, wireless power transfer and identification and diagnosis of renewable sources and storage systems. He is Associate Editor of *IEEE Journal of Photovoltaics and of IET Power Electronics. He is member of the editorial board of MPDI Applied Sciences journal*. Since 2017, he is a senior member of IEEE. Prof. Petrone is co-author of more than 150 papers published in international journals and conference proceedings. He is also co-author of two books, two IEEE e-Learning library courses and five patents in the field of power electronics for photovoltaic applications.

Part I
Modelling and Simulation

Modelling of Photovoltaic Systems for Real-Time Hardware Simulation

Harshavardhan Palahalli, Yujia Huo, and Giambattista Gruosso

Abstract The real-time simulation is a valid help to test electrical systems when a physical device is not available. This is significantly evident when used in hardware and software co-simulation environment, where it is possible to connect the emulator to a real subsystem to test or validate it. In this paper, a model of the photovoltaic system is presented that can be implemented within a hardware simulator to be able to interface it with a real circuit, the hardware simulator used is the National Instruments RIO system.

1 Introduction

Real-time simulation (RT) is one of the frontiers of electrical system analysis. Its task is to control or react to events that occur in the real world, with the same speed as the real world clock. With the aim of having systems that can be interfaced with other software or real systems in order to simulate increasingly complex systems and on a large scale, also taking communication infrastructure into account, as in the case of smart grids [1–4].

It is necessary to be aware of the execution time of the simulation system to ensure that it is shorter or equivalent to the selected time step especially in simulating scenarios that cannot normally be performed, for example, to model the high switching characteristics of the converters, a fast processor is required to reach the minimum step size of up to microseconds [5, 6]. To simulate very quick phenomena in the order of microseconds, it is necessary to think of architectures with more than one processor, and based on FPGAs, that allow high sampling rates, a faster and parallel computing capacity [7]. As the penetration of PV generation increases, its impact on stability and security of the power system will become more and more significant, due to the characteristic of randomness and volatility

H. Palahalli · Y. Huo · G. Gruosso (✉)
Politecnico di Milano - DEIB, Milano, Italy
e-mail: harshavardhan.palahalli@mail.polimi.it; Yujia.Huo@polimi.it;
giambattista.gruosso@polimi.it

W. Zamboni, G. Petrone (eds.), *ELECTRIMACS 2019*, Lecture Notes
in Electrical Engineering 615, https://doi.org/10.1007/978-3-030-37161-6_1

[8]. Modeling and simulation are the basic technologies to study the impact on the power grid in which, large-scale PV generation systems are integrated. Simulations in the usual platform may give good results, but they are not able to deliver the results for dynamic change in input as present in real world in run time, the model may not respond for such a change. When we try to simulate to know the long-term performance of a system, the normal simulation requires a very long time to deliver the results and the accuracy of the results may also get compromised. While simulating the complex model like PV system interaction with the grid in real-time, we may encounter many problems. The important ones to mention are,

1. The need for a PV mathematical model that can deliver results faster to keep the real-time simulation properties during execution. Solving the algebraic loop in the PV model is an important task, as algebraic loops are not supported in the real-time hardware.
2. There is a need of cost-effective test bench/platform for simulating PV systems in real-time that can be used for control validation, studies of the storage system, and integration of PV system to the power drive train or grid.
3. Model-based design of process and systems is very popular, there are tools available, for automatic code generation for the developed model, it is required to use these tools, that can deliver C code from the model, which can be used for the cost-effective target hardware.
4. The memory of the real-time digital simulator (RTDS) is the main constraint while simulating a complex model, like the grid, this memory is used for storing and executing the compiled C code in real-time. It may be necessary to split the model into two or more separate systems and bridge them using an appropriate interface.
5. Interfacing the two models using respective interfacing algorithm introduces some errors in the execution, that results in, instability of the system during run time and also the accuracy of the results varies according to the interfacing algorithm used.

The main objective of this work is to conduct the real-time simulation of the PV considering as hardware under test (HUT), with the MPPT (Maximum Power Tracking Point) system, storage and the load connected to DC bus acting as a rest of the system (ROS) of hardware-in-the-loop (HIL) simulation methodology. The idea is that the photovoltaic system and the rest of the system are executed in parallel in real-time and the maximum power is monitored for irradiation variation in the real world and the extracted power is supplied to the DC bus to which storage element and the fixed load resistance are connected. The modeling of these systems is done in Simulink [9]. The modeled systems are compiled and deployed in NI myRIO's FPGA [10] for real-time simulation.

The rest of this article is organized as follows: Sect. 2 deals with the modeling of photovoltaic and MPPT systems, Sect. 3 explains the real-time simulation architecture used in this work, Sect. 4 tells the real-time simulation with the results and conclusions are drawn in Sect. 5.

2 PV System Modelling

The mathematical model of the PV system is drawn accordingly with [11, 12]. The equivalent described in Fig. 1 consists of a current source in parallel with the diode and R_{sh} and R_s are the intrinsic shunt and series resistances of the cell, respectively. The I–V characteristic of the cell is determined by the diode. PV cells are grouped into larger units to form PV modules, these modules are connected in series or parallel to create a PV array to obtain required electrical power.

$$I_{ph} = \left[I_{sc} + K_i(T_{cell} - T_{ref})\right]\lambda \tag{1}$$

$$I_{rs} = \frac{I_{sc}}{\left[\exp\left(\frac{qV_{oc}}{N_{cell}KnT_{cell}}\right) - 1\right]} \tag{2}$$

$$I_o = I_{rs}\left[\frac{T_{cell}}{T_{ref}}\right]^3 \exp\left\{\frac{qE_g}{nK}\left(\frac{1}{T_{ref}} - \frac{1}{T_{cell}}\right)\right\} \tag{3}$$

$$V_t = \frac{KT_{cell}}{q} \tag{4}$$

$$I_{sh} = \frac{V_{pv}\frac{N_p}{N_{cell}N_s} + I_{pv}R_s}{R_{sh}} \tag{5}$$

$$I_{pv} = N_pI_{ph} - N_pI_o\left[\exp\left(\frac{\frac{V_{pv}}{N_{cell}N_s} + \frac{I_{pv}R_s}{N_p}}{nV_t}\right) - 1\right] - I_{sh} \tag{6}$$

where

I_{pv}	Cell current
V_{pv}	Array terminal voltage
I_{ph}	Photo current
I_{sc}	Short-circuit current at 25 °C and 1 kW/m^2
K_i	Short-circuit current temperature coefficient
T_{cell}, T_{ref}	PV cell and reference temperature in K

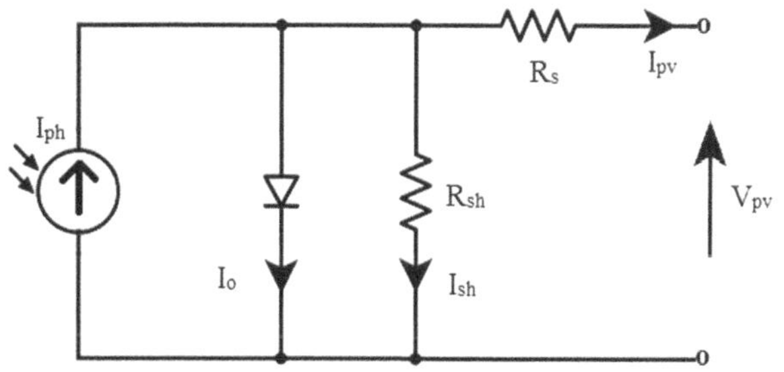

Fig. 1 PV cell equivalent circuit

λ	Ratio of solar irradiation to reference 1 kW/m^2
I_{rs}	Module reverse saturation current
q	Charge on electron = 1.6×10^{-19} C
N_{cell}	Number series cells in a module
n	Diode ideal factor
k	Boltzmann's constant = 1.3805×10^{-23} J/K
I_o	Module saturation current
E_g	Energy band gap of semiconductor material
V_t	Diode thermal voltage
I_{sh}	Shunt current
N_s	Number of series modules connected in string
N_p	Number of parallel string connected in array
V_{oc}	Open circuit voltage
R_s,R_{sh}	Series and shunt resistance

The method to find the resistances of PV cell can be found in [13], the above mentioned Eq. (6) is solved using Newton Raphson method with the error tolerance of 10^{-5} for better accuracy. In this work Mitsubishi Electric PV-UD180MF5 PV module is selected as reference, the parameters of the module are given in Table 1. The IV curve of the modeled PV array for different irradiation is given in Fig. 2, later the same model is used for PV real-time simulation, fixed step discrete solver with the sample time of 1 ms is used in both normal and real-time simulation.

The PV panels are connected to the load using a DC/DC converter governed by an MPPT controller, based on perturb and observe ($P\&O$) method. The choice of a DC–DC converter is based on many factors such as load resistance, resistance at the maximum power point, DC output voltage [14]. Figure 3 shows the block diagram of PV with MPPT system.

Table 1 Mitsubishi electric PV-UD180MF5 PV Module details and parameters for 100 kW array

Parameter	Value
Open circuit voltage (V_{oc})	30.4 V
Short-circuit current (I_{sc})	8.03 A
Maximum power rating ($Pmax$)	180 W
Voltage @ maximum power	24.2 V
Current @ maximum power	7.45 A
Number of series connected cells in module (N_{cell})	50
Diode ideal factor (n)	1.7
Band Energy Gap (E_g)	1.121 eV
I_{sc} temperature co-efficient k_i	0.0032 K^{-1}
Shunt resistance (R_{sh})	3.767 Ω
Series resistance (R_s)	0.00019 Ω
Number of series modules used N_s	11
Number of parallel strings used N_p	51

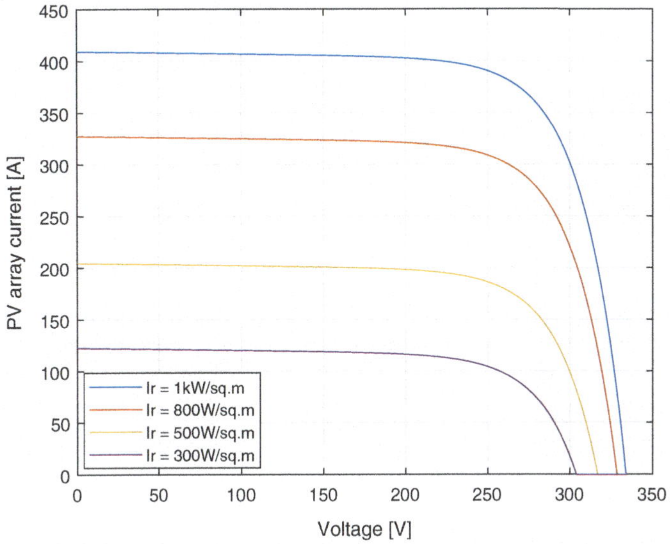

Fig. 2 *IV* curve of the 100 kW array for different irradiation

Fig. 3 Block diagram of PV system with MPPT

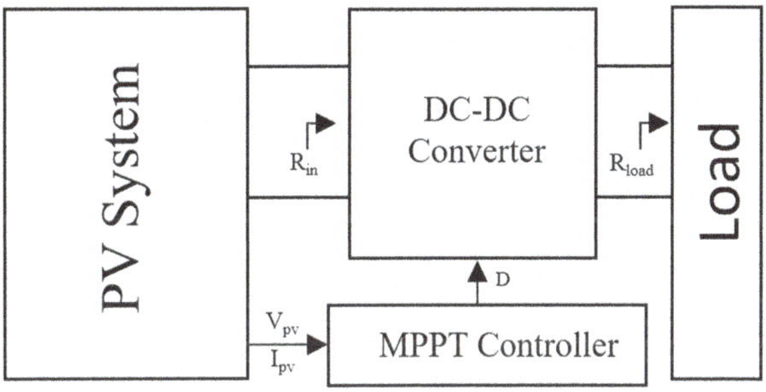

3 Real-Time Simulation Architecture

In this section the test benches created for conducting HIL experiment will be presented. The mathematical model will be simulated in myRIO in real-time. FPGA IO's are used to integrate the model in-ports and out-ports to the real world measurement systems (Fig. 4).

The model will be implemented in MATLAB-Simulink [9] and translated in into C code with its internal C code generating tool using the NI VeriStand compilers which adds on to the MATLAB and it can generate the file ".SO" which defines our model with the libraries. NI VeriStand Engine has to be deployed in the RT target to make it compatible to run with the host computer. .lvbitx file defines the custom FPGA personality used for the simulation in the RT target. The main operation is shown in Fig. 5. National Instruments myRIO-1900 is a Linux based device, it

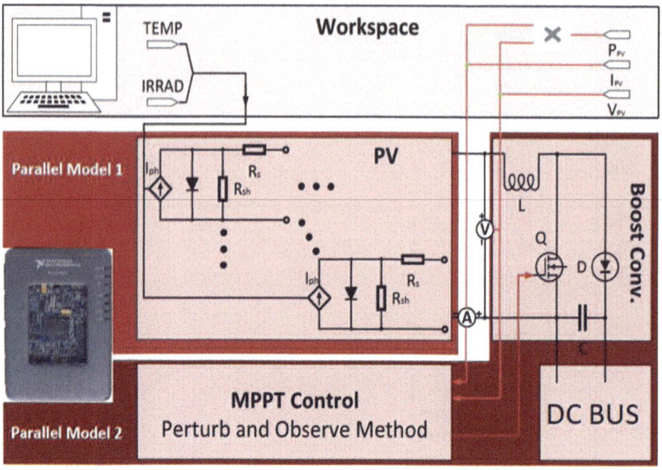

Fig. 4 PV with MPPT system in myRIO

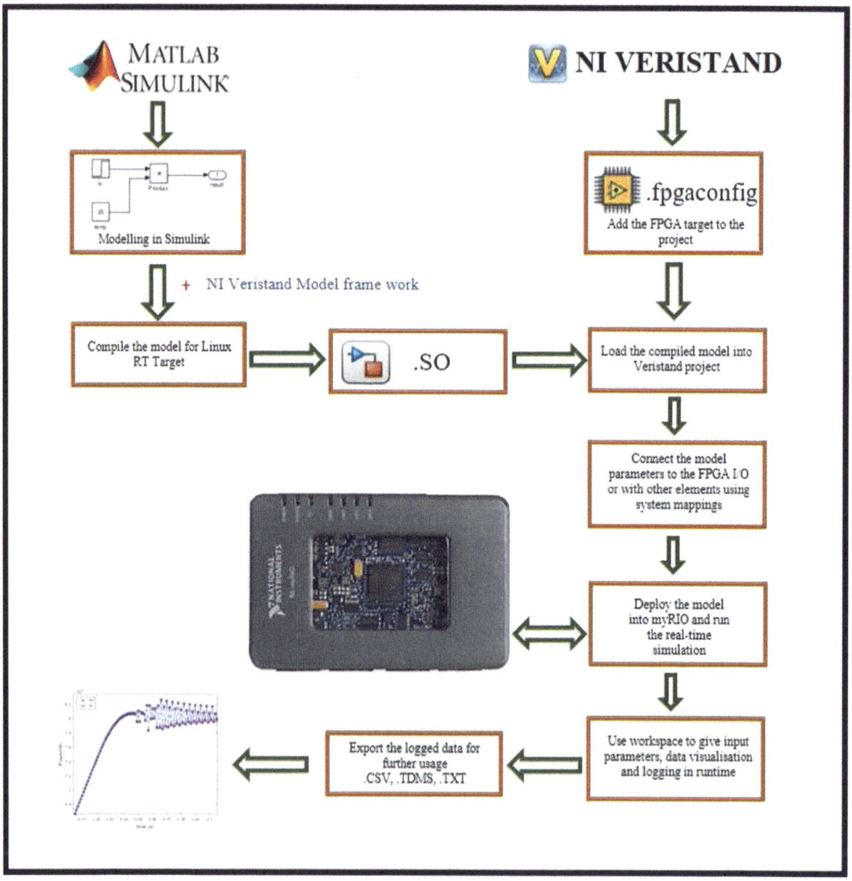

Fig. 5 Functional scheme of the procedure

comes with built-in Xilinx Z-7010 FPGA [10] and it also allows us to build custom FPGA applications with real-time capabilities. In this study, the real-time simulation of PV system is done in myRIO using NI VeriStand [10], it is a configuration-based testing software, specifically designed to run the real-time testing application, thus allowing us to develop and test control system using hardware I/O and simulation models.

4 Simulation and Results

This test bench consists of a 100 kW PV array model as HUT, whose interaction with the ROS for the change in temperature and the irradiation is studied. The ROS consists of MPPT, DC–DC converter, DC bus is created where the load and the storage are connected with the breaker and the bi-directional intelligent converter, respectively. The temperature and the irradiation data are measured using a temperature sensor and a small 5 W PV cell, respectively. The model used to simulate using this test bench is shown in Fig. 6.

The power rating and other parameters of the proposed RT simulated system are given in Table 2, the voltage of DC bus is maintained at 250 V all the time with the help of storage and the bi-directional DC-DC converter, which allows the power flow in both the directions. PID control is used for the voltage control of the converter. When the power required by the load is more than the generation, DC bus voltage drops. Power will be compensated by the energy stored and the voltage is brought regulated. When the power generated is more than the load requirement, the DC bus voltage increases, and the controller takes action to bring back the voltage to the set reference of 250 V. In this experiment, the fixed resistive load of 1 Ω is connected to the DC bus at 5th s, the outputs are driven to the real world using FPGA IO and read using the digital oscilloscope, the test bench setup is shown in Fig. 7.

The irradiation and the temperature measured in real-time are given in Fig. 8. They both are measured by myRIO and fed to the model during run time.

The load is connected to the DC bus at 5th s, the DC bus voltage and its variations during run time is given in Fig. 9. It can be observed that the controller is able to maintain the bus voltage constant even after connecting load at 5th s.

To ensure the simulation running in real-time the bus voltage is also measured using a digital oscilloscope as shown in Fig. 10. The real-time simulation started at −9 s, this corresponds to 0th s of the model time, the load is configured to get connected to the DC bus at 5th s of the model time, and in the oscilloscope the voltage variation due to load application is recorded at -4th s, this proves that the simulation is running in real-time, just the voltage is scaled down by 100 V as 1 V to drive output through DAC, finally the simulation is stopped at 1.6 s as seen in oscilloscope.

The maximum power delivered by PV and the power consumed by the load after connecting it to the DC bus is shown in Fig. 11. The power delivered by PV before connecting the load is used to charge the storage system of 10 Ah. The state of

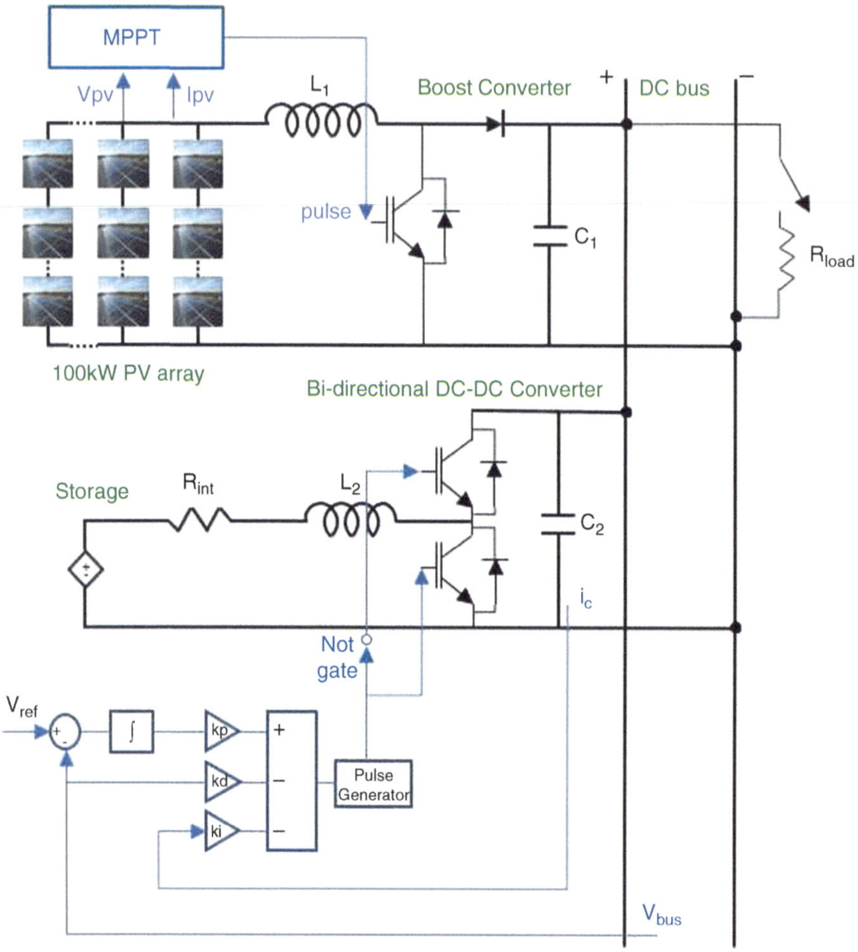

Fig. 6 Block diagram of the simulated model

charge measured in the storage system and the current flowing to the batteries are given in Fig. 12. Initial SOC is about 50% and raised rapidly till connecting the load to the DC bus, later the slope of SOC decreases since the flow of current is decreased to the storage system. The voltage seen across the storage system throughout the simulation is given in Fig. 13.

Table 2 Parameters of electric system

Parameter	Value
L_1	11 mH
C_1	10 mF
MPPT switching frequency (f_{s1})	1 kHz
R_{int}	0.14 Ω
L_2	15 mH
C_2	10 mF
Controller gains, kp	0.055
ki	0.001
kd	0.0003
Bi-directional converter (f_{s2})	5 kHz
Nominal voltage of battery	140 V
Initial SOC of battery	50%
DC Bus voltage V_{bus}	250 V
R_{load}	1 Ω
Sample time used in RT simulation	1 ms

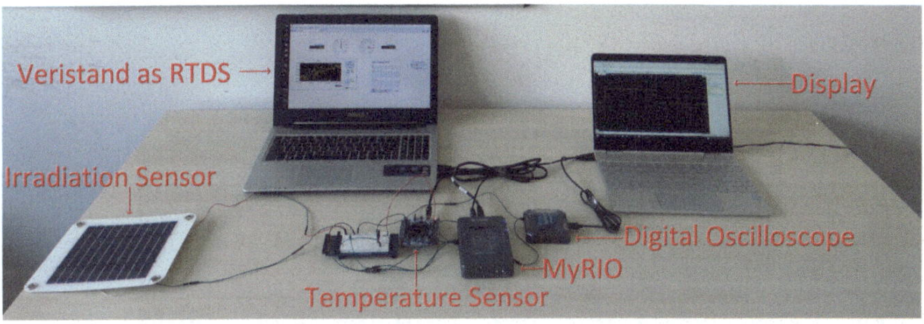

Fig. 7 Test bench used for HIL simulation

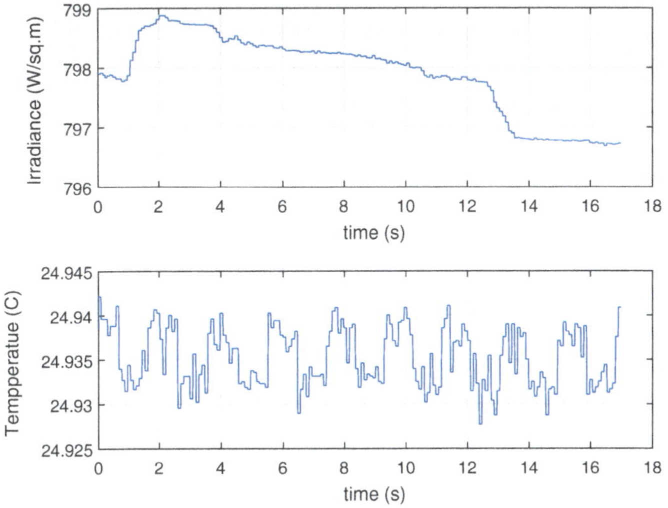

Fig. 8 Irradiation and temperature measured in real-time

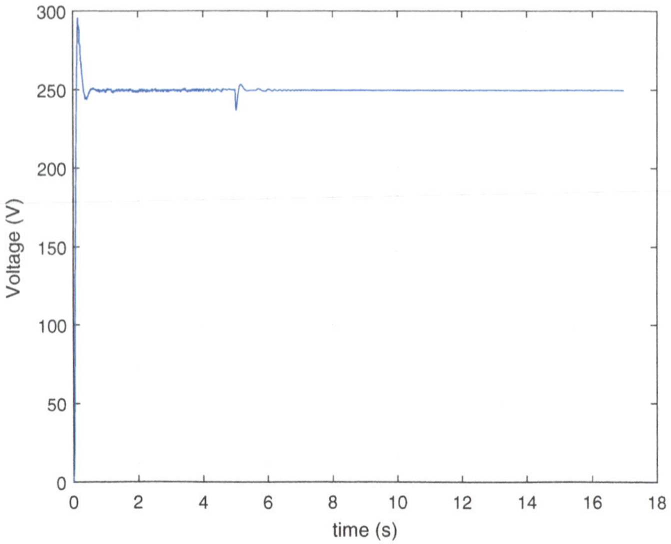

Fig. 9 Voltage measured across DC bus in real-time

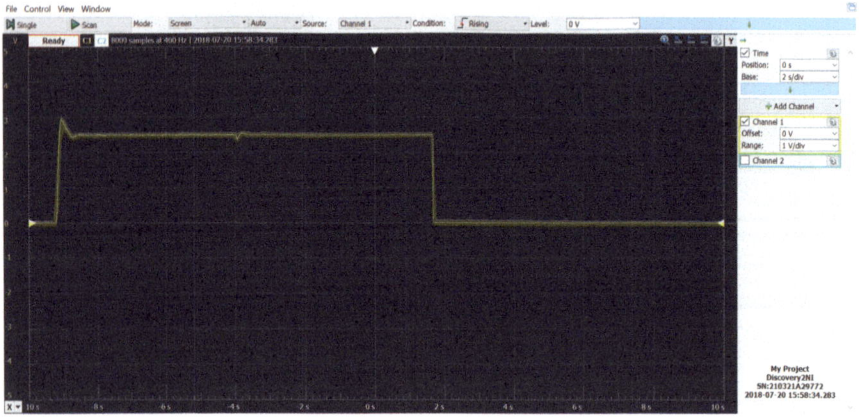

Fig. 10 Voltage measured across DC bus using oscilloscope

5 Conclusion

This article presents a methodology for real-time simulation, based on a model created in Simulink environment and then downloaded to the hardware of National Instruments. The step is done through VeriStand tool, to demonstrate that, it is possible to create small real-time simulators for laboratory or educational use. In the article, the methodology is illustrated in detail and accompanied by the results.

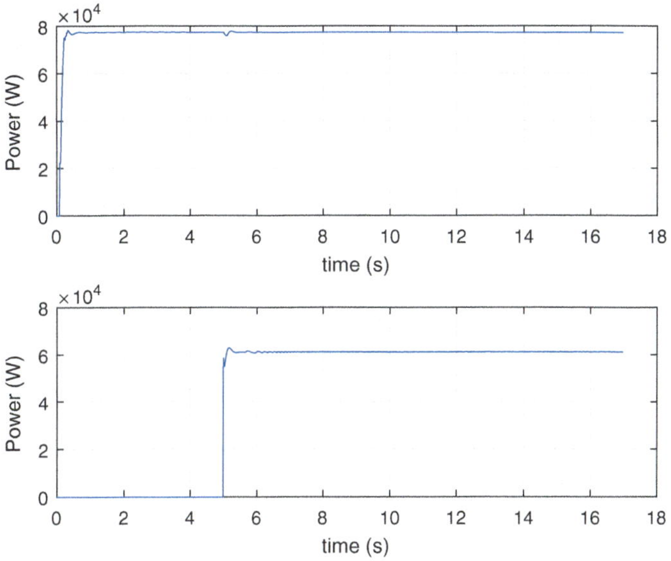

Fig. 11 Power delivered by PV and consumed by load

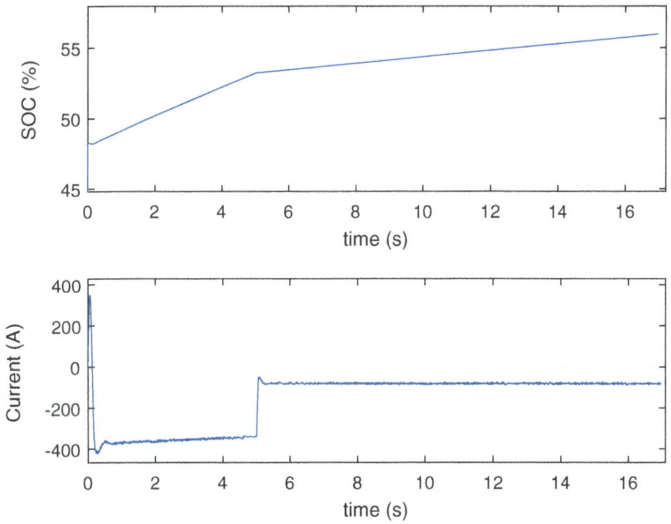

Fig. 12 SOC and the battery current

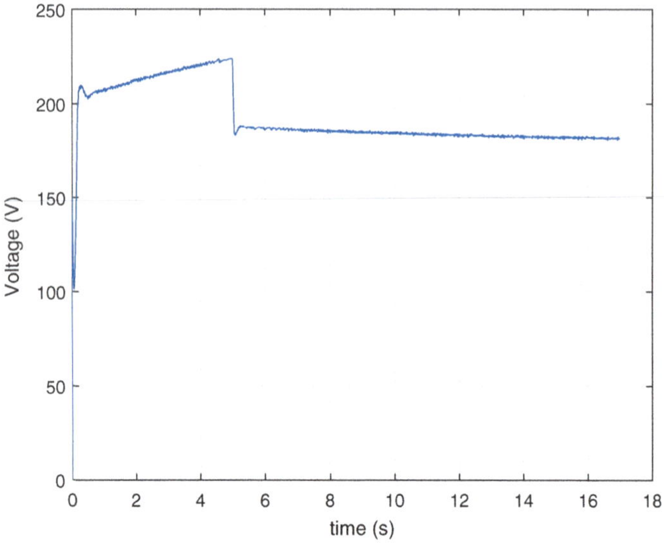

Fig. 13 Battery terminal voltage

References

1. Y. Huo, G. Gruosso, L. Piegari, Power hardware in the loop simulator of photovoltaic plant for smart grid interation analysis, in *2017 IEEE International Conference on Environment and Electrical Engineering and 2017 IEEE Industrial and Commercial Power Systems Europe (EEEIC / I CPS Europe)*, June, pp. 1–5
2. M.D. Omar Faruque, T. Strasser, G. Lauss, V. Jalili-Marandi, P. Forsyth, C. Dufour, V. Dinavahi, A. Monti, P. Kotsampopoulos, J.A. Martinez et al. Real-time simulation technologies for power systems design, testing, and analysis. IEEE Power Energy Technol. Syst. J. **2**(2), 63–73 (2015)
3. A. Ingalalli, H. Satheesh, M. Kande, Platform for hardware in loop simulation, in *2016 International Symposium on Power Electronics, Electrical Drives, Automation and Motion (SPEEDAM)* (IEEE, New York, 2016), pp. 41–46
4. W. Li, G. Joós, J. Bélanger, Real-time simulation of a wind turbine generator coupled with a battery supercapacitor energy storage system. IEEE Trans. Ind. Electron. **57**(4), 1137–1145 (2010)
5. P.M. Menghal, A. Jaya Laxmi, Real time control of electrical machine drives: a review, in *2010 International Conference on Power, Control and Embedded Systems (ICPCES)* (IEEE, New York, 2010), pp. 1–6
6. B.L. Schenkman, D.G. Wilson, R.D. Robinett, Photovoltaic distributed generation for lanai power grid real-time simulation and control integration scenario, in *2010 International Symposium on Power Electronics Electrical Drives Automation and Motion (SPEEDAM)* (IEEE, New York, 2010), pp. 154–157
7. M. Dagbagi, L. Idkhajine, E. Monmasson, I. Slama-Belkhodja, FPGA implementation of power electronic converter real-time model, in *2012 International Symposium on Power Electronics, Electrical Drives, Automation and Motion (SPEEDAM)* (IEEE, New York, 2012), pp. 658–663

8. J.V. Paatero, P.D. Lund, Effects of large-scale photovoltaic power integration on electricity distribution networks. Renew. Energy **32**(2), 216–234 (2007)
9. The Mathworks. http://www.mathworks.com/
10. National Instruments. http://www.ni.com
11. F. Bizzarri, A. Brambilla, G. Gruosso, C. Guardiani, A. Sangiovanni Vincentelli, G. Storti Gajani, Modeling and estimating yield and efficiency of photovoltaic solar parks, in *2013 IEEE International Conference on Industrial Technology (ICIT)* (2013), pp. 734–739
12. F. Bizzarri, M. Bongiorno, A. Brambilla, G. Gruosso, G.S. Gajani, Model of photovoltaic power plants for performance analysis and production forecast. IEEE Trans. Sustain. Energy **4**(2), 278–285 (2013)
13. S. Sumathi, L. Ashok Kumar, P. Surekha, Application of MATLAB/SIMULINK in solar PV systems, in *Solar PV and Wind Energy Conversion Systems* (Springer, New York, 2015), pp. 59–143
14. M. Pisano, F. Bizzarri, A. Brambilla, G. Gruosso, G.S. Gajani, Micro-inverter for solar power generation, in *International Symposium on Power Electronics Power Electronics, Electrical Drives, Automation and Motion*, June (2012), pp. 109–113

Hardware-in-the-Loop Simulation of High-Power Modular Converters and Drives

Christian Dufour, Karthik Palaniappan, and Brian J. Seibel

Abstract This paper explains several industrial cases involving the HIL simulation of MW-range drives and inverters using CPU cores with FPGAs to compute model equations. The use of HIL simulators is common today in the industry to accelerate design cycles, mitigate financial and human risks and support software updates throughout the product life cycle.

The first case presented is a 2-level inverter scheme in which increasing power specifications are met by adding parallel IGBT-modules. The second case is a multi-level motor drive with low harmonic injection on the AC-side. The third case is a modular multi-level converter in a grid application. We also discuss a new T-type inverter model that uses an industry PV-to-grid power converter.

In each case, all power system modelling was done using Simulink and Sim-PowerSystems in conjunction with the SSN solver from the ARTEMiS blockset in addition to code generation for CPU execution at time steps in the 20–50 μs range, with an exception for MMC models on FPGA. In all cases the firing accuracy of the IGBTs remains in the nanosecond range using time-stamping techniques and an FPGA board. In the case of the parallel 2-level inverters, there is significant difficulty regarding the small firing delays (typically <500 ns) between modules that create circulating currents. These circulating currents are rendered correctly on the HIL bench.

Also discussed in the paper are the various optimisations, solvers and methods that enable these performances.

The original version of this chapter was revised: Line 13 of page 17 has been corrected. The correction to this chapter is available at https://doi.org/10.1007/978-3-030-37161-6_57

C. Dufour (✉)
Opal-RT Technologies, Montréal, QC, Canada
e-mail: christian.dufour@opal-rt.com

K. Palaniappan · B. J. Seibel
Rockwell Automation Inc., Milwaukee, WI, USA
e-mail: kpalani@ra.rockwell.com; bjseibel@ra.rockwell.com

W. Zamboni, G. Petrone (eds.), *ELECTRIMACS 2019*, Lecture Notes in Electrical Engineering 615, https://doi.org/10.1007/978-3-030-37161-6_2

1 Introduction

Real-time simulation is an important part of power system and industrial drive development as it enables engineers to test system controllers in the lab before field commissioning. This helps to reduce overall costs by providing early detection of issues during the design process. After deployment, the real-time simulator continues to be used to test controller software updates before release in the laboratory, at safe power levels, without the need to maintain a real drive.

This paper presents some of the most challenging HIL simulation cases for drives and converters undertaken by Opal-RT, as of 2019. HIL simulation of these drives and converters is made possible through constant advancements in the field of multi-core CPU and FPGA technologies combined with clever algorithms and simulation methods.

2 Some Requirements for HIL Simulation of Drives

The main objective of the HIL test system is to verify the functionality of the real system in controlled laboratory conditions and at low power levels, before actual release, including software upgrades and commissioning.

In particular, the HIL test system must:

- Check the drive start-up and shutdown sequences. Models must be able to output truly null currents before the starting sequence for this to be checked correctly in HIL mode. Otherwise an error occurs and the controller goes into safe mode.
- Check the circulating currents between IGBT inverter modules in parallel configurations. Small circulating currents are normal in the real system and are due to variations in firing caused by wiring and element tolerances. The drive closely monitors these currents and shuts them down if they exceed a pre-determined level, which is an indication of system malfunction.
- The HIL system must be able to adequately reproduce the PWM inverter characteristics on a CPU-based simulator running the model within a 25–50 μs range. This is achieved using the time-stamping technique [1] and special inverter models called Time-Stamped Bridges or TSB. Direct connection of controller PWM pulses to the HIL simulator is mandatory (i.e., use of averaged models is not possible).
- These TSBs must be able to work correctly in natural rectifying mode; this is especially important for drives with Active-Front-End rectifiers.

3 Some MW Converter Topologies Simulated in HIL

3.1 Two-Level Parallel IGBT Modular Motor Drive and Active-Front-End Rectifier

This topology comes from the Rockwell PowerFlex 750-Series products for the Low-Voltage market (from 160 to 6000 kW) and is depicted in Fig. 1 This topology provides a scalable power level by using parallel 2-level IGBT modules up to 6 MW. More details can be found in [2].[1]

An OPAL-RT Hardware-In-the-loop (HIL) system was chosen to perform a wide variety of product software and hardware verification and validation during the product design phase and will be used for regression testing over the life of the product. As with any simulation, fidelity and accuracy of the simulation must adequately match the product itself to guarantee usefulness and confidence in any testing and verification. This section shows the PowerFlex 755TM Common Bus Inverter. The inverter PWM frequency is 1.33 kHz and the complete model runs at 50 μs in HIL mode. In HIL tests, close to 2000 I/Os were required in the simulator to interface and test the actual controllers in the real-time simulated drive system, with up to ten converters at any given time during operation.

IGBT modules are connected together through interphase reactance to smooth out any small difference of voltage output between the parallel inverters.

The drive controller closely monitors this current difference between the same phases of parallel inverters and puts the system in fault mode if it rises above a certain threshold.

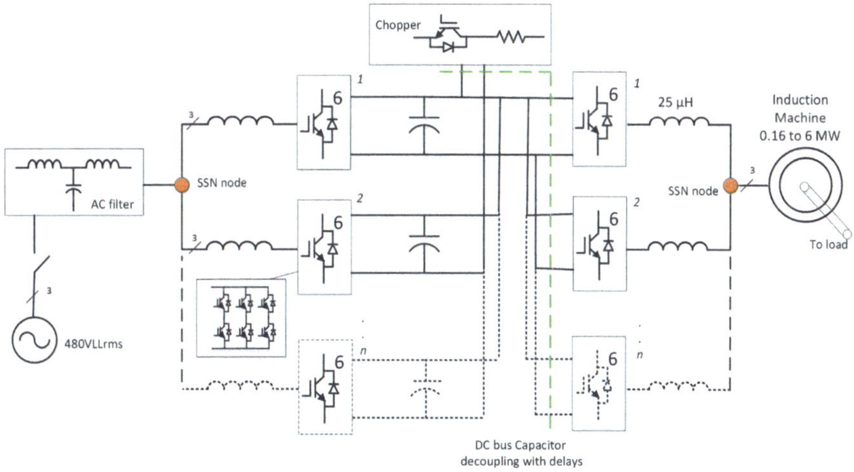

Fig. 1 Parallel 2-level IGBTs drive with AFE rectifier

[1]The authors acknowledge the re-use of some of their own verbatim of this reference.

3.1.1 Modelling Techniques

This inverter topology poses several difficulties, such as the accurate computation of inter module currents and the need for accurate current levels at low power. Furthermore, the model must be decoupled in small parts, running on different cores of the CPU to maintain the simulation time step within an acceptable range.

The system has a large DC-link, which makes the separation of the model into separate parts easier. DC-bus voltages and currents can be transmitted between separated parts with an inserted delay, with negligible errors.

The model uses modified interpolating inverter models (Time-Stamped Bridge, TSB); these inverter models are able to incorporate I/O gating signal time-stamps and to interpolate voltage outputs, resulting in very accurate simulation, mainly limited by the simulator FPGA sampling rate (in this instance, 5 ns).

Of special interest is the treatment of cases in which dead-time occurs near zero-crossing. In this case, the standard TSB model may actually output opposite voltages on different modules of the same phase.

For example, for a DC voltage of 640 V and an interphase inductance of 25 μH, the circulating current rises at the rate of $640/(2 \times 25e{-}6) \times \tau$ (for two parallel IGBT inverters), where τ is the firing delay between parallel IGBTs. For a 500 ns delay, the current would rise to about 7 A, within the controller's tolerance level.

As a simple test, we ran the model offline at 50 μs, in a simple motor start-up sequence with two parallel IGBT modules, with a high-resolution PWM modulator and variable inserted delay between the parallel IGBTs. The result is depicted in Fig. 2 and is consistent with the theory.

The requirement for null current during blocked modes, as well as accurate natural rectification modes, led us to model the diode as binary switches. This in turn caused the total number of connected diodes to be very high (60 diodes for ten parallel modules), dramatically increasing the total number of possible electric modes and permutations of state-space equations (possibly 2^{60}). By using SSN, each TSB module, containing six binary diodes each, was set into a different SSN group, solving this potential issue for real-time calculation.

Finally, the induction motor model used was an SSN, meaning that no minimum load or stabilisation snubber was required to run the model stably, even at very low power levels [3].

3.2 *Multi-level Motor Drive with Low Input Harmonic*

The multi-level motor drive depicted in Fig. 3 is designed to provide low harmonic AC-voltage to the load, as well as low input current harmonic to the feeding grid. This is achieved by arranging the multiple diode rectifiers with zig-zag transformers with varying phase shifts. The system runs in HIL with attached industrial controller under test at a time step of 25 μs.

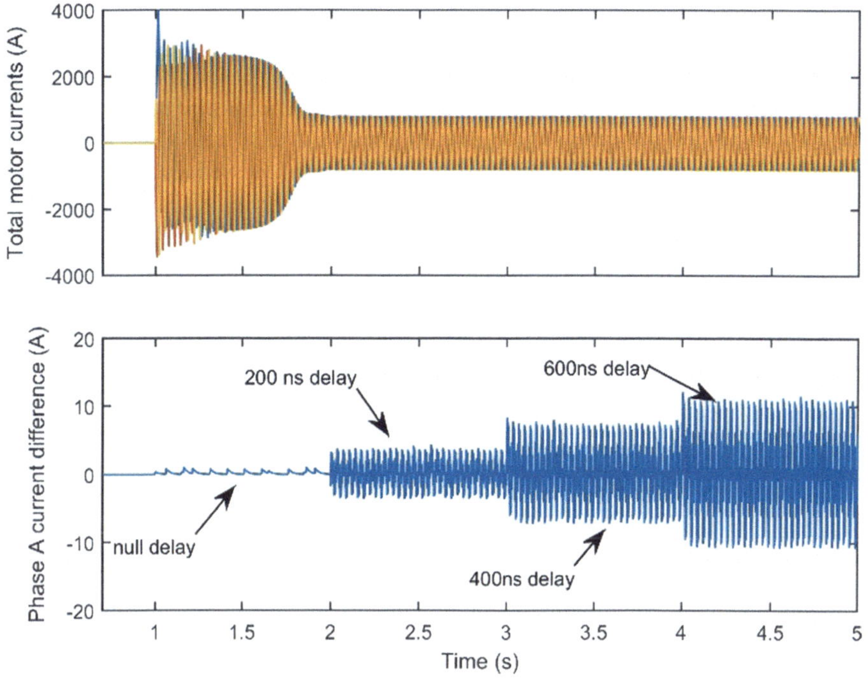

Fig. 2 Motor start-up with variable firing delay between two IGBT modules. Top: motor current; bottom: current difference between the parallel inverters

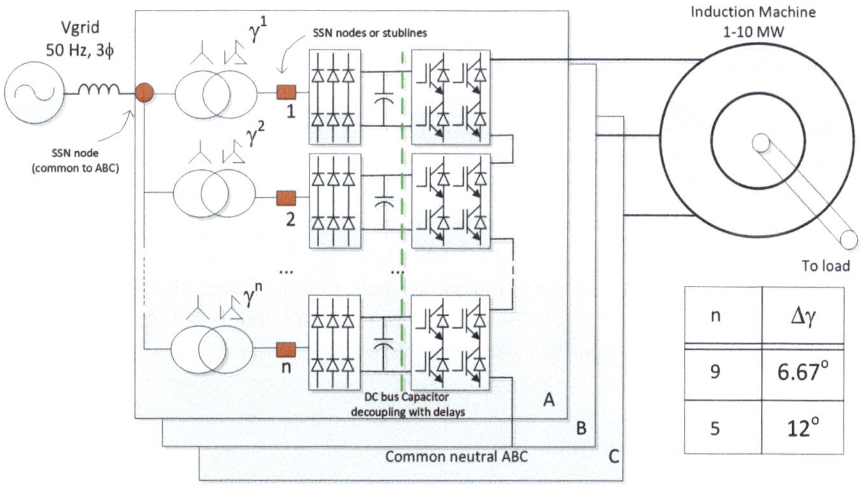

Fig. 3 Multi-level motor drive with low input harmonic

3.2.1 Modelling Techniques

This inverter topology presents several difficulties for HIL simulation. One of these is the common inductive connection of all zig-zag transformer primary windings; the SSN solver makes it possible to break down this huge state-space system of equations into several smaller ones, *without delays*. In one version of the model, where the $3 \times n$ 6-pulse rectifier (here n is the number of inverter stages) was separated using SSN nodes, SSN was also required to decouple the $3 \times n$ 6 pulse rectifiers. In the final implementation, stublines were used to provide a better decoupling at the expense of slightly lower precision.

Also, the presence of a DC-bus at each 6-pulse rectifier allowed us to simulate the rectifiers and the inverters in different cores of the simulator.

The SSN node location allows for the creation of three large SSN groups, one for each phase of zig-zag transformers, which allowed us to compute these groups on different cores.

Stublines are artificial one-time-step-delay transmission lines and are used to completely decouple a circuit from one end of the stubline to the other. As they add a little capacitance to the circuit, they are less precise than SSN, which does not approximate or add delays to the simulated equations.

3.3 *Multi-level Modular Converter (MMC)*

Modular Multi-level Converter (MMC)-based High-Voltage Direct Current (HVDC) is a rapidly emerging technology for DC current transport and high-power converters in general. MMC, by its modular nature, has the advantage of being very reliable, easily maintainable and scalable: the inverters can continue to work correctly if a module fails, and the defective module can be replaced without completely shutting down the converter. This is an important issue with sites difficult to access, such as offshore windfarms. The converter described in this section is a 271-level MMC system in a power grid, configured for STATCOM operation, as specified for the factory acceptance test of an OPAL-RT client in the Zhoushan region of China. The MMC-STATCOM is part of a multi-terminal HVDC system and, in this case, its DC breaker is open, disconnecting the MMC station from the other stations [4, 5]. In this configuration, depicted in Fig. 4, the MMC is used as a STATCOM to stabilize the AC voltage level.

3.3.1 Modelling Techniques

To achieve HIL simulation of this MMC system with 3240 IGBTs and 1620 capacitors, the modelling approach is different from the previous cases: all IGBT switching and capacitor voltage calculations are done directly on the FPGA, and a

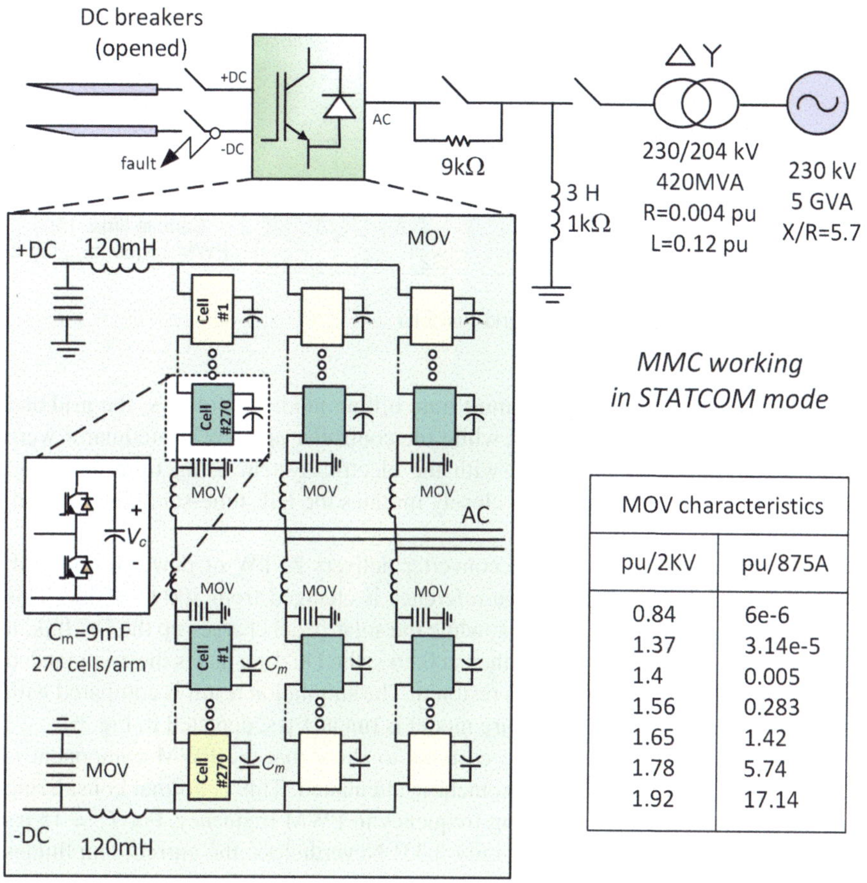

MOV characteristics	
pu/2KV	pu/875A
0.84	6e-6
1.37	3.14e-5
1.4	0.005
1.56	0.283
1.65	1.42
1.78	5.74
1.92	17.14

Fig. 4 MMC converter in a power grid

Thevenin equivalent is sent to the CPU for the SSN solver updates with iterative surge arresters. This model runs in HIL at a time step of 25 μs [6].

3.4 T-Type 3-Level Inverter for PV-to-Grid Converter

Among recent developments is the creation of a TSB-type model for the T-type 3-level inverter in ARTEMiS-SSN. The topology is tested in this section for a 25 kW industrial PV converter for grid, depicted in Fig. 5.

The converter's PWM switching frequency is 20 kHz. This is typical for PV inverters rated below 30 kW for residential and commercial PV applications where audible noise is not acceptable.

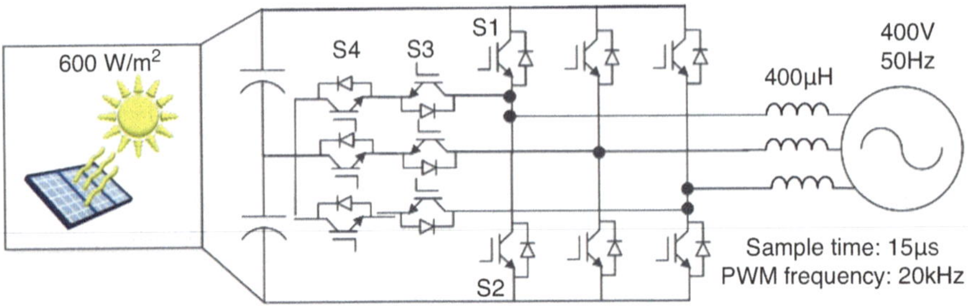

Fig. 5 T-type 3-level inverter in PV-to-grid converter

The model was simulated in a multi-rate offline model as follows: the grid and converter were simulated at 15 μs while the controller and PWM modulator were simulated at 1 μs and interfaced with the electric system using time-stamping. This time-stamping approach very closely imitates the HIL time-stamping method, described in Sect. 4.1.

This test case is one where the converter delivers 25 kW of power to the grid. Then at 0.25 s, the DC-link voltage reference is changed from 800 to 850 V. This results in a small gap in the currents; while the solar panel charges up the DC link, it does not deliver power to the grid; then at 0.26 s, the DC-link meets the commanded values and power flow to the grid is restored. The simulation result is compared with a reference model in which the entire model is run at 1 μs, depicted in Fig. 6.

The figure is zoomed in on the currents to show that the PWM component is still present in the simulation but somehow attenuated. This is normal considering the low ratio of simulation sampling frequency to PWM frequency. For $T_s = 15$ μs and $F_{PWM} = 20$ kHz, this ratio is only 3.33! Nevertheless, the current amplitudes are very accurate and free of amplitude jitter, which is sufficient to fully test the controller.

This also clears up some common confusion about TSBs: they are NOT averaged models, because the PWM component is included in the simulation. Rather, TSBs could be called 'Per time step averaging models', while standard averaged models are, by definition, 'Per fundamental period averaging models'.

4 Miscellaneous Topics

4.1 Time-Stamping Technique and Real-Time Simulators

Interpolating IGBT models are very important to obtain accurate simulation on a CPU-based computational platform. Interpolating data for power inverters is obtained by time-stamping the gate transitions on a high-frequency FPGA; logic and transition time values are then used by the interpolating model (TSB) on the

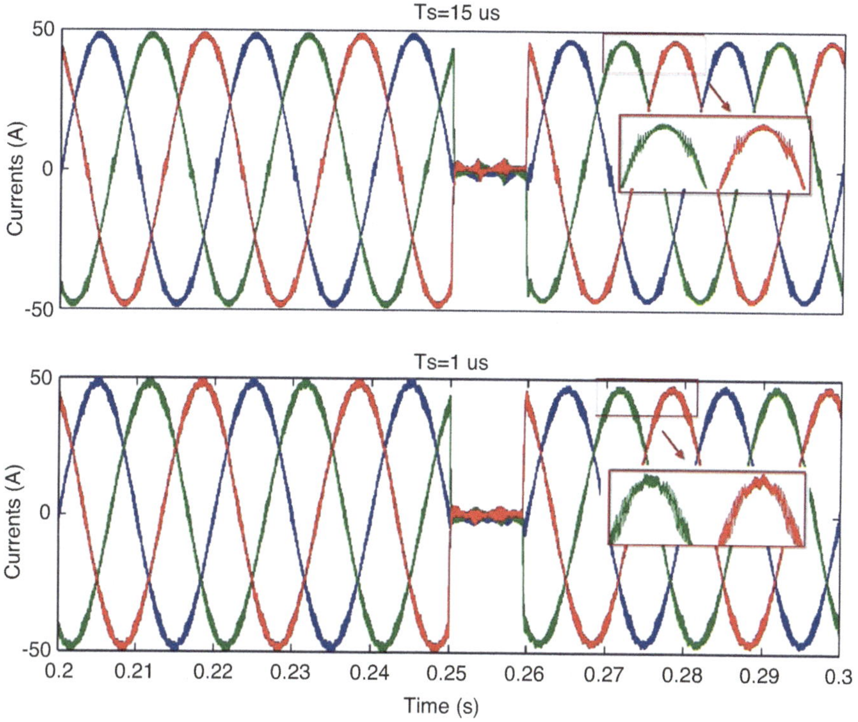

Fig. 6 Comparison of simulation at 15 μs using time-stamping method (upper) vs. reference model running at 1 μs (lower)

HIL simulator, as shown in Fig. 7 [7, 8]. A similar approach is done without I/Os by calculating the intra-step zero-crossing of PWM modulators.

In certain cases, such as MMC, some parts of the model can be computed on the FPGA.

4.2 *Machine Models and SSN*

All the machine models used in these models were SSN machines; that is, machine models developed using the nodal admittance method of SSN [3, 9, 10]. The main reason for this choice was to ensure stability without any parasitic loads or stabilisation snubbers at the machine terminals.

This was especially important for the Two-Level Parallel IGBT Modular Motor Drive because of the inductive connection of the motor and the interface inductor of the different modules. Standard SPS machine models use the current injection technique, which is prone to numerical instability, and is actually unstable when injected into such inductive node.

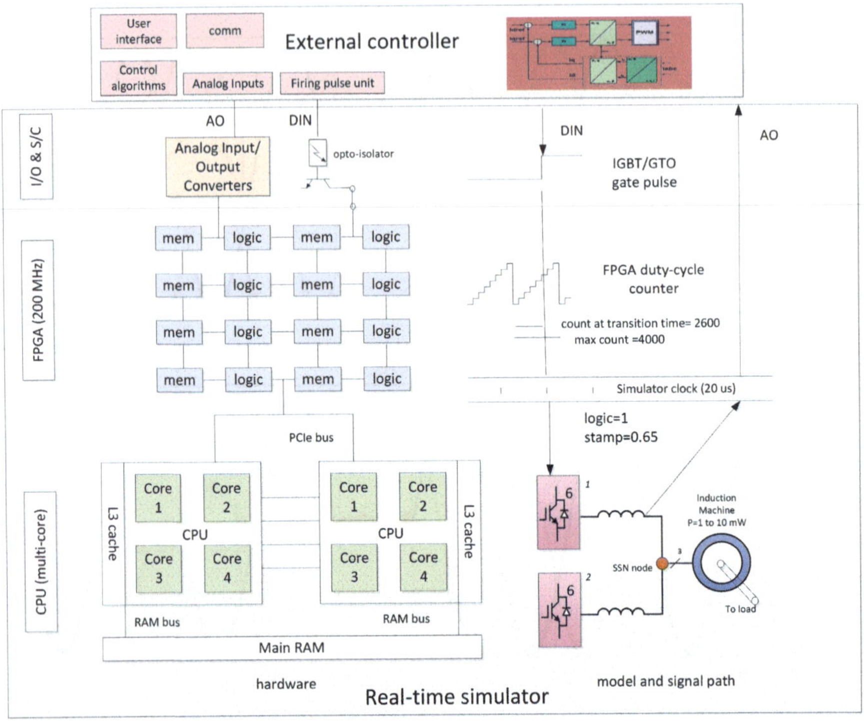

Fig. 7 RT-LAB simulator hardware structure and time-stamping

4.3 *Decoupling for Power Circuits and Drives*

Figure 8 summarizes the various techniques to decouple a power system or electric drive. Long transmission lines (where the transport delay is larger than the simulation time step) are the best way to decouple a power system into small parts; however this technique relates more to power systems than drives.

More common in drives and converters, DC links are also an excellent choice to decouple a drive with a delay between fractioned parts. Stublines are also a good choice when substituted for large inductor or transformer leakage inductance. As stublines are an approximation of the real circuit, this decoupling method should be validated when used.

SSN is also an excellent choice to parallelise the calculation of a grid or drive. In this case, the parallelisation is made during the calculation itself, similar to computing a 'parallel for' loop in MATLAB. For example, *SSN introduces neither approximation nor delays* into the calculation of the equations, but the parallelisation effectiveness is less than full task decoupling created by DC link delays and stublines.

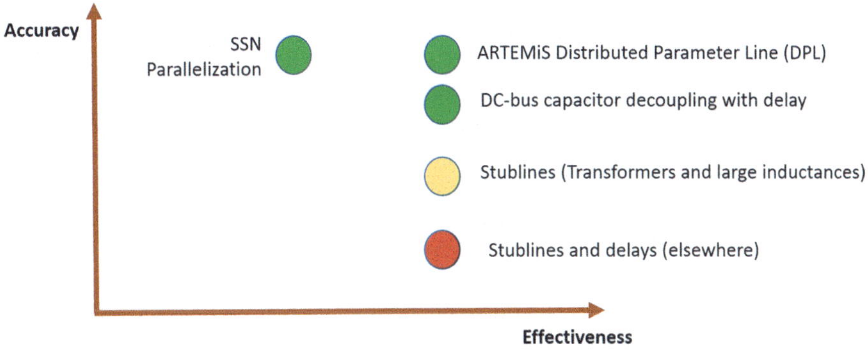

Fig. 8 Comparison of accuracy vs. effectiveness of common decoupling techniques

4.4 CPU vs. FPGA Modelling

FPGA chips can be powerful enough to compute drive equations in HIL simulators.

Special advantages of FGPA include:

- Smaller time steps, which are typically under 1 μs.
- Lower input–output latency in HIL systems (considering that I/Os and models are together on the FPGA).
- A natural way to sample high-frequency PWM drives.

Recent advances in the field of electric circuit simulation on FPGA [11, 12] now enable the simulation of variable topology, variable parameter circuits, all without the need for bitstream generation, which until recently has been a powerful disincentive in FPGA technology since it can take hours to place and route (the equivalent of compiling with a CPU). Today, FPGAs are still considered as a specialist's domain with several limitations, in particular the coding of complex and flexible algorithms such as SSN. This is the main reason that CPU-based HIL simulation is still preferred today.

The best approach here is to use both technologies in a pragmatic approach: unless very low-latency (<5 μs) or PWM components are absolutely required in the simulation (e.g., resonant converters), CPUs are preferable. When the model is run on a CPU, there is a typical two-time step latency involved, caused by data transfer to and from I/Os at each time step.

The case of the MMC is particularly interesting in its pragmatism: the MMC topology is indeed better suited to FPGA simulation because of its extremely high modularity and number of devices. FPGA makes it possible to simulate this topology efficiently, using pipelining methods. It also allows for direct connectivity of thousands of I/O points. The more complex model parts of the complete MMC system, such as iterative surge arresters, are then computed on the simulator's CPU using the SSN solver.

4.5 *The Case for CPU-Based High-PWM-Frequency Drives*

It is worth noting that high PWM frequency is NOT always a good reason to run a model on FPGA. With the time-stamping techniques described in this paper, accuracy is excellent even at large simulation time steps. Even when the sampling frequency is close to the PWM frequency, this technique still provides good accuracy by filtering out the PWM component in the simulation. It happens that most converters are designed so that PWM components are filtered out (with some exceptions such as resonant converters); therefore this PWM filtering effect often has no impact on tests. Also, the time-stamping method has an obvious advantage over average models: in HIL mode, power electronic devices can be directly driven by the controller firing pulses.

In this paper, we have shown a good example using a 3-level T-type inverter in which the model sampling frequency was only 3.33 times higher than the PWM frequency where accuracy remains high. PWM modulation is simply a way to convert DC to AC and the PWM component is not normally tested (again, there are exceptions, such as resonant converters). So, for control testing purposes, it is often acceptable to have this PWM component damped.

The inverter used in the models of this paper are of a novel generation (called TSB-RD), a kind of hybrid between interpolated switching-functions, for the active mode (similar to older generation of TSB) and binary switches (in SimPowerSystems, these are 2-state switches with very low resistance if ON or very high resistance if OFF) for the natural rectification modes. It is worth noting that older generations of TSB, in which the rectifying modes were simulated using current-nulling feedback loops, can still be used because they are sometimes more computationally efficient than using TSB-RD with SSN. That was actually the case in the *Multi-level Motor Drive with Low Input Harmonic* (Sect. 3.2).

5 Conclusions

This paper discussed the different modelling techniques used to simulate several large motor drives and converters. All these topologies have been successfully commissioned at industrial client sites.

Here the motto of HIL simulation is 'Make it work!', and this motto is additionally the reason for providing all these various models, options, decoupling techniques and solvers in ARTEMiS: to meet all the demanding client objectives.

References

1. P. Terwiesch, T. Keller, E. Scheiben, Rail vehicle control system integration testing using digital hardware-in-the-loop simulation. IEEE Trans. Cont. Syst. Technol. **7**(3), 352–362 (1999)
2. K. Palaniappan, B.J. Seibel, C. Dufour, Real time hardware in the loop validation of common bus inverter low voltage drives, in *IEEE Applied Power Electronics Conference & Exposition (APEC-2019)*, Anaheim, California, 17–21 Mar 2019

3. C. Dufour, Highly stable rotating machine models using the state-space-nodal real-time solver, in *COMPENG-2018 conference*, Florence, Italia, 10–12 Oct 2018
4. G. Li, Y. Dong, J. Tian, W. Wang, W. Li, J. Bélanger, Factory acceptance test of a five-terminal MMC control and protection system using hardware-in-the-loop method, in *Proc. of 2015 IEEE Power & Energy Society General Meeting*, Denver, USA, 26–30 July 2015
5. Z. Zhu, X. Li, H. Rao, W. Wang, W. Li, Testing a complete control and protection system for multi-terminal MMC HVDC links using hardware-in-the-loop simulation, in *IECON 2014 - 40th Annual Conference of the IEEE Industrial Electronics Society*, pp. 4402–4408
6. C. Dufour, W. Li, X. Xiao, J.-N. Paquin, J. Bélanger, Fault studies of MMC-HVDC links using FPGA and CPU on a real-time simulator with iteration capability, in *11th International Conference on Compatibility, Power Electronics and Power Engineering (IEEE CPE-POWERENG 2017*, Cadiz, Spain, 4–6 Apr 2017
7. C. Dufour, S. Abourida, J. Belanger, Real-time simulation of electrical vehicle motor drives on a PC cluster, in *10th European Conference on Power Electronics EPE*, Sep 2003
8. M. Harakawa, H. Yamasaki, T. Nagano, S. Abourida, C. Dufour, J. Bélanger, Real-time simulation of a complete PMSM drive at 10 μs time step, in *Proceedings of the 2005 International Power Electronics Conference - Niigata (IPEC-Niigata 2005)*, 2005
9. C. Dufour, J. Mahseredjian, J. Bélanger, A combined state-space nodal method for the simulation of power system transients. IEEE Trans. Power Deliv. **26**(2), 928–935 (2011). (ISSN 0885–8977)
10. C. Dufour, J. Mahseredjian, J. Bélanger, J.L. Naredo, An advanced real-time electro-magnetic simulator for power systems with a simultaneous state-space nodal solver, in *IEEE/PES T&D 2010 - Latin America*, São Paulo, Brazil, 8–10 Nov 2010
11. C. Dufour, S. Cense, J. Bélanger, An induction machine and power electronic test system on a field-programmable gate array. Math. Comput. Simul., 112–123 (2016). https://doi.org/10.1016/j.matcom.2016.03.014
12. T. Ould Bachir, C. Dufour, J. Bélanger, J. Mahseredjian, J.P. David, A fully automated reconfigurable calculation engine dedicated to the real-time simulation of high switching frequency power electronic circuits. Math. Comput. Simul. **91**, 167–177 (2013)

Concept and Implementation of a Rotor Position Detection Method for Permanent Magnet Synchronous Machines Based on Linear Halls

Yuchen Wang, Kai Liu, Wei Hua, Xiaofeng Zhu, and Baoan Wang

Abstract The purpose of this paper is to propose and implement a novel rotor position detection method for permanent magnet synchronous machines (PMSMs) based on linear Halls, which are embedded inside of stator of PMSMs. A three-phase 9-slots/8-poles PMSM is exampled to verify the method. Firstly, a special point located in stator yoke (back-iron) is found by two-dimensional finite element analysis (2D-FEA), where the open-circuit flux-density due to permanent magnets versus rotor position (B_{PM}) shows a high amplitude and good linearity, while the armature-reaction flux-density ($B_{armature}$) due to armature currents exhibits a low amplitude and good linearity versus armature currents. Then, an analytical model is built and the analytical relationship between armature currents and the $B_{armature}$ is derived. Based on the analytical mode, B_{PM} can be obtained by separating the $B_{armature}$ from the synthetic magnetic field ($B_{Synthetic}$). Thereafter, the resultant B_{PM} can be used to detect the rotor position information with differential-type piecewise-linear analytical method. The feasibility of the proposed detection method is verified by co-simulations and experiments. The simulation results show that the novel linear Hall-based angle sensor can achieve the accuracy equivalent to 3000-line. The experimental results indicate that compared with an encoder, the maximum error of electric angle position at different speeds is less than 0.3%.

1 Introduction

Owing to the requirements of high-power density, high-power efficiency, large output torque, and relatively simple control, permanent magnet synchronous machines (PMSMs) have been widely used [1, 2], where accurate rotor position detection is essential for stable and reliable operation [3]. However, in some special occasions, it is difficult to mount rotary rotor sensors, e.g., encoder, directly to the shafts of

---

Y. Wang · K. Liu · W. Hua (✉) · X. Zhu · B. Wang
EE—Southeast University, Nanjing, China
e-mail: kliu@seu.edu.cn; huawei1978@seu.edu.cn; wbaseu@seu.edu.cn

W. Zamboni, G. Petrone (eds.), *ELECTRIMACS 2019*, Lecture Notes in Electrical Engineering 615, https://doi.org/10.1007/978-3-030-37161-6_3

PMSMs. In addition, for ultra-precision position control, a sensor-less algorithm is unsuitable [4]. Compared with encoder and resolver, which are expensive and have complex coupling structure, linear Hall-based magnetic encoder can be embedded in stator and free of oil, moisture, dust, obstacle, vibration, and shock.

A linear Hall magnetic encoder-based rotor position detection scheme is to extend the rotor shaft and install a magnetized magnetic ring with sinusoidal field distribution [5, 6], which inevitably take up extra volume. Another type of linear Hall magnetic encoder [7–9] detects the magnetic leakage field of rotor-PMs at the end of the shaft, utilizing two linear Halls orthogonal embedded in stator slot or on a PCB located in the end-face of housing directly. However, the leakage field of rotor-PMs is relatively low, leading to a weak output of linear Halls and reduction of the position detection accuracy. Meanwhile, the temperature variation cannot be compensated when only two orthogonally embedded linear Halls are employed, unless two additional linear Halls are symmetrically installed to obtain the opposite signals versus the initial signals from two former linear Halls. Additional installation causes extra cost and the inaccurate installation results in unnecessary phase-shift error. In [10], three 120°-symmetric-distribution linear Halls are embedded in slots, but an extended Kalman Filter is used to eliminate the impact from temperature variation, which weakens the dynamic response due to a large amount of calculation. In general, none of the above methods consider the effect of armature current.

Hence, to resolve the contradictions above, a novel magnetic encoder with a linear Hall embedded in stator yoke is proposed in this paper, taking the influence of armature reaction into account.

2 The Principle of Angle Solving System

2.1 Selection of Magnetic Field Detection Points

Ideally, the magnetic field detected by the linear Hall should be in the form of an analytical expression versus rotor position, which can enhance the accuracy of angle decoding. Firstly, two-dimensional finite element analysis (2D-FEA) is carried out by Ansoft Maxwell to find suitable mounting locations for the linear Hall.

From the perspective of magnetic field amplitude, the amplitude of the B_{PM} should be as large as possible, while the amplitude of the B_{armature} should be as small as possible. From the perspective of solving accuracy, both the B_{PM} versus rotor position and B_{armature} versus armature currents should be linear.

A 9-slot/8-pole PMSM is exampled as shown in Fig. 1. Considering the stator is symmetrical in the circumference, nine special points are selected as magnetic field detection points within a tooth slot range.

Secondly, produced separately by PMs and armature currents, both the tangential and radial components of the flux-densities of the nine detection points are analysed. Among them, the tangential component of the B_{PM} versus rotor position is shown in

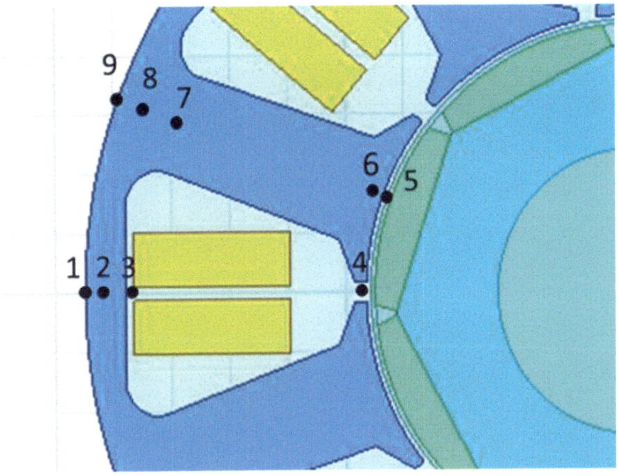

Fig. 1 Distributions of particular points in a 9-slot/8-pole PMSM

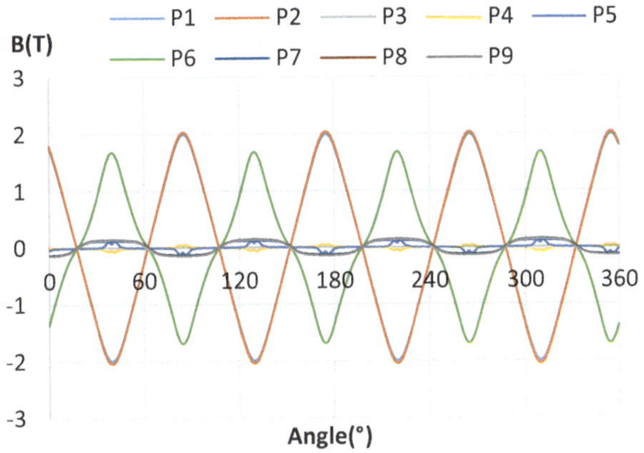

Fig. 2 The tangential component of the B_{PM} versus rotor position

Table 1 Comparison of flux-densities at points P1–P9

Magnetic field source	Amplitude of magnetic field	Waveform of magnetic field
PM	Large (567/126)	Linear (7/12)
Armature-reaction	Small (1234/356)	Sinusoidal (137/12789)

Note: In the brackets, the numbers 1–9 represent points P1–P9, radial components/tangential components

Fig. 2. The simulation results are shown in Table 1. In summary, point 2 is selected as the magnetic detection point due to its high linearity for both B_{PM} and $B_{armature}$. For convenience, 'point D' represents 'point 2' in the following.

Moreover, considering the geometrical periodic of the 9-slot/8-pole PMSM, there are other eight 'point Ds' on the stator, i.e., totally nine 'point Ds' with the detailed locations listed in Table 2. Moreover, the nine 'point Ds' are divided into three groups, e.g., D_1, D_4, and D_7 for Group 1. The three $B_{armature}$ versus time waveforms in same group are distributed symmetrically by 120° in electrical degrees, and the

Table 2 The locations of points D_1–D_9

Group	D	Locations
1	D_1	(43.8, 180°, 0)
	D_4	(43.8, 60°, 0)
	D_7	(43.8, −60°, 0)
2	D_2	(43.8, 140°, 0)
	D_5	(43.8, 20°, 0)
	D_8	(43.8, −100°, 0)
3	D_3	(43.8, 100°, 0)
	D_6	(43.8, −20°, 0)
	D_9	(43.8, −140°, 0)

Note: The radial distance from detection point to the axis is 43.8 mm

waveforms at D_1, D_2, and D_3 are sequentially different by 15° in phase. Hence, D_1, D_4, and D_7 are chosen as a group of signal detection points in the following process.

2.2 *Mathematical Model of Armature-Reaction Field*

The waveform of B_{armature} versus time of point D is essentially sinusoidal with a small amount of harmonics and the same frequency as the armature current. According to the nature of the trigonometric function, the B_{armature} can be linearly combined by two phase armature currents.

The three-phase armature currents are as follows:

$$\begin{cases} I_a = I_m \cos\theta \\ I_b = I_m \cos\left(\theta - 120^\circ\right) = I_m\left(\frac{\sqrt{3}}{2}\sin\theta - \frac{1}{2}\cos\theta\right) \\ I_c = I_m \cos\left(\theta + 120^\circ\right) = I_m\left(-\frac{\sqrt{3}}{2}\sin\theta - \frac{1}{2}\cos\theta\right) \end{cases} \tag{1}$$

where I_a, I_b, and I_c are three-phase armature currents; I_m is the amplitude of armature currents; $\theta = \omega t$; ω is the electrical angular velocity (rad/s).

Set the tangential B_{armature} of point D as:

$$\begin{aligned} B_T &= B_m \sin(\theta + \varphi) \\ &= B_m (\sin\theta\cos\varphi + \cos\theta\sin\varphi) \\ &= B_m\left(\frac{I_b - I_c}{\sqrt{3}I_m}\cos\varphi - \frac{I_b + I_c}{I_m}\sin\varphi\right) \\ &= EI_b + FI_c + G \end{aligned} \tag{2}$$

where, $E = -\frac{2B_m}{\sqrt{3}I_m}\sin\left(\varphi - 30^\circ\right)$, $F = -\frac{2B_m}{\sqrt{3}I_m}\sin\left(\varphi + 30^\circ\right)$, and G represents the error term.

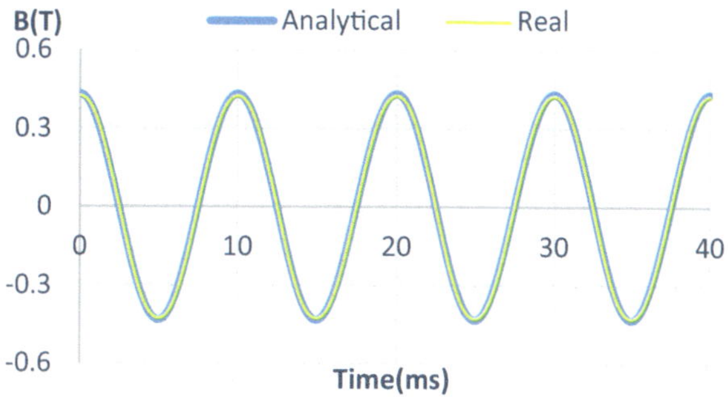

Fig. 3 Analytical result and actual FEA-predicted value of B_{armature}

The B_{armature} with armature currents of point D_1 can be fit by using the curve fitting toolbox in Matlab and the following result is obtained:

$$B_{\text{T}} = 0.004981 I_{\text{b}} - 0.004981 I_{\text{c}} - 2.7 \times 10^{-5} \tag{3}$$

The comparison between the analytical result and FEA-predicted value of B_{armature} is shown as Fig. 3, showing that the B_{armature} can be expressed linearly by the armature currents within 1.2% error.

2.3 Angle Solving Algorithm

Assuming that a series of sinusoidal signals have been obtained by separating the B_{armature} from $B_{\text{Synthetic}}$, a reasonable signal processing algorithm should be selected to generate the rotor position from these signals. The commonly used signal processing algorithms are based on (1) Arc tangent, (2) Look-up table, (3) Linear analytical, and (4) Adaptive notch filter (ANF) and phase-locked loop (PLL).

Algorithm 1 is simple to execute, but it requires a long operation time and cannot meet the requirements of dynamic response [3, 6, 7]. Algorithm 2 greatly simplifies the calculation; however, the sine function has a low resolution accuracy in the portion with a small slope, which seriously reduces the resolution accuracy [11]. Meanwhile, a large amount of index data for high accuracy will reduce dynamic response and lead to over-fitting. Algorithm 3 makes use of the linear segment at the zero crossing of a sine wave to estimate the rotor position [12]. Algorithm 4 requires additional filters, which decreases the dynamic response [9].

This paper proposes a novel differential-type piecewise-linear analytical method. After subjected to approximate differential process, the processed signal is divided into several quadrants. In each quadrant, one of the signal with the highest linearity is used to solve the rotor position and the linear analytical expressions are obtained by least squares.

For the three symmetrical signals distributed by 120°, the same DC offset ΔU is generated due to the environment, U_{m} is the amplitude of the waveform and U_{OQ} is the static output voltage from linear Hall, then:

$$
\begin{cases} U_{\mathrm{a}} = U_{\mathrm{m}} \sin(\theta) + U_{\mathrm{OQ}} + \Delta U \\ U_{\mathrm{b}} = U_{\mathrm{m}} \sin\left(\theta - 120^{\circ}\right) + U_{\mathrm{OQ}} + \Delta U \\ U_{\mathrm{c}} = U_{\mathrm{m}} \sin\left(\theta + 120^{\circ}\right) + U_{\mathrm{OQ}} + \Delta U \end{cases} \tag{4}
$$

$$
\begin{cases} U_{\mathrm{ab}} = U_{\mathrm{a}} - U_{\mathrm{b}} = \sqrt{3}U_{\mathrm{m}} \sin\left(\theta + 30^{\circ}\right) \\ U_{\mathrm{bc}} = U_{\mathrm{b}} - U_{\mathrm{c}} = \sqrt{3}U_{\mathrm{m}} \sin\left(\theta - 90^{\circ}\right) \\ U_{\mathrm{ca}} = U_{\mathrm{c}} - U_{\mathrm{a}} = \sqrt{3}U_{\mathrm{m}} \sin\left(\theta + 150^{\circ}\right) \end{cases} \tag{5}
$$

where U_{a}, U_{b}, and U_{c} are signals from three linear Halls located at points D_1, D_4, and D_7, as shown in Fig. 4; U_{ab}, U_{bc}, and U_{ca} are signals after the approximate differential process, as shown in Fig. 5.

As for Eq. (5), a six-quadrant symmetric partition is performed according to Fig. 5. In each quadrant region, the signal with highest linearity (marked with red line) will be utilized by linear analytical solving, which treats the waveform as a line segment, accelerating the dynamic response.

This novel method eliminates the DC offset without adapting addition differential linear Hall and increases the signal amplitude to $\sqrt{3}$ times, improving the

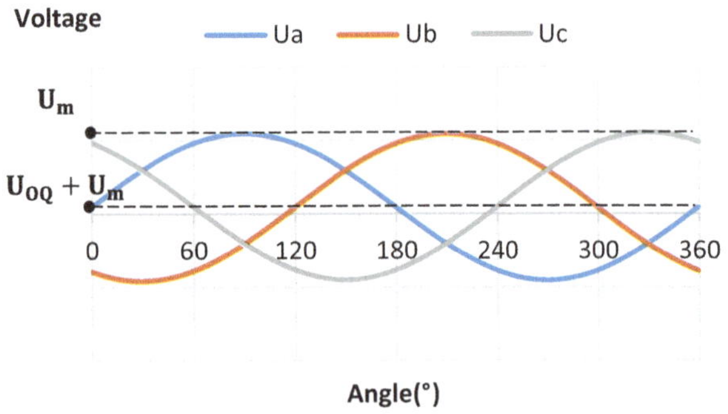

Fig. 4 The initial signals

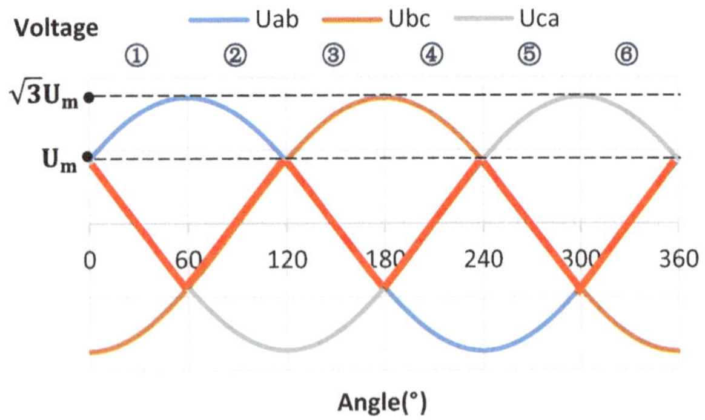

Fig. 5 The quadrant diagram of processed signals

solving accuracy correspondingly. It coordinates the contradiction between dynamic response, accuracy, and cost.

2.4 Angle Solving System Diagram

The rotor position can be obtained, according to the flow chart (Fig. 6), where two sets of raw data are needed in angle solving system

1. Three-phase armature currents are obtained by high-precision current sensor detection.
2. The voltage signals of Hall group are derived from three linear Halls embedded in the stator.

Then,

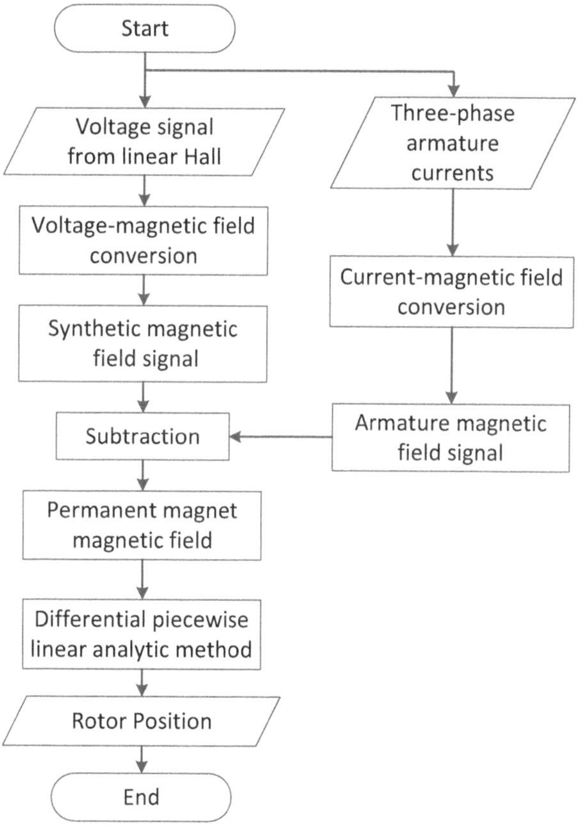

Fig. 6 Angle solving system diagram

3 Simulation and Analysis

In this simulation, the mechanical speed of the PMSM is 1500 rpm, which means the electrical angular velocity $\omega_e = 200\pi$ rad/s; the amplitude of armature currents $I_m = 50$ A; for three-phase symmetrical armature current, $\theta = 0\ or \pm 2\pi/3$; $t_{simulation}$ means simulation time; *step* means simulation time step.

3.1 Fitting Coefficients Acquisition

Before solving the rotor position, fitting coefficients of $B_{armature}$ versus armature currents is required, i.e., the current-magnetic field conversion shown in Fig. 6.

The 2D-FEA is carried out by Ansoft Maxwell under the condition: (1) $t_{simulation} = 0.01$ s; (2) $step = 0.00001$ s; (3) $I = (1 - e^{-t/0.04})I_m \sin(\omega_e t + \theta)$; (4) The remanence of PM is 0 T.

The waveforms of $B_{armature}$ at D_1, D_4, and D_7 are shown in Fig. 7. The $B_{armature}$ at nine points are separately fit to armature currents, obtaining a coefficient matrix as shown in Table 3. E and F represent the fitting coefficients mentioned in Eq. (2).

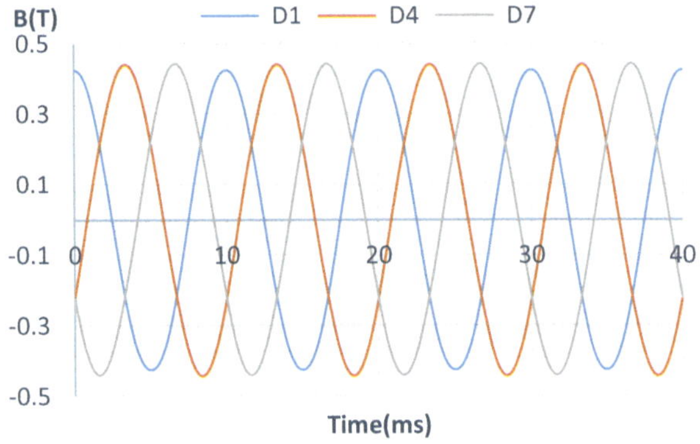

Fig. 7 $B_{armature}$ at D_1, D_4, and D_7

Table 3 Fitting coefficients of $B_{armature}$ versus armature currents

Number of points	E	F
1	0.004981	−0.004981
2	−0.01015	−0.00214
3	0.008197	−0.00192
4	−0.01012	−0.004981
5	0.007979	0.009931
6	−0.009931	−0.007979
7	0.004982	0.01012
8	0.001919	−0.008198
9	0.00214	0.01015

3.2 Angle Solving Process

The 2D-FEA is carried out by Ansoft Maxwell under the condition: (1) $t_{\text{simulation}} = 0.04$ s; (2) $step = 0.00005$ s; (3) $I = (1 - e^{-t/0.04})I_m \sin(\omega_e t + \theta)$.

Then, two sets of raw data can be obtained: (1) synthetic flux-density and (2) three phase armature currents.

After confirming the value of B_{armature} by the expression of armature currents, which is mentioned earlier, B_{PM} is obtained by separating the B_{armature} from the $B_{\text{Synthetic}}$.

Via differential piecewise linear analytical method, the comparison between angle decoding result and the actual rotor position is shown in Figs. 8 and 9.

The waveform in Fig. 9 can be regarded as the combination of basic waveform with same amplitude and spike waveform with increasing amplitude. There are eight pairs of basic waveforms in Fig. 9, which are independent of the armature currents. The spike waveform is related to the incremental current value, which is caused by the error of fitting coefficient in Table 3. The simulation error between decoding result and actual value of rotor angle is 0.12 degrees, reaching a resolution level of 3000 lines.

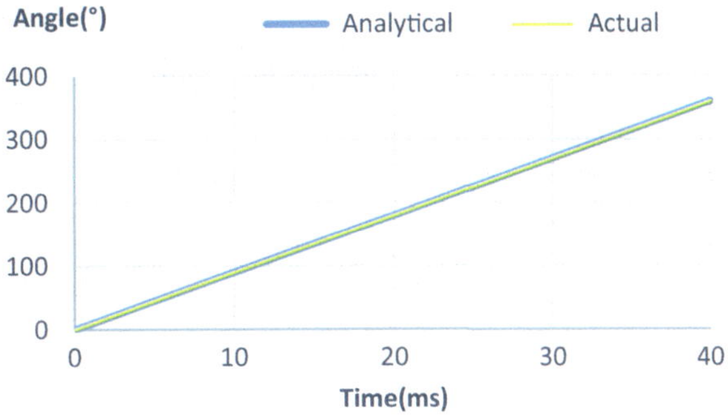

Fig. 8 The comparison between decoding result and actual value of rotor position

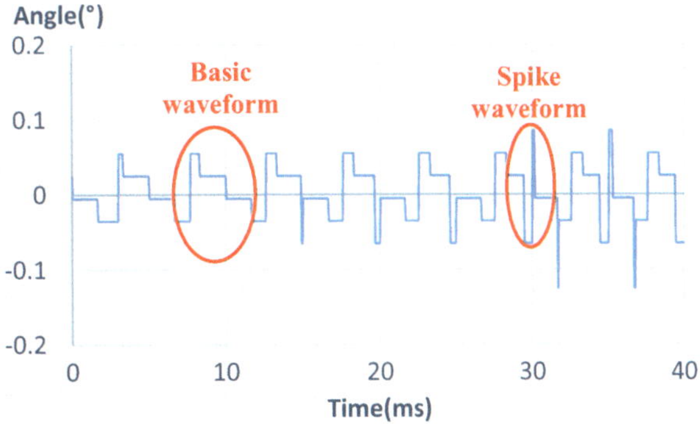

Fig. 9 The rotor angle error between decoding result and actual value

4 Experiment

4.1 Experiment Platform

In order to verify the feasibility of this novel angle detection system, the PMSM, the permanent magnet DC motor and the rotary encoder are coaxially connected and mounted on a bench, as shown in Fig. 10, and the assembled view of linear Halls is shown in Fig. 11.

(1) 24 V regulated power supply module, (2) OMRON rotary encoder with the accuracy of 2000 lines, (3) Permanent magnet DC motor (Drives PMSM to imitate no-load operation or serves as the load of PMSM), (4) Three-phase 6-slots/4-poles PMSM (Three linear Halls are embedded in the stator yoke which are marked with dashed ellipse), (5) Permanent magnet DC motor driver, (6) Digital signal processor and PMSM driver.

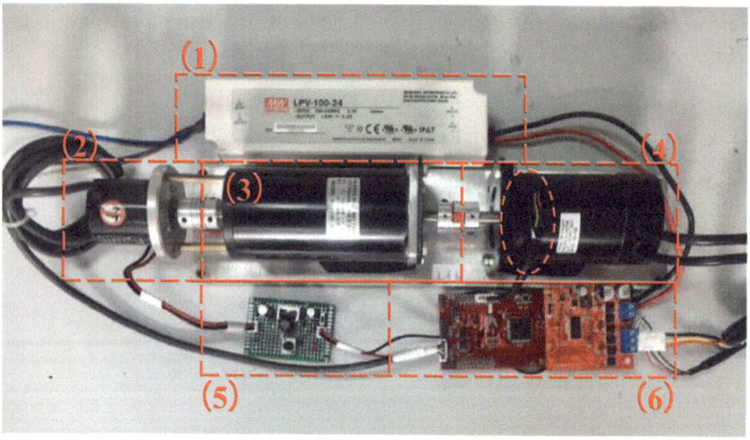

Fig. 10 Experiment platform

Fig. 11 Assembled view of linear Halls at the stator yoke

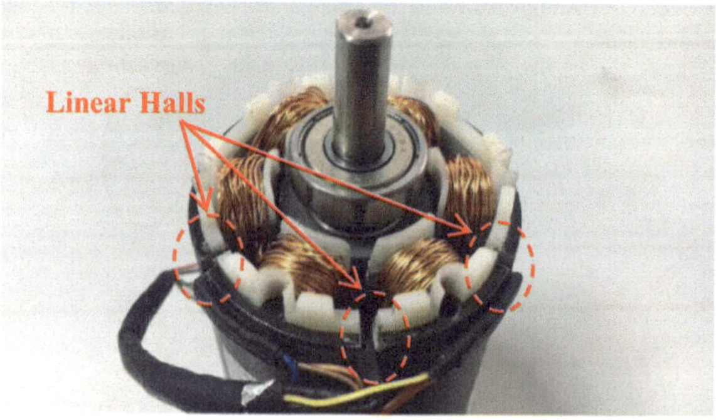

4.2 *Experiment Results*

Power the permanent magnet DC motor and drag the PMSM to intimate the no-load operation of PMSM, observing the voltage signals from three linear Halls (Fig. 12), angular output from rotary encoder, and novel angle solving system (Fig. 13).

Since the install holes of the linear Halls are manually punched, there exists mechanical asymmetry of three mounting holes, resulting in asymmetry of the three-phase waveform, shown in Fig. 12. The width deviation of mounting holes causes waves' difference in amplitude, and the radial deviation of the mounting holes causes waves' difference in phase. Considering that the peak-to-peak value of signal is up to 1.415 V, the decoding accuracy can be improved by precise installation and replacing small-scale linear Halls.

This novel encoder prototype has a maximum electrical angle error of 0.3%, which is mainly derived from the asymmetry of the mounting holes. Although the error at switch point is inevitable, it can be eliminated as much as possible by setting the calibration angle between encoder and this novel encoder properly. Furthermore, the dynamic response is almost the same as using a rotary encoder.

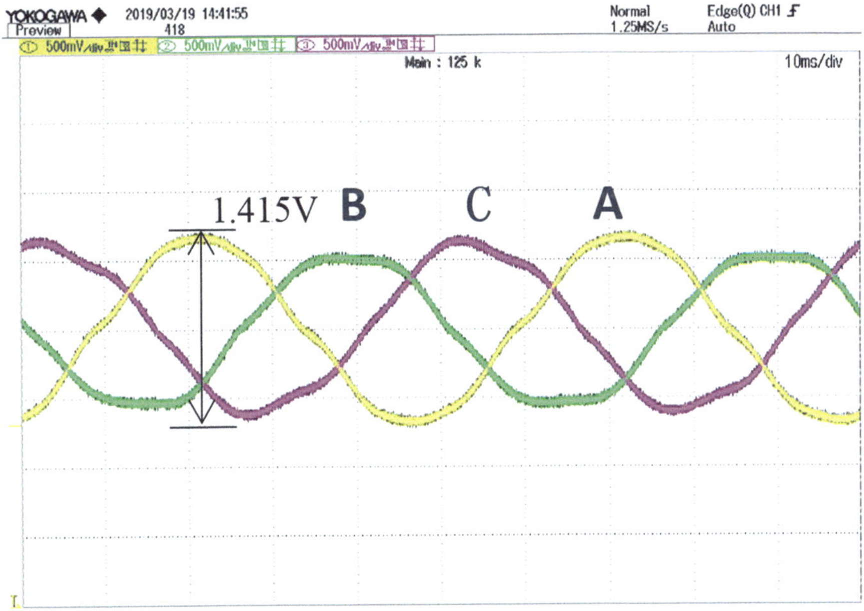

Fig. 12 Linear Hall group's output under no-load operation

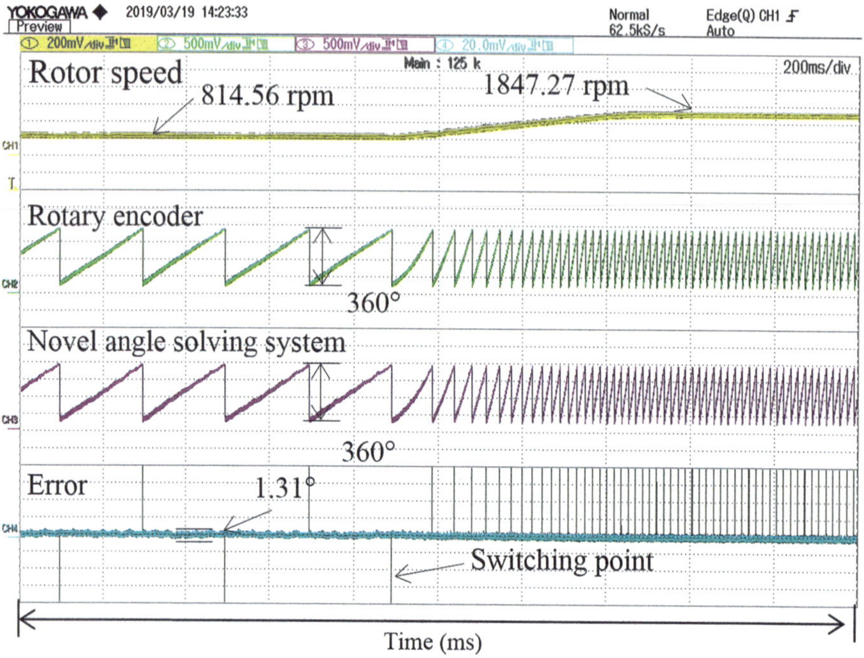

Fig. 13 The comparison of angular outputs between rotary encoder and novel angle solving system at different speeds

5 Conclusions

The linear Hall-based embedded angle sensor proposed in this paper is verified by simulation and experiment, showing that the simulation decoding resolution reaches 3000 lines and the novel encoder prototype has a maximum electric angle error of 0.3% under no-load operation. It reduces the cost of the angle sensor and saves the sensor installation space while maintaining a certain level of resolution. This preliminary experiment does not take the effect of armature reaction currents into account. So the novel linear Hall-based angle sensor should be validated by further experiments under various operating conditions.

Acknowledgment This work was supported in part by the National Natural Science Foundation of China under Grant 51825701 and Key R&D Program of Jiangsu Province under Grant BE2019073.

References

1. B.-G. Gu, J.-H. Choi, I.-S. Jung, Development and analysis of interturn short fault model of PMSMs with series and parallel winding connections. IEEE Trans. Power Electron. **29**(4), 2016–2026 (2014)
2. K. Jezernik, J. Korelic, R. Horvat, PMSM sliding mode FPGA-based control for torque ripple reduction. IEEE Trans. Power Electron. **28**(7), 3549–3556 (2013)
3. X. Song, F. Jiancheng, H. Bangcheng, High-precise rotor position detection for high-speed surface PMSM drive based on linear hall-effect sensors. IEEE Trans. Power Electron. **31**, 4720–4731 (2015)
4. C.G. Anisha, A. Parthan, Position and speed control of BLDC motor using Hall sensor. Int. J. Eng. Res. **3**(32), 3 (2015)
5. Q. Lin, T. Li, Z. Zhou, Error analysis and compensation of the orthogonal magnetic encoder, in *2011 First International Conference on Instrumentation, Measurement, Computer, Communication and Control, Beijing, China*, (2011), pp. 11–14
6. J. Hu, J. Zou, F. Xu, Y. Li, Y. Fu, An improved PMSM rotor position sensor based on linear Hall sensors. IEEE Trans. Magn. **48**(11), 3591–3594 (2012)
7. S. Jung, B. Lee, K. Nam, PMSM control based on edge field measurements by Hall sensors, in *2010 Twenty-Fifth Annual IEEE Applied Power Electronics Conference and Exposition (APEC), Palm Springs, CA, USA*, (2010), pp. 2002–2006
8. Y.F. Shi, Z.Q. Zhu, D. Howe, EKF-based hybrid controller for permanent magnet brushless motors combining Hall sensors and a flux-observer-based sensorless technique, in *IEEE International Conference on Electric Machines and Drives, 2005, San Antonio, TX, USA*, (2005), pp. 1466–1472
9. S.-T. Lee, Y.-K. Kim, J. Hur, Pseudo-sensorless control of PMSM with linear Hall-effect sensor, in *2017 IEEE Energy Conversion Congress and Exposition*, (ECCE, Cincinnati, OH, 2017), pp. 1896–1900
10. A. Simpkins, E. Todorov, Position estimation and control of compact BLDC motors based on analog linear Hall effect sensors, in *Proceedings of the 2010 American Control Conference, Baltimore, MD*, (2010), pp. 1948–1955
11. W.U. Jie, C. Feihong, W. Tianlei, et al., Permanent magnetic synchronous motor drives control system based on magnetic encoder. Large Elec. Mach. Hydraulic Turbine. **31**, pp. 18–22 (2017)
12. Liu ZT. Researches on the single-pole magnetic encoder based on HALL sensor. (2016)

A Discrete-Time Robust MRAC Applied on Grid-Side Current Control of a Grid-Connected Three-Phase Converter with LCL Filter

Paulo Jefferson Dias de Oliveira Evald, Rodrigo Varella Tambara, and Hilton Abílio Gründling

Abstract This paper proposes a discrete-time control strategy of grid-side currents of a three-phase grid-connected converter with LCL filter. The implemented control system, a RMRAC (Robust Model Reference Adaptive Control) by state feedback, presents robustness to parametric uncertainties and rejection of periodic disturbances. To demonstrate the controller performance, numerical simulation results, considering parameters of a real plant, are presented.

1 Introduction

Converters are devices of great importance, with a wide applicability in grid-connected systems, being the main integration form among renewable energy sources and electrical grids [1–3]. In order to connect the energy source to the grid, L or LCL filters are often used. Comparing both filters, the connection with an LCL filter is more advantageous due to its greater attenuation of the high frequency harmonics, from the pulse width modulation at the converter output voltage [4], which is −60 dB/decade, while the L filter attenuates −20 dB/decade. Furthermore, although the LCL filter has a more complex topology, it is built with smaller reactive elements, which results in a lower cost filter, weight, switching frequency and reactive power. However, the use of this filter may induce the current control loop to instability [5], especially when the grid impedance at PCC (Point of Common Coupling) is unknown and may cause a high resonance peak [6]. Consequently, to ensure equilibrated grid-side currents, it is indispensable to use a reliable current controller.

Several works dedicated to the control of grid-connected converter dynamics are reported in literature. In [7], a robust adaptive control, in discrete time, was implemented for controlling the grid-side currents. In the experiments, only same sensors were needed, due to the use of a modified least squares algorithm integrated

P. J. D. de O. Evald (✉) · R. V. Tambara (✉) · H. A. Gründling (✉)
Federal University of Santa Maria, Santa Maria, Rio Grande do Sul, Brazil

W. Zamboni, G. Petrone (eds.), *ELECTRIMACS 2019*, Lecture Notes in Electrical Engineering 615, https://doi.org/10.1007/978-3-030-37161-6_4

in the control system. Moreover, it was considered the time-delay in the control project. The experimental results demonstrated high performance.

Already in [8], an optimal digital control strategy was implemented to compensate time-delay related to analogical-digital conversions and computations of control algorithm. The proposed method is based on the estimation of future values of the plant states. Experimental results were presented to validate the proposed time delay compensation strategy. Other approaches to control grid-side currents of grid-tied converters are found in the literature, such as: PI controller, predictive control and more.

In this bias, the main contribution of this work is the proposal of a RMRAC (Robust Model Reference Adaptive Control) by state feedback to regulate the grid-side currents of a three-phase grid-connected converter with LCL filter. This controller, because of its adaptive nature, performs better than fixed-gain controllers that are designed for a specific operation point, which have their performance degraded when the grid inductance varies. Moreover, this controller was improved with adaptive rejection to periodic disturbances, which fixed-gain controllers cannot be able to handle in variables operation points.

The organisation of this work is given as follows: in Sect. 2 is presented the model of three-phase grid-connected converter with LCL filter, filter in $\alpha\beta 0$-coordinates, followed by the RMRAC mathematical background and its implementation in Sects. 3 and 4, respectively. Finally, in Sect. 5, simulation results are discussed and the final considerations are presented in Sect. 6.

2 Modelling of Three-Phase Grid-Connected Converter with LCL Filter

To facilitate the modelling of this system, some hypotheses are assumed as true, they are:

- (H1) the electric grid is assumed to be predominantly inductive, modelled by a sinusoidal source V_d in series with an inductance L_{g2} and resistance R_{g2}
- (H2) the input bus is assumed to be stabilised and represented by a direct voltage source;
- (H3) the filter output voltage and PCC voltage have ensured synchronism;
- (H4) the converter switches are considered ideal.

The electrical diagram of this system is shown in Fig. 1. It is worth noting that the equivalent LCL circuit is represented by a Thevenin equivalent to the PCC. Moreover, R_c and L_c are the resistance and inductance of converter, respectively. Besides, C is the LCL filter capacitance, $L_g = L_{g1} + L_{g2}$ and $R_g = R_{g1} + R_{g2}$ are the total inductance and total resistance of the grid, respectively. Furthermore, i_c, v_c and i_g are converter current, capacitor voltage and grid current, respectively. These three variables, in each phase (a, b and c), are measured (indicated in grey colour in

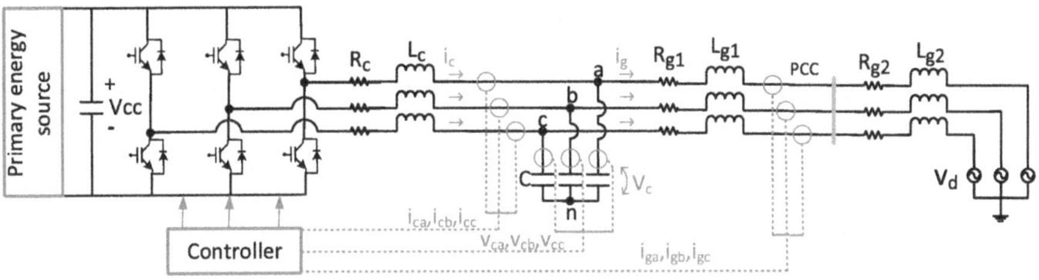

Fig. 1 Electrical diagram of the three-phase grid-connected converter with LCL filter

the diagram) and form the states vector, $\boldsymbol{x}_{abc}$, in abc-coordinates.

$$\boldsymbol{x}_{abc} = [i_{ca}\ i_{cb}\ i_{cc}\ v_{ca}\ v_{cb}\ v_{cc}\ i_{ga}\ i_{gb}\ i_{gc}]^T. \tag{1}$$

As the model in abc-coordinates, obtained by Kirchhoff's laws, is highly coupled, the control project becomes a complex task. To simplify the control project, the system in abc-coordinates has been transformed to $\alpha\beta0$-coordinates, which results in two identical single-phase decoupled systems, one associated with the α axis and other associated with the β axis. As the system is considered equilibrated, there is no way to drive current in the 0 axis, being able to disregard it. The new state vector for each coordinate is formed by the following variables:

$$\boldsymbol{x}_\alpha = [i_{c_\alpha}\ v_{c_\alpha}\ i_{g_\alpha}]^T \quad \text{and} \quad \boldsymbol{x}_\beta = [i_{c_\beta}\ v_{c_\beta}\ i_{g_\beta}]^T. \tag{2}$$

The transfer function of the decoupled single-phase system, in coordinates α (and also in β, because they are identical), considering $v_d = 0$, is shown as follows [7]:

$$\frac{i_g(s)}{u(s)} = \frac{\frac{1}{L_g L_c C}}{s^3 + \frac{(R_g L_c + R_c L_g)C}{L_g L_c C} s^2 + \frac{L_c + L_g + R_g R_c C}{L_g L_c C} s + \frac{R_g + R_c}{L_g L_c C}}, \tag{3}$$

where $u(s)$ is the voltage synthesised by converter. In the next section is given the mathematical background for the grid-side current control.

3 Robust Model Reference Adaptive Control

In this section is presented the control theory and it is emphasised that vectors and matrices are in bold. The RMRAC evolved from the classical model reference control [9]. This control strategy uses a reference model defined with the same relative degree as the plant. The reference model is designed according to the dynamics that are desired to impose on the controlled plant. Thus, the error is given between reference model output and plant output.

The main differences between RMRAC and its predecessor are that it has adaptive gains and imposes robustness to the system. Thus, the parametric uncertainties and non-modelled dynamic are properly handled. For more information about this control strategy and its stability proofs, is indicated the reading of [9].

Since the plant to be controlled consists of two identical single-phase decoupled systems, it is possible to write it as two identical single-input single-output systems, as follows:

$$y_p(z) = G(z)u_p(z) : \tag{4}$$

where

$$G(z) = G_0(z)[1 + \mu\Delta_m(z)] + \mu\Delta_a(z) \tag{5}$$

and

$$G_0(z) = k_p \frac{Z_0(z)}{R_0(z)}, \tag{6}$$

where $G(z)$ is the system transfer function, $G_0(z)$ is the modelled part of plant, $\mu\Delta_a$ and $\mu\Delta_m$ are the additive and multiplicative non-modelled dynamics, respectively. Furthermore, k_p is a gain, $Z_0(z)$ and $R_0(z)$ are monic polynomials with degree m_0 and n_0, respectively. The relative degree of the modelled plant part is $n_0^* = n_0 - m_0$.

The modelled part of the plant G_0(z), must satisfy the following conditions [9]:

1. The signal k_p and the values of m_0 and n are known and with no loss of generality $k_p > 0$;
2. $Z_0(z)$ is a Schur polynomial with degree $m_0 \leq n_0 - 1$;
3. $\Delta_m(z)$ is a stable transfer function and $\Delta_a(z)$ is a stable transfer function and is strictly proper ($m_0 < n_0$);
4. The lower bound $p_0 > 0$ is marginally stable, whose poles of $\Delta_a(z - p)$ and $\Delta_m(z - p)$ are stable and known.

Therefore, the control objective is defined as: given a reference model, shown in (7), from a defined control project, so that, for some $\mu^* > 0$ and any $\mu \in [0, \mu^*]$, the closed-loop control will be globally stable and plant output, $y_p(z)$, will track the reference model output, $y_m(z)$, as close as possible, even if the plant is subject to non-modelled dynamics, $\Delta_m(z)$ and $\Delta_a(z)$, taking into account the presented hypotheses [9].

$$y_m(z) = W_m(z)r(z) = k_m \frac{Z_m(z)}{R_m(z)} r(z), \tag{7}$$

where $W_m(z)$ is the reference model transfer function, $Z_m(z)$ and $R_m(z)$ are the monic Schur polynomials with degree m_m and n_m, respectively. The relative degree of the $W_m(z)$ is $n_m^* = n_m - m_m$ and is equal to the relative degrees of $W_m(z)$.

Moreover, k_m is a gain and $r(z)$ is a uniformly limited reference signal, which defines the control signal $u(z)$.

In the structure of RMRAC by state feedback, the states must be measurable or estimated. These variables form the $\boldsymbol{\omega}$ vector as follows:

$$\boldsymbol{\omega}_i(k) = \boldsymbol{x}_i(k) \ , \ i = 1, \ldots, n, \tag{8}$$

where n is the plant order and $\boldsymbol{x}_i(k)$ are plant states.

The control action $u_p(k)$ is shown in (9).

$$u_p(k) = \boldsymbol{\theta}(k)^T \boldsymbol{\omega}(k), \tag{9}$$

where $\boldsymbol{\theta}(k) = [\boldsymbol{\theta}_1(k), \boldsymbol{\theta}_2(k), \theta_3(k), \theta_r(k)]^T$ represents the control vector parameters. The algorithm for gains adaptation, considering $v_d = 0$, is presented in the next section.

3.1 Parametric Adaptation Algorithm

The parametric adaptation algorithm must be able to ensure that system remains stable, even in the presence of external disturbances and non-modelled dynamics [9].

In this work, the gradient algorithm was implemented, because it presents high performance and due to its simple structure, which implies on low computational cost. This algorithm uses as parameters: the control action, plant output, states and tracking error, to perform adjustment in the adaptive gains vector $\boldsymbol{\theta}(k)$. The adaptation law of this vector is shown as follows:

$$\boldsymbol{\theta}(k+1) = \boldsymbol{\theta}(k) - T_s \sigma(k) \Gamma \boldsymbol{\theta}(k) - T_s sign(C_0) \frac{\varepsilon(k) \Gamma \boldsymbol{\zeta}(k)}{m^2(k)} \tag{10}$$

$$sign(C_0) = sign\left(\frac{k_m}{k_p}\right) \tag{11}$$

$$m^2(k) = 1 + \boldsymbol{\zeta}(k)^T \boldsymbol{\zeta}(k), \quad m(0) = 1 \tag{12}$$

$$\boldsymbol{\zeta}(k) = W_m(z) \boldsymbol{\omega}(k) \tag{13}$$

$$e_1(k) = y_p(k) - y_m(k) \tag{14}$$

$$\nu(k) = W_m(z) u_p(k) \tag{15}$$

$$\varepsilon(k) = e_1(k) + \boldsymbol{\theta}(k)^T \boldsymbol{\zeta}(k) - \nu(k), \tag{16}$$

where $e_1(k)$ is the tracking error, given between the plant output and the reference model output, and $\varepsilon(k)$ is the augmented error. The function $sign$ is used to obtain

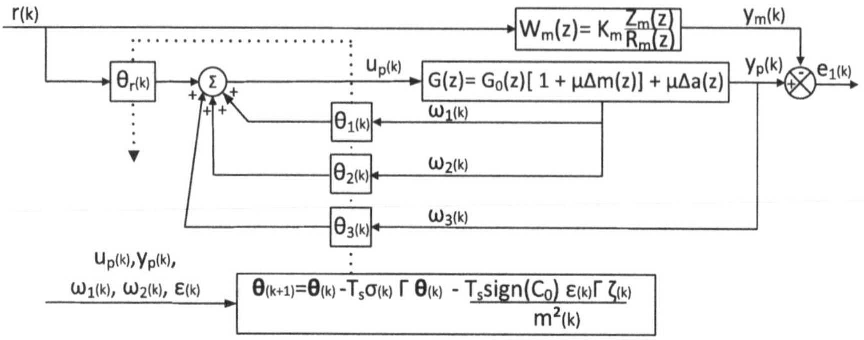

Fig. 2 RMRAC block diagram

the signal of high frequency gains ratio. Moreover, $\nu(k)$, $\boldsymbol{\zeta}(k)$ and $\boldsymbol{\omega}$ are the control signal (u_p) filtered by $Wm(z)$, a regressor filter and state vector, respectively.

The parameter Γ, which must be positive, defines the convergence speed of the plant response with respect to the reference model output. However, if it is too high, it may occur over-stressing or even inducing the system to instability. In addition, the normaliser $m^2(k)$ can be modified according to the project need. Also, it was incorporated into the parametric adaptation algorithm the σ-modification, to increase the robustness of this algorithm [9]. This modification is shown in (17).

$$\sigma(k) = \begin{cases} 0 & \text{if} \quad \|\boldsymbol{\theta}(k)\| < M_0 \\ \sigma_0 \left(\frac{\|\boldsymbol{\theta}(k)\|}{M_0} - 1 \right) & \text{if} \quad M_0 \leq \|\boldsymbol{\theta}(k)\| < 2M_0 \\ \sigma_0 & \text{if} \quad \|\boldsymbol{\theta}(k)\| \geq 2M_0, \end{cases} \tag{17}$$

where $M_0 > \|\boldsymbol{\theta}^*\|$ is the superior bound of norm of $\boldsymbol{\theta}$, overdimensioned due to the nonknowledge of $\|\boldsymbol{\theta}^*\|$ and σ_0 is the maximum value of modification function.

To aggregate periodic disturbance rejection, the control law was increased, incorporating these dynamics into the control action, as discussed in the next subsection. The RMRAC block diagram is shown in Fig. 2. Note that the dotted arrows represent the adaptation of the parameters. Within the dashed box is the gradient parametric adaptation algorithm, used in this work.

3.2 *Rejection of Periodic Disturbances*

In real applications, the presence of periodic external disturbances on the plant is often identified. These disorders can impair the control law performance if not properly treated. One of the ways to attenuate these disturbances is through their measurement and incorporation into control law.

Considering a measurable sinusoidal disturbance $d(k)$, given by $d(k) = A\sin(\omega_d k T_s + \phi)$, where A, ω_d and ϕ are the amplitude, frequency and phase of the disturbance, respectively. Then, the disturbance can be rewritten as:

$$d(k) = A_c\cos(\omega_d k T_s) + A_s\sin(\omega_d k T_s) \tag{18}$$

which is compensated by the control law shown in (19). Note that the gains from this control action are also adaptive and calculated in the same way as the other $\boldsymbol{\theta}$ gains.

$$u_d(k) = \theta_c(k)\cos(\omega_d k T_s) + \theta_s(k)\sin(\omega_d k T_s). \tag{19}$$

Thus, this control action is incorporated to the control law u_p, in each coordinate α and β, resulting in:

$$u(k) = u_p(k) + u_d(k). \tag{20}$$

4 Control Project

In this section, the control project will be presented. Firstly, the reference model was defined. As the plant, in continuous-time, has relative degree equal to 3, the chosen model has also relative degree equal to 3. The reference model is shown below.

$$W_m(z) = \frac{(1-0.15)^3}{(z-0.15)(z-0.15)(z-0.15)}. \tag{21}$$

The plant was discretised, with Z transformation, considering a zero order holder (ZOH), with sampling time of 1/4800 s. As previously mentioned, the model in the α coordinate is identical to the β coordinate model. So, the control design for both is the same, as well as its transfer function shown in (22). The model, considering time-delay due digital systems processing, and controller parameters are presented in Tables 1 and 2, respectively.

$$G_0(z) = 0.05342\frac{z^2 + 3.4743z + 0.984}{z(z^3 - 0.9843z^2 + 0.9806z - 0.9672)} \tag{22}$$

Table 1 Plant parameters

Symbol	Parameter	Value
L_c	Converter-side inductance	1 mH
R_c	Converter-side resistance	0.05 Ω
L_g	Grid-side inductance	0.5 mH
R_g	Grid-side resistance	1 mΩ
C	LCL filter capacitance	60 μF

Table 2 Controller parameters

Parameter	Value	Parameter	Value
Γ	500	$\boldsymbol{\theta}(0)$	$[0\,0\,0\,0\,0\,0]^T$
σ_0	0.1	$\boldsymbol{\zeta}(0)$	$[0\,0\,0\,0\,0\,0]^T$
M_0	5	$\boldsymbol{\omega}(0)$	$[0\,0\,0\,0\,0\,0]^T$
$m^2(0)$	1	$A_c = A_s$	15

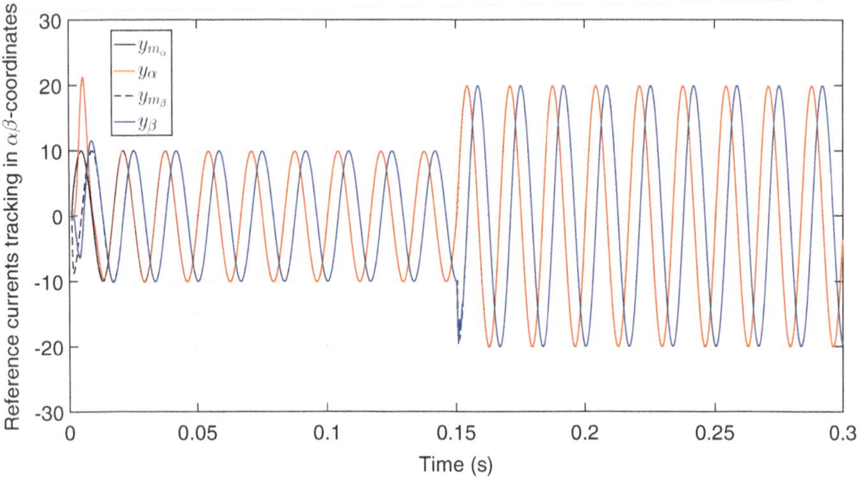

Fig. 3 Currents tracking in $\alpha\beta$-coordinates

5 Simulation Results

In this section we present the simulation results of the projected control strategy. First, in Fig. 3 are shown the grid-side currents in $\alpha\beta$-coordinates, tracking their respective reference model outputs. Initially, it is considered that system is free of parametric variations and without disturbances, that is, disconnected from the grid. The reference current for α-coordinate has amplitude equal to 10 A and 60 Hz. Then, at 0.15 s, the amplitude is changed to 20 A. The references for β-coordinate are the same as α-coordinate references, however, with 90° in phase delay. For the other simulations, the reference amplitude is maintained at 10 A.

As can be seen in Fig. 3, both grid-side currents quickly tracked their respective reference model outputs, presenting a very small error in steady state in less than 0.1 s. The only high errors observed in the simulation are the initial errors and in the reference change, which presented a considerable overshoot. However, these errors converged fast to zero.

Furthermore, note that, error in the reference change was less than initial error. It occurred due to $\boldsymbol{\theta}$ gains initialisation, which are null in the first instants, so far from ideal $\boldsymbol{\theta}$ gains, and in the reference change they are closer to the ideal $\boldsymbol{\theta}$ gains. This adaptation is clearly observed in Fig. 4, where there is initially a large movement of

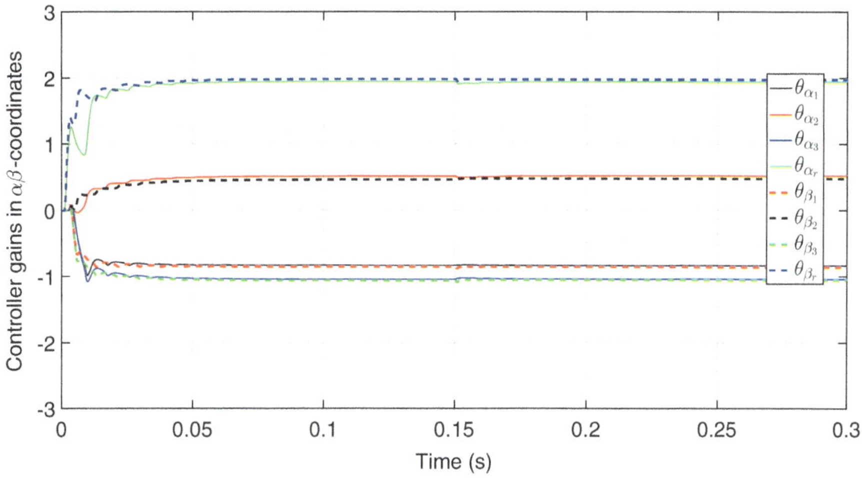

Fig. 4 Gains adaptation in $\alpha\beta$-coordinates

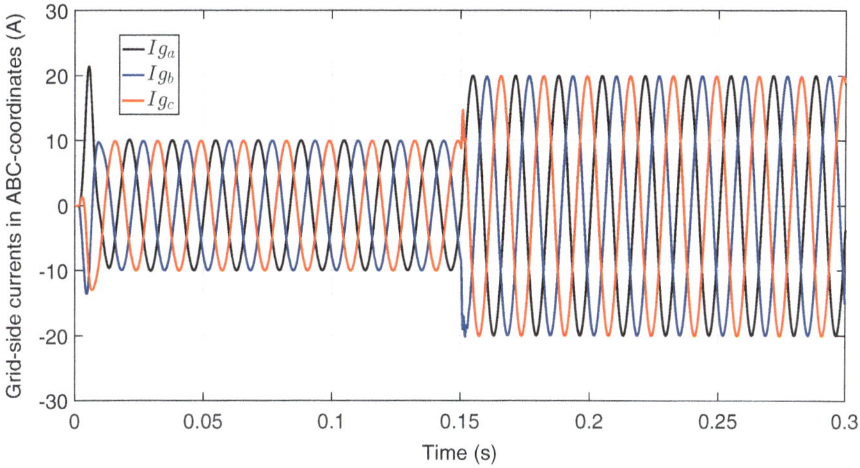

Fig. 5 Grid-side currents in *abc*-coordinates

$\boldsymbol{\theta}$ gains, however, after they enter steady state, only a brief adaptation occurs for the system tracking the new references. The grid-side currents in *abc*-coordinates are shown in Fig. 5, where it is notable that the reference was adequately tracked.

To evaluate the controller performance, simulations were carried out imposing parametric variations to the model. The parametric variation was imposed to the grid-side inductance, with ten times the nominal value, starting at 0.15 s. In the physical plant, the total grid-side inductance is uncertain, due to the grid inductance variations.

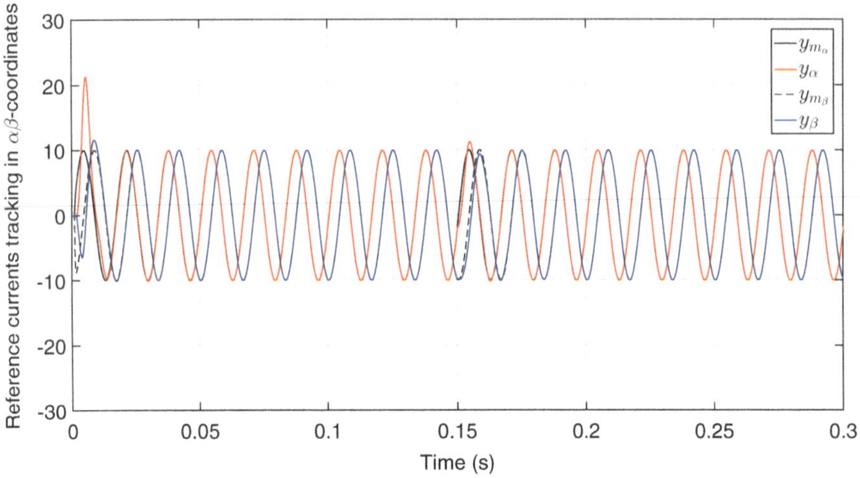

Fig. 6 Currents tracking in $\alpha\beta$-coordinates, considering parametric variation in grid inductance

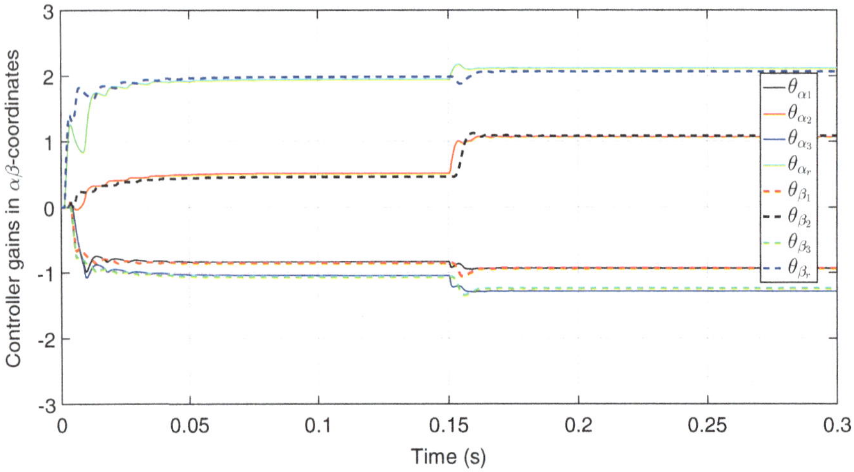

Fig. 7 Gains adaptation in $\alpha\beta$-coordinates, considering parametric variation in grid inductance

In Fig. 6 are shown the grid-side currents in $\alpha\beta$-coordinates. Besides, Fig. 7 shows controller gains adaptation. Note that, although it was considered a high value in the parametric variation of grid inductance, even though the control maintained a good performance. Next, in Fig. 8 are shown the grid-side currents in *abc*-coordinates.

Finally, simulations results considering periodic disturbances were presented. The disturbances are grid-side current perturbation are sinusoidal signals, with frequency f equal to 60 Hz, phase delay equal to 120° each other, in *abc*-coordinates and v_{pcc} amplitude equal to 311 V. The amplitude is derived from the

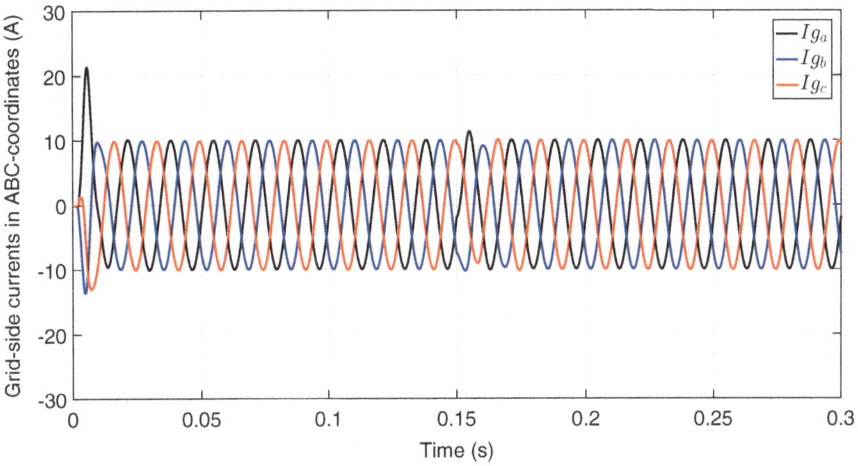

Fig. 8 Grid-side currents in *abc*-coordinates, considering parametric variation in grid inductance

voltage at point of common coupling of the plant with the grid. In order to obtain such a voltage, the presence of a transformer with variable taps connected to the grid was considered. Thus, the disturbance amplitude was gradually elevated (10 V each 0.005 s during the 0.015 s initial instants, next 30 V each 0.005 s and the last change was 11 V to reach the 311 V), avoiding an excessive overshoot in the grid-side current. The disturbance, with $f = 60\,\text{Hz}$, $\omega_d = 2\pi f$ and $t = kT_s$, is described by

$$v_{abc} = v_{pcc}\left[\sin(\omega_d t) \quad \sin\left(\omega_d t - \frac{2\pi}{3}\right) \quad \sin\left(\omega_d t + \frac{2\pi}{3}\right)\right]^T. \tag{23}$$

Figure 9 shows currents tracking in $\alpha\beta$-coordinates and the gains adaptation is shown in Fig. 10. Note that, although the initial transient response is oscillatory, the maximum amplitude did not reach 20 A and the disturbances were quickly compensated. This fast compensation can be seen in the gains accommodation to each increase in the value of the disturbances in Fig. 10. The grid-side currents in *abc*-coordinates are shown in Fig. 11.

6 Conclusion

In this work, the implementation of a Robust Model Reference Adaptive Control by state feedback, in discrete time, for grid-side current reference tracking of a three-phase grid-connected converter with LCL filter was presented. Simulation results shown that this controller is practicable. Its performance presented very small steady-state current tracking error and is robust to parametric variations in

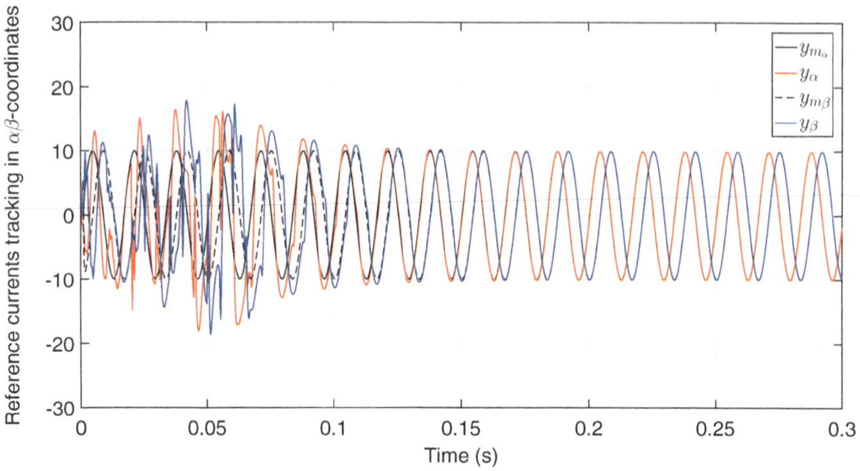

Fig. 9 Currents tracking in $\alpha\beta$-coordinates, considering periodic disturbances

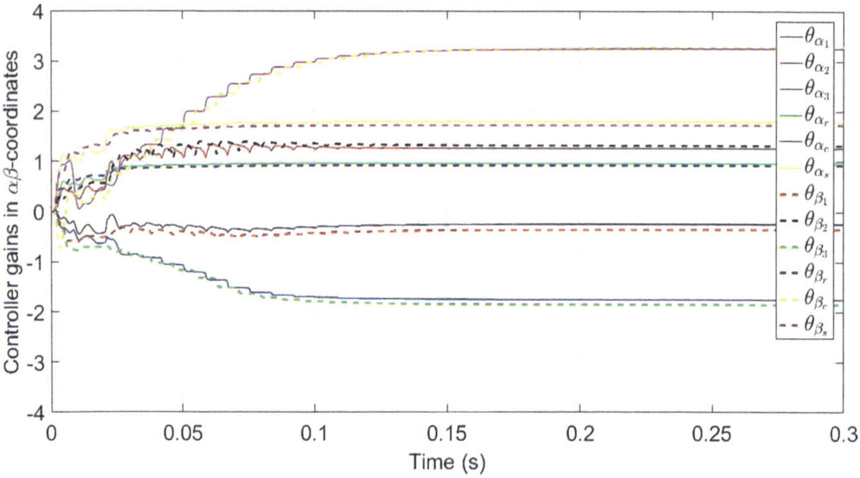

Fig. 10 Gains adaptation in $\alpha\beta$-coordinates, considering periodic disturbances

grid inductance and periodic disturbances rejection. The next step of this study is to implement the controller in a DSP (Digital Signal Processor) for experimental validation.

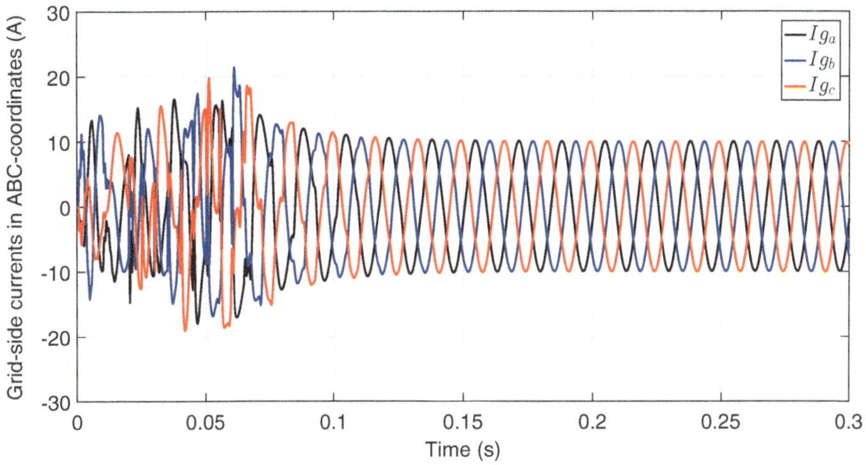

Fig. 11 Grid-side currents in *abc*-coordinates, considering periodic disturbances

Acknowledgements This study was financed in part by the Coordenação de Aperfeiçoamento de Pessoal de Nível Superior - Brasil (CAPES/ PROEX) - Finance Code 001.

References

1. Z. Zeng, H. Yang, S. Tang, R. Zhao, Objective-oriented power quality compensation of multifunctional grid-tied inverters and its application in microgrids. IEEE Trans. Power Electron. **30**, 1255–1265 (2015)
2. Y. Tang, P.C. Loh, P. Wang, F.H. Choo, F. Gao, Exploring inherent damping characteristic of LCL-filters for three-phase grid-connected voltage source inverters. IEEE Trans. Power Electron. **27**, 1433–1443 (2012)
3. Y. Wang, X. Wang, F. Blaabjerg, Z. Chen, Harmonic instability assessment using state-space modeling and participation analysis in inverter-fed power systems. IEEE Trans. Ind. Electron. **64**, 806–816 (2017)
4. R. Teodorescu, M. Liserre, P. Rodríguez, *Grid Converters for Photovoltaic and Wind Power Systems* (Wiley-IEEE, Hoboken, 2011)
5. R. Teodorescu, M. Liserre, P. Rodriguez, *Grid Converters for Photovoltaic and Wind Power Systems* (Wiley, Hoboken, 2011)
6. M. Liserre, R. Teodorescu, F. Blaabjerg, Stability of photovoltaic and wind turbine grid-connected inverters for a large set of grid impedance values. IEEE Trans. Power Electron. **21**, 263–272 (2006)
7. R.V. Tambara, J.M. Kanieski, J.R. Massing, M. Stefanello, H.A. Gründling, A discrete-time robust adaptive controller applied to grid-connected converters with LCL filter. Springer J. Control Autom. Electrical Syst. **28**, 371–379 (2017)
8. M. Semasa, T. Kato, K. Inoue, A simple and effective time delay compensation method for grid-connected inverter with an LCL filter: application to active damping method, in *Proceedings of IEEE 18th Workshop on Control and Modeling for Power Electronics (COMPEL)* (2017), pp. 1–7
9. P. Ioannou, J. Sun, *Robust Adaptive Control* (Courier Corporation, Mineola, 2012)

Electrothermal Characterization of Double-Sided Cooling Si Power Module

Sébastien Sanchez, C. Nguyen, Claudia Cadile, Jean-Pierre Fradin, Patrick Tounsi, and Jean-Michel Reynes

Abstract This paper presents an electrothermal characterization of a prototype double-sided cooling power module. The junction temperature T_j is an important parameter of power devices. Different methods exist for junction temperature measurement. In this work, an electrical method based on temperature sensitive electrical parameter (TSEP) is conducted to estimate the junction temperature of the power module. A 3D thermal model was built to better comprehend thermal behavior within the module. A comparison between simulation and measurement results is performed and analyzed. Results have shown that 3D numerical modeling help understanding several manufacturing defects (soldering, sintering, die defaults, etc.).

1 Introduction

Nowadays, most applications like electric drive transportation require higher and higher power densities and efficiency for power electronics. In the case of power modules, there is a trend towards higher switching frequencies and power levels, while decreasing the total volume. The increased power densities demand an

S. Sanchez (✉) · C. Cadile · J.-P. Fradin
Icam - Site de Toulouse, Toulouse, France
e-mail: sebastien.sanchez@icam.fr

C. Nguyen (✉)
IRT Saint-Exupery, Toulouse, France
Univ. De Toulouse, INSA, LAAS, Toulouse, France
e-mail: quang-chuc.nguyen@irt-saintexupery.fr

P. Tounsi
Univ. De Toulouse, INSA, LAAS, Toulouse, France
e-mail: patrick.tounsi@laas.fr

J.-M. Reynes
IRT Saint-Exupery, Toulouse, France
e-mail: jean-michel.reynes@irt-saintexupery.fr

W. Zamboni, G. Petrone (eds.), *ELECTRIMACS 2019*, Lecture Notes in Electrical Engineering 615, https://doi.org/10.1007/978-3-030-37161-6_5

improved thermal management of power modules. By the way, the junction temperature (T_j) has a large impact on performance and reliability of power devices. This parameter is really important for thermal characterizations (thermal resistance R_{th} and thermal impedance Z_{th}) or for offline/online monitoring of power modules. Optical, physical contact or electrical methods are several technical solutions to measure the temperature of power modules [1, 2].

Optical methods, such as IR camera, photo-luminescence, Raman effect, thermo-reflectance, and refraction index, require direct visual access of power chips [2–4]. In most cases, these measurement methods are not used.

Physical contact methods, e.g., thermocouple, scanning thermal microscopy, liquid crystals, or optical fibers [5, 6], not only need to have physical access to the module, but also might be intrusive.

Electrical methods allow us to have an indirect estimation of the junction temperature without having access inside the power module. However, a drawback of this method is that the measured temperature is only an average temperature of the die. Some electrical methods give the possibility of online temperature measurement monitoring of transistors. For example, if the converter encountered a problem, an immediate temperature measurement would give an indication about the relationship between the fault and the junction temperature. These methods might be employed, for example, in aeronautical and automotive applications for fault detection.

Hence, electrical methods are based on the temperature dependence of electrical parameters (voltage, current) of power devices. By measuring the temperature sensitive electrical parameter (TSEP) and using a known temperature dependence of this parameter, the junction temperature can be estimated. Different TSEPs are investigated for temperature measurements of Si devices [1, 2]. The forward voltage V_F of a diode under low current (10–100 mA) is widely used to measure junction temperature [7]. For IGBTs temperature measurement, different TSEPs can be used: saturation voltage V_{CESAT}, threshold voltage V_{TH}, gate-emitter voltage V_{GE}, saturation current I_{SAT}, or gate current I_G [4].

The content of this work consists in applying an electrothermal characterization applied to a prototype double-sided cooling Si power module. One of the goals is to be able to identify manufacturing defects by means of numerical modeling performed by 6SigmaET software. In the first part, the prototype power module is presented (physical and numerical modeling). Then, a non-intrusive T_j measurement method and its associated test bench are detailed. Experimental and numerical results about junction temperature and thermal resistance are compared and discussed.

2 Wire-Bondless, Double-Sided Cooling Power Module

2.1 *Overview of the Power Module*

The device under test (DUT) used in this work is a double-sided cooling power module.

As already demonstrated by many authors, cooling power modules on both sides of the active switching devices reduce the operational junction temperature compared to conventional single-sided cooling [8]. The outstanding facts are the increased current carrying capabilities by 100–200% and an improvement around 50% of the equivalent thermal resistance by using double-sided cooling embedded Si power packages.

The DUT is a Si IGBT/Diode power module (600 V–400 A) designed by aPSI^{3D} and IRT Saint-Exupery. It is composed of both upper and lower switches. Each switch includes a Si IGBT and an anti-parallel Si diode (see Fig. 1b). As one can see in Fig. 1a, gate driver connections are implemented within the module. The module plastic case is designed so that numerous modules can be mechanically connected in series.

Instead of using conventional wire-bonding, copper bumps are used to interconnect the dice. This technology will help reduce parasite inductances within the

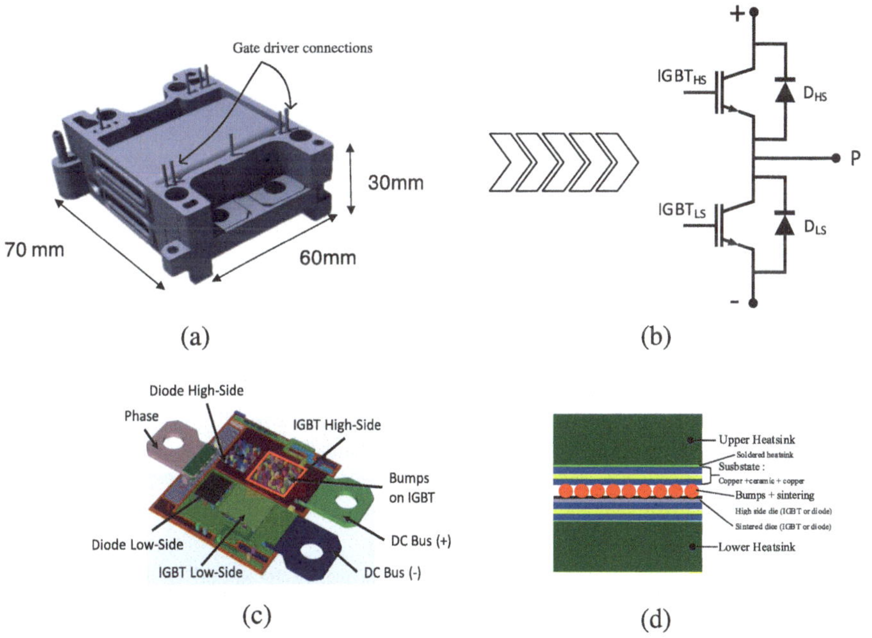

Fig. 1 DUT: double-sided cooling Si IGBT/diodes power module. (**a**) 3D view. (**b**) Electrical circuit. (**c**) Inside. (**d**) Cross section die assembly

module. The upper switch dice are directly mounted on the lower Active Metal Brazed (AMB) substrate. The front side of these dies is connected to the upper AMB substrate through copper bumps. Inversely, the lower switch dice are mounted directly on the upper AMB substrate and connected through copper bumps to the lower AMB substrate. These interconnections within the power module are presented in Fig. 1c, d. Hence, the first advantage of this wire-bondless module is that a double-sided cooling can be performed. The second one is to reduce parasitic inductance within the module by eliminating wire-bonding. Other simulations and measurement showed good values of module inductances (under 8 nH) [9].

Afterwards, an optimized heatsink, leads reducing the pressure losses and increasing the thermal capability at constant flow rate, are soldered on both sides of the power module. Several geometry parameters were optimized, such as base plate thickness, cooling fin shape, heatsink position, and symmetry. To illustrate these facts, computational fluid dynamics (CFD) simulations and experimental tests were performed to optimize the pressure losses of the heatsink (test bench presented in Sect. 4). Figure 2 shows that the optimized heatsinks allow lower pressure losses compared to classical solutions (heatsink cooling fins) at several water cooling temperatures. The gain gap between 0 and 60 °C is due to the liquid cooling viscosity, which decreases with increasing temperature.

The whole module is packaged inside a plastic case which allows liquid cooling.

2.2 *Numerical Modeling*

A numerical modeling, performed by 6SigmaET software, was set up to analyze the thermo-fluidic behavior of the Si IGBT power modules. 6SigmaET is really suitable for electronics applications and it brings a good compromise between accuracy and quick simulations. The whole power module is covered in our simulations (packaging, substrate, die, bumps, etc.). More than 700 elements were included. A regular unstructured meshing was implemented, finer for smaller dimensions, including 31 millions of cells.

Thermo-fluidic coupling was taken into account by means of a simplified model (wall laws, turbulence model k–ε, etc.). Coolant was used as a cooling fluid in double-sided cooling configurations.

All simulations are treated by an Eulerian approach in stationary steady-state [10], with a k–ε turbulence model at constant density [11].

The turbulent kinetic energy k [$\mathrm{m^2\,s^{-2}}$] is mainly carried by large structures, with any interaction of fluid viscosity: formation vortices.

The dissipation ε [$\mathrm{m^2\,s^{-3}}$] is mainly due to the small structures, which are destroyed by the fluid viscosity effect, corresponding to speed of large vortices destruction.

Eulerian phase around the module at T_a was carried out in the air environment. The T_j was evaluated by volume average temperature of the die. Table 1 summarizes fluid and simulation parameters.

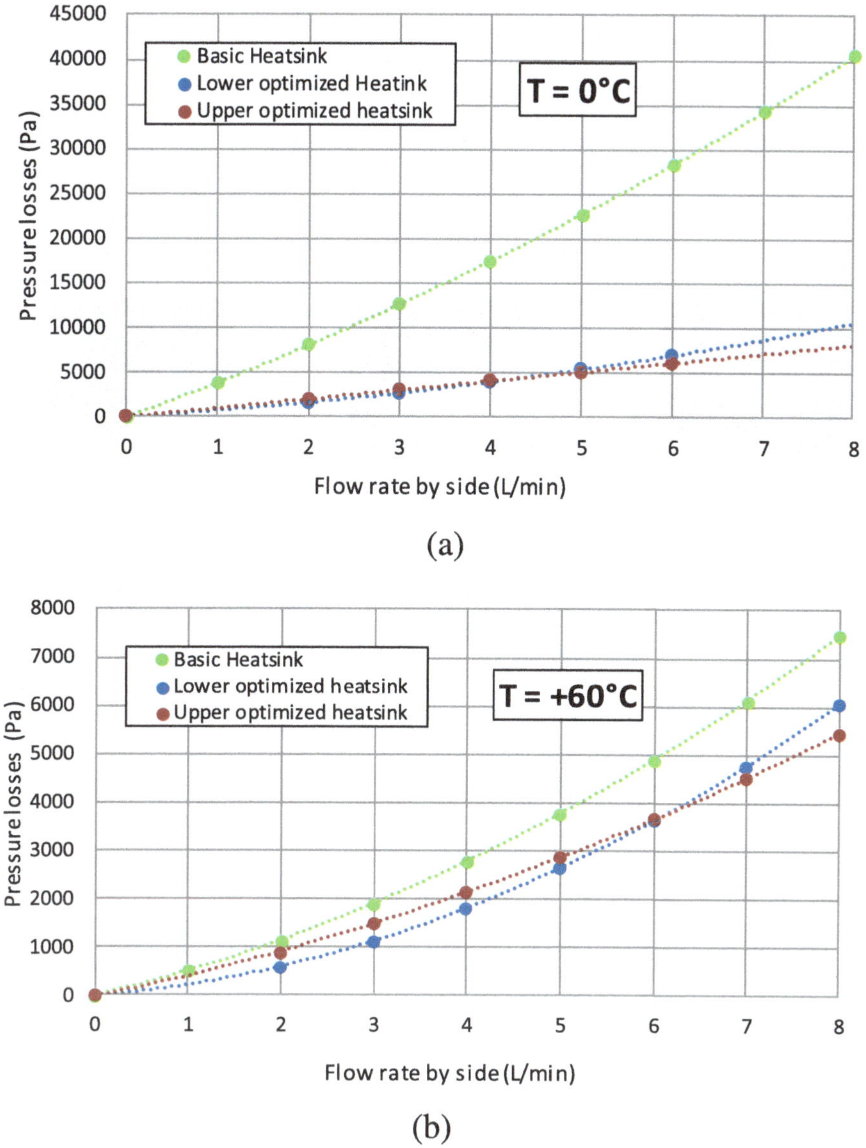

Fig. 2 Experimental results of pressure drop at different fluid temperature. (**a**) At 0 °C. (**b**) At 60 °C

Figure 3 shows several surface temperature fields of the power module for biased high side Si IGBT. Three fluidic configurations were taken into account: double side cooling, upper or lower single side cooling. During double side cooling, a vertical distribution of the heat flow is clearly illustrated. Heat extraction is around

Table 1 Simulation parameters (fluidic and electric considerations)

Parameters	Values
Fluid temperature T_{fluid}	65 °C
Ambient temperature T_a	20 °C
Fluid density ρ_f	1045 kg/m^3
Fluid dynamic viscosity	1.4 mPa s
Thermal conductivity λ	0.42 W/m K
Flow rate ṁ	6 L/min
Dissipated power by die P	100 W

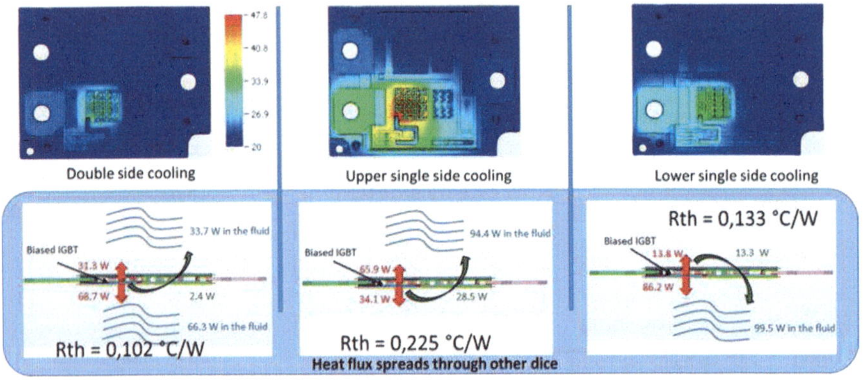

Fig. 3 Surface temperature fields of the power module for biased high side Si IGBT

2/3 through the die back side and 1/3 through the front side by means of copper bumps.

Heat extraction is more efficient through the back-side joint than through the bumps on the front side of the die. Based on a surface average temperature, the thermal resistance was estimated at 0.102 K/W.

A comparison between simulations and experimental results concerning equivalent thermal resistance is presented in Sect. 4.

3 Non-intrusive T_j Measurement

3.1 Principle

As presented above, several methods are known to measure the junction temperature of power modules. In our study, an electrical method using a TSEP was chosen and applied to IGBT/diode half bridge (forward voltage V_F and saturation voltage V_{CESAT}, respectively, to the diode and the IGBT). The TSEP is used to indirectly estimate junction temperature of the power module and to benefit from a precise and linear thermometric parameter (around −2 mV/°C), an easy implementation of

Fig. 4 Law of variation between the TSEP and the temperature under a weak current. (**a**) $V_f = f(T°)$. (**b**) $Vce = f(T°)$

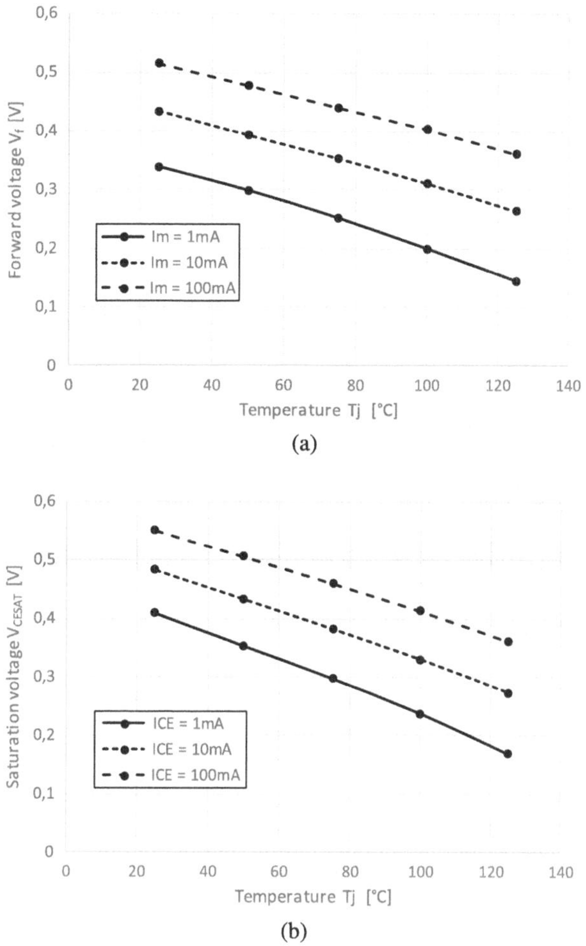

the simple measurement as well as a calibration. It was shown [12] that estimated temperatures with V_F or V_{CESAT} are close to average junction temperatures of the power module.

Basically, there are two different steps. The first step, which is a calibration step, consists in establishing a law of variation between the TSEP and the temperature at low current (below 100 mA) of excitation, in order not to produce conduction losses. Figure 4 shows the TSEP evolution under DC-biased current while power modules are located inside a controlled oven.

Once the TSEP evolution at low current is reached, an electrical method is carried out. Figure 5 shows its measurement principle. It consists in injecting a power current (I_P) flowing through the DUT until it reaches the thermal steady-state. Power current is then replaced by measurement current I_m used in previous calibrations (see Fig. 4). Cooling curve of the DUT is recorded through voltage

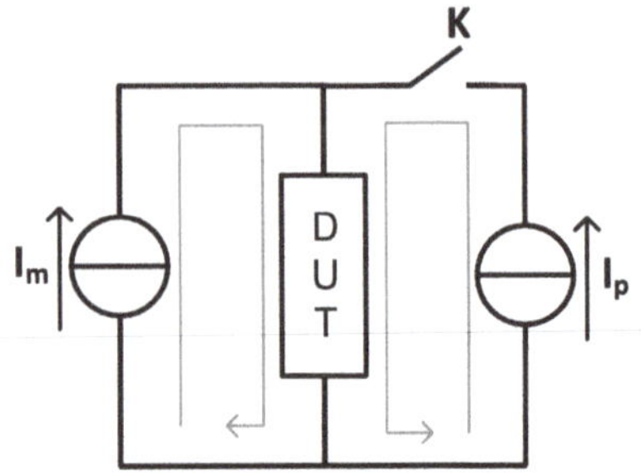

Fig. 5 Schematic of the junction temperature measurement method

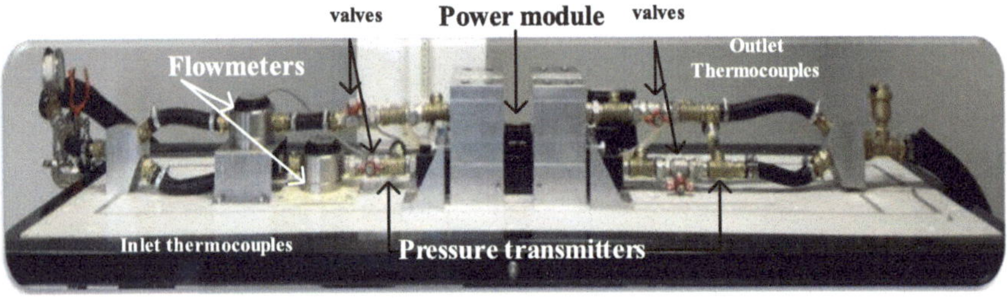

Fig. 6 Overview of the hydraulic system

drop variations. The initial temperature T_{j0} at the end of power current injection is then calculated using square root time based extrapolation on the linear part of the cooling curve. Authors in [13] showed that the strong dynamics observed during short times (close to 40 μs) do not correspond to a thermal response but correspond to electrical effects due to parasitic capacitive and inductive elements of the measuring circuit.

Therefore, the estimated junction temperature allows us to calculate the average thermal resistance of the DUT:

$$R_{th} = \frac{T_j - T_{amb}}{P}. \tag{1}$$

In this formula, P is the power absorbed by the DUT, corresponding to the conduction losses and T_{amb} is the inlet coolant temperature.

The method described and its associated test bench (see Sect. 3.2) shall be extended to a transient thermal impedance measurement, based on pulsed heating curve technique [14].

3.2 Test Bench

In order to estimate the thermal resistance, a complete electrothermal fluidic bench was developed, as shown in Fig. 6.

3.2.1 Hydraulic Parts

The coolant circuit is divided into two parallel channels in order to supply the coolant to both faces of the module. DUT is located in the middle of the coolant circuit. A cryostat is used in order to control the coolant temperature. Several valves are located in the upper and lower coolant circuit to regulate the flow rate.

In addition, different measurements are provided at the inlet and the outlet of each module:

- Two volumetric flowmeters (MACNAUGH, model MX12, accuracy of ±0.5%, repeatability < 0.03%) are used to measure the coolant flow rate flowing into each module face. The measuring range is between 1 and 16 L/min, for coolant temperature from −30 to +90 °C.
- Four high accuracy pressure transmitters (WIKA, S20, accuracy of ±1.25 mbar). It allows us to measure the pressure drop of each module side, in the range of coolant flow rate from 1 to 8 L/min, for temperatures from −20 to +80 °C.
- Four type T thermocouples (accuracy of ±0.5 °C), in the widest operating temperature range: −30 to +90 °C. They quantify the impact of the module heating on the coolant temperature. Other thermocouples of type K (accuracy of ±1 °C) are used to measure, respectively, the reference outside ambient temperature, the power terminal, and the casing surface module.

Several studies have been conducted to validate the very low pressure losses of the cooling circuit and the DUT [9] by means of experimental and CFD simulations.

3.2.2 Electric Parts

Figure 7 shows the electrical circuit of the I_m and I_p currents control.

High current pulses I_P are created by storing energy in an inductor (L). Two MOSFET transistors (T1 and T2) are used to drive the current into the DUT. The injected power current is measured using a current probe with a large bandwidth. A TSEP current (I_M), adjusted by a low current source, is injected permanently into the DUT and checked by using a shunt resistor (R_m). The Schottky diode

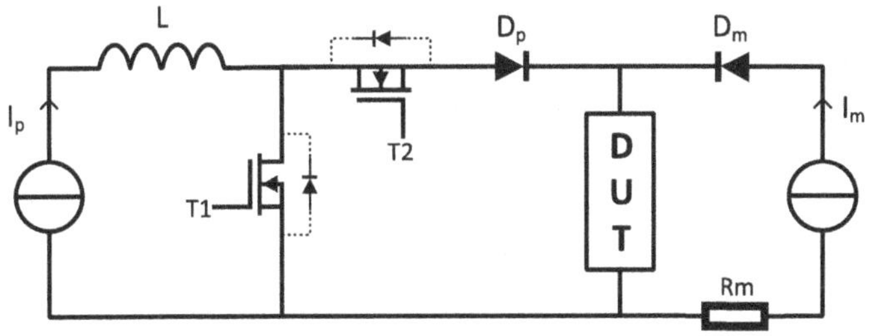

Fig. 7 Electrical current switcher

D_p prevents the current I_M from passing through the MOS2 body diode and MOS1 during the indirect junction temperature measurement of the DUT by voltage acquisition under low current. The diode D_m prevents from the influence of current I_P on the I_M current source. In our case, the current I_M is set to 10 mA, which is low enough to avoid self-heating during the temperature measurements and high enough to get a good sensitivity on the whole temperature range from 0 °C up to 150 °C, as shown in Fig. 4.

4 Experimental and Numerical Results: Comparison of Thermal Resistance R_{th}

In this section, the thermal resistance R_{th} of the power module IGBT is measured under certain conditions:

- Liquid temperature: (20 °C)
- Flow rate: 6 L/min/channel
- Dissipated power 65 W up to 160 W
- cooling configurations: double-sided cooling, single high-sided cooling, single low-sided cooling
- DC-biased current I_M: 10 mA

Figure 8 demonstrates the evolution of junction temperature of high side IGBT under varying dissipated power and different types of cooling. The junction temperature increases linearly with the dissipated power of the die.

Figure 9 shows that thermal resistance of high side IGBT is invariant with different dissipated power. It is evident that double-sided cooling results have a lowest thermal resistance. Thermal resistance of the IGBT under single low sided

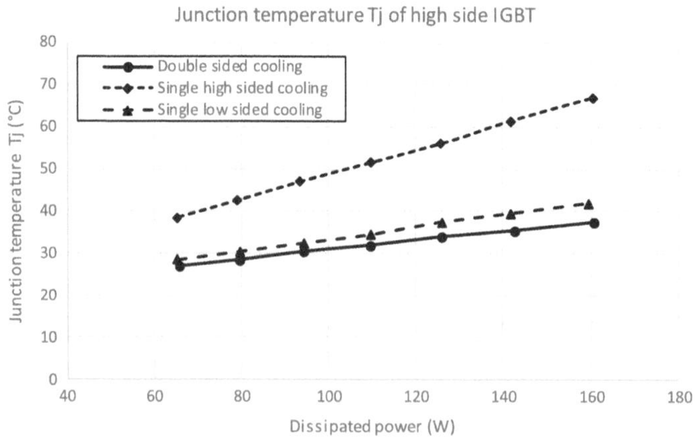

Fig. 8 Variation of junction temperature of high side IGBT with varying dissipated power

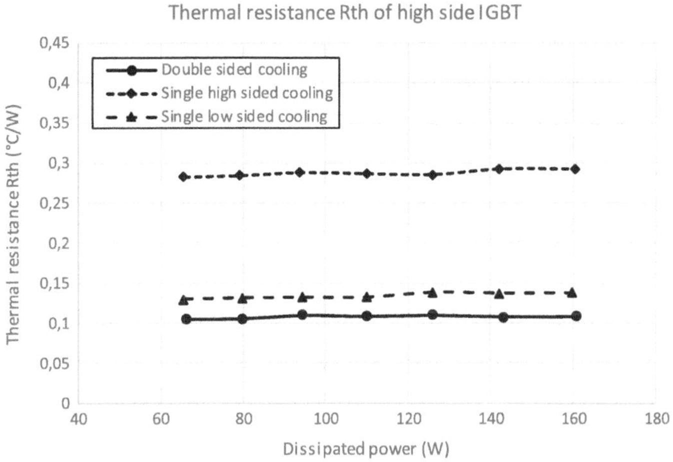

Fig. 9 Thermal resistance of high side IGBT with varying dissipated power

Table 2 Simulated and measured thermal resistance R_{th} of the power module's IGBT under double and single-sided cooling

Cooling	Simulation	Measurement	Difference
Double-sided cooling	0.106	0.108	2%
Single high-sided cooling	0.220	0.288	31%
Single low-sided cooling	0.130	0.134	3%

cooling (heat transfers through under side of the die) is much lower than the one under single high-sided cooling.

Table 2 summarizes measured (average value) and simulated thermal resistances of the IGBT and the relative difference between these values.

As mentioned, under double-sided cooling, the measured thermal resistance is very close to the simulated one (only 2% difference). The measured and simulated thermal resistances under single low sided cooling (heat is cooled through back side of the die) are also only 3% difference. However, the measured thermal resistance under single high-sided cooling is 31% higher than the simulated one. This means the heat transfer through front sides (and so through copper bumps) is not as good as expected on the basis of the numerical results.

5 Conclusions

An electrothermal characterization of a double-sided cooling Si power module using an electrical method is presented in this paper. In addition, a numerical model helps to have a thermal performance reference for the power module. Experimental results show good thermal performance of the power module under double-sided cooling and single back-sided cooling. However, a difference of 30% is found between experimental and simulated thermal resistance under single front-sided cooling. It is important to notice that the value of the thermal resistance under single-sided cooling dramatically depends on the contact between the bumps and the sintering layers. The difference between computer-aided design (CAD) modeling and real module may explain such differences. Further work are being conducted in order to explain these facts and improve the assembly technology of the power module and in using MOSFET SiC technology.

Acknowledgements Great thanks to aPSI^{3D} company for giving us Si power module samples in order to perform a real case study.

References

1. D.L. Blackburn, Temperature measurements of semiconductor devices a review, in *Semiconductor Thermal Measurement and Management Symposium* (IEEE, Piscataway, 2004)
2. Y. Avenas, L. Dupont, Z. Khatir, Temperature measurement of power semiconductor devices by thermo-sensitive electrical parameters—a review. IEEE Trans. Power Electron. **27**(6), 3081–3092 (2012)
3. D. Werber, G. Wachutka, Interpretation of laser absorption measurements on 4 H-SiC bipolar diodes by numerical simulation, in *Proceedings of International Conference on Simulation of Semiconductor Processes and Devices*, Hakone, September 2008, p. 8992
4. Y. Avenas, L. Dupont, Comparison of junction temperature evaluations in a power IGBTs module using an IR camera and three thermosensitive electrical parameters, in *Applied Power Electronics Conference and Exposition*, Orlando, FL, February 5–9, 2012
5. D. Bin, J.-L. Hudgins, E. Santi, A.-T. Bryant, P.-R. Palmer, H.-A. Mantooth, Transient electro thermal simulation of power semiconductor devices. IEEE Trans. Power Electron. **25**(1), 237–248 (2010)
6. A. Majumdar, Scanning thermal microscopy. Annu. Rev. Mater. Sci. **29**, 505–585 (1999)
7. B. Thollin, L. Dupont, Z. Khatir, Y. Avenas, J.C. Crebier, P.O. Jeannin, Partial thermal impedance measurement for die interconnection characterization by a microsecond pulsed heating curve technique, in *Power Electronics and Applications (EPE), 2013 15th European Conference* (2013), pp. 1–10
8. B.C. Charboneau, F. Wang, J.D. van Wyk, D. Boroyevich, Z. Liang, E.P. Scott, C.W. Tipton, Double-sided liquid cooling for power semiconductor devices using embedded power packaging. IEEE Trans. Ind. Appl. **44**(5), 1645–1655 (2008)
9. J.-P. Fradin, D. Elzo, C. Cadile, J. de Martres, G. Ramecourt, J.-M. Reynes, B. du Trieu, P. Monfraix, Characterisation of highly dissipative power modules: experimental fluidic tests CFD thermal modelling, in *12th European ATW on Micropackaging and Thermal Management-IMAPS*, 2016
10. I.G. Currie, *Fundamental Mechanics of Fluids* (CRC Press, New York, 2002)
11. W. Rodi, *Turbulence Models and Their Application in Hydraulics* (CRC Press Book, New York, 2017)
12. L. Dupont, Y. Avenas, Preliminary evaluation of thermo-sensitive electrical parameters based on the forward voltage for online chip temperature measurements of IGBT devices. IEEE Trans. Ind. Appl. **51**(6), 4688–4698 (2015)
13. D.L. Blackburn, A review of thermal characterization of power transistors, in *Semiconductor Thermal and Temperature Measurement Symposium, Semi-Therm*, 1988
14. D.L. Blackburn et al., Transient thermal response measurements of power transistors. IEEE Trans. Ind. Electron. Control Instrum. **IECI-22**(2), 134–141 (1976)

Integrated Design Process and Sensitivity Analysis of a Hybrid Electric Propulsion System for Future Aircraft

Matthieu Pettes-Duler, Xavier Roboam, and Bruno Sareni

Abstract This paper is situated in the framework of future hybrid electric aircraft in which embedded weight minimization and maximization of power efficiency are the key challenges to address fuel reduction and environmental constraints. In the first part, the integrated design process aiming the overall power train optimization is described. The second part presents models specifically oriented towards the integrated design. Finally, a sensitivity analysis is carried out at the power train system level to study the influence of both electric components-specific powers and efficiencies on the Maximum Take Off Weight (MTOW) and on the fuel burn of the hybrid propulsion aircraft.

1 Introduction

Nowadays, transport is the fastest growing sector worldwide [1], which significantly contributes to environmental degradation. Finding sustainable solutions is a key challenge to solve this issue especially for the aircraft sector, which represents about 2% of the global CO_2 emissions. The Clean Sky (H2020 EU) project assists the aircraft manufacturers in finding aircraft cleaning/noiseless solutions. Currently hybrid electric or all electric aircraft are not commercialized but several aircraft manufacturers explore the future of electric flight. Airbus has teamed up with Rolls Royce and Siemens to build a 100-passenger hybrid electric technology flight demonstrator, E-Fan X, scheduled for 2020 [2]. The same year, backed by Boeing and Jetblue technology ventures, Zunum Aero, a hybrid electric aircraft manufacturer startup, plans to fly a 12-passenger hybrid electric prototype in 2020 [3]. The hybrid electric technology is taking off and the studies too.

M. Pettes-Duler (✉) · X. Roboam · B. Sareni
Laplace Lab., Université de Toulouse, CNRS, INPT, UPS, Toulouse Cedex 7, France
e-mail: pettes@laplace.univ-tlse.fr; roboam@laplace.univ-tlse.fr; sareni@laplace.univ-tlse.fr

W. Zamboni, G. Petrone (eds.), *ELECTRIMACS 2019*, Lecture Notes
in Electrical Engineering 615, https://doi.org/10.1007/978-3-030-37161-6_6

There exist several electrified propulsion architectures: Full electric, Turbo-electric and Hybrid Electric. A complete state of the art has been conducted by the National Aeronautics and Space Administration (NASA) [4]. In order to project the new aircraft generation, several electrified propulsion architectures (series hybrid/parallel hybrid/series-parallel partial hybrid/turbo-electric) have been studied. Many assessments have been done to estimate fuel consumption. The University of Michigan proposes different architectures [5], following the same dynamic, analyzing future electric components. The following step in the Technology Readiness Level (TRL) is the validation of concept through models. The NASA develops a mission analysis coupled with a regional aircraft sizing [6] by means of very simple models to estimate the weight of a parallel hybrid electric aircraft. In the same idea, a complete study has been done in Airbus with ICA and LAPLACE labs [7] about hybrid electric propulsive architectures for a regional aircraft. In those works, the overall aircraft architecture has been studied taking into account both energy and aerodynamics benefits. Hybridization with electric auxiliary sources (batteries or fuel cells) may also offer the opportunity to redesign the Internal Combustion Engines by limiting mechanical power demand constraints in particular flight sequences thus allowing the optimization of the engine weight [8] and fuel consumption. The obtained results are the same for all of these hybrid architectures: hybrid electric power trains are heavier than conventional thermal systems. But the global gain, due to aerodynamics, engine redesign and energy optimization are often positive in terms of fuel burn. These studies clearly emphasize the technological sensitivity; the lower the MTOW, the lower the fuel burn. The key challenge is then to minimize embedded weights by optimizing efficiencies and specific powers.

This objective is the main target of the HASTECS (Hybrid Aircraft Academic reSearch on Thermal and Electrical Components and Systems) EU project that aims at coupling thermal and electrical studies for a regional aircraft. Only a series hybrid electric architecture has been studied in this project, as it leads to huge power constraints on the electric power train, the thrust being fully provided by electric devices. From an input data set and given environmental conditions (temperature, pressure, aircraft speed, etc.), we use different "surrogate" models to simplify assessments of efficiencies and masses from each device to the whole power train. Unlike the previous references, a fixed aircraft structure is considered. Only the propulsive system is refined through a looped process linking weight variations and thrust consequences as described in Sect. 2: this integrated design approach allows assessing energy efficiency and mass benefits. As in some examples described in Sect. 3, the main issue of this study is to provide surrogate models in view of an integrated design allowing to efficiently progress (with moderate computation cost) towards the optimal trade-off of the whole hybrid electric power train. Finally, after setting up the integrated design process, a sensitivity analysis is carried out in the last section showing the effects of variations on both specific power and efficiency of the main electric devices at the system level.

2 Integrated Design Process

A series hybrid electric propulsion system (see Fig. 1) is composed of gas turbines as main power sources that drive electric generators with rectifiers supplying a uHVDC[1] Electric Power Distribution Unit (EPDU). This latter is also powered by auxiliary sources composed of batteries or fuel cells. This uHVDC power distribution unit supplies inverter-fed electric motors (here four motors), which drive propellers through gearbox. For each component of the hybrid electric propulsive device, a "system-oriented design model", also called "surrogate model", must be integrated into a global looped process estimating the MTOW at the aircraft level and the fuel consumption from Turboshaft power demand. In this framework, "surrogate models" based on simplified analytic derivation, scaling, regression laws or response surfaces are necessary to address the integrated design process objectives with moderate computation cost.

Both efficiency and mass are assessed from each component model at the end of each iteration of the looped process; masses are added in the whole power train to be included in the MTOW. After an iteration, based on the new (after redesign) MTOW ($MTOW_{new}$) and its ratio versus the weight reference value ($MTOW_{ref}$) at the initial point of the process, a new thrust must be derived:

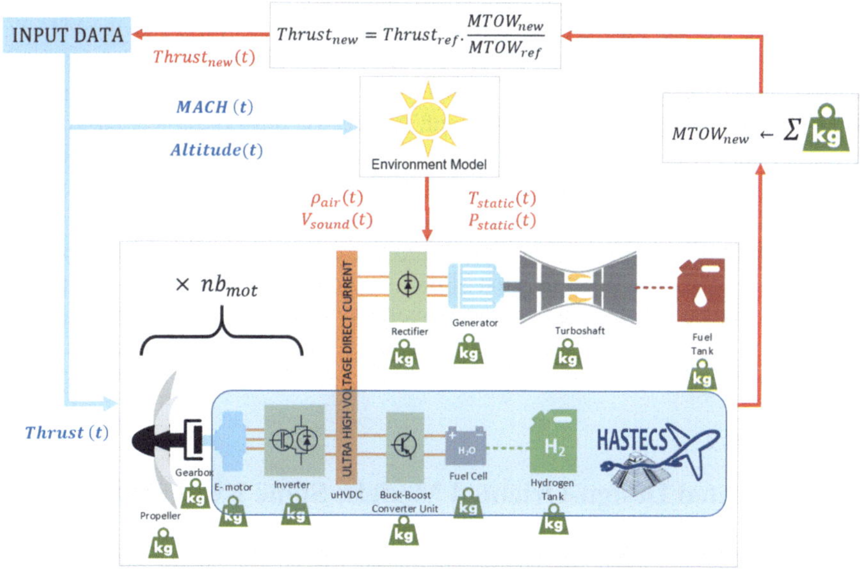

Fig. 1 Integrated design process

[1]uHVDC means "ultra-high voltage DC" standards beyond HVDC bus (±270 V), leading to bus voltages in the range of kVs.

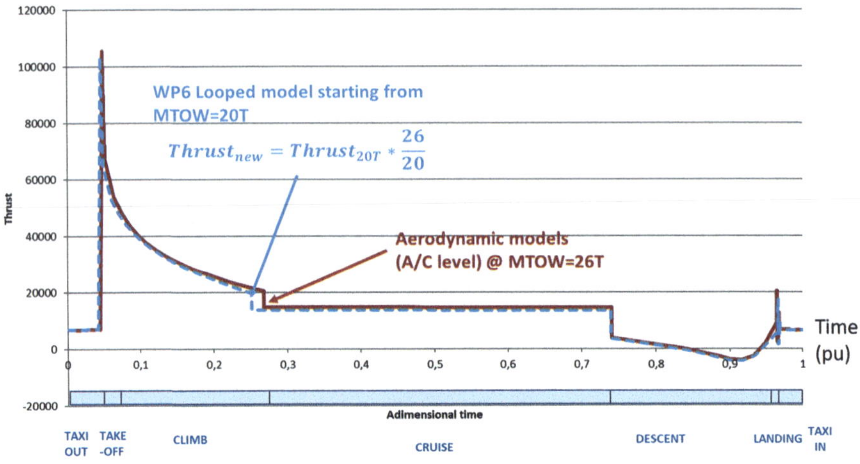

Fig. 2 Comparison of aerodynamic and looped models

$$\text{Thrust}_{\text{new}} = \text{Thrust}_{\text{ref}} \times \frac{\text{MTOW}_{\text{new}}}{\text{MTOW}_{\text{ref}}} \tag{1}$$

Thus, by considering the evolution of "new" variables displayed in Eq. (1) at each iteration, an implicit looped process is achieved, which is stopped when the MTOW error between two successive iterations is lower than 50 kg. The process is initialized with a full thermal reference aircraft with an MTOW_{ref} of 20 T. Thus, adding the embedded weights corresponding with the electric devices (power electronics, electric motors, auxiliary source, cables), implies new MTOW and consequently new thrust values. Our looped simplified model has been validated with reference to a complete aerodynamic model described in [7]. For a final MTOW of 26 T, the results obtained by both models are quite similar (see Fig. 2); the accordance of both thrust shapes over the flight mission allows validating simplification assumptions with the looped process. Note that this implicit loop converges within ten iterations, which corresponds to an acceptable CPU cost in view of an integrated design by optimization.

3 Integrated Design-Oriented Models: Some Examples

From AIRBUS side, input data have been given:

- Environmental aircraft data: Altitude(t)
- Mach speed along the mission profile: MACH(t)

Given these inputs, some examples of surrogate models are proposed to design the hybrid electric power train of the studied aircraft. Readers will find more detailed models in [7].

3.1 Environment Model

An environment model is required to design the different components of the power train (turboshaft, propeller) over the flight mission. The main environment variables are derived as follows:

$$P_{\text{static}} = 101,325 \times \left(1 - \frac{\text{Altitude}}{44,331}\right)^{5.256} \tag{2}$$

$$T_{\text{static}} = 288.15 - 0.0065 \times \text{Altitude} + \text{DISA} \tag{3}$$

$$\rho_{\text{air}} = 0.0034837 \times \frac{P_{\text{static}}}{T_{\text{static}}} \tag{4}$$

$$P_{\text{dynamic}} = P_{\text{static}} \times \left(1 + 0.2 \times \text{MACH}^2\right)^{3.5} \tag{5}$$

$$T_{\text{dynamic}} = T_{\text{static}} \times \left(1 + 0.2 \times \text{MACH}^2\right) \tag{6}$$

$$V_{\text{sound}} = \sqrt{401.8 \times T_{\text{static}}} \tag{7}$$

where Altitude based on the input flight mission is in meter, DISA is the temperature difference from the International Standard Atmosphere (ISA). ρ_{air} is the air density in kg/m^3. Finally, static and dynamic pressures ($P_{\text{static}}, P_{\text{dynamic}}$) and temperatures ($T_{\text{static}}$, T_{dynamic}) are respectively assessed in [Pa] and in [K]. These variables will be used in particular in turboshaft and propeller models.

3.2 Propeller Model

A six-bladed variable-pitch propeller is considered. The propeller geometry is sized by two equations. The same tip speed limit and the same maximum shaft power are considered as for the ATR 72 propeller used in [7]. The propeller diameter can be sized from the maximum shaft power ($P_{\text{max}_{\text{shaft}}}$) giving the maximum rotation speed (N_{prop}) and the propeller mass (M_{prop}):

$$D_{\text{prop}} = 3.93 \times \sqrt{\frac{P_{\text{max}_{\text{shaft}}}}{2.4 \times 10^6}} \tag{8}$$

$$N_{\text{prop}} = 1200 \times \frac{3.93}{D_{\text{prop}}} \tag{9}$$

$$M_{prop} = 185 \times \left(\frac{D_{prop}}{3.93}\right)^2 \tag{10}$$

where D_{prop} is the corresponding diameter. The model is based on the Disk Actuator Theory (Eq. 16) with a saturation function at low speed, which is implemented in Eq. (15). In Eq. (13), the maximum thrust T_{max} has been estimated. The use of the actuator disk theory during low-speed phases tends to overpredict the propeller thrust, that is why we have refined the thrust T_{vir} in a saturation function (Eq. 15) from the real thrust T_{real}. T_{real} comes from the input data.

$$kP = \frac{1}{\rho_{air}} \quad 0.1 \leq kP \leq 1.5 \tag{11}$$

$$k = 1.6 + \frac{\frac{V_{A/C}}{3.93 \times 1200}}{60} \tag{12}$$

$$T_{max} = 10^{f(kP)}.P_{max_{shaft}} \tag{13}$$

$$f(kP) = \alpha_{prop} + \beta_{prop} \times \log{(kP)}^{\gamma} \tag{14}$$

$$T_{vir} = \left(\frac{2.k.T_{max}}{\pi}\right) \times \tan\left(\frac{T_{real}\pi}{2.k.T_{max}}\right) \tag{15}$$

$$P_{shaft} = \frac{1}{2k_{eff}}\left(V_{A/C}T_{vir} + \sqrt{\left(V_{A/C}.T_{vir}\right)^2 + \frac{2T_{vir}^3}{\rho_{air}A_{prop}}}\right) \tag{16}$$

where kP is the input value of the linear regression of the model Eq. (14). ρ_{air} is the air density varying during the flight mission in kg/m^3. $V_{A/C}$ is the aircraft speed in m/s also defined by the mission profile. We have defined an advance ratio k for the saturation function. $f(kP)$ is the linear regression function used by the model where α_{prop}, β_{prop} and γ are constant values. After that, we can use the disk actuator theory formula with a constant value $k_{eff} = 0.9$ and with the propeller disk area: $A_{prop} = \frac{\pi.D_{prop}^2}{4}$.

3.3 Gearbox Model

For the gearbox, the linear regression model proposed by NASA in [6] is considered for estimating the mass:

$$\text{Gbox} = \left(\frac{P_{\text{propHP}}}{\text{RPM}_{\text{prop}}}\right)^{0.75} \times R_{\text{gbox}}^{0.15} \tag{17}$$

$$M_{\text{Gbox}_{\text{LBS}}} = -37.43 + 116.33 \times \text{Gbox} \tag{18}$$

$$\eta_{\text{Gbox}} = 0.985 \tag{19}$$

where P_{propHP} is the maximum propeller shaft power in horsepower [HP], RPM_{out} the propeller rotation speed in [rpm] and R_{gbox} is the gearbox ratio. A linear function allows estimating the gearbox weight in [lbs] (Eq. 18). A constant efficiency value is finally assessed in Eq. (19).

3.4 Turboshaft Model

The Turboshaft model is based on response surfaces. We consider three inputs: the static power at sea level P_{SLS}, the ratio between total and static pressures at sea level δ [Pa] and the ratio between total and static temperatures at sea level θ [K] previously calculated in Eqs. (5) and (6). From these inputs, a one–column vector is derived in Eq. (22) to assess the maximum mechanical turboshaft power $P_{\text{max}_{\text{shaft}_{\text{GT}}}}$. α_1 and β_1 are two constant values of the linear regression. A convenient P_{SLS} value must be designed in order to supply conveniently the hybrid electric propulsion during the whole flight mission.

$$\delta = \frac{P_{\text{dynamic}}}{101,325} \tag{20}$$

$$\theta = \frac{T_{\text{dynamic}}}{288.15} \tag{21}$$

$$x_{\text{C}} = \begin{bmatrix} \log\left(P_{\text{SLS}_{\text{kW}}}\right) \\ \log(\delta) \\ \log(\theta) \\ \log\left(P_{\text{SLS}_{\text{kW}}}\right)^2 \\ \log(\delta)^2 \\ \log(\theta)^2 \\ \log\left(P_{\text{SLS}_{\text{kW}}}\right) \times \log(\delta) \\ \log\left(P_{\text{SLS}_{\text{kW}}}\right) \times \log(\theta) \\ \log(\delta) \times \log(\theta) \end{bmatrix} \tag{22}$$

$$P_{\text{max shaftGT}} = 10^{\alpha_1 + \beta_1 . x_C} \tag{23}$$

With the same vector (x_C), the maximum Power Specific Fuel Consumption $\text{PSFC}_{\max}$ is assessed (in [kg/s/kW] in Eq. (24)); it corresponds to the turboshaft PSFC when this engine provides its maximum power leading to a minimum consumption. For the real turboshaft fuel consumption, we have considered a ratio ($\text{Part}_{\text{load}}$) between the maximum mechanical power and the required power function of the environmental data δ and θ in Eq. (25). Another linear regression is used to estimate the actual PSFC (Eq. 27) from which the fuel flow rate is defined given a required power, P_{required}.

$$\text{PSFC}_{\max} = 10^{\alpha_2 + \beta_2 . x_C} \tag{24}$$

$$\text{Part}_{\text{load}} = \frac{P_{\text{required}}}{P_{\text{max}_{\text{shaftGT}}} \times \delta \times \sqrt{\theta}} \tag{25}$$

$$y = \begin{bmatrix} \log\left(\text{Part}_{\text{load}}\right) \\ \log\left(\text{Part}_{\text{load}}\right)^2 \\ \cdots \\ \log\left(\text{Part}_{\text{load}}\right)^{10} \end{bmatrix} \tag{26}$$

$$\text{PSFC} = 10^{\alpha_3 + \beta_3 . y} \times \frac{\text{PSFC}_{\max}}{\theta} \tag{27}$$

$$D_{\text{m}_{\text{fuel}}} = \text{PSFC} \times \frac{P_{\text{required}}}{1000} \tag{28}$$

In this model, the specific fuel consumption variation is taken into account: As displayed in Fig. 3, the turboshaft consumption is minimum at 100% of the rating power [7]. Contrarily, at low rating, the SFC evolution may be referred to a hyperbolic function (see Fig. 3). In that way, using secondary energy source in taxi and descent phases switching off the main thermal source could be interesting in a hybrid electric aircraft. An efficient, reliable and fast starting system is mandatory.

The turboshaft mass is difficult to estimate. For that purpose, a typical specific power of 9.86 kW/kg is assessed from experimental data set [7]. Finally, the fuel mass is obtained by the time integral of the mass flow rate (Eq. 29):

$$M_{\text{fuel}} = \int D_{\text{m}_{\text{fuel}}}(t)\,\mathrm{d}t \tag{29}$$

$$M_{\text{turboshaft}} = P_{\text{SLS}} \times P_{\text{SPE}} \tag{30}$$

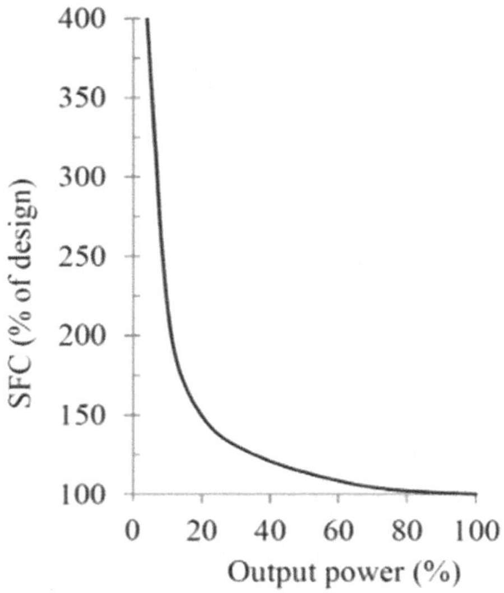

Fig. 3 Specific fuel consumption evolution versus rating power

Table 1 Electric component assessments

	2025 target	2035 target	20xx target
Emotor/Egenerator			
SP + cooling	5 kW/kg	10 kW/kg	15 kW/kg
Efficiency	96%	98.5%	99%
Power electronics			
SP + cooling	15 kW/kg	25 kW/kg	35 kW/kg
Efficiency	98%	99.5%	99.8%
Fuel cell—liquid H_2			
H_2 + tank SE	3.3 kWh/kg		
Auxiliary SP	1.3 kW/kg		
Stack SP	4 kW/kg		
DC bus			
Ultra HVDC	2000 V		

3.5 Electric Component Models

For the electric components, technological models with corresponding design are currently achieved in the HASTECS project. However, in this paper mainly oriented towards system level sensitivity analysis, very simple models based on assessments with fixed efficiency and specific power/energy targets have been considered. In the HASTECS project, two targets have been challenged for 2025 and for 2035 as displayed in Table 1. In this latter table of assessments, future 20xx target has been added with more aggressive assumptions. In our case, specific powers of the

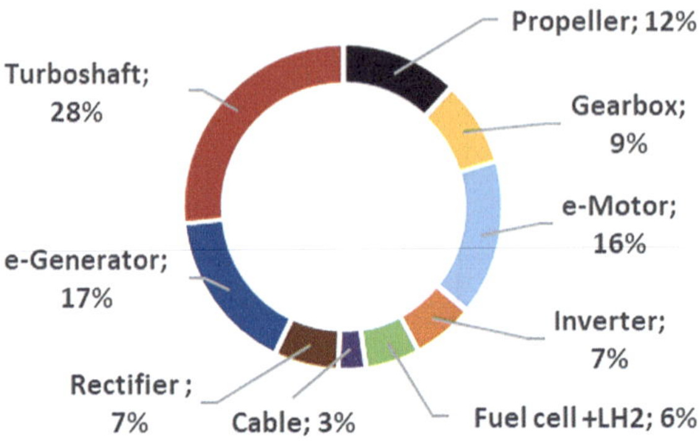

Fig. 4 Hybrid power train mass sharing

electric motors, (respectively electric generators) and power electronics (rectifiers and inverters) include the cooling system. A liquid hydrogen storage with fuel cell stack is considered in this study with corresponding assessments in Table 1 [9].

$$\text{SP} = \text{Specific Power, SE} = \text{Specific Energy}$$

Aggressive targets have been chosen, but certain targets are already achieved; in particular, Siemens [10], with the electric motor SP260D at 5.2 kW/kg. The University of Illinois is designing and building a permanent magnet synchronous motor to exceed a specific power of 13 kW/kg and efficiency of 96%, showing that these targets can be reached [11]. Concerning the inverters, General Electric [12] has reached a specific power by 19 kW/kg for its product.

Other simple models have been developed to assess cable weights and losses, following the transferred power, the voltage level and the cable type (AC or DC).

Based on this model set, the mass distribution of the hybrid power train is displayed in Fig. 4, with the particular assessments dealing with the "2035 Target".

4 Sensitivity Analysis at System Level

We consider a hybrid electric aircraft composed of two same-sized turboshafts, which turn generators to supply the electric power train. The power distribution over the flight sequences is displayed in Fig. 5; in this management strategy, fuel cells are used during full electric taxi and descent phases, which may correspond to a "light hybridization scenario". For the other phases, the power is equally shared between turboshafts. The sensitivity analysis consists in varying both the efficiencies and specific powers of two device classes (power electronics and electric machines) of the hybrid electric conversion chain with regard to the 2025 and 2035 targets.

In the first analysis (see Figs. 6, 7, 8, and 9), the influence of specific power and efficiency variations are separately estimated for each class of device.

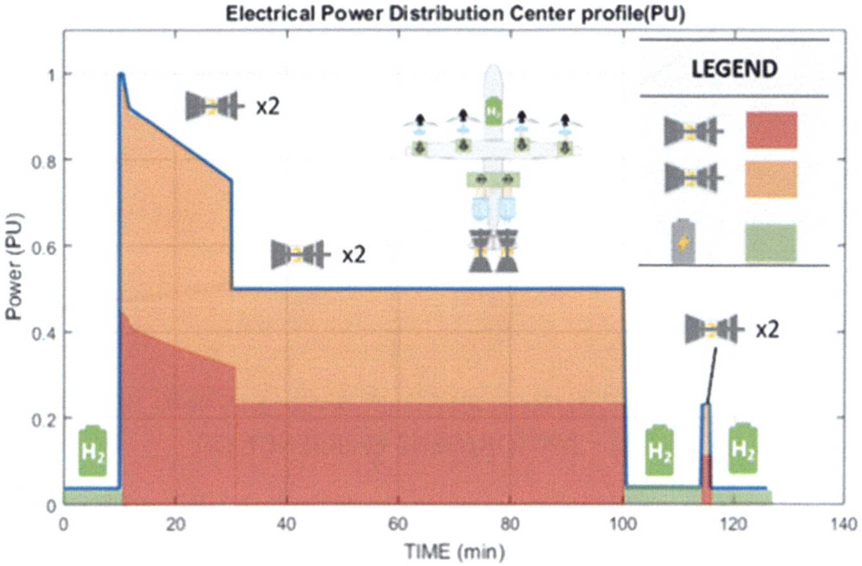

Fig. 5 Power distribution between sources

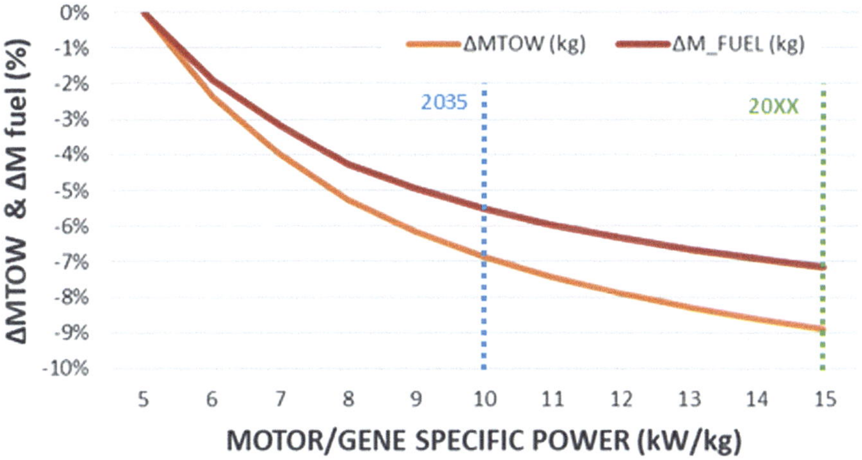

Fig. 6 Impact of Specific Power variations of electric motor/generator on the MTOW and the fuel mass from the 2025 target

It can be noticed that the impact of technological progress both in terms of specific powers and efficiency is significant on the fuel burn, even if this impact is slightly lower for power electronics than for electromechanical converters. It can also be seen that both progress axes (efficiency and specific power) are really influent.

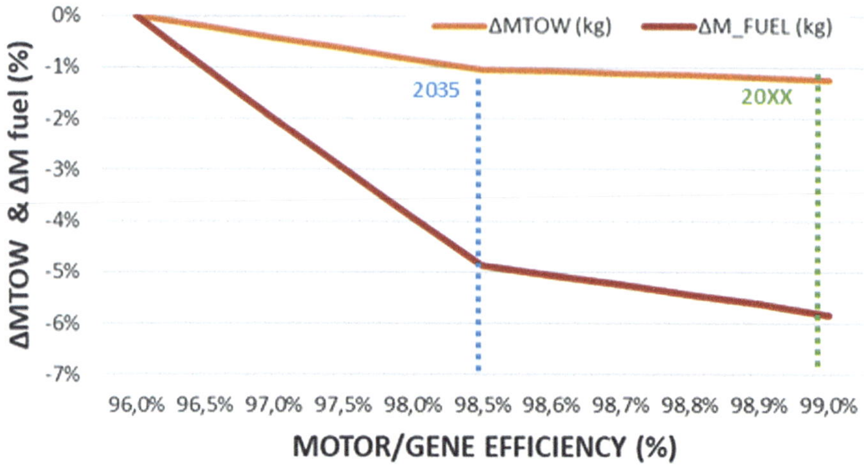

Fig. 7 Impact of Efficiency variations of electric motor/generator on the MTOW and the fuel mass from the 2025 target

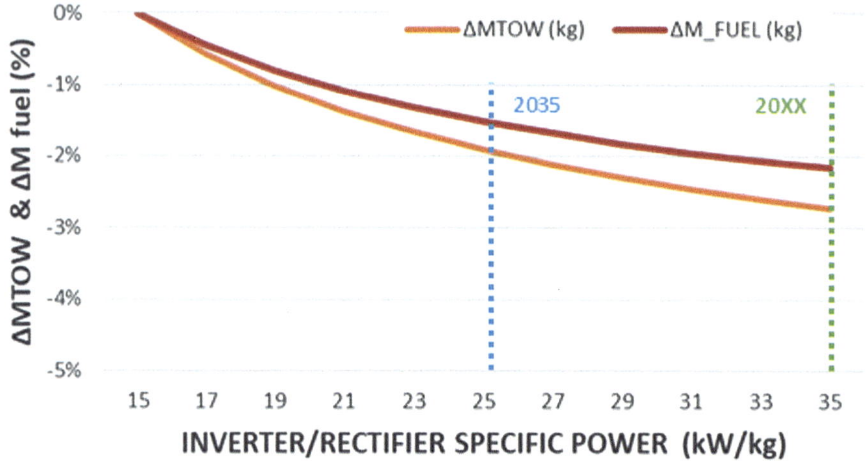

Fig. 8 Impact of Specific Power variations of inverters/rectifiers on the MTOW and the fuel mass from the 2025 target

The following analysis deals with the sensitivity of the same factors on the whole hybrid electric system by coupling all the elements of the system in the same figures.

In the final figures, the two influence factors have been coupled (see Figs. 10 and 11). In this last sensitivity analysis, it is interesting to gather the two influent factors (specific powers and efficiencies) that characterize the technological and conceptual advance in terms of power integration and efficiency. With reference

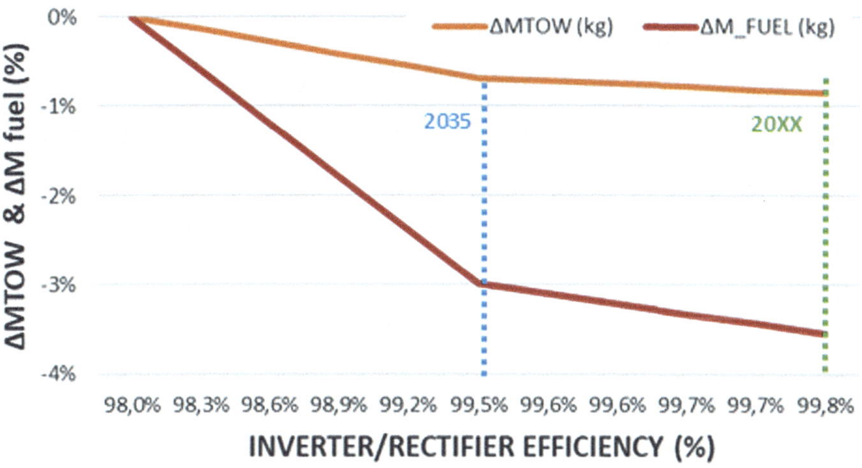

Fig. 9 Impact of efficiency variations of inverters/rectifiers on the MTOW and the fuel mass from the 2025 target

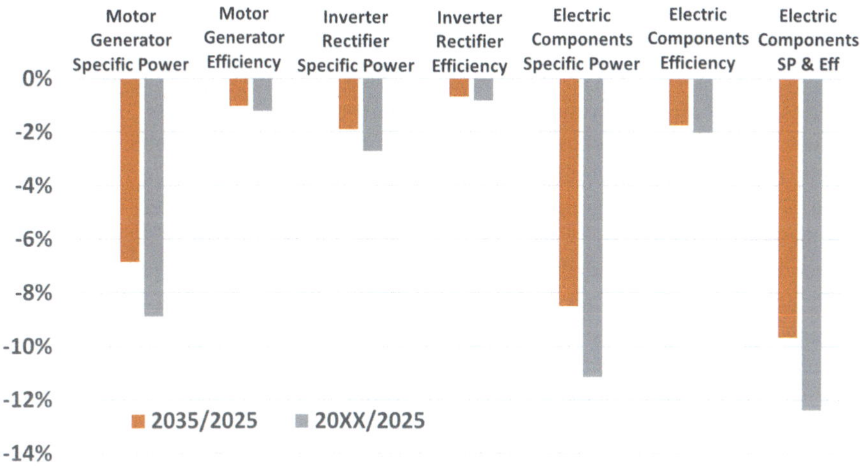

Fig. 10 MTOW variations versus technological improvement of the electric conversion chain from the 2025 target

to the less-aggressive assumptions (target 2025) and regarding the targets 2035 then 20XX assessments, the whole weight would be reduced by more than 12% (see Fig. 10) when the fuel burn is reduced by 17% (see Fig. 10). To conclude, electric components-specific power is more sensitive to lower the MTOW. But the efficiency effect is not negligible; a trade-off between both specific powers and efficiencies has to be found.

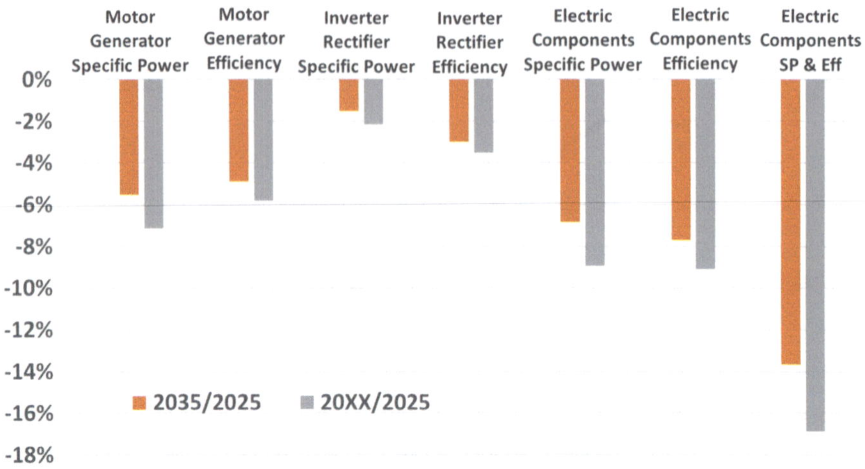

Fig. 11 Fuel mass variations versus technological improvement of the electric conversion chain from the 2025 target

5 Conclusions

A series hybrid electric aircraft architecture performance is "strongly" linked with specific power and efficiency of each electric component. The whole embedded weight (MTOW) of a hybrid structure is a priori higher than for a full thermal chain. Two different progress axes exist to minimize the embedded weight: the specific power of each device as well as its efficiency. When adding supplementary masses of electric devices, there is a snowball effect with the MTOW that leads to the enhancing of the aircraft thrust, consequently enhancing fuel consumption. Fortunately, energy gains offered by a higher efficiency of the electric chain, especially at low-power operation (taxi and descent phases), compensate the weight increase. But this paper emphasizes why both specific powers and device efficiency have a huge impact on the MTOW and consequently the fuel burn.

Future works in the HASTECS project consist in a more precise sensitivity analysis based on the technical models of power electronics and electromechanical converters. The aim of this prospect is to find the more sensitive input parameters which may be the decision variable of our future system design optimization. For that purpose, sizing models of electric machines and power electronics have to be refined, but sufficiently simplified to be included in an integrated design process involving the whole power train.

Acknowledgements This project has received funding from the European Union Horizon 2020 (cleansky 2 JTI) research and innovation programme, 2014–2024 under grant agreement No 715483.

References

1. X. Roboam, B. Sareni, A. DeAndrade, More electricity in the air: toward optimized electrical networks embedded in more-electrical aircraft. IEEE Indus. Electron. Mag. **6**(4), 6–17 (2012). ISSN 1932-4529
2. Rolls Royce, The E-FAN X programme (2018), https://www.rolls-royce.com/media/our-stories/insights/2018/paul-stein-talks-about-e-fan-x.aspx
3. Zunum Aero (2018), https://www.jetblueventures.com/portfolio/zunum-aero/
4. R.H. Jansen, C. Bowman, A. Jankovsky, R. Dyson, J. Felder, Overview of NASA electrified aircraft propulsion research for large subsonic transports, in *AIAA Propulsion and Energy 2017 Forum*, Atlanta, GA, United States, 10–12 July 2017
5. B.J. Brelje, J.R.R.A. Martins, Electric, hybrid, and turboelectric fixed-wing aircraft: a review of concepts, models, and design approaches. Prog. Aerosp. Sci. **104**, 1–19 (2019)
6. K.R. Antcliff, M.D. Guynn, T. Marien, D.P. Wells, S.J. Schneider, M.J. Tong, Mission analysis and aircraft sizing of a hybrid-electric regional aircraft, in *54th AIAA Aerospace Sciences Meeting, American Institute of Aeronautics and Astronautics*, San Diego, California, USA, Jan 2016
7. J. Thauvin, Exploring the design space for a hybrid-electric regional aircraft with multidisciplinary design optimisation methods, PhD Université de Toulouse, France, Oct 2018
8. A. Sgueglia, P. Schmollgruber, N. Bartoli, O. Atinault, E. Bénard, J. Morlier, Exploration and sizing of a large passenger aircraft with distributed electric ducted fans, in *AIAA - Scitech 2018*, Kissimmee, United States, 8 Jan 2018–12 Jan 2018
9. F. Michel, H. Fieseler, L. Allidieres, Liquid hydrogen technologies for mobil use, in *WHEC*, Lyon, France, 16 June 2006
10. Siemens, Electric propulsion components with high power densities for aviation (2015), https://nari.arc.nasa.gov/sites/default/files/attachments/Korbinian-TVFW-Aug2015.pdf
11. X. Yi, A. Yoon, K. Haran, Multi-physics optimization for high-frequency air-core permanent-magnet motor of aircraft application, in *2017 IEEE International Electric Machines and Drives Conference (IEMDC)*, Miami, FL, USA, pp. 1–8
12. D. Zhang, NASA SiC light-weight inverter for MW-power (SLIM)—phase I, in *55th AIAA Aerospace Sciences Meeting, AIAA*, Reston, VA, 2017

Direct Torque Control Applied to DFIG Supplied via a Three-Level Inverter Under Random Behavior Wind Speed

Salah Tamalouzt, Nabil Benyahia, Mariama Said Mohamed, Angel Scipioni, and Bernard Davat

Abstract A doubly fed induction generator (DFIG) driven by a variable speed wind turbine (WT) is presented in this paper. The DFIG is partially interfaced with its rotor via a back-to-back converter. This one is supplied via a three-level inverter in the rotor side and controlled with a flexible algorithm based on DTC technique, to ensure mastery of this generator. The main aim of this contribution is to analyze the performances and robustness of the proposed control technique. The aim of this structure is to obtain at the generator output AC sine waveforms signals with a constant frequency and a low THD, as well as minimum output voltages ripples, regardless of the variation of the wind speed. Indeed, the main objective of this work is the performance analysis of the DTC applied to a three-level inverter in the rotor side of the DFIG considering some constraints that reflect the real operation of the wind turbine generator, such as the randomness behavior of the wind speed, allowing all operation modes of this generator. These operation modes are carried out in a successive and continuous manner, while showing synchronous and overspeed modes. Simulation results, performed under Matlab/Simulink, are presented and analyzed.

1 Introduction

The wind power generation systems are among the most effective systems available to generate electricity. Particularly, the DFIG remains widely used in the case of a

S. Tamalouzt (✉)
Laboratoire LTII, Département de Génie Electrique, Faculté de Technologie, Université A/Mira de Bejaia, Bejaia, Algeria

N. Benyahia
Laboratoire LATAGE, Université de M/Mammeri Tizi-Ouzou, Tizi-Ouzou, Algeria

M. Said Mohamed · A. Scipioni · B. Davat
Groupe de Recherche en Electrotechnique et Electronique de Nancy (GREEN), Université de Lorraine, Nancy, France

W. Zamboni, G. Petrone (eds.), *ELECTRIMACS 2019*, Lecture Notes in Electrical Engineering 615, https://doi.org/10.1007/978-3-030-37161-6_7

grid-connected structure. This particularity is due to the potential of delivering high yields in regards to energy production and allowing the independent control of active and reactive power generation [1]. The main advantage of this generator is the size of the power electronic converters that are smaller than those of conventional full-size stator converters. In order to develop an efficient conversion of wind energy connected to the grid, several structures and topologies of converters and control techniques associated with generators of production are proposed in the literature [1–18]. Indeed, the structures based on multilevel inverters improve efficiency and maximize the power delivered to the grid. A multilevel inverter has several advantages over a conventional two-level inverter that uses high switching frequency PWM. Multilevel inverters continue to receive more and more attention because of their high voltage operation capability, low switching losses, high efficiency, and low output of electromagnetic interference. Also, it is possible to verify the good performance of these structures, when supplying the electric grid, by sinusoidal current with less total harmonic distortion (THD), and an adjustable output power factor. Multilevel converters not only can generate the output voltages with very low distortion but also can reduce the dV/dt stress; therefore, electromagnetic compatibility problems can be reduced. They can operate at both fundamental switching frequency and high PWM switching frequency [18–22].

To highlight the advantages of the chosen generator and to develop a wind energy conversion connected to an efficient and economical micro-grid, the control technique selection to adopt is indispensable. One competitive control technique is a direct torque control because of its numerous advantages as a simpler structure with high dynamic performances. Several works have been done in this field [7–15, 18].

Nevertheless, the combination of the WT-DFIG with its converters' advantages and those of DTC applied to a three-level inverter in all operation modes of the DFIG (subsynchronous, supersynchronous, and synchronous mode) in a successive and continuous manner has never been reported in the literature. The evolution of these modes can take long time intervals in a successive and continuous manner, it is why, its study becomes a necessity. Especially, if we want to consider the randomness behavior of the wind speed profile, which is never reported or studied except in [9–12], in which theses modes were studied in a successive and continuous manner, using two-level inverter.

In the field of systems control, the two main handicaps of control techniques can be summed up as follows: The pursuit of references to abrupt changes and the robustness to external and internal disturbances. In this context, the main objectives of this paper are to show the performances and the robustness of the proposed DTC, applied to a WT-DFIG supplied via a three-level inverter, considering the randomness variations of wind speed, allowing the DFIG operation in all modes, while showing synchronous and overspeed modes. That is to show clearly the main advantages of the DFIG with its converter. In addition, the insertion of a three-level inverter to the DFIG will allow the generated signals of very acceptable waveforms and very low THD. Moreover, the quality and performances of this technique are verified to reproduce some constraints that reflect the real operation of the wind, such as the random variation of the wind speed in the most interesting operating zones of the WT characteristics.

The present work is started with the presentation and description of the proposed system. After that, the models of the WT operation and the DFIG with their control technique are discussed and developed. The effectiveness and the performances of the proposed system are verified by simulation under MATLAB/SIMULINK environment.

2 Description of the Proposed System

The proposed overall system is based on a DFIG dedicated to a wind turbine partially interfaced with the grid, via an indirect frequency converter (AC/AC), as illustrated in Fig. 1. This frequency converter consists of two converters separated by an intermediate DC bus, serving as a DC-grid. The first converter (I) is a three-level inverter connected to the rotor of this generator and controlled by the proposed control technique (DTC). Whereas, the connection to the AC-grid is provided via a second converter (II) controlled so as to master its output at an adjustable power factor operation with DC-grid voltage regulation, and to guarantee sinusoidal signals with a constant frequency of 50 Hz.

3 Direct Torque Control of the Three-Level Inverter

The objectives of wind turbine controller depend on the operating zone, as described in [1].

The DFIG model expressed in the rotor frame is written as follows in [1, 12].

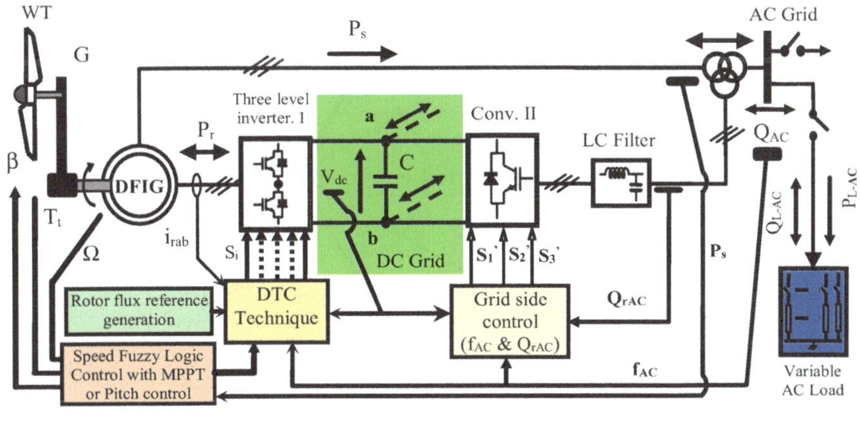

Fig. 1 Global control scheme

The DTC is based on the rotor flux magnitude regulation and the generator electromagnetic torque value. Where the rotor flux magnitude is estimated from its components along the α and β axes, as given in [11]:

$$\begin{cases} \Phi_{r\alpha}(t) = \int_0^t (v_{r\alpha} - R_r i_{r\alpha})\, dt \\ \Phi_{r\beta}(t) = \int_0^t \left(v_{r\beta} - R_r i_{r\beta}\right)\, dt \\ \Phi_r = \sqrt{\Phi_{r\alpha}{}^2 + \Phi_{r\beta}{}^2} \end{cases} \quad (1)$$

The electromagnetic torque is given in [1, 11] by the following expression:

$$T_{em} = p\left(\Phi_{r\alpha} i_{r\beta} - \Phi_{r\beta} i_{r\alpha}\right) \quad (2)$$

In order to analyze the output voltage from this three-level inverter, every leg is illustrated by three switches, which permit to independently connect the rotor inputs to the following source voltages ($V_{dc}/2$, 0, and $-V_{dc}/2$).

The transformation into $\alpha\beta$ (or dq) subspace obtains a voltage vector that can be associated with the spatial position of the rotor flux. The number of the different states of this vector is 19, because the 8 other combinations produce the same voltage vector. By using the diagonal between the adjacent medium and long vector, it can be seen that the space voltage vector diagram for the three-level inverter is divided into 12 sectors. In Fig. 2, we illustrate an optimal voltage vector, which is selected from the 19 available vectors according to the errors that are set on electromagnetic torque and the rotor flux linkage.

Fig. 2 Space voltage vectors presentation

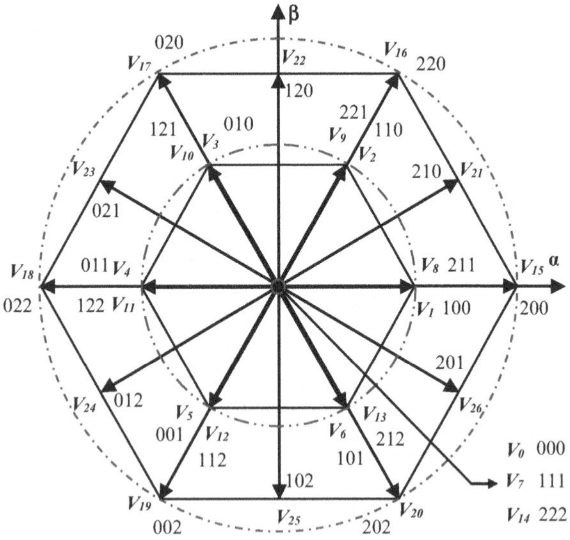

The operation of DFIG's direct torque control is characterized by two steps: in one step, the appropriate rotor voltage vector to be applied to the three-stage inverter NPC structure is selected. The other step is characterized by the estimation of the rotor flux and the electromagnetic torque. These estimated values are compared to the respective references, and the errors are used through hysteresis controllers.

The direct torque control of the DFIG topology is based on the hysteresis controller output relating to the variable flux (C_{flx}) and the variable torque (C_{trq}) associated to the sector N corresponding to the rotor flux vector position.

In this paper, we consider that the rotor flux control is achieved by using a three-level hysteresis comparator, while the electromagnetic torque is controlled by using 5-level hysteresis. The inverter state is considered as high if the output of the torque comparator is high or equal to two. Otherwise, the state is low, as given in equations (3) and (4).

$$\begin{cases} \text{If } \Delta tr > \varepsilon_{\text{tr}_2} & \text{Then; } C_{\text{trq}} = +2 \\ \text{If} -\varepsilon_{\text{tr}_1} \le \Delta tr \le \varepsilon_{\text{tr}_2} \ \& \ \frac{\Delta tr}{dt} > 0 & \text{Then; } C_{\text{trq}} = +1 \\ \text{If} -\varepsilon_{\text{tr}_1} \le \Delta tr \le \varepsilon_{\text{tr}_2} \ \& \ \frac{\Delta tr}{dt} < 0 & \text{Then; } C_{\text{trq}} = +2 \\ \text{If } 0 < \Delta tr \le \varepsilon_{\text{tr}_1} \ \& \ \frac{\Delta tr}{dt} > 0 & \text{Then; } C_{\text{trq}} = 0 \\ \text{If } 0 < \Delta tr \le \varepsilon_{\text{tr}_1} \ \& \ \frac{\Delta tr}{dt} < 0 & \text{Then; } C_{\text{trq}} = +1 \\ \text{If} -\varepsilon_{\text{tr}_1} \le \Delta tr \le 0 \ \& \ \frac{\Delta tr}{dt} > 0 & \text{Then; } C_{\text{trq}} = -1 \\ \text{If} -\varepsilon_{\text{tr}_1} \le \Delta tr \le 0 \ \& \ \frac{\Delta tr}{dt} > 0 & \text{Then; } C_{\text{trq}} = 0 \\ \text{If} -\varepsilon_{\text{tr}_2} \le \Delta tr \le -\varepsilon_{\text{tr}_1} \& \ \frac{\Delta tr}{dt} > 0 & \text{Then; } C_{\text{trq}} = -2 \\ \text{If} -\varepsilon_{\text{tr}_2} \le \Delta tr \le -\varepsilon_{\text{tr}_1} \& \ \frac{\Delta tr}{dt} > 0 & \text{Then; } C_{\text{trq}} = -1 \\ \text{If } \Delta tr < -\varepsilon_{\text{tr}_2} & \text{Then; } C_{\text{trq}} = -2 \end{cases} \tag{3}$$

$$\begin{cases} \text{If } \Delta\varphi_{\text{r}} > \varepsilon_{\text{flx}} & \text{Then; } C_{\text{flx}} = +1 \\ \text{If } 0 \le \Delta\varphi_{\text{r}} \le \varepsilon_{\text{flx}} \ \& \ \frac{\Delta\varphi_{\text{r}}}{dt} > 0 & \text{Then; } C_{\text{flx}} = 0 \\ \text{If } 0 \le \Delta\varphi_{\text{r}} \le \varepsilon_{\text{flx}} \ \& \ \frac{\Delta\varphi_{\text{r}}}{dt} < 0 & \text{Then; } C_{\text{flx}} = +1 \\ \text{If} -\varepsilon_{\text{flx}} < \Delta\varphi_{\text{r}} < 0 \ \& \ \frac{\Delta\varphi_{\text{r}}}{dt} > 0 & \text{Then; } C_{\text{flx}} = -1 \\ \text{If} -\varepsilon_{\text{flx}} < \Delta\varphi_{\text{r}} < 0 \ \& \ \frac{\Delta\varphi_{\text{r}}}{dt} < 0 & \text{Then; } C_{\text{flx}} = 0 \\ \text{If } \Delta\varphi_{\text{r}} < -\varepsilon_{\text{flx}} & \text{Then; } C_{\text{flx}} = -1 \end{cases} \tag{4}$$

The electromagnetic torque control is ensured by using a five-level hysteresis comparator, which allows having both directions of machine rotation, the corrector output variable is a Boolean type C_{trq}, it indicates if the electromagnetic torque amplitude must be increased, decreased or maintained constant ($C_{\text{trq}} = 1, 2, 0, -2, -1$), as modeled by Eq. (3). A three-level hysteresis comparator is used to control the rotor flux. The comparator output variable is a Boolean type C_{flx}, it indicates if the rotor flux amplitude must be increased ($C_{\text{flx}} = 1$), decreased ($C_{\text{flx}} = -1$), or

Table 1 DTC Commutation table of the three-level inverter

C_{flx}	C_{trq}	*N* 1	2	3	4	5	6	7	8	9	10	11	12
+1	+2	V_{21}	V_{16}	V_{22}	V_{17}	V_{23}	V_{18}	V_{24}	V_{19}	V_{25}	V_{20}	V_{26}	V_{15}
	+1	V_{21}	V_2	V_{22}	V_3	V_{23}	V_4	V_{24}	V_5	V_{25}	V_6	V_{26}	V_1
	0	Zero vector											
	−1	V_{26}	V_1	V_{21}	V_2	V_{22}	V_3	V_{23}	V_4	V_{24}	V_5	V_{25}	V_6
	−2	V_{26}	V_{15}	V_{21}	V_{16}	V_{22}	V_{17}	V_{23}	V_{18}	V_{24}	V_{19}	V_{25}	V_{20}
−1	+2	V_{17}	V_{23}	V_{18}	V_{24}	V_{19}	V_{25}	V_{20}	V_{26}	V_{15}	V_{21}	V_{16}	V_{22}
	+1	V_3	V_{23}	V_4	V_{24}	V_5	V_{25}	V_6	V_{26}	V_1	V_{21}	V_2	V_{22}
	0	Zero vector											
	−1	V_5	V_{25}	V_6	V_{26}	V_1	V_{21}	V_2	V_{22}	V_3	V_{23}	V_4	V_{24}
	−2	V_{19}	V_{25}	V_{20}	V_{26}	V_{15}	V_{21}	V_{16}	V_{22}	V_{17}	V_{23}	V_{18}	V_{24}
0	+2	V_{22}	V_{17}	V_{23}	V_{18}	V_{24}	V_{19}	V_{25}	V_{20}	V_{26}	V_{15}	V_{21}	V_{16}
	+1	V_{22}	V_3	V_{23}	V_4	V_{24}	V_5	V_{25}	V_6	V_{26}	V_1	V_{21}	V_2
	0	Zero vector											
	−1	V_{25}	V_6	V_{26}	V_1	V_{21}	V_2	V_{22}	V_3	V_{23}	V_4	V_{24}	V_5
	−2	V_{25}	V_{20}	V_{26}	V_{15}	V_{21}	V_{16}	V_{22}	V_{17}	V_{23}	V_{18}	V_{24}	V_{19}

maintained constant ($C_{flx} = 0$) so as to maintain: $|\Phi_{r\_ref} - \Phi_r| \leq \Delta\varphi_r$, as given by Eq. (4).

Table 1 represents the commutation table, which is considered as one of the solutions adapted to choose the optimally selected voltage vector for each sector. Where, the rotor flux and the electromagnetic torque are achieved by using, respectively, three levels and five levels hysteresis comparator.

In analyzing the effect of each available voltage vector, it can be seen that the vector affects the electromagnetic torque and the rotor flux linkage with the variation of the module and direction of the selected vector. The nulls voltages vectors (V_0, V_7, and V_{14}) are chosen alternately, so as to minimize the number of commutations in the arms of the inverter [20].

4 Simulation Results and Discussion

The effectiveness of the proposed control strategy that leads to better performances of the WT-DFIG system is carried out. Simulations have been realized under wind speed variations, in the aim to allow all the DFIG operation modes in a successive and continuous manner. The WT-DFIG simulation parameters are given in [1]. The DTC strategy presented in the previous paragraph is then introduced, to control the three-level inverter, under MATLAB/SIMULINK environment. The weather data in terms of wind speed profiles are shown in Fig. 3a.

The DFIG mechanical speed is presented in Fig. 3a. The tracking of the references is shown by the mechanical speed, electromagnetic torque, and the rotor

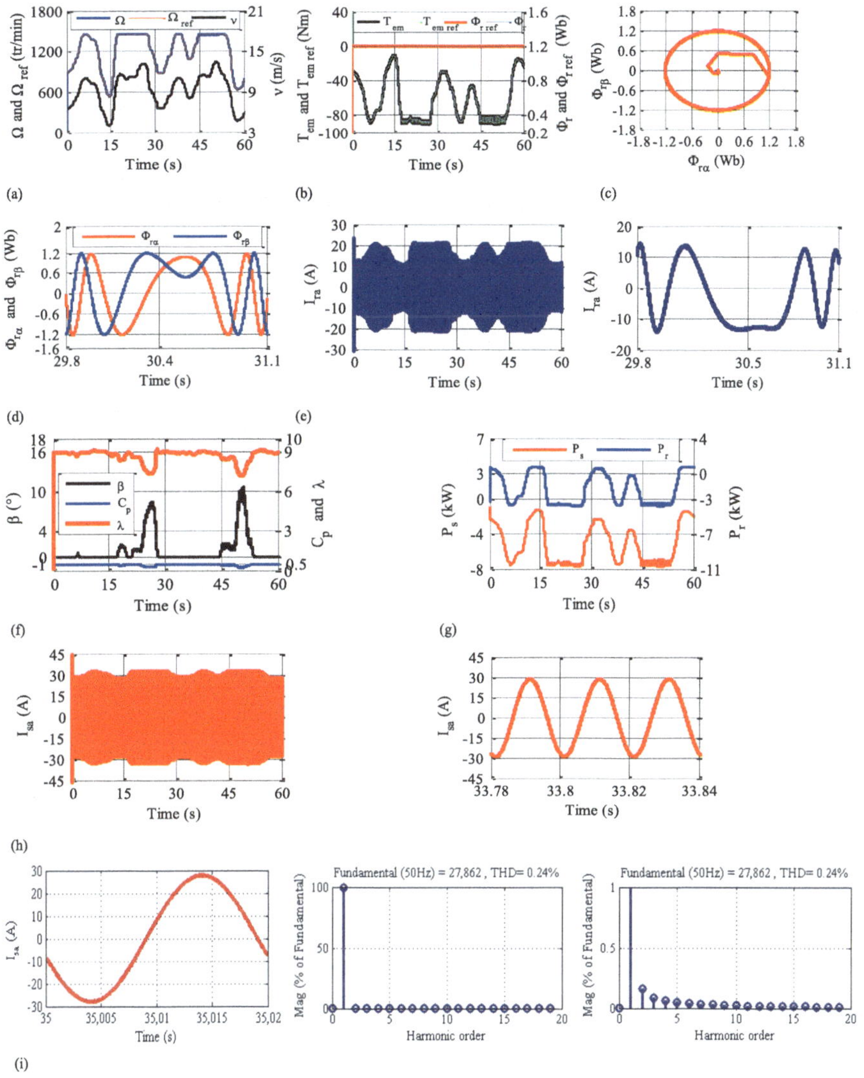

Fig. 3 Waveforms of the simulation results for a variable wind speed: (**a**) Ω, Ω_{ref} and ν; (**b**) Φ_{r}, $\Phi_{\text{r ref}}$, T_{em} and $T_{\text{em ref}}$; (**c**) $\Phi_{\text{r}\beta} = f(\Phi_{\text{r}\alpha})$; (**d**) β, λ and C_{p}; (**e**) P_{r} and P_{s}; (**f**) Zooms of the $\Phi_{\text{r}\alpha}$, $\Phi_{\text{r}\beta}$ in the three modes DFIG operation; (**g**) I_{ra} and its zoom in the three modes DFIG operation; (**h**) I_{sa} and its zoom; (**i**) Generated phase current (stator current, I_{sa}) and its harmonic spectrum in supersynchronous mode operation of DFIG

flux magnitude waveforms, as shown, respectively, in Fig. 3a–c. Hence, these results confirm better performances and robustness of the proposed control. Moreover, the torque ripples are about 60% less than the results given in [1, 11].

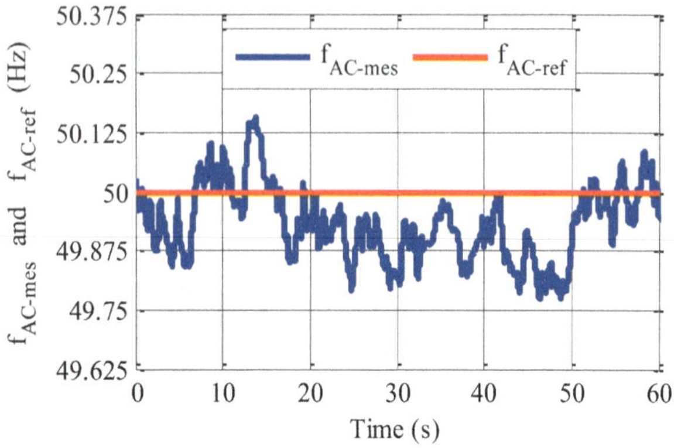

Fig. 4 The output frequency f_{AC} and its reference $f_{AC\text{-ref}}$

Figure 3c shows the rotor flux waveform, which is circular and kept constant at 1.2 Wb. This form leads to a sinusoidal fluxes $\Phi_{r\alpha}$ and $\Phi_{r\beta}$ behavior (Fig. 3d) in both the sub- and supersynchronous modes. However, in the synchronous mode the rotor fluxes have continuous behavior. Also, this operation mode is proved by a direct form of the rotor phase currents, as presented in Fig. 3e. In these figures, we illustrate the sinusoidal evolution of the rotor currents, in the sub- and supersynchronous operation modes. Besides, the current waveform in Fig. 3e shows good signal quality.

Figure 3f illustrates the β, λ, and C_p (blade pitch angle, tip speed ratio, and power coefficient respectively), where one can see that when the power reached its maximum value, the pitch angle control is activated, as presented in Fig. 3g, which shows all the exchanged powers between the grid and DFIG. The P_s depends on the wind speed evolution, whereas, P_r changes its direction, marked by its sign, according the generator slip.

Finally, the sinusoidal form of the generated current shown in Figs. 3h, i and 4 show the quality of the power injected in the AC-grid from the DFIG stator. This remains true also for the three operating modes with a constant 50 Hz frequency. Moreover, the variable amplitude is a consequence of the wind speed variation.

Compared with the results obtained in [1, 11], the proposed DTC associated to the three-level inverter, offers a significant improvement to the stator currents supplied to the AC-grid, with a total harmonic distortion (THD) largely below the limits imposed by IEEE std 519, as illustrated in Fig. 4. This confirms a better quality of energy generated in the AC-grid.

5 Conclusions

The performances of the DTC applied to a three-level inverter have been presented in this paper; taking into account the mains advantages of the DFIG when associated with its converter. In this work, the wind speed random variation is imposed in order to check the quality and performances of this control. Thus, the randomness behavior of the wind speed allowed the utilization of the DFIG in its different operation modes. The study in this paper is focused on the DTC dynamic performances. The effectiveness and validity of this control are verified by numerical simulation using Matlab/Simulink.

The obtained simulation results show that the application of the DTC significantly improves the performance of the conversion system, which contributes to reducing the harmonic distortion rate in the output voltage and current waves and the torque ripples (the generated power).

The results obtained confirm an acceptable robustness of the proposed control, and the performances are satisfactory. The three-level inverter inserted in the DFIG rotor also contributes to improving the quality of the energy supplied to the AC-grid.

References

1. S. Tamalouzt, N. Benyahia, T. Rekioua, D. Rekioua, R. Abdessemed, Performances analysis of WT-DFIG with PV and fuel cell hybrid power sources system associated with hydrogen storage hybrid energy system. Int. J. Hydrog. Energy **41**(45), 21006–21021 (2016)
2. K.C. Wong, S.L. Ho, K.W.E. Cheng, Direct torque control of a doubly-fed induction generator with space vector modulation. Elec. Power Comp. Syst. **36**(12), 1337–1350 (2008)
3. M. Rahimi, M. Parniani, Dynamic behavior analysis of doubly-fed induction generator wind turbines, the influence of rotor and speed controller parameters. Int. J. Elec. Power Energy Syst. **32**(5), 464–477 (2010)
4. K. Kerrouche, A. Mezouar, K. Belgacem, Decoupled control of doubly fed induction generator by vector control for wind energy conversion system. Energy Procedia **42**, 239–248 (2013)
5. M. Adjoudj, M. Abid, A. Aissaoui, Y. Ramdani, H. Bounoua, Sliding mode control of a doubly fed induction generator for wind turbines. Rev. Roum. Sci. Techn. Electrotechn. et Energ. **56**(1), 15–24 (2011)
6. A. Abdelfetteh, M. Abid, Hybrid fuzzy sliding mode control of a doubly fed induction generator in wind turbines. Rev. Roum. Sci. Techn. Electrotechn. et Energ. **57**(4), 15–24 (2012)
7. J. Arbi, M.J. Ben Ghorbal, I. Slama-Belkhodja, L. Charaabi, Direct virtual torque control for doubly fed induction generator grid connection. IEEE Trans. Ind. Electron. **56**(10), 4163–4173 (2009)
8. J. Ben Alaya, A. Khedher, M.F. Mimouni, Speed-sensorless DFIG wind drive based on DTC using sliding mode rotor flux observer. Int. J. Renew. Energy Res. **2**(4), 735–745 (2012)
9. A.F. Payam, M.N. Hashemnia, J. Faiz, Robust DTC control of doubly fed induction machines based on input-output feedback linearization using recurrent neural networks. J. Power Electron. **11**(5), 719–725 (2011)
10. S. Tamalouzt, T. Rekioua, R. Abdessamed, K. Idjdarene, Direct torque control of grid connected doubly fed induction generator for the wind energy conversion, in *International Renewable Energy Congress IREC'2012*, Sousse, Tunisia, 19–22 Dec. 2012, pp. 1–6
11. S. Tamalouzt, K. Idjdarene, T. Rekioua, R. Abdessamed, Direct torque control of wind turbine driven doubly fed induction generator. Rev. Roum. Sci. Techn. Electrotechn. et Energ. **61**(3), 244–249 (2016)

12. S. Tamalouzt, T. Rekioua, R. Abdessemed, Direct torque and reactive power control of grid connected doubly fed induction generator for the wind energy conversion, in *International Conference on Electrical Sciences and Technologies in Maghreb (CISTEM)*, Tunis, 3–6 Nov. 2014, pp. 1–7
13. J. Hu, X. Yuan, VSC-based direct torque and reactive power control of doubly fed induction generator. Renew. Energy **40**, 13–23 (2012)
14. M. Sandhu, T. Thakur, Multilevel inverters: literature survey – topologies, control techniques & applications of renewable energy sources - grid integration. J. Eng. Res. Appl. **04**(03 (Version 1)), 644–652 (2014)
15. Y. Djeriri, A. Meroufel, B. Belabbes, A. Massoum, Three-level NPC voltage source converter based direct power control of the doubly fed induction generator at low constant switching frequency. Rev. Energ. Renouv. **16**(01), 91–103 (2013)
16. A. Djoudi, S. Bacha, H. Iman-Eini, T. Rekioua, Sliding mode control of DFIG powers in the case of unknown flux and rotor currents with reduced switching frequency. Electr. Power Energy Syst. **96**, 347–356 (2018)
17. L. Gidwani, H. Tiwari, R.C. Bansal, Improving power quality of wind energy conversion system with unconventional power electronic interface. Electr. Power Energy Syst. **44**, 445–453 (2013)
18. N. Taïb, B. Metidji, T. Rekioua, Performance and efficiency control enhancement of wind power generation system based on DFIG using three-level sparse matrix converter. Int. J. Electr. Power Energy Syst. **53**, 287–296 (2013)
19. T. Ghennam, E.M. Berkouk, B. Francois, DC-link voltage balancing algorithm using a space-vector hysteresis current control for three-level VSI applied for wind conversion system, in *2007 European Conference on Power Electronics and Applications*, Aalborg, 2007, pp. 1–10
20. A. Nami, J. Liang, F. Dijkhuizen, G.D. Demetriades, Modular multilevel converters for HVDC applications: review on converter cells and functionalities. IEEE Trans. Power Electron. **30**(1), 18–36 (2015)
21. A. Nami, F. Zare, *Multilevel Converters in Renewable Energy Systems* (In-Tech, Cooperstown, NY, 2010)
22. A. Lega, Multilevel converters: dual two-level inverter scheme, Ph.D thesis in Electrical Engineering, University of Bologna, pp. 39–51, 2007

Sensorless Control Strategy for Switched Reluctance Traction Drive Based on High Frequency Injection

Pavel Sovicka, Giacomo Scelba, Pavol Rafajdus, and Vladimir Vavrus

Abstract This paper deals with the implementation of a sensorless control strategy devoted to Switched Reluctance Motor Drives used in the traction drive of a small truck. The sensorless technique operates at low and zero speed. In the proposed approach an additional high frequency magnetic field is injected into the machine and a suitable demodulation algorithm is exploited to extract useful information on the rotor position and speed. The feasibility of the implementation is verified by simulations.

1 Introduction

The anticipated proliferation of electric vehicles in upcoming years leads to a development focused on electric traction drives. Thanks to its inherent robustness, simplicity, and low manufacturing cost the Switched Reluctance Motor (SRM) is a perspective machine. For efficient operation it requires a suitable converter and accurate position information [1, 2]. An example of this focus can be seen in [3, 4]. To increase reliability and fault-tolerance, the position sensor must be backed up by a sensorless alternative. This paper focuses on a development and simulation analysis of a new sensorless method for low speed and low load operating conditions. For high speed and high load torque conditions other methods have to be used, e.g., [5–7].

P. Sovicka (✉) · P. Rafajdus · V. Vavrus
KVES – University of Zilina, Univerzitna 8215/1, Zilina, Slovakia
e-mail: pavel.sovicka@fel.uniza.sk; pavol.rafajdus@uniza.sk; vladimir.vavrus@kves.uniza.sk

G. Scelba
DIEEI – Universit degli Studi di Catania, Catania (CT), Italy
e-mail: giacomo.scelba@dieei.unict.it

W. Zamboni, G. Petrone (eds.), *ELECTRIMACS 2019*, Lecture Notes in Electrical Engineering 615, https://doi.org/10.1007/978-3-030-37161-6_8

There are two main approaches for low speed sensorless control of an SRM—Active probing (e.g., [8, 9]) and high frequency injection. Compared to other machines, the use of high frequency approach with the SRM provides additional challenges. Most of these are connected to the square waveform of motoring current, which is composed of a significant amount of higher harmonics. This complicates position signal separation. As was shown in [10], continuous phase excitation can improve the harmonic spectrum at the cost of generating negative torque. Alternatively synchronous frames can be used for current filtering [11]. Another challenge is the usage of asymmetric H-bridge converters that do not enable controlled application of negative voltage to the phase. Therefore a DC bias might be necessary. These issues can be mitigated significantly by using separate sensing coils, but this approach requires additional hardware [12]. Use of advanced algorithms to process information, such as Artificial Neural Networks, is also being investigated [13]. While complex, these methods can account for non-linearities and disturbances that are difficult to implement using classical mathematical modeling.

Differently than [11], the proposed solution extracts the information on the rotor position through a demodulation algorithm using a stationary high frequency sine wave voltage injection. A combination of band-pass and Low-Pass Filters, coordinate transformations, and synchronous frames is used to extract position error information. This position error is used by a Phase Locked Loop (PLL) to calculate rotor speed and position values for the control structure. The delay caused by this algorithm is actively compensated.

2 Switched Reluctance Drive Description

Switched Reluctance Motor Drive (SRMD) used in this study is composed of a 3 phase 18/12 SRM and an asymmetric half-bridge converter. Both the motor and converter were developed at Department of Power Electrical Systems on University of Zilina for a small electric truck. Basic parameters of the machine are shown in Table 1 [14]. Inductance profile is shown in Fig. 1. The converter uses FF400R06KE3 IGBT modules. Sampling frequency is set to 20 kHz.

Table 1 SRM parameters

Parameter	Value
Nominal voltage	300 V
Nominal phase current (RMS)	22.4 A
Maximum allowed instantaneous phase current amplitude	200 A
Rated power	16 kW
Rated speed	5000 rpm
Rated torque	30 Nm
Maximum static torque	250 Nm

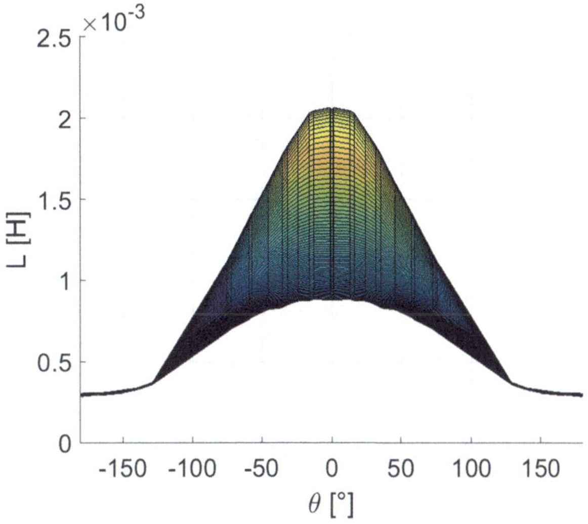

Fig. 1 Inductance profile of analyzed SRM. Note that peak inductance is saturated from 2.06 to 0.89 mH as the current increases from 0 to 200 A

3 Signal Injection

Performance and accuracy of any high frequency voltage sensorless method is based on a correct signal injection. This is especially true in the SRMD for several reasons:

- Square shape of main phase excitation current produces a significant amount of higher harmonics.
- The asymmetric half-bridge converter does not allow injection of negative voltage. Therefore a DC bias is required to inject a harmonic waveform correctly.
- An unipolar switching strategy might be preferred to reduce current ripple and converter losses. However, this approach distorts the waveform shape.
- By injecting into an active phase, the signal gets distorted by inductance saturation. On the other hand, not using the excited phase for injection removes at least one source of information for position estimation.

A 2 kHz stationary field voltage injection was selected for this work. While higher injection frequencies are desirable as they allow easier separation from motoring current harmonics, the selected frequency is a maximum practical value achievable for a 20 kHz switching and sampling frequency. Bipolar switching is used for the injection in unexcited phases and in the active phase when the demanded current is lower than 40 A. To lower current ripple during motoring an unipolar switching scheme is used for the active phase above this current threshold. Certain load dependence can therefore be expected due to inductance saturation.

While this approach can deteriorate accuracy and stability estimation during motoring, it decreases current ripple significantly. Figure 2 shows a block diagram of injection. The PWM block not only calculates correct duty cycle for the converter

Fig. 2 Block diagram of voltage injection

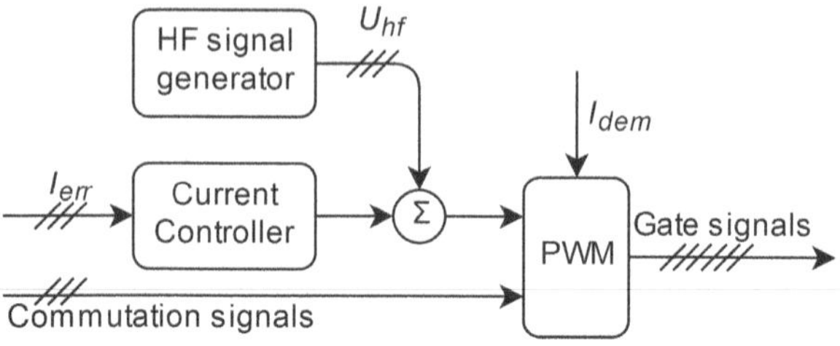

transistors, but also selects the appropriate switching scheme depending on the demanded current value I_{dem} and commutation signals.

4 Low Speed Sensorless Method

As it was shown in [11], the reverse inductance profile $1/L$ of a SRM phase can be decomposed by a series of cosine terms. In the following study the reciprocal value of the inductances is represented by approximating it with a DC term plus a cosinusoidal term. Used $1/L$ waveform equations for each phase are shown in (1)–(3).

$$\frac{1}{L_a} = A + B \cdot \cos(\omega_r t) \tag{1}$$

$$\frac{1}{L_b} = A + B \cdot \cos\left(\omega_r t - \frac{2\pi}{3}\right) \tag{2}$$

$$\frac{1}{L_c} = A + B \cdot \cos\left(\omega_r t - \frac{4\pi}{3}\right) \tag{3}$$

Constants "*A*" and "*B*" represent the inductance profile constants. As will be shown later in this paper, their exact value is irrelevant for this analysis. A stationary voltage (4) is injected into all phases. By neglecting the voltage drop in stator resistance and using a Band-Pass Filter, the resulting current response to the injection can be calculated according to (5)–(7). For better clarity during principle explanation, these equations assume that all current components related to motoring current have been filtered out.

$$U_{hf} = U_{amp} \cdot \sin\left(\omega_{hf} t\right) \tag{4}$$

$$i_{aBPF} = \frac{U_{amp}}{\omega_{hf}} \left\{ A \cdot \sin\left(\omega_{hf} t\right) + \frac{1}{2} B \cdot \sin \right.$$

$$\left. \cdot \left[\left(\left(\omega_{hf} + \omega_r \right) t \right) + \left(\left(\omega_{hf} - \omega_r \right) t \right) \right] \right\} \tag{5}$$

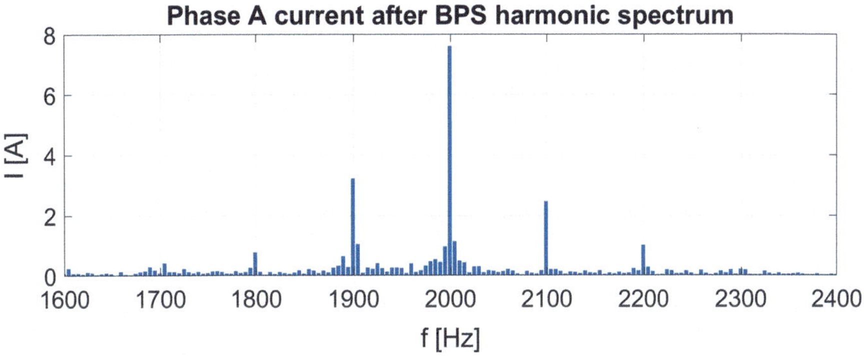

Fig. 3 Harmonic spectrum of the filtered motor phase A

$$i_{bBPF} = \frac{U_{amp}}{\omega_{hf}}\left\{A \cdot \sin\left(\omega_{hf}t\right) + \frac{1}{2}B\sin \cdot\left[\left(\left(\omega_{hf}+\omega_r\right)t - \frac{2\pi}{3}\right) + \left(\left(\omega_{hf}-\omega_r\right)t + \frac{2\pi}{3}\right)\right]\right\} \tag{6}$$

$$i_{cBPF} = \frac{U_{amp}}{\omega_{hf}}\left\{A \cdot \sin\left(\omega_{hf}t\right) + \frac{1}{2}B\sin \cdot\left[\left(\left(\omega_{hf}+\omega_r\right)t - \frac{4\pi}{3}\right) + \left(\left(\omega_{hf}-\omega_r\right)t + \frac{4\pi}{3}\right)\right]\right\} \tag{7}$$

As it can be seen in the high frequency current responses, three harmonic waveforms are expected to be present in the harmonic spectrum as a result of the high frequency voltage injection. Two of these waveforms, which contain rotor position information, form sidebands in respect to the main injected frequency. This can clearly be seen in harmonic spectrum shown in Fig. 3, which was captured from a simulation. The simulation will be described later. The rotor speed of 520 rpm corresponds to a frequency of 104 Hz. Sidebands containing double of the rotor position can also be seen in this figure, but they are suppressed by the demodulation algorithm. To improve figure clarity during principle explanation, no-load conditions were selected as they are not distorted by interference from the motoring current.

Because the studied machine has three phases a regular Clarke transformation (8) can be used to obtain currents in the orthogonal stationary frame. As can be seen in (9) and Fig. 4, the main high frequency current response is canceled out in this transformation.

$$i_{\alpha\beta} = \begin{bmatrix} 1 & -\frac{1}{2} & -\frac{1}{2} \\ 0 & +\frac{\sqrt{3}}{2} & -\frac{\sqrt{3}}{2} \end{bmatrix} \begin{bmatrix} i_a \\ i_b \\ i_c \end{bmatrix} \tag{8}$$

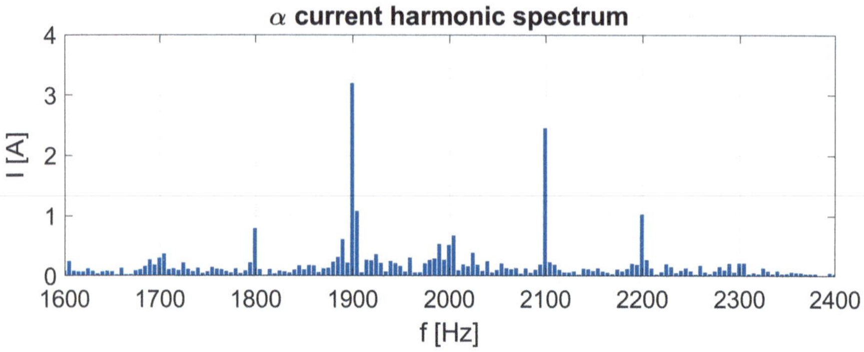

Fig. 4 Harmonic spectrum of α current component

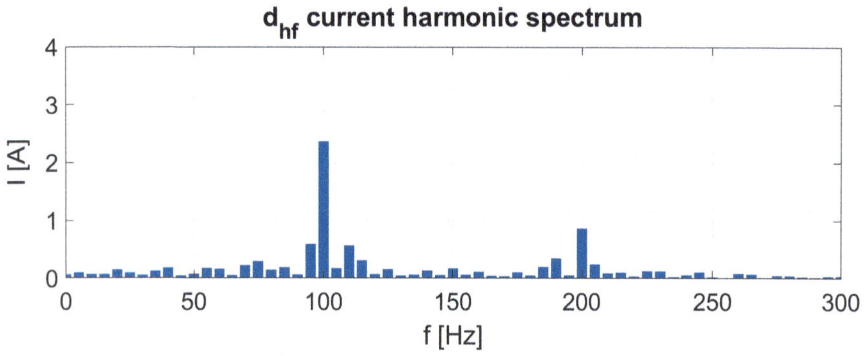

Fig. 5 Harmonic spectrum of d_{hf} current component

$$i_{\alpha\beta} = \frac{U_{amp}}{\omega_{hf}} \left[\frac{3}{4} B \cdot \left(e^{j\frac{\pi}{2} - j(\omega_{hf} - \omega_r)t} + e^{j\frac{\pi}{2} - j(\omega_{hf} + \omega_r)t} \right) \right] \tag{9}$$

To further simplify the equations an amplitude of remaining high frequency current response is defined in (10).

$$I = \frac{3}{4} B \cdot \frac{U_{amp}}{\omega_{hf}} \tag{10}$$

A further reference frame transformation, which is synchronous to the injected high frequency, is exploited next in (11). As it can be seen in Fig. 5 the 104 Hz rotor information is easily distinguishable.

$$i_{dqhf} = i_{\alpha\beta} \cdot e^{j\omega_{hf} t} = I \left[e^{j\left(\frac{\pi}{2} + \omega_r t\right)} + e^{j\left(\frac{\pi}{2} - \omega_r t\right)} \right] \tag{11}$$

Another step rotates the signal using estimated rotor speed $\hat{\omega}_r$ from the last known step according to (12). The first of the two remaining terms is $\hat{\omega}_r + \omega_r$.

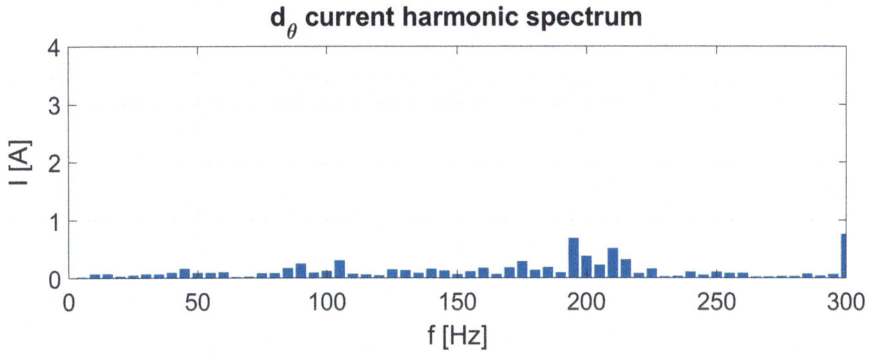

Fig. 6 Harmonic spectrum of d_θ current component

Assuming correct estimation this term rotates with double of the actual rotor speed. Above certain speed the following Low-Pass Filter mitigates this term. It can be seen in Fig. 5 as a small amplitude at a frequency of 208 Hz. The second term $\hat{\omega}_r - \omega_r$ is effectively a direct representation of position error. Decreasing cut-off frequency of the Low-Pass Filter limits impact of the first term, but also limits the dynamics of the entire structure because the position error cannot be higher than this cut-off frequency (13) (Fig. 6).

$$i_{dq\theta} = i_{dqhf} \cdot e^{j\left(-\frac{\pi}{2}+\hat{\omega}_r t\right)} = I\left[e^{j(\hat{\omega}_r+\omega_r)t} + e^{j(\hat{\omega}_r-\omega_r)t}\right] \tag{12}$$

$$i_{dqLPF} = LPF\left(i_{dq\theta}\right) = Ie^{j(\hat{\omega}_r-\omega_r)t} \tag{13}$$

Using both d and q components to extract the position error information according to (16) effectively cancels the amplitude represented by I introduced in (10). Therefore this method is dependent only on the inductance waveform shape and not its absolute values. Of course, the signal has to have a high enough Signal to Noise Ratio to be processed correctly.

$$\theta_{err} = \left(\omega_r - \hat{\omega}_r\right)t = \theta_r - \hat{\theta}_r \tag{14}$$

$$i_{dqLPF} = Ie^{j(-\theta_{err})} \tag{15}$$

$$\theta_{err} = -\text{atan2}\left(i_{dLPF}, i_{qLPF}\right) \tag{16}$$

The High Frequency Demodulator (HFD) is shown in Fig. 7.

A Phase Locked Loop (PLL) is used to estimate speed and position based on the position error information provided by the HFD. Speed output is connected to the speed controller. Position information is used not only for the commutation control, but also for the reference frame transformation (12) in the HFD. Note that the position for the commutation control is compensated for the processing delay.

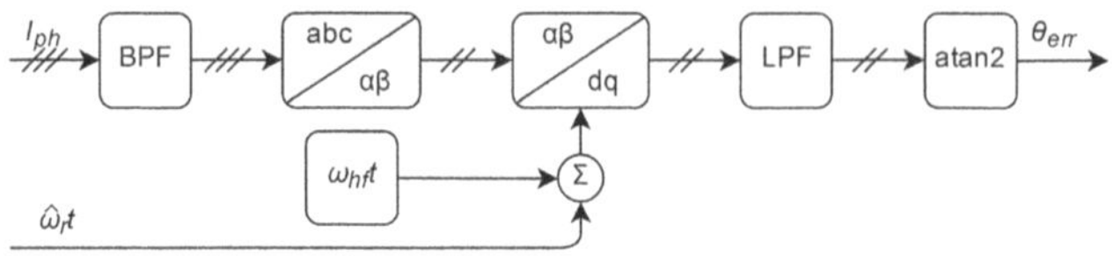

Fig. 7 Block diagram of High Frequency observer

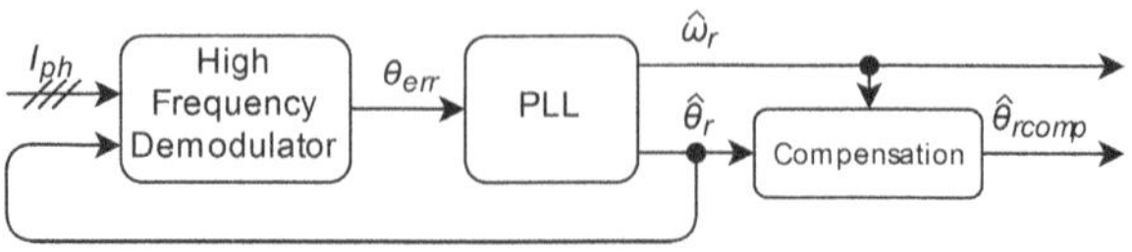

Fig. 8 Block diagram of the proposed method

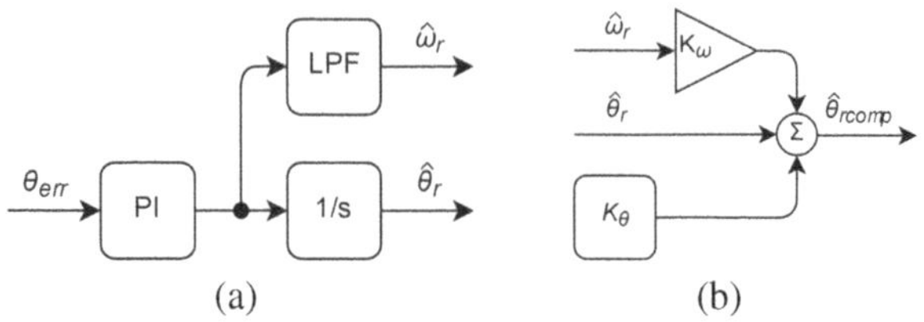

Fig. 9 Block diagrams of Phased Locked Loop and Compensation block. (**a**) Phase Locked Loop. (**b**) Compensation block

The compensated position is not used for the HFD as it would introduce interference with the internal dynamics of the demodulator (Fig. 8).

The PLL consists of a PI controller, an integrator, and another Low-Pass Filter, which is connected according to Fig. 9a. Compensation block, shown in Fig. 9b, corrects errors made by the delay in signal processing. This includes not only the calculation time itself, but also the phase delay caused by the multiple filters used in the algorithm. It consists of two parts. Firstly the static delay is compensated by adding a constant to the estimated position. Secondly a compensation of delay for the given speed is added. As the delay of position calculation is a linear function of speed a simple multiplication of estimated speed and an appropriate constant is sufficient. Estimation delay due to acceleration is not added because it would require speed difference information, which can be affected significantly by noise. Also if the SRMD dynamic is high, increasing the bandwidth of PLL PI controller can effectively reduce this particular estimation delay.

5 Performance Evaluation

5.1 Modeling of the SRM Drive

Block diagram of simulation implemented in Simulink is shown in Fig. 10. It is composed from the SRM itself, an accurate converter model (using models of actual IGBTs) created with Simscape Power Systems toolbox, speed controller, current controllers, phase commutation control, and the proposed method estimation block.

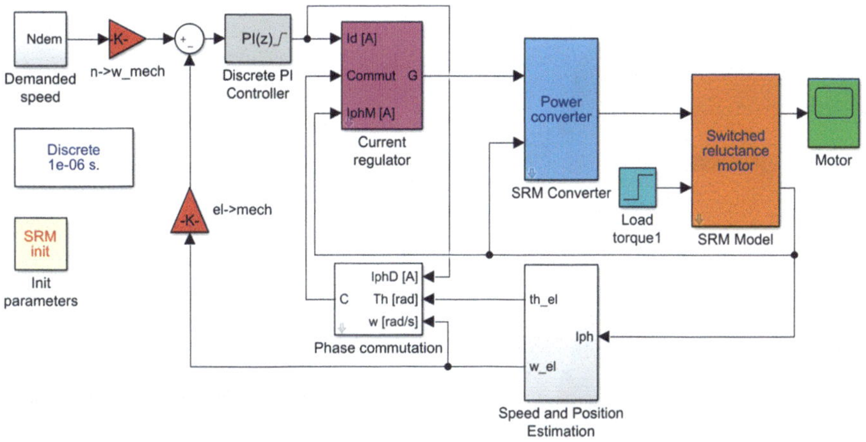

Fig. 10 Sensorless SRMD implementation in Simulink

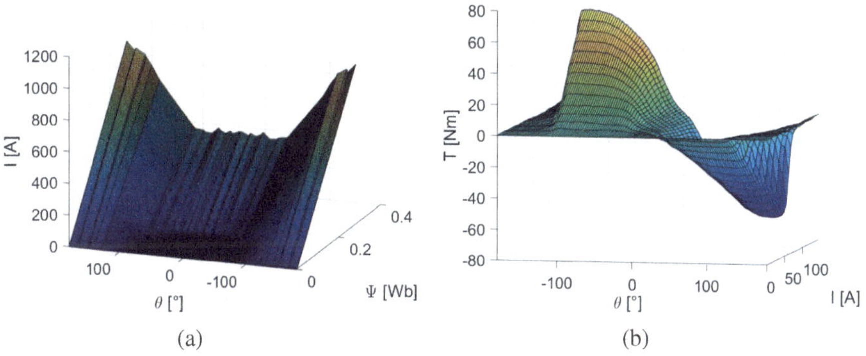

Fig. 11 Parameter maps used in analysis. (**a**) Reverse flux-linkage map. (**b**) Torque map

During the entire simulation the control algorithm uses only estimated speed and position. While the simulation runs at a step of 1 μs, the control and sensorless blocks are set to 50 μs according to the 20 kHz sampling frequency of the actual drive.

The SRM model is based on a reverse flux-linkage map $i = f(\theta, \Psi)$ and the torque map $T_e = f(\theta, i)$ [14], which are shown in Fig. 11a, b.

5.2 Simulation Results

As can be seen in Fig. 12, the simulation run consists of five distinct parts:

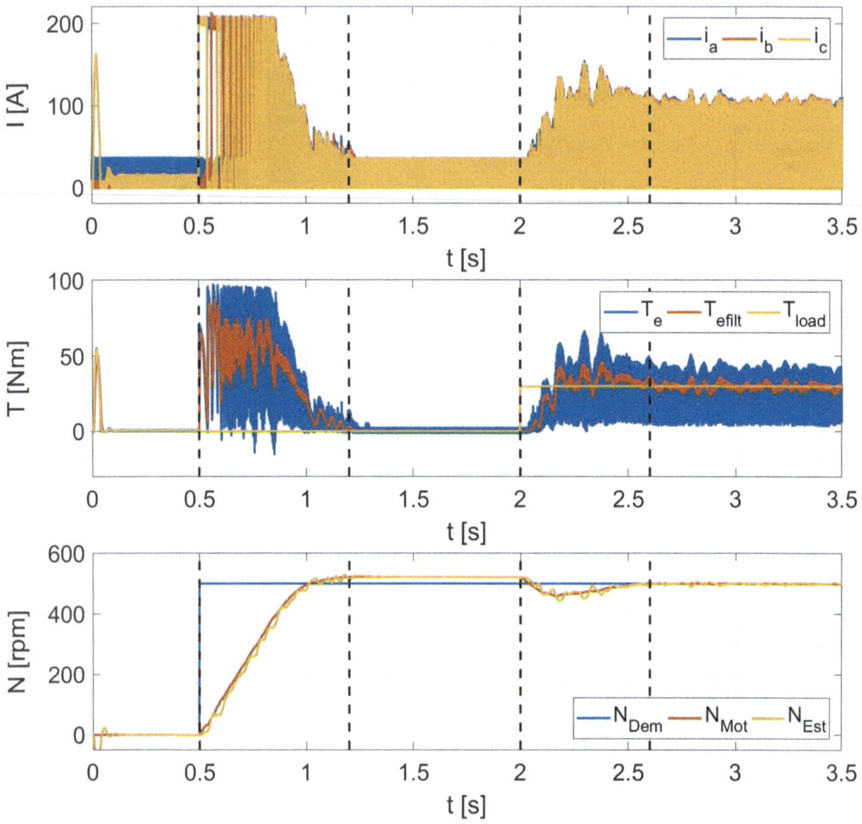

Fig. 12 Phase currents (top), torques (middle) and speed (bottom) of the SRMD operation using only estimated values

- $t \in \langle 0; 0.5)$—Demanded speed is set to zero and rotor position is deliberately set to 180°. The rotor is braked in this position and cannot move. This simulates the worst case of an unknown starting position.
- $t \in \langle 0.5; 1.2)$—A speed of 500 rpm (10% of nominal speed) is demanded with zero load torque. Because of this and the control structure dynamics, the speed overshoots slightly to approximately 520 rpm.
- $t \in \langle 1.2; 2)$—Demanded current is zero because of the overshoot and no load torque.
- $t \in \langle 2; 2.6)$—A step change of load torque to nominal value of 30 Nm is applied to demonstrate robustness of the algorithm. It should be noted that this sudden loading is not expected in a traction application.
- $t \in \langle 2.6; 3.5)$—This last interval shows steady state accuracy under nominal load.

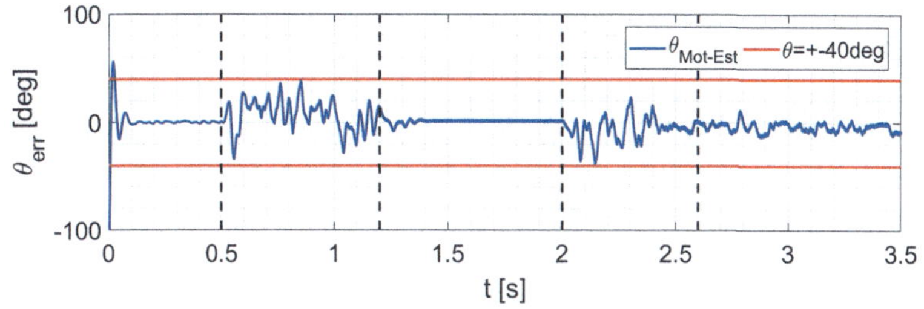

Fig. 13 Electrical position estimation error during the simulation

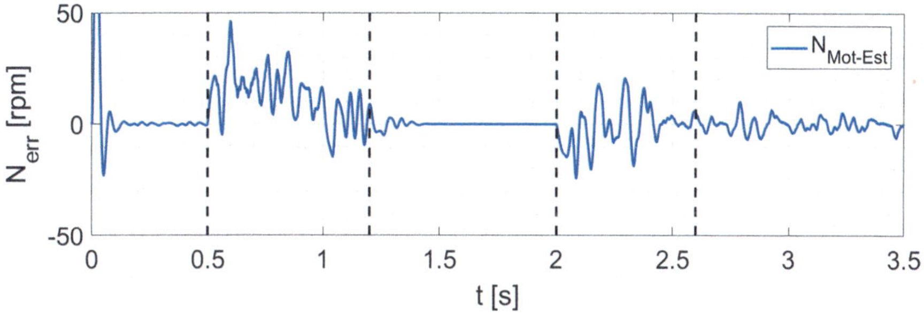

Fig. 14 Speed error during the simulation

Electrical position error is shown in Fig. 13. The red boundary marks the ±40° of error, or 3.33° mechanical. If the rotor position stays within these limits, torque generation capability should not be reduced by more than 25%. This is acceptable for a low speed and low load operation and is comparable to [10]. The algorithm correctly converges to the starting rotor position within a very short period. During the acceleration to demanded speed, the position accuracy is negatively affected by interference from the motoring current, but stays within the acceptable limits. The no-load interval again reaches very good accuracy and stability due to the lack of motoring currents. During the load torque step change transient, the position error is again increased. It can be seen that during stable nominal load operation, the position error ripple is higher than during no load and also the mean value of position error is not zero. However, the error is still within specified limits.

Speed error, recalculated to rpm, is shown in Fig. 14. In regions without motoring current, the error is negligible, similarly to position error. In the presence of motoring current there is a noticeable ripple even with the Low-Pass Filter in the PLL. Design of speed controller must account for this ripple. During the initial transient to correct position, the negative estimated speed might cause the speed controller to apply motoring current, causing the rotor to move if adequate braking is not applied.

It must be noted that the high torque ripple seen in Fig. 12 is not caused only by the position error, but also by the requirement to use only one phase for motoring at any given moment. Because the studied SRM is a 3 phase machine, this causes severe ripple. The filtered torque values stay high, verifying correct estimation.

6 Conclusions

A new approach for the estimation of position and speed of a Switched Reluctance Motor Drive based on stationary high frequency voltage injection was shown. This method improves signal processing, which separates the necessary information from the measured current signal. The principle was described thoroughly and verified by a simulation using an accurate model of the motor and converter.

Using both bipolar and unipolar switching scheme under appropriate conditions significantly reduces current ripple and improves converter efficiency. The peak injected current reaches values up to 35 A (corresponding to 17.5% of allowed instantaneous current amplitude), but given the SRM characteristics the resulting negative torque is negligible. The inductance saturation at this current level is also not significant, but during motoring section of the period. It was shown that even during high-power motoring the accuracy, decreased by inductance saturation from motoring current, is still within required limits. High amplitude of injected voltage is also permissible, as this method is considered for low speed operation, where there is sufficient supply voltage available compared to the Back EMF.

Simulation results show that the method is capable of correct startup from an unknown position to at least 10% of the nominal speed of the machine and sustain this speed both in no-load and nominal load conditions. Robustness of the algorithm was verified by the acceleration to speed and load torque step change transients. Experimental verification will be published in a following paper.

Acknowledgements This work was supported by VEGA 1/0774/18 and by ITMS: 26210120021.

References

1. T.J.E. Miller (ed.), *Electronic Control of Switched Reluctance Machines*. Newnes Power Engineering (Reed Educational and Professional Publishing Ltd., London, 2001). ISBN: 0-7506-50737
2. M. Cacciato et al., A switched reluctance motor drive for home appliances with high power factor capability, in *2008 IEEE Power Electronics Specialists Conference*, June 2008, pp. 1235–1241
3. N. Sakr et al., A combined switched reluctance motor drive and battery charger for electric vehicles, in *IECON 2015 - 41st Annual Conference of the IEEE Industrial Electronics Society*, November 2015, pp. 001770–001775. https://doi.org/10.1109/IECON.2015.7392357
4. A. Peniak, J. Makarovic, P. Rafajdus, Replacing of DC motor in the first Slovak electric car by an optimized switched reluctance motor, in *2016 ELEKTRO*, May 2016, pp. 350–354

5. M.S. Islam et al., Design and performance analysis of sliding-mode observers for sensorless operation of switched reluctance motors. IEEE Trans. Control Syst. Technol. **11**(3), 383–389 (2003). ISSN: 1063-6536
6. Y. Saadi et al., Sensorless control of switched reluctance motor with unknown load torque for EV application using extended Kalman filter and second order sliding mode observer, in *2018 IEEE International Conference on Industrial Technology (ICIT)*, February 2018, pp. 522–528
7. P. Brandstetter, O. Petrtyl, J. Hajovsky, Luenberger observer application in control of switched reluctance motor, in *2016 17th International Scientific Conference on Electric Power Engineering (EPE)*, May 2016, pp. 1–5
8. J. Bekiesch et al., A simple excitation position detection method for sensorless SRM drive, in *2007 European Conference on Power Electronics and Applications*, September 2007, pp. 1–8
9. A. Sarr et al., Sensorless control of switched reluctance machine, in *IECON 2016 - 42nd Annual Conference of the IEEE Industrial Electronics Society*, October 2016, pp. 6693–6698
10. E. Kayikci, M.C. Harke, R.D. Lorenz, Load invariant sensorless control of a SRM drive using high frequency signal injection, in *Conference Record of the 2004 IEEE Industry Applications Conference, 2004. 39th IAS Annual Meeting*, vol. 3, October 2004, pp. 1632–1637
11. R. Hu et al., Sensorless control of switched reluctance motors based on high frequency signal injection, in *17th International Conference on Electrical Machines and Systems (ICEMS)*, October 2014, pp. 3558–3563
12. S.M. Ahmed, P.W. Lefley, A new simplified sensorless control method for a single phase SR motor using HF signal injection, in *2007 42nd International Universities Power Engineering Conference*, September 2007, pp. 1075–1078
13. P. Kappes, I. Krueger, G. Griepentrog, Self-sensing method for a switched reluctance motor using delta-sigma modulators and neural networks, in *2017 IEEE International Symposium on Sensorless Control for Electrical Drives (SLED)*, September 2017, pp. 55–60
14. P. Sovicka et al., Current controller with slope compensation for a switched reluctance motor, in *2018 ELEKTRO*, May 2018, pp. 1–6

Optimization of Offshore Wind Farms with HVAC and HVDC Transmission Networks

Asma Dabbabi, Salvy Bourguet, Rodica Loisel, and Mohamed Machmoum

Abstract Recently, offshore wind farms have attracted more and more attention because of their greater energy capacity. To get the best performances of a wind farm park, a technical and economic compromise between energy yields and overall investment must be established. In this paper, a study was done on Borssele I & II offshore wind farm with HVAC and HVDC transmission technologies to compare their performances with different transmission distances.

1 Introduction

The European Union has set targets to reduce the amount of CO_2 through the integration of renewable energy sources. By 2020, 20% of European energy supplied should come from renewable sources with a 20% reduction in CO_2 emissions compared to 1990 [1].

Offshore wind energy is one among renewable energies, which has several advantages over land-based wind, such as stronger and more regular wind at sea. The exploitation of offshore energy involves investments in energy transmission to the terrestrial network. Nevertheless, it is important to consider cheaper transmission solutions with less energy loss. There are two main transmission technologies: HVAC and HVDC. In the literature, studies have been done by P. Monjean [2] comparing several topologies of internal connections of offshore wind farms according to various criteria such as technological feasibility and availability. Thus, this comparison led to favor HVDC transport over HVAC or MVAC transport under certain conditions (significant distance from the coast). In addition, S. Gasnier [3] has developed a technical and economic evaluation tool for different connection architectures. His work has focused on the economic study of DC topologies by

A. Dabbabi (✉) · S. Bourguet · R. Loisel · M. Machmoum
IREENA, Institut de Recherche en Energie Electrique de Nantes Atlantique, CRTT, University of Nantes, Saint Nazaire, Cedex, France
e-mail: asma.dabbabi@univ-nantes.fr; Salvy.Bourguet@univ-nantes.fr; Rodica.Loisel@univ-nantes.fr; Mohamed.Machmoum@univ-nantes.fr

W. Zamboni, G. Petrone (eds.), *ELECTRIMACS 2019*, Lecture Notes in Electrical Engineering 615, https://doi.org/10.1007/978-3-030-37161-6_9

calculating the Levelized Cost of Energy (LCOE), which involves electrical losses and overall investment costs of the wind farm. Besides, O. Dahmani focused on the modeling and optimization of topologies of offshore wind farms with AC distribution and transmission networks, and studying their reliability [4].

The objective of this paper is to make a comparison between HVAC and HVDC technologies for a large offshore wind farm such as Borssele I & II. The paper is divided into five sections, the first presents the difference between the HVAC and HVDC architecture, the second and third parts respectively present the electric and economic models of an offshore wind farm, the fourth part presents the algorithm developed for offshore wind farms optimization, and the last part is dedicated to the study and comparison of the energy and economic performances between both technologies for Borssele I & II park.

2 Transmission Technologies

Figure 1 shows two different architectures of an offshore wind farm. The first topology consists of wind turbines exploiting wind energy, a MV distribution

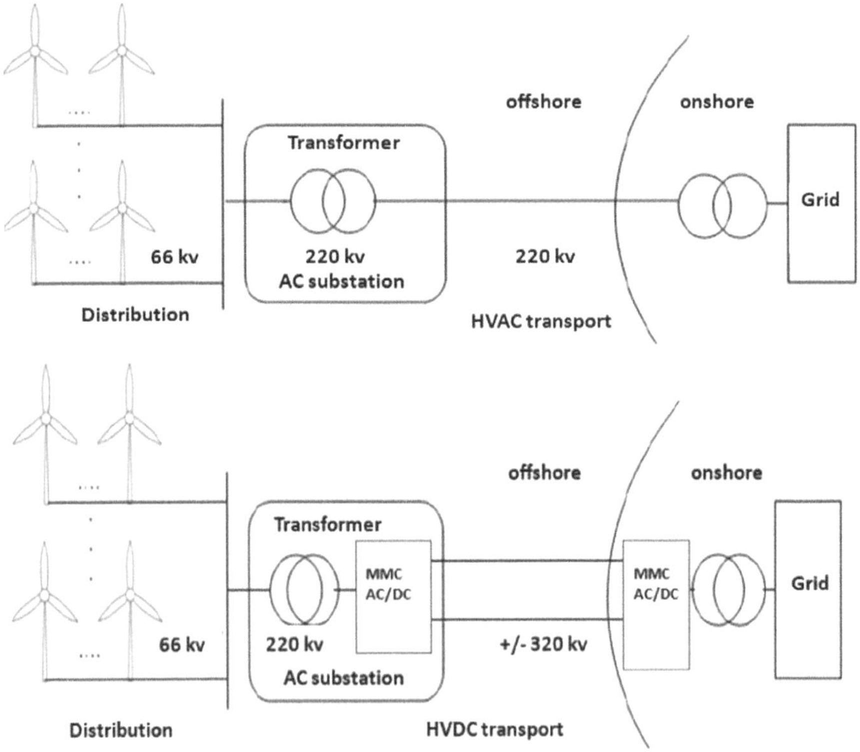

Fig. 1 Transmission technologies

network in which the wind turbines are connected and the AC offshore substations are founded under a medium voltage 66 kV and finally a 220 kV HVAC transmission network that is directly connected to the onshore point called the delivery point (DP). The second topology has the same distribution network as the first, but here the transport is in HVDC under a voltage of $\pm$ 320 kV; therefore, the DC offshore substations contain AC/DC converters and the DP must contain a DC/AC converter in order to inject the energy transmitted to the terrestrial network.

3 Offshore Wind Farm Electrical Models

3.1 Turbines

The study of the energy losses of the different topologies of offshore wind farms is carried out over a total operating life estimated between 20 and 30 years. Since the analysis is done for a long term, the modeling of the various components of the electrical network is made for a balanced, permanent, and sinusoidal operating regime.

Borssele park zone I & II is composed of 116 turbines with 6 MW each. The turbines used for this study are General Electric Haliade with a rotor diameter (RD) of 150 m and a hub height of 100 m [5]. Each turbine is modeled by an active power generation curve as a function of wind speed. The typical production curve of the Haliade wind turbine is shown in Fig. 2.

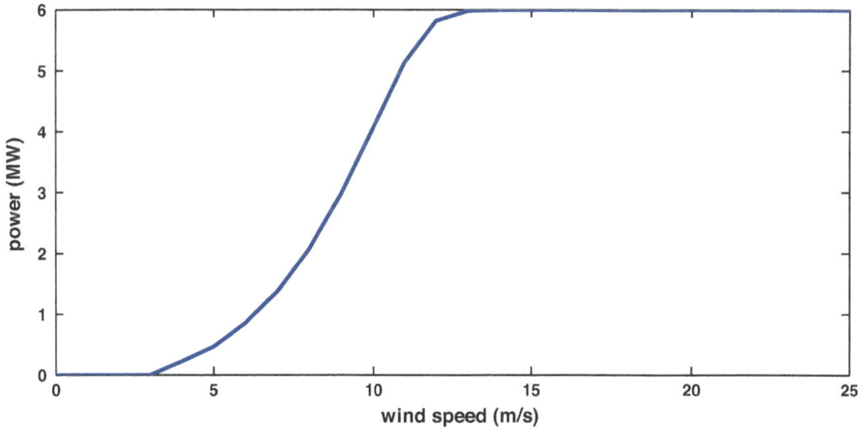

Fig. 2 Power curve of the Haliade 6 MW wind turbine

3.2 Submarine Cables

All cables are modeled using a π model that takes into account the type of soil surrounding cables and the cable geometric dimensions in accordance with IEC 60228 and 60287 [6]. Table 1 shows the properties of all the cables used for Borssele I & II.

3.3 Multilevel Modular Converters (MMCs)

In order to ensure the energy conversion in the case of HVDC transmission, a multilevel modular converter is chosen. This structure easily adapts to very high voltage and power levels. Figure 3a shows the VSC HVDC converter model, while Fig. 3b presents the equivalent single-phase power flow model of a converter connected to the AC network. The converter is represented as a controllable voltage source U_c behind the phase reactor, represented as a complex impedance Z_c. The

Table 1 Submarine cables parameters of Borssele park [7]

	MVAC		HVAC			HVDC			
Rated voltage (kV)	66		220			±320			
Section (mm^2)	240	630	500	800	1000	240	630	1200	2000
Rated power (MW)	54.9	81.7	250	295	330	446	797	1147	1540
Rated current (A)	480	715	655	775	866	697	1246	1791	2406
R at 20 °C (mΩ/km)	85	41.4	39.1	24	21	75	27.3	15.1	9
C (nF/km)	220	320	140	170	190	–	–	–	–
L (mH/km)	0.38	0.33	0.43	0.4	0.39	–	–	–	–

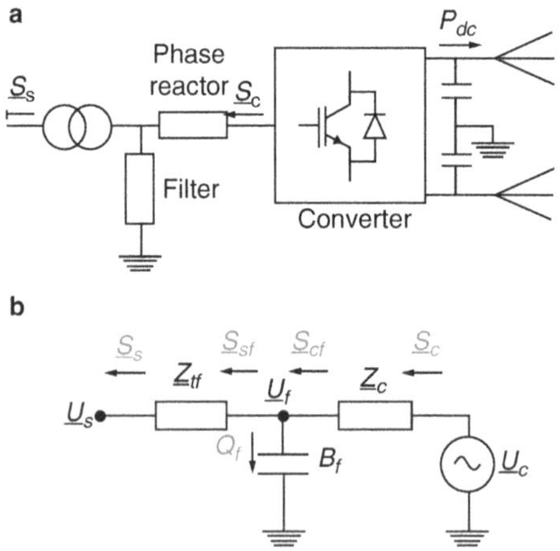

Fig. 3 (**a**) VSC HVDC converter model [8]. (**b**) Equivalent single-phase power flow model of a VSC HVDC converter [8]

Table 2 *a*, *b*, and *c* parameters values [9]

VSC	*a*	*b*	*c*
Rectifier	11.033e−3	3.464e−3	4.400e−3
Inverter	11.033e−3	3.464e−3	6.667e−3

filter is represented as a susceptance B_f at system frequencies and Z_{tf} is the complex impedance of the transformer.

In our case, the nominal power of the MMC is the sum of the nominal powers of the turbines that are interconnected. Concerning the calculation of the MMC losses, we used a generalized model of losses which is a model of ABB Corporate Research Sweden based on the data of the HVDC link Light of Södra Länken, evaluated with 600 MW and ±300 kV [9]. The VSC HVDC link is assumed to operate at a power factor of about 1.

The losses of the converter are written according to the converter current I_c determined by the active and reactive power exchanged with the AC system. The overall converter losses P_{loss} can be divided into three loss components, discussed further in [9]: no load losses, linear losses, and quadratic losses as shown in Eq. (1).

$$P_{loss} = a + b\, I_c + c\, I_c^2 \tag{1}$$

With:

$$I_c = \frac{\sqrt{P_c{}^2 + Q_c{}^2}}{\sqrt{3}\, U_c} \tag{2}$$

P_c, Q_c: active and reactive power to be injected/absorbed by the converter and U_c: converter voltage.

Table 2 shows the per unit values of coefficients *a*, *b*, and *c* for the Södra Länken model.

a, *b*, and *c* values depend on the converter nominal power and its voltage level. So for different values of power and voltage other than those of the Södra Länken model, a calculation of *a*, *b*, and *c* must be made (more details are given in [10]). Note that the inverter losses are greater than rectifier losses because the inverter uses the IGBTs more frequently, while the rectifier uses the diodes more frequently.

3.4 Transformers

The transformers are modeled by a T-model presented in the Fig. 4.

The impedance of the transformer depends on its rated power. The per unit values of the transformer parameters are:

$R_{TS} = 0.007$ p.u, $X_{TS} = 0.1$ p.u, $R_F = +\infty$, and $X_m = 50$ p.u [6].

Fig. 4 Transformer model in per unit

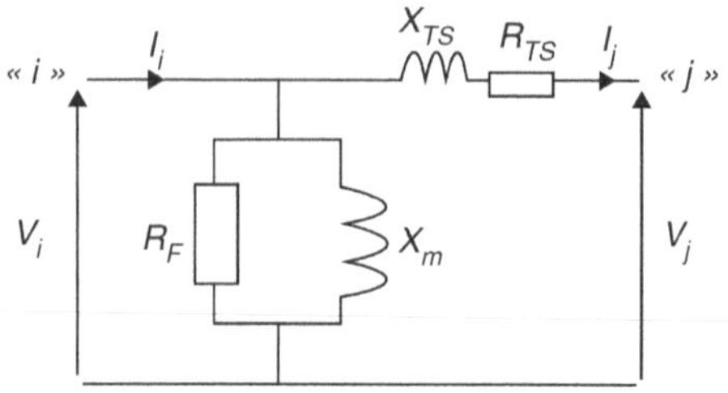

Table 3 Cost of submarine DC cables [7]

	HVDC			
Rated voltage (kV)	±320	±320	±320	±320
Section (mm^2)	240	630	1200	2000
Rated power (MW)	446	797	1147	1540
Cost (k€/km)	502	677	1019	1405

4 Offshore Wind Farm Economic Models

4.1 Submarine Cables

The cost model for an AC cable is represented by the following equation:

$$C_{\text{cable},ij} = C_{\text{SI}} + \left(\alpha_{\text{c}} + \beta_{\text{c}} \exp\left(\gamma_{\text{c}}.10^{-8} S_{\text{C},n,ij}\left(C_{\text{section}}\right)\right)\right) \tag{3}$$

This function depends on the initial investment cost, the transport, and installation cost C_{SI}, which are proportional to the cable length and the cable nominal power $S_{\text{C}n,ij}$. The coefficients α_{c}, β_{c}, and γ_{c} are dependent on the cable voltage level. The reference model of S. Lundberg [4] was used for AC submarine cables costs. The DC cables cost is modeled with the following equation [11]:

$$C_{\text{cable pair HVDC}} = 0.652 + 0.00098 P_{\text{ratedcabpair}} - 0.002363 U_{\text{HVDC}} \tag{4}$$

Table 3 presents the DC cable cost in k€/km in function of different sections and rated powers.

4.2 Offshore Substations

The AC offshore substations' AC cost is given by the following equation [11]:

$$C_{\text{ss}}^{\text{HVAC}} = 2.534\ \text{Me} + 0.0887\ P_{\text{ss}} \tag{5}$$

where P_{ss} is the sum of the nominal power of the turbines connected to the substation (MW) and C_{ss}^{HVAC} is the cost of the substation (M€). The same expression is used for DC substations, but with an increase of 85% to account for additional components [11].

4.3 Transformers

The expression of the transformer cost is presented by Eq. (6) [4]:

$$C_{\text{transfo}} = 0.03327 P_{\text{transfo}}^{0.7513} \tag{6}$$

where P_{transfo} is the transformer nominal power (MVA) and C_{transfo} is the cost (M€). The minimum transformer nominal capacity is estimated at 40 MW.

4.4 MMCs

The MMC cost [11] is given by the following expression:

$$C_{\text{MMC}} = 54.985 \text{ M€} + 0.0589\ P_{\text{MMC}} \tag{7}$$

where P_{MMC} is the MMC rated power (MVA) and C_{MMC} is the MMC cost (M€).

4.5 Reactive Power Compensation

The reactive power compensation cost is the sum of the compensation equipment cost C_{ec} (€) and the increment of the platform cost C_{incremt} (€) [4].

$$C_{\text{comp}} = C_{\text{ec}} + C_{\text{incremt}} \tag{8}$$

The compensation equipment cost C_{ec} is proportional to the total reactive power of the network (consumed or produced) Q_{comp}.

$$C_{\text{ec}} = \alpha_{\text{ec}} + \beta_{\text{ec}} Q_{\text{comp}}^{\gamma_{\text{ec}}} \tag{9}$$

C_{incremt} is proportional to reactive power installed on each offshore substation $Q_{\text{comp,sub}}$.

$$C_{\text{incremt}} = 608.Q_{\text{comp,sub}}^{0.765} \tag{10}$$

G. Guidi's model is used to estimate both the parameters α_{ec}, β_{ec}, γ_{ec}, and the term C_{incremt} [4].

5 Offshore Wind Farm Optimization (Borssele I & II)

Optimal structure of an offshore wind farm is a structure that verifies certain constraints such as minimizing the cost of energy transmitted to the grid and minimizing power losses. Indeed, each electrical topology is characterized by a number of offshore substations as well as connections of wind turbines in the distribution network and a transmission network configuration HVDC or HVAC.

The genetic algorithm provides the first topologies of connections such as the connection between the wind turbines and the connection between offshore substations. Next, the Prim's algorithm is used to complete the connection between each group of wind turbines and the nearest substation as well as between each group of substations to DP (search for the shortest path) [4]. The optimization algorithm flowchart is depicted in Fig. 5.

5.1 *Economic Functions*

The Borssele I & II park is optimized according to an economic function (mono objective).

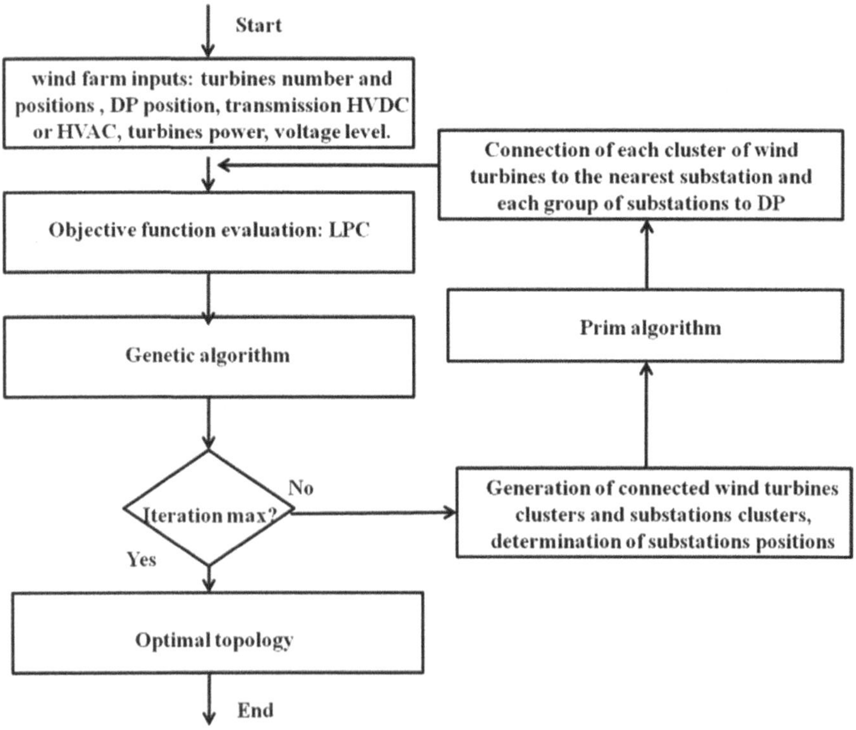

Fig. 5 The optimization algorithm flowchart

The objective function is the levelized production cost (LPC (€/kWh)), which is the ratio between the total investment cost and the average energy delivered to the grid.

$$\mathrm{LPC} = \frac{C_{\mathrm{invest}}}{E_{\mathrm{d}}} \tag{11}$$

With:

$$C_{\mathrm{invest}} = \frac{r\,(1+r)^T\,T}{(1+r)^T - 1}\,\frac{1}{1-\mathrm{PR}}\,C_0 \tag{12}$$

$$E_{\mathrm{d}} = P_{\mathrm{sortie}}\,N_{\mathrm{t}}T = \left[(n_{\mathrm{eol}}P_{\mathrm{eol}}) - P_{\mathrm{pertes}} - P_{\mathrm{L}}\right]N_{\mathrm{t}}\,T \tag{13}$$

C_0: total initial investment, T: offshore wind farm lifetime set at 20 years, r: bank interest rate of 4%, PR: banks profit 2%, n_{eol}: number of wind turbines, P_{eol}: wind turbine power, P_{losses}: power losses, P_{L}: total active power consumed, $N_{\mathrm{t}} = 8760$ (number of hours of wind farm operation per year).

5.2 *Power Flows*

Power flows aim to calculate the nodal voltage modules and phases. The Mat AC/DC tool [12] implemented with Matlab® allows this calculation.

- Load flow AC

  In the AC wind farm case, the turbines and the transformers (66 kV/220 kV) are modeled as PQ nodes. In addition, in the terrestrial network, a last transformer (220 kV/380 kV) is added to increase the voltage and to inject it directly to the grid. Finally, the transformer output must be a reference node to balance the power flow.
- Load flow AC/DC

  In this case, the studied system is hybrid (AC distribution and HVDC transport), so the calculation of load flow must take into account the two regimes.

In fact, the turbines are considered as PQ nodes. The AC side of the MMCs and the terrestrial network node are defined as reference nodes. Therefore, the network is divided into different AC zones where each zone contains a reference node, and between these zones there are DC zones which present the HVDC transmission lines. Transformers are considered as PQ nodes.

For the calculation of the AC/DC load flow, the sequential method is used. It consists in calculating the state variables of the AC and DC systems and iterating

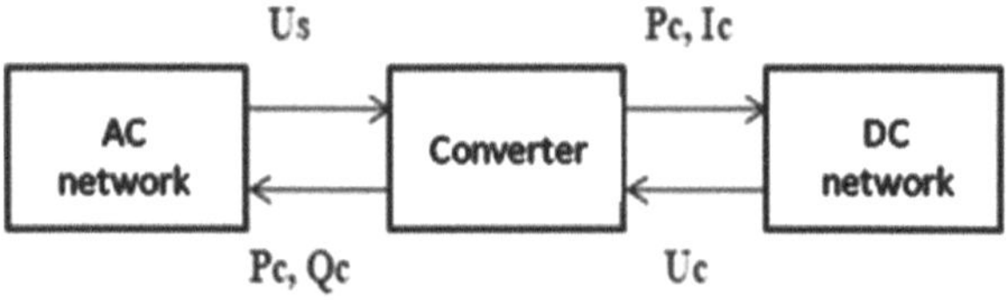

Fig. 6 The calculation interface of the AC/DC load flow

them one by one, until all state variables converge. During this process, the AC energy flow equations and the DC equations are solved separately.

To calculate the variables of the AC system, the converter can be considered as an equivalent PQ node connecting to an AC bus. Then, to solve the DC equations, the AC system is considered as a constant voltage of the converter bus. Figure 6 shows the calculation interface of the AC/DC load flow by the sequential method.

For converters control, there are two different modes:

1. Active power injection control which can be divided into three types:
   (a) Constant P: the converter injects a constant active power to the AC network.
   (b) Constant U_{DC}: the active power changes to keep the DC bus voltage constant.
   (c) DC droop: the active power changes linearly with its power.
2. Reactive power injection control which can be divided into two types:
   (a) Constant Q: the converter injects a constant reactive power to the AC network.
   (b) U_s constant: reactive power changes to maintain constant AC bus voltage.

In our case, each terminal of the AC zone (reference node) contains a converter, which must be controlled in constant active and reactive power "converter PQ," the last converter connecting all the transmission lines must be controlled by the imposition of a constant AC and DC voltage to keep the stability of HVDC lines, it is a "DC reference converter." Injections of active and reactive powers should be limited by a PQ diagram [8] defining the converter steady state operating points.

6 Results

6.1 *Performance Evaluation of Borssele I & II with Genetic Algorithm Optimization*

In this part, a comparison between the reference topology of Borssele I & II and the topology found with the optimization algorithm was done. For the genetic algorithm, four substations are fixed with variable positions. Figures 7 and 8 show, respectively, the basic architecture of Borssele I & II and the topology found after optimization.

The performances of reference topology with a transmission distance equal to 50 km are: LPC [c€/kWh] = 0.6893, active losses P_{ac} [MW] = 45.806, and reactive

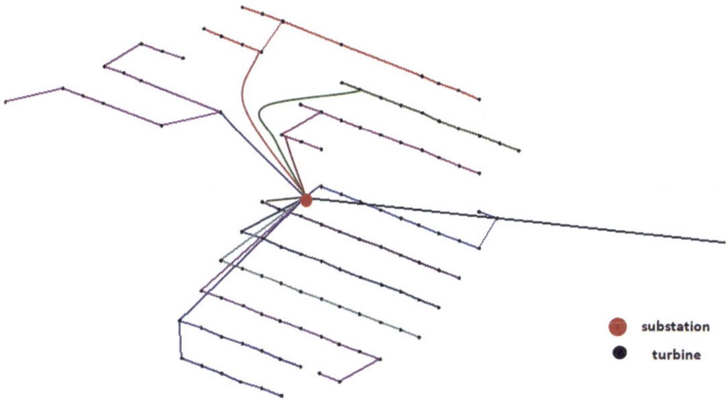

Fig. 7 The reference topology of Borssele I & II

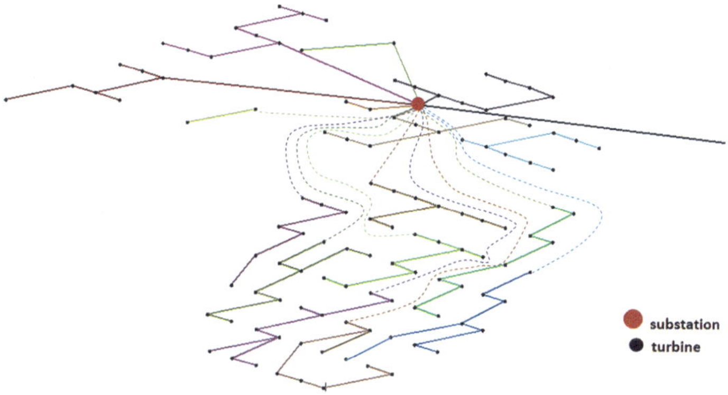

Fig. 8 Borssele I & II topology obtained by optimization

losses Q_{ac} [MVar] = 171.41. For the same transmission distance, the performances of the topology obtained by optimization algorithm are: LPC [c€/kWh] = 0.6889, active losses P_{ac} [MW] = 47.796, and reactive losses Q_{ac} [MVar] = 170.80.

The results found with the genetic algorithm are almost the same as the reference one, which validates that the reference topology is already optimized. In this case study, the minimization of the LPC is privileged than that of the losses. Furthermore, for other network topologies the optimization tool optimizes both the LPC and the losses.

6.2 Comparison Between HVAC and HVDC

In this part, Borssele park I & II is optimized for both transmission technologies. A comparison of the LPC is achieved depending on the transmission distance. Borssele I & II contains 116 turbines each producing 6 MW, one substation for both cases AC and DC is fixed, its position is determined by the genetic algorithm. For a DC network, the losses are divided into three categories: transmission losses (HV network), converters losses, and distribution losses (MV network). Q_{dc} are the reactive losses for the distribution network in DC case. Table 4 shows the comparison between transmission technologies for Borssele I & II for different transmission distances.

Between the distances 50 and 103 km, the HVAC technology presents more advantages than the HVDC regarding the LPC but the total DC losses remain less than the AC losses by a factor about 1/4. In fact, the DC transmission lines contain no reactive losses and the only losses are the resistive losses, which cause the reduction of the overall losses. From the distance 140 km, the HVDC is more efficient than the HVAC technology; it presents a lower LPC, and it is explained by the fact that for large transmission distances the total AC structure cost is greater than structure cost of wind farms in DC. The DC total losses are still lower than AC losses network for large transmission distances. Table 5 shows the detailed components costs in both cases of HVAC and HVDC at a distance 168 km.

Table 4 Comparison between HVAC and HVDC

Distance (km)	Losses HVAC MW MVA	Losses HVDC MW MVAR	LPC_{ac} (c€/kWh)	LPC_{dc} (c€/kWh)
50	$P_{ac} = 47.79$ $Q_{ac} = 170.80$	$P_{dc} = L_{transHV} + L_{conv} + L_{distrMV}$ $P_{dc} = 3.41 + 22.80 + 11.57 = 37.78$ $Q_{dc} = 110.27$	0.688	1.019
80	$P_{ac} = 63.450$ $Q_{ac} = 186.13$	$P_{dc} = 5.17 + 22.83 + 12.231 = 40.23$ $Q_{dc} = 111.34$	0.895	1.116
103	$P_{ac} = 75.535$ $Q_{ac} = 198.03$	$P_{dc} = 7.14 + 22.90 + 11.936 = 41.97$ $Q_{dc} = 111.39$	1.008	1.182
140	$P_{ac} = 96.260$ $Q_{ac} = 218.73$	$P_{dc} = 9.15 + 22.96 + 12.269 = 44.37$ $Q_{dc} = 112.15$	1.272	1.252
168	$P_{ac} = 109.98$ $Q_{ac} = 232.33$	$P_{dc} = 11 + 23.05 + 11.525 = 45.57$ $Q_{dc} = 112.68$	1.481	1.253
199	$P_{ac} = 126.26$ $Q_{ac} = 248.70$	$P_{dc} = 13.19 + 23.09 + 12.228 = 48.50$ $Q_{dc} = 112.89$	1.734	1.359

Table 5 Components costs for HVAC and HVDC cases

	HVAC		HVDC	
	M€	%	M€	%
MV cables	317.82	22.01	289.38	23.41
HV cables	998.79	69.16	270.55	21.88
Transformers	14.037	0.972	13.84	1.12
Platforms	99.97	6.92	363.49	29.40
MV switchgears	9.29	0.64	9.29	0.75
HV switchgears	3.67	0.25	1.22	0.099
Compensations	0.396	0.0274	–	–
Converters	–	–	288.25	23.32

According to Table 5, at a distance of 168 km, the cost of HV cables represents 69.16% of the total cost. However, the HVDC cables cost represents only 21.88% of the total cost. This cost evolves according to the transmission distance, which proves the increase of the LPC_{ac} compared to LPC_{dc} for large distances. The costs of converters and substations platforms in the DC case present the greatest costs, but it does not vary with the increase in transmission distance.

7 Conclusion

This research work highlights the comparison between two transmission technologies HVAC and HVDC. Investment costs and energy losses regarding different transmission distances are studied for an offshore wind farm topology Borssele I & II. HVDC has lower energy losses compared to HVAC, but AC network can be more efficient regarding LPC for distances lower than 103 km. As a result, imposing an optimization framework where needs and constraints are clarified is essential to find the best technology according to economical and energetic compromise. In future work, the maintenance cost (OPEX) will be integrated in the economic function for a better evaluation of offshore wind farm performances.

Acknowledgement This work was carried out within the framework of the WEAMEC, West Atlantic Marine Energy Community, and with funding from the Pays de la Loire Region. It was also funded with the French National Research Agency (ANR), with the collaboration of France Energies Marines Institute (FEM).

References

1. International Renewable Energy Agency, IRENA, *Renewable Energy Prospects for the European Union* (IRENA, Abu Dhabi, 2018)
2. P. Monjean, Optimisation de l'architecture et des flux énergétiques de centrales à énergies renouvelables offshore et onshore équipées de liaisons en continu, PhD Thesis in Electrical Engineering, National School of Arts and Crafts, 2012
3. S. Gasnier et al., Technical economic assessment tool for offshore wind generation connection scheme: application to comparing 33 kV and 66 kV AC collector grids, in *the 18th European Conference on Power Electronics and Applications (EPE'16 ECCE Europe)*, 2016
4. O. Dahmani, Modélisation, optimisation et analyse de fiabilité de topologies électriques AC de parcs éoliens offshore, PhD Thesis in Electrical Engineering, University of Nantes, 2014
5. General Electric Haliade 150-6MW, https://en.wind-turbine-models.com/turbines/1446-general-electric-haliade-150-6mw
6. S. Fragoso Rodrigues, A multi-objective optimization framework for the design of offshore wind farms, PhD Thesis in Electrical Engineering, Delft University of Technology, 2016
7. ABB, *XLPE Submarine Cable Systems - Attachment to XLPE Land Cable Systems - User's Guide* (ABB, Lyckeby, Sweden, 2010)
8. J. Beerten, Modeling and control of DC grids, Dissertation presented in partial fulfillment of the requirements for the degree of Doctor in Engineering Science, ARENBERG Doctoral School, 2013

9. J. Beerten, S. Cole, R. Belmans, A sequential AC/DC power flow algorithm for networks containing multi-terminal VSC HVDC systems, IEEE, Power and Energy Society General Meeting, 2010
10. J. Beerten, S. Cole, R. Belmans, Generalized steady state VSC MTDC model for sequential AC/DC power flow algorithms. IEEE Trans Power Syst, 27, 821–829 (2012)
11. R.L. Joaquin, Power transmission systems for offshore wind farms: technical-economic analysis, Bachelor Thesis, Polytechnic University of Catalunya, 2015
12. J. Beerten, R. Belmans, Development of an open source power flow software for HVDC grids and hybrid AC/DC systems: MatACDC. IET Gener. Transm. Distrib. **9**(10), 966–974 (2015)

Low-Side MOSFET Current Sensing Technique for Automotive Applications

Patrik Varecha, Giacomo Scelba, Mario Cacciato, Pavol Makys, and Matej Pacha

Abstract This paper deals with the analysis and simulation of a current sensing technique based on the estimation of the current flowing in the low-side MOSFETs of an inverter. In this case the power MOSFET is utilized as a current sensor, by estimating its internal on-state resistance. Due to the temperature dependency of the internal on-state resistance, the temperature of MOSFET die has been also estimated by measuring the internal body diode forward voltage of the power MOSFET. The proposed low-side MOSFET current sensing method will be used as current sensing in three-phase inverters for automotive applications.

1 Introduction

The power electronics area is divided into many sectors, such as automotive area, industrial area and aerospace industry. Each of the mentioned sectors of power electronics industry has to fulfil specific criterions about quality, reliability, price, etc. Automotive application area requires to fulfil the high reliability of the device with the lowest price for the device possible. Therefore, the development of many automotive applications is focused on the increasing of device reliability and decreasing the price of devices and the costs of production [1]. This paper is focused on the development and utilization of a different approach for current sensing technique, in order to reduce the price and increase the reliability and efficiency of power electronic devices. The proposed current sensing technique utilizes the on-state resistance R_{dson} of MOSFET as a current sensor and internal body diode of MOSFET as a temperature sensor. The idea of current or temperature estimation by utilization of MOSFET power device was developed and published

P. Varecha (✉) · P. Makys · M. Pacha
Department of Power Systems and Electric Drives, University of Zilina, Zilina, Slovakia
e-mail: patrik.varecha@fel.uniza.sk; pavol.makys@fel.uniza.sk; matej.pacha@ieee.org

G. Scelba · M. Cacciato
DIEEI, University of Catania, Catania, Italy
e-mail: giacomo.scelba@dieei.unict.it; mario.cacciato@dieei.unict.it

W. Zamboni, G. Petrone (eds.), *ELECTRIMACS 2019*, Lecture Notes in Electrical Engineering 615, https://doi.org/10.1007/978-3-030-37161-6_10

many years ago, also a few workarounds were patented [2–5]. However, the results of the latest published technique did not achieve the sufficient accuracy of the estimated current or temperature, mainly for lower currents, therefore it has been utilized just for monitoring purposes [2]. Firstly, the sensorless current sensing idea has been patented for synchronous Buck DC-DC converter with MOSFET power device. The estimation process was based on the measurement of body diode forward voltage of MOSFET and drain-source voltage of MOSFET. The temperature and current has been estimated by means of look-up tables [5]. Later, the similar idea of sensorless current sensing technique has been patented for the utilization in high-power applications. In this case, the IGBT power devices have been used for control purpose [4]. The latest patented technique related just to the temperature estimation by means of MOSFET power devices utilization, mainly the internal body diode forward voltage. The temperature was estimated by means of the measurement of diode forward voltage and load current. This technique has been primarily established for motor control application [3]. Due to the requirements on high reliability and low cost of automotive devices, the current is usually sensed by shunt resistors. Utilization of on-state internal resistance of power MOSFET as a current sensor is difficult, even impossible, to implement due to variation of many parameters and mainly the operating temperature [6, 7]. The widely used shunt resistor fulfils many requirements to be used as a current sensor. The temperature stability of each shunt resistor current sensor is one of the most important requirements. In the case of MOSFET power devices, the internal on-state resistance changes with temperature significantly. This fact makes it difficult to use MOSFETs as current sensors. On the other hand, it enables the utilization of the temperature dependence of on-state resistance for device temperature monitoring [6, 8]. In the case of the utilization of internal on-state resistance as a current sensor, the characteristic of resistance as a function of temperature must be known in advance. The R_{dson} as a function of temperature is part of a datasheet of power MOSFET. However, this characteristic can be calculated or measured directly in more details and in various conditions [2, 9]. The temperature of power MOSFET usually changes heavily during the operation of power converter. The automotive applications are designed to work within a temperature range of −40–125 °C (grade 1) or to 150 °C (grade 0). The R_{dson} of the power MOSFET can be subjected to significant temperature variations [10, 11]. During the normal operation, the real value of R_{dson} can be identified as far as the temperature of power MOSFET die is known. The temperature of power MOSFET can be obtained directly from temperature sensor, or it can be obtained indirectly by calculation. Indirect methods of temperature estimation utilize the parameters of power MOSFET, which are temperature dependent, i.e. on-state resistance, forward voltage of body diode, and gate-source threshold voltage. Monitoring of the temperature of power MOSFET by means of R_{dson} is a reliable way, due to linear dependence of on-state resistance on temperature [8, 12]. Another method of power MOSFET temperature monitoring utilizes the internal body diode, where the linear temperature dependence of forward voltage is utilized [2, 13–15]. This method can be used only in applications in which the freewheeling operation exists, i.e.

inverters. A similar method has been published in [2]. The main disadvantage of this previous published method was related with poor precision of current estimation for low currents. The proposed method utilizes a more precise mathematical model for current estimation, in which the wider operation range of current with high precision of estimation is reached. This method of low-side MOSFET current sensing technique has been simulated in LTspice. It utilizes two important temperature and current characteristics of power MOSFET, which are expressed by mathematical equations. The first characteristic describes the R_{dson} as a function of temperature and the second characteristic describes forward voltage of MOSFET body diode as a function of temperature and diode current. The characteristic of body diode forward voltage as a function of temperature for different forward current is described by two equations, where the slope of characteristic and *y*-intersection as a function of forward current are mathematically expressed. For the sake of higher precision of temperature estimation, the *y*-intersection calculation is divided for three particular equations. The points of characteristic of internal R_{dson} as a function of temperature are interpolated by quadratic equation, which is subsequently used for current estimation. The mathematical model of the system is described in detail in the second section. Algorithm is primarily focused on high precision achievement for whole temperature and current operation range. The high precision current sensing is an important achievement for the utilization in automotive applications.

2 Mathematical Model of the Power MOSFET

2.1 General Information and Mathematical Model of Power MOSFET

The power MOSFET can work in three main operating regions called the cut-off region, saturation region and triode region [16]. The cut-off region defines the non-conductive operating state of MOSFET, where the excitation voltage V_{GS} does not reach the threshold value of voltage V_T. Saturation region defines the conduction operating state of MOSFET, but the value of drain-source voltage V_{DS} is higher than the difference of gate-source value of voltage V_{GS} and threshold voltage V_T. In the triode region, a constant value of R_{dson} can be observed. If in case the power MOSFET is used for power electronic purposes, the operation of MOSFET in triode region is necessary. The drain current in the triode region is given by:

$$I_d = k\left[(V_{GS} - V_T)\,V_{DS} - \frac{{V_{DS}}^2}{2}\right] \tag{1}$$

where *k* represents the transconductance of MOSFET, [16]. Transconductance and threshold voltage are highly temperature dependent. The transconductance parameter is given by:

$$k = \mu_0 C_{\text{ox}} \frac{W}{L_{\text{eff}}} \tag{2}$$

where μ_0 represents electron mobility, C_{ox} represents oxide capacitance per unit area, W represents width of channel and L_{eff} represents effective length of channel. The behaviour of drain current (Eq. 2) influences the character of internal on-state resistance of power MOSFET. The behaviour of diode forward current is given by the Shockley equation:

$$I_{\text{d}} = I_{\text{s}} \left(e^{\frac{V_{\text{D}}}{nV_{\text{T}}}} - 1 \right) \tag{3}$$

where the current I_{s} is the reverse bias saturation current, V_{D} is voltage across diode, n is the emission coefficient and V_{T} is the thermal voltage. The diode forward voltage (Eq. 4) is determined starting from Eq. (3):

$$V_{\text{d}} = \frac{kT}{q} \ln \left(\frac{I_{\text{d}}}{I_{\text{s}}} - 1 \right) \tag{4}$$

The parameter k is the Boltzmann constant, T is the temperature and q is the magnitude of charge [17]. The thermal voltage parameter is expressed through the Boltzmann constant, the temperature and the magnitude of charge. The expressed equation of thermal voltage is given by:

$$V_{\text{T}} = \frac{kT}{q} \tag{5}$$

In the case of temperature estimation based on diode forward voltage, the value of diode current has to be known. The mathematical model of MOSFET helps to understand the behaviour of the key parameters of MOSFET when the temperature is changed. The proposed low-side MOSFET current sensing algorithm utilizes the temperature-dependent characteristics of power MOSFET, meanwhile the analytical approach for the proposed technique will be presented in the full paper.

2.2 *Characteristics of Power MOSFET Intended for Low-Side MOSFET Current Sensing Technique*

In the case of implementation of the proposed low-side MOSFET current sensing method, the characteristics of the on-state resistance as a function of temperature and forward voltage of internal body diode as a function of temperature have to be known. These two characteristics have been obtained from the spice model of MOSFET transistor BUK7Y3R5-40E. Datasheet of this MOSFET is given in Ref. [18]. R_{dson} varies following a quasi-linearly function of the temperature. Figure 1

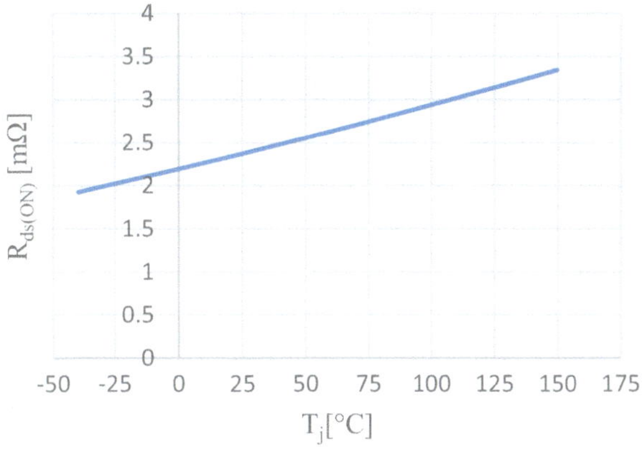

Fig. 1 Internal on-state resistance of power MOSFET as a function of temperature

shows the characteristic of the R_{dson} as a function of temperature, where the points of characteristic can be interpolated by using the quadratic equation (6).

$$R_{\mathrm{ds(ON)}} = 5T^2 10^{-6} + 0.0069T + 2.1972 \tag{6}$$

If the temperature of the MOSFET die is known, the internal on-state resistance of MOSFET can be calculated. The precision of the calculated value depends on the accuracy of the interpolating function R_{dson}. The second important characteristic defines the internal diode forward voltage as a function of temperature. Similarly, as in the previous case, this characteristic was obtained from the spice simulation model of the considered MOSFET transistor. The forward voltage of the diode linearly depends on temperature. This assumption allows to utilize the semiconductor diode as a simple temperature sensor [14, 15]. Figure 2 shows the internal diode forward voltage characteristic as a function of temperature for some values of forward currents. These characteristics for each current can be expressed by a matrix of line equations with the same pattern but different parameters (Eq. 7). The y-intersection represents the forward voltage V_{f0} at 0 °C while the slope k_{T} is a temperature-dependent parameter.

$$V_{\mathrm{f}} = V_{\mathrm{f0}} - k_{\mathrm{T}}T \tag{7}$$

The temperature can be expressed from the previous equation (7):

$$T = \frac{V_{\mathrm{f0}} - V_{\mathrm{f}}}{k_{\mathrm{T}}} \tag{8}$$

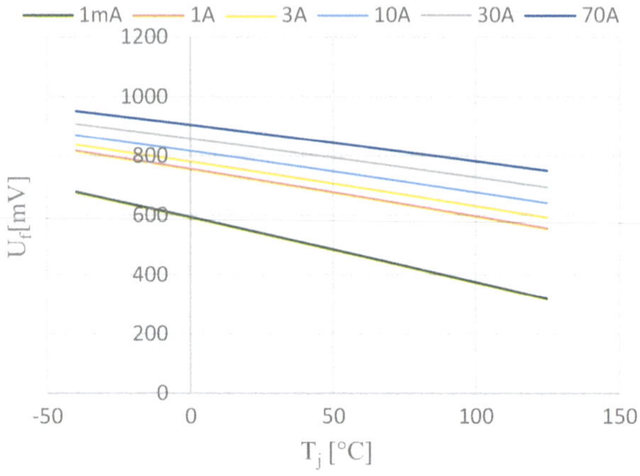

Fig. 2 Forward voltage of internal body diode of MOSFET as a function of temperature for different diode currents

The slope of each characteristic is influenced by the diode current, in which this change has non-linear behaviour. For the sake of simplicity, the characteristic of the slope as a function of current and characteristic of *y*-intersection as a function of current have been expressed according to the characteristics of Fig. 2. If the forward diode current is known, the slope of voltage–temperature characteristic can be calculated by:

$$k_{\mathrm{T}} = -0.086 \ln (I_{\mathrm{d}}) + 1.5866 \tag{9}$$

In the case of *y*-intersection, also called diode forward voltage for 0 °C, the characteristic was expressed by three equations in order to achieve a better accuracy in the estimation of the forward voltage. In practice, when the diode currents are lower than 10 A, V_{f0} is obtained by using Eq. (10), while if the diode forward currents are in the range of 10 A and 50 A, the V_{f0} is calculated by the relationship (Eq. 11). Finally, when the diode forward currents are higher than 50 A, the V_{f0} is obtained by Eq. (12).

$$V_{\mathrm{f0(0A\text{-}9A)}} = 0.0256 \ln(I) + 0.7576 \tag{10}$$

$$V_{\mathrm{f0(10A\text{-}50A)}} = I^3 10^{-7} - 2I^2 10^{-5} + 0.0027I + 0.7942 \tag{11}$$

$$V_{\mathrm{f0(50A\text{-}100A)}} = 2I^3 10^{-8} - 7I^2 10^{-6} + 0.0017I + 0.8139 \tag{12}$$

These equations utilize the diode forward current as a key parameter. The forward diode current can be obtained from the current sensor or it can be estimated. In our case, the forward current is carried out by the knowledge R_{dson} as a function of temperature and sensed drain-source voltage of MOSFET during the turn-on.

3 Proposed Current Sensing Technique

The proposed current sensing technique has been evaluated in a H-Bridge converter realized with the Si MOSFETs model BUK7Y3R5-40E and simulated in LTspice [18]. The converter has been controlled in current closed loop with the switching frequency 20 kHz and a dead time equal to 700 ns. The current is controlled by PI controller on the constant value of current and the load is represented by R-L components. Figure 3 shows block diagrams of full-bridge converter and proposed low-side MOSFET current sensing algorithm.

The drain-source voltage of MOSFET is sensed by a second differential amplifier, whose gain is adjusted on different value than differential amplifier intended for diode forward voltage measurement. Figure 4 shows the flowchart adopted by the proposed algorithm to estimate the temperature and current. In the first step, the algorithm checks the state of low side MOSFET. In the case of deadtime (DT) occurrence, the diode forward voltage is measured. In the case of low side switched on state occurrence, the drain-source voltage is measured. When the diode forward voltage was measured, the estimation of MOSFET die temperature is carried out based on the Eqs. (8–12). Algorithm checks the value of estimated current and chooses the proper equation for y-intersection calculation. When the temperature is estimated, and the low-side MOSFET is going to be switched on, the on-state

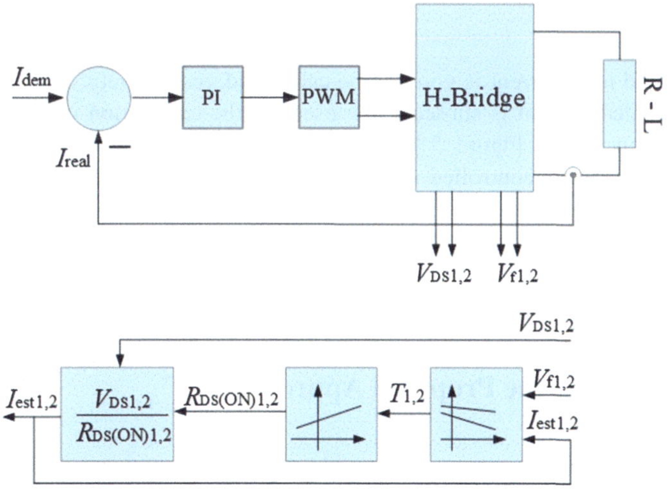

Fig. 3 Block diagram of the current control loop (up) and proposed low-side MOSFET current sensing algorithm (down)

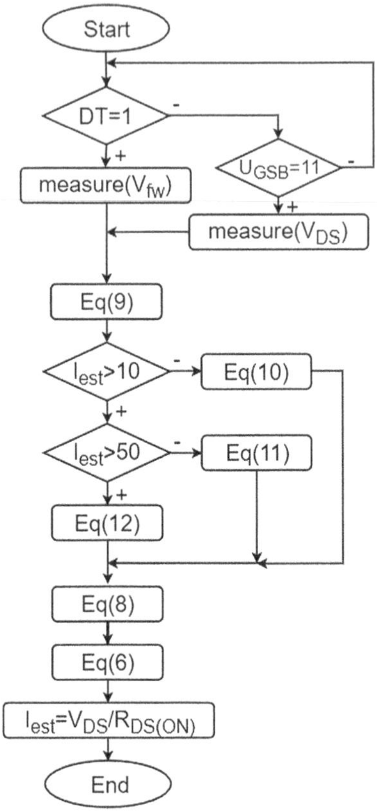

Fig. 4 Flowchart of proposed low-side MOSFET current sensing algorithm

resistance and load current is going to be calculated, respectively. The mean value of the estimated current is subsequently used for the calculation of the following value of temperature. Figure 5 shows the waveforms of voltages and current, when the converter is controlled in current closed loop. The estimation of current works in closed loop, where the inaccuracy of estimated temperature or current is compensated to each other. This assumption is based on the opposite thermal factor of internal on-state resistance and diode forward voltage.

4 Validation of the Proposed Approach

The low-side MOSFET current sensing algorithm was simulated in LTspice program, as was mentioned before. The converter was controlled by PWM modulation with 20 kHz switching frequency. The load was represented by RL components, where the resistance part was 50 mΩ and the inductance part was 30 μH. The

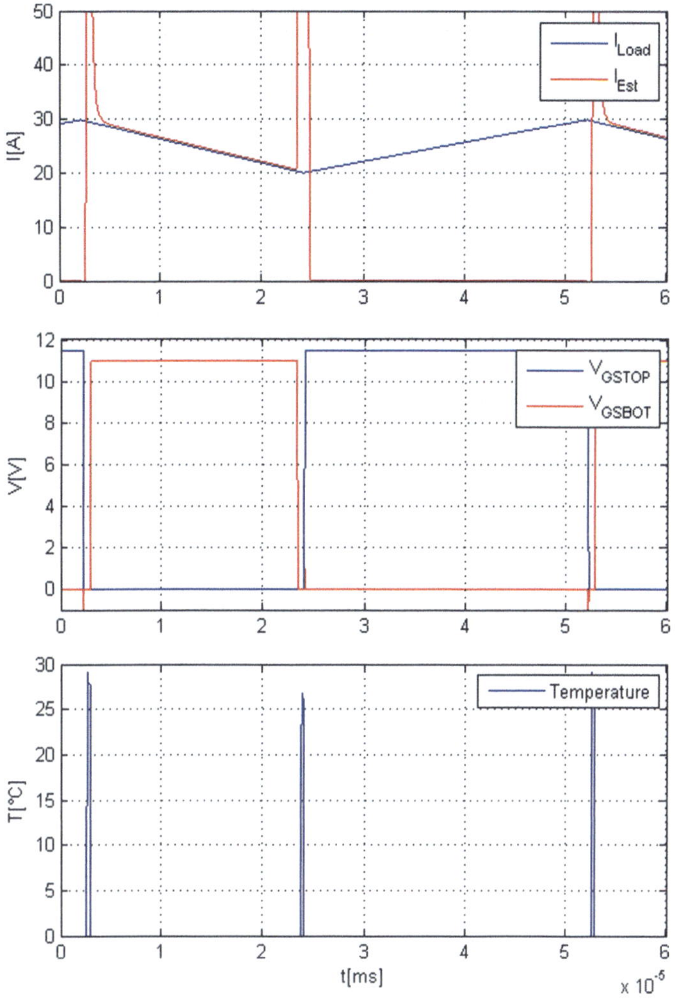

Fig. 5 Waveforms of real and estimated current, excitation voltage of MOSFETs and estimated value of temperature of MOSFET die for proposed low-side MOSFET current sensing algorithm

spice model of power MOSFET allows to make the simulation more realistic for experimental result comparison. The proposed algorithm has been tested in open loop, where the estimated current has not been used as a feedback signal for closed loop control. The temperature has been changed from −40 °C up to 150 °C and load current has been changed from 1 A up to 80 A.

Figure 6 shows the waveform of deviation between real and estimated current (top), real and estimated current (middle) and estimated temperature (bottom) for 3 A load current and 40 °C. The temperature has been measured and calculated during deadtime period and the value has been used for calculation of internal on-

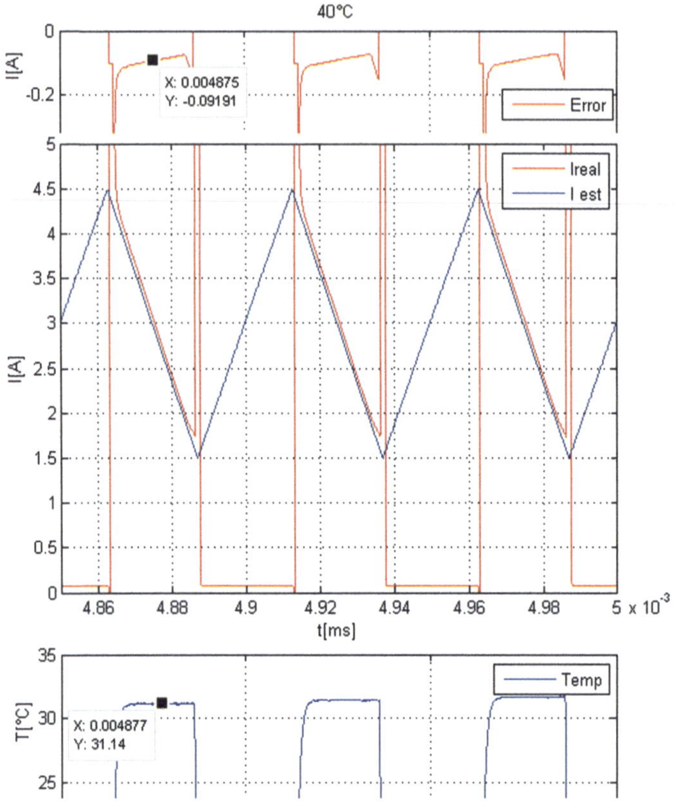

Fig. 6 Waveforms of estimated current, real load current, error between real and estimated current and estimated temperature of MOSFET die for 3 A load current and 40 °C temperature

state resistance and estimation of current, respectively. The deviation between real and estimated current is small, and the algorithm is able to estimate current with high accuracy. The deviation of estimated temperature is higher, but this inaccuracy does not influence the accuracy of estimated current. Figure 7 shows the waveforms of deviation between the real current and estimated current (top), real and estimated current (middle) and estimated value of MOSFET die temperature (bottom) for 30 A of load current and −40 °C of temperature. In this case, the temperature has been calculated by another group of Equations (6, 7, 9). The accuracy of estimated temperature is higher compared with the previous set of equations. The accuracy of estimated current is around 1%, which can be supposed as a sufficient result. Figure 8 shows the waveforms of deviation between the real current and estimated current (top), real and estimated current (middle) and estimated value of MOSFET die temperature (bottom) for 80 A of load current and 135 °C. The deviation of estimated temperature is higher compared with the two previous examples, but

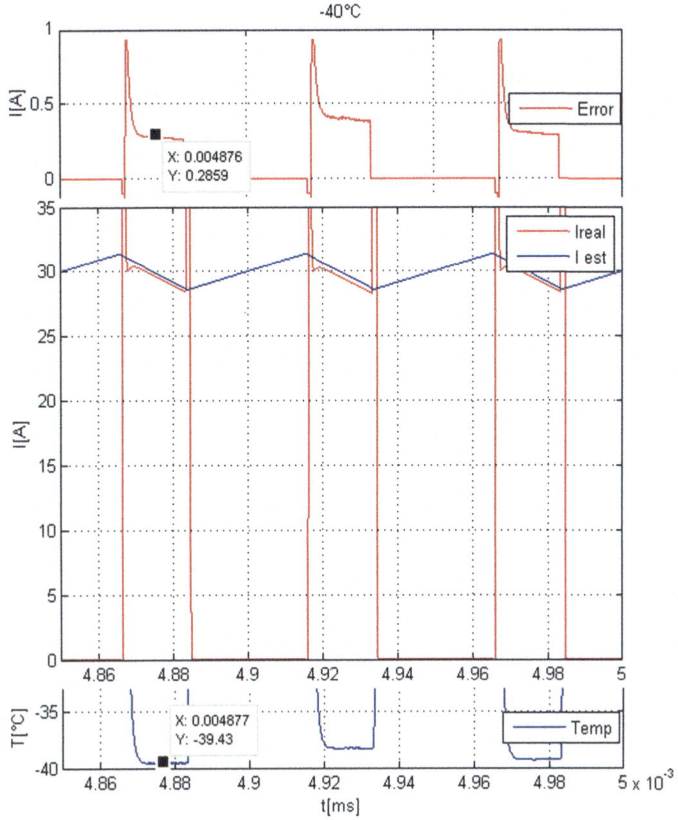

Fig. 7 Waveforms of estimated current, real load current, error between real and estimated current and estimated temperature of MOSFET die for 30 A load current and −40 °C temperature

the accuracy of estimated current is quite high. The deviation between real and estimated current is smaller than 0.5%, which can be considered as a great result.

Figure 9 shows the map of percental deviations as a function of temperature and current. The proposed algorithm is able to reach great accuracy in high current range and also in the middle current. The accuracy of estimated current for low currents can be supposed as sufficient, where the percental value of error is lower than 5%. Inaccuracy can be caused by short switched on time of low side MOSFET together with low signal–to-noise ratio. Due to the small value of internal on-state resistance together with low current, the recognized voltage drop is too small for following postprocessing. This assumption brings inaccuracy to the following calculation of estimated current and temperature, respectively.

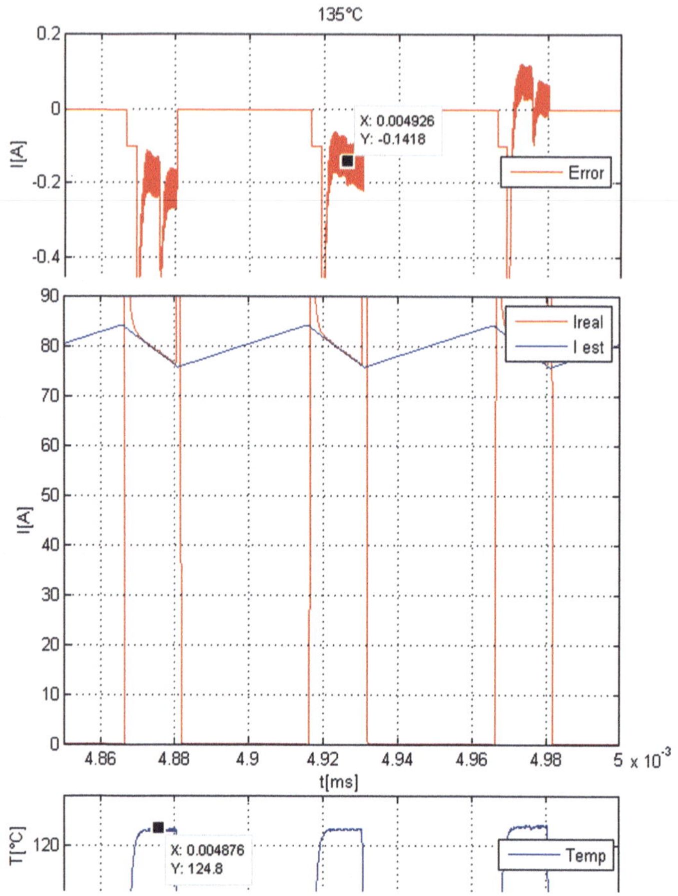

Fig. 8 Waveforms of estimated current, real load current, error between real and estimated current and estimated temperature of MOSFET die for 80 A load current and 135 °C temperature

5 Conclusion

This paper deals with the analysis and simulation of a different approach for current sensing technique primarily intended for automotive applications. The proposed method for current sensing technique utilizes the MOSFET as a current and temperature sensor. The proposed technique utilizes two characteristics: diode forward voltage as a function of temperature for different forward currents and R_{dson} as a function of temperature. The characteristics have been interpolated and the equations have been utilized for estimation purposes. The accuracy of estimated current achieved was quite high, also for lower currents. The main contribution of the proposed technique is the improvement in accuracy of estimated current, mainly for lower current compared with the previously published method [2]. This

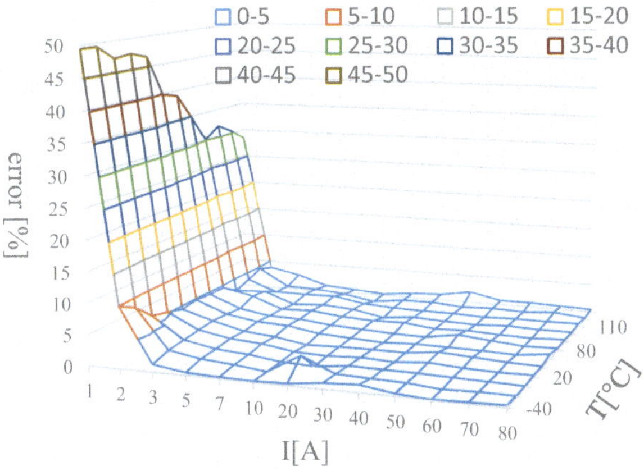

Fig. 9 Deviation between real and estimated current for whole temperature and current range

high accuracy has been achieved by using a different approach of temperature and current estimation, which is based on using a more precise mathematical model. Moreover, due to the opposite thermal coefficient of forward voltage and on-state resistance, the inaccuracy has also been particularly compensated, and the stability of the proposed algorithm has been ensured. The improvement in accuracy is hard to compare and quantify due to the missing simulation results in the previous publication. The comparison of experimental results with the simulated results of the proposed technique will be published later.

Acknowledgements The authors of this paper would like to thank for the support to the project VEGA 1/0774/18 and by project ITMS: 26210120021, co-funded from EU sources and European Regional Development Fund. Authors would also like to thank the NXP Semiconductors in Czech Republic for their support.

References

1. S.M. Agren, E. Knauss, R. Heldal, P. Pelliccione, G. Malmqvist, J. Boden, The manager perspective on requirements impact on automotive systems development speed, in *2018 IEEE 26th International Requirements Engineering Conference (RE), Banff, AB*, (2018), pp. 17–28
2. B. Cyril, B. Dominique, M. Herv, A. Bruno, E. Rene, B. Pascal, Towards a sensorless current and temperature monitoring in MOSFET-based H-bridge, in *IEEE 34th Annual Conference on Power Electronics Specialist, 2003, PESC '03, Acapulco, Mexico*, vol. 2, (2003), pp. 901–906
3. M. Bogus, C.H. Heiling, *Calculation of MOSFET Switch Temperature in Motor Control* (Infineon Technologies AG, 2017), US9608558B1
4. N.B. Singh, *Sensorless Current Sensing Method for Power Electronic Converters* (Deere&Company, 2015), US20150168462A1
5. G.L. Pearce, Technique for Measuring Temperature and Current via a MOSFET of a Synchronous Buck Voltage Converter (Charles E Wands ESQ, Allen, Dyer, Doppelt, Milbrath&Gilchrist, P.A., 2005), US20050088863A1

6. D.L. Blackburn, D.W. Berning, Power MOSFET temperature measurements, in *1982 IEEE Power Electronics Specialists Conference*, Cambridge, MA, USA, 1982, pp. 400–407
7. A. Ferrara et al., Comparison of electrical techniques for temperature evaluation in power MOS transistors, in *2013 IEEE International Conference on Microelectronic Test Structures (ICMTS), Osaka, Japan*, (2013), pp. 115–120
8. A. Koenig, T. Plum, P. Fidler, R.W. de Doncker, On-line junction temperature measurement of CoolMOS devices, in *2007 7th International Conference on Power Electronics and Drive Systems, Bangkok*, (2007), pp. 90–95
9. Vishay Siliconix AN-957: Measuring Power MOSFET Characteristics, https://www.vishay.com/docs/90715/an957.pdf. Accessed Nov 2010
10. B. Nguyen, X. Zhang, A. Ferencz, T. Takken, R. Senger, P. Coteus, Analytic model for power MOSFET turn-off switching loss under the effect of significant current diversion at fast switching events, in *2018 IEEE Applied Power Electronics Conference and Exposition (APEC), San Antonio, TX*, (2018), pp. 287–291
11. C. Leonardi, A. Raciti, F. Frisina, R. Letor, S. Musumeci, A new power MOSFET model including the variation of parameters with the temperature, in *Proceedings of the 1998 Second IEEE International Caracas Conference on Devices, Circuits and Systems, ICCDCS 98, On the 70th Anniversary of the MOSFET and 50th of the BJT.* (Cat. No.98TH8350), Isla de Margarita, Venezuela, 1998, pp. 261–266
12. Y.-S. Kim, S.-K. Sul, On-line estimation of IGBT junction temperature using on-state voltage drop, in *Conference Record of 1998 IEEE Industry Applications Conference*, Thirty-Third IAS Annual Meeting (Cat. No. 98CH36242), St. Louis, MO, USA, 1998, pp. 853–859, vol. 2
13. J.N. Davidson, D.A. Stone, M.P. Foster, D.T. Gladwin, Measurement and characterization technique for real-time die temperature prediction of MOSFET-based power electronics. IEEE Trans. Power Electron. **31**(6), 4378–4388 (2016)
14. R.C.S. Freire, S. Daher, G.S. Deep, A highly linear single p-n junction temperature sensor. IEEE Trans. Instrum. Meas. **43**(2), 127–132 (1994)
15. M.L. Walters, R.D. Lorenz, Circuit modeling methodology for isolated, high bandwidth junction temperature estimation, in *The 2010 International Power Electronics Conference - ECCE ASIA -*, Sapporo, 2010, pp. 970–974
16. H.M. Rashid, *Power Electronics Handbook, Devices, Circuits and Applications* (Elsevier, Amsterdam, 2011), p. 1362. ISBN 978-0-12-982036-5
17. B. Dobrucký, V. Ráček, P. Špánik, R. Gubric, Výkonové polovodičové štruktúry, Vysoká škola dopravy a spojov, 1995, 180 p. ISBN 9788071002840
18. Nexperia BUK7Y3R5-40H Datasheet, https://assets.nexperia.com/documents/data-sheet/BUK7Y3R5-40E.pdf. Accessed June 2015

Input-State Feedback Linearization of a Boost DC/DC Converter

Andrea Cervone and Gianluca Brando

Abstract The paper presents a procedure to achieve an input-state feedback linearization on a bidirectional Boost DC/DC converter connected to a passive load. The system linearization is achieved by a proper state-space/output transformation performed on a non-dimensional form of the analytical model. The resulting system is then controlled through a standard linear regulator. An online load estimation technique is also provided to overcome the transformations parameter dependency. The proposed approach has been numerically tested and compared with a standard two-loop controller.

1 Introduction

Power conversion is a key problem for electrical engineering, needed to adequate the voltage and current waveform from the energy sources to the loads [1, 2]. The DC/DC conversion, in particular, plays a major role in several applications like photovoltaic sources [3], battery-based energy storages [4] or High Voltage DC (HVDC) systems [5].

Among the several existing architectures, the Boost converter topology is one of the most widespread for industrial applications [1, 2]. It is particularly suited as a voltage stabilizer, where a variable input voltage has to be transformed into a higher and almost constant output one, independently from the supplied load [1, 2, 6].

The mathematical model of the system, even after a state-space averaging transformation, is non-linear with respect to the state variables. Standard controllers either rely on the linearized small-signal model around an equilibrium point [6] or use a cascaded control structure with an inner current control loop and an outer voltage control loop [6, 7].

A. Cervone · G. Brando (✉)
DIETI, University of Naples Federico II, Naples, NA, Italy
e-mail: andrea.cervone@unina.it; gianluca.brando@unina.it

W. Zamboni, G. Petrone (eds.), *ELECTRIMACS 2019*, Lecture Notes in Electrical Engineering 615, https://doi.org/10.1007/978-3-030-37161-6_11

However, in the last decades, the non-linear control theory has spread out, thanks both to the many analytical results, which guarantee the enhancement of the control performances, and to the availability of fast and cheap digital controllers, which make it possible to easily implement even complex control algorithms [8, 9].

One of the main approaches is the Feedback Linearization (FBL) [6–9], aimed to transform the original non-linear model into a globally linear one, thus allowing the controlled design to be approached with the well-established linear control theory.

The Input–Output Feedback Linearization [8] requires the definition of an equivalent control input, which directly controls a desired output variable through a linear relationship. This technique, well-suited when the equivalent input–output differential equation has the same order of the original system, introduces non-observable state variables when the system order is higher. Their evolution, named Zero-Dynamics [8], may lead to an unstable behaviour. It can be verified that the voltage stabilizer boost converter belongs to this kind of systems [6].

This problem can be avoided by employing the Input-State Feedback Linearization, which performs a linearizing transformation in the state-space. This technique requires the original model to satisfy some differential geometry properties [8, 9] and is usually hard to implement, since the desired transformation can be analytically found only through the resolution of a partial differential equation.

The present paper finds a state-space linearizing transformation of the considered system by properly manipulating the non-dimensional averaged model of the converter. The resulting system is then controlled through a simple linear regulator.

Since the feedback linearization approach largely depends on the converter load, an additional online estimation process is provided to overcome the intrinsic parameter dependence.

With respect to similar approaches covering the same topic, like the one described in [10], the proposed state transformation is independent from the control goals and theoretically allows the full decoupling between the control and the parameter estimation.

The control is then validated through numerical simulations in the Matlab/Simulink environment and compared with a standard cascaded two-loop control, thus highlighting some of its main benefits.

2 System Architecture and Mathematical Model

Given the discontinuous nature of power electronic converters, the mathematical model of the examined architecture is a piece-wise smooth system with respect to the switching devices control signals. In order to obtain a suitable form for the feedback linearization approach, it first needs to be transformed into a globally smooth model and then explicitly put in a control-affine form [8]:

$$\dot{\mathbf{x}} = \mathbf{f}(\mathbf{x}) + \mathbf{g}(\mathbf{x}) \cdot u \tag{1}$$

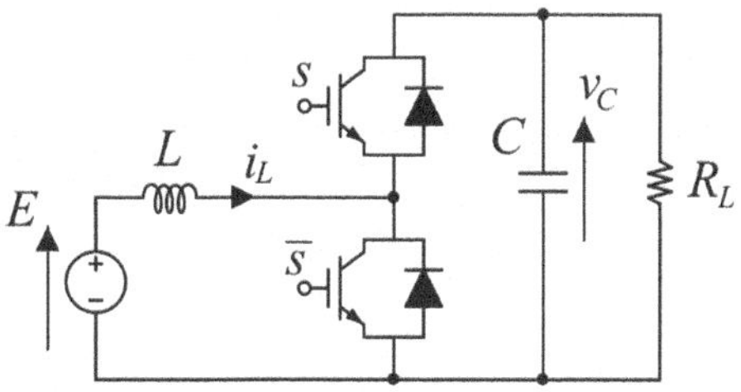

Fig. 1 Current-bidirectional Boost Converter circuit

2.1 Piece-Wise Smooth Dynamical Model

The examined structure, schematically represented in Fig. 1, is a current-bidirectional Boost Converter feeding a passive load. Its mathematical model can be easily derived once some simplifying hypotheses are met [7]:

- The load is modelled as a simple resistor.
- The supplying voltage is slowly variable with respect to the dynamics of interests (i.e. it can be modelled as a constant voltage source).
- The switching devices are ideal ones (i.e. it is possible to neglect their voltage drops, leakage currents and switching behaviour).
- The power losses on the filtering devices can be neglected.

The mathematical model is represented by a set of two differential equations [7, 11]:

$$\begin{cases} L\frac{\mathrm{d}i_{\mathrm{L}}}{\mathrm{d}t} = E - v_{\mathrm{C}} \cdot s \\ C\frac{\mathrm{d}v_{\mathrm{C}}}{\mathrm{d}t} = i_{\mathrm{L}} \cdot s - \frac{v_{\mathrm{C}}}{R_{\mathrm{L}}} \end{cases} \tag{2}$$

where L is the input filter's inductance, C is the DC-bus capacitance, E is the input voltage source, R_{L} is the load equivalent resistance and $s \in \{0; 1\}$ is the top-leg device switching signal (the bottom device switching signal is controlled through the complementary signal $\overline{s} = 1 - s$).

2.2 Averaged Model

For control purposes, it is convenient to transform the piece-wise smooth model (Eq. 2) into a globally smooth one through a standard averaging approach [11–14]: each state and input variable $x(t)$ is decomposed into a smoothly varying contribution $\langle x \rangle_{T_{\mathrm{s}}}(t)$, defined as its moving average over a modulation period with T_{s} width:

$$\langle x \rangle_{T_s}(t) = \frac{1}{T_s} \int_{t-T_s}^{t} x(\tau)\, d\tau \tag{3}$$

and into a rapidly varying contribution

$$\delta x(t) = x(t) - \langle x \rangle_{T_s}(t) \tag{4}$$

By neglecting the high-frequency contributions given by the product of two rapidly varying terms, the average model of (Eq. 2) is found:

$$\begin{cases} L \frac{d\langle i_L \rangle_{T_s}}{dt} = \langle E \rangle_{T_s} - \langle v_C \cdot s \rangle_{T_s} \cong E - \langle v_C \rangle_{T_s} \cdot d \\ C \frac{d\langle v_C \rangle_{T_s}}{dt} = \langle i_L \cdot s \rangle_{T_s} - \frac{\langle v_C \rangle_{T_s}}{R_L} \cong \langle i_L \rangle_{T_s} \cdot d - \frac{\langle v_C \rangle_{T_s}}{R_L} \end{cases} \tag{5}$$

where $d = \langle s \rangle_{T_s}$ is the duty-cycle of the top switching device and, contrary to the discrete nature of corresponding switching signal, can be chosen continuously in its feasible interval [0; 1].

2.3 *Non-dimensionalized Affine Model*

The non-dimensionalized model can be found from Eq. (5) by applying the variables transformations:

$$\begin{matrix} t' = t/\sqrt{LC} & ; & x_1 = \langle i_L \rangle_{T_s} / \left(E \cdot \sqrt{C/L} \right) \\ G = \sqrt{L/C}/R_L & ; & x_2 = \langle v_C \rangle_{T_s} / E \end{matrix} \tag{6}$$

leading to the compact form:

$$\begin{cases} \dot{x}_1 = 1 - x_2 \cdot d \\ \dot{x}_2 = x_1 \cdot d - G \cdot x_2 \end{cases} \tag{7}$$

In this form, the model (Eq. 7) represents a second-order control-affine system in the canonical form (Eq. 1), whose input is the duty-cycle d and whose dynamical functions are:

$$\mathbf{f}(\mathbf{x}) = \begin{bmatrix} 1 \\ -G \cdot x_2 \end{bmatrix} \quad ; \quad \mathbf{g}(\mathbf{x}) = \begin{bmatrix} -x_2 \\ x_1 \end{bmatrix} \tag{8}$$

3 Input-State Feedback Linearization

The input-state feedback linearization [8, 9] is based on the search of a proper state transformation $\mathbf{z} = \mathbf{T}(\mathbf{x})$ and of a proper input transformation $u = \alpha(\mathbf{x}) + \beta(\mathbf{x}) \cdot v$ (with v an equivalent control input) able to put the n-th order system (Eq. 1) in an equivalent linear form:

$$\dot{\mathbf{z}} = \mathbf{A} \cdot \mathbf{z} + \mathbf{B} \cdot v \tag{9}$$

A standard approach for this purpose is to find a scalar state variable $z = \phi(\mathbf{x})$ whose dynamics is in the Brunovsky canonical form [15]:

$$z^{(n)} = f_z\left(z, \dot{z}, \cdots, z^{(n-1)}\right) + b_z\left(z, \dot{z}, \cdots, z^{(n-1)}\right) \cdot u = v \tag{10}$$

where $z^{(n)} = \mathrm{d}^n z/\mathrm{d}(t')^n$.

This transformation can be found if [8]:

- The set $\left\{\mathbf{g}(\mathbf{x}), \mathrm{ad}_{\mathbf{f}(\mathbf{x})}(\mathbf{g}(\mathbf{x})), \cdots, \mathrm{ad}_{\mathbf{f}(\mathbf{x})}^{(n-1)}(\mathbf{g}(\mathbf{x}))\right\}$ is linearly independent
- The set $\left\{\mathbf{g}(\mathbf{x}), \mathrm{ad}_{\mathbf{f}(\mathbf{x})}(\mathbf{g}(\mathbf{x})), \cdots, \mathrm{ad}_{\mathbf{f}(\mathbf{x})}^{(n-2)}(\mathbf{g}(\mathbf{x}))\right\}$ is involutive

where $\mathrm{ad}_{\mathbf{f}(\mathbf{x})}(\mathbf{g}(\mathbf{x})) = \frac{\partial \mathbf{g}}{\partial \mathbf{x}} \cdot \mathbf{f}(\mathbf{x}) - \frac{\partial \mathbf{f}}{\partial \mathbf{x}} \cdot \mathbf{g}(\mathbf{x})$ is the Lie Bracket differential operator [8].

3.1 Linearizing Transformation Search

Since the model (Eq. 7) is a second-order dynamical system, the involutivity condition is not required. The linear independence is easily verified since, from Eq. (8), it results:

$$\mathbf{g}(\mathbf{x}) = \begin{bmatrix} -x_2 \\ x_1 \end{bmatrix}; \quad \mathrm{ad}_{\mathbf{f}(\mathbf{x})}(\mathbf{g}(\mathbf{x})) = \begin{bmatrix} G \cdot x_2 \\ G \cdot x_1 + 1 \end{bmatrix} \tag{11}$$

The state transformation operator $\phi(\mathbf{x})$ needs to verify the partial differential equation $\nabla\phi \cdot \mathbf{g}(\mathbf{x}) = 0$, resulting in [8]:

$$x_1 \cdot \frac{\partial \phi}{\partial x_2} = x_2 \cdot \frac{\partial \phi}{\partial x_1} \tag{12}$$

A non-trivial solution of Eq. (12) is given by:

$$\phi(\mathbf{x}) = x_1^2 + x_2^2 \tag{13}$$

and it is easy to verify that it satisfies the additional condition for the Input-State Feedback Linearization [6, 8]:

$$\nabla\phi \cdot \mathrm{ad}_{\mathbf{f}(\mathbf{x})}\left(\mathbf{g}\left(\mathbf{x}\right)\right) = 2\, x_2 \left(G \cdot x_1 + 1\right) + 2\, G \cdot x_1 \cdot x_2 \neq 0 \tag{14}$$

The complete transformation is, then, given by:

$$\begin{bmatrix} z \\ \dot{z} \end{bmatrix} = \begin{bmatrix} z_1 \\ z_2 \end{bmatrix} = \mathbf{z} = \mathbf{T}\left(\mathbf{x}\right) = \begin{bmatrix} \phi\left(\mathbf{x}\right) \\ \nabla\phi\left(\mathbf{x}\right) \cdot \mathbf{f}\left(\mathbf{x}\right) \end{bmatrix} = \begin{bmatrix} x_1^2 + x_2^2 \\ 2x_1 - 2Gx_2^2 \end{bmatrix} \tag{15}$$

The transformation (Eq. 15), although invertible, has a polydromic nature linked to the square function of the normalized capacitor voltage x_2. Nevertheless, since the considered converter is unidirectional with respect to the DC-bus voltage, this effect can be neglected by restricting the state-space to $x_2 > 0$.

It can also be noted that $z = z_1$, which represents the Brunovsky equivalent state, is the sum of the normalized square of the original state variables and, therefore, represents the normalized energy stored into the system.

3.2 Feedback Linearized System Model

From the state transformation (Eq. 15), the dynamics of the converter can be effectively expressed as:

$$\ddot{z} = f_z\left(z, \dot{z}\right) + b_z\left(z, \dot{z}\right) \cdot d = f_z'\left(\mathbf{x}\left(z, \dot{z}\right)\right) + b_z'\left(\mathbf{x}\left(z, \dot{z}\right)\right) \cdot d \tag{16}$$

with:

$$f_z'\left(\mathbf{x}\right) = 4\, G^2\, x_2^2 + 2; \quad g_z'\left(\mathbf{x}\right) = -2\, x_2 - 4\, G\, x_1\, x_2 \tag{17}$$

The globally linear form (Eq. 10) is then achieved by setting:

$$d = \frac{v - f_z'\left(\mathbf{x}\right)}{b_z'\left(\mathbf{x}\right)} = \frac{4\, G^2\, x_2^2 + 2 - v}{2\, x_2 + 4\, G\, x_1\, x_2} \tag{18}$$

resulting in the desired form $\ddot{z} = v$.

4 Controller Design

Once the feedback linearization transformation has been found, it is possible to design a proper feedback controller, aimed to stabilize the DC-Bus voltage v_C (corresponding to the non-dimensionalized state x_2) to a desired constant reference value v_C^* (corresponding to a reference state x_2^*).

4.1 Equivalent Reference Computation

The linearized system (Eq. 16) controllable variable is the equivalent state $z = x_1^2 + x_2^2$: to keep into account the desired control of the original state variable x_2, its reference needs to be properly adapted.

A simple, but effective, solution consists in finding, for each x_2^*, the correspondent steady state value of the normalized input inductor current x_1^*. From Eq. (7), by nullifying the time derivatives, a second-order algebraic system in the (d, x_1) variables is found. The corresponding solution is:

$$d^* = 1/x_2^*; \quad x_1^* = G\left(x_2^*\right)^2 \tag{19}$$

As a result, the equivalent reference input for the feedback linearized system can be chosen as:

$$z^* = \left(x_1^*\right)^2 + \left(x_2^*\right)^2 = \left(x_2^*\right)^2 + G^2\left(x_2^*\right)^4 \tag{20}$$

4.2 System Controller Design

Once the feedback linearization has been performed, the dynamic behaviour of the equivalent state (Eq. 16) with the control transformation (Eq. 18) is a simple double integrator, whose transfer function in the Laplace domain is:

$$\mathfrak{G}(s) = 1/s^2 \tag{21}$$

A feedback controller acting on the error $e_z = z - z^*$ can be designed through any standard linear control design procedure. A simple solution is to implement a Proportional-Derivative (PD) controller, whose transfer function in the Laplace domain is:

$$\mathfrak{R}(s) = k_P + s \cdot k_D = k_P \cdot (1 + s \cdot \tau_{PD}) \tag{22}$$

The introduction of the zero $\tau_{PD} = k_D/k_P$ is able to compensate for the double pole in 0 of the equivalent feedback-linearized boost converter transfer function.

The effective open-loop transfer function of the system can be found by taking into account the time delay introduced by the PWM modulation technique. For a digital controller with a sampling period of T_s (the same referred in Eq. (3) for the averaging procedure) and with a synchronous discrete update of the reference duty-cycle d, the average non-dimensional time delay is $\tau_s = 1.5\ T_s/\sqrt{LC}$ and can be represented in the Laplace domain through the transfer function:

$$\mathfrak{A}(s) = \mathrm{e}^{-s\ \tau_s} \tag{23}$$

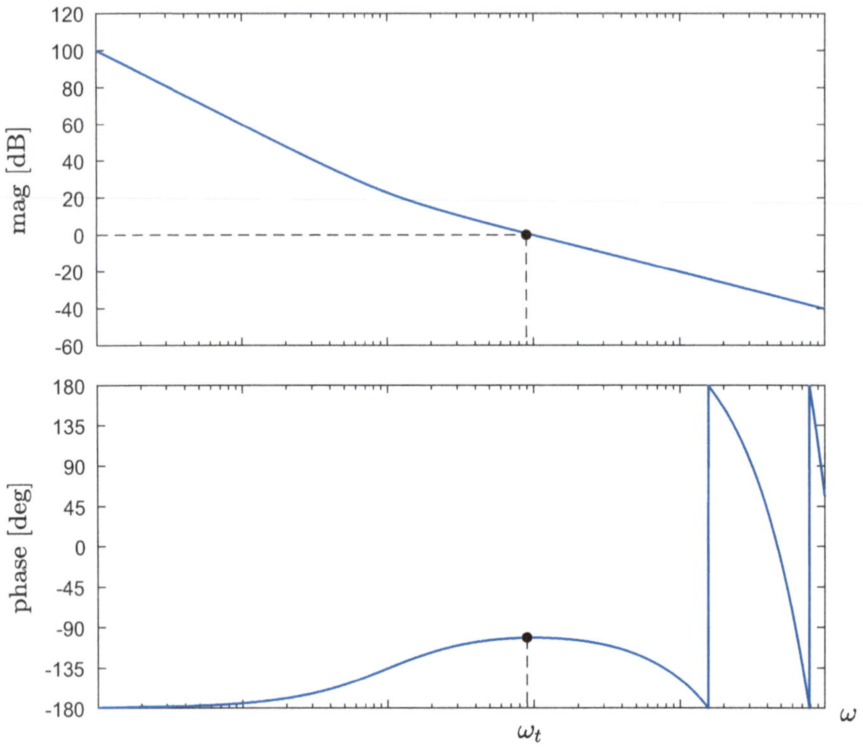

Fig. 2 Qualitative Bode diagrams of the open-loop transfer function

leading to the open-loop transfer function:

$$\mathfrak{L}(s) = \mathfrak{R}(s) \cdot \mathfrak{G}(s) \cdot \mathfrak{A}(s) = \frac{k_{\mathrm{P}} \cdot (1 + s \cdot \tau_{\mathrm{PD}})}{s^2}\, \mathrm{e}^{-s\, \tau_{\mathrm{s}}} \tag{24}$$

The robustness of the control is achieved by guaranteeing a proper phase margin through a proper choice of the open-loop cut-off frequency ω_{t} (and its corresponding non-dimensional one). The resulting Bode diagram is depicted in Fig. 2.

It should be noted that, since the measurable equivalent state already represents the time derivative of the equivalent output, no explicit computation is required for the input of the controller derivative component and, therefore, no additional high frequency poles are needed. Then, the PD regulator behaves as a state-feedback controller.

Moreover, since the proposed approach does not use any integrator, no anti wind-up strategy is required.

4.3 Online Load Estimation

The feedback linearized controller needs both to non-dimensionalize the state variables (in order to find the equivalent state z) and to compute the duty-cycle as per Eq. (18). For both the operations there is the explicit need for the knowledge of the system parameters.

While the inductance L and the capacitance C, being chosen during the hardware converter design, can be known with a sufficiently high precision, the equivalent load resistance R_L and the corresponding non-dimensional conductance G cannot be known a priori. This parameter dependence is found in almost every feedback linearization technique [6, 8, 9] and might affect the whole controller performance, leading to high errors and stability issues.

Standard two-loop controllers compensate for parameters' inaccuracy through the DC-bus voltage controller with the integral contribution (in steady state) and with a proper sizing of its bandwidth (in transient operations).

Contrarily, the implementation of an additional integral contribution in the controller (Eq. 22) would not compensate for parameter inaccuracies, since the reference state z^* would be affected too. This drawback might be solved through an external control loop acting on v_C, but this approach might give satisfactory results only in steady state and would not guarantee a stable behaviour of the inner loop acting on z.

A way to overcome this effect is to provide an online estimation of the load parameter. With reference to the second equation in the non-dimensional model (Eq. 7), a simple estimation rule is immediately found as:

$$\hat{G} = (x_1 \cdot d - \dot{x}_2)/x_2 \tag{25}$$

To avoid the estimation through the time derivative $\dot{x}_2$ (which, in an experimental setup, might be noisy) it is possible to integrate the same equation in a time interval of width T_{est} (corresponding to a non-dimensional width $T'_{\text{est}} = T_{\text{est}}/\sqrt{LC}$) while assuming the load as constant. The estimation rule is then modified as:

$$\hat{G} = \left(\int_{t'-T'_{\text{est}}}^{t'} x_1 \cdot d \, \mathrm{d}\tau' - x_2\left(t'\right) + x_2\left(t' - T'_{\text{est}}\right) \right) \Big/ \int_{t'-T'_{\text{est}}}^{t'} x_2 \, \mathrm{d}\tau' \tag{26}$$

The optimal T_{est} width value needs to be chosen as a trade-off between the speed of the desired estimation response and the noise filtering capabilities.

In case the converter is needed to supply a load varying among a known set of parameters (which is reasonable in most industrial applications), the estimation method can be further improved through a proper introduction of a feed-forward action.

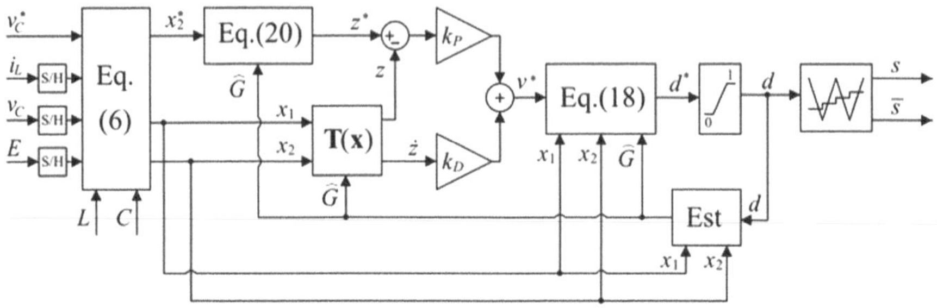

Fig. 3 Feedback Linearization-based control scheme

4.4 Control Scheme

The proposed control algorithm is schematically represented in Fig. 3.

The input voltage E and the state variables i_L and v_C are sampled and, since the sampling instants are synchronized with the PWM carrier signal, this operation is almost equivalent to the discretization of the averaged model state variables $\langle E\rangle_{T_s}$, $\langle i_L\rangle_{T_s}$ and $\langle v_C\rangle_{T_s}$ of system (Eq. 5).

The sampled variables, together with the reference DC-Bus voltage, are then non-dimensionalized through Eq. (6).

Then, the feedback linearization technique is applied to the state variables x_1 and x_2, which are transformed into the equivalent state variables $z_1 = z$ and $z_2 = \dot{z}$ by Eq. (15), while the equivalent state reference z^* is computed by Eq. (20).

The PD controller, acting on the feedback linearized state variables, computes the reference equivalent control v^*, which is then used to find the actual reference duty-cycle d^* through Eq. (18).

The reference duty-cycle is then saturated in its feasible interval [0; 1] and compared with a standard triangular carrier signal in order to produce the effective discrete signals s and $\bar{s}$ to control the switching devices.

The online load estimation is performed in parallel with the main controller and, from the knowledge of the non-dimensional state variables x_1, x_2 and the applied duty-cycle d, computes through Eq. (26) the value $\hat{G}$, implemented into the state and output transformations.

5 Numerical Validation

The proposed control strategy has been numerically tested in the MATLAB/Simulink environment and compared with a standard two-loop controller in different scenarios.

Table 1 Rated parameters of the numerical simulations

Parameter	Symbol	Value
Rated power	P_R	10 kW
Rated input voltage	E_R	80 V
Rated output voltage	V_{CR}	200 V
Rated switching frequency	f_{swR}	5 kHz
Inductance	L	1 mH
Capacitance	C	2 mF
Rated load equivalent resistance	R_L	4 Ω
Sampling frequency	f_s	10 kHz

The hardware system parameters are reported in Table 1.

The feedback linearization-based controller has been set with an open-loop cut-off frequency $\omega_t \cong 500$ Hz, while the time window for the online load estimation has been set to $T_{est} = 5$ ms (i.e. 50 sampling periods).

The standard two-loop controller has been implemented though an inner Proportional (P) current controller with ancillary current prediction and an outer Proportional-Integral (PI) voltage controller acting on the DC-bus voltage. To guarantee a fair comparison between the two approaches, the open-loop cut-off frequency for the voltage control has been set to the same value $\omega_t \cong 500$ Hz at rated conditions.

The comparisons are shown with respect to the sampled variables of the controllers: since the sampling is synchronous with the pulse width modulation, the fast varying contributions to the state variables due to the switching transitions are filtered out via aliasing and an easier visualization of the results is facilitated.

Moreover, in order to prove the robustness of the proposed control with respect to parametric uncertainties, the inductance and capacitance have been simulated with a random uncertainty with respect to their rated value in the interval [95%; 105%].

5.1 Reference Voltage Transitions

The test is aimed to verify the output voltage response with respect to a reference transition. In Fig. 4 the reference step assumes low values with respect to the rated output voltage (5%) and is performed at the rated load, while in Fig. 5 the step is half of the rated voltage at half of the rated load.

In each test, the DC-bus voltage, before seeking the new reference value, shows an initial decrease from its initial state, thus exhibithing a non-minimum phase behaviour (as can be analytically shown through the small-signal analysis of the mathematical model (Eq. 2)). This is a direct consequence of the fast increase of the inductor current.

In both cases the standard two-loop controller manifests an higher initial voltage drop and a slower setting time, in addition to higher peak transient currents. This behaviour can be justified by the phase delay introduced on the outer voltage loop both by the inner current control and by the presence of the integral component.

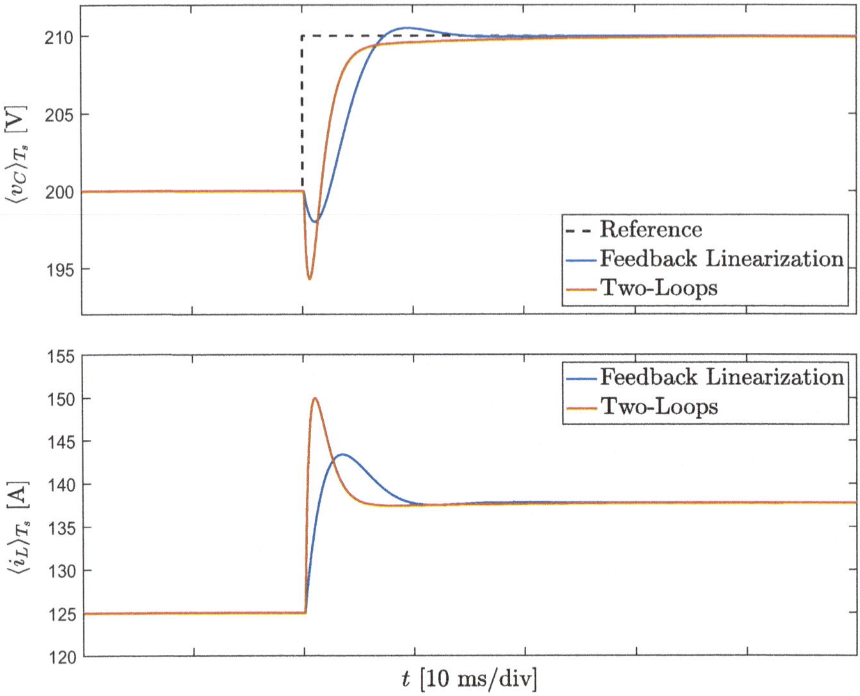

Fig. 4 Small reference voltage step transition at full load

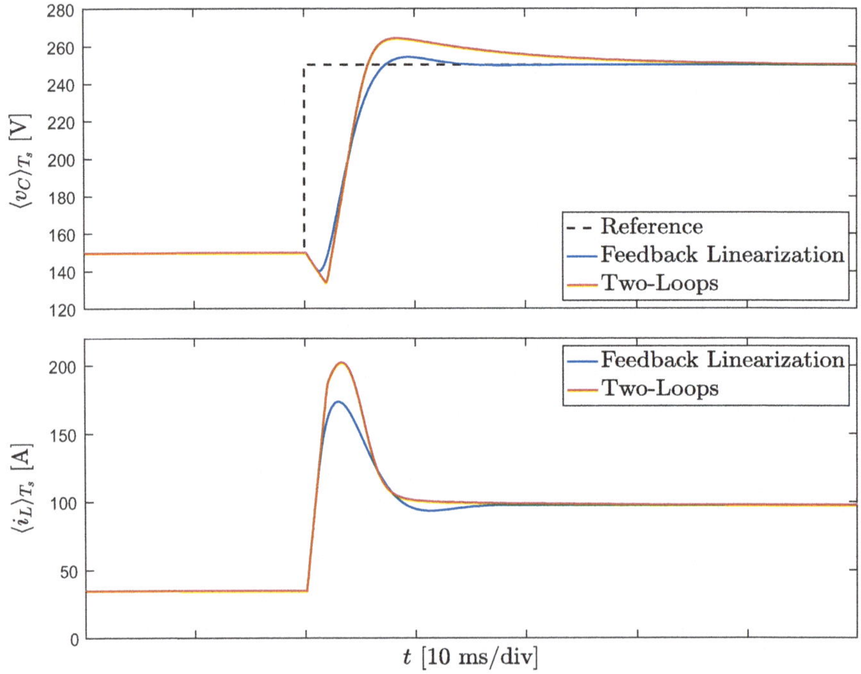

Fig. 5 Big reference voltage step transition at half of the rated load

5.2 Load Transition

The test verifies the system response with respect to a load step variation from no-load to rated load condition. The behaviour of both the DC-bus voltage and the inductor current (see Fig. 6) is similar for the two different controllers. However, the feedback linearization-based one, despite showing a slightly higher initial voltage drop, manifests a faster setting time. The transient behaviour of the two-loop controller is governed by the setting time of the PI voltage regulator, while in the feedback linearization the transient depends on the online load estimation, which can be designed almost independently from the main controller.

5.3 Input Voltage Variation

The test verifies the system response with respect to an input voltage variation by performing a step variation of E from 80 V to 60 V. The results, shown in Fig. 7, show, again, a similar behaviour. However, contrary to the previous test, the feedback linearization controller is slower. Indeed, the online load estimation block,

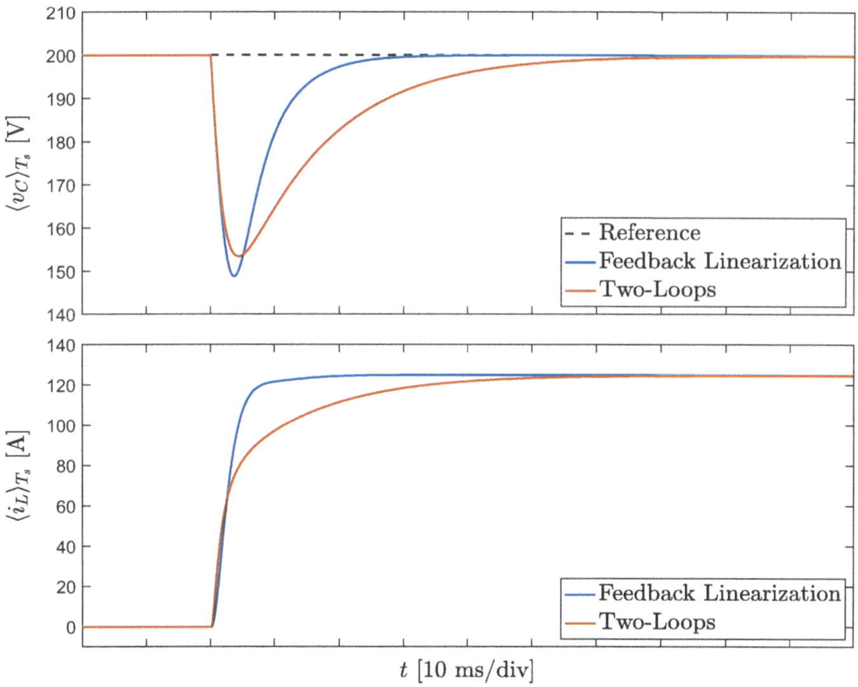

Fig. 6 Step transition from no-load to rated load

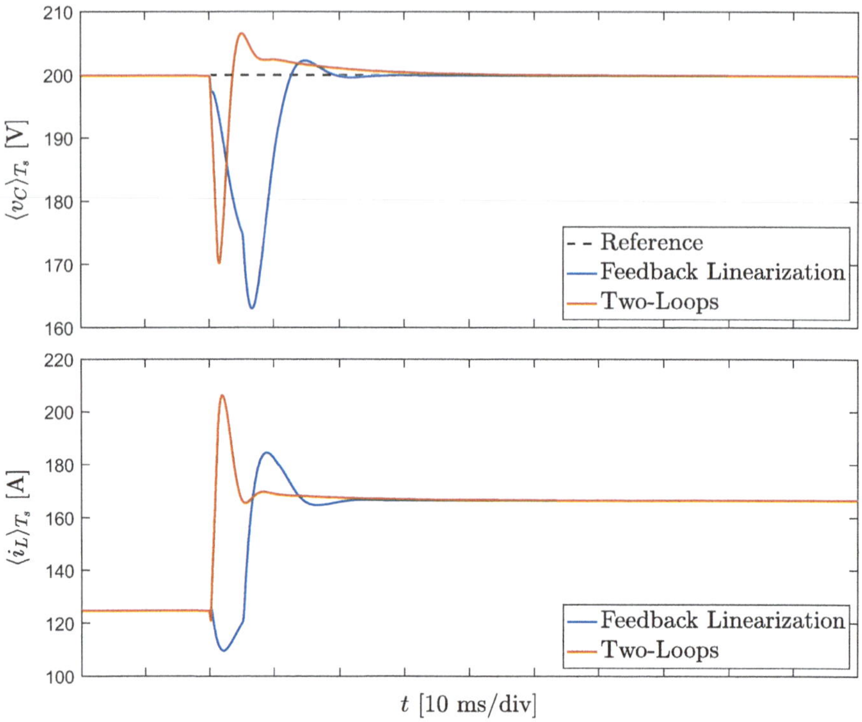

Fig. 7 Step decrease of the input voltage

perturbed by the resulting step variation in the non-dimensional state variables, wrongly drives the state and output transformations accordingly by assuming a load transition. The effect becomes negligible with a proper reduction of the estimation time. A more refined approach, which considers the real state variables v_C and i_L, might improve this aspect further.

6 Conclusions

An input-state feedback linearization approach has been applied to a current-bidirectional boost converter. From a non-dimensional mathematical model, through the application of some differential geometry tools, a partial differential equation is obtained and analytically solved, leading to a state transformation able to globally linearize the system. Then, a feedback PD controller and an online load estimation procedure have been designed to meet the control goals. The resulting system has been numerically tested and compared with a standard two-loop controller: in most of the testing scenarios the proposed controller is able to achieve better performances, therefore representing a good alternative to standard approaches.

References

1. N. Mohan, T.M. Undeland, W.P. Robbins, *Power electronics, converters, applications and design* (Wiley, Hoboken, 2003)
2. M.H. Rashid, *Power electronics: circuits, devices & applications*, 4th edn. (Pearson, London, 2014)
3. W. Li, X. He, Review of nonisolated high-step-up DC/DC converters in photovoltaic grid-connected applications. IEEE Trans. Ind. Electron. **58**, 1239–1250 (2011)
4. A. Khaligh, Z. Li, Battery, ultracapacitor, fuel cell, and hybrid energy storage systems for electric, hybrid electric, fuel cell, and plug-in hybrid electric vehicles: state of the art. IEEE Trans. Veh. Technol. **59**, 2806–2814 (2010)
5. S. Kedia, H.J. Bahirat, DC-DC converter for HVDC grid application, in *2017 National Power Electronics Conference (NPEC)*, Pune 2017
6. S. Bacha et al., Power electronic converters modeling and control: with case studies, in *Advanced Textbooks in Control and Signal Processing*, (Springer, London, 2014)
7. L. Dixon, Average current mode control of switching power supplies, Unitrode Power Supply Design Seminar Manual
8. J.-J.E. Slotine, W. Li, *Applied Nonlinear Control* (Prentice Hall, Englewood Cliffs, NJ, 1991)
9. H.K. Khalil, *Nonlinear Systems*, 3rd edn. (Prentice Hall, Upper Saddle River, NJ, 2002)
10. H. Zheng, D. Shuai, Nonlinear control of Boost converter by state feedback exact linearization, in *2012 24th Chinese Control and Decision Conference*, (CCDC, Taiyuan, 2012), pp. 3502–3506
11. C.T. Rim, G.B. Joung, G.H. Cho, A state-space modeling of nonideal DC-DC converters, in *PESC '88 Record., 19th Annual IEEE Power Electronics Specialists Conference*, Kyoto, Japan, 1988, pp. 943–950, vol. 2
12. G.V. Wester, R.D. Middlebrook, Low-frequency characterization of switched DC-DC converters. IEEE Trans. Aerosp. Electron. Syst. **AES-9**, 376–385 (1973)
13. R.D. Middlebrook, S. Cuk, A general unified approach to modelling switching converter power stages, in *Proceedings of the IEEE Power Electronic Specialists Conference*, Cleveland, Ohio, USA
14. G.C. Verghese, C.A. Bruzos, K.N. Mahabir, Averaged and sampled-data models for current mode control: a reexamination, in *Proceedings of the 20th Annual IEEE Power Electronics Specialists Conference – PESC* 1989
15. P. Brunovský, A classification of linear controllable systems. Kybernetika **06**(3), 173–188 (1970)

2-D Analytical Model of Conventional Switched Reluctance Machines

Walid Belguerras, Youcef Benmessaoud, Fréderic Dubas, Kamel Boughrara, and Hilairet Mickael

Abstract In this paper, we present a two-dimensional (2-D) analytical model of conventional switched reluctance machines (SRMs). This model has been applied to an 8/6 conventional SRM supplied by conventional excitation (viz., standard asymmetric H-bridge). The goal is to determine the electromagnetic performances. The proposed analytical model is based on solving the partial differential equations (PDEs) due to Maxwell's equations in each domain of the studied machine (viz., air-gap, rotor and stator slots). A parametric study by using the developed analytical model has been compared with that obtained by numerical computations in linear and non-linear conditions. The results showed that the analytical and numerical results are in good agreements in linear conditions. However, in non-linear conditions, the developed model overestimates the performances. Indeed, to predesign the machine, this model can be incorporated in optimization environments where savings in computation time are needed.

1 Introduction

The SRMs present many benefits for high-speed applications (e.g., electric compressor) compared with other types of machines. They can be operated at a very high-speed as they have no sliding contacts [1, 2], and can be operated in extreme temperature conditions. They are a competitor for the permanent-magnet synchronous machines in electric vehicle applications because of their simplicity and low cost, and their ability to operate at high speed with low maintenance [3–5].

In the literature, we find different methods of electromagnetic modeling of electric machines; semi-analytical modeling based on the magnetic equivalent

W. Belguerras (✉) · Y. Benmessaoud · F. Dubas · H. Mickael
FEMTO-ST, CNRS, Univ. Bourgogne Franche-Comté, Belfort Cedex, France
e-mail: walid.belguerras@femto-st.fr; mickael.hilairet@univ-fcomte.fr

K. Boughrara
Ecole Nationale Polytechnique (LRE-ENP), Alger, Algeria
e-mail: kamel.boughrara@g.enp.edu.dz

W. Zamboni, G. Petrone (eds.), *ELECTRIMACS 2019*, Lecture Notes in Electrical Engineering 615, https://doi.org/10.1007/978-3-030-37161-6_12

circuit (or permeance network) [6, 7], subdomain method in linear conditions (i.e., infinite permeability of the iron parts) [8–10], and the exact subdomain method, which takes into account the iron permeability [11–15]. In addition, we find analytical methods based on multilayers [16–19] or elementary subdomains for the local saturation effect [20, 21].

The analytical methods cited previously give an accurate electromagnetic result compared with numerical calculations, with reduced computation time. In this paper, we will present a comparison study of 2-D electromagnetic performances between the developed linear model based on the subdomain method in linear conditions. In order to analyze the validity of the developed model, the results have been compared with those computed numerically using FEMM [22] in linear and non-linear conditions.

2 Analytical Model

The analytical model based on the subdomain method in linear conditions is given in [8–10]. In order to simplify the model, we have considered the following assumptions:

- End-effects are neglected, i.e., $A = \{0;\ 0;\ A_z\}$.
- Eddy-currents effects in all materials are neglected.
- Current density in the stator slots has only one component along the z-axis, i.e., $J = \{0;\ 0;\ J_z\}$.
- The slots have a radial sides.
- The relative permeability is considered infinite for the iron parts (i.e., the saturation effect is neglected).

The schematic representation of the studied 8/6 conventional SRM is shown in Fig. 1.

In developing the 2D analytical model, a magnetic vector potential formulation is used in polar coordinates. It consists of solving the partial differential equation [23] due to

Maxwell's equations in each domain of the studied machine. The equations to be solved in each region are given by

$$\frac{\partial^2 A_{zI}}{\partial r^2} + \frac{1}{r} \cdot \frac{\partial A_{zI}}{\partial r} + \frac{1}{r^2} \cdot \frac{\partial^2 A_{zI}}{\partial \theta^2} = 0 \quad \text{(Region I)}, \tag{1a}$$

$$\frac{\partial^2 A_{zj}}{\partial r^2} + \frac{1}{r} \cdot \frac{\partial A_{zj}}{\partial r} + \frac{1}{r^2} \cdot \frac{\partial^2 A_{zj}}{\partial \theta^2} = 0 \quad (\text{Region } j\text{th}). \tag{1b}$$

$$\frac{\partial^2 A_{zi}}{\partial r^2} + \frac{1}{r} \cdot \frac{\partial A_{zi}}{\partial r} + \frac{1}{r^2} \cdot \frac{\partial^2 A_{zi}}{\partial \theta^2} = -\mu_0 \cdot J_{zi} \quad (\text{Region } i\text{th}). \tag{1c}$$

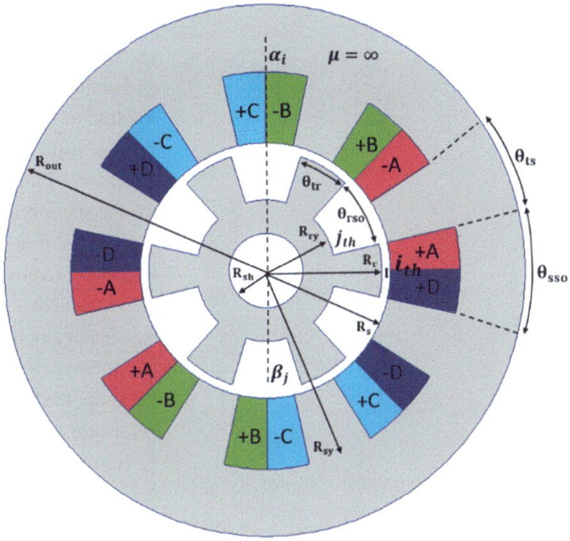

Fig. 1 Example of 8/6 conventional SRM

Solving of PDEs given by Eqs. (1a–1c) allows to obtain the general solution of A_z in each domain [8–10]. The integration constants are determined by using a Fourier series expansion of A_z in each region and the boundary conditions (BCs) [8–21]. The linear systeme can be written as

$$[A] \cdot [X] = [B], \tag{2}$$

where:

- $[A]$ is the square matrix of the integration constants obtained by BCs of dimension $Q \times Q$ with $Q = 4N + Q_r \cdot (1 + M) + Q_s \cdot (1 + K)$ in which Q_s and Q_r represent, respectively, number of the stator and rotor slots, and N, M, and K represent, respectively, the finite number of spatial harmonics terms in various regions.
- $[X]$ is the vector of unknowns (integration constants to determine) with dimension $Q \times 1$.
- $[B]$ is the vector of electromagnetic sources terms with dimension $Q \times 1$.

The vector $[X]$ can divided into three parts as follows:

- Part 1 is the air-gap (i.e., Region I) with dimension $4N \times 1$.
- Part 2 is the rotor slots (i.e., Region *j*th) with dimension $Q_r \cdot (1 + M) \times 1$.
- Part 3 is the stator slots (i.e., Region *i*th) with dimension $Q_s \cdot (1 + K) \times 1$.

Using the geometrical and physical parameters given in Table 1, we have calculated the computing time necessary for obtaining the vector $[X]$ (viz., $t = 58.85$ s for 31 position) and the computing time necessary for making the mesh and analyze on FEMM (viz., $t = 151.33$ s for 31 positions. The auto-mesh isused).

Table 1 Parameters of 8/6 conventional SRM

Parameter	Value
Q_s Number of stator slots	8
Q_r Number of rotor slots	6
R_{sh} Radius of the shaft	11.5 mm
R_{ry} Radius of the rotor yoke	20.9 mm
R_r Radius of the rotor surface	33.9 mm
R_s The stator bore radius	34.3 mm
R_{sy} Radius of the stator yoke	53.4 mm
R_{out} Outer radius of the stator	71.5 mm
L Axial length	125 mm
θ_{sso} Stator slot opening	25.2°
θ_w Opening of a slot coil	12.6°
θ_{ts} Stator teeth opening	19.8°
θ_{rso} Rotor slot opening	39.3°
θ_{tr} Rotor teeth opening	20.6°
N Number of harmonic in air-gap	150
M Number of harmonic in rotor slots	150
K Number of harmonic in stator slots	150
P Pitch used in calculation	12°

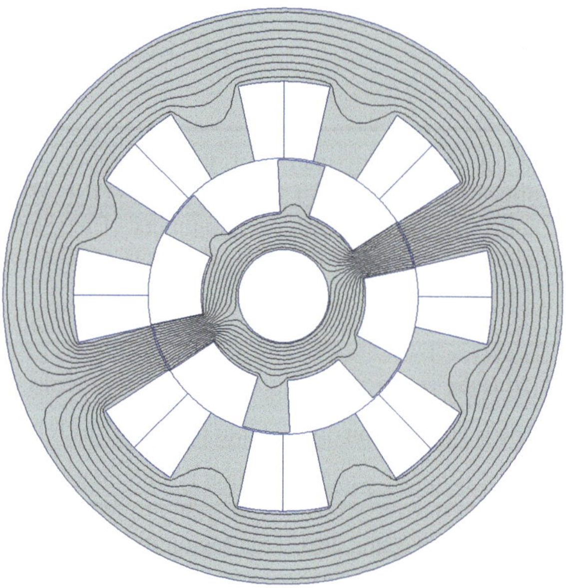

Fig. 2 Equipotential lines of A_z due to phase A for $I_A = 20$ A and 180° rotor position

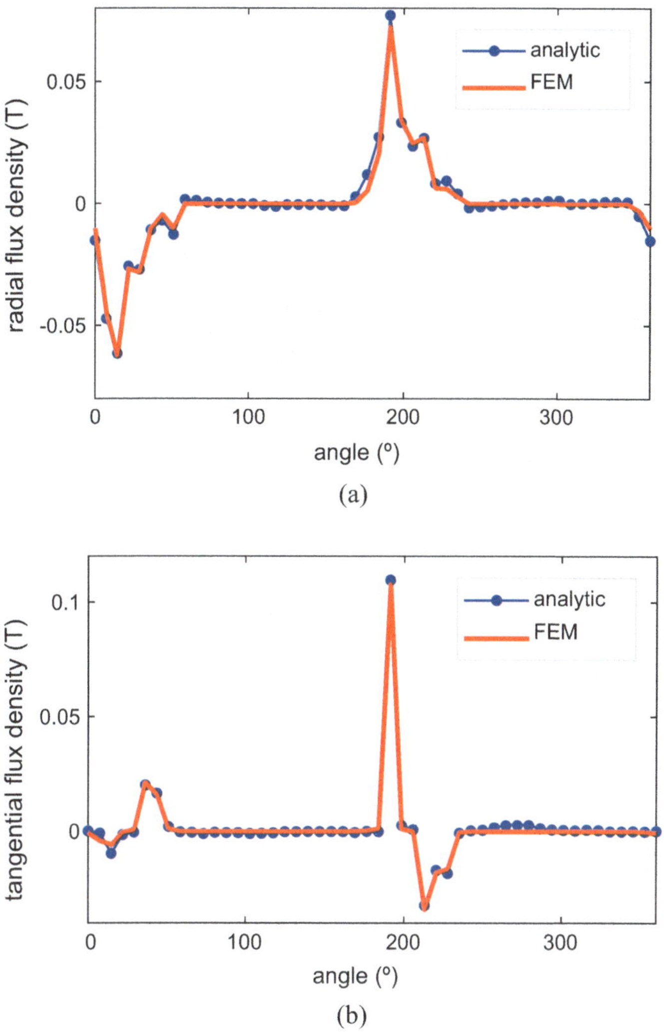

Fig. 3 Radial (**a**) and tangential (**b**) components of magnetic flux density in the air-gap for $I_A = 10$ A and rotor position 45°

3 Simulation Results

The analytical expression of electromagnetic torque and the method for calculating the flux are given in [9, 10, 13–21]. We have used the parameters of SRM given in [24] for our comparison study.

Figure 2 shows the equipotential lines of A_z in the machine due to phase A ($I_A = 20$ A) at 180° rotor position obtained by numerical model. Figure 3 shows the

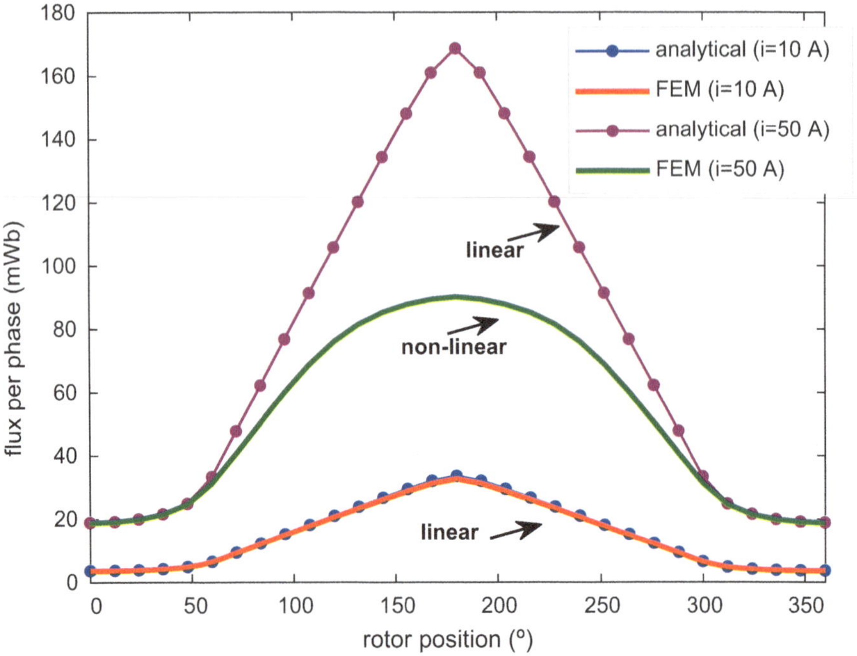

Fig. 4 Waveform of flux per phase due to phase A

radial and tangential magnetic flux density in the middle of the air-gap by feeding only phase A ($I = 10$ A) for rotor position 45°.

Figure 4 shows a comparison between the numerical and analytical result of flux per phase due to phase A. In linear ($I = 10$ A), the numerical and analytical results are well in agreement. In non-linear, the relative error between analytical and numerical results is 46.56%. The mutual flux between phase A and others phases is shown in Fig. 5. In non-linear, the analytical and numerical results present 49.12% relative error. The mutual flux between phases A–C is null because the opening between this phases is $\pi/2$.

Figure 6 shows the comparison between analytical and numerical results of static electromagnetic torque due to phase A. It can be seen that the analytical model gives the accurate results in linear. In non-linear, we have a 42.15% relative error between the analytical and numerical results. This important relative error due to no considering the relative permeability of iron parts in the analytical model.

Figure 7 shows the comparison between analytical model and non-linear numerical results of the maximum electromagnetic torque due to phase A for different values of current (0–100 A). The results of the max self-inductance due to phase A for different values of current (0–100 A) obtained by analytical model and non-linear numerical model are shown in Fig. 8. It can be seen that the analytical model gives a good result and the non-linear numerical results in limited current rang ($I \leq 30$ A for max torque, and $I \leq 20$ A for max self-inductance).

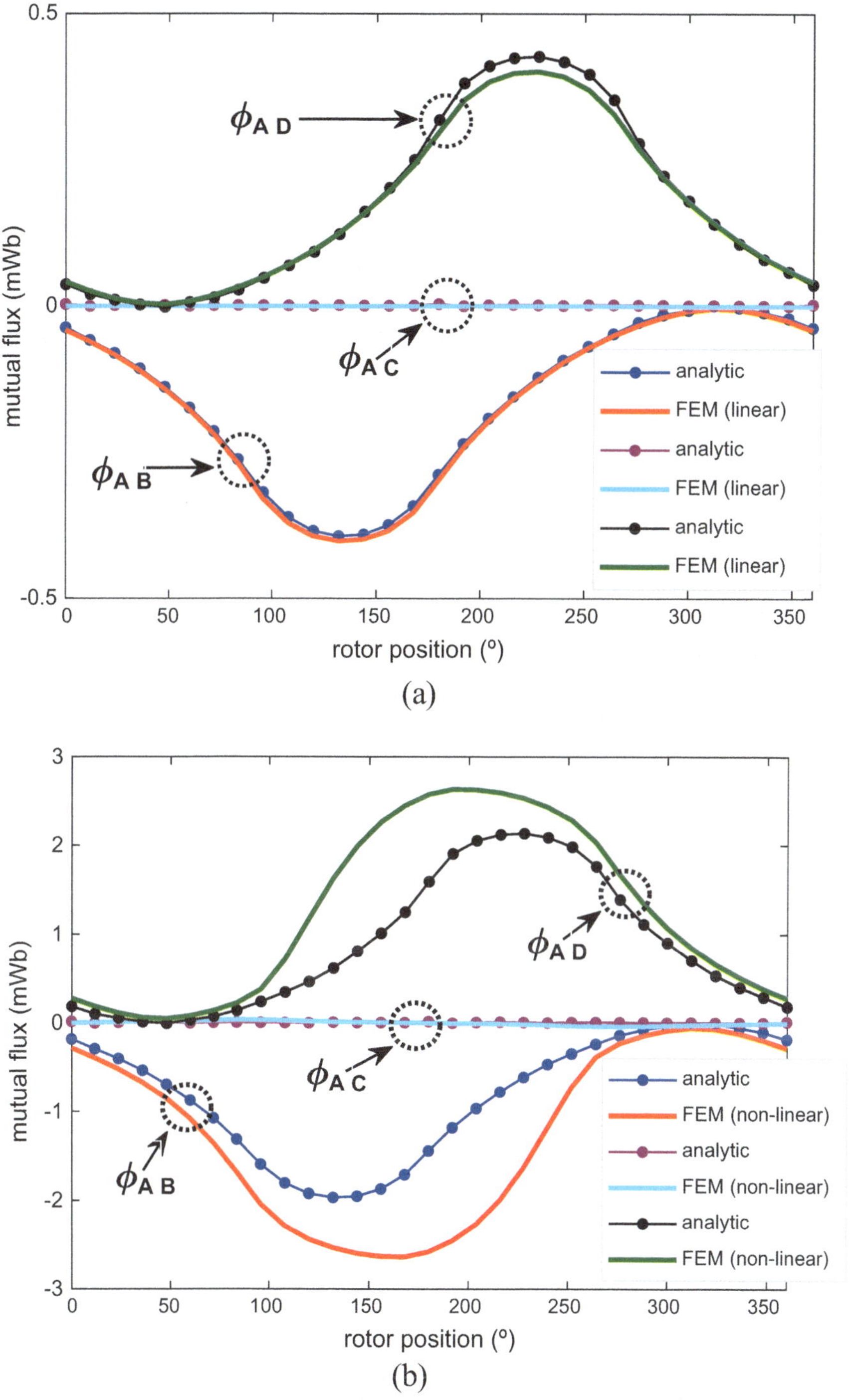

Fig. 5 Waveform of mutual flux obtained by feeding phase A in: (**a**) linear ($I_A = 10$ A) and (**b**) non-linear ($I_A = 50$ A)

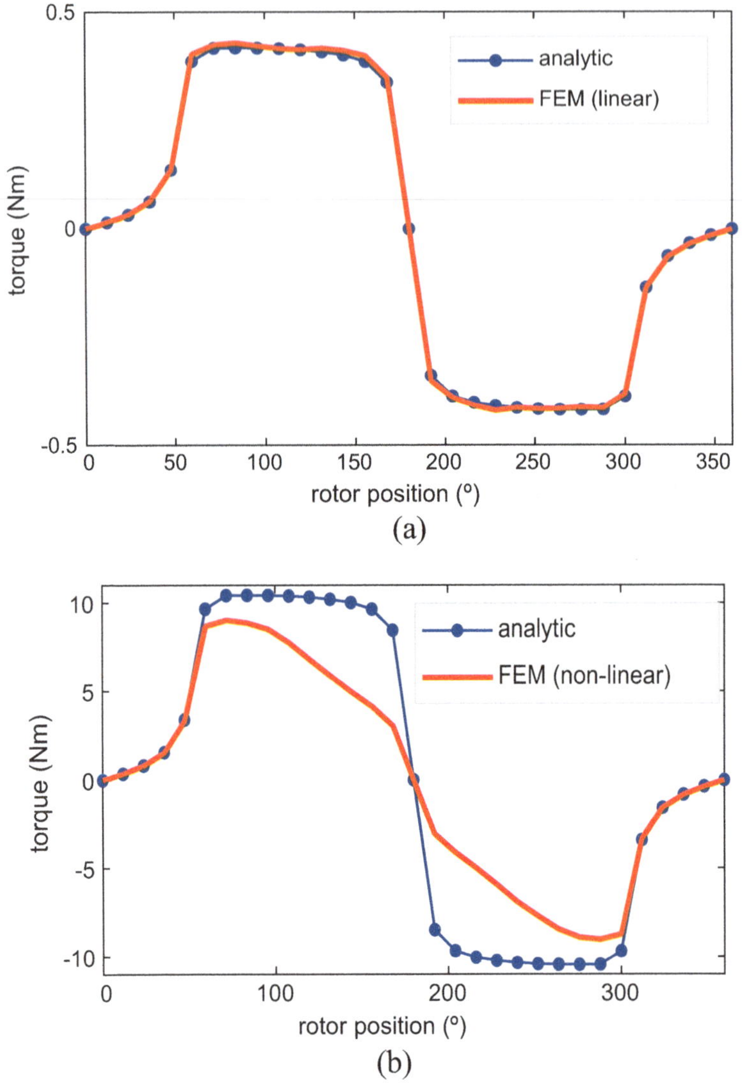

Fig. 6 Waveform of the static electromagnetic torque due to phase A in: (**a**) linear ($I_A = 10$ A) and (**b**) non-linear ($I_A = 50$ A)

4 Conclusion

In this paper, we have presented an analytical model based on the subdomain method of 8/6 conventional SRM in linear conditions. The analytical results are in excellent agreement with numerical ones. However, in non-linear conditions ($I = 50$ A), the developed model overestimates the electromagnetic performances

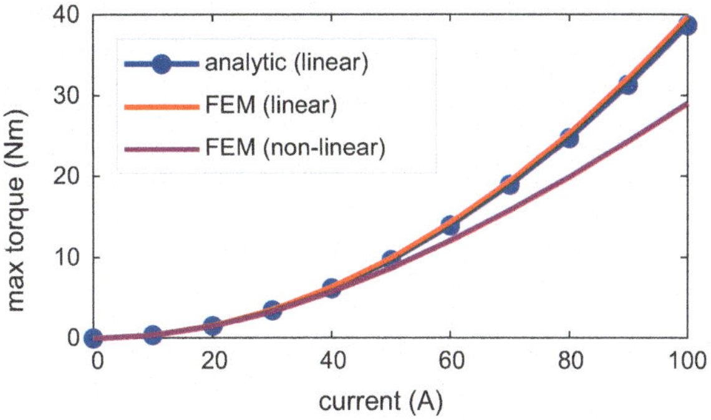

Fig. 7 The maximum electromagnetic torque due to phase A for different values of current

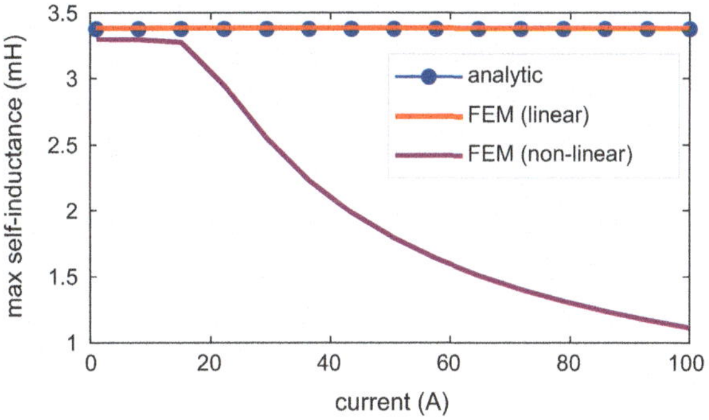

Fig. 8 The maximum self-inductance due to phase A for different values of current

with a maximum error of about 46%. However, by taking this error into account, this model can be effectively used in the optimization procedure, in which saving of computation time is required. In order to obtain more accurate performances in non-linear conditions, this model can be extended to the saturation case (taking into account the characteristic of relative permeability of iron parts). It will be the object of our future works.

References

1. C. Gong, T. Habetler, A novel rotor design for ultra-high speed switched reluctance machines over 1 million rpm, in *2017 IEEE International Electric Machines and Drives Conference (IEMDC)*, (2017), pp. 1–6
2. P.T. Hieu, D. Lee, J. Ahn, Design of a high speed 4/2 switched reluctance motor for blender application, in *2017 IEEE Transportation Electrification Conference and Expo, Asia-Pacific (ITEC Asia-Pacific)*, (2017), pp. 1–5
3. M. Takeno, A. Chiba, N. Hoshi, S. Ogasawara, M. Takemoto, A. Rahman, Test results and torque improvement of the 50-kW switched reluctance motor designed for hybrid electrical vehicles. IEEE Trans. Ind. Appl. **48**(4), 1327–1334 (2012)
4. X. Rain, Contributions à la commande et à la conception des machines à réluctance variable à double saillance, Thesis, Paris 11, 2013

5. M.-W. Arab, Étude et conception d'un groupe motopropulseur électrique à faibles niveau vibratoire et sonore pour véhicule électrique. Aspects 'contrôle - commande, Thesis, Paris 11, 2015
6. R. Benlamine, F. Dubas, S. Randi, D. Lhotellier, C. Espanet, Modeling of an axial-flux interior PMs machine for an automotive application using magnetic equivalent circuit, in *2015 18th International Conference on Electrical Machines and Systems (ICEMS)*, (2015), pp. 1266–1271
7. R. Benlamine, Y. Benmessaoud, F. Dubas, C. Espanet, Nonlinear adaptive magnetic equivalent circuit of a radial-flux interior permanent-magnet machine using air-gap sliding-line technic, in *2017 IEEE Vehicle Power and Propulsion Conference (VPPC)*, (2017), pp. 1–6
8. F. Dubas, C. Espanet, Analytical solution of the magnetic field in permanent-magnet motors taking into account slotting effect: no-load vector potential and flux density calculation. IEEE Trans. Magn. **45**(5), 2097–2109 (2009)
9. T. Lubin, S. Mezani, A. Rezzoug, 2-D Exact analytical model for surface-mounted permanent-magnet motors with semi-closed slots. IEEE Trans. Magn. **47**(2), 479–492 (2011)
10. T. Lubin, S. Mezani, A. Rezzoug, Exact analytical method for magnetic field computation in the air gap of cylindrical electrical machines considering slotting effects. IEEE Trans. Magn. **46**(4), 1092–1099 (2010)
11. F. Dubas, K. Boughrara, New scientific contribution on the 2-D subdomain technique in Cartesian coordinates: taking into account of iron parts. Math. Comput. Appl. **22**(1), 17 (2017)
12. F. Dubas, K. Boughrara, New scientific contribution on the 2-D subdomain technique in polar coordinates: taking into account of iron parts. Math. Comput. Appl. **22**(4), 42 (2017)
13. L. Roubache, K. Boughrara, F. Dubas, R. Ibtiouen, New subdomain technique for electromagnetic performances calculation in radial-flux electrical machines considering finite soft-magnetic material permeability. IEEE Trans. Magn. **54**(4), 8103315 (2018)
14. M. Ben Yahia, L. Roubache, Z. Djelloul-Khedda, K. Boughrara, F. Dubas, R. Ibtiouen, A 2-D exact subdomain technique in switched reluctance machines taking into account of finite soft-magnetic material permeability, in *2018 International Conference on Electrical Sciences and Technologies in Maghreb (CISTEM)*, (2018), pp. 557–562
15. M. Ben Yahia, K. Boughrara, F. Dubas, L. Roubache, R. Ibtiouen, Two-dimensional exact subdomain technique of switched reluctance machines with sinusoidal current excitation. Math. Comput. Appl. **23**(4), 59 (2018)
16. R.L.J. Sprangers, J.J.H. Paulides, B.L.J. Gysen, J. Waarma, E.A. Lomonova, Semi-analytical framework for synchronous reluctance motor analysis including finite soft-magnetic material permeability. IEEE Trans. Magn. **51**(11), 8110504 (2015)
17. R.L.J. Sprangers, J.J.H. Paulides, B.L.J. Gysen, E.A. Lomonova, Magnetic saturation in semi-analytical harmonic modeling for electric machine analysis. IEEE Trans. Magn. **52**(2), 8100410 (2016)
18. Z. Djelloul-Khedda, K. Boughrara, F. Dubas, R. Ibtiouen, Nonlinear analytical prediction of magnetic field and electromagnetic performances in switched reluctance machines. IEEE Trans. Magn. **53**(7), 8107311 (2017)
19. Z. Djelloul-Khedda, K. Boughrara, F. Dubas, A. Kechroud, B. Souleyman, Semi-analytical magnetic field predicting in many structures of permanent-magnet synchronous machines considering the iron permeability. IEEE Trans. Magn. **54**(7), 8103921 (2018)
20. L. Roubache, K. Boughrara, F. Dubas, R. Ibtiouen, Elementary subdomain technique for magnetic field calculation in rotating electrical machines with local saturation effect. COMPEL (2018). https://doi.org/10.1108/COMPEL-11-2017-0481.
21. L. Roubache, K. Boughrara, F. Dubas, R. Ibtiouen, Technique en Sous-Domaines Élementaires dans les Machines Asynchrones à Cage d'Écureuil: Saturation Magnétique Locale & Courants de Foucault dans les Barres, in *Symposium du Génie Électrique (SGE)*, (2018), pp. 1–7
22. D. Meeker, Finite element method magnetics, http://www.femm.info
23. S.J. Farlow, *Partial Differential Equations for Scientists and Engineers* (Courier Corporation, North Chelmsford, 1993)
24. H. Hannoun, Etude et mise en oeuvre de lois de commande de la machine à réluctance variable à double saillance, Thesis, Paris 11, 2008

Effects of Synchronous Motors Parameters Variations on Efficiency Maps

Amina Bensalah, Habibou Lawali Ali, Abed Al Kader Al Asmar, Yacine Amara, and Georges Barakat

Abstract This contribution presents a study of the effect of synchronous motors losses parameters variations on the machines behaviour. In particular, the effects on efficiency maps are investigated. The goal is to identify parameters sets allowing improving energy efficiency for a given application.

1 Introduction

Energy efficiency is a very significant issue nowadays, thanks to its direct link to pollution and green gazes' emissions reduction. Any system or device should be designed not only according to envelopes requirements but also considering its behaviour along its lifetime [1].

Electrical traction vehicles are a typical application in which energy efficiency is very important. Indeed, up–to-date batteries still suffer from a relatively low energy density [2]. It is then very important to have the most efficient traction system in order to increase the autonomy of such vehicles, and increase their attractiveness.

First, the model used to conduct this study is presented. Parameters variations are then discussed. The way efficiency maps are estimated will be clearly explained, and the parametric studies are conducted. Finally, some conclusions are drawn based on the effects of losses parameters variations on the efficiency maps.

A. Bensalah · H. Lawali Ali · A. A. K. Al Asmar · Y. Amara (✉) · G. Barakat
GREAH Groupe de Recherche en Electrotechnique et Automatique du Havre, EA 3220, Université Le Havre Normandie, Le Havre, France
e-mail: amina.bensalah@g.enp.edu.dz; habibou.lawali-ali@doct.univ-lehavre.fr; abed-al-kader.al-asmar@univ-lehavre.fr; yacine.amara@univ-lehavre.fr; georges.barakat@univ-lehavre.fr

W. Zamboni, G. Petrone (eds.), *ELECTRIMACS 2019*, Lecture Notes in Electrical Engineering 615, https://doi.org/10.1007/978-3-030-37161-6_13

2 Synchronous Motors Model

The conventional first harmonic model defined in the synchronous *d–q* reference frame, including iron loss [3–5], is used in this study. The magnetic saturation is neglected. Nevertheless, the effect of magnetic saturation could be studied via its impact on machines inductances. Hybrid excitation synchronous machines (HESM) model being the more general synchronous machines model is used as the base of this study. The models of other synchronous machines, PMSM (permanent magnet synchronous machine), electrically excited machines and synchronous reluctance machines, could all be easily derived from the HESM model.

Figure 1a–c show *d–q* equivalent circuits of the armature windings, and the wound field excitation equivalent circuit, respectively.

More details about this model could be found in [5]. Symbols in Fig. 1 are defined as:

i_d, i_q	*d* and *q* components of armature current
I_e	current in excitation windings
i_{fd}, i_{fq}	*d* and *q* components of iron loss current
v_d, v_q	*d* and *q* components of terminal voltage
V_e	excitation voltage
R_a	armature windings resistance per phase
R_f	iron loss resistance
R_e	excitation winding resistance
Φ_a	permanent magnet flux linkage
Φ_{exc}	total excitation flux linkage
k_e	mutual inductance between the armature and the excitation windings
L_d, L_q	*d* and *q* components of synchronous inductance

2.1 Per-Unit System

The per-unit system defined in [5] is briefly recalled in this section. It allows a better understanding of parameters effect on machines performance. It also constitutes a powerful tool for electric machines classification [6–8]. Base values of EMF, current and mechanical speed Ω_b are chosen as the reference values. More details could be found in Ref. [5]. Per-unit formula for the armature current and the terminal voltage are given by

$$I_n = \frac{\sqrt{i_d^2 + i_q^2}}{I_m} \tag{1}$$

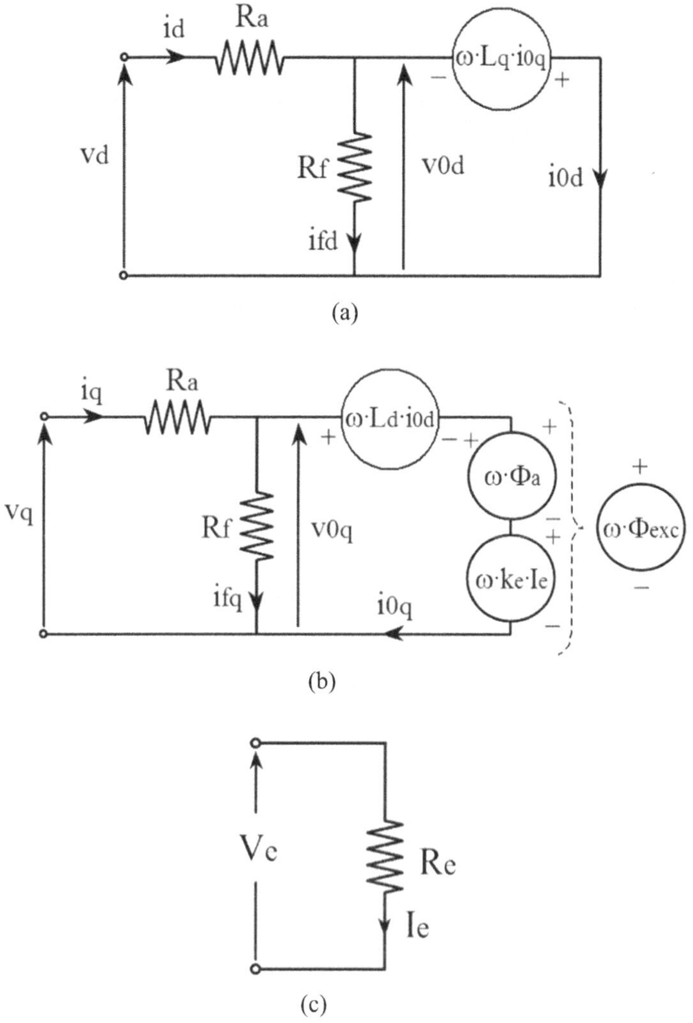

Fig. 1 Synchronous machines equivalent circuits. (**a**) Equivalent circuit in d axis. (**b**) Equivalent circuit in q axis. (**c**) Equivalent circuit of the wound field excitation

$$V_n = \frac{\sqrt{v_d^2 + v_q^2}}{\Phi_{\mathrm{e\,max}} \cdot p \cdot \Omega_{\mathrm{b}}} \tag{2}$$

where,

I_{m}: armature current maximum value (d–q referential),
p: number of pole pairs,

Ω_b: base speed (rated speed),
$\Phi_{e\,max}$: total excitation flux maximum value.

Per-unit values are defined using subscript n. Per-unit expressions of some quantities are defined by Eq. (3):

$$\Omega_n = \frac{\Omega}{\Omega_b},\ P_n = \frac{P}{V_m \cdot I_m} \text{ and } T_n = \frac{P_n}{\Omega_n} \tag{3}$$

where V_m is the maximum applied armature voltage (in d–q referential).

Per-unit values of the excitation current and the mutual inductance between the excitation and the armature windings are given by

$$I_{en} = \frac{I_e}{I_{em}} \text{ and } k_{en} = \frac{k_e \cdot I_{em}}{\Phi_{e\,max}} \tag{4}$$

where,

I_{em}: maximum excitation current,

The saliency ratio ρ has also to be defined ($=L_q/L_d$). Furthermore, the excitation coefficient k_f, linking the total excitation flux with the maximum flux linkage, is defined by

$$\Phi_{exc} = \Phi_a + k_e \cdot I_e = k_f \cdot \Phi_{e\,max} \tag{5}$$

The excitation coefficient varies ideally between 0 and 1. However, the thermal constraints, demagnetization limits and/or the magnetic saturation could limit the possibility of cancelling the PM excitation flux ($k_{f\,min} > 0$).

Hybridization ratio α, which is a parameter specific to HESM and which offers an additional degree of freedom, is given by

$$\alpha = \frac{\Phi_a}{\Phi_{e\,max}} \tag{6}$$

Using Eqs. (4), (5) and (6), k_f could be expressed as a function of the normalized excitation current:

$$k_f = \alpha + k_{en} \cdot I_{en} \tag{7}$$

2.2 *Parameters variations*

Table 1 gives normalized parameters variations intervals. These intervals are bounded by values between which the different parameters could reasonably vary.

Table 1 Normalized parameters variations intervals

Parameter	Variations interval
L_{dn}	[0, 5]
ρ	[0, 5]
R_{an}	[0, 0.5]
R_{fn}	[5, +∞]
R_{en}	1
k_{en}	[0, 1]
β	[5, +∞[

Main parameters of the various types of synchronous motors are given by

$$\begin{cases} \text{Wound field excited motors} & \Rightarrow (\alpha = 0, k_{en} \neq 0) \\ \text{PM excited motors} & \Rightarrow (\alpha = 1, k_{en} = 0) \\ \text{Synchronous reluctance motors} & \Rightarrow (\alpha = 0, k_{en} = 0) \\ \text{Hybrid excited motors} & \Rightarrow (\alpha \neq 0, k_{en} \neq 0) \end{cases} \tag{8}$$

β the power rating ratio of converters connected to both the armature windings and excitation coils, needs to be defined for wound field and hybrid excitation machines. It is given by

$$\beta = \frac{V_{\mathrm{m}} \cdot I_{\mathrm{m}}}{V_{\mathrm{em}} \cdot I_{\mathrm{em}}} \tag{9}$$

where, V_{em} is the maximum excitation coil terminal voltage. More details are provided in [5].

Machines where losses are neglected have $R_{\mathrm{a}} = 0\ \Omega$ and $R_{\mathrm{f}} \rightarrow +\infty\ \Omega$. For PM excited and synchronous reluctance machines $\beta = +\infty$.

3 Efficiency Maps Computation

Efficiency maps represent a practical way to evaluate motor designs [5, 9–12]. Efficiency maps estimation is based on the algorithm shown in Fig. 2. Per-unit value of speed varies between 0 and $\Omega_{n\max} > 1$, and T_n could vary between 0 and 1. The algorithm helps identify (k_{f}, I_n, ψ) combination, which maximizes efficiency for each (Ω_n, T_n) combination. ψ is the phase shifting between the EMF and the armature current. For each operating point, three imbricated loops allow identifying (k_{f}, I_n, ψ) combinations, which meet the (speed, torque) demand and the current and voltage limits. Then, the combination of (k_{f}, I_n, ψ) that maximize the efficiency is selected. For the operating points with unfounded combination of (k_{f}, I_n, ψ) that respect the voltage limit, the efficiency is set to 0. Efficiency maps are obtained by displaying iso-lines of efficiency. It should be highlighted that mechanical losses are neglected.

Fig. 2 Efficiency maps computation algorithm

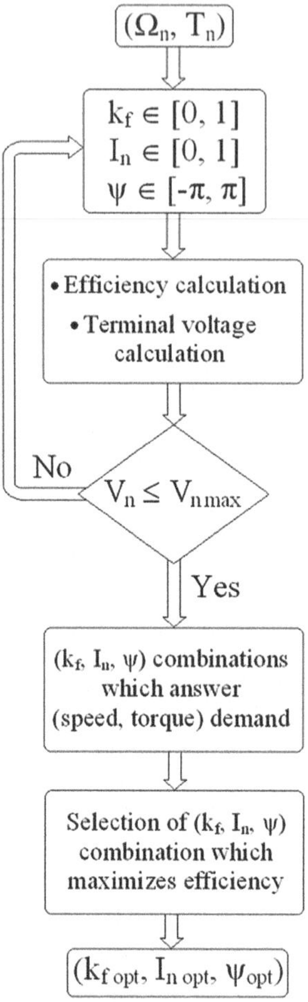

Figure 3 shows efficiency maps for a wound field synchronous motor (Fig. 3a), a PM motor (Fig. 3b) and a hybrid excited motor (Fig. 3c). Only excited synchronous machines are considered because, if well designed, they should have higher efficiencies as compared to non-excited synchronous machines, i.e., synchronous reluctance machines. Efficiency estimation in these figures do not include mechanical and the static converter losses.

The values of the different parameters are given as follows: $L_{dn} = 0.5$; $\rho = 1$; $R_{an} = 0.1$; and $R_{fn} = 20$. For the wound field and the hybrid excited machines, additional parameters are given as follows: $k_{en} = 1$; $R_{en} = 1$; $\beta = 27$, and $\alpha = 1$ for the hybrid excited machine. These parameters are derived from an existing prototype [5, 13].

As could be expected, the PM and hybrid excited motors have higher efficiency as compared to the wound field motor. While the high efficiency zones are located

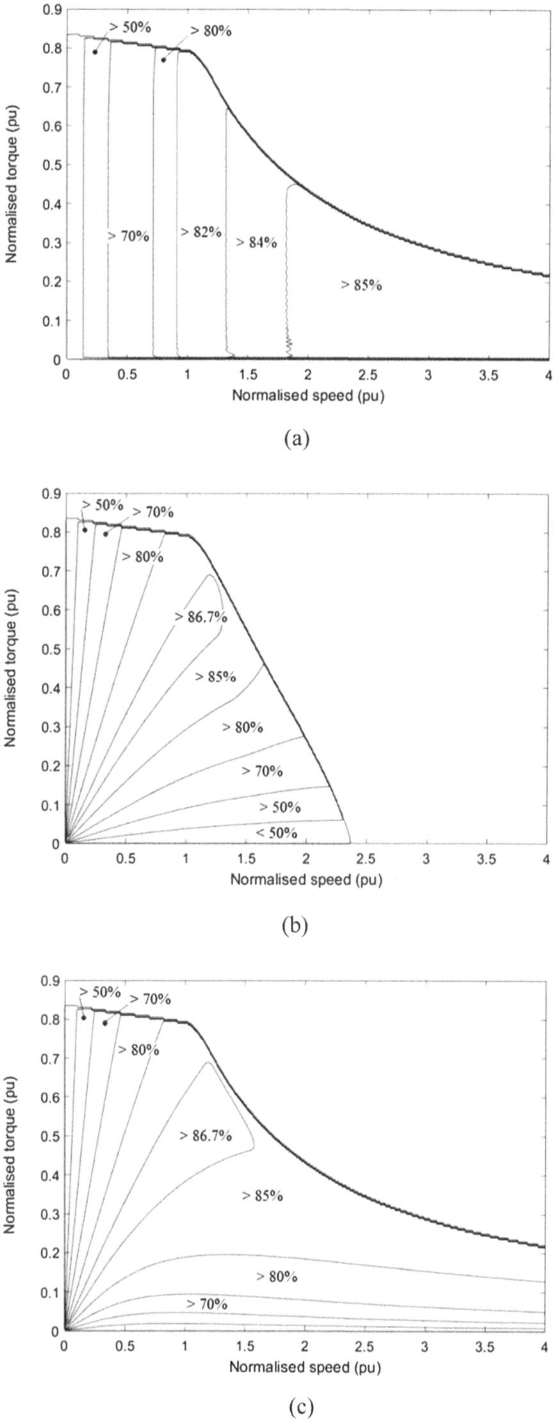

Fig. 3 Efficiency maps of the different machines. (**a**) Wound field machine. (**b**) PM machine. (**c**) Hybrid excited machine

at high speed for the wound field motor, they are located around the base speed for the PM and the hybrid excited motors.

4 Parametric Study

The parametric study presented in this section is conducted in order to establish the separate effects of each parameter on the efficiency maps. At this stage, only losses parameters are considered, i.e., R_{an} and R_{fn}.

Figure 4 shows the efficiency maps when R_{an} is null, all other parameters are kept equal to previously defined values. As can be seen the efficiency is obviously increasing for same operating conditions as compared to maps shown in Fig. 3. In high-excitation flux requiring regions, the wound field machine has a lower efficiency, as could be expected, since the excitation flux is obtained at the price of excitation joule losses. For the PM machine, the lowest efficiency zones are located in regions where the flux weakening is demanding. The HES machine combines the advantages of both the previous structures. For all machines, the power capability at relatively low speed is increased.

Figure 5 shows efficiency maps when $R_{an} = 0.5$, all other parameters being kept equal to previously defined values. The maximum efficiency is decreasing as expected. It is also the case of the maximum power capability at relatively low speed for all machines. However, it has been observed that the power capability at higher speed has been enlarged for the PM machine [8]. As previously observed, while the high-efficiency zones are located at high speed for the wound field motor, they are located around the base speed for the PM and the hybrid excited motors.

Figure 6 shows efficiency maps when $R_{fn} = 10{,}000$, all other parameters being kept equal to previously defined values ($R_{an} = 0.1$). Having such a high value of R_{fn} means that the iron loss is neglected. It could be observed that if the iron losses are neglected, or insignificant, the torque envelope at low speed is independent of speed. The efficiency is of course higher as compared to maps shown in Fig. 3. The maximum power capability is slightly improved. The reachable speed for the PM machine is not significantly affected.

Figure 7 shows efficiency maps when $R_{fn} = 5$, all other parameters are kept equal to previously defined values ($R_{an} = 0.1$). R_{fn} has a significant effect on the torque envelope slope at low and high speeds. The maximum torque increases at very low speed when R_{fn} decreases. The maximum efficiency zones are shifted to low speed regions. The maximum reachable speed for PM machine decreases as the value of R_{fn} decreases (increasing iron loss). The effect of R_{fn} on this speed seems less important as compared to the effect of R_{an}.

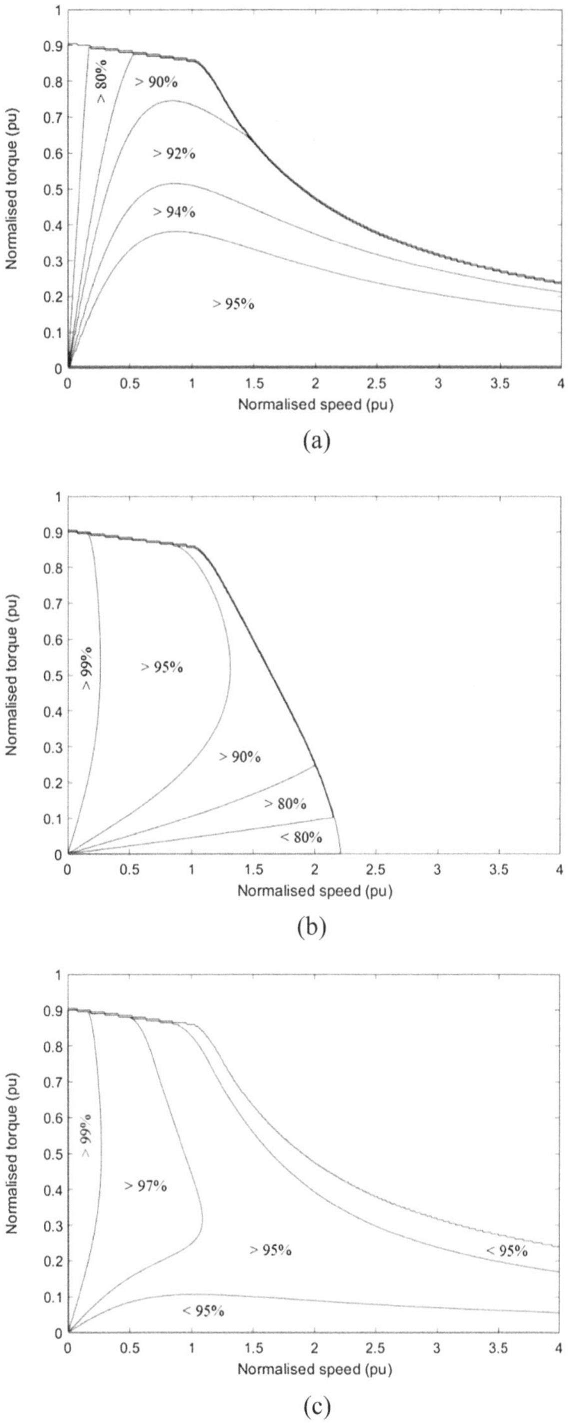

Fig. 4 Efficiency maps of the different machines when $R_{an} = 0$. (**a**) Wound field machine. (**b**) PM machine. (**c**) Hybrid excited machine

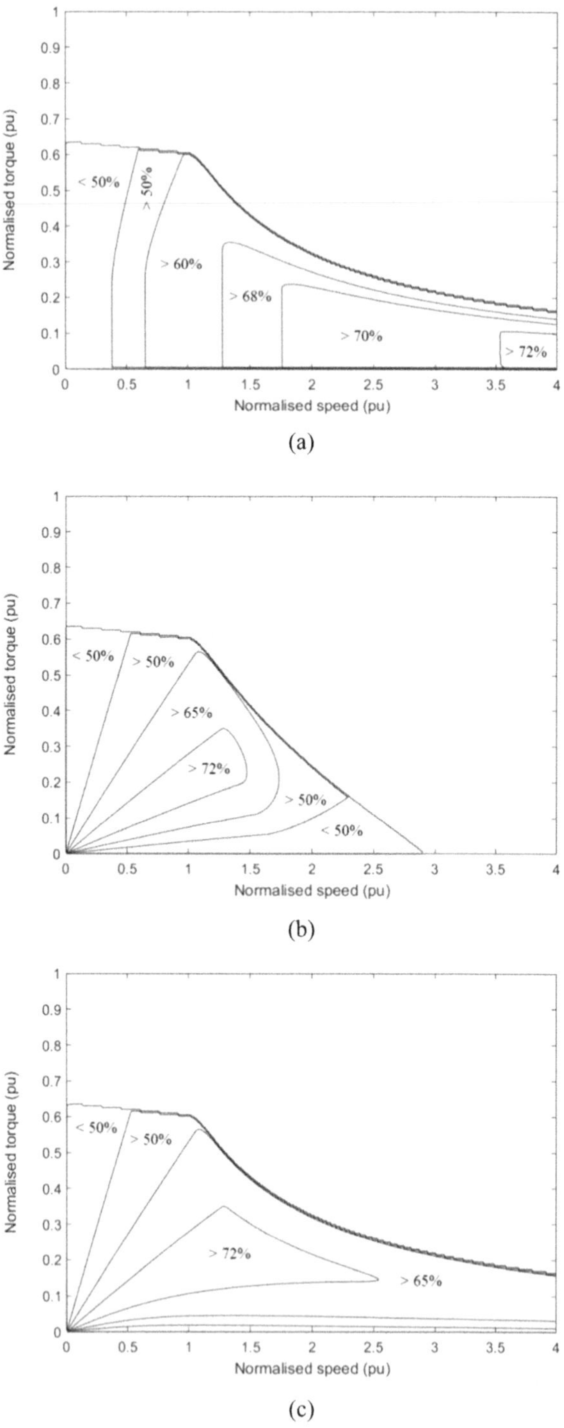

Fig. 5 Efficiency maps of the different machines when $R_{an} = 0.5$. (**a**) Wound field machine. (**b**) PM machine. (**c**) Hybrid excited machine

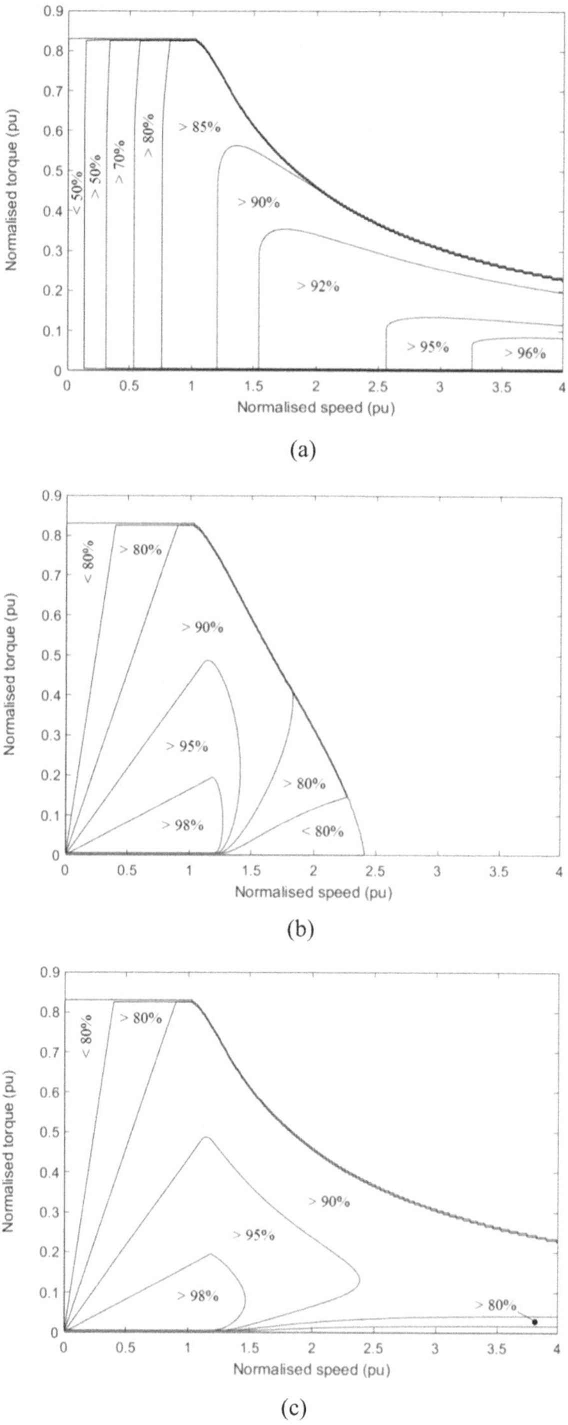

Fig. 6 Efficiency maps of the different machines when $R_{fn} = 10{,}000$. (**a**) Wound field machine. (**b**) PM machine. (**c**) Hybrid excited machine

Fig. 7 Efficiency maps of different machines when $R_{fn} = 5$. (**a**) Wound field machine. (**b**) PM machine. (**c**) Hybrid excited machine

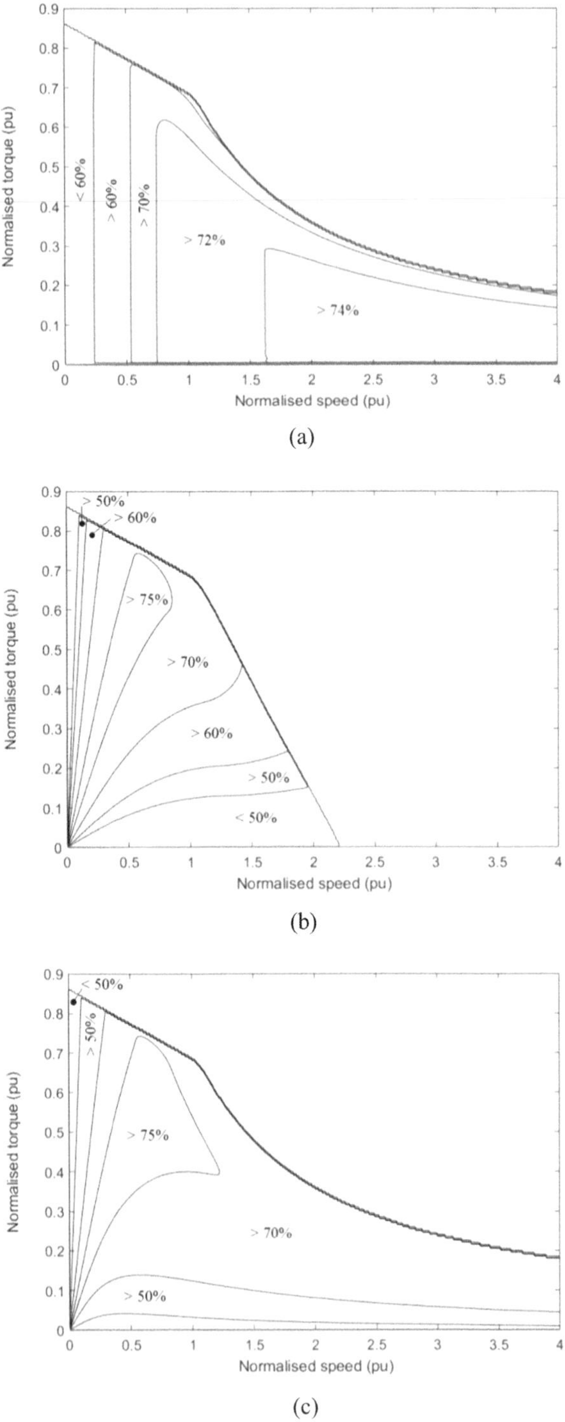

5 Conclusions

A parametric study has been conducted in order to evaluate the effects of losses parameters, i.e., R_{an} and R_{fn}, on the efficiency maps of excited synchronous motors. Interesting features have been identified concerning the behaviour of the PM machine operation at high speeds. It has been observed that the maximum reachable speed increases when the value of R_{an} increases. The maximum power capability is obviously reduced at relatively low speed when R_{an} increases. The maximum reachable speed for the PM machine when the iron loss increases (R_{fn} decreasing) is decreasing. The power capability at low speed is less affected by the variations of R_{fn}, as compared to variations of R_{an}.

All observations show that the established tool is consistent with scientific literature. It constitutes therefore a very interesting tool to explore different characteristics and extend the parametric study.

This study should be regarded as a general theoretical analysis of the effect of synchronous machines parameters on efficiency maps. The use of normalized parameters, and quantities, was aiming at serving this goal. An experimental validation was not pursued, since only tendencies were sought. Nevertheless, the used tool was previously compared to a corresponding experimental study in [5]. In order to link this theoretical study to real design considerations, here are described the effects of some general design considerations on the two considered parameters, i.e., R_{an} and R_{fn}:

1. The reduction of armature resistance could be sought by the use of concentrated tooth windings with reduced end-windings. The use of an adapted cooling can also help keep the windings temperature as low as possible to avoid increasing the windings resistivity.
2. The reduction of iron loss could be achieved by using low iron loss ferromagnetic materials grades. In contrast to joule losses, the temperature increase induces a reduction of iron loss.

References

1. G. Choi, T.M. Jahns, Design of electric machines for electric vehicles based on driving schedules, in *Proceedings of 2013 International Electric Machines & Drives Conference (IEMDC2013)*, Chicago, IL, USA, 12–15 May 2013, pp. 54–61
2. M. Yilmaz, P.T. Krein, Review of battery charger topologies, charging power levels, and infrastructure for plug-in electric and hybrid vehicles. IEEE Trans. Power Electron. **28**(5), 2151–2169 (2013)
3. S. Morimoto, Y. Tong, Y. Takeda, T. Hirasa, Loss minimization control of permanent magnet synchronous motor drives. IEEE Trans. Ind. Electron. **41**(5), 511–517 (1994)
4. F. Fernandez-Bernal, A. Garcia-Cerrada, R. Faure, Determination of parameters in interior permanent-magnet synchronous motors with iron losses without torque measurement. IEEE Trans. Ind. Appl. **37**(5), 1265–1272 (2001)

5. Y. Amara, L. Vido, M. Gabsi, E. Haong, A.H. Ben Ahmed, M. Lécrivain, Hybrid excitation synchronous machines: energy-efficient solution for vehicles propulsion. IEEE Trans. Veh. Technol. **58**(5), 2137–2149 (2009)
6. R.F. Schiferl, T.A. Lipo, Power capability of salient pole permanent magnet synchronous motors in variable speed drive applications. IEEE Trans. Ind. Appl. **26**(1), 115–123 (1990)
7. W.L. Soong, T.J.E. Miller, Field-weakening performance of brushless synchronous AC motor drives. IEE Proc. Electr. Power Appl. **141**(6), 331–340 (1994)
8. J. Soulard, B. Multon, Maximum power limits of small single-phase permanent magnet drives. IEE Proc. Electr. Power Appl. **146**(5), 457–462 (1999)
9. S.S. Williamson, S.M. Lukic, A. Emadi, Comprehensive drive train efficiency analysis of hybrid electric and fuel cell vehicles based on motorcontroller efficiency modeling. IEEE Trans. Power Electron. **21**(3), 730–740 (2006)
10. Y. Sugii, M. Yada, S. Koga, T. Ashikaga, Applicability of various motors to electric vehicles, in *Proc. 13th Int. EVS*, Osaka, Japan, Oct 1996, pp. 757–764
11. A. Fonseca, C. Chillet, E. Atienza, A.-L. Bui-Van, J. Bigeon, New modelling methodology for different PM motors for electric and hybrid vehicles," in *Proc. IEEE Int. Electr. Mach. Drives Conf.*, Cambridge, MA, June 2001, pp. 173–178
12. M. Menne, J. Reinert, R.W. De Doncker, Energy-efficiency evaluation of traction drives for electric vehicles," in *Proc. 15th EVS*, Brussels, Belgium, Oct 1998, 7 pages, CD-ROM
13. Y. Amara, E. Hoang, M. Gabsi, M. Lécrivain, A.H. Ben Ahmed, S. Dérou, Measured performances of a new hybrid excitation synchronous machine. EPE J. **12**(4), 42–50 (2004)

Structural Analysis and Modular Control Law for Modular Multilevel Converter (MMC)

Pierre-Baptiste Steckler, Jean-Yves Gauthier, Xuefang Lin-Shi, and François Wallart

Abstract This paper proposes an in-depth analysis from the control point of view of dynamic models of a modular multilevel converter (MMC) for high-voltage direct current (HV-DC) application. Firstly, a generic method of analysis is presented for a natural arm-level state-space model. Its structural analysis highlights the decoupled nature of the MMC. Secondly, the well-known sum and difference of the upper and lower arm state and control variables is considered to obtain a (Σ/Δ) model. This transformation leads to a coupling between state and control variables and to an increase of the system complexity. Using the analysis results of the natural model and the (Σ/Δ) model, an original arm-modular control is finally proposed. The simulation results show the effectiveness of the proposed control, which is simpler to design compared to a conventional (Σ/Δ) control.

1 Introduction

With the constant development of renewable energy and interconnections between countries, the high-voltage direct current (HVDC) transmission technology is in expansion, thanks to its lower footprint and greater controllability. The reference conversion structure today is the modular multilevel converter (MMC), which solves most of the problems of the former technologies, providing a good efficiency and an easy voltage scalability. MMC [1] has been invented and patented by Rainer Marquardt in 2001, and published for the first time in 2003. Even if different configurations of submodules are available, the most used today is the "half-bridge", shown in Fig. 1. The SM configuration has little or no influence on the normal

P.-B. Steckler (✉) · J.-Y. Gauthier · X. Lin-Shi
SuperGrid Institute SAS, Villeurbanne, France

Laboratoire Ampère, UMR CNRS 5005, Université de Lyon, INSA Lyon, Villeurbanne, France
e-mail: pierre-baptiste.steckler@supergrid-institute.com; jean-yves.gauthier@insa-lyon.fr

F. Wallart
SuperGrid Institute SAS, Villeurbanne, France

W. Zamboni, G. Petrone (eds.), *ELECTRIMACS 2019*, Lecture Notes in Electrical Engineering 615, https://doi.org/10.1007/978-3-030-37161-6_14

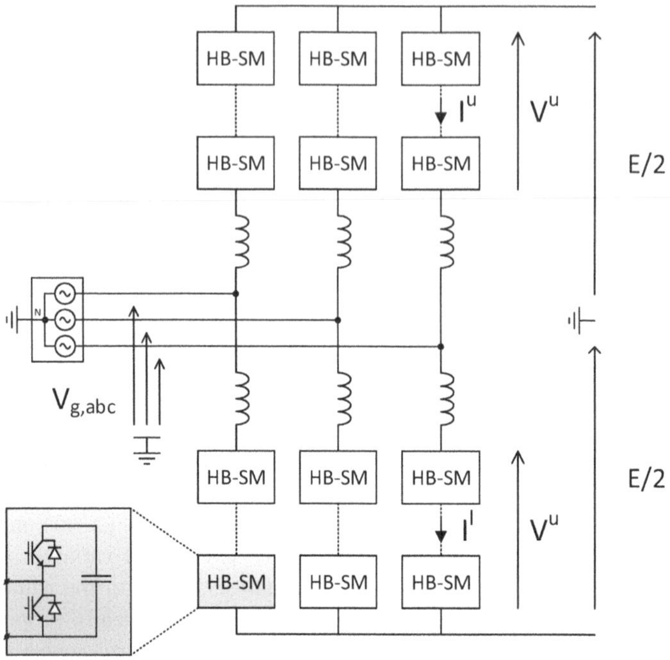

Fig. 1 Half-Bridge MMC schematic

operation, and is essentially important for the fault condition, especially when the DC link is shorted.

Many works have been made for the MMC modelling, such as switching models, arm averaged models, non-linear state-space models or linearized small-signal state-space models [2, 3]. Averaged type modelling is generally used for control design. After examining the prior art in modelling analysis and control design of MMC, it was found that on the one hand, not most models were enough exploited from the control point of view, and on the other hand, most controls were based on (Σ/Δ) transformation [4, 5] by using the sum and difference of the upper and lower arm state and control variables. Compared to the previous published work on MMC model analysis, the first contribution of this paper is to propose a generic analysis method, which demonstrates the decoupled nature of the averaged model in natural coordinates contrary to the (Σ/Δ) model where the state and control variables are coupled between arms leading to a complex control system. This approach comes with inherent benefits, like the use of the output variables as state variables or the partial separation of AC and DC quantities, allowing the use of very classical control means (e.g., Park transform for AC-grid-side and PI control for DC grid side). However, additional conditions are necessary to make the (Σ/Δ) model decoupled for its control design. When these assumptions are not verified, the control quality can be degraded, hence leading to an interest of arm-based control. Thus, the second contribution of this paper is to propose a new MMC control based on the averaged model in natural coordinate.

The paper is organized as follows: in Sect. 2, the description of average equivalent schematic of the MMC is given. Based on a unified formulation, a global non-linear state-space model is proposed both for natural coordinate system and for (Σ/Δ) transformation. In Sect. 3, a structural analysis is performed for both models. A new control is developed and validated in simulation in Sect. 4.

2 MMC Modelling

For MMC modelling, the "average" approach [3] is often used. Relying on the assumption that every capacitor in a given stack has the same voltage level thanks to a lower level control [6], it reduces strongly the model dimension. Each stack is simply replaced by a DC transformer with a controllable ratio, feeding an equivalent capacitor. This controllable ratio is called the modulation index of the stack. For an half-bridge stack, it has the range $m \in [0, +1]$, and it becomes $m \in [-1, +1]$ for a full-bridge stack. The resulting schematic is shown in Fig. 2.

The modelling will be made under the following assumptions:

- All switching elements are considered to be ideal
- Each element of the equivalent schematic is linear

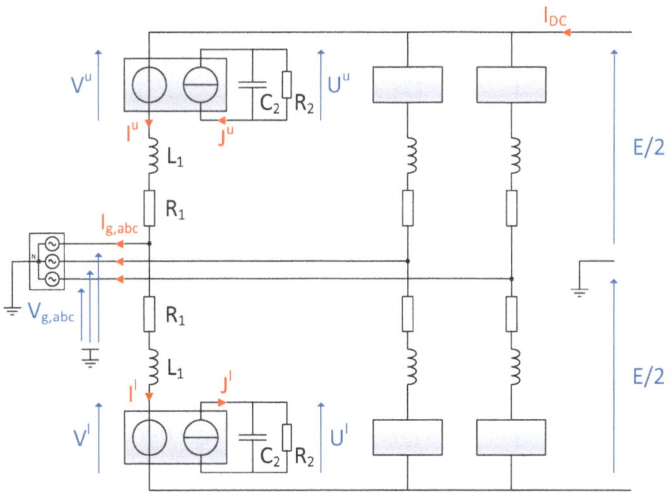

Fig. 2 MMC—equivalent average schematic

- Each storage element is associated with a dissipative term,[1] as shown in Fig. 2
- The saturation of control variables is not modelled

The three-phase MMC has 12 state variables, 6 arm currents (noted $I^{u,l}_{abc}$) and 6 equivalent capacitor voltages (noted $U^{u,l}_{abc}$). It also has six control variables, the modulation indexes of the six stacks, noted $m^{u,l}_{abc}$. The state vector is expressed in (1), and the control vector in (2). The arm inductor is defined by its inductance L_1 and its series resistance R_1, whereas the stack capacitor is defined by its capacitance C_2 and its parallel resistance R_2. The DC voltage E and the AC voltages $V_{g,abc}$ appear as exogenous disturbances on the model.

$$\begin{aligned} x &= \left[I^u_a, U^u_a, I^l_a, U^l_a, I^u_b, U^u_b, I^l_b, U^l_b, I^u_c, U^u_c, I^l_c, U^l_c\right]^T \\ &:= [x_{1a}, x_{2a}, x_{3a}, x_{4a}, x_{1b}, x_{2b}, x_{3b}, x_{4b}, x_{1c}, x_{2c}, x_{3c}, x_{4c}]^T \end{aligned} \tag{1}$$

$$\begin{aligned} u &= \left[m^u_a, m^l_a, m^u_b, m^l_b, m^u_c, m^l_c\right]^T \\ &:= [u_{1a}, u_{2a}, u_{1b}, u_{2b}, u_{1c}, u_{2c}]^T \end{aligned} \tag{2}$$

A non-linear average state-space model of the MMC, under the form $\dot{x} = f(x, u)$, is presented in (3), and the state and control space over which it is defined is shown in (4). Note that the absence of $_{abc}$ and/or ^{u,l} indexes indicates that the result is relevant whatever the considered phase or arm.

$$f(x,u) = \begin{bmatrix} 1/L_1 \cdot \left(E/2 - V_{ga} - R_1 \cdot x_{1a} - x_{2a} \cdot u_{1a}\right) \\ 1/C_2 \cdot (x_{1a} \cdot u_{1a} - x_{2a}/R_2) \\ 1/L_1 \cdot \left(E/2 + V_{ga} - R_1 \cdot x_{3a} - x_{4a} \cdot u_{2a}\right) \\ 1/C_2 \cdot (x_{3a} \cdot u_{2a} - x_{4a}/R_2) \\ 1/L_1 \cdot \left(E/2 - V_{gb} - R_1 \cdot x_{1b} - x_{2b} \cdot u_{1b}\right) \\ 1/C_2 \cdot (x_{1b} \cdot u_{1b} - x_{2b}/R_2) \\ 1/L_1 \cdot \left(E/2 + V_{gb} - R_1 \cdot x_{3b} - x_{4b} \cdot u_{2b}\right) \\ 1/C_2 \cdot (x_{3b} \cdot u_{2b} - x_{4b}/R_2) \\ 1/L_1 \cdot \left(E/2 - V_{gc} - R_1 \cdot x_{1c} - x_{2c} \cdot u_{1c}\right) \\ 1/C_2 \cdot (x_{1c} \cdot u_{1c} - x_{2c}/R_2) \\ 1/L_1 \cdot \left(E/2 + V_{gc} - R_1 \cdot x_{3c} - x_{4c} \cdot u_{2c}\right) \\ 1/C_2 \cdot (x_{3c} \cdot u_{2c} - x_{4c}/R_2) \end{bmatrix} \tag{3}$$

$$x_{1,3} \in \mathbb{R}\,,\ x_{2,4} \in \mathbb{R}^+\,,\ u_{1,2} \in [0, +1] \tag{4}$$

[1]A rough idea of the converter losses is necessary to take into account all the "first-order" damping phenomena.

As shown in [4, 5, 7, 8], the most common method for MMC analysis and control is based on both a state and a control change of coordinates, using sums and differences of top/bottom state and control variables in a given leg.

This corresponds to state and control variable transformations defined in (5).

$$\begin{aligned} \left[z_1, z_2, z_3, z_4\right]^T &= \left[x_1 + x_3, x_2 + x_4, x_1 - x_3, x_2 - x_4\right]^T \\ \left[v_1, v_2\right]^T &= \left[u_1 + u_2, u_1 - u_2\right]^T \end{aligned} \tag{5}$$

The transformed model becomes $\dot{z} = \bar{f}(z, v)$, whose components are detailed in (6).

$$\bar{f}(z, v) = \begin{bmatrix} 1/L_1 \cdot (E - R_1 \cdot z_{1a} - v_{1a} \cdot z_{2a}/2 - v_{2a} \cdot z_{4a}/2) \\ 1/C_2 \cdot (v_{1a} \cdot z_{1a}/2 - z_{2a}/R_2 + v_{2a} \cdot z_{3a}/2) \\ 1/L_1 \cdot \left(-2V_{ga} + R_1 \cdot z_{3a} - v_{2a} \cdot z_{2a}/2 - v_{1a} \cdot z_{4a}/2\right) \\ 1/C_2 \cdot (v_{2a} \cdot z_{1a}/2 - z_{4a}/R_2 + v_{1a} \cdot z_{3a}/2) \\ 1/L_1 \cdot (E - R_1 \cdot z_{1b} - v_{1b} \cdot z_{2b}/2 - v_{2b} \cdot z_{4b}/2) \\ 1/C_2 \cdot (v_{1b} \cdot z_{1b}/2 - z_{2b}/R_2 + v_{2b} \cdot z_{3b}/2) \\ 1/L_1 \cdot \left(-2V_{gb} + R_1 \cdot z_{3b} - v_{2b} \cdot z_{2b}/2 - v_{1b} \cdot z_{4b}/2\right) \\ 1/C_2 \cdot (v_{2b} \cdot z_{1b}/2 - z_{4b}/R_2 + v_{1b} \cdot z_{3b}/2) \\ 1/L_1 \cdot (E - R_1 \cdot z_{1c} - v_{1c} \cdot z_{2c}/2 - v_{2c} \cdot z_{4c}/2) \\ 1/C_2 \cdot (v_{1c} \cdot z_{1c}/2 - z_{2c}/R_2 + v_{2c} \cdot z_{3c}/2) \\ 1/L_1 \cdot \left(-2V_{gc} + R_1 \cdot z_{3c} - v_{2c} \cdot z_{2c}/2 - v_{1c} \cdot z_{4c}/2\right) \\ 1/C_2 \cdot (v_{2c} \cdot z_{1c}/2 - z_{4c}/R_2 + v_{1c} \cdot z_{3c}/2) \end{bmatrix} \tag{6}$$

3 Structural Analysis

The underlying idea of the proposed analysis method is to obtain simple and if possible non-parametric conclusions from the non-linear models presented above. These conclusions play a role into the comprehension of the behaviour of MMC, thus making the control law development easier. The difference between structural and parametric properties will be highlighted.

3.1 Natural Model Analysis

3.1.1 Local Differentiation (Jacobian Analysis)

Formally, the Jacobian matrix $\mathscr{J}_f$ of the vector field $f(\cdot)$ is defined over the entire argument vector, $\alpha := [x|u]$. We introduce the partial Jacobian matrix $\mathscr{J}_{f,x}$ and $\mathscr{J}_{f,u}$ such that $\mathscr{J}_{f,\alpha} = \left[\mathscr{J}_{f,x}|\mathscr{J}_{f,u}\right]$

These matrix are presented, respectively, in (7) and (8).

$$\mathcal{J}_{f,x} = \begin{bmatrix} -\frac{R_1}{L_1} & -\frac{u_{1a}}{L_1} & 0 & 0 & 0 & 0 & 0 & 0 & 0 & 0 & 0 & 0 \\ \frac{u_{1a}}{C_2} & -\frac{1}{R_2 \cdot C_2} & 0 & 0 & 0 & 0 & 0 & 0 & 0 & 0 & 0 & 0 \\ 0 & 0 & -\frac{R_1}{L_1} & -\frac{u_{2a}}{L_1} & 0 & 0 & 0 & 0 & 0 & 0 & 0 & 0 \\ 0 & 0 & \frac{u_{2a}}{C_2} & -\frac{1}{R_2 \cdot C_2} & 0 & 0 & 0 & 0 & 0 & 0 & 0 & 0 \\ 0 & 0 & 0 & 0 & -\frac{R_1}{L_1} & -\frac{u_{1b}}{L_1} & 0 & 0 & 0 & 0 & 0 & 0 \\ 0 & 0 & 0 & 0 & \frac{u_{1b}}{C_2} & -\frac{1}{R_2 \cdot C_2} & 0 & 0 & 0 & 0 & 0 & 0 \\ 0 & 0 & 0 & 0 & 0 & 0 & -\frac{R_1}{L_1} & -\frac{u_{2b}}{L_1} & 0 & 0 & 0 & 0 \\ 0 & 0 & 0 & 0 & 0 & 0 & \frac{u_{2b}}{C_2} & -\frac{1}{R_2 \cdot C_2} & 0 & 0 & 0 & 0 \\ 0 & 0 & 0 & 0 & 0 & 0 & 0 & 0 & -\frac{R_1}{L_1} & -\frac{u_{1c}}{L_1} & 0 & 0 \\ 0 & 0 & 0 & 0 & 0 & 0 & 0 & 0 & \frac{u_{1c}}{C_2} & -\frac{1}{R_2 \cdot C_2} & 0 & 0 \\ 0 & 0 & 0 & 0 & 0 & 0 & 0 & 0 & 0 & 0 & -\frac{R_1}{L_1} & -\frac{u_{2c}}{L_1} \\ 0 & 0 & 0 & 0 & 0 & 0 & 0 & 0 & 0 & 0 & \frac{u_{2c}}{C_2} & -\frac{1}{R_2 \cdot C_2} \end{bmatrix} \tag{7}$$

$$\mathcal{J}_{f,u} = \begin{bmatrix} -\frac{x_{2a}}{L_1} & 0 & 0 & 0 & 0 & 0 \\ \frac{x_{1a}}{C_2} & 0 & 0 & 0 & 0 & 0 \\ 0 & -\frac{x_{4a}}{L_1} & 0 & 0 & 0 & 0 \\ 0 & \frac{x_{3a}}{C_2} & 0 & 0 & 0 & 0 \\ 0 & 0 & -\frac{x_{2b}}{L_1} & 0 & 0 & 0 \\ 0 & 0 & \frac{x_{1b}}{C_2} & 0 & 0 & 0 \\ 0 & 0 & 0 & -\frac{x_{4b}}{L_1} & 0 & 0 \\ 0 & 0 & 0 & \frac{x_{3b}}{C_2} & 0 & 0 \\ 0 & 0 & 0 & 0 & -\frac{x_{2c}}{L_1} & 0 \\ 0 & 0 & 0 & 0 & \frac{x_{1c}}{C_2} & 0 \\ 0 & 0 & 0 & 0 & 0 & -\frac{x_{4c}}{L_1} \\ 0 & 0 & 0 & 0 & 0 & \frac{x_{3c}}{C_2} \end{bmatrix} \tag{8}$$

3.1.2 Coupling Analysis

In the proposed analysis method, both input-state and state-state coupling will be studied. Whereas the former describes the access of energy inside the system, the latter corresponds to its propagation inside it. To study the behaviour of the global converter, both are significant. For the first point, it is clear from the block structure of $\mathcal{J}_{f,u}$ that the arms are decoupled from their inputs, because each modulation index only affects its own arm. On the other hand, the analysis of $\mathcal{J}_{f,x}$ shows that it is purely block-diagonal. Consequently, there is no coupling paths between the phases of the converter, neither between the upper and lower arms in each phase. The natural representation of the MMC is consequently purely decoupled.

The analysis performed above shows that the converter can effectively be decomposed in three phases and each phase in two arms, and that its whole behaviour can be described by one latter only. By definition, this arm will be defined as the irreducible element of the converter. After linearization and by defining $x_i := X_i + \tilde{x}_i$ (resp. $u_i := U_i + \tilde{u}_i$), the small variations $\tilde{x}_i$ dynamics around X_i (resp. U_i) are described in (9).

$$\begin{bmatrix} \dot{\tilde{x}}_1 \\ \dot{\tilde{x}}_2 \end{bmatrix} = \begin{bmatrix} -\frac{R_1}{L_1} & -\frac{U}{L_1} \\ \frac{U}{C_2} & -\frac{1}{R_2 \cdot C_2} \end{bmatrix} \cdot \begin{bmatrix} \tilde{x}_1 \\ \tilde{x}_2 \end{bmatrix} + \begin{bmatrix} -\frac{X_2}{L_1} \\ +\frac{X_1}{C_2} \end{bmatrix} \cdot \tilde{u} \tag{9}$$

It plays a very significant role in the converter analysis, since all the important properties (like stability, observability and controllability) can be studied directly on this element, whose size is much reduced compared to the full-order model.

3.2 *Transformed (Σ/Δ) Model Analysis*

3.2.1 Local Differentiation (Jacobian Analysis)

Using the same formalism, the two partial Jacobian matrix of the transformed system are shown in (10) and (11).

$$\mathcal{J}_{\bar{f},z} = \begin{bmatrix}
-\frac{R_1}{L_1} & -\frac{v_{1a}}{2L_1} & 0 & -\frac{v_{2a}}{2L_1} & 0 & 0 & 0 & 0 & 0 & 0 & 0 & 0 \\
\frac{v_{1a}}{2C_2} & -\frac{1}{R_2C_2} & \frac{v_{2a}}{2C_2} & 0 & 0 & 0 & 0 & 0 & 0 & 0 & 0 & 0 \\
0 & -\frac{v_{2a}}{2L_1} & -\frac{R_1}{L_1} & -\frac{v_{1a}}{2L_1} & 0 & 0 & 0 & 0 & 0 & 0 & 0 & 0 \\
\frac{v_{2a}}{2C_2} & 0 & \frac{v_{1a}}{2C_2} & -\frac{1}{R_2C_2} & 0 & 0 & 0 & 0 & 0 & 0 & 0 & 0 \\
0 & 0 & 0 & 0 & -\frac{R_1}{L_1} & -\frac{v_{1b}}{2L_1} & 0 & -\frac{v_{2b}}{2L_1} & 0 & 0 & 0 & 0 \\
0 & 0 & 0 & 0 & \frac{v_{1b}}{2C_2} & -\frac{1}{R_2C_2} & \frac{v_{2b}}{2C_2} & 0 & 0 & 0 & 0 & 0 \\
0 & 0 & 0 & 0 & 0 & -\frac{v_{2b}}{2L_1} & -\frac{R_1}{L_1} & -\frac{v_{1b}}{2L_1} & 0 & 0 & 0 & 0 \\
0 & 0 & 0 & 0 & \frac{v_{2b}}{2C_2} & 0 & \frac{v_{1b}}{2C_2} & -\frac{1}{R_2C_2} & 0 & 0 & 0 & 0 \\
0 & 0 & 0 & 0 & 0 & 0 & 0 & 0 & -\frac{R_1}{L_1} & -\frac{v_{1c}}{2L_1} & 0 & -\frac{v_{2c}}{2L_1} \\
0 & 0 & 0 & 0 & 0 & 0 & 0 & 0 & \frac{v_{1c}}{2C_2} & -\frac{1}{R_2C_2} & \frac{v_{2c}}{2C_2} & 0 \\
0 & 0 & 0 & 0 & 0 & 0 & 0 & 0 & 0 & -\frac{v_{2c}}{2L_1} & -\frac{R_1}{L_1} & -\frac{v_{1c}}{2L_1} \\
0 & 0 & 0 & 0 & 0 & 0 & 0 & 0 & \frac{v_{2c}}{2C_2} & 0 & \frac{v_{1c}}{2C_2} & -\frac{1}{R_2C_2}
\end{bmatrix} \tag{10}$$

$$
\mathscr{J}_{\bar{f},v} = \begin{bmatrix}
-\frac{z_{2a}}{2L_1} & -\frac{z_{4a}}{2L_1} & 0 & 0 & 0 & 0 \\
\frac{z_{1a}}{2C_2} & \frac{z_{3a}}{2C_2} & 0 & 0 & 0 & 0 \\
-\frac{z_{4a}}{2L_1} & -\frac{z_{2a}}{2L_1} & 0 & 0 & 0 & 0 \\
\frac{z_{3a}}{2C_2} & \frac{z_{1a}}{2C_2} & 0 & 0 & 0 & 0 \\
0 & 0 & -\frac{z_{2b}}{2L_1} & -\frac{z_{4b}}{2L_1} & 0 & 0 \\
0 & 0 & \frac{z_{1b}}{2C_2} & \frac{z_{3b}}{2C_2} & 0 & 0 \\
0 & 0 & -\frac{z_{4b}}{2L_1} & -\frac{z_{2b}}{2L_1} & 0 & 0 \\
0 & 0 & \frac{z_{3b}}{2C_2} & \frac{z_{1b}}{2C_2} & 0 & 0 \\
0 & 0 & 0 & 0 & -\frac{z_{2c}}{2L_1} & -\frac{z_{4c}}{2L_1} \\
0 & 0 & 0 & 0 & \frac{z_{1c}}{2C_2} & \frac{z_{3c}}{2C_2} \\
0 & 0 & 0 & 0 & -\frac{z_{4c}}{2L_1} & -\frac{z_{2c}}{2L_1} \\
0 & 0 & 0 & 0 & \frac{z_{3c}}{2C_2} & \frac{z_{1c}}{2C_2}
\end{bmatrix} \tag{11}
$$

3.2.2 Coupling Analysis

In the (Σ/Δ) coordinates, the system is no longer decoupled. Whereas the three phases remain independent, a structural coupling appears between the two arms of each one through the "difference" quantities, in this case the equilibrium points of v_2, z_3 and z_4, as shown in (10) and (11). To illustrate these phenomena, it is possible to consider an identical current flowing through both arms in the presence of two different modulation indexes ($v_2 \neq 0$). As this current will charge differently both capacitors, it contributes to the apparition of a differential voltage: the sum current z_1 contributes to the differential voltage z_4 through the differential modulation index v_2.

From the coupling analysis presented before, it is obvious that the irreducible element is now a whole phase. The small-signal model of one phase around an arbitrary $[V, Z]$ operating point is shown in (12).

$$
\begin{bmatrix} \dot{\tilde{z}}_1 \\ \dot{\tilde{z}}_2 \\ \dot{\tilde{z}}_3 \\ \dot{\tilde{z}}_4 \end{bmatrix} = \begin{bmatrix}
-\frac{R_1}{L_1} & -\frac{V_1}{2L_1} & 0 & -\frac{V_2}{2L_1} \\
\frac{V_1}{2C_2} & -\frac{1}{R_2C_2} & \frac{V_2}{2C_2} & 0 \\
0 & -\frac{V_2}{2L_1} & -\frac{R_1}{L_1} & -\frac{V_1}{2L_1} \\
\frac{V_2}{2C_2} & 0 & \frac{V_1}{2C_2} & -\frac{1}{R_2C_2}
\end{bmatrix} \cdot \begin{bmatrix} \tilde{z}_1 \\ \tilde{z}_2 \\ \tilde{z}_3 \\ \tilde{z}_4 \end{bmatrix}
+ \begin{bmatrix}
-\frac{Z_2}{2L_1} & -\frac{Z_4}{2L_1} \\
\frac{Z_1}{2C_2} & \frac{Z_3}{2C_2} \\
-\frac{Z_4}{2L_1} & -\frac{Z_2}{2L_1} \\
\frac{Z_3}{2C_2} & \frac{Z_1}{2C_2}
\end{bmatrix} \cdot \begin{bmatrix} \tilde{v}_1 \\ \tilde{v}_2 \end{bmatrix} \tag{12}
$$

3.2.3 Parametric Condition for Decoupling

To obtain a decoupling configuration comparable with the initial one, the conditions $V_2 = Z_3 = Z_4 = 0$ should be respected. In steady-state, v_2, z_3 and z_4 are essentially sinusoidal, corresponding indeed to the AC component of the manipulated quantities. Even if the former assumptions are relevant from the control point of view (if $v_2(t) = \sin(\omega t)$ then $V_2 =< \sin(\omega t) >= 0$), they are questionable because the so-called sine waves inside m^u, m^l, I^u and I^l are comparable in magnitude to their DC components. In other words, the Σ/Δ transformation changes a structural arm-arm decoupling into a questionable parametric arm-arm decoupling.

4 Proposed Control and Simulation Results

Making use of the arm-decoupled behaviour of the MMC, the proposed controller contains six control systems, almost identical except the feedforward/modulating terms. Its structure is shown in Fig. 3. Each control system includes two standard single-input, single-output dynamic controllers (arm current controller and stack voltage controller), making each state variable controlled explicitly one time. The proposed scheme allows both active, reactive and circulating current control.

The DC-grid power is explicitly controlled through the DC arm current, which is equal between top and bottom arms. To maintain the DC stack voltage, the associated controller generates an individual AC power set point, converted into an instantaneous active current reference. By adding the latter, the former DC component, a reactive current reference and possibly a $2f$ circulating current reference (not shown in Fig. 3), the global arm current reference is obtained. It is then fed to the dedicated controller with generates the corresponding arm voltage.

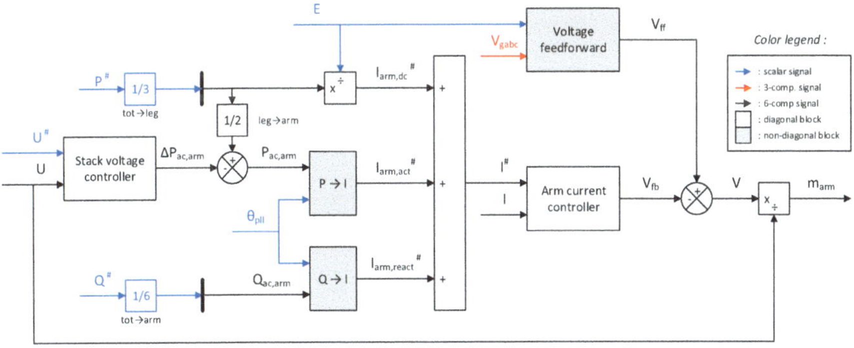

Fig. 3 Block diagram of the proposed control law

As explained in the colour legend, all the white blocks have a diagonal behaviour, meaning that a given output does only depend on the input(s) which share the same index. For instance, the arm current controller first[2] output $V_{fb}[1]$ depends freely on $I[1]$ and $I^{\#}[1]$, but not on $I[2]$ or $I^{\#}[2]$. For the stack voltage controller, the same principle applies excepted that the reference is common to all the six arms.

The non-diagonal (grey) blocks behaviour is detailed in (13) and (14). The Q-to-I block is obtained by replacing $\cos(\cdot)$ by $\sin(\cdot)$ in (14).

$$V_{ff} = \left[\tfrac{E}{2} - V_{ga},\ \tfrac{E}{2} + V_{ga},\ \tfrac{E}{2} - V_{gb},\ \tfrac{E}{2} + V_{gb},\ \tfrac{E}{2} - V_{gc},\ \tfrac{E}{2} + V_{gc}\right]^T \tag{13}$$

$$I^{\#}_{arm,act} = \frac{2}{\hat{V}_g} \times \begin{bmatrix} +P_{ac,arm}[1] \times \cos(\theta_{PLL} - 0\pi/3) \\ -P_{ac,arm}[2] \times \cos(\theta_{PLL} - 0\pi/3) \\ +P_{ac,arm}[3] \times \cos(\theta_{PLL} - 2\pi/3) \\ -P_{ac,arm}[4] \times \cos(\theta_{PLL} - 2\pi/3) \\ +P_{ac,arm}[5] \times \cos(\theta_{PLL} - 4\pi/3) \\ -P_{ac,arm}[6] \times \cos(\theta_{PLL} - 4\pi/3) \end{bmatrix} \tag{14}$$

To validate the proposed control law, simulations based on a three-phase MMC average model have been made with Matlab/Simulink. Its parameters, corresponding to Fig. 2, are summarized in Table 1

The power references have a rise time of 100 ms,[3] and an amplitude of ±700 MW and ±300 MVAR. Both controllers of each control system are polynomial, linear controllers. The global simulation results are shown in Fig. 4, with a zoom on a steady-state condition ($P = 700$ MW, $Q = 0$ MVAR) in Fig. 5. In general two slow active and reactive power controllers are used for HVDC-VSC control, but they are not shown here since they would hide the proposed, inner control dynamics. These results exhibit both good transient and steady-state performance, with good dynamics, little or no overshoot and low harmonic distortion.

Table 1 Simulation parameters

Parameter	Value	Parameter	Value
E	640 kV	L_1	50 mH
$\hat{V}_g$	250 kV	R_1	1 Ω
S_n	1 GVA	C_2	25 μF
f_n	50 Hz	R_2	1 MΩ

[2]The $\beta[i]$ notation here refers to the i-st component of the vector β.

[3]10PU/s is considered as very fast from a power transmission point of view, and corresponds to the fastest response that may be required in real HVDC point-to-point applications.

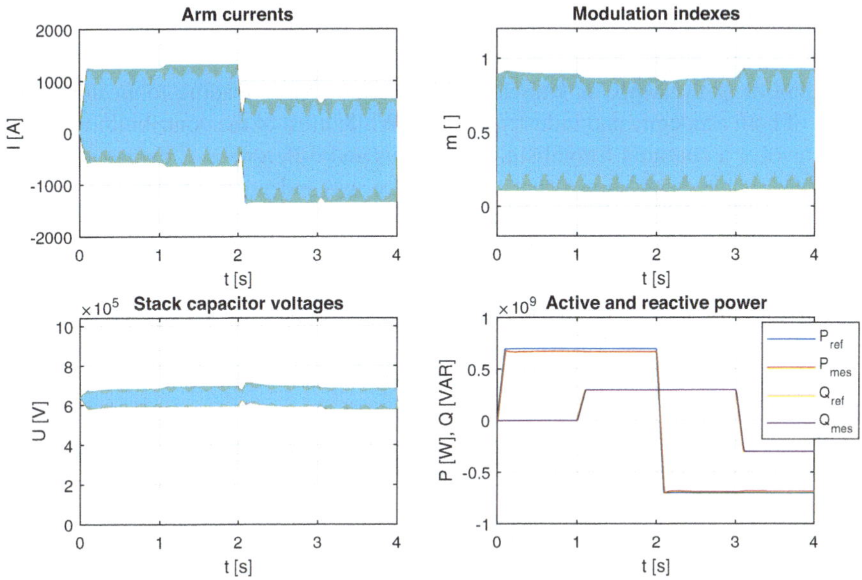

Fig. 4 Simulation results: global overview

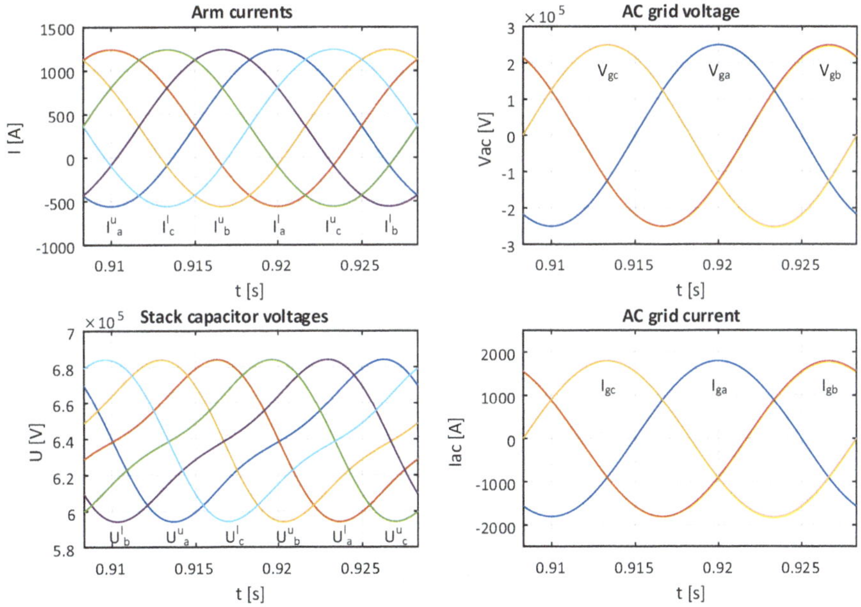

Fig. 5 Simulation results: steady-state operation

5 Conclusion

The control of the MMC is known as a challenging task, leading to an important work of both academic and industrial actors. While most of the contributions in this field involve a common formalism, using sums and differences of the normal state variables, it has been shown that this practice was indeed increasing the system complexity, introducing coupling between state and control variables, leading to hard to interpret phenomena. After the computation of the non-linear model of the MMC, it has been analysed and the independence of arms has been proved. Based on this concept, a purely independent control based on arm-modularity with few parameters and straightforward tuning was proposed, and its validity was shown in simulation. Moreover, this control law lends itself well to distributed control, and to MMC-type converters with an arbitrary number of phases, for example, for HVDC DC/DC [9] applications.

The study of the consequence of the HVDC link configuration (monopolar/bipolar) and the station transformer coupling is considered as a perspective, such as the design of different controllers for increased performance.

Acknowledgements This work was supported by a grant overseen by the French National Research Agency (ANR) as part of the Investissements d'Avenir Program (ANE-ITE-002-01).

References

1. A. Lesnicar, R. Marquardt, An innovative modular multilevel converter topology suitable for a wide power range, in *2003 IEEE Bologna Power Tech Conference Proceedings*, vol. 3, June 2003, p. 6
2. L. Zhang, Y. Zou, J. Yu, J. Qin, V. Vittal, G.G. Karady, D. Shi, Z. Wang, Modeling, control, and protection of modular multilevel converter-based multi-terminal HVDC systems: a review. CSEE J. Power Energy Syst. **3**(4), 340–352 (2017)
3. A. Antonopoulos, L. Angquist, H. Nee, On dynamics and voltage control of the modular multilevel converter, in *2009 13th European Conference on Power Electronics and Applications*, September 2009, pp. 1–10
4. E.N. Abildgaard, M. Molinas, Modelling and control of the modular multilevel converter (mmc). Energy Procedia **20**, 227–236 (2012). Technoport 2012 - Sharing Possibilities and 2nd Renewable Energy Research Conference (RERC2012). Available: http://www.sciencedirect.com/science/article/pii/S1876610212007539
5. K. Shinoda, A. Benchaib, J. Dai, X. Guillaud, Dc voltage control of mmc-based HVDC grid with virtual capacitor control, in *2017 19th European Conference on Power Electronics and Applications (EPE'17 ECCE Europe)*, September 2017, pp. P.1–P.10
6. V. Hofmann, M. Bakran, A capacitor voltage balancing algorithm for hybrid modular multilevel converters in HVDC applications, in *2017 IEEE 12th International Conference on Power Electronics and Drive Systems (PEDS)*, December 2017, pp. 691–696
7. G. Bergna-Diaz, J.A. Suul, S. D'Arco, Energy-based state-space representation of modular multilevel converters with a constant equilibrium point in steady-state operation. IEEE Trans. Power Electron. **33**(6), 4832–4851 (2018)

8. Y. Hsieh, F.C. Lee, Modeling of the modular multilevel converters based on the state-plane analysis and $\sigma\delta$ coordinate transformation, in *2018 IEEE 18th International Power Electronics and Motion Control Conference (PEMC)*, August 2018, pp. 993–999
9. J.D. Pez, D. Frey, J. Maneiro, S. Bacha, P. Dworakowski, Overview of DC-DC converters dedicated to HVDC grids. IEEE Trans. Power Deliv. **34**(1), 119–128 (2019)

Simulation of an Islanded DC Microgrid Using Instantaneous and Average Modeling Approaches

Elie Hleihel, Maurice Fadel, and Hadi Y. Kanaan

Abstract With the proliferation of renewable energy sources and the adoption of several policies to reduce environmental risks caused by traditional polluting sources, the concept of microgrids, especially DC microgrids, is currently gaining interest. In fact, most renewable energy sources (RESs) and loads are inherently DC type. Moreover, DC microgrids offer many merits over AC ones in terms of ease of control and efficiency. While most of researches address the control hierarchy and strategy in DC microgrids, this paper focuses on the modeling and simulation aspect. A typical configuration of an islanded DC microgrid is modeled in MATLAB/Simulink, and a primary-level control strategy is adopted where two approaches of converters modeling are tested: instantaneous and average model. The two approaches of modeling are compared in terms of precision of losses modeling, dynamic response of the system, simulation time, and computational burden. Simulation tests are conducted, and the results show that, despite its accuracy, the instantaneous model can be applied only for short-term simulations due to many limitations, whereas average converter modeling presents a better solution for long-time simulations, since it ensures a tradeoff between model accuracy and simulation time, which makes the application of the three levels of hierarchical control in DC Microgrids valid in one simulation model.

1 Introduction

In the recent decades, there has been renewed interest in renewable energy sources, which can be an alternative to traditional sources of polluting energy such as coal, oil, and natural gas, which are being depleted and the cause of global warming.

E. Hleihel (✉) · H. Y. Kanaan
Faculty of Engineering—ESIB, Saint-Joseph University of Beirut, Beirut, Lebanon
e-mail: hleihel@laplace.univ-tlse.fr; hadi.kanaan@usj.edu.lb

M. Fadel
LAPLACE, Université de Toulouse, Toulouse, France
e-mail: maurice.fadel@laplace.univ-tlse.fr

W. Zamboni, G. Petrone (eds.), *ELECTRIMACS 2019*, Lecture Notes
in Electrical Engineering 615, https://doi.org/10.1007/978-3-030-37161-6_15

The use of renewable energy sources is growing rapidly, which allowed to adopt a new approach of distributed architectures and to stand out from the conventional approach of centralized generation [1]. The distribution and connection of distributed sources to the conventional utility grid has been the subject of several studies over the past 10 years, especially when these generators based on renewable energy sources are of different and intermittent nature, which poses a problem of energy quality when considering energy variations, voltage stability problems (voltage drop, rapid fluctuation, interruption, harmonic distortions . . .) [2].

The concept of microgrid was proposed more than a decade ago as a solution to electrify remote areas and integrate renewable energy sources (RESs) with energy storage systems (ESSs) and different types of sensitive and non sensitive loads [3]. Microgrids could operate in islanded mode as well as in grid connected mode, and based on the type of connection between its different components, one can identify three types of microgrids: AC, DC, and hybrid AC-DC microgrids [4, 5]. Several researches have been carried out in AC microgrids [6]. In fact, AC distribution systems are the most adopted all over the world and offer many benefits in terms of ease of voltage transformation, especially for high-voltage transmissions, and maturity of standardized technology [6]. On the other hand, DC microgrids are currently gaining interest and have many merits over AC grids due to the increased number of DC type loads such as: computers, lighting, electric vehicle, home electronics, etc... so as most of RESs et ESSs which are inherently DC type [7]. Thus, adopting a DC type connection between all these components will result in less number of conversion stages and consequently a higher system efficiency [8]. Moreover, the control of a DC microgid is less complex than an AC one because there are no issues with power quality, grid synchronization, frequency, and reactive power regulation.

Mainly, the prime control priority in a DC microgid is the regulation of the common DC bus voltage; in other terms, the achievement of a balance between the power generated by RESs and load demands and ensuring a proper load sharing between source converters [9, 10].

However, the topology and the configuration of a microgrid define its possible operating modes and the best strategy of control to be adopted. A typical islanded DC microgrid configuration is depicted in Fig. 1. It consists of RESs: a solar PV array, a wind turbine, a diesel generator for emergency cases, an ESS, and different types of DC loads. While most of studies and surveys address the control strategies and hierarchy in DC microgrids [11–14], this paper focuses on the modeling of an islanded DC microgrid and the problems related to the simulation side. It compares two approaches of converters modeling: instantaneous and average models in terms of precision of losses modeling, dynamic response of the system, simulation time, and computational burden. The simulations are performed on MATLAB/Simulink. Since the main objective of this paper is to compare the two model approaches in a DC microgrid simulation, a simple control strategy of microgrids based on first-level control functions, such as MPPT, current, and voltage regulation loops, is adopted; advanced functionalities of secondary and tertiary levels of control are not applied and are out of scope of this work. This paper is organized as follows:

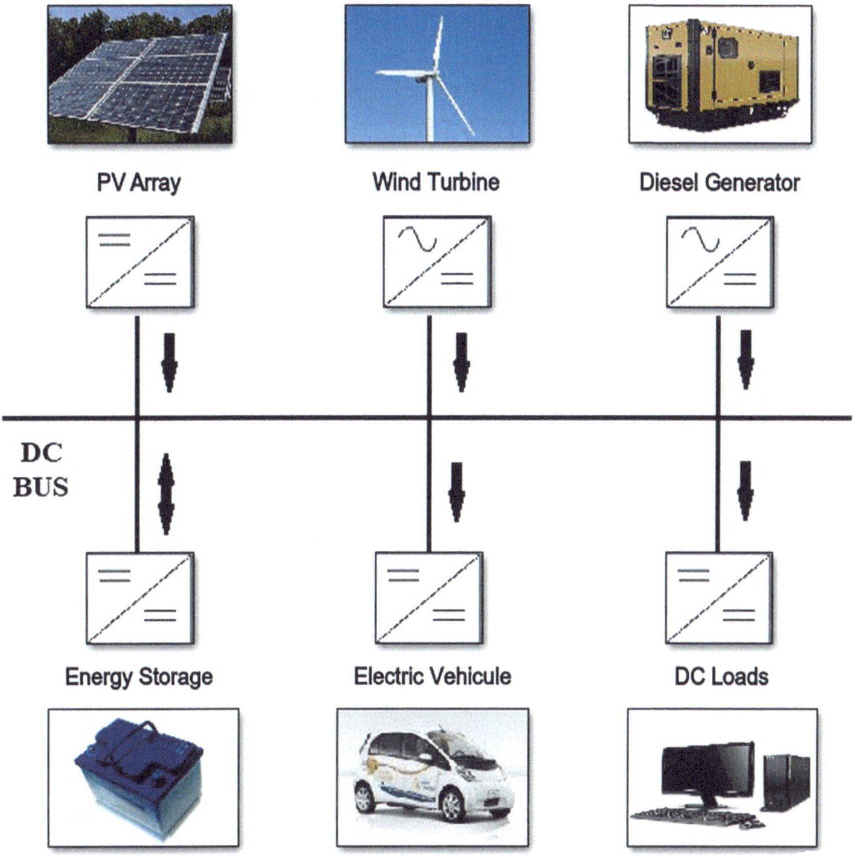

Fig. 1 Islanded DC microgrid configuration

In Sect. 2, the microgrid configuration and the strategy of control of each unit is represented. In Sect. 3, the instantaneous and the average model of converters are discussed. Simulation tests are performed in Sect. 4, and a comparison between the two models approaches is carried out. Finally, Sect. 5 concludes the paper.

2 Microgrid Configuration and Strategy of Control

Figure 1 shows the microgrid configuration adopted in the frame of this work: each unit is connected to the common DC bus through its own converter controlled locally. In order to model required RESs, ESSs and DC Microgrid power electronic interfaces, "Simpower Systems" library in MATLAB/Simulink is used. As a global strategy of control, all RESs are functioning in MPPT mode, while the battery

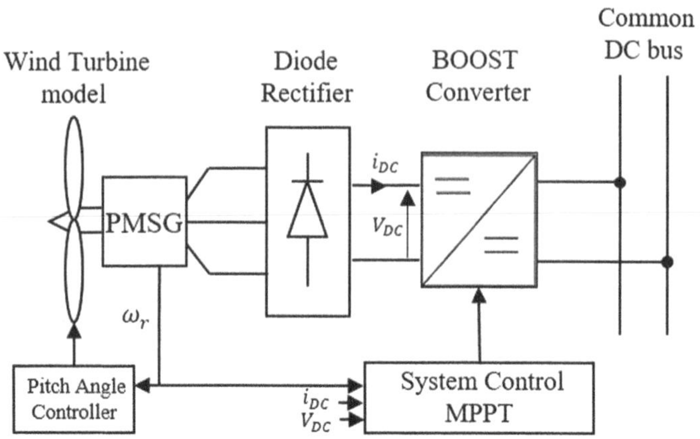

Fig. 2 Wind Turbine Configuration

takes charge of regulating the DC bus voltage by charging/discharging processes. However, the diesel generator, as a polluting energy source, intervenes only in emergency cases to generate power and feed the loads. In the following list, each microgrid unit is represented individually including its own configuration and converter's strategy of control:

1. *PV Array.*

   In order to model the PV solar generation, the block "PV array" is used. Knowing that the generated power from a PV is DC type, the PV is connected to the DC common bus through a DC/DC Boost converter, which is functioning in MPPT mode to extract the maximum available power.
2. *Wind Turbine.*

   The wind turbine configuration is shown in Fig. 2. It is a typical configuration [16] constituted by a wind turbine model, a permanent magnet synchronous generator (PMSG), a Diode rectifier, and a Boost converter connected to the common DC bus. Regarding the control, a pitch angle controller block regulates the pitch angle of the turbine model and a system control block implements the MPPT algorithm to extract the maximum available wind power.
3. *Diesel Generator.*

   Different mathematical models of diesel generator (DG) are mentioned in [15]. The dynamic model adopted in this study addresses the electrical aspects of a DG rather than the mechanical ones. As shown in Fig. 3, it consists of a diesel engine governor, which includes a diesel engine and governor system, an excitation system block which implements a synchronous machine (SM) voltage regulator, and exciter. The output is the field voltage to apply to the input of a SM block. The AC output of the SM is connected to the DC common bus through a three-phase universal bridge rectifier. As previously mentioned, the DG

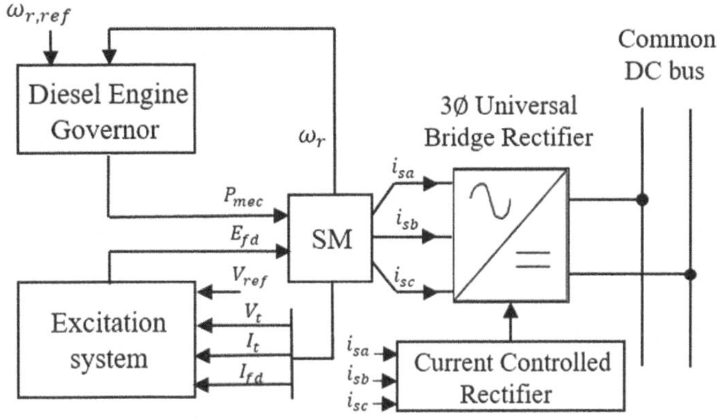

Fig. 3 Diesel Generator Configuration

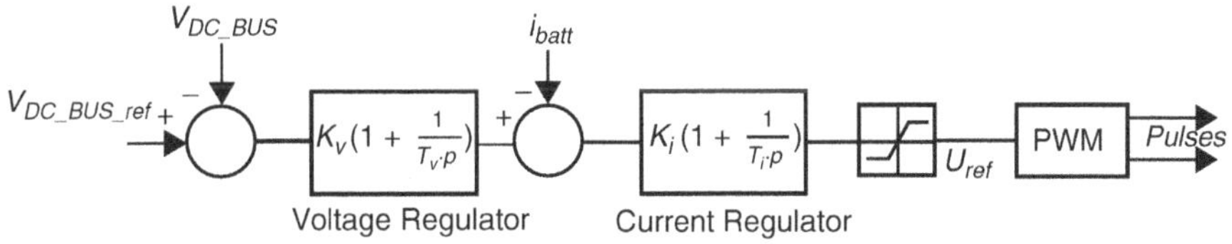

Fig. 4 Battery cascaded loops regulation

intervenes only in emergency cases, e.g., if the generated power is less than the load demands and the battery is already discharged, then the DG is connected to the microgrid to balance the power flow and prevent battery excess discharging. Once the power deficit is no longer persistent, the DG is disconnected from the microgrid. Hence, the DG converter is current/power controlled through a current loop in order to provide a specific amount of power when needed for a limited period of time.

4. *Battery and DC Loads.*

   The "battery" block is used to implement a Lithium-Ion 500 Ah–300 V battery model. It is connected to the DC bus voltage through a DC/DC bidirectional converter, which functions as a Boost in discharging process and as a Buck in charging process. The battery main control objective is the regulation of the common DC bus voltage. A cascaded voltage and current loops regulation technique is adopted as shown in Fig. 4: the DC bus voltage is regulated using a P.I. controller, which generates the reference current to be injected/absorbed from the battery. Then the battery current is regulated using another P.I (*all P.I.s are equipped with anti-windup systems*). Variable DC current sources are utilized to model the DC loads.

3 Instantaneous and Average Modeling of Converters

Based on the hierarchy, control objective, and level in a microgrid, the timescale in which the microgrid should be simulated varies. In turn, the simulation time is well limited by the model adopted in MATLAB/Simulink, which induces many constraints in terms of precision, sampling time, and computational burden. When the instantaneous converters models are used with real switching devices, the system behavior is modeled with best accuracy in terms of dynamic response, losses modeling, etc., whereas long-time simulations cannot be performed due to the reduced sampling time, related to the switching frequency of converters, which results in a slower simulation. In addition, the complexity of the model contributes to high computational burden requiring a large memory space in the hardware, which can block the simulation if the PC does not have a high computing capacity. Hence, instantaneous converters modeling can be utilized to simulate few seconds of the microgrid behavior and is not adequate for power flow analysis and optimization, especially while sources and loads profiles are realistic and spread over wide times. Thus to cope with the tradeoff between the model accuracy and the time range of the simulation, average converters model solution is adopted.

3.1 Average Model Converter Configuration

To adopt the average model of converters, "average model" converters blocks in Simscape/Simulink are used. As a principle, the real switching devices are omitted and converters are controlled directly by duty cycles instead of firing pulses, which allows to eliminate the pulse width modulation (PWM) block, increase the sampling time, and so speeding up the simulation. The average model configuration of a Boost converter is shown in Fig. 5a: voltage- and current-controlled sources are used to model the boost converter. First, the voltage of the controlled source (V_a) is computed based on the boost equations mentioned below:

$$E = L\frac{\mathrm{d}i_a}{\mathrm{d}t} + V_s.(1-\alpha) \tag{1}$$

$$i_a.(1-\alpha) = C\frac{\mathrm{d}V_s}{\mathrm{d}t} + \frac{V_s}{R} \tag{2}$$

The input current i_a and the output voltage V_s are measured instantaneously in order to compute the values of the voltage and current controlled sources. From Eq. (1), it can be concluded that $V_a = V_s.(1-\alpha)$, where V_a, V_s, and α are, respectively, the voltage-controlled source value at the input side of the converter, the output voltage of the converter, and the duty cycle. In order to compute the value of the current-controlled source (i_s), power equality equation is used:

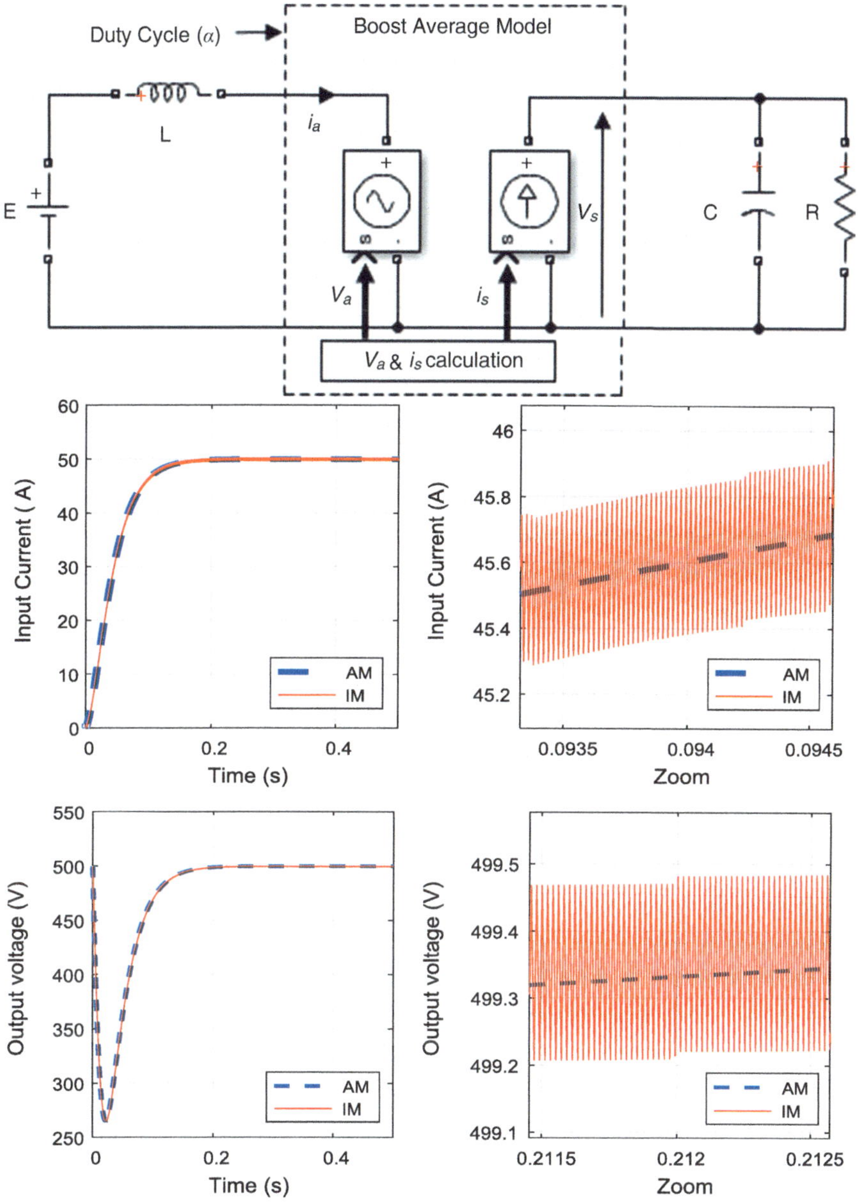

Fig. 5 (**a**) Boost Average model configuration. (**b**) Input current and output voltage of Boost converter in instantaneous and average model simulations

$$P = V_a . i_a = V_s . i_s \tag{3}$$

$\rightarrow i_s = \frac{V_a . i_a}{V_s}$. By substituting V_a with its value we obtain: $i_s = i_a . (1 - \alpha)$ which leads to Eq. (2) of the boost converter. Hence by imposing V_a and i_s, the configuration represented in Fig. 5 corresponds precisely to the equations of the boost. It can be seen from Fig. 5a that boost components L and C still exist with this model configuration. Moreover, a snubber resistance could be added in parallel to the current-controlled source to model losses in the diode and the switch. Thus, the adopted average model configuration does not affect the accuracy of the converter model and exhibits the same behavior as the instantaneous model. Similarly, the DC/DC bidirectional converter and the three-phase universal bridge rectifier are modeled with the same strategy, based on their own equations.

3.2 Sampling Time Selection

The selection of the sampling time (T_s) in each model is directly influenced by the time constants of the system and the switching time of converters. In instantaneous model, the switching time of each converter (T_{sw_i}) should be lower than the smallest time constant (τ_i) of the system related to its own unit ($T_{sw\_i} \ll \tau_i$) then, the sampling time of the simulation is selected based on the lowest switching time of converters. It can be seen that a ratio of 100 between the sampling and the switching time ($T_{s\_IM} = T_{sw}/100$) presents a good tradeoff between the quality of the obtained curves and the duration of the simulation. On the other hand, the sampling time in average model is no more influenced by the switching time of converters, since switches are omitted in the average approach. So, the sampling time is chosen to be lower than the lowest time constant of the system ($T_{s\_AM} \ll \tau_{lowest}$). In order to obtain accurate results, the sampling time is chosen to be equal to the switching time of the instantaneous model ($T_{s\_AM} = T_{sw}$). By this, good quality results are guaranteed and a gain of 100 in sampling time is ensured between the two simulations models: ($T_{s\_AM} = 100.\ T_{s\_IM}$), which makes the average model simulation much faster than the instantaneous one. To test the validity of the average model, the boost converter configuration of Fig. 5a is simulated with both models. A voltage and a current regulation loop are realized. The input voltage is set to $E = 200$ V, the load resistance $R = 25\ \Omega$, and the reference output voltage $V_{s\_ref} = 500$ V. A switching frequency $f_{sw} = 50$ kHz is selected $\rightarrow T_{s_IM} = 0.2\ \mu$s and $T_{s_AM} = 20\ \mu$s. The results are shown in Fig. 5b. It can be seen that the output voltage and the input current are well regulated in both simulations, and the curves of instantaneous and average converter models are typically identical in transient and in steady state, which validate the average model.

4 Simulation Tests and Results

Since the instantaneous model is the most accurate, it is used as a reference to validate the viability of the average model on the overall microgrid through many simulation tests. Then, a comparison between the two modeling approaches is conducted. Simulations are performed on a Core i5 2.6 GHz CPU processor and a 6 GB RAM memory. All microgrid parameters and data are listed in Table 1. First, the instantaneous model is tested apart with a specific scenario to validate the strategy of control of the system. Figure 6a shows the results of the simulation. In this scenario, RESs generate variable power and function in MPPT mode. Since the generated power of RESs is lower than the load demand, the battery discharges to stabilize the DC bus voltage and ensure the power balance in the microgrid. After a while (at $t = 2.9$ s), the battery state of charge (SOC) decreases below its admissible limit, so the diesel generator is connected to prevent battery excess discharging and balance the power in the DC microgrid. It is shown in Fig. 6b that the DC bus voltage is stabilized on its reference value (730 V) within the allowable variation limits. In addition, the SOC of the battery stops its fast decrease and is almost stabilized after the connection of the diesel generator. Despite the accuracy of this model, it requires a lot of time to be simulated (*2 min 40 s*), and for higher simulation time ($t > 6$ s in our case), the software is blocked due to the limitation in the hardware computational capacity. In order to test the validity of the average model, we apply the same scenario of Fig. 6 in both simulations models.

Figure 7a, b show, respectively, the instantaneous and the average model curves of diesel generator, PV, wind turbine, and battery power flow. It can be seen that the two models' curves are almost identical for PV and diesel generator power, whereas wind turbine curves present a slight delay in the first transient state ($T_d = 0.19$ s), then the delay is reduced to 7 ms in the subsequent transients. An error of 1.3% is noticed between the two models in steady state. The wind curves delay and steady state error appear in battery power curves, which have a delay in transients ($T_d = 0.2$ s) and a maximum steady state error of 5.91%. It can be seen that the

Table 1 DC Microgrid parameters and data

Common DC bus rated voltage	730 V
Allowable DC bus voltage deviation	±5%
Power electronic switches type	IGBTs
Switching frequency for all converters	5 kHz
Wind turbine maximum rated power	55 kW
PV array maximum rated power	100.7 kW
Diesel generator maximum rated power	55 kW
Battery rated capacity	300 Ah
Battery rated voltage	200 V
Simulation time	6 s
Sampling time (instantaneous model)	$T_{s\_IM} = 2\ \mu s$
Sampling time (average model)	$T_{s\_AM} = 200\ \mu s$

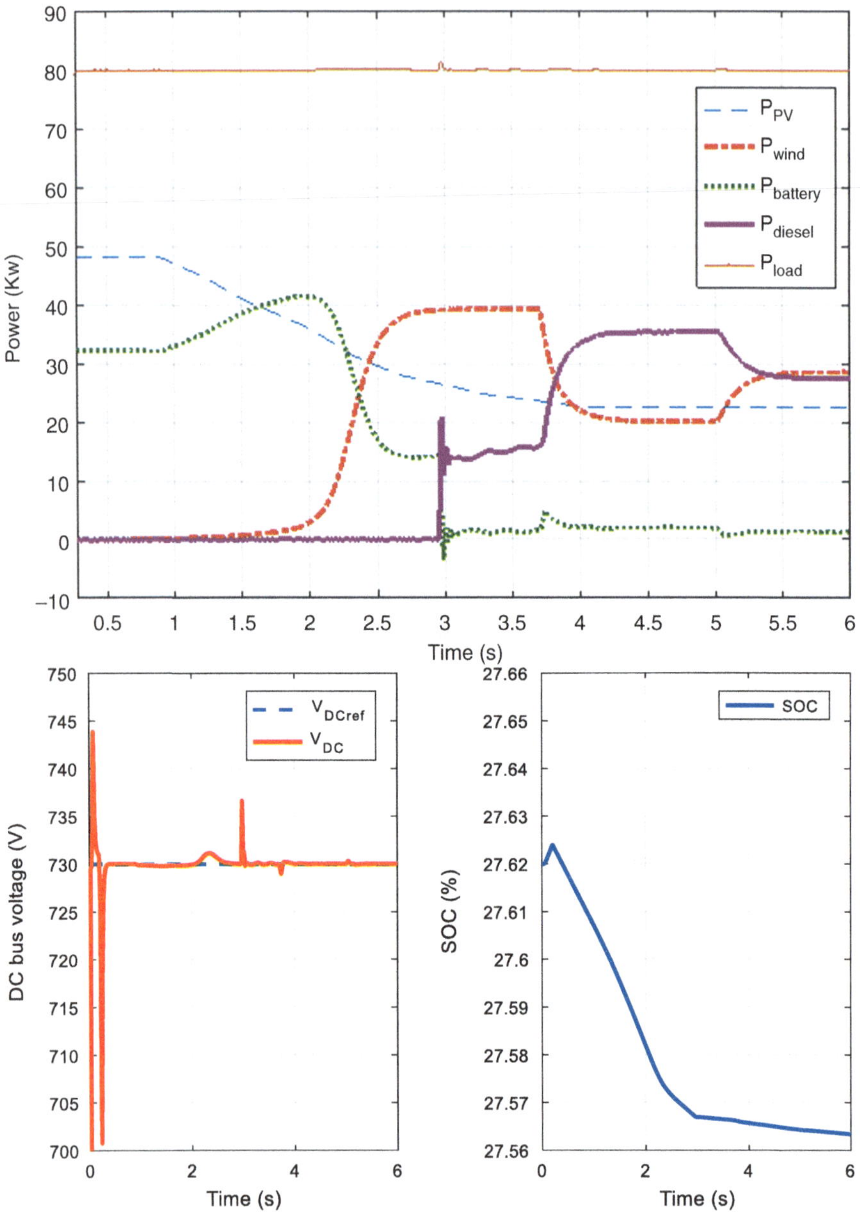

Fig. 6 (**a**) DC microgird power flow. (**b**) DC bus voltage regulation and SOC of the battery

average model reflects precisely the behavior of the system in transients and in steady state. However, a notable reduction in simulation time is obtained with the average approach (*6 s are simulated in 15 s*). Hence, average model simulation is

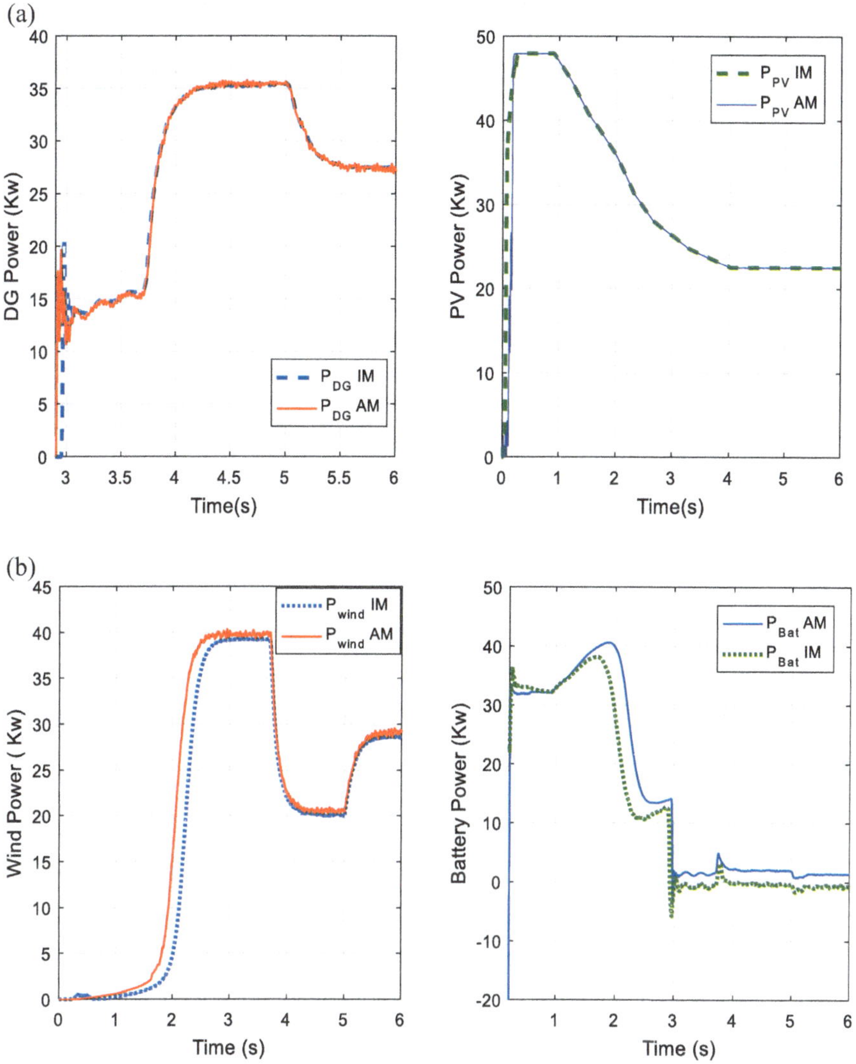

Fig. 7 (**a**) DG and PV power curves in average and instantaneous model. (**b**) Wind and battery power curves in average and instantaneous model

11 times faster than the instantaneous one. Moreover, for higher simulations time ($t > 6$ s), we no longer encounter software blocking problems as in the case of the instantaneous model.

Figure 8 shows losses in wind turbine rectifier in both models. It can be seen that losses are modeled in both simulations with an error of (1.3%). So, adopting the average model does not affect the modeling of losses in converters. Hence, losses

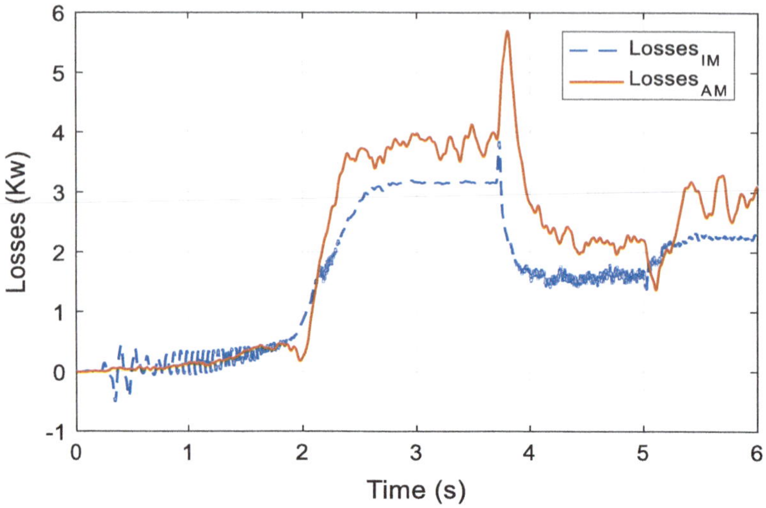

Fig. 8 Losses in wind turbine rectifier

reduction algorithms in microgrids can always be applied to the average model simulation.

Finally, the average model is tested apart with a long-time simulation (600 s) and slower dynamic variation of sources profiles. In this scenario, RESs are generating variable power and functioning in MPPT mode, a constant power load of $P_{\text{load}} = 84.68$ kW is demanded.

Results are shown in Figs. 9, 10, and 11. It can be seen from Fig. 9 that the battery (*Blue line curve*) is discharging to stabilize the DC bus voltage since ($P_{\text{RESs}} < P_{\text{load}}$). At $t = 198$ s, SOC of the battery decreases to 30% (Fig. 10), hence the diesel generator intervenes (*Green line curve*) to prevent battery excess discharging and generate power ($P_{\text{Diesel}} = P_{\text{load}} - P_{\text{RESs}}$) to ensure power balance in the microgrid. It is shown that during the operation of the diesel generator (198 s < t < 382 s) the battery power is set to zero, which stabilizes the SOC on 30%. At $t = 382$ s, the diesel generator is disconnected since ($P_{\text{RESs}} \cong P_{\text{load}}$), then the RESs generated power increases and surpasses the load demand, so the surplus power is used to recharge the battery ($P_{\text{batt}} = P_{\text{RESs}} - P_{\text{load}}$) and thus the SOC increases again. Figure 11 shows a stable operation of the DC bus voltage during all the simulation with acceptable transient.

5 Conclusion

In this paper, an islanded DC microgrid configuration is modeled and simulated using MATALB/Simulink software. Two converters modeling approaches were

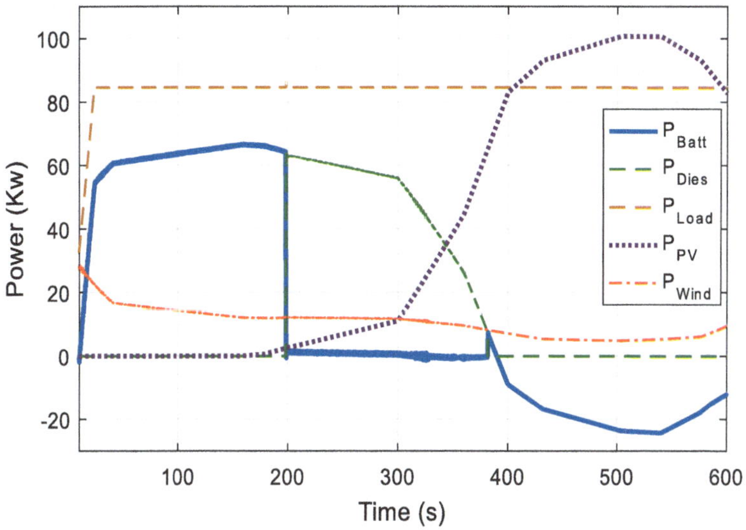

Fig. 9 DC Microgrid power flow in average model simulation

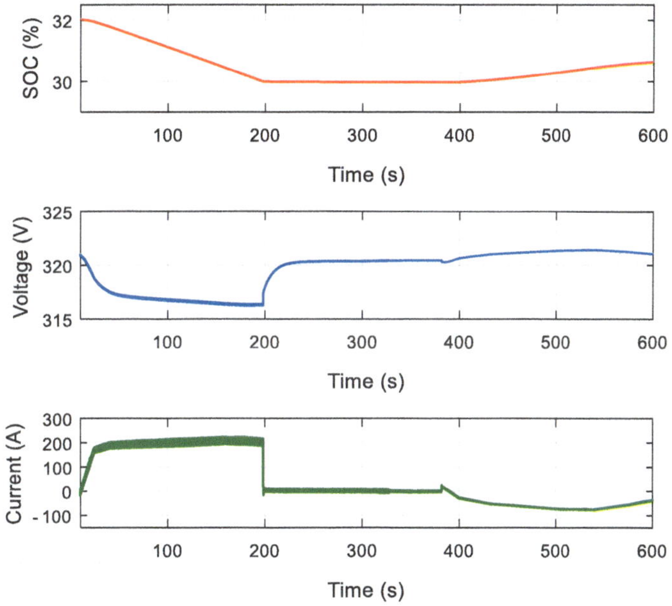

Fig. 10 Battery State of charge (%), Voltage (V) and Current (A)

investigated: instantaneous and average model. The instantaneous model is used as a reference to validate the effectiveness of the average model in terms of accuracy, dynamic response of the system, simulation time, and computational burden.

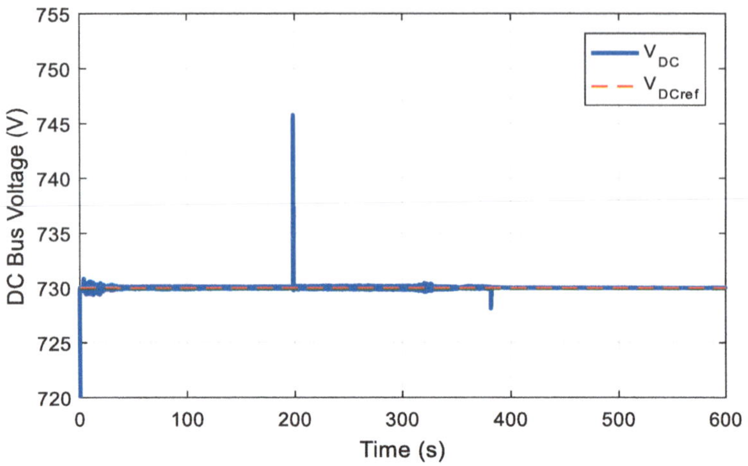

Fig. 11 DC bus voltage regulation in average model simulation

Results show that despite its accuracy, instantaneous model cannot be applied for real power flow analysis simulation due to the limitation in real switching devices resulting in long-time simulation and high computational burden. On the other hand, the average converters approach brings a reduction in simulation time (ratio of 11) and computational burden by canceling high frequencies phenomena of the system (high switching frequencies of converters are omitted). However, middle and low frequencies phenomena are always modeled, which keep the accuracy of the system and make the application of traditional primary-level control strategies (MPPT, Droop Control, Current, voltage regulations, . . .) always valid. Consequently, the average model approach is a better platform to adopt hierarchical control in DC microgrids, in which the three levels of control can be applied simultaneously in one simulation with an acceptable accuracy and an improved time simulation and model complexity.

Acknowledgments The authors gratefully thank the Agence Universitaire de la Francophonie (AUF), the Lebanese National Council for Scientific Research (CNRS-L) and the Research Council of Saint-Joseph University of Beirut for their financial support.

References

1. F. Katiraei, R. Iravani, N. Hatziargyriou, A. Dimeas, Microgrids management. IEEE Power Energy Mag. **6**(3), 54–65 (2008)
2. A. Ipakchi, F. Albuyeh, Grid of the future. IEEE Power Energy Mag. **7**(2), 52–62 (2009)
3. R. Lasseter, A. Akhil, C. Marnay, J. Stevens, J. Dagle, R. Guttromson, A.S. Meliopoulous, R. Yinger, J. Eto, *The Certs Microgrid Concept—White Paper on Integration of Distributed Energy Resources* (U. S. Department of Energy, Washington, DC,. Technical Rep., 2002)

4. R.H. Lasseter, Microgrids and distributed generation. J. Energy Eng. **133**(3), 144–149 (2007)
5. J.M. Guerrero, J.C. Vasquez, J. Matas, L.G. de Vicuna, M. Castilla, Hierarchical control of droop-controlled AC and DC microgrids—a general approach toward standardization. IEEE Trans. Ind. Electron. **58**(1), 158–172 (2011)
6. J.C. Vasquez, J.M. Guerrero, A. Luna, P. Rodriguez, R. Teodorescu, Adaptive droop control applied to voltage-source inverters operation in grid-connected and islanded modes. IEEE Trans. Ind. Electron. **56**(10), 4088–4096 (2009)
7. D.J. Hammerstrom, AC versus DC distribution systems Did we get it right? in *2007 IEEE Power Engineering Society General Meeting*
8. T. Dragicevic, J.C. Vasquez, J.M. Guerrero, D. Skrlec, Advanced LVDC electrical power architectures and microgrids: a step toward a new generation of power distribution networks. IEEE Electrif. Mag. **2**(1), 54–65 (2014)
9. S. Parhizi, H. Lotfi, A. Khodaei, S. Bahramirad, State of the art in research on microgrids: a review. IEEE Access. **3**, 890–925 (2015)
10. T. Dragičević, X. Lu, J.C. Vasquez, J.M. Guerrero, DC microgrids—Part I: A review of control strategies and stabilization techniques. IEEE Trans. Power Electron. **31**, 4876–4891 (2016)
11. S.K. Sahoo, A.K. Sinha, N.K. Kishore, Control techniques in AC, DC and hybrid AC-DC microgrid: a review. IEEE J. Emerg. Select. Top. Power Electron. **6**(2), 738–759 (2018)
12. S. Anand, B.G. Fernandes, J.M. Guerrero, Distributed control to ensure proportional load sharing and improve voltage regulation in low-voltage DC microgrids. IEEE Trans. Power Electron. **28**(4), 1900–1913 (2013)
13. Q. Shafiee, T. Dragicevic, J.C. Vasquez, J.M. Guerrero, Hierarchical control for multiple DC-microgrids clusters. IEEE Trans. Energy Convers. **29**(4), 922–933 (2014)
14. T. Dragicevic, J.M. Guerrero, J.C. Vasquez, A distributed control strategy for coordination of an autonomous LVDC microgrid based on power-line signaling. IEEE Trans. Ind. Electron. **61**(7), 3313–3326 (2014)
15. S. Benhamed et al., Dynamic modeling of diesel generator based on electrical and mechanical aspects, in *2016 IEEE Electrical Power and Energy Conference (EPEC)*, (2016), pp. 1–6
16. H. Xu, J. Hui, D. Wu, W. Yan, H. Xu, Implementation of MPPT for PMSG-based small-scale wind turbine, in *2009 4th IEEE Conference on Industrial Electronics and Applications*, (2009), pp. 1291–1295

Part II
Control and Power Management

Digital Control of an FFC NMR Relaxometer Power Supply

Rúben J. A. Lopes, Pedro J. Sebastião, Duarte M. Sousa, António Roque, and Elmano Margato

Abstract The fast field cycling (FFC) experimental technique allows to overcome a technical difficulty associated with the nuclear magnetic resonance (NMR) signal-to-noise ratio (SNR) at low frequency spin-lattice relaxation measurements when using conventional NMR spectrometers. Constituting a step forward than the classical analog approaches, in this paper, a digital control system for an FFC-NMR relaxometer power supply was developed. The hardware and software were designed to allow for the modulation of the Zeeman field as required by this technique. Experimental results show that under digital control the system performs fast transitions between the high and low magnetic flux density levels, i.e., the switching times obtained are in the millisecond range, and, assures a good stability of the field during the steady states. Comparative proton relaxometry measurements in two compounds (liquid crystal 5CB and ionic liquid [BMIM]BF4) were made to assess the digital control system performance.

R. J. A. Lopes · P. J. Sebastião
Department of Physics, Instituto Superior Técnico, Universidade de Lisboa, Lisbon, Portugal
e-mail: pedro.jose.sebastiao@tecnico.ulisboa.pt

D. M. Sousa
Department of Electrical and Computer Engineering, Instituto Superior Técnico & INESC-ID, Universidade de Lisboa, Lisbon, Portugal
e-mail: duarte.sousa@tecnico.ulisboa.pt

A. Roque (✉)
Department of Electrical Engineering, Escola Superior de Tecnologia de Setúbal/Instituto Politécnico de Setúbal, Setúbal, Portugal
e-mail: antonio.roque@estsetubal.ips.pt

E. Margato
CEI, ISEL-Instituo Superior de Engenharia de Lisboa, Instituto Politécnico de Lisboa, and INESC-ID, Lisbon, Portugal
e-mail: efmargato@isel.ipl.pt

W. Zamboni, G. Petrone (eds.), *ELECTRIMACS 2019*, Lecture Notes in Electrical Engineering 615, https://doi.org/10.1007/978-3-030-37161-6_16

1 Introduction

NMR spectroscopy is used to study molecular order and dynamics in different materials, as for instance, organic compounds like ionic liquids and liquid crystals [1–11].

There are different NMR techniques used to study molecular dynamics by measuring relaxation times T_1 spin-lattice and T_2 spin-spin. The main difficulty associated with the NMR relaxometry at different magnetic fields is the signal-to-noise ratio (SNR) of the measured NMR signal, since the NMR SNR decreases with the magnetic field [2–4]. Therefore, conventional NMR spectrometers, which operate at fields above 0.1 T, are not suitable to perform studies of resonance frequencies associated with low magnetic fields [5]. The FFC-NMR apparatus is used to overcome this difficulty, by applying different magnetic fields at distinct times, but always measuring the NMR induction signal when the sample is submitted to a high magnetic field.

In an FFC-NMR experiment to measure the T_1 spin-lattice relaxation time, the sample is submitted to different Zeeman fields B_0 at different times, allowing measurement of the magnetization decay with time at a low magnetic flux density, but detecting the NMR signal when the sample is submitted to a high magnetic flux density, which provides good SNR measuring conditions.

In general, the magnetic flux density varies cyclically, as it is illustrated in Fig. 1.

At first, the sample is submitted to a strong magnetic flux density $B_{0\mathrm{P}}$ to be polarized. Following this, the magnetic flux density switches down to a lower value $B_{0\mathrm{E}}$ that is applied during a time Δt_{E}. Next, the field switches up to a stronger magnetic flux density $B_{0\mathrm{D}}$ during Δt_{E} and finally the magnetic flux density switches back to the initial value $B_{0\mathrm{P}}$ after Δt_{D}. Since the detection of the NMR signal is made when the sample is submitted to $B_{0\mathrm{D}}$, which is larger than $B_{0\mathrm{E}}$, the sensitivity of the quality of the NMR signal detected is independent of the evolution occurring during Δt_{E}. As it is essential to cycle the magnetic flux density accurately, after Δt_{D} the magnetic flux density changes to $B_{0\mathrm{P}}$ being the transition time Δt_{on2} [2–4].

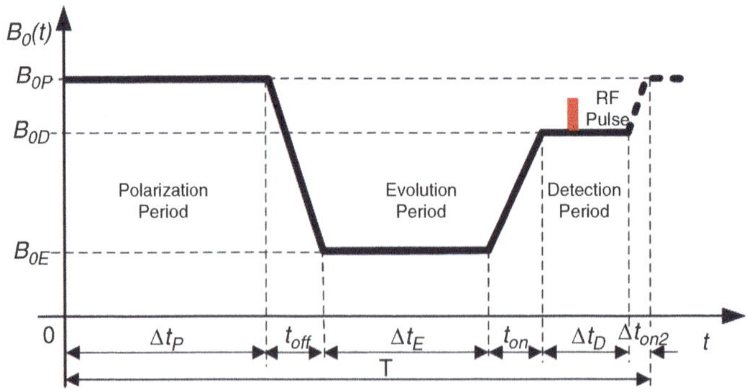

Fig. 1 Standard time diagram of the magnetic flux density of the FFC-NMR technique

In order to implement the cycle represented in Fig. 1, the current supplied to the magnet coils that produce the magnetic flux density B_0 is controlled in such a way that the transitions between the different magnetic flux densities are fast, but the magnetic field is stable during the steady states. This means that the power source of an FFC apparatus requires a control system that guarantees steady and stable currents during Δt_P, Δt_E, and Δt_D and regulated transitions of the current between these steady periods of the cycle.

Classical solutions use an analog PID regulator controlling the Zeeman field in FFC-NMR experiments. A step forward in the development of FFC relaxometers is to implement a digital controller taking advantage of the features offered by low cost microcontrollers. The digital approach of a PID controller is based on a discrete algorithm.

The real implementation of the digital control is used in a pre-production NMR relaxometer prototype. This embedded system is going to replace part of the analog solution already implemented and tested in the power supply control system. With this solution, the digital control system modulates the Zeeman field of the setup as required by the FFC-NMR technique. In order to modulate the magnetic flux density, the magnet current is controlled throughout the semiconductors of the relaxometer power supply. This is achieved using the commercial Microchip dsPIC30F4013 microcontroller, which also requires the usage of additional filters and driving on-chip peripherals interfacing sensors and power semiconductors.

The developed digital solution is tested performing FFC-NMR experiments for the liquid crystal 5CB and the ionic liquid [BMIM]BF4.

2 The Apparatus

The FFC-NMR technique is a technique for applications requiring molecular dynamics characterization [2–4]. Cycling the magnetic flux density using mechanical systems was the earliest form of the field cycling technique.

In the latest NMR relaxometers, the magnetic flux density is controlled using up-to-date power supplies based on modern power semiconductors, allowing to control the electric current flowing in the FFC magnets [12–19]. A recent design of an FFC power supply uses IGBT semiconductors and a direct control of their gate voltage, which allows regulating the magnet current within the limits of the FFC NMR specifications. One common aspect of the control units of the power supplies developed so far is the use of analog electronics in their implementation [18–20].

In this paper, a digital power control unit was developed and integrated in the power supply of a new relaxometer prototype using an FFC magnet with $R_M = 3\ \Omega$ and $L_M = 270$ mH [14]. The architecture of this relaxometer is presented in Fig. 2.

In this solution, the magnet is excited by two power sources (U_0 and U_{aux}). The control system of the main power source of the FFC relaxometer is represented in Fig. 3. The main power source (U_0, 24 V DC power supply) supplies the magnet accordingly to perform the magnetic field cycle. In addition, a high voltage

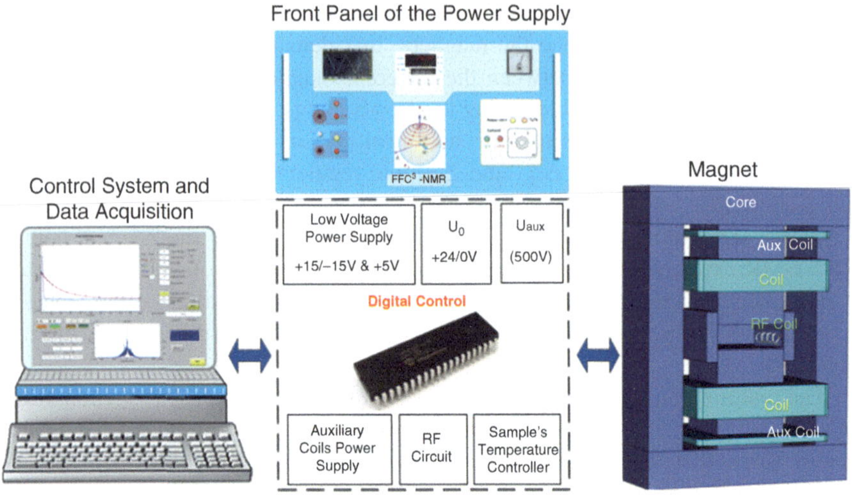

Fig. 2 Main parts of an FFC relaxometer [13]

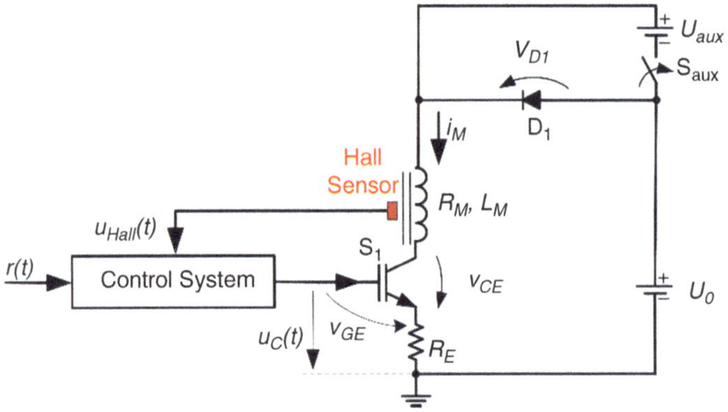

Fig. 3 Main power source control circuit of the FFC equipment

($U_{\text{aux}} = 500$ V DC) power source is switched on by the control system during the up-down (Δt_{on}) magnetic flux density transitions. The time interval Δt_{on} is in the milliseconds range, allowing to obtain short switching times from a low magnetic flux density level to the high polarization magnetic flux density level.

This spectrometer directly controls the magnetic flux density feedbacking the output signal of a Hall effect sensor ($u_{\text{Hall}}(t)$) placed in the sample's cavity. The other input of the control system is the reference magnetic flux density signal $r(t)$. The output of the control chain ($u_{\text{C}}(t)$) corresponds to the command signal of the semiconductors in order to get the accurate magnetic flux density level B_0 corresponding to the reference imposed.

Without the contribution of the auxiliary power supply U_{aux}, the behavior of the circuit is given by

$$\frac{\mathrm{d}i_{\mathrm{M}}}{\mathrm{d}t} = \frac{1}{L_{\mathrm{M}}}(U_0 - v_{\mathrm{CE}}) - \frac{i_{\mathrm{M}}}{\tau_{\mathrm{L}}} \tag{1}$$

where $\tau_{\mathrm{L}} = \frac{L_{\mathrm{M}}}{R_{\mathrm{E}}+R_{\mathrm{M}}}$.

As an IGBT semiconductor is used, the incremental changes in the control voltage (gate-emitter voltage) lead to a change in the semiconductor current as follows:

$$\Delta v_{\mathrm{CE}} = \beta\,(R_{\mathrm{E}}\Delta i_{\mathrm{M}} - \Delta u_{\mathrm{C}}) \tag{2}$$

Being the factor β set according to the characteristics of the IGBT.

So that, the magnet current can be set by changing the gate command voltage u_{C}. As referred before, the transitions between magnet current levels, i.e., magnetic flux density levels, need to be fast, but should be long enough so that the net magnetization follows the Zeeman field reorganization. Clearly, these limits depend upon the spin system in consideration. In order to fulfill the requirements of the FFC technique, transition times in the milliseconds range (≈3 ms and less) are acceptable for a wide range of compounds [5–15, 21].

To reach the required switching times during a down-up transition (Δt_{on}), the auxiliary power supply is turned on. During this transition, the IGBT semiconductor operating point is forced to the saturation region as the magnet current rises, reaching its steady-state value. A PI controller is the typical solution to control the Zeeman field in FFC experiments. The PI controller changes the command voltage $u_{\mathrm{C}}(t)$ of the IGBT minimizing the error $e(t)$ between the Hall sensor output $u_{\mathrm{Hall}}(t)$ and the reference signal $r(t)$. The command voltage $u_{\mathrm{C}}(t)$ is the sum of the proportional and integral correcting terms:

$$u_{\mathrm{C}} = k_{\mathrm{P}}\left[e(t) + \frac{1}{T_{\mathrm{I}}}\int_0^t e(\tau)\,\mathrm{d}\tau\right] \tag{3}$$

where k_{P} is the proportional gain, T_{I} is the integral time, and

$$e(t) = r(t) - u_{\mathrm{Hall}}(t) \tag{4}$$

Being the gain of the integral component given by $k_{\mathrm{I}} = \frac{k_{\mathrm{P}}}{T_{\mathrm{I}}}$, as usual [20, 22].

A standard option is to use analog PI controllers to achieve fast switching of the current, i.e., the magnetic flux density [13]. This project aims to develop a digital PID controller, which can replace the analog version adding more flexibility in setting the control of the power supply of the relaxometer. Nowadays, it is possible to use user-friendly programmable solutions in order to change the settings of a digital control loop instead of changing the physical parts of its analog counterpart. A digital solution can be a strong contribution for spanning the specifications of the fast field cycling technique. The most promising enhancements behind the digital approach are to change the switching times and defining different magnetic field sequences, which are usually unpractical to implement with the analog approach.

3 Digital PID Controller

The implementation of a digital PID controller is based on a discrete algorithm that uses as reference the continuous time behavior of a PID controller [22, 23]:

$$u(t) = k_{\mathrm{P}}\left[e(t) + \frac{1}{T_{\mathrm{I}}}\int_0^t e(\tau)\,\mathrm{d}\tau + T_{\mathrm{D}}\frac{\mathrm{d}}{\mathrm{d}t}e(t)\right] \tag{5}$$

Considering that the continuous time PID controller behavior can be expressed by finite differences approximations, the behavior of discrete time PID controller can be expressed by:

$$u(t_k) = u(t_{k-1}) + k_{\mathrm{P}}\left[e(t_k) - e(t_{k-1})\right] + k_{\mathrm{I}}e(t_k) + k_{\mathrm{D}}\left[e(t_k) - 2e(t_{k-1}) + e(t_{k-2})\right] \tag{6}$$

Equation (6) is the *Velocity Algorithm* for a PID controller [22, 23], derived by approximating the first-order derivatives via backward finite differences, where the parameters of the controller are the gains $k_{\mathrm{I}} = \frac{k_{\mathrm{P}}}{T_{\mathrm{I}}}$ and $k_{\mathrm{D}} = k_{\mathrm{P}}T_{\mathrm{D}}$.

3.1 Digital Control Chain

The digital control chain for an FFC solution needs to perform the following functionalities:

- To read the output signal of a Hall sensor (magnetic flux density measurement).
- To react to pulse signals (0–5 V TTL) that set the instant of the magnetic field transitions.
- To generate the pulse signal that sets on the auxiliary power supply during the up-down magnetic flux density transition (Δt_{on}).
- To generate the IGBT gate signals suitable for the different operating modes of the power supply.

In this project, the PID controller is based on the Microchip ® dsPIC30F4013 *digital signal controller*. This processing unit incorporates several useful peripheral components, including 5 timers (16-bit resolution), 12-bit analog-to-digital converter (ADC), and an output compare module. For example, the ADC is used to acquire the Hall sensor output voltage. On the other hand, the dsPIC30F4013 does not incorporate a digital-to-analog converter (DAC) necessary to generate the analog

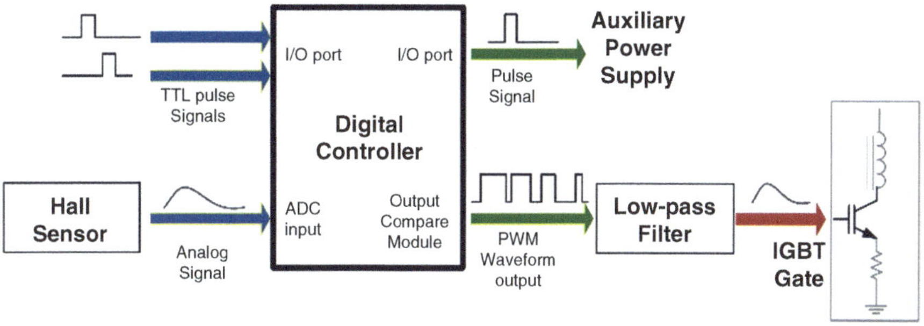

Fig. 4 Digital control system configuration

output signal to command the IGBT. Therefore, a pulse width modulation (PWM) DAC will be implemented to overcome this issue.

However, this approach requires removing the high frequency components of the PWM signal by the implementation of an analog low-pass filter. Therefore, the control variable is the PWM duty cycle, and the analog output signal corresponds to a fraction of the pulse voltage. The digital control system configuration (with the inputs and outputs referred before) is represented in Fig. 4.

The digital implementation requires the predetermination of the digital value of the reference signal $r(t)$. The reference signal changes once a magnetic flux density transition is launched throughout the TTL pulses. When a TTL pulse corresponding to a down-up magnetic flux density transition is launched, the digital control system also generates the pulse signal that switches on the auxiliary power source (U_{aux}). The power supply U_{aux} is switched off when the magnetic flux density reaches the reference signal (measuring and AD converting the output signal of the Hall sensor feedback signal (u_{Hall})). During this process, the digital algorithm adjusts the PID output and therefore the IGBT gate voltage, minimizing the error between the signal u_{Hall} and the magnetic flux density reference value r. Once an up-down TTL pulse occurs, the IGBT command voltage is set accordingly by the digital PID in order to observe a fast up-down transition and the required stability and accuracy for low magnetic field levels.

3.2 Control Operations

Generally speaking, a digital PID controller operates continuously reading signals from sensors, sampling and converting them to digital form by means of an analog-to-digital converter. The digital control signal is computed and then converted to an analog signal. The sequential operation is:

1. Wait for a clock interrupt.
2. Read the signal of the sensor.

3. Compute the control signal.
4. Update controller variables.
5. Send output to the actuator.
6. Repeat from 1.

The computation of the controller parameters must occur out of the main loop, to minimize the time delay.

The relaxometer power supply control system is implemented as an embedded application based on on-chip peripheral interrupts that initiate processes on CPU to deal with the events. For example, when an A/D conversion is completed, a signal is sent to the processor indicating that this event has occurred, demanding the interruption of the current code that the processor is executing. A timer is used to start the sampling and A/D conversion process at a desired frequency (25 kHz). Upward and downward transitions are determined by the change of state on specific input ports (RB4 and RB5), as illustrated in Fig. 5.

When a TTL pulse is detected on the input ports RB4 or RB5, a change notification (CN) interrupt request is generated and the interrupt service routine (ISR) is triggered, as represented in Fig. 6. In this interrupt handler the I/O ports are read, determining which pin was driven high. If PORTBbits.RB4=1, a down-up transition has been commanded (the magnetic flux density reference is set to high level) and the state of the RB6 pin changes to high in order to switch on the auxiliary power supply U_{aux}. The magnetic flux density reference is set as a value corresponding to a high magnetic field. If PORTBbits.RB5=1, an up-down transition has been commanded and the output signal of the controller corresponds to a low magnetic flux density reference value.

A transition to a high polarization field is dictated by a TTL pulse, which originates a change of state on input port RB4. Change notification ISR is executed, and it determines the digital reference value that is used in the control loop, by setting the value of γ. When RB4 pin is driven high (reading PORTBbits.RB4=1), its value is set to $\gamma = 1$.

Variable γ determines not only the control reference value used but also the PID parameters in the control calculations. CN ISR starts Timer5 and RB6 pin is driven high, leading the IGBT to the saturation region and turning on the auxiliary voltage supply. This way, a transition to a high polarization field can be accomplished within milliseconds.

Timer5 also limits the duration of the down-up transition, there is, the time interval during which high voltage is applied to the magnet in order to avoid damaging the IGBT.

Furthermore, an A/D sampling is enabled in the Timer4 ISR and the conversion of the Hall sensor analog signal starts. This interrupt handler compares the A/D converted signal to the high magnetic flux density reference value. PID control of the IGBT voltage gate starts when the control error is less than a threshold value ε_1. Otherwise, the control output signal is maintained at a high level. Consequently, the PWM duty-cycle is near 100% in the first control cycles during the upward transition.

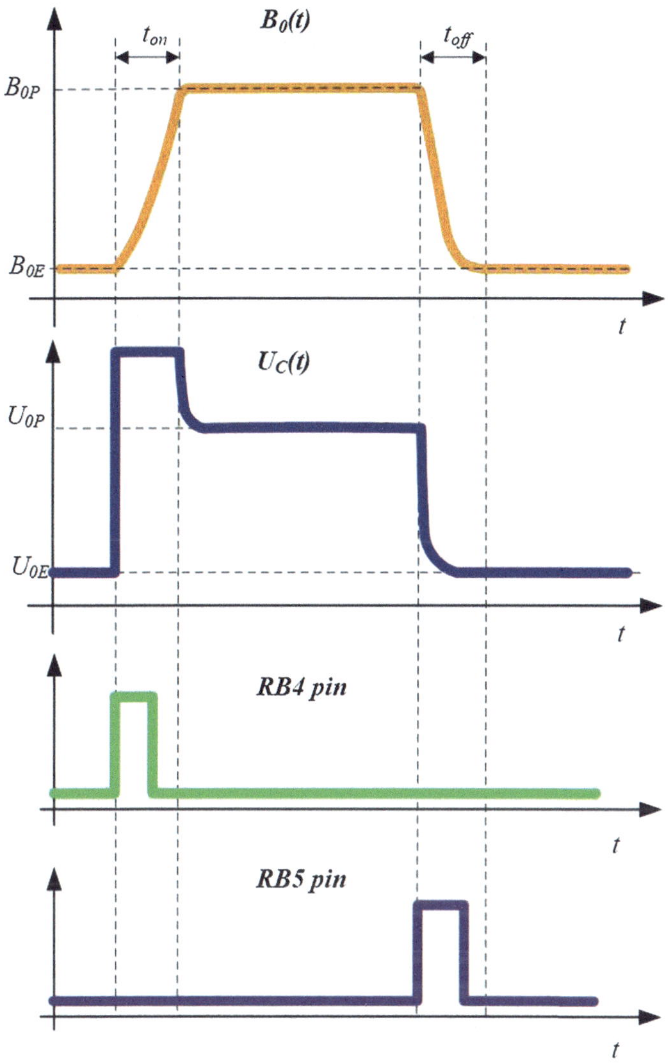

Fig. 5 Field cycling sequence. Shown are the magnetic flux density $B(t)$, the IGBT gate voltage $u_C(t)$, and the states on RB4 and RB5 pins

Once the error is less than ε_1, the RB6 pin changes to the low status, switching off the auxiliary power supply U_{aux}, and, from this point on, the PID algorithm takes into account and the control system output (i.e. the PWM duty cycle) is the output of the PID algorithm.

From the moment the high power supply is switched off, the control action is the result of the PID calculations. The control signal is the error between the magnetic flux density reference and the digital value returned by the ADC

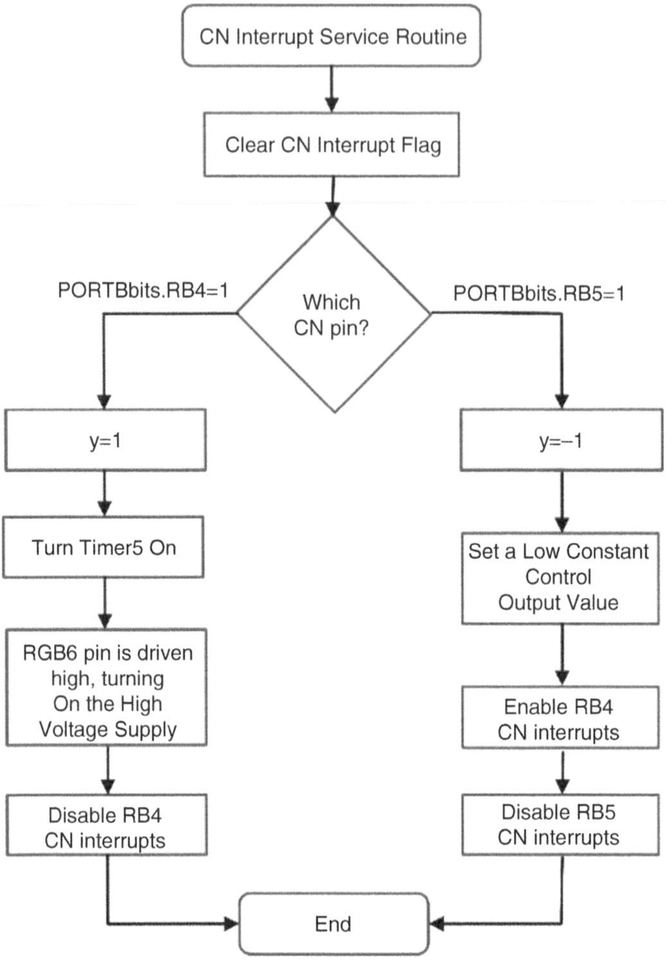

Fig. 6 Change notification (CN) interrupt service routine (ISR) flow chart

module, corresponding to the last measurement of the Hall sensor analog signal. PID controller is responsible for the adjustment of the PWM duty cycle. This action on the IGBT command signal must guarantee a fast-settling time, after an down-up field transition.

A transition to a low magnetic flux density level is dictated by a TTL pulse, detected on pin RB5. Changing the RB5 state (reading PORTBbits.RB5=1) initiates a change notification interrupt request to the processor. In the interrupt handler, variable γ is set as -1 and a constant low value is kept as the control output.

From the instant the processor executes the CN ISR, every single ADC measurement of the Hall sensor signal is compared to a low field reference value in the ADC interrupt handler.

If the value read from the Hall sensor is far from the low reference value, then the control output is kept a constant low value, allowing a fast up-down field transition. For this reason, in the first control cycles of the transition to a low relaxation field, the PWM duty cycle is constant and near the minimum value allowed in the IGBT conduction state.

Once the error is smaller than a threshold value ε_2, PID controller starts adjusting the PWM duty cycle, in order to settle the field on its final value.

After the downward transition, the control signal is formed from the error between the field reference and the digital value returned by the ADC module, corresponding to the last measurement of the Hall sensor analog signal. PID controller action on the IGBT gate voltage must guarantee a fast-settling time, after a downward field transition.

3.3 Experimental Setup

Figure 7 shows a conceptual scheme of the control system, with different conditioning stages and low pass filters necessary for this application.

First, the PWM waveform is converted to an analog signal throughout a low-pass filter, followed by a conditioning stage, where amplification of the signal is done in such a way that it meets the requirements of the plant (IGBT gate). A Bessel low-pass filter was designed with a cut-off frequency of $f_c = 2.5$ kHz for this purpose (*Filter 1*). On the other hand, an anti-alias low pass filter and a conditioning stage are placed between the Hall sensor and the A/D converter (*Filter 2*). This low-pass allows to reduce the higher-frequency noise components in the PWM analog signal. Noise components with a frequency much higher than the control system bandwidth can be aliased down, so that the closed-loop system responds to noise.

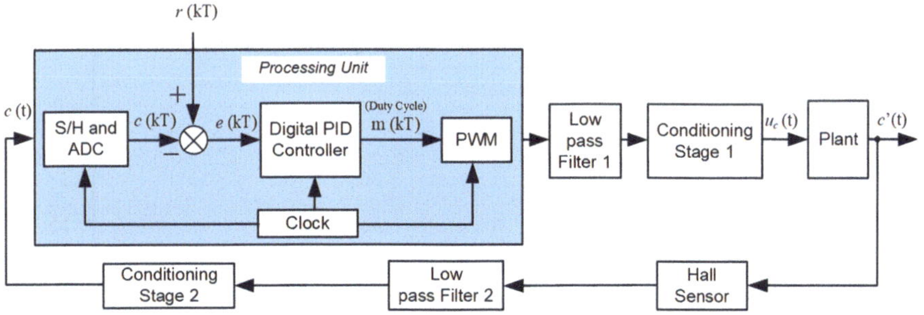

Fig. 7 Conceptual scheme simplified of the control system implemented

4 Experimental Results

The digital controller was tested replacing the main analog controller of the magnetic flux density in an FFC prototype relaxometer [24, 25].

As explained earlier, the duration of the upward and downward magnetic flux density transients needs to be faster compared with the relaxation times of the spin systems to reduce the energy transfer as much as possible during these time intervals. On the other hand, the duration of these transitions need to be adequately set in order to keep the alignment of the magnetization and avoid any transverse component in the magnetic field. Both conditions depend on the sample in study, but in terms of technical realization, the upward and downward switching times should be in the order of milliseconds (t_{on}, $t_{off} \approx 3$ ms) [26].

In Fig. 8 is presented the evolution of the PWM waveform and magnetic flux density during a downward transition between high magnetic flux density level and a low magnetic flux density level.

As expected, the controller adjusts the PWM cycle in order to minimize the error between the magnet current level digital reference voltage and the Hall sensor feedback voltage. Figure 8 corresponds to the time instant that a downward transition is triggered and the digital reference of the controller changes to a predetermined lower value. PWM duty cycle is immediately reduced, stabilizing in a lower value corresponding to the IGBT gate voltage required, so that the magnetic flux density reaches the reference value.

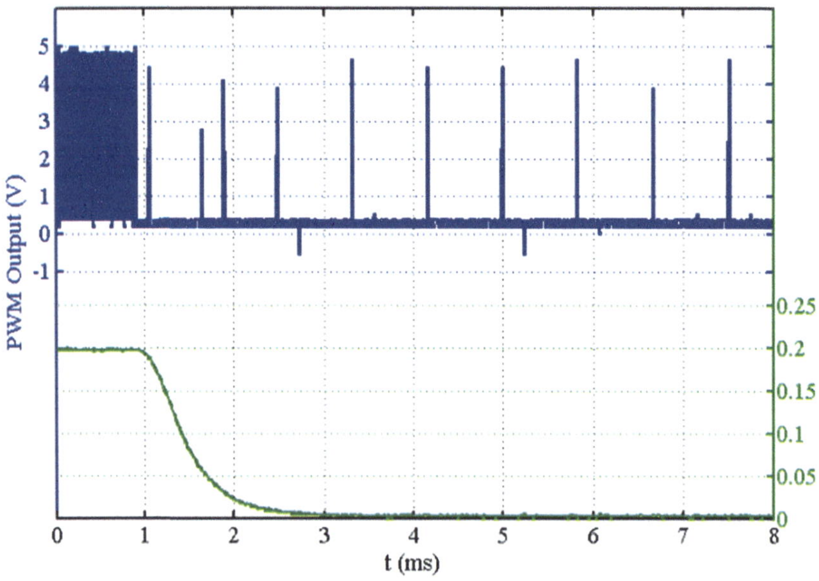

Fig. 8 Magnetic flux density and PWM waveform during a downward transition

Table 1 PID parameters used for controlling the upward and downward transitions

PID parameters	Upward transition	Downward transition
k_P	2800	3000
k_I	33	20
k_{DP}	0	0

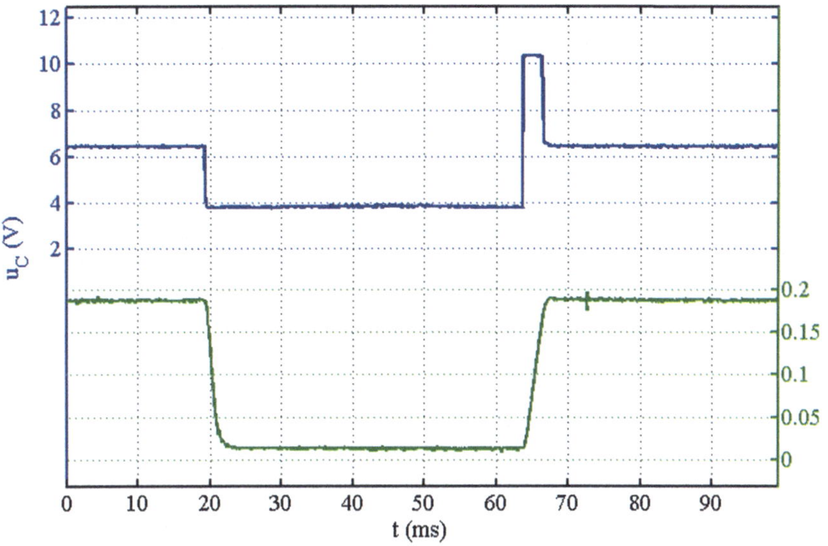

Fig. 9 IGBT command voltage and magnetic flux density in a complete field cycle

Several sets of PID parameters were tested on the control system to obtain fast adiabatic field transitions to perform spin-lattice relaxation measurements. Table 1 presents the PID controller parameters for the upward and downward transitions. These parameters allowed obtaining fast and smooth magnetic flux density transitions, without changes in the magnetic flux density values in the steady-state regimes as showed in Fig. 9.

Several transitions of the magnet current and the magnetic field strength are presented in Fig. 10.

Experimental results for a down-up and an up-down magnetic flux density transitions obtained with the proposed digital controller are shown in Figs. 11 and 12, respectively. A Bessel low-pass filter dimensioned for a cut-off frequency of 10 kHz was used in the PWM-DA converter to obtain these results (*Filter 2*). In these figures it is also possible to observe the IGBT gate signals during transitions between the B_{0P} and B_{0E} magnetic flux density levels.

These results are consistent for controlling the relaxometer power supply during the transitions. Besides being fast, the magnetic flux density transitions have to assure the repeatability of the measurements. By the results obtained (Fig. 10), these conditions are satisfied. The stability of the magnetic field and the reproducibility of

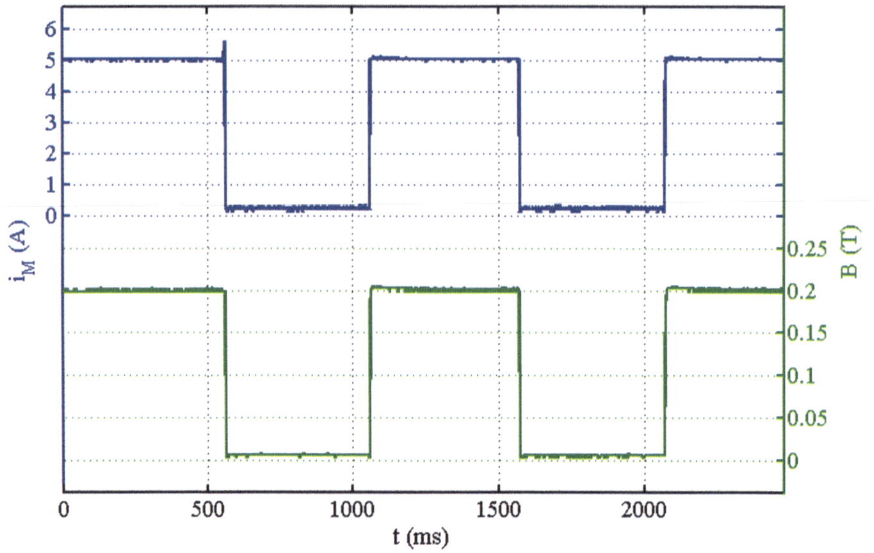

Fig. 10 Magnet current and magnetic flux density during several transitions

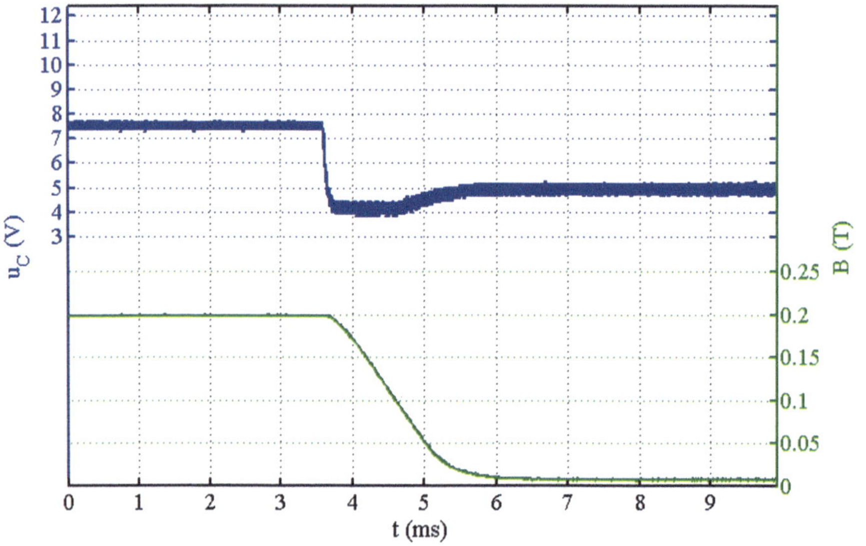

Fig. 11 Downward magnetic flux density transition

the transitions are demonstrated when performing FFC experiments by observing a repetitive free induction decay signal (FID).

In order to assess the digital control system performance, the proton spin-lattice relaxation time T_1 is measured for two distinct samples using the developed solution

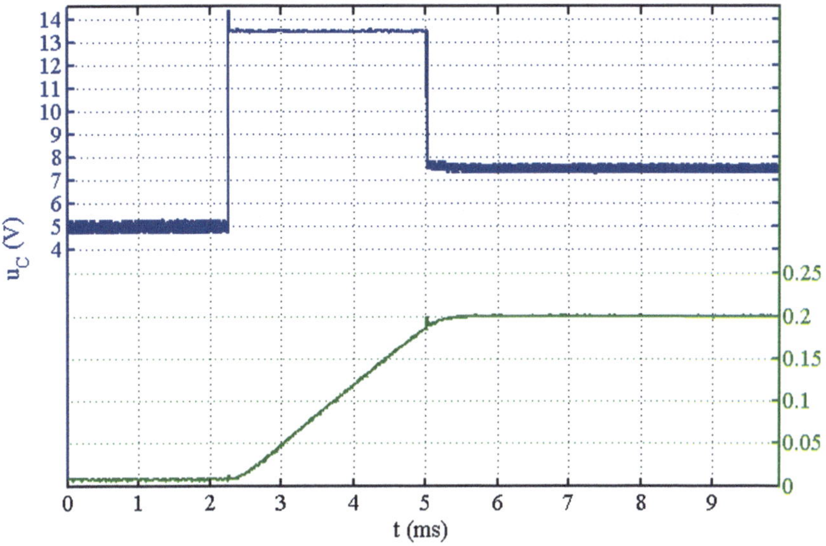

Fig. 12 Upward magnetic flux density transition

Table 2 Results of the spin-lattice relaxation time measurements using digital and analog control systems

Sample	f (kHz)	Digital control		Analog control	
		T (°C)	T_1 (ms)	T (°C)	T_1 (ms)
5CB	300	26.0	43.4 ± 4.3	25.0	42.9 ± 4.3
	300	26.0	46.6 ± 4.7	25.8	42.5 ± 4.3
	1000	20.0	40.8 ± 4.1	23.8	48.6 ± 4.9
	1000	26.0	59.2 ± 5.9	24.2	49.8 ± 5.0
[BMIM]BF4	300	26.0	80.3 ± 8.0	26.4	78.6 ± 7.9

to control the magnet current. These experiments were performed with the liquid crystal 5CB and the ionic liquid compound [BMIM]BF4. The liquid crystal 5CB is commonly used as a reference for testing and comparing the performance of relaxometer experimental setups [12]. The ionic liquid was used because it is one of the systems recently studied with this technique.

The FFC-NMR measurements quality depends on many factors other than the control system efficiency. A good adjustment of all the system components is crucial for the NMR detected signal and, overall, the FFC-NMR measurements quality. Therefore, measurements under the same conditions were performed with the same relaxometer, using both the analog and digital control systems to the power supply.

Table 2 compares the results obtained with each control system for a better assessment of the digital system performance. The analysis of the spin-lattice relaxation time measurements obtained with the analog and the digital control solutions shows that the results obtained with the digital approach are in line with the measurements performed with the analog controller.

In the case of the 5CB liquid crystal, the slight differences are due to small frequency and temperature discrepancies between measurements with the digital control system than with the analog controller. These instabilities can produce transverse nuclear magnetization, lowering the magnitude measured, which should be the result of the longitudinal relaxation process only. Anyway, these results are compatible within the experimental error suggesting that the digital system can provide the magnetic flux density control as efficient as the analog implementation.

The experimental results obtained for the ionic liquid compound [BMIM]BF4 at a frequency of 300 kHz using the analog and digital controllers are similar, showing that the digital solution modulates the Zeeman Field as required by the FFC-NMR experimental technique.

5 Conclusions

FFC-NMR is a relaxometry technique that allows to perform spin-lattice relaxation studies in a wide range of frequencies. This technique involves having the sample in different fields B_0 at different times, which allows measuring the relaxation time T_1 related to a low field B_{0E} with a NMR signal sensitivity of a common high field measurement.

This technique has been developed setting strictly the dynamic characteristics of experimental setup. It requires to perform repetitive magnetic flux density cycles between with accurate steady-state regimes and with switching times within the 1–3 ms range a wide range of compounds. With this work, it is clearly demonstrated that a digital controller is capable of modulating the Zeeman field fulfilling the FFC-NMR requirements. This was accomplished using a programmable Microchip ® dsPIC microcontroller driving on-chip peripherals interfacing sensors and power electronic devices and designing additional filters. The program developed configures the microcontroller unit core and several peripherals required for the application. The sequence of control operation starts with a timer interrupt, enabling the analog-to-digital conversion of the signal from the Hall sensor placed near the sample inside the magnet. The result of this conversion is compared with a predetermined digital reference. The error difference between these two values is then minimized, as the PID algorithm adjusts the control system output.

The software and hardware designed under the scope of this work allow to fulfill the system requirements, mainly the required fast transitions between magnetic field levels. The experimental results consistently show that time duration of the transitions between magnetic field levels are within the milliseconds range.

Furthermore, the digital system developed was used to modulate the Zeeman field while measuring the relaxation time T_1 of the liquid crystal 5CB and the ionic liquid [BMIM]BF4. The exponential decay of the nuclear magnetization was observed and the related time constant was determined for each case. The results obtained show that the digital approach can be integrated in an FFC relaxometer power supply system.

Acknowledgements This work was supported by national funds through Instituto Politécnico de Setúbal, ISEL/Instituto Politécnico de Lisboa, Fundação para a Ciência e a Tecnologia (FCT) with reference UID/CEC/50021/2019 and UID/CTM/04540/2019.

References

1. M. Levitt, *Spin Dynamics: Basics of Nuclear Magnetic Resonance* (Wiley, Chichester, 2001)
2. F. Noack, NMR field-cycling spectroscopy: principles and applications. Prog. NMR Spectrosc. **18**, 171–276 (1986)
3. R. Kimmich, E. Anoardo, Field-cycling NMR relaxometry. Prog. NMR Spectrosc. **44**, 257–320 (2004)
4. E. Anoardo, G. Galli, G. Ferrante, Fast-field-cycling NMR: applications and instrumentation. Appl. Magn. Reson. **20**(3), 365–404 (2001)
5. T. Farrar, E. Becker, *Pulse and Fourier Transform NMR: Introduction to Theory and Methods* (Academic Press, New York, 1971)
6. I. Matter, G. Scott, T. Grafendorfer, A. Macovski, S. Conolly, Rapid polarizing field cycling in magnetic resonance imaging. IEEE Trans. Med. Imaging **25**(1), 84–93 (2006)
7. P. Ross, L. Broche, D. Lurie, Rapid field-cycling MRI using fast spin-echo. Magn. Reson. Med. **73**(3), 1120–1124 (2015)
8. J. Pine, R. Davies, J. Lurie, Field-cycling NMR relaxometry with spatial selection. Magn. Reson. Med. **63**(6), 1698–1702 (2010)
9. J. Lurie et al., Fast field-cycling magnetic resonance imaging - Imagerie de resonance magnétique en champ cyclé, Multiscale NMR and relaxation/RMN et relaxation multi-échelles. C. R. Phys. **11**(2), 136–148 (2010)
10. F. Bonetto, E. Anoardo, Proton field-cycling nuclear magnetic resonance relaxometry in the smectic Amesophase of thermotropic cyanobiphenyls: effects of sonication. J. Chem. Phys. **121**(1), 554–561 (2004)
11. J. Lurie, R. Davies, A. Foster, J. Hutchison, Field-cycled PEDRI imaging of free radicals with detection at 450 mT. Magn. Reson. Imaging **23**(2), 175–181 (2005)
12. M. Sousa, P. Fernandes, G. Marques, A. Ribeiro, P. Sebastião, Novel pulsed switched power supply for a fast field cycling NMR spectrometer. Solid State Nucl. Magn. Reson. **25**(1–3), 160–166 (2004)
13. D. Sousa, G. Marques, J. Cascais, P. Sebastião, Desktop fast-field cycling nuclear magnetic resonance relaxometer. Solid State Nucl. Magn. Reson. **38**(1), 36–43 (2010)
14. A. Roque, D. Sousa, E. Margato, J. Maia, Current source of a FFC NMR relaxometer linearly controlled, in *EPE13-15th European Conference on Power Electronics and Applications*, Sept 2013
15. D. Sousa, G. Marques, P. Sebastiao, A. Ribeiro, New isolated gate bipolar transistor two-quadrant chopper power supply for a fast field cycling NMR spectrometer. Rev. Sci. Instrum. **74**(3), 4521–4528 (2003)
16. R. Seitter, R. Kimmich, Magnetic resonance: relaxometers, in *Encyclopedia of Spectroscopy and Spectrometry*, (Academic Press, London, 1999), pp. 2000–2008
17. J. Constantin, J. Zajicek, F. Brown, Fast field-cycling nuclear magnetic resonance spectrometer. Rev. Sci. Instrum. **67**(6), 2113–2122 (1996)

18. E. Rommel, K. Mischker, G. Osswald, K. Schweikert, F. Noack, A powerful NMR field-cycling device using GTO's and MOSFET's for relaxation dispersion and zero-field studies. J. Magn. Reson. **70**(2), 219–234 (1986)
19. D. Sousa, E. Rommel, J. Santana, F. Silva, P. Sebastião, A. Ribeiro, Power supply for a fast field cycling NMR spectrometer using IGBTs operating in the active zone, in *7th European Conference on Power Electronics and Applications (EPE97)*, Trondheim, Norway, pp. 2.285–2.290, 1997
20. A. Roque, D. Sousa, E. Margato, J. Maia, G. Marques, Power source of a relaxometer - topology and linear control of the current, in *Annual Seminar on Automation, Industrial Electronics and Instrumentation (SAAEI'13)*, Madrid – Espanha, 2013
21. K. Gilbert et al., Design of field-cycled magnetic resonance systems for small animal imaging. Phys. Med. Biol. **51**(11), 2825–2841 (2006)
22. G. Franklin, J. Powell, A. Emami-Naeini, *Feedback Control of Dynamic Systems* (Pearson Education, Upper Saddle River, NJ, 2011)
23. K. Ogata, *Modern Control Engineering, Instrumentation and Controls Series* (Prentice Hall, Boston, 2010)
24. A. Redfield, W. Fite, H. Bleich, Precision high speed current regulators for occasionally switched inductive loads. Rev. Sci. Instrum. **39**(5), 710–715 (1968)
25. I. Nathaniel et al., Noise performance of a precision pulsed electromagnet power supply for magnetic resonance imaging. IEEE Trans. Med. Imaging **27**(1), 75–86 (2008)
26. A. Roque, S. Pinto, J. Santana, D. Sousa, E. Margato, J. Maia, Dynamic behavior of two power supplies for FFC NMR relaxometers, in *IEEE International Conference on Industrial Technology – ICIT*, Athens, Greece, 2012

Blended Antilock Braking System Control Method for All-Wheel Drive Electric Sport Utility Vehicle

Andrei Aksjonov, Valery Vodovozov, Klaus Augsburg, and Eduard Petlenkov

Abstract At least two different actuators work in cooperation in regenerative braking for electric and hybrid vehicles. Torque blending is an important area, which is responsible for better manoeuvrability, reduced braking distance, improved riding comfort, etc. In this paper, a control method for electric vehicle blended antilock braking system based on fuzzy logic is promoted. The principle prioritizes usage of electric motor actuators to maximize recuperation energy during deceleration process. Moreover, for supreme efficiency it considers the battery's state of charge for switching between electric motor and conventional electrohydraulic brakes. To demonstrate the functionality of the controller under changing dynamic conditions, a hardware-in-the-loop simulation with real electrohydraulic brakes test bed is utilized. In particular, the experiment is designed to exceed the state-of-charge threshold during braking operation, what leads to immediate switch between regenerative and friction brake modes.

1 Introduction

One of the advantageous features of the electric vehicles (EVs) is their ability to recuperate energy during a deceleration process. In EVs, friction braking (FB) cooperates with regenerative braking (RB), which opens a need to efficient torque control between two separate actuators (i.e. torque blending), which are characterized by different dynamics. In some cases, RB is simply not enough to achieve the requested

A. Aksjonov (✉)
ŠKODA Auto a.s., Mladá Boleslav, Czech Republic
e-mail: andrei.aksjonov@outlook.com; andrei.aksjonov@aalto.fi

V. Vodovozov · E. Petlenkov
Tallinn University of Technology, Tallinn, Estonia
e-mail: valery.vodovozov@taltech.ee; eduard.petlenkov@taltech.ee

K. Augsburg
Technische Universität Ilmenau, Ilmenau, Germany
e-mail: klaus.augsburg@tu-ilmenau.de

W. Zamboni, G. Petrone (eds.), *ELECTRIMACS 2019*, Lecture Notes
in Electrical Engineering 615, https://doi.org/10.1007/978-3-030-37161-6_17

braking torque; therefore, the FB system is activated in parallel or in series. In other cases, the battery conditions (e.g. temperature, battery's state of charge (SOC)) must be considered. For instance, when the battery is fully charged, the recuperation is no longer useful and even dangerous [1].

The SOC is a ratio of the remaining battery capacity to the fully charged one. It is one of the most important parameters in EVs. Its feature is used not only in battery management to estimate potential driving range before the next recharge but also in vehicle traction (e.g. hybrid EV) and braking (e.g. blended braking system) control strategies [1]. For example, to avoid electric battery overcharge, and consequent damage, the regeneration by electric motors is usually limited to a specific upper bound, 80–90% [2]. Therefore, the SOC must be always involved in a blended antilock braking system (ABS).

Nowadays, the fuzzy logic controllers (FLCs) are widely used in automotive engineering to solve various problems [3]. For instance, in [4], the effectiveness and strong robustness of a fuzzy sliding mode control over conventional proportional-integral-derivative (PID) and Mamdani's type FLC in energy recuperation for EV in simulation environment was demonstrated. The complexity of vehicle dynamics in deceleration process, especially during emergency braking, was not integrated in the study.

An FLC-based RB strategy integrated with series RB was developed in [5]. The FLC received the driver's force command, vehicle speed, battery's SOC and the temperature to determine distribution between FB and RB to improve energy recuperation efficiency. In [6], the FLC involved SOC and a ratio between the brake torque and the biggest brake torque to determine the factual FB and RB brake torques. An RB control strategy applying FLC was presented in [7]. The simulation results demonstrated that the developed method is able to recover energy and distribute power flow to maintain SOC around target value. A PID in combination with FLC ensured efficient RB strategy of the EV [8]. The SOC was taken as an input of the FLC. Despite impressive results, all these works only focused on the base brake case. The ABS function was not considered.

In [9], the authors applied genetic algorithm in EV stability control logic using RB of the rear wheels motor and FB of electrohydraulic brake (EHB). The simulation results showed that the optimal recuperation strategy is able to provide an increase in recuperation energy. However, neither SOC in torque allocation nor ABS performance were under investigation. The brake force distribution strategy for EVs based on estimation of tire–road friction coefficient was provided in [10]. The road condition estimation was also based on fuzzy theory. An efficient torque blending was demonstrated in [11]. The experiment was conducted on a real vehicle braking on low-friction road surface. In [12], the FLC was used to adjust braking torque between RB and FB. However, in these works torque blending or force distribution did not consider the SOC of a battery.

Scholars in [13] integrated sliding mode controller with FLC for an ABS control to maintain optimal wheel slip ratio deceleration. The SOC was reckoned in torque blending in this instant. However, the ABS controller inputed fixed slip ratio. Thus, wheel slip adaptability to various road conditions was not considered. Advanced control allocation with energy recuperation for EV was introduced in [14]. The authors also involved the battery's SOC. Both works did not study the situation when SOC exceeds its bound during braking manoeuvre.

In [15], the authors proposed the EV torque blending with recuperation capabilities taking into account the SOC. It was integrated with three types of controllers, namely PID, tabular, and FLC. In this paper, the attention is once again focused on the SOC's influence on the EV's blended ABS. To this aim, the intelligent FLC control method previously developed by the authors [16] was applied in the hardware-in-the-loop (HIL) simulation with real EHB system. The HIL testbed accompanied with a hardware delay is exploited to represent actual EHB dynamics, making the simulation experiment more valued for real-life application. The deceleration test is designed in a way that SOC reaches its maximum threshold in the middle of the braking process. Consequently, blended braking system rapidly switches from RB to FB.

In RB, the recovered energy is not stored directly in the battery, but in the ultracapacitor. From the latter, the recuperated energy is transmitted slowly to the battery or is used for vehicle acceleration. Thus, the SOC shall also consider the capacity of an ultracapacitor. Furthermore, the electronic power converters play an essential role in energy recuperation in EVs, as they are an intermediate connection between energy sources and motors. In this paper, it is assumed that the energy is transferred directly to the battery; thus, the SOC may surpass its maximum limit during the EV deceleration. However, the power electronics losses are neglected in the powertrain model. Nevertheless, possible consideration of the SOC of ultracapacitor in torque blending was also proposed by the authors in [15].

This paper is organized as follows. The next section (Sect. 2) stresses the HIL simulation environment together with vehicle modelling. Section 3 describes the blended ABS control method. In Sect. 4, the HIL simulation results are delivered. A brief conclusion is provided in Sect. 5.

2 Vehicle Model and Experimental Set-Up

2.1 Single-Wheel Model

A simplified schematic single-wheel brake diagram is presented in Fig. 1. The rolling resistance and lateral dynamics are neglected, because only the straight braking manoeuvre was studied in this work. The torque balance about a wheel axis is expressed as:

$$J_W \dot{\omega}_W = T_d - r_W \cdot F_x - T_b, \qquad (1)$$

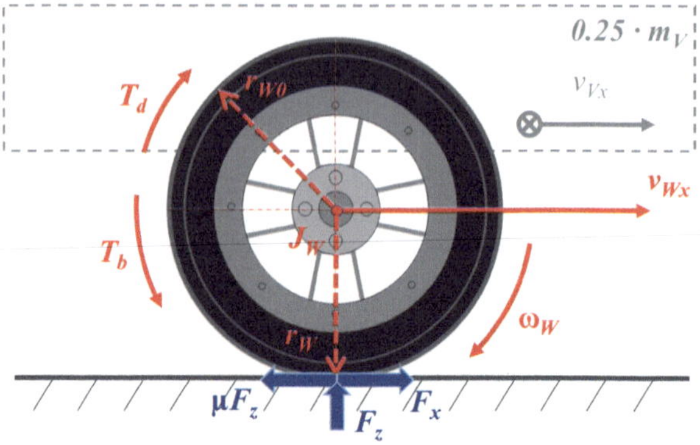

Fig. 1 A schematic drawing of a braked wheel for a single-wheel model

where J_{w}—moment of inertia of wheel; ω_W—angular velocity of wheel; T_{b}—braking torque; T_{d}—driving torque; r_{w}—radius of deformed tire; F_x—longitudinal force of tire.

A distinctive feature of the EV: its braking torque T_{b} is a summation of the RB T_{RB} and FB T_{FB} braking torques [16]:

$$T_{\mathrm{b}} = T_{\mathrm{FB}} + T_{\mathrm{RB}}. \tag{2}$$

In practice, T_{RB} and T_{FB} are not measured by the sensors directly. They change proportionally to phase current of a switched reluctance motor (SRM) and line pressure of an EHB, accordingly. Those states are measured by on-board sensors available in modern vehicles. In this paper, the variables are represented as torques directly.

2.2 State Estimation

An essential characteristic of an ABS is tire–road friction coefficient μ. Straight direction braking manoeuvre neglects lateral dynamics; hence, μ is calculated as a ratio of longitudinal F_x and normal F_z forces:

$$\mu = \frac{F_x}{F_z}. \tag{3}$$

The proposed control method uses μ to understand the road surface under the tires of the EV. In this regard, μ is assumed to be proportional to the EV body deceleration rate [16]:

$$\mu^* = \frac{F_x}{F_z} = \frac{m_{\mathrm{V}} \cdot a_{\mathrm{V}x}}{m_{\mathrm{V}} \cdot g} = \frac{a_{\mathrm{V}x}}{g}, \tag{4}$$

where m_V—mass of vehicle; a_{Vx}—longitudinal acceleration of vehicle; g—gravitational acceleration.

The maximum vehicle deceleration achieved during the first period of heavy braking manoeuvre is related to road surface conditions and is used as the road recognizer in the proposed control method. The variable is expressed as μ^* [16].

Another important state for the control method is longitudinal wheel slip λ, which is estimated from vehicle v_{Vx} and wheel v_{Wx} longitudinal velocities:

$$\lambda = \frac{v_{Wx} - v_{Vx}}{v_{Vx}} \cdot 100\%. \tag{5}$$

Vehicle longitudinal velocity is derived from the vehicle body deceleration signal:

$$v_{Vx} = \int a_{Vx} \mathrm{d}t. \tag{6}$$

Wheel longitudinal speed is found as:

$$v_{Wx} = r_W \cdot \omega_W. \tag{7}$$

2.3 Electric Vehicle Model

The EV model is completed in IPG CarMaker® 6.0 (Germany) software. The 14 degrees-of-freedom model is interacted with MATLAB® from MathWorks, Inc. (USA), allowing users for rapid control algorithm development and testing. The software's integration in the HIL systems opens a great possibility for advance prototypes testing and concepts engineering, which sensitively saves development time and cost.

The sport utility EV model with all-wheel drive powertrain represents a vehicle under investigation. The specifications of the vehicle parameterizations are provided by the EV's manufacturer (e.g. mass, dimensions, electric propulsion system) or are collected experimentally (e.g. suspension, tire model).

Each of the four wheels is equipped with SRM. In-wheel motor transmission type is a two-stage reducer with helical gear and half-shaft. Considering the SRM's peak torque (i.e. 200 Nm at 800 V) together with overall SRM–gear ratio (i.e. 1:10.5), the maximum torque achieved on single wheel reaches 2100 Nm. The motor's behaviour is defined by the first-order transfer function. More information about vehicle model together with its parameters is available in [16].

The braking linings' coefficient of friction was modelled by means of a dynamic model, which was validated against data collected on the brake dynamometric test bed at Technische Universität Ilmenau (Germany). This model considers the influence of speed, pressure, and temperature on the brake linings' coefficient of friction [17].

The tire dynamics are approximated with Pacejka's "Magic Formula" with experimentally obtained coefficients. The tire–road model is a relevant element for the control method design. Particularly, it is important to know the most efficient workspace for the λ with various road surfaces. Deceleration with the optimal λ results in maximum braking manoeuvre efficiency that impacts the deceleration distance. Moreover, when the wheel slip is equal or smaller than its corresponding peak (so-called stable region), the EV presumes steerability. On the contrary, deceleration with the λ exceeding its optimal one (i.e. unstable zone) leads to wheels' lockage and lateral control aggravation. The ABS's task is to avoid wheel slip unstable region.

2.4 Electrohydraulic Brake System Test Bed

The EHB with control unit test bed (Fig. 2) was provided by Technische Universität Ilmenau. The test rig is developed by the ZF TRW Automotive GmbH (Germany). The EHB set-up is used in vehicle braking dynamics studies for reproduction of the real pressure dynamics of the brake circuit.

The vehicle model sends demanded braking pressure for each wheel to the EHB control unit. The dSPACE® (Germany) platform is utilized as an intermediate

Fig. 2 Electrohydraulic brake system test bed

connection between the vehicle numerical model and the EHB. The requested braking pressure received from the vehicle model activates the valves that generate corresponding braking pressure between the wheels and the calipers. Finally, after measuring with the appropriate sensors, the line braking pressure on each wheel is returned to the vehicle model.

3 Blended Anti-lock Braking System Control Method

3.1 Control Method

The ABS control method supplies an appropriate braking torque to decelerate the vehicle with optimal wheel slip for each wheel. Different road surfaces are taken into account in control method design. A detailed description of control method and its design can be found in [16]. Only a brief introduction is presented here.

The control area network bus provides vehicle longitudinal deceleration together with wheel velocity (Fig. 3). In the next steps, applying Eqs. (4)–(7), two commanded variables, λ and μ^*, are obtained. They are used by the FLCs to generate the required torque for the actuators.

Two FLCs are designed separately for SRM and EHB control for each wheel. The fuzzy system accepts the information about vehicle body deceleration during the first step of heavy braking. Its maximum value is fed as a constant crisp input to the FLC to recognize road surface. This crisp input is uncertain. Hence, computational intelligence methods, such as a fuzzy set theory, are capable of dealing with such ill-defined and vague data. Thanks to the method's robustness, precise mathematical modelling may be avoided. The second input is wheel slip, which is used to decide whether to decrease or increase the requested torque $T^{(\mathrm{req})}$.

Both the FLC inputs have symmetrically dispersed over the whole universe of discourse triangular membership functions, five for λ and seven for μ^*. Overlapping membership functions ensure equal sensitivity of the inputs. The λ is bounded in [0 18] and μ^*—in [0 10]. Sugeno's inference method was exploited in this study. In Table 1, the rule base for the front and rear wheels in regenerative braking mode is provided. Considering the motor's peak torque limits, the output torque for the SRM has 11 linguistic values from 0 to 200 Nm. The T_{FB} is between 0 and 150 bar and has the same design principle as the RB (Table 1). The FLCs are designed based on the given tire model and the expert's knowledge concerning efficient plant control.

The modus ponens rules (**If** premise **Then** consequence) are the common expression of fuzzy input–output fit. As done in this work, the tabular representation is often acquired by trial and error. The main criterion for Table 1 design is to achieve wheel slip for each tire as close as possible to the optimal one. An example of input–output linguistic mapping is as follows: **If** wheel "*slip is 3% (S3)*" and road surface is "*Dry*", **Then** request from the SRM "*200*" Nm for the front wheels and "*120*"

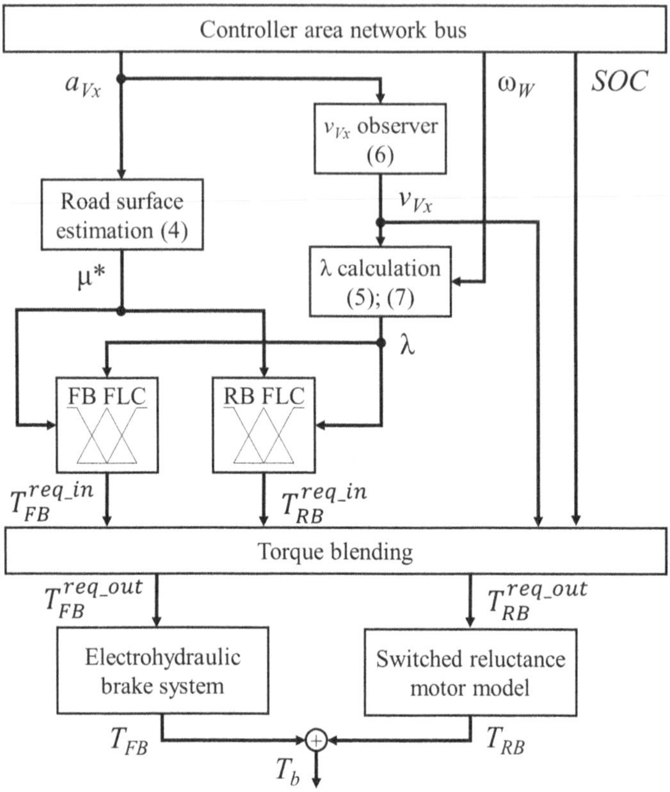

Fig. 3 Control block scheme for a single wheel of the EV: superscript req stands for "requested"; *RB FLC* regenerative braking fuzzy logic controller; *FB FLC* friction braking fuzzy logic controller

Table 1 FLC rule base for front/rear wheels in regenerative mode

T_{RB} (Nm)		μ^* Zero	Icy	Wet	Damp	Dry
λ (%)	S_0	60	80	160	200/120	200/140
	S_3	40	60	140	200/100	200/120
	S_6	20	40	120	200/80	200/100
	S_9	0	20	100	180/40	200/80
	S_{12}	0	0	60	160/20	200/40
	S_{15}	0	0	20	140/0	180/20
	S_{18}	0	0	0	120/0	160/0

Nm for the rear wheels. The final step is to translate the output linguistic variables back to crisp numbers. To achieve this, the center of gravity method is considered.

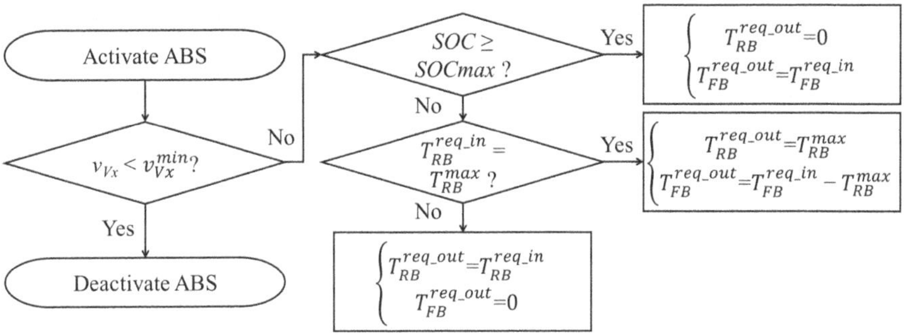

Fig. 4 Control flowchart of torque blending for a single wheel

3.2 *Torque Blending*

Torque blending is realized with simple logic rules. It requires several inputs, in particular, requested input RB and FB torques $T_{\mathrm{RB}}^{\mathrm{req\_in}}$ and $T_{\mathrm{FB}}^{\mathrm{req\_in}}$, vehicle longitudinal velocity $v_{\mathrm{V}x}$, and SOC of the battery SOC. Torque blending block outputs are requested RB torque $T_{\mathrm{RB}}^{\mathrm{req\_out}}$ for the SRM and requested FB torque $T_{\mathrm{FB}}^{\mathrm{req\_out}}$ for the EHB (Fig. 3). The approach flowchart is presented in Fig. 4. It is developed to prioritize the usage of the SRMs, yet without battery damage due to overcharge.

Firstly, the algorithm checks the velocity of the vehicle. When vehicle longitudinal speed is slower than a desired minimum threshold $v_{\mathrm{V}x}{}^{\min}$ (typically 8–15 km/h), the ABS control is deactivated, because the distance travelled with very low speed with locked wheels is not critical.

Secondly, when the SOC reaches the maximum allowed threshold $\mathrm{SOC}_{\max}$ (e.g. 90%), the braking switches to pure FB mode, where the torque for the SRM is equal to zero:

$$\begin{cases} T_{\mathrm{RB}}^{\mathrm{req\_out}} = 0 \\ T_{\mathrm{FB}}^{\mathrm{req\_out}} = T_{\mathrm{FB}}^{\mathrm{req\_in}} \end{cases}. \tag{8}$$

Thirdly, the blended ABS considers the SRM's peak performance. Specifically, when peak torque $T_{\mathrm{RB}}^{\max}$ of the SRM is requested by the FLC, the block supplies the peak torque request to the SRM and calculates additional torque for the FB actuator to ensure optimal λ deceleration as:

$$\begin{cases} T_{\mathrm{RB}}^{\mathrm{req\_out}} = T_{\mathrm{RB}}^{\max} \\ T_{\mathrm{FB}}^{\mathrm{req\_out}} = T_{\mathrm{FB}}^{\mathrm{req\_in}} - T_{\mathrm{RB}}^{\max} \end{cases}. \tag{9}$$

Finally, when none of the previous conditions are true, the EV decelerates only with SRMs as the ABS actuators:

$$\begin{cases} T_{\mathrm{RB}}^{\mathrm{req\_out}} = T_{\mathrm{RB}}^{\mathrm{req\_in}} \\ T_{\mathrm{FB}}^{\mathrm{req\_out}} = 0 \end{cases}. \tag{10}$$

4 Results

The results of the EV heavy braking with activated ABS on a dry asphalt road ($\mu \approx 1$) is presented in Fig. 5. At the experiment time 3.5 s, the SOC of the EV's battery is assumed to exceed its upper bound. At this moment, the torque blending control is easily noticeable, because the EV switches from RB to pure FB mode. Energy recuperation is no longer conducted. The vehicle speed the wheels' speeds are presented in Fig. 5a.

Before the controller intervention, the wheels' slip rates grow due to exceeded torque requested by the driver (i.e. the braking pedal is instantly pressed to its maximum) (Fig. 5b). Nevertheless, after the ABS activation, the wheel slip rates drop down to their optimal values for a given road surface. The optimal slips for every wheel are also depicted.

In Fig. 5c, the road surface estimation method is scoped. At the beginning of the braking manoeuvre, the controller measures the maximum deceleration rate of the EV body. Its peak value is mapped with an appropriate road surface. The blue line symbolizes the crisp input for the estimation road conditions. The variable together with wheel slip is therefore processed by the FLC to estimate the relevant braking torques.

In Fig. 5d, braking torques for each in-wheel SRM of the EV are presented. Until the SOC makes an impact on blended ABS, it is seen that the SRM supplies its maximum available torque for the front wheels. As a result, the torque blending requests additional torque from the EHB (Fig. 5e) to lead λ as close as possible to their theoretical optimal values in accordance to Eq. (9). For the rear wheels, however, the generated torques by the SRMs (Fig. 5d) are enough to reach optimal rate. Thus, the FB torques are not required (Fig. 5e) as stated by Eq. (10).

When the SOC overshoot steps in (i.e. $t = 3.5$ s), the energy regeneration stops, and the SRMs are not used as braking actuators any more (Fig. 5d). Consequently, the RB torques for all wheels drop to zero. On the contrary, the system moves to the pure FB mode. Now, only the EHB's torques are applied to decelerate the transport (Fig. 5e), applying Eq. (8).

Moreover, the FB torques are not as smooth as RB ones. Furthermore, the optimal wheel slip achievability is not as precise as in the case when the SRMs affect vehicle deceleration (Fig. 5b). This phenomenon is mainly due to the EHB's significant delay as well as the plant complexity (i.e. the wheel tire's highly non-linear behaviour). It was also studied by the authors in the previous work [16].

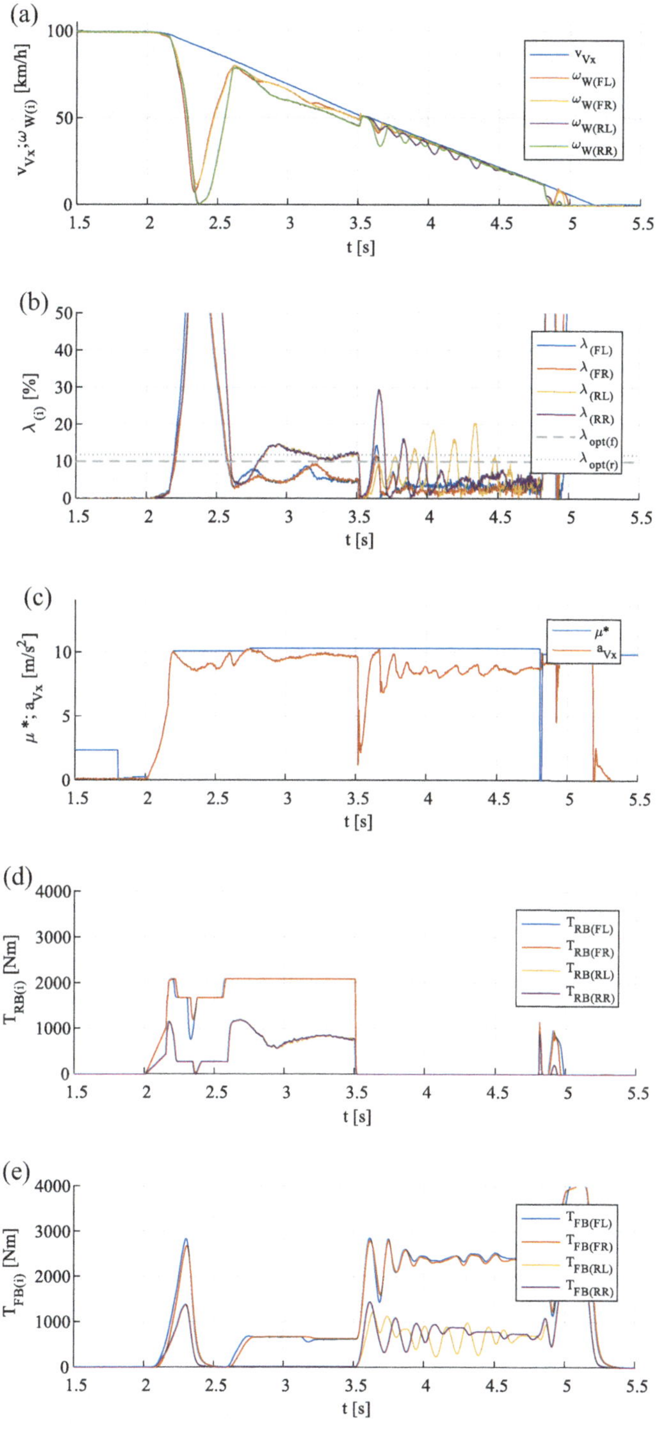

Fig. 5 Experimental results from vehicle braking on a high-μ ($\mu \approx 1$) road surface: (**a**) vehicle and wheels speeds; (**b**) wheels longitudinal slips; (**c**) road recognition with vehicle body deceleration rate (**d**) RB torques; (**e**) FB torques; (*FL* front left, *FR* front right, *RL* rear left, *RR* rear right)

The conclusion was made that, thanks to the electric drives' fast response, the control method accomplishes fast and more accurate control. As a result, the EV's RB braking affords noticeably more efficient EV deceleration performance under the ABS operation. It allows for vehicle stopping distance diminishment. However, the electric motors are not always fully available as the braking systems, as for instance in the case of deceleration on high-μ surfaces or in case with the exceeded SOC threshold.

5 Conclusions

In this paper, the blended ABS control method for all-wheel drive sport utility EV is described. The SOC and the requested torques from both actuators, namely in-wheel SRMs and EHB, are taken into consideration for providing a sufficient braking torque to presume maximum deceleration efficiency for every wheel independently. The efficiency is guaranteed by the optimal wheel slip ratio braking for each wheel. In combination with the intelligent FLC, the blended ABS control method provides high efficiency and robustness against varying road conditions and changing system states.

The proposed solution is verified against HIL simulation, in which the experimentally validated EV is coupled with EHB test bed, which provides real brakes dynamics followed by significant hardware delay. The presented experimental results are dedicated to heavy braking conditions on a high-μ road surface, during which the upper SOC threshold is achieved. As a result, the blended ABS switches from the RB mode to the pure FB, and the vehicle continues the deceleration process with optimal slip ratio without motors' impact and performance degradation.

Despite good optimal slip control for various road surfaces, the proposed FLC-based control method has several limitations. Firstly, the method depends on tire model, whose behaviour is also different for other types of vehicles characterized by different centres of gravity, masses, etc. Hence, for other vehicle and tire types, the control method must be slightly modified. Secondly, to design an FLC for complex control system, like ABS, an expert's multidisciplinary knowledge is essential.

Acknowledgements This project has received funding from the European Union's Horizon 2020 research and innovation program under grant agreement No. 675999 and from the Estonian Research Council grant No. PRG658.

References

1. M. Ehsani, Y. Gao, S.E. Gay, A. Emadi, *Modern Electric, Hybrid Electric, and Fuel Cell Vehicles* (CRC Press, Boca Raton, FL, 2005)
2. A. Emadi, *Handbook of Automotive Power Electronics and Motor Drives* (Taylor and Francis Group, Boca Raton, FL, 2005)

3. V. Ivanov, A review of fuzzy methods in automotive engineering applications. Eur. Transp. Res. Rev. **7**, 1–10 (2015)
4. X. Zhang, Y. Wang, G. Liu, X. Yuan, Robust regenerative charging control based on T–S fuzzy sliding-mode approach for advanced electric vehicle. IEEE Trans. Transp. Electrif. **2**, 52–65 (2016)
5. G. Xu, W. Li, K. Xu, Z. Song, An intelligent regenerative braking strategy for electric vehicles. Energies **4**, 1461–1477 (2011)
6. J. Zhang, B. Song, S. Cui, J. Zhang, D. Ren, Fuzzy logic approach to regenerative braking system, in *2009 Intern. Conf. on Intel. Human–Machine Syst. and Cybern.*, (2009), pp. 451–454
7. X. Li, L. Xu, L. Hua, J. Li, M. Ouyang, Regenerative braking control strategy for fuel cell hybrid vehicles using fuzzy logic, in *2008 Intern. Conf. on Elect. Machines and Syst.*, (2008), pp. 2712–2716
8. X. Nian, F. Peng, H. Zhang, Regenerative braking system of electric vehicle driven by brushless DC motor. IEEE Trans. Ind. Elect. **61**, 5798–5808 (2014)
9. D.-H. Kim, J.-M. Kim, S.-H. Hwang, H.-S. Kim, Optimal brake torque distribution for a four-wheel-drive hybrid electric vehicle stability enhancement. Proc. Inst. Mech. Eng. D J. Automob. Eng. **221**, 1357–1366 (2007)
10. D. Paul, E. Velenis, D. Cao, T. Dobo, Otpimal μ-estimation based regenerative braking strategy for an AWD HEV. IEEE Trans. Transp. Electrif. **3**, 249–258 (2017)
11. D. Savitski, V. Ivanov, B. Shyrokau, T. Pütz, J. de Smet, J. Theunissen, Experimental investigation on continuous regenerative anti-lock braking system of full electric vehicle. Int. J. Automotive Technol. **17**, 327–338 (2016)
12. P. Dong, Z. Jianwu, Y. Chengliang, Regenerative braking control system improvement for parallel hybrid electric vehicle, in *Intern. Technol. Innov. Conf. 2016 (ITIC 2006)*, (2006), pp. 1902–1908
13. J. Guo, X. Jian, G. Lin, Performance evaluation of an anti-lock braking system for electric vehicles with a fuzzy sliding mode controller. Energies **7**, 6459–6476 (2014)
14. B. Shyrokau, D. Wang, D. Savitsky, V. Ivanov, Vehicle dynamics control with energy recuperation based on control allocation for independent wheel motors and brake system. Int. J. Powertrain **2**, 153–181 (2013)
15. A. Aksjonov, V. Vodovozov, Z. Raud, Improving energy recovery in blended antilock braking systems of electric vehicles, in *16th Inter. Conf. of Indust. Informat. (INDIN'2018)*, (2018), pp. 589–594
16. A. Aksjonov, V. Vodovozov, K. Augsburg, E. Petlenkov, Design of regenerative anti-lock braking system controller for 4-in-wheel-motor drive electric vehicle with road surface estimation. Int. J. Automotive Technol. **19**, 727–742 (2018)
17. V. Ricciardi, D. Savitski, K. Augsburg, V. Ivanov, Estimation of brake friction coefficient for blending function of base braking control. SAE Int. J. Passeng. Cars – Mech. Syst. **10**, 774–785 (2017)

Voltage Boost by Neutral Point Supply of AC Machine

Jean-Yves Gauthier and Xuefang Lin-Shi

Abstract A new concept of AC motor drive is proposed in this paper. It allows to boost voltage of an AC machine without supplementary components. The main idea is to wisely connect the neutral point of the AC machine to the DC power supply. With some modifications on the control algorithm, the proposed solution allows to supply the AC machine with a higher voltage than with classical inverters. The concept is general for different AC machines and different topologies of inverters. The case study of an induction motor driven by a three-phase two-level inverter is illustrated. In steady state, a factor gain up of 1.7 of maximum RMS voltage can be obtained with the proposed solution compared to a classical scheme. Experimental validation on an induction motor test bench shows the effectiveness of the proposed concept.

1 Introduction

The systems non-connected to the grid such as embedded systems, electric vehicles, or micro-grids need the use of energy storage systems (battery, super-capacitor, etc.). For AC machines, such as synchronous motors, induction motors, or variable reluctance motors, the output power depends on the supply level delivered by the energy storage systems. The most classical scheme for an AC machine drive is the use of a three-phase two-level inverter as described on Fig. 1 because of its simplicity and robustness. However, one of its drawbacks is that the required DC voltage should be a sufficiently high value to supply the machine to obtain the nominal AC voltage. Unfortunately, due to constraints of cost, size, or safety, the supply level of the energy storage systems is limited. Moreover, with the classical

J.-Y. Gauthier (✉) · X. Lin-Shi
Université de Lyon – INSA Lyon – CNRS – Laboratoire Ampère, Villeurbanne, France
e-mail: jean-yves.gauthier@insa-lyon.fr; xuefang.shi@insa-lyon.fr

W. Zamboni, G. Petrone (eds.), *ELECTRIMACS 2019*, Lecture Notes in Electrical Engineering 615, https://doi.org/10.1007/978-3-030-37161-6_18

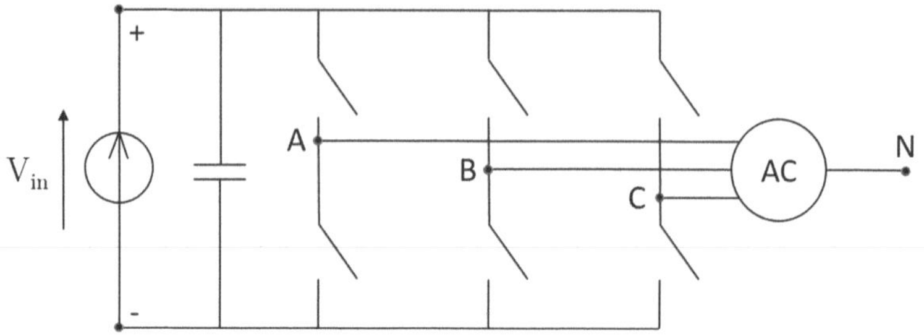

Fig. 1 Classical scheme for supplying AC machine by using an inverter. The neutral point of the AC machine is not connected to supply

scheme, DC supply voltage is fixed. This induces a very low modulation index at low speed. Indeed, at low speed, the required AC voltage for the motor is low too. So the voltage viewed by the switches (IGBT or MOSFET) is high (i.e., the DC voltage) but the generated AC voltage is low. How to obtain the nominal AC voltage power with a limited lower supply and how to adjust the supply voltage according to operation conditions are two interesting issues for AC machine drives. Generally, a DC/DC converter between the DC supply and the inverter can be used (Fig. 4). The DC/DC converter can boost the DC supply on the one hand, and can adjust the DC voltage in front of the inverter on the other hand. The latter can reduce the voltage of the switches when a low voltage is required on the machine. However, this solution requires to add other passive and semiconductor elements for the DC/DC converter. Therefore the size and the cost will increase and the robustness of the whole system can decrease.

In this paper, we propose a new and simple way to increase or adjust the DC voltage by using the stator neutral point of AC motors. The use of the neutral point of an AC machine when it was star-connected can be found in many cases for fault-tolerant control applications and others [1, 2]. The most famous application is, for example, the addition of a fourth leg in the inverter to control the voltage value of the neutral point [3]. In each of these papers, some added semiconductor devices can be found. In [4, 5], a voltage source is moreover added in series in the fourth leg. This second DC power supply is added as a battery.

Different from existing works, the new concept proposed in this paper is to supply the stator neutral point of AC motor to increase the DC voltage without adding passive or semiconductor components. The paper is organized as follows. In Sect. 2, the main idea and scheme of the proposed concept are described. In Sect. 3, the modelling of an AC machine with the proposed concept is presented. The possibility to control the DC voltage is underlined. A steady-state behavior is studied to show the gain compared to a classical star-connected machine configuration. In Sect. 4, the proposed control scheme is given followed by current analysis. Finally, experimental validation of the proposed concept on an induction motor is illustrated in Sect. 5.

2 Main Idea and Scheme

2.1 Scheme of the Proposed Concept

As underlined before, the approach proposed in this paper is generalizable for different topologies of inverters. For the sake of simplicity, a three-phase two-level inverter will be considered in the following. Figure 1 presents the classical way to supply an AC machine: a DC voltage source V_{in} supplies an inverter which is connected to an AC machine. The neutral point of the AC machine is disconnected.

On Fig. 2, the proposed scheme is shown: the DC power source is directly connected between the neutral point of the machine and the lowest voltage point ($-$) of the inverter. The capacitor is still connected. Note that the highest voltage point of the inverter ($+$) is not connected to the power supply. In the following, the name of neutral point supplied (NPS) will be used to refer to this scheme.

2.2 Power Flow

With NPS, the power can flow from the source V_{in} through the AC machine coils in order to charge the capacitor and to maintain its voltage V_{DC}. By only considering the inductors of the machine, it looks like a multi-phase DC/DC boost converter. An additional DC current component is flowing through each of the three phases of the machine corresponding to a third of the DC current delivered by the source. A controlled capacitor voltage can therefore be reached by tuning duty cycles of the inverter acting as a multi-phase boost converter.

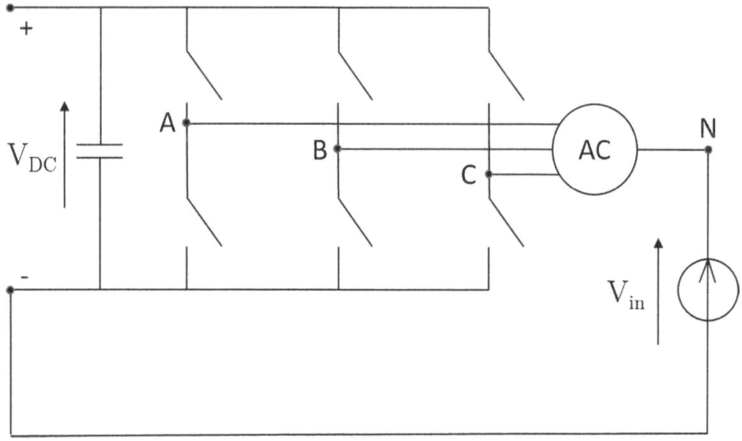

Fig. 2 Proposed Neutral Point Supply (NPS) scheme. The DC (+) point is not connected

Then the classical AC motor controls can be applied to supply the AC machine from the capacitor voltage V_{DC} leading to AC currents in the machine.

In a first analysis, the main advantage of the proposed scheme is the voltage boost effect between the DC source and the capacitor voltage (see Sect. 3). Moreover, the capacitor voltage can be controlled and tuned depending on the operating point of the machine.

3 AC Machine Drive Modelling and Steady-State Study

3.1 AC Machine Drive Modelling

By considering a simple model of an AC machine (with phase back-electromotive forces E_k, resistor R, and inductor L), the relation between phase voltage V_{kN} and phase current i_k is as follows:

$$V_{kN}(t) = L\frac{di_k(t)}{dt} + R \cdot i_k(t) + E_k(t) \quad k \in \{A, B, C\} \tag{1}$$

As the machine is assumed to be balanced which is generally the case with healthy machines:

$$E_A(t) + E_B(t) + E_C(t) = 0 \tag{2}$$

From the inverter point of view, an average model is considered (i.e., the switching frequency is sufficiently high to neglect the current ripple). The expressions linking duty cycles α_k, capacitor voltage V_{DC}, source voltage V_{in}, and phase voltages are thus:

$$\alpha_k(t)V_{DC}(t) = V_{kN}(t) + V_{in} \quad k \in \{A, B, C\} \tag{3}$$

Therefore:

$$\begin{aligned}&(\alpha_A(t) + \alpha_B(t) + \alpha_C(t))\, V_{DC}(t)\\ &= L\frac{d(i_A + i_B + i_C)}{dt} + R(i_A(t) + i_B(t) + i_C(t)) + 3V_{in}\end{aligned} \tag{4}$$

By defining homopolar current $i_H(t) = \frac{1}{3}(i_A(t)+i_B(t)+i_C(t))$ and homopolar duty cycle $\alpha_H(t) = \frac{1}{3}(\alpha_A(t) + \alpha_B(t) + \alpha_C(t))$, the following equation can be obtained:

$$\alpha_H(t)V_{DC}(t) = L\frac{di_H}{dt} + Ri_H(t) + V_{in} \tag{5}$$

This expression corresponds to the average model of a DC/DC converter. In our case, the voltage source is V_{in} and the output voltage is V_{DC}: the behavior is there related to a boost converter. Note that in a real boost converter, the duty cycle corresponds to the duty cycle of the bottom switch in the leg of the converter. But in this paper the considered duty cycle is related to the top switch of the leg because it is the classic approach in inverter modelling.

3.2 Capacitor Voltage Control

A simple way to control this system is to separate the control of the capacitor voltage from the control of the machine. We therefore choose in this paper the following duty cycles:

$$\begin{aligned} \alpha_A(t) &= \alpha_H(t) + \frac{V_A(t)}{V_{DC}} \\ \alpha_B(t) &= \alpha_H(t) + \frac{V_B(t)}{V_{DC}} \\ \alpha_C(t) &= \alpha_H(t) + \frac{V_C(t)}{V_{DC}} \end{aligned} \tag{6}$$

$V_A(t)$, $V_B(t)$, and $V_C(t)$ are the three-phase voltages containing only the AC component of the voltages, i.e., $V_A(t) + V_B(t) + V_C(t) = 0$. Thus, it suits the previous definition of α_H.

In the particular case where $\alpha_H = 0.5$, the shape of the duty cycles corresponds to the sinusoidal pulse width modulation.

The voltages $V_A(t)$, $V_B(t)$, and $V_C(t)$ can be generated by a classical control law either in open loop (U/f control, for example) or in closed loop with phase current control (field-oriented control, for example).

Concerning the capacitor voltage control, by neglecting the voltage drop due to the resistance R, the steady-state average relation between the input voltage and the capacitor voltage coming from (5) becomes:

$$V_{DC} = V_{in}/\alpha_H \tag{7}$$

It can be seen that the homopolar duty cycle α_H controls the capacitor voltage value. For the sake of simplicity, in this paper only an open-loop control is applied to control the voltage V_{DC}, but a feedback controller can be used as it is usually done for DC/DC converters.

3.3 Steady-State Behavior

In steady state, α_H tends to a constant value. Moreover, $V_A(t)$, $V_B(t)$, and $V_C(t)$ are sinusoidal functions with equal RMS value V. The values of the duty cycles $\alpha_k(t)$ are restricted between 0 and 1, therefore V is limited. By including these limits in (6), we obtain:

$$-\alpha_H V_{\rm DC} \leq V_k(t) \leq (1-\alpha_H)V_{\rm DC} \quad k \in \{A, B, C\} \tag{8}$$

With (7), we obtain:

$$-V_{in} \leq V_k(t) \leq \frac{1-\alpha_H}{\alpha_H}.V_{in} \tag{9}$$

With α_H lower than 0.5, the limiting value to create a sinusoidal wave is the lower one ($-V_{in}$) and so the maximum reachable amplitude is V_{in}. The maximum RMS value for the sinusoidal voltage is therefore $V_{MAX} = V_{in}/\sqrt{2}$.

With α_H larger than 0.5, the limiting value to create a sinusoidal wave is the larger one $\left(\frac{1-\alpha_H}{\alpha_H} V_{in}\right)$ and the maximum reachable RMS value is thus:

$$V_{MAX} = \frac{1-\alpha_H}{\alpha_H\sqrt{2}} V_{in} \tag{10}$$

Figure 3 shows, respectively, the reachable RMS value of the sine voltage applied to the machine and the capacitor voltage according to α_H in steady state. The higher the value α_H, the lower the reachable AC voltage and the lower the capacitor voltage. With α_H lower than 0.5, the maximum AC voltage is fixed to $V_{in}/\sqrt{2}$ but the capacitor voltage is higher than twice V_{in}. This leads to higher losses in the semiconductor devices and induces more design constrains. Therefore, it is not interesting to have a homopolar duty cycle lower than 0.5 and it will be chosen between 0.5 and 1 in the following.

3.4 Comparisons Between the NPS Scheme and a Classical Star-Connected Machine

In this section, the proposed NPS architecture is compared with the classical star-connected machine configuration presented in Fig. 1.

Firstly, the reachable RMS voltage will be discussed. The maximum RMS voltage depends on the modulation algorithm. Generally, a space vector pulse width modulation is used. With this modulation, for the NPS structure, the reachable maximum RMS voltage is $V_{in}/\sqrt{2} \simeq 0.707 \cdot V_{in}$. For the classical scheme, the

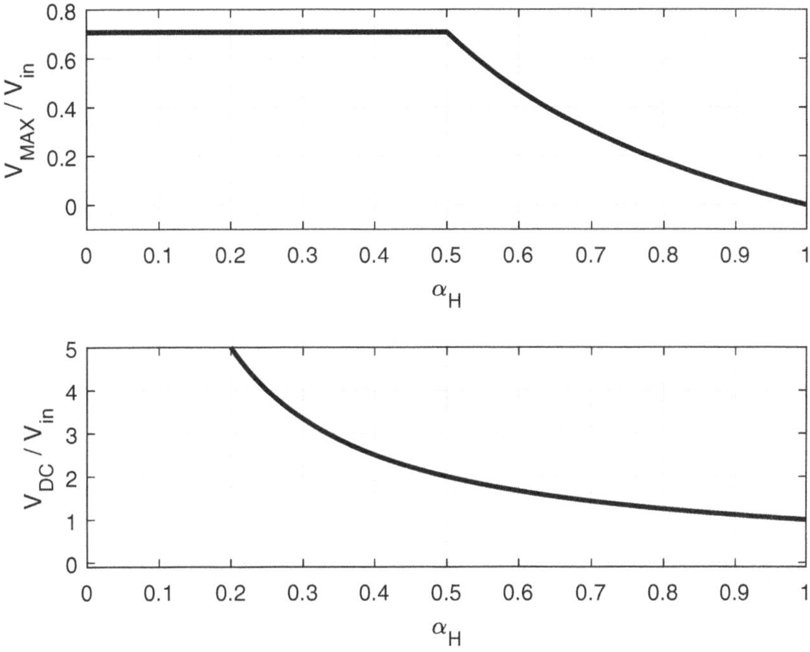

Fig. 3 Maximum reachable AC RMS voltage (top) and DC capacitor voltage (bottom) in steady state according of homopolar duty cycle

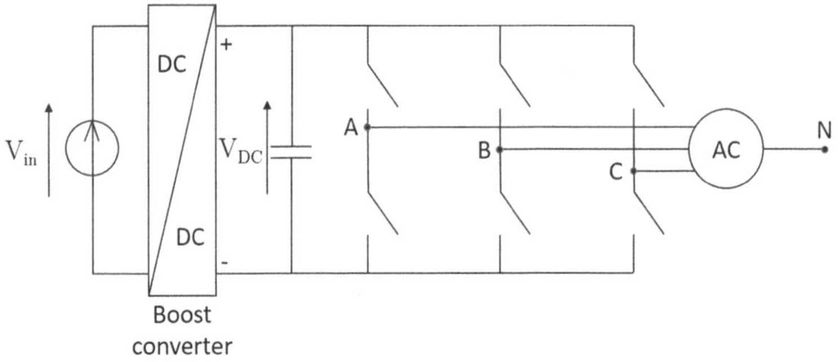

Fig. 4 Inverter with a boost converter to increase the capacitor voltage

reachable maximum RMS voltage is $V_{in}/\sqrt{6} \simeq 0.408V_{in}$. To reach the same maximum RMS voltage as the NPS scheme, a boost converter must be added (see Fig. 4). It is clear that this increases the number of components, the size, and the cost of the whole system. Its reliability can be decreased. Consequently, the advantages of the NPS structure in comparison with a classical star-connected one are clear.

By considering a RMS line current I in a classical scheme, the RMS line current in the NPS structure I_{line} contains both the current I and an additional homopolar current I_H:

$$I_{\text{line}} = \sqrt{I_H^2 + I^2} \tag{11}$$

By neglecting the converter losses, the added homopolar current is directly related to the power consumption P of the AC connected machine:

$$I_H = \frac{P}{3V_{in}} \tag{12}$$

The addition of DC current into the machine and the converter can lead to some losses. This is the main drawback of the NPS structure. However, at low speed, this drawback disappears (see Sect. 4.2).

4 Control

4.1 Proposed Control Scheme

The control scheme for NPS structure is presented in Fig. 5. A classical field-oriented control (FOC) [6] is used to control the AC phase currents by way of $V_A(t)$, $V_B(t)$, and $V_C(t)$. An open-loop control of the capacitor voltage V_{DC} is made by way of α_H. V_{DC} is chosen as low as possible but sufficiently high to reach the required voltage from the FOC controller. By inverting the expression in (10), we obtain α_H as follows:

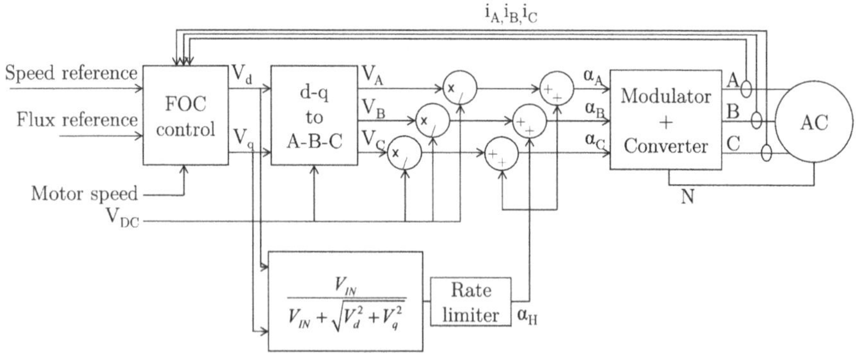

Fig. 5 Control scheme: a classical field-oriented control with a homopolar duty cycle open-loop control

$$\alpha_H = \frac{V_{in}}{V_{in} + \sqrt{2}V_{MAX}} \tag{13}$$

By using Park transformation, the voltages in the d–q reference frame V_d and V_q lead to:

$$\alpha_H = \frac{V_{in}}{V_{in} + \sqrt{V_d^2 + V_q^2}} \tag{14}$$

A rate limiter is added for α_H to have a low response time since only steady state is considered in this paper.

4.2 RMS Current Analysis

By using the proposed control, the RMS voltage is maintained at its maximum value V_{MAX}. By noting, respectively, I, P, and $\cos\phi$, the RMS AC current per winding, the power, and the power factor of the machine, we have

$$P = 3V_{MAX} I \cos\phi \tag{15}$$

By using (10)–(12), the RMS value of line current can be written as:

$$\frac{I_{\text{line}}}{I} = \sqrt{\left(\cos\phi \frac{1-\alpha_H}{\alpha_H\sqrt{2}}\right)^2 + 1} \tag{16}$$

The expression (16) is drawn on Fig. 6 for $\alpha_H = 0.5$. It can be seen that the ratio between the RMS line current and the AC machine phase current is larger than 1 due to the DC component, but it depends on the power factor $|\cos\phi|$. For low $|\cos\phi|$, the line current reaches the coil current because less power is transmitted and so less DC current flows into the coils. Considering the line and the converter RMS current, the NPS scheme is more interesting for low values of power factor.

Considering the special case where a near zero voltage is applied to the machine. Only V_{in} is applied on the capacitor (see Fig. 3 for $\alpha_H = 1$). For AC machines, this case appears at low speed because of the low electromotive force. At low speed, the machine has a low power too. From (12), it can be seen that the homopolar current has therefore a small value. But the mechanical torque and the AC current can be large. In this case:

$$I_{\text{line}} \simeq I \tag{17}$$

$$V_{\text{DC}} \simeq V_{in} \tag{18}$$

A small value for voltage and current on the switches of the converter leads to less power losses.

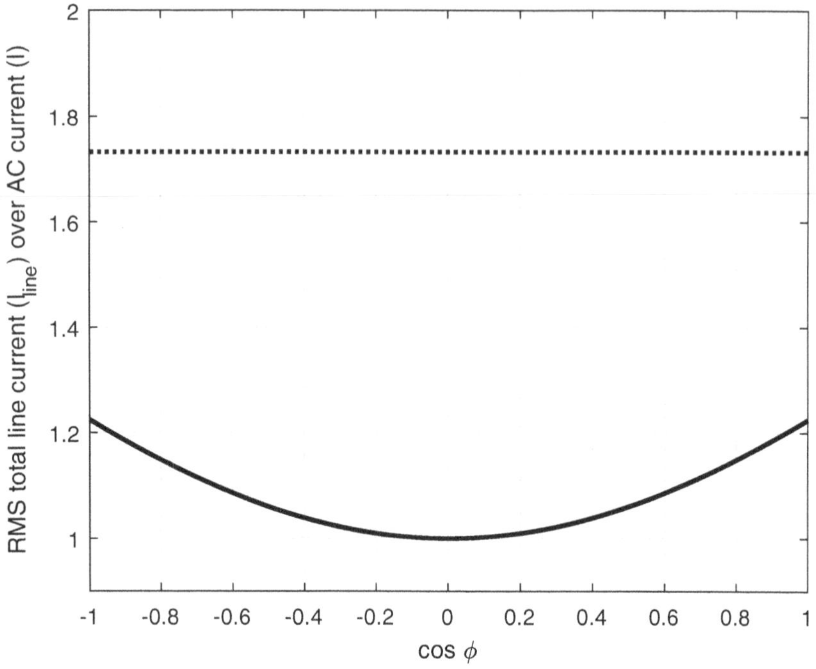

Fig. 6 Ratio between the RMS line current and the AC machine phase current in the NPS structure. $\sqrt{3}$ value is also plotted in dotted line as comparison

5 Experimental Validation

5.1 Test Bench Description

The NPS scheme (Fig. 2) has been tested with an induction motor drive.

The parameters of the motor are described in Table 1. A DC regulated voltage power supply Xantrex is used with $V_{in} = 300\,\text{V}$. An IGBT based ARCEL three-phase two-level inverter drives the motor with a 6.6 mF capacitor.

The control scheme of Fig. 5 is implemented on a rapid control prototyping system dSPACE DS1104 by using Matlab/Simulink. The motor is connected to a powder brake system which can deliver a mechanical load torque on the motor.

5.2 Experimental Results

Figure 7 shows the experimental time evolution of the applied duty cycles and the capacitor voltage when the machine speed increases slowly from 0 to 1500 rpm. It can be seen that firstly at low speed, the voltage is at low value. It increases with

Table 1 Induction motor parameters

Parameter	Value
Rated voltage	230 V/400 V
Nominal current	6.9 A/4 A
Nominal power	1.8 kW
$\cos\phi$	0.81
2 pole pairs	50 Hz

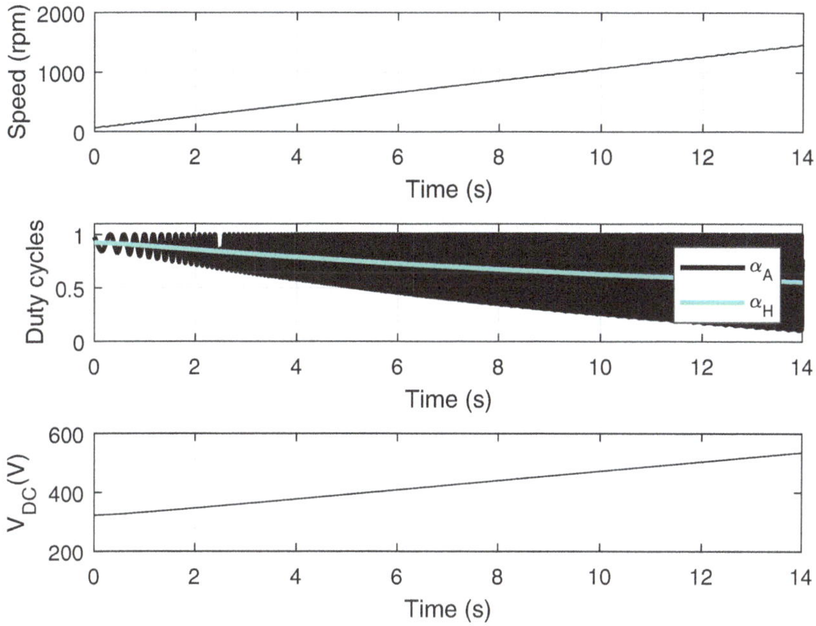

Fig. 7 Experimental time evolution of the applied duty cycles and the capacitor voltage when the machine speed increases slowly without load

the speed due to the decrease of homopolar duty cycle α_H tuned by the open-loop control. Secondly, the applied duty cycle α_A on the leg A increases with the speed due to the fact that the applied voltage in steady state increases too. Thirdly, the maximum value of α_A is kept as one due to the decrease of homopolar duty cycle α_H with respect to the increase of the amplitude of α_A.

Figure 8 presents the time evolution of the phase current i_A with its mean in an experimental test at 200 rpm with an load torque. The nominal flux is applied and the measured quadratic current i_q is at 5.4 A. The obtained capacitor voltage is $V_{\mathrm{DC}} = 364$ V. Moreover, from the duty cycles α_A and α_H shown in Fig. 8, it can be seen that the homopolar duty cycle α_H is near to its theoretical value obtained with a simplified boost converter model $V_{in}/V_{\mathrm{DC}} = 0.824$.

Figure 9 presents another experiment time evolution of the phase current i_A with its mean at 1500 rpm without load torque. The nominal flux is applied and

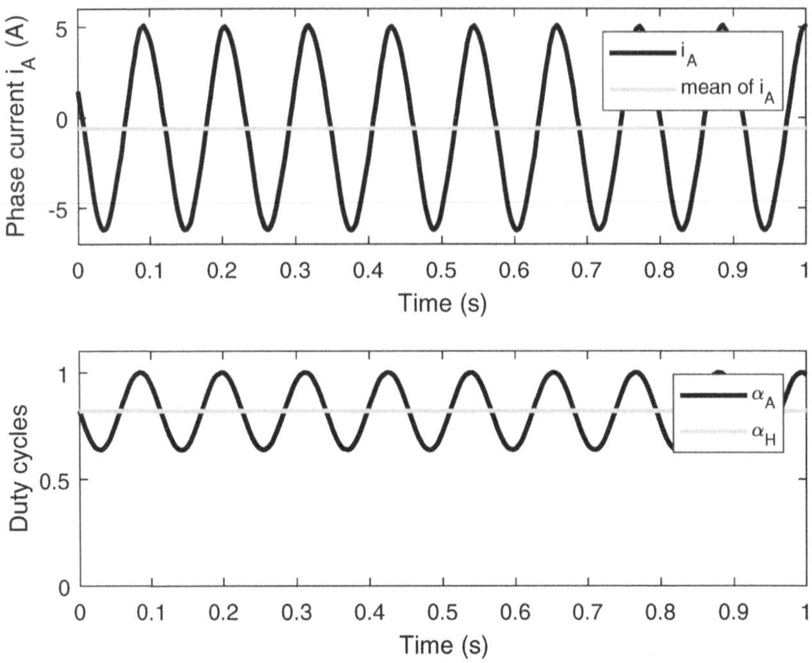

Fig. 8 Experimental time evolution of the phase current and duty cycles at 200 rpm with a load ($I_d = 4.1$ A, $I_q = 5.4$ A, $V_{DC} = 364$ V)

the quadratic current i_q is measured at 0.3 A. The obtained capacitor voltage is $V_{DC} = 546$ V. Moreover, from the duty cycles α_A and α_H shown in Fig. 9, it can be seen that the homopolar duty cycle α_H is near to its theoretical value obtained with a simplified boost converter model $V_{in}/V_{DC} = 0.55$.

6 Conclusion

A new concept for supplying AC machine drive was proposed and validated experimentally. It is based on an original connection of the neutral point of the machine. It allows for a voltage boost effect between the input voltage source and the capacitor voltage without the use of an additional boost converter. Another advantage of the proposed NPS concept is related to its flexibility. Even if it is not the best structure when a nominal operation condition is required, when a low speed or a low torque is required, this structure becomes more attractive. One particular interesting application can be the possibility to create a high current at low speed with a converter scaled for a low current with high speed value. Some studies are still necessary to validate definitively the performances of the proposed concept, such as a study of the impact of the DC component on the machine coils (self-heating, useful

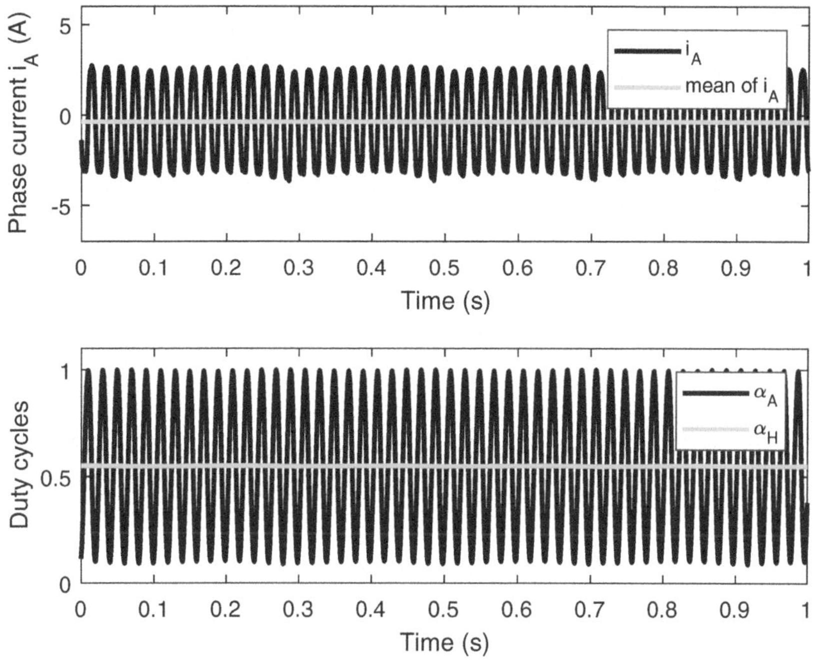

Fig. 9 Experimental time evolution of the phase current and duty cycles at 1500 rpm without load ($I_d = 4.1$ A, $I_q = 0.3$ A, $V_{DC} = 546$ V)

life, etc.). However, the proof of concept was made with a simple associated control algorithm and opens many research ways and some perspectives. For example, the comparison study with a Δ-connected machine is to be considered. First results show that with the addition of a DC current in the machine coils, the advantages of NPS are the possibility to have a lower RMS line current and lower current ripple in the motor. Another perspective concerns the extension of the NPS concept to other inverter architectures, such as multilevel ones for examples. The experimental tests were made with an induction machine in this paper. Other purposes can be found to supply other AC machines or other AC loads. Moreover, the use of the neutral point of a machine is often used to create a fault-tolerant supply system: a study of the NPS concept for this application looks promising.

References

1. B.A. Welchko, T.A. Lipo, T.M. Jahns, S.E. Schulz, Fault tolerant three-phase AC motor drive topologies: a comparison of features, cost, and limitations. IEEE Trans. Power Electron. **19**(4), 1108–1116 (2004)
2. M. Farhadi, M.T. Fard, M. Abapour, M.T. Hagh, DC–AC converter-fed induction motor drive with fault-tolerant capability under open- and short-circuit switch failures. IEEE Trans. Power Electron. **33**(2), 1609–1621 (2018)

3. B. Mirafzal, Survey of fault-tolerance techniques for three-phase voltage source inverters. IEEE Trans. Ind. Electron. **61**(10), 5192–5202 (2014)
4. G.T. Chiang, J. Itoh, A high efficiency three-phase AC motor drive converter that utilized the neutral point of a motor, in *2009 International Conference on Power Electronics and Drive Systems (PEDS)*, Taipei, 2009, pp. 783–788
5. G.T. Chiang, J. Itoh, DC/DC boost converter functionality in a three-phase indirect matrix converter. IEEE Trans. Power Electron. **26**(5), 1599–1607 (2011)
6. G. Lefebvre, J.-Y. Gauthier, A. Hijazi, X. Lin-Shi, V. Le Digarcher, Observability-index-based control strategy for induction machine sensorless drive at low speed. IEEE Trans. Ind. Electron. **64**(3), 1929–1938 (2017)

Performance Analysis of a Micro-Grid System Composed of Renewable Energy Sources with Hybrid Energy Storage System

S. Tamalouzt, N. Benyahia, A. Tounzi, and K. Idjdarene

Abstract The present paper focuses on the study of the generated power smoothing performances of hybrid renewable energy sources dedicated to the application of a micro-grid. The latter is constituted of a variable speed wind turbine (VSWT) based on a DFIG, a PV system and a hybrid energy storage system. The DFIG rotor side is powered by a three-level inverter and controlled by a direct reactive power control (DRPC) technique. This topology is chosen to improve the quality of the output power and current injected into the grid. The hybrid storage system is made of a combination of battery banks (BBs) and supercapacitors (SCs). The main purpose of the work is to manage the energy used by the system and to control the active and reactive powers flowing through the micro-grid. Also, the use of the DFIG with its converter as a local reactive power compensator is considered. In addition, to mitigate fluctuations due to random changes in wind speed and solar radiation, an energy management algorithm is introduced. The latter must also ensure the smoothing of the output power, the control of the discharge depth of the batteries, and the continuity of service. Simulation results, performed under Matlab/Simulink, are presented and analyzed.

1 Introduction

Strong penetrations of renewable energy sources (RESs) in electrical grids generate numerous problems in terms of reliability, energy quality, and stability of power

S. Tamalouzt · K. Idjdarene
Laboratoire LTII, Département de Génie Electrique, Faculté de Technologie, Université A/Mira de Bejaia, Bejaia, Algeria

N. Benyahia
Laboratoire LATAGE, Université de M/Mammeri Tizi-Ouzou, Tizi-Ouzou, Algeria

A. Tounzi (✉)
Arts et Metiers ParisTech, Centrale Lille, HEI, EA 2697 - L2EP - Laboratoire d'Electrotechnique et d'Electronique de Puissance, University of Lille, Lille, France
e-mail: abdelmounaim.tounzi@univ-lille.fr

W. Zamboni, G. Petrone (eds.), *ELECTRIMACS 2019*, Lecture Notes in Electrical Engineering 615, https://doi.org/10.1007/978-3-030-37161-6_19

grids. To avoid these problems, micro-grids, with energy storage systems (ESSs), constitute a valuable solution. Thus, the quality of energy can be improved while reducing power variations and energy imbalances [1]. Indeed, a suitable electrical energy storage system (EESS) along with the adopted approach allows dealing with the unpredictability of the RES generations and their intermittency [1–3]. The EESS is a process that enables electricity to be stored at the time of either low demand and low generation cost or from intermittent energy sources and to be used at the time of high demand and high generation cost or when no other generation means is available [1, 4, 5].

The various energy storage systems can be classified according to two criteria: function and form. In terms of function, EESS technologies are divided into the following: (1) those intended for high power ratings with a relatively small energy content, making them suitable for power quality or uninterruptible power supply (UPS), such as supercapacitors, and (2) those designed for energy management for continuity of service, such as the large-scale batteries [4, 5]. Electrical energy storage technologies can also be classified according to the form of energy to store. For this classification, we can mention several cases, which we can point out as (1) electrical storage, as the case of supercapacitors, and (2) chemical storage, as the conventional batteries [1, 5]. In general, the ESS is desirable with high power density in order to make face power fluctuations, whereas a high energy density is determinant to ensure micro-grid autonomy. For these reasons, hybrid energy storage systems (HESSs) are recommended [1–3].

Economically, it is admitted that hybrid systems are more efficient with battery bank ESSs (BBESSs) [6]. However, the lifetime is strongly affected by the battery charge and discharge cycles [3]. Therefore, a third energy source is needed to improve the supply reliability of these systems. The choice of the third system is according to the role of the storage system and its form. So, for power quality and balance, the electrical form is the best suited, as the case of supercapacitor ESS (SCESS) [3].

Recently, numerous micro-grid configurations based on RESs associated with HESSs have been proposed [2, 3, 7–10]. However, little attention has been paid to the integration of DFIG into micro-grids as compared to other types of machines, therefore not considering the benefits and usefulness of DFIG. In [1–3], authors proposed some micro-grids made up of hybrid energy sources with better performances and efficiency. In this context, a three-level inverter is used instead of the classical two-level one in the DFIG rotor. This converter can not only generate the output voltages with very low distortion but also reduce dv/dt stresses [11].

The significant benefit of the proposed configuration is an easier accessibility of DC and AC grids. This topology permits an optimal and efficient control of the different sources, better AC grid voltage, and frequency regulation with a low rate of total harmonic distortion (THD), continuous service, and local reactive power compensation, ensuring better power quality.

The layout of the present paper is as follows: A description of the overall proposed system is first given. Then, the operation principle and the modeling of the system components are presented. The global control and the power management strategy which is the main contribution of this work are developed. Finally, through several simulation tests, carried out under different conditions such as power demand and meteorological changes, the effectiveness of the proposed system is verified and highlighted.

2 Description of the Proposed Micro-Grid System

The proposed micro-grid topology is shown in Fig. 1; it combines a hybrid energy source (WT-DFIG and PV system) associated with the hybrid energy storage system. The DFIG is partially interfaced with the grid, via an indirect frequency converter (AC/DC/AC). The latter consists of two converters separated by an intermediate DC bus assumed as a DC grid. The first converter (I) is a three-level inverter connected to the rotor of this machine and controlled by the proposed control technique (DRPC). The connection to the AC grid is provided via a second converter (II) controlled so as to maintain its output at unity power factor operation and to guarantee sinusoidal signals with a constant frequency of 50 Hz. Both HESS and PV generator are of course connected to the DC bus. The PV system control is detailed in Ref. [1]. To extract the maximum power available in the photovoltaic generator, it is necessary to control the boost converter associated with the PV by the maximum power point tracking (MPPT) algorithm. The classical MPP technique of disturbance and observation is used; it is a simple technique but has the advantage of a very easy implementation. The WT control is insured in two different stages. The MPPT algorithm is applied below the nominal turbine speed, in the zone I and II of the WT control characteristic, as illustrated in [1]. Beyond rated speed, the pitch angle control is activated such as the power extraction is maintained at its

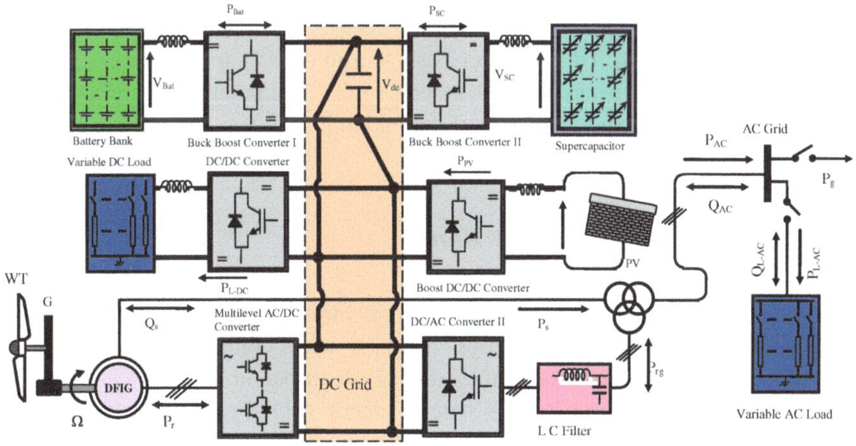

Fig. 1 The proposed micro-grid scheme

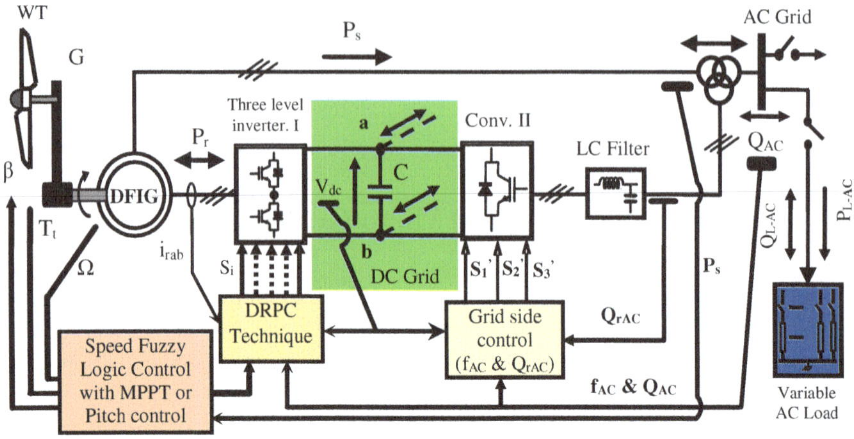

Fig. 2 Synoptic scheme of the proposed control

rate value. The description of control strategy (DRPC) is developed in [1, 3]. This control combines the Direct Torque Control (DTC) and fuzzy logic technique; the reactive and active powers are controlled with the rotor flux and electromagnetic torque, respectively. These parameters are considered to be the outputs of the fuzzy controllers performed to reactive power and the mechanical DFIG speed control, as given in Fig. 2.

2.1 Supercapacitor Model

The supercapacitor model used in this paper is based on a simplified (RC) circuit; it is more suitable for applications where the energy stored in the capacitor is of a primary importance [10]. This model consists of a nonlinear capacitance $C(v_{sc0})$ and an equivalent series resistance (ESR), as given in [3]. Its parameters are identified using the constant current test method described in [3, 10].

2.2 Modeling of the Battery Bank

A standard battery model presented in [1, 8] is implemented in this study. Thus, the battery bank (BB) is modeled using a simple series connected to a controlled voltage source with a constant resistive value, as described in [1].

In the studied case, the battery's state of charge (SOC) range is kept between 25% and 75%. In this work, the DC-grid voltage is controlled at its reference value ($V_{\text{dc-ref}}$) throughout the buck-boost DC-DC converter associated with the battery banks. This converter has the advantage of using a smaller number of batteries that must be connected in series. In the proposed micro-grid system, this voltage is maintained around 444 V with 37 batteries connected in series.

3 Description of the Power Management and V_{dc} Control

The BBESS voltage can be kept lower than the DC-grid voltage reference ($V_{dc\text{-}ref}$) by using a buck-boost converter, the reason for which is that several parallel–/series-connected batteries are often necessary. In the present case, the voltage across the BBESS as SCESS is kept around 420 V.

In the case of modern lead-acid batteries, charge acceptance is very high; the battery charge/discharge is considered with a SOC range between 25% and 75% [1, 3]. According to weather conditions and power requirement, BBESS or SCESS is able to act either as a power supply or as system storage; they should discharge/charge within specified limits in case of deficit/surplus of hybrid power. Of lower dynamic, the BBESS cannot feed the power instantaneously and hence unable to stabilize the DC-grid voltage during the transient state; for this reason, a SCESS is inserted using a second Buck-Boost DC Converter (BBDC) (connected between SC and DC grid), in order to increase the power of the BBESS which is decreasing.

The power difference between the generation sources and the required power demand is calculated as given in Eq. (1), where P_s and P_r are the DFIG stator and rotor powers. The P_{PV} is the generated PV power. $P_{L\text{-}AC}$ and $P_{L\text{-}DC}$ are the AC and DC load powers, P_{rAC} is the power transferred through DC/AC converter II, P_{SC} and P_{Bat} are the SCESS and BBESS powers, and P_{AC} and P_g are the powers exchanged between the proposed system and the AC grid.

$$\begin{cases} P_{AC} = P_{L\text{-}AC} + P_g \\ P_{AC} = P_{rAC} + P_s \\ P_{Net} = P_r + P_{PV} + P_{rg} \\ P_{rg} = P_{L\text{-}DC} + P_{rAC} \\ P_{Sc} = P_{Net} - P_{Bat} \end{cases} \tag{1}$$

By managing the power available according to the flowchart, Fig. 3, the DC-grid voltage (V_{dc}) control is achieved by the buck-boost DC converter associated to the BBESS.

The DC/AC converter II is controlled to operate at unity power factor, whereas the reactive power exchange between the micro-grid and the AC side (grid) (Q_{AC}) is provided by the DFIG.

$$\begin{cases} Q_{AC} = Q_s + Q_{rAC} \\ Q_{AC} = Q_g + Q_{L\text{-}AC} \\ Q_{L\text{-}AC} = P_{L\text{-}AC}.tg\left(\varphi_{L\text{-}AC}\right) \end{cases} \tag{2}$$

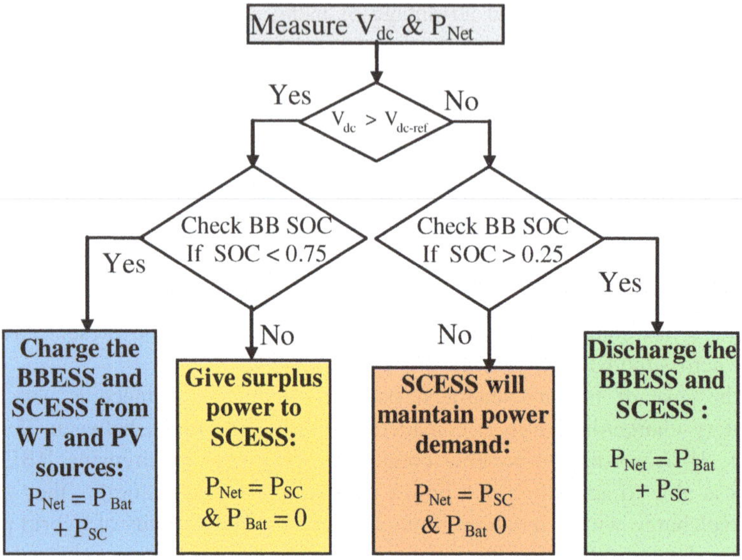

Fig. 3 Illustration of a flowchart for DC-grid voltage control

Q_s represents the reactive power exchange between the DFIG stator and the AC grid. $Q_{L\text{-}AC}$ and Q_{AC} are the reactive powers exchanged with the AC load and the AC grid together, respectively. Q_{rAC} is the reactive power transferred through DC/AC converter II, and Q_g is the reactive power in the grid. Finally, $\varphi_{L\text{-}AC}$ is the local load phase.

4 Simulation Results and Discussion

The proposed system associated with the HESS is connected to the micro-grid. This configuration can be used as an uninterruptible power supply to ensure micro-grid autonomy and to make face power fluctuations in order to ensure continuous and smooth supplying.

The proposed system is simulated under Matlab/Simulink environment. The simulation parameters are given in [1, 10]. In order to verify the system performances, the simulations have been realized under variable power requirement and weather data (wind speed profile, solar irradiance, and air temperature). The simulation tests are given below, in Fig. 4a–k. The purpose of these tests is to highlight the contribution and the effects of this hybrid system on the injected or requirement energy quality and efficiency.

Figure 4a shows the variable wind speed and irradiation profiles. The wind speed profile is chosen such that the DFIG operation area is in different zones (I, II, and III), as shown in [1]. The simulation results represented in Fig. 4a–c for

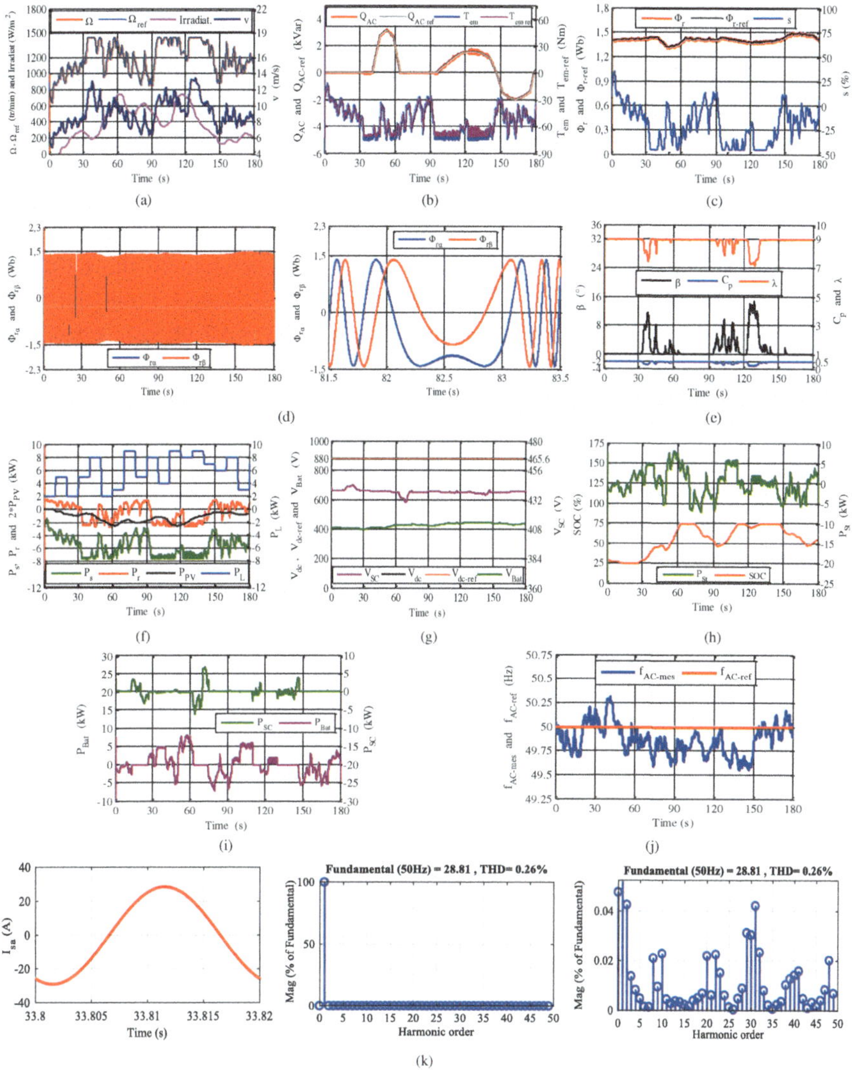

Fig. 4 Waveforms of the simulation results for a variable wind speed and a reactive power requirements profiles: (**a**) Ω, Ω_{ref}, and υ; (**b**) Q_{AC}, $Q_{\text{AC-ref}}$, T_{em}, and $T_{\text{em-ref}}$; (**c**) Φ_{r}, $\Phi_{\text{r-ref}}$, and s; (**d**) $\Phi_{\text{r}\alpha}$, $\Phi_{\text{r}\beta}$, and their zooms; (**e**) β, λ, and C_{p}; (**f**) P_{r}, P_{s}, P_{PV}, and P_{L}; (**g**) V_{dc}, $V_{\text{dc-ref}}$, V_{Bat}, and V_{SC}; (**h**) SOC and P_{Bat}; (**i**) P_{Bat} and P_{SC}; (**j**) f_{AC} and $f_{\text{AC-ref}}$; (**k**) generated phase current (stator current, I_{sa}) and its harmonic spectrum, in synchronous mode operation of DFIG

the mechanical speed, electromagnetic torque, reactive power compensation, and rotor flux, respectively, show that the proposed DRPC ensure a correct reference tracking and robustness under random behavior of wind speed. In zones I and II [1], the power coefficient is always optimized, by adapting the electromagnetic torque according to the wind speed. However, in the zone III, that torque is constant to keep constant the power generation. Moreover, it can be seen that the reactive power follows the rotor flux variations, ensuring a WT optimal operation and, therefore, a possible contribution of the DFIG for active and reactive power management.

In Fig. 4d, a sinusoidal waveform of the rotor flux is obtained for both components during the sub- and super-synchronous modes. In the synchronized asynchronous mode, these flux components have continuous forms, and the resulting flux magnitude is always kept constant, as shown in the zoom of these fluxes (Fig. 4d). Consequently, a continuous and successive operation of the DFIG is obtained in the three modes under random wind speed change. From the sub-synchronous ($s > 0$) to the super-synchronous mode ($s < 0$) or vice versa, the transition is realized through the synchronous mode, as shown in Fig. 4a, c, d.

The blade pitch angle, tip speed ratio, and the power coefficient are illustrated in Fig. 4e. As expected, at maximum power ($P_{s\text{-max}}$), in zone III, the pitch angle control is activated to generate the variation of the tip speed ratio and the power coefficient. Finally, in this zone, the mechanical speed is kept constant at its limit value, as presented in Fig. 4a. However, in zones I and II, the power is maximized by the MPPT algorithm.

Figure 4f illustrates the powers generated, absorbed, and required by the load (P_L). This figure clearly shows that when there is an excess of power in the DC bus ($P_{Net} < 0$), the storage of this power is performed at BBESS and SCESS, thus increasing their voltages. On the other hand, when there is a power deficit ($P_{Net} > 0$), it is the BBESS and SCESS storage system that fills the lack of energy.

In Fig. 4g, the DC bus and storage element voltages are represented. The DC-grid voltage (V_{dc}) does not present abrupt changes, and it is kept at 880 V. In Fig. 4h, i, when the WT-DFIG and PV powers are below the power demand, with $SOC \leq SOC_{min}$, the supercapacitor supplies the missing energy. In case of power excess from the WT-DFIG and PV according to demand, with the battery fully charged ($SOC \geq SOC_{max}$), the excess energy is used to charge the supercapacitor.

Finally, the sinusoidal form of the generated current shown in Fig. 4j, k shows the quality of the power injected in the AC grid from the DFIG stator. This remains true also for the three operating modes with a constant 50 Hz frequency. Moreover, the variable amplitude is a consequence of the wind speed variation.

Compared with the results obtained in [1, 3], the proposed DRPC associated to the three-level inverter offers a significant improvement to the stator currents supplied to the AC grid, with a total harmonic distortion (THD) largely below the limits imposed by IEEE std 519, as illustrated in Fig. 4k. This confirms a better quality of energy generated in the AC grid.

5 Conclusions

In this paper, a hybrid micro-grid has been studied; it consists of a combination of WT-DFIG and PV generators with a battery (BBESS) and supercapacitor (SCESS) energy storage system. This hybridization acts as a dump load using all the excess power available for the BESS-SCESS load, as well as backup generation and system power in case of power failure while the WT-DFIG and PV generators represent the main source of energy production. The DFIG is controlled by a three-level inverter in its rotor by a proposed DRPC technique. In addition, global energy control and management strategies are also proposed to optimize the quality of the energy produced.

From the simulation tests carried out for different situations, in particular, those relating to the generation and variation of the energy demand, the behaviors expected for the entire system have been verified. Indeed, the hybrid storage system has shown its effectiveness to ensure a continuous and regular supply while ensuring a rapid response to regulate the difference between production and energy demand. In addition, the DFIG can act as a synchronous compensator and can also operate in an asynchronous mode as a synchronous generator. These results allowed us to confirm the robustness behavior and satisfactory performances of the proposed control. The three-level inverter inserted in the DFIG rotor also contributes to improving the quality of the energy supplied to the alternative grid.

References

1. S. Tamalouzt, N. Benyahia, T. Rekioua, D. Rekioua, R. Abdessemed, Performances analysis of WT-DFIG with PV and fuel cell hybrid power sources system associated with hydrogen storage hybrid energy system. Int. J. Hydrog. Energy **41**(45), 21006–21021 (2016)
2. S. Tamalouzt, N. Benyahia, T. Rekioua, R. Abdesemed, A doubly fed induction generator wind turbine and fuel cell hybrid power sources system for micro-grid applications, in *Proc. of 3rd International Conference on Control, Engineering & Information Technology, CEIT'2015, Tlemcen, Algeria, May 25–27*, (2015), pp. 1–6
3. S. Tamalouzt, F. Hamoudi, T. Rekioua, D. Rekioua, Variable speed wind generator associated with hybrid energy storage system-application for micro-grids, in *Proc. of 5th International Renewable Sustainable Energy Conference, IRSEC'2017, Marrakech & Ouarzazate, Morocco, December 4–7*, (2017), pp. 1–6
4. X. Luo, J. Wang, M. Dooner, J. Clarke, Overview of current development in electrical energy storage technologies and the application potential in power system operation. Appl. Energy **137**, 511–536 (2015)
5. H. Chen, T.N. Cong, W. Yang, C. Tan, Y. Li, Y. Ding, Progress in electrical energy storage system: a critical review. Prog. Nat. Sci. **19**(3), 291–312 (2009)
6. H. Yang, L. Lu, W. Zhou, A novel optimization sizing model for hybrid solar-wind power generation system. J. Solar Energy **81**, 76–84 (2007)
7. S. Obara, Y. Morizane, J. Morel, Economic efficiency of a renewable energy independent microgrid with energy storage by a sodium-sulfur battery or organic chemical hydride. Int. J. Hydrog. Energy **38**, 8888–8902 (2013)
8. C. Wang, M. Hashem Nehrir, Power management of a stand-alone wind/photovoltaic/fuel cell energy system. IEEE Trans. Energy Convers. **23**(3), 957–967 (2008)
9. S.G. Malla, C.N. Bhende, Voltage control of stand-alone wind and solar energy system. Electr. Power Energy Syst. **56**, 361–373 (2014)

10. N. Benyahia, H. Denoun, M. Zaouia, S. Tamalouzt, M. Bouheraoua, N. Benamrouche, T. Rekioua, S. Haddad, Characterization and control of supercapacitors bank for stand-alone photovoltaic energy. Energy Procedia **42**, 539–548 (2013)
11. A. Nami, J. Liang, F. Dijkhuizen, G.D. Demetriades, Modular multilevel converters for HVDC applications: review on converter cells and functionalities. IEEE Trans. Power Electron. **30**(1), 18–36 (2015)

Management and Control of Smart Transformer-Fed LV Distribution Networks During Grid Faults

Rongwu Zhu, Giovanni De Carne, and Marco Liserre

Abstract The reliability and service continuity in the distribution grid are the basic requirements, but are threatened by various grid faults. Compared to the conventional power transformer, a smart transformer (ST) can flexibly switch between current control mode and voltage control mode, and thus has high capability on power flow control and to mesh the electric grid in secondary substations. Consequently, the ST can significantly improve the reliability and service continuity in secondary networks under grid faults. This paper presents the possible solutions to manage faults in MV and LV distribution grid, respectively, and the corresponding control strategies to improve system performances.

1 Introduction

Due to the continuously increasing penetration of distributed generators (DGs) and electric vehicle (EV) charging stations, the distribution grid experiences more challenges in terms of voltage limits violation [1], low power quality [2, 3], stability issues [4], and overload of network (transformers and cables) [5]. Moreover, since it has been proved that the dc network has higher efficiency as compared to the ac network, it has drawn more attentions to develop the dc network to integrate EV charging stations, renewable energies, and household dc loads, which can avoid power losses in the extra inverse stage [6].

In order to improve the aforementioned challenges, grid operators generally update the electric grid by installing extra power converter-based devices, e.g., STATCOM [7], active power filter (APC) [8], energy storage system (ESS) [9], and reinforcing the network assets, e.g., larger size of cables and higher power transformers. However, such upgrades are time-consuming, costly, and may lead to disruptive works.

R. Zhu (✉) · G. De Carne · M. Liserre
Christian-Albrechts-Universität zu Kiel, Kiel, Germany
e-mail: rzh@tf.uni-kiel.de; gdc@tf.uni-kiel.de; ml@tf.uni-kiel.de

W. Zamboni, G. Petrone (eds.), *ELECTRIMACS 2019*, Lecture Notes in Electrical Engineering 615, https://doi.org/10.1007/978-3-030-37161-6_20

In contrast, smart transformer (ST), which is based on a solid-state transformer, can provide advanced services, e.g., voltage regulation, harmonic compensation, power flow control, low voltage ride through to support MV grid and can decouple MV and LV grid by the dc-link in terms of harmonics, reactive power, and faults, as shown in Fig. 1. On the other hand, the ST can optimize the capability in the secondary substations by flexibly meshing the electric distribution grid, reducing or postponing the reinforcement in the electric grid. Moreover, ST can provide dc connectivity [5].

The configuration of conventional power transformer (CPT)-based secondary distribution grids, generally include radial networks, spot networks, and grid networks, and the capability to mesh electric grid in secondary substations is also limited due to uncontrollable power flow in CPT. In contrast, the ST can actively control its power flow and smoothly switch between current control mode (CCM) and voltage control mode (VCM), which provides extra degree of freedom to mesh electric grid.

When the faults locate in the upstream ac grid, the CPT cannot provide any solutions to improve the reliability and service continuity of the electric grid. In contrast, the ST can flexibly mesh the electric grid in secondary substations and provide advanced service and control to manage grid faults and to ensure the post-fault operation capability of electric grid. When asymmetrical faults locate in the

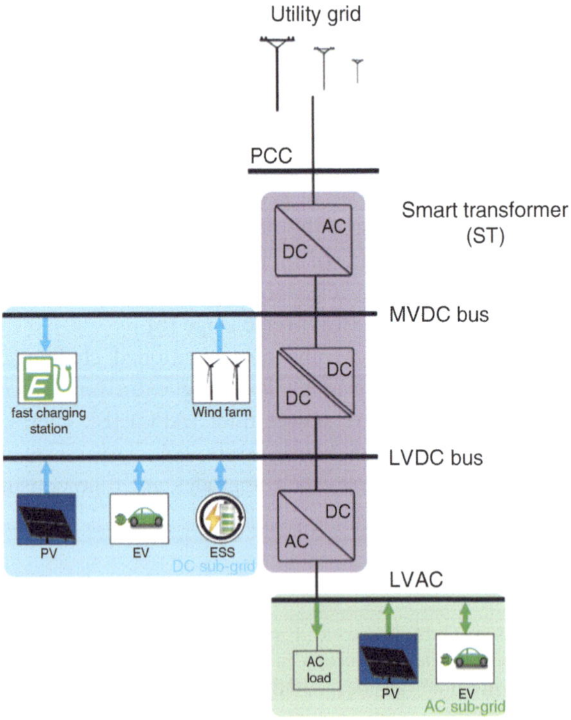

Fig. 1 Smart Transformer-fed distribution network

downstream LVac grid, the CPT will experience high neutral current and thermal stress, limiting the continuous operation capability of CPT. In the ST scenario, since the LVac grid voltage is established by ST, various optimal objectives focusing on LVac grid can be achieved by implementing the improved voltage control strategies in ST LVac side inverter. Under single-phase to ground faults, the two-phase operation of ST with phase shift angle control to optimize the system performances between the suppression of dc active power oscillation and neutral current amplitude will be introduced as well.

The reset of this paper is organized as follows. Section 2 presents the post-fault management of ST under the faults in upstream ac grid of the distribution transformer. Section 3 presents the post-fault management of ST under the faults in downstream ac grid of the distribution transformer and the possible optimization of system performance by the phase shift angle control. Section 4 summarizes this paper.

2 Post-fault Management in MVAC Grid

2.1 Basic Configuration of Three-Stage Smart Transformer

As shown in Fig. 1, the ST includes three stages (MVac/dc stage, MV/LVdc/dc stage, and LVdc/ac stage) and four terminals (MVac and dc, as well as LVac and dc), the MVac terminal is connected to the MVac distribution network, the MVdc connectivity is suitable for integrating the EV fast charging station and large wind/PV farm. The LVdc connectivity is suitable for DGs, e.g., PV system and micro-wind turbine, EV charging station and ESS connection. In the LVac grid, besides the conventional loads, the DGs and EV charging stations are also suitable to be integrated.

2.2 Configuration of Radial Networks

The radial operation, which is the most widespread network in LV system, as shown in Fig. 2, can provide sufficient reliability and service continuity for most customers, e.g., in the resident area. It is a cost-efficient configuration as compared to the spot network, which needs two or more CPTs and each is fed by an independent MV feeder.

In radial configuration, under normal operation, each LV feeder is independently fed by a MV/LV distribution transformer, but under the maintenance of the MV/LV distribution transformer, the secondary distribution feeders will be meshed by solid connection of the normally open point (NOP), (NOP1 or NOP 2), as shown in Fig. 2. Consequently, one MV/LV distribution transformer will feed two feeders, which

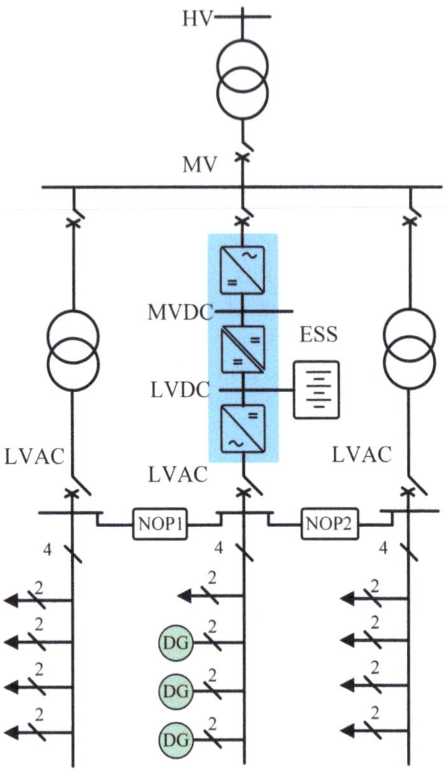

Fig. 2 Smart Transformer-fed distribution network

cannot provide the full power demand in both feeders due to the limit capacity of the distribution transformer.

Unlike the CPT, the ST can actively regulate ac grid voltage amplitude, frequency, and phase angle and smoothly switch between the CCM and VCM. Such characteristics allow the ST to operate in parallel with the CPT even under normal conditions [10]. The ST can control the active power flow in real-time, which allows to optimize the capacity in the substations, avoiding high voltage violation and overloading.

2.3 Dual-Microgrids and Power Loop Controller

When faults occur in the upstream ac grid of the distribution transformer, the ST can continue to support LVac grid by a ESS, which is integrated through the LVdc link, as shown in Fig. 3. Meanwhile, if the CPT-fed feeder is switched to the LVac busbar of ST via the solid connection of a NOP, the of ST LV side inverter cannot provide enough power for both feeders. Instead of connecting two feeders to ST LVac busbar, the CPT-fed feeder can be connected to the MV side inverter of the

Fig. 3 Smart transformer meshed LV distribution grid of (**a**) dual-microgrid and (**b**) loop power controller

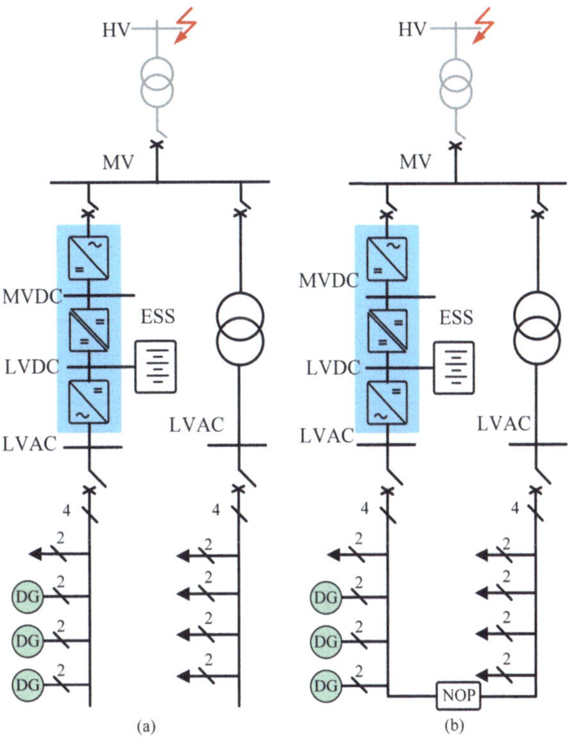

ST through the CPT and common MVac busbar. In this case, ST will not experience overload, because two feeders are connected to MVac/dc converter and LVdc/ac inverter, respectively, and two feeders work as a dual-microgrids [11], as shown in Fig. 3a.

Due to the change of grid configuration, the control system of ST is also needed to be adapted accordingly. Under normal operation, the control block diagram of ST is shown in Fig. 4a, where the MVac/dc converter is controlled as an active front end (AFE) rectifier, the MV/LVdc/dc stage regulates the LVdc link voltage and the LVdc/ac inverter adopts the VCM to establish the LVac grid voltage. Under dual-microgrid operation, the control system of ST will switch to Fig. 4b. Compared to Fig. 4a, the LVdc-link link voltage is controlled by the ESS converter, MV dc-link voltage is controlled by the MV/LVdc/dc converter, and the LVdc/ac as well as MVac/dc converter both work at VCM to establish ac voltage for both feeders, respectively.

The concept of power loop controller which connects the ends of two independent LVac feeders by a back-to-back converter, can avoid power unbalance and reduce power losses [12, 13]. In the ST application, the ends of two LVac feeders can be connected by solid connection of NOP instead of extra back-to-back converter, as shown in Fig. 3b. In this cases, the control block diagram of the ST is shown

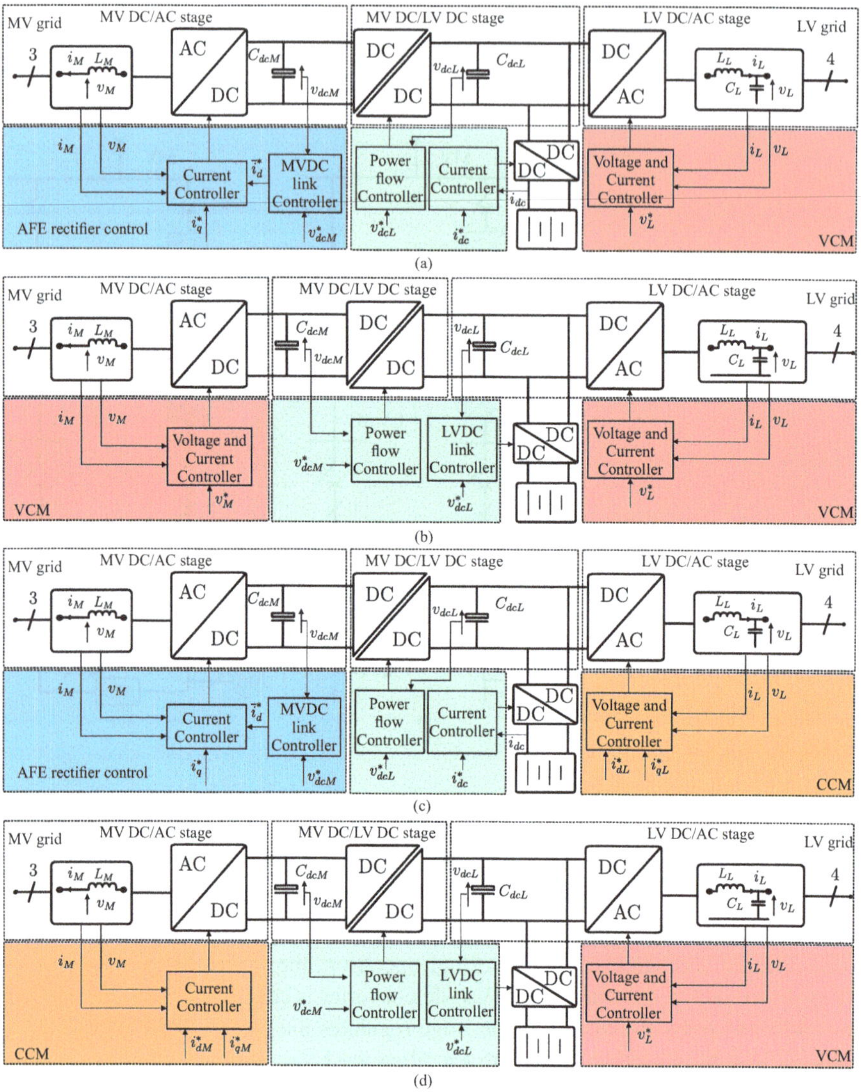

Fig. 4 ST control block diagram under (**a**) normal operation, (**b**) dual-microgrid operation during faults, (**c**) loop power control during normal operation (**d**) loop power control during faults

in Fig. 4c, where the LVdc/ac inverter operates at the CCM instead of the VCM as shown in Fig. 4a, because ST will share the power between CPT and ST.

When grid faults occurring in the upstream ac grid of the distribution transformer, the control block diagram of ST will switch from Fig. 4c to d, because the LVdc/ac converter will be in charge of establishing LVac grid voltage and the MV side inverters will regulate the power flow between them in real-time by the CCM.

3 Post-fault Management in LVac Grid

When a single-phase fault exists in the downstream feeder of CPT, the faulty phase will be disconnected directly to isolate the faulty area. The continuous operation capability of the CPT is limited by the thermal stress in CPT and high neutral voltage variation caused by high neutral current under extreme asymmetrical two-phase operating conditions due to the uncontrollable terminal voltages in CPT. Consequently, the interruption of CPT due to the protection leads to outage in a large area, potentially increasing economical losses.

In contrast, the ST can fast block voltage output in the faulty phase and continue the power supply in the left two healthy phases. The neutral current can be effectively suppressed by ST and therefore the neutral potential variation is reduced as well. Moreover, unlike the CPT, the ST does not suffer from high thermal stress under two-phase operation. The detailed management of the LV distribution grid during single-phase to ground faults is introduced as follows.

According to the location of faults, the operation model of ST can be divided into three-phase and two-phase operation mode, as shown in Fig. 5. When the faults locate at the main distribution feeder as shown in Fig. 5a, highlight as red (e.g., point A), ST will work in two-phase operation mode, in contrast, when the faults locate at the terminal of feeder, e.g., point B, the ST will continue working in three-phase operation. The aim of working mode division is to reduce the outage area as small as possible.

Fig. 5 Three-phase and two-phase operation mode

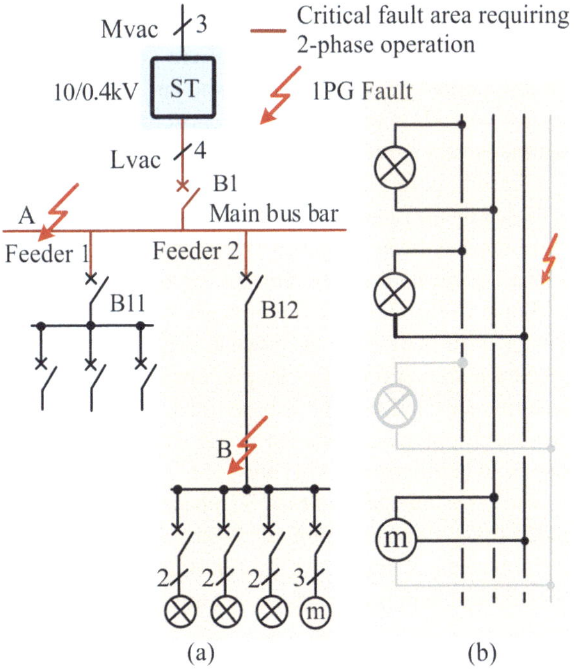

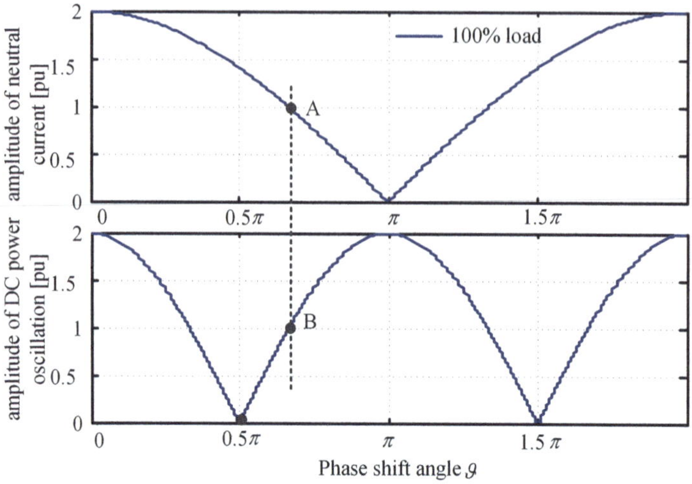

Fig. 6 Control diagram of Smart Transformer under two-phase operating model

Compared to three-phase symmetrical operation, under two-phase operation, active power oscillation will appear in LVdc-link and high current will be aroused in neutral line as well. The oscillatory dc power will lead to LVdc-link voltage oscillation and thus potentially reduce the lifetime of LVdc-link capacitors. On the other hand, higher neutral current increases power losses in neutral conductor and high neutral potential variation, reducing the reliability and safety of precise devices.

In order to solve aforementioned challenges, the profile of active power oscillation in LVdc link and neutral current amplitude versus phase shift angle, which is between two-healthy phases is shown Fig. 6. It shows that when the phase shift angle $\vartheta = \pi/2$ or $3\pi/2$ active power oscillation is disappeared, and when $\vartheta = \pi$ the neutral current is zero, which shows the possibility to improve system performances by controlling the phase shift angle.

Based on the phase shift angle, an improved voltage control strategy of two-phase operation has been proposed to either suppress the LVdc-link voltage oscillation or reduce the neutral line current, or obtain the tradeoff between them [14, 15], as shown in Fig. 7. In the three-phase operation mode, the reference of three-phase voltage is given directly, as shown in Fig. 7a, while in the two-phase operation mode, an extra phase shift control loop is involved which can be used to optimize system performances and the SOGI-based PLL is adopted to measure the current angle in two-healthy phases, as shown in Fig. 7b. The detail of voltage and current control loops can be referred to [14] and [15].

The experimental results of ST phase voltages, active power, and neutral current under two-phase operation mode are shown in Fig. 8. It clearly shows that the performances of the active power oscillation and neutral current oscillation are effected by changing the phase shit angle. When the phase shift angle is controlled

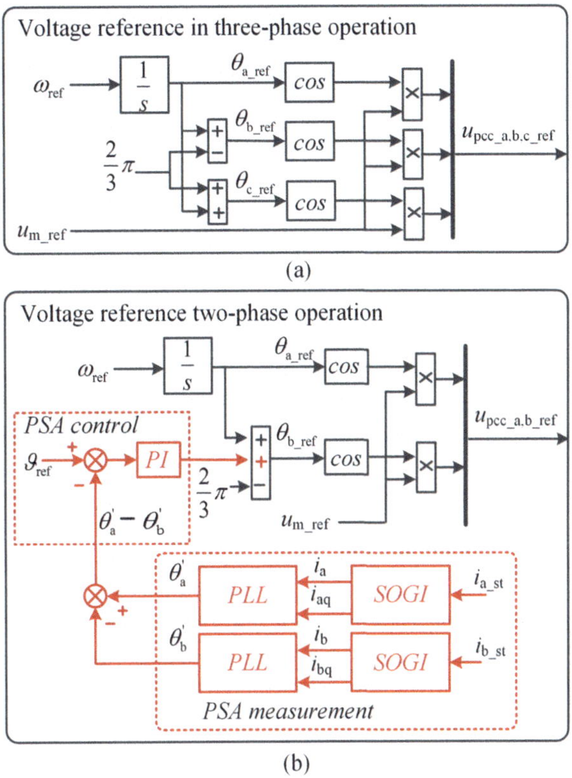

Fig. 7 Voltage reference under (**a**) three-phase operation mode and (**b**) two-phase operating model

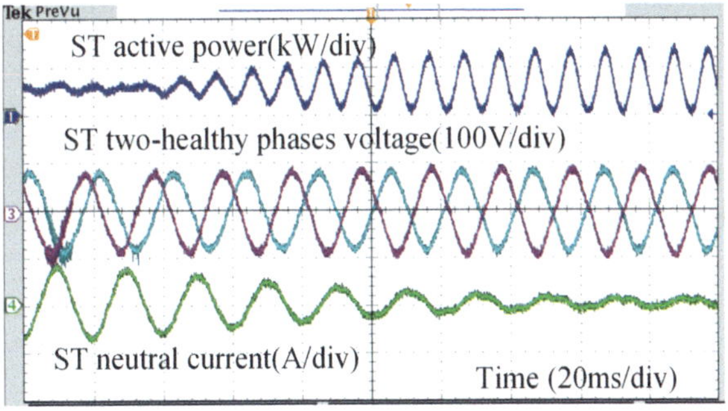

Fig. 8 Experimental results under two-phase operation

at $\pi/2$, the active power oscillation is almost zero and when the phase shift angle is controlled as π, the neutral current is almost zero, matching well with Fig. 6.

4 Conclusions

In this paper, the post-fault management of smart transformer used in the secondary distribution grid has been introduced. Based on the radial networks, under the grid faults in the upstream ac grid of the secondary substation, the smart transformer can mesh the LV feeders as a dual-microgrid and power loop controller as well, which can improve the reliability and service continuity in the smart transformer-fed two LV feeders without unnecessary outage. The corresponding control system design of smart transformer under normal operation, dual-microgrid operation, and power loop controller is presented. Under single-phase to ground faults in the downstream of the secondary distribution transformer, the smart transformer can switch from the normal three-phase operation mode to the two-phase operation mode with the aim to reduce the outage area as small as possible. The phase shift angle-based control strategy can improve the system performances by reducing either active power oscillation or neutral current amplitude under two-phase operation. The experimental results validate the correctness of the phase shift angle control. Thus, smart transformer has high potential to improve LVac distribution grid performances and improve the of reliability and service continuity for the customers.

Acknowledgements The research leading to these results in this paper has received funding from the European Research Council under the European Union's Seventh Framework Programme (FP/2007-2013)/ERC Grant Agreement n.[616344]—HEART, and from the German Federal Ministry of Education and Research (BMBF) within the Kopernikus Project ENSURE "New ENergy grid StructURes for the German Energiewende" (03SFK1I0).

References

1. P. Li, C. Zhang, X. Fu, G. Song, C. Wang, J. Wu, Determination of local voltage control strategy of distributed generators in active distribution networks based on Kriging metamodel. IEEE Access 1–1 (2019, early access)
2. J. Enslin, P. Heskes, Harmonic interaction between a large number of distributed power inverters and the distribution network. IEEE Trans. Power Electron. **19**(6), 1586–1593 (2004)
3. R. Zhu, Z. Zou, M. Liserre, High power quality voltage control of smart transformer-fed distribution grid, in *IECON-2018* (2018), pp. 5547–5552
4. N. Pogaku, M. Prodanovic, T. Green, Modeling, analysis and testing of autonomous operation of an inverter-based microgrid. IEEE Trans. Power Electron. **22**(2), 613–625 (2007)
5. LV-Engine, [available]. https://www.spenergynetworks.co.uk/pages/lv_engine.aspx
6. R. Zhu, G. Buticchi, M. Liserre, Investigation on common-mode voltage suppression in smart transformer-fed distributed hybrid grids. IEEE Trans. Power Electron. **33**(10), 8438–8448 (2018)

7. T. Lee, S. Hu, Y. Chan, D-STATCOM with positive-sequence admittance and negative-sequence conductance to mitigate voltage fluctuations in high-level penetration of distributed-generation systems. IEEE Trans. Ind. Electron. **60**(4), 1417–1428 (2013)
8. M. Hashempour, M. Savaghebi, J. Vasquez, J.M. Guerrero, A control architecture to coordinate distributed generators and active power filters coexisting in a microgrid. IEEE Trans. Smart Grid **7**(5), 2325–2336 (2016)
9. J. Choi, Y. Shin, M. Choi, W. Park, I. Lee, Robust control of a microgrid energy storage system using various approaches. IEEE Trans. Smart Grid 1–1 (2018, early access)
10. R. Zhu, G. De Carne, F. Deng, M. Liserre, Integration of large photovoltaic and wind system by means of smart transformer. IEEE Trans. Ind. Electron. **64**(11), 8928–8938 (2017)
11. C. Kumar, M. Liserre, A new prospective of smart transformer application: dual microgrid (DMG) operation, in *IEEE IECON-2015* (2015), pp. 4482–4487
12. C. Kumar, X. Gao, M. Liserre, Smart transformer based loop power controller in radial power distribution grid, in *IEEE ISGT-Eupro-2018* (2018), pp. 1–6
13. X. Xing, J. Lin, C. Wan, Y. Song, Model predictive control of LPC-looped active distribution network with high penetration of distributed generation. IEEE Trans. Sustain. Energy **8**(3), 1051–1063 (2017)
14. R. Zhu, M. Liserre, Continuous operation of smart transformer-fed distribution grid with single-phase faults, in *IEEE ECCE2018* (2018), pp. 458–464
15. G. De Carne, M. Langwasser, R. Zhu, M. Liserre, Smart Transformer-based single phase-to-neutral fault management. IEEE Trans. Power Deliv. 1–1 (2019, early access)

Efficient Power Management Strategies for High-Energy-Demanding Industrial Plants

Libero Paolucci, Mostafa Abdollahi, Francesco Grasso, Alberto Reatti, and Fabio Corti

Abstract In the present work, the authors present a comparative study of an electrical power distribution system. The work aims to study efficient and cost-effective ways to manage power in high-energy-demanding plants. By doing this, power losses can be reduced with economic and environmental benefits. As a case study, a galvanization plant is taking under consideration. Such plants indeed need a high quantity of electrical energy for the galvanizing process; hence, also a slight increase of the overall efficiency can lead to consistent economic benefits. The proposed plant in addition has a PV solar roof which is meant to integrate the grid for the process supply. In the work, the present situation is analyzed and a different solution is hence proposed. The plant indeed is now supplied with a medium voltage (MV) connection and transformed into three-phase 400 V AC low voltage (LV). Galvanization pools and auxiliary utilities are supplied with this voltage. Concerning the PV solar roof, the strips output 800 V DC. In order to exploit the PV produced energy, a DC-AC inverter connects the PV plant to the AC 400 V line. The presented structure involves several power losses which can be reduced by exploiting a more efficient power management strategy.

1 Introduction

In the present work, the authors present a comparative study of an electrical power distribution system. The work aims to study efficient and cost-effective ways to manage power in high-energy-demanding plants. By doing this, power losses can be reduced with economic and environmental benefits. As a case study, a galvanization plant is taking under consideration. Such plants indeed need a high quantity of electrical energy for the galvanizing process; hence, also a slight increase of the overall efficiency can lead to consistent economic benefits. The proposed plant in

L. Paolucci · M. Abdollahi · F. Grasso (✉) · A. Reatti · F. Corti
DINFO, University of Firenze, Florence, Italy
e-mail: libero.paolucci@unifi.it; mostafa.abdollahi@unifi.it; francesco.grasso@unifi.it; alberto.reatti@unifi.it; fabio.corti@unifi.it

W. Zamboni, G. Petrone (eds.), *ELECTRIMACS 2019*, Lecture Notes in Electrical Engineering 615, https://doi.org/10.1007/978-3-030-37161-6_21

addition has a PV solar roof which is meant to integrate the grid for the process supply. In the work, the present situation is analyzed and a different solution is hence proposed. The plant indeed is now supplied with a medium voltage (MV) connection and transformed into three-phase 400 V AC low voltage (LV). Galvanization pools and auxiliary utilities are supplied with this voltage. Concerning the PV solar roof, the strips output 800 V DC. In order to exploit the PV produced energy, a DC-AC inverter connects the PV plant to the AC 400 V line. The presented structure involves several power losses which can be reduced by exploiting a more efficient power management strategy.

The proposed solution involves the use of a DC bus for the supply of the galvanization pools, while, for the auxiliary utilities which absorb less power, 400 V AC power is supplied. By doing this, the PV plant can be directly connected to the DC bus through a DC-DC converter. This solution is meant to reduce conduction losses due to low voltage in the power distribution lines along the plant. Concerning the PV plant power production, the advantage of avoiding a DC-AC conversion is due to reduced inverter losses and to a more continuous exploitation. Hence, due to phase discrepancy and mismatches between the grid frequency and generated one, in traditional operation, the PV plants are often disconnected from the grid. Furthermore, phase discrepancies induce frequency variations on grid voltage which degrade power quality.

Losses and disturbances degrade the overall efficiency of the plant. High current on the electrical distribution circuit requires larger cable sizes to minimize conduction losses, because reduction of power quality causes electrical stress in the connected appliances, which need power factor correction systems, and, therefore, further power losses. These factors are widely underestimated in industrial plants where standardization and functionality are privileged. However, the economic losses due to low efficiency can be relevant. According to [1], economical costs due to bad power quality accounts to up 4% of a company turnover, while conduction losses could be up to 9% of the dissipated power in Europe and up to 17% in Latin America and the Caribbean area [2]. Hence, the adoption of power management strategies that could decrease these losses could be of great interest especially for these industries which have high-energy needs.

Clearly, these strategies can be costly and difficult to implement; hence, an accurate study is needed to define the most reliable and convenient ones. The main criteria are clearly the maximum energetic efficiency and power quality but also ease of integration with the actual system in order to simplify these improvement strategies [3–11].

In this work, several configurations are simulated in order to minimize power transport and conversion losses. Simulation results and conclusions are presented together with a possible power distribution plant setup.

2 Power Management in Traditional Galvanizing Plants

A common power distribution setup in galvanizing plants is illustrated in Fig. 1. It is possible to recognize the MV connection to the grid through the MV-LV transformer. The power is hence distributed along the plant in order to supply the utilities in AC. Concerning the pool supply, each of them has an AC-DC converter as schematically shown in Fig. 1. The power along the plant is thus distributed with a 400 V AC voltage for every utility. This guarantees a relatively safe and highly standardized operation of the plant, but at the same time, if high distances are covered and high loads are attached, considerable conduction and conversion losses occur.

Concerning the PV solar roof, it generates an 800 V DC voltage which must be converted to 400 V AC in order to be exploited. This is done through an AC-DC converter connected in parallel with the LV line. These converters are largely exploited in most of household and industrial application because they are highly integrable in existing plants [3]. However, in addition to the conversion losses that this system introduces, they cannot continuously match the grid frequency and phase [4]. This means that the PV plant will be periodically disconnected from the grid causing a discontinuous exploitation of PV source and overall power quality reduction [5]. Clearly, by avoiding DC-AC conversion, a better exploitation of PV produced energy could be performed.

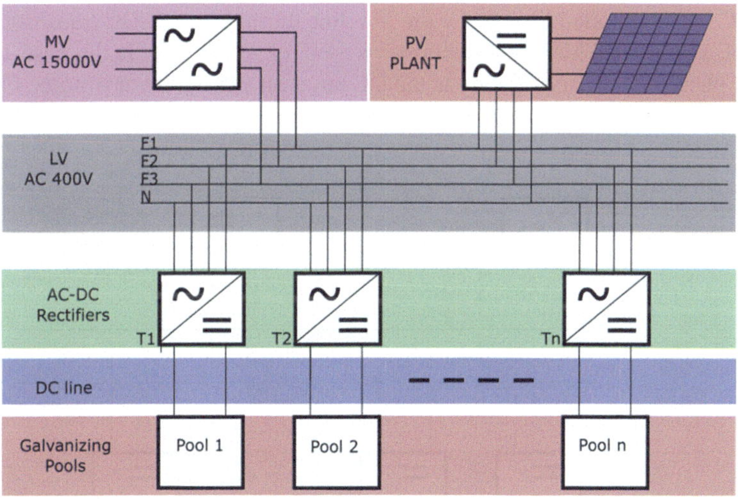

Fig. 1 Power distribution scheme, traditional setup

3 Proposed Power Management Setup for Galvanizing Plant

In the previous paragraph, some of the problems related to power losses in traditional power distribution of industrial plants and PV integration are listed. The study and implementation of new, more efficient solutions is slowed down for several reasons. The phenomena are often underestimated, and existing systems guarantee high levels or reliability and standardization. Hence, when developing new strategies to reduce power losses and disturbances, these elements need to be taken under consideration. The proposed solution has been developed starting from the factors that cause power losses which are mainly the following:

- Conduction losses
- Conversion/switching losses
- Distortion losses

In this preliminary stage, the work is focused on conduction loss reduction strategies because the operating scenario presents an extended distribution line. In order to minimize these losses, a different architecture is proposed as shown in Fig. 2.

In this case, a DC bus is distributed along the plant. Optimal bus voltage is a parameter that had to be studied in order to minimize conduction losses while permitting an optimal integration of the new setup with the existing system. Consequently, several voltages have been simulated to achieve the optimal trade-off.

By using this strategy, it is necessary to use two different supply lines: a DC line for the supply of the galvanizing pools and a LV AC line for auxiliary utilities. The DC bus is supplied directly by the MV line through an AC-DC transformer. In parallel, a traditional MV-LV transformer is used for AC LV line. In this case, hence, the PV plant is connected directly to the DC bus with a DC-DC transformer. Finally, the galvanizing pools are supplied by the DC bus through a DC-DC converter. In the

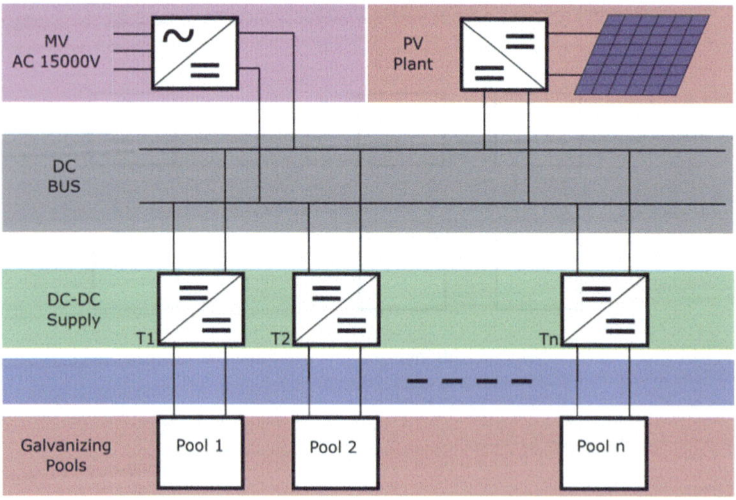

Fig. 2 Power distribution scheme, proposed setup

previous configuration, each pool was supplied with an AC-DC isolated transformer composed of a rectifier stage and a current regulator. In this case, the rectifier stage is not necessary, and line voltage is fed directly to the pool current regulator.

4 Modeling of the Plant and Simulation Study

The modeling methodology of both the traditional setup (AC plant) and the proposed one (DC plant) is described in this section. Then, the results of the analysis in Matlab/Simulink are presented. The specific data of the plants are reported in Table 1.

Table 1 Data of grid, PV system, and AC and DC subunits

Power grid
AC-DC plant: 1000 MVA, 15 kV, $X/R = 20$, 50 Hz Transformer: 650 kVA, 15 kV/400 V, 50 Hz $X_{\text{trans}} = 0.05$ pu, $R_{\text{trans}} = 0.004$ pu LC filter in DC plant: $L_f = 600$ μH, $C_f = 50$ mF, $R_d = 0.05$ Ω
PV system
AC plant: Power converter: 30 kW, $V_{ac} = 400$ V, $V_{dc} = 700$ V, 10 kHz LC filter: $L_f = 30$ μH, $C_f = 3.3$ mF, $R_d = 9.5$ mΩ Control system: PI active controller: $K_P^P = 0.5$, $K_I^P = 150$ PI reactive controller: $K_P^Q = 0.5$, $K_I^Q = 150$ PR current controller: $K_P^I = 0.65$, $K_R^I = 1$
DC plant: Buck DC-DC converter: 30 kW, $V_{ac} = 400$ V, $V_{dc} = 700$ V LC filter: $L_f = 0.6$ mH, $C_f = 50$ mF, $R_d = 0.05$ Ω Control system: PI active controller: $K_P^{Pdc} = 0.5$, $K_I^{Pdc} = 150$ PI current controller: $K_P^{Idc} = 0.65$, $K_R^{Idc} = 1$
AC and DC subunits (10 kW, 3 kW)
AC plant: Three-phase AC cable, π model, 50 Hz $R_{ac} = 5.88$ Ω/km, $L_{ac} = 0.2957$ mH/km, $C_{ac} = 10^{-12}$ F/km Internal transformer: 10 kW Subunit: 10 kVA, 400 V/16 V, 50 Hz, $X_t = 0.05$ pu, $R_t = 0.004$ pu 3 kW Subunit: 3 kVA, 400 V/12 V, 50 Hz, $X_t = 0.05$ pu, $R_t = 0.004$ pu Diode rectifier and LC filter (10 kW, 3 kW): $L_f = 600$ μH, $C_f = 50$ mF, $R_d = 0.05$ Ω
DC plant: DC bipolar cable: $R_{dc} = 5.88$ Ω/km DC-DC buck converter: 10 kW Subunit: 10 kW, 400 V/16 V 3 kW Subunit: 3 kW, 400 V/12 V LC filter (10 kW, 3 kW): $L_f = 600$ μH, $C_f = 50$ mF, $R_d = 0.05$ Ω

4.1 Modeling Methodology

A schematic representation of the AC plant is presented in Fig. 3. The power grid supplies the main bus of the galvanizing pools through a step-down power transformer. Moreover, the main bus is supplied by a PV system as a secondary supplier. There are two different pool sizes of 10 kW and 3 kW which are schematically divided into two subunits.

The PV system is modeled as shown in Fig. 4. The voltage source converter is controlled in current mode, and it is equipped with a proper LC filter.

The control system of the PV system is designed using a well-known technique as presented in Fig. 5 [12]. According to this method, the outer layer of the control system manages the active and reactive power using proportional-integral (PI) controllers. The current references are subsequently calculated according to Eq. (1). After that, a proportional resonant (PR) current controller is used as inner control layer to control the converter in current mode. The PR controller transfer function is expressed in Eq. (2) [13]. Finally, the resulting voltage references are converted into a three-phase system using the Clark transformation as described in Eq. (3).

$$\begin{bmatrix} I_\alpha \\ I_\beta \end{bmatrix} = \frac{1}{V_\alpha^2 + V_\beta^2} \begin{bmatrix} V_\alpha & V_\beta \\ V_\beta & -V_\alpha \end{bmatrix} \begin{bmatrix} P \\ Q \end{bmatrix} \tag{1}$$

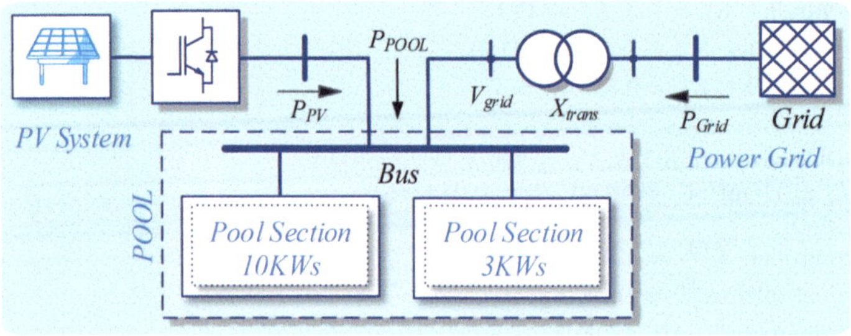

Fig. 3 Schematic of the plant modelling

Fig. 4 PV system configuration in AC plant

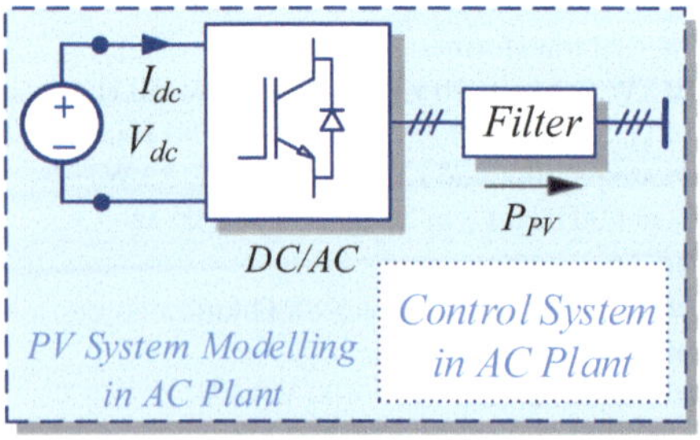

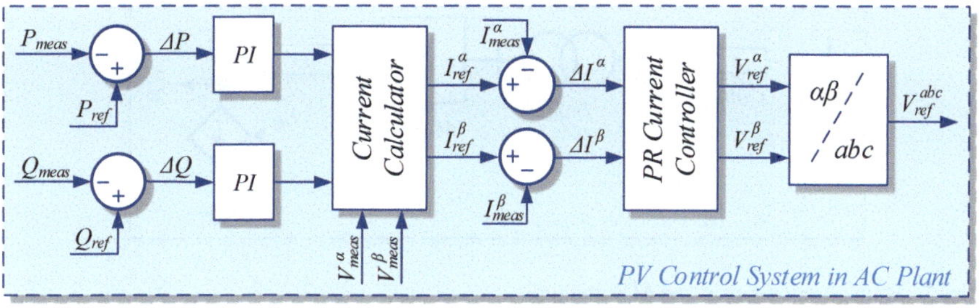

Fig. 5 Control mechanism of the PV system in AC plant

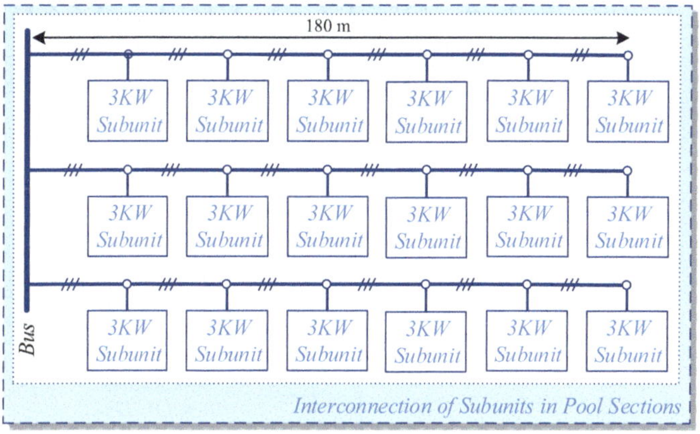

Fig. 6 Internal configurations of 3 kW subunits in poll section

$$G^{\mathrm{I}}_{\mathrm{PR}}(s) = K^{\mathrm{I}}_{\mathrm{P}} + K^{\mathrm{I}}_{\mathrm{R}} \frac{s}{s^2 + \omega_0^2} \tag{2}$$

$$\begin{bmatrix} V_a \\ V_b \\ V_c \end{bmatrix} = \sqrt{\frac{2}{3}} \begin{bmatrix} 1 & 0 \\ -\frac{1}{2} & \frac{\sqrt{3}}{2} \\ -\frac{1}{2} & -\frac{\sqrt{3}}{2} \end{bmatrix} \begin{bmatrix} V_\alpha \\ V_\beta \end{bmatrix} \tag{3}$$

As shown in Fig. 3, the plant contains two main subsections. Each subsection (10 kWs and 3 kWs) contains 18 subunits that are arranged as depicted in Fig. 6. In each pool section, there are three parallel corridors 180 m long, and each corridor is supplied by the main bus.

All the subunits are supplied through the main bus by an AC-DC power converter which includes isolated transformer, full wave rectifier bridges, and LC filter as depicted in Fig. 7. The high-power subunits (10 kW) operate at 0–16 V DC output voltage and 600 A DC output current. The low-power subunits (3 kW) operate at 0–

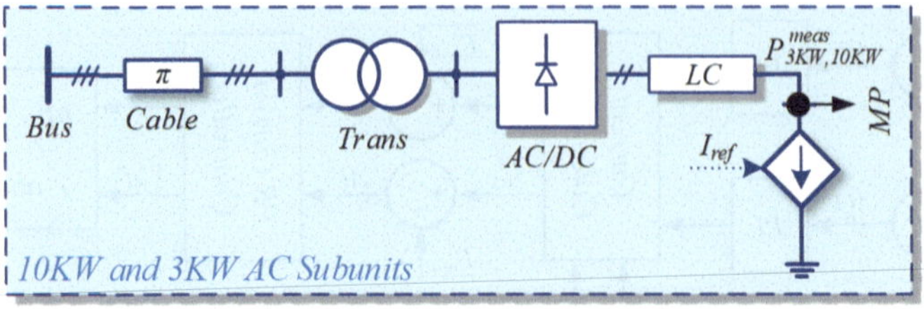

Fig. 7 Main components in the AC subunits

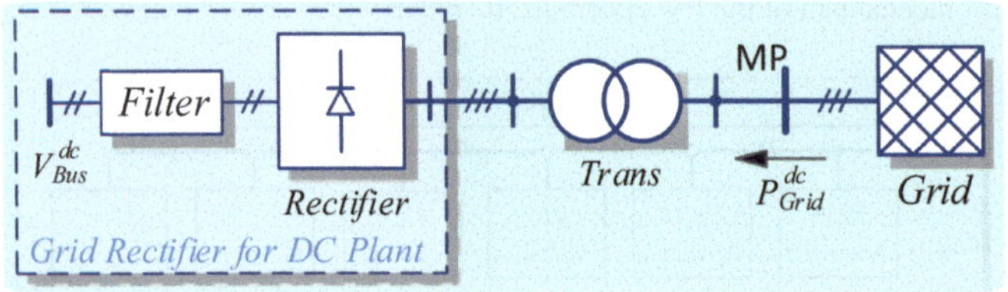

Fig. 8 Grid side high power rectifier for DC plant

Fig. 9 Buck converter and control of PV system in DC plant

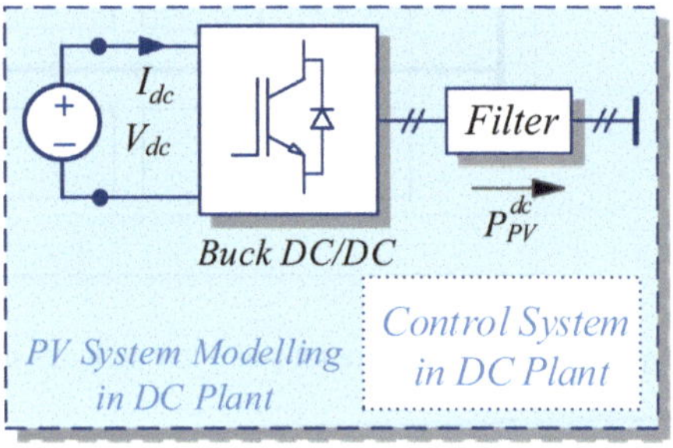

12 V DC output voltage and 300 A DC output current. More details of the subunits are reported in Table 1.

In the present work, an alternative solution based on the scheme of Fig. 2 is proposed with the aim of increasing the overall system efficiency. Three main configurations have been studied. In this case, the entire plant is supplied in DC; hence, a high-power rectifier is needed beside the LC filter to supply the main DC bus as shown in Fig. 8.

Concerning the PV system, the DC-AC power converter used to supply the main AC bus has to be replaced with a DC-DC buck converter with a similar power rating as shown in Fig. 9. Also, the LC filter should be modified to obtain proper output DC power. Concerning the DC-DC control strategy in this case, PI controllers are used for both voltage and current control as shown in Fig. 10.

Finally, as described in Fig. 11, the AC-DC controller which supplies each pool subunit has to be replaced with a DC-DC buck converter.

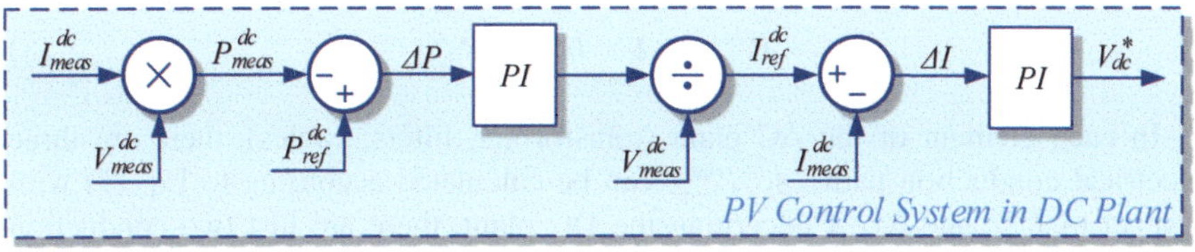

Fig. 10 Control of PV system in dc plant

Fig. 11 Main components of the DC subunits

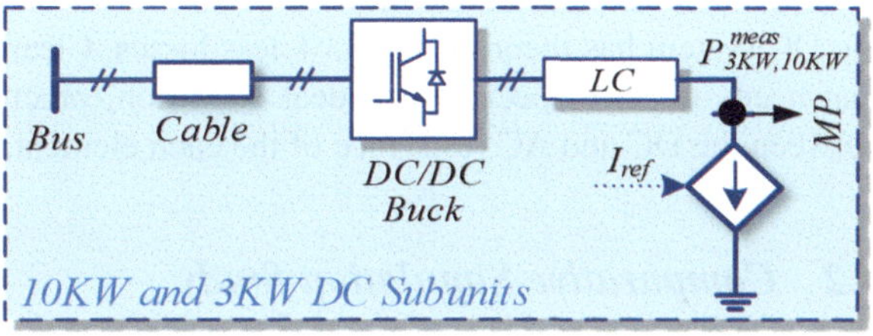

The π model is used to model the cables in both AC and DC plants. Concerning AC plant, a five-core, 4 mm AC cable has been modeled, while the DC system is modeled with a unipolar 4 mm cable (with operational voltage 0/+400 V).

The conduction losses P_{Losses} in both AC and DC plants can be calculated according to Eq. (4). The total absorbed power of all the subunits P_{Loads} (Eq. 5) is subtracted from $P_{\text{Generation}}$ which includes grid and PV power (Eq. 6). The measurement points used for power estimation of the subunits in AC and DC plants are shown in Figs. 7 and 11, respectively, and they are marked with "MP." Power from the grid instead is measured at the high-voltage side of the coupling transformer as shown in Fig. 8. Moreover, the power generated by the PV plant is measured at the entrance of the LC filter in both (AC and DC) plants.

$$P_{\text{Losses}} = P_{\text{Generation}} - P_{\text{Loads}} \tag{4}$$

$$P_{\text{Loads}} = \sum_{i=1}^{18} P_{3\,\text{kW}(i)} + \sum_{j=1}^{18} P_{10\,\text{kW}(j)} \tag{5}$$

$$P_{\text{Generation}} = P_{\text{PV}} + P_{\text{Grid}} \tag{6}$$

The resistance of the interconnected cables, isolator transformers, and LC filters also contribute to the conduction losses of the system. Hence, the selection of the correct measurement points is an essential task to obtain an accurate comparison of conduction losses. Equations (7) and (8) describe the basic equations for conduction loss calculation in each element for the AC and DC systems, respectively [14].

$$P^{\text{ac}}_{\text{Loss}} = 3 \times I^{\text{rms}2}_{\text{ac}} \times R_{\text{ac}} \tag{7}$$

$$P_{\mathrm{Loss}}^{\mathrm{dc}} = 2 \times {I_{\mathrm{dc}}}^{2} \times R_{\mathrm{dc}} \quad (8)$$

In each element of the AC plant (transformer, filters, cables), there are three electrical conduction paths, so $P_{\mathrm{Loss}}^{\mathrm{ac}}$ can be calculated according to Eq. (7) with respect to $I_{\mathrm{ac}}^{\mathrm{rms}}$ and R_{ac}. Concerning the DC plant, there are just two conduction paths for the DC current, and the conduction loss $P_{\mathrm{Loss}}^{\mathrm{dc}}$ can be calculated according to Eq. (8) with respect to I_{dc} and R_{dc}. The comparison between Eqs. (7) and (8) confirms that, in equivalent condition of both the systems ($R_{\mathrm{dc}} = R_{\mathrm{ac}}$, $I_{\mathrm{dc}} = I_{\mathrm{ac}}^{\mathrm{rms}}$) the DC system has theoretically 33% less losses. Clearly, this is an ideal condition that needs to have specific and ideal condition; practically, there are differences between the DC and AC resistance of the each element.

4.2 Comparative Simulation Study

In the present work, the AC plant and the DC one are modeled with Matlab/Simulink. In the next sections, where it is not specified, authors will refer to AC quantities; otherwise, for DC quantities, it will be specified with an index. The details of plants parameters are given in Table 1.

The power injection from the PV system is considered as an operational scenario for both models with the aim to analyze its impact in both the AC and DC systems (see Figs. 12 and 13). From Figs. 14, 15, 16, 17, 18, 19, 20, 21, 22, 23, 24, 25, 26, 27, 28 and 29, the following signals are shown:

(a) Grid- and PV-generated power (P_{Grid}, P_{PV}, $P_{\mathrm{Grid}}^{\mathrm{dc}}$, $P_{\mathrm{pv}}^{\mathrm{dc}}$).
(b) Total pool absorbed power (P_{Pool}, $P_{\mathrm{Pool}}^{\mathrm{dc}}$) including power losses.
(c) Current, voltage, and power signals of a 3 kW and a 10 kW subunit.
(d) Total power losses (P_{Losses}, $P_{\mathrm{Losses}}^{\mathrm{dc}}$).

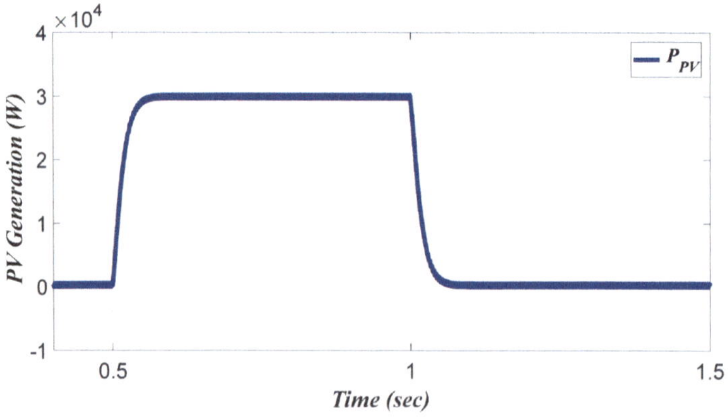

Fig. 12 PV system power generation profile in the AC model

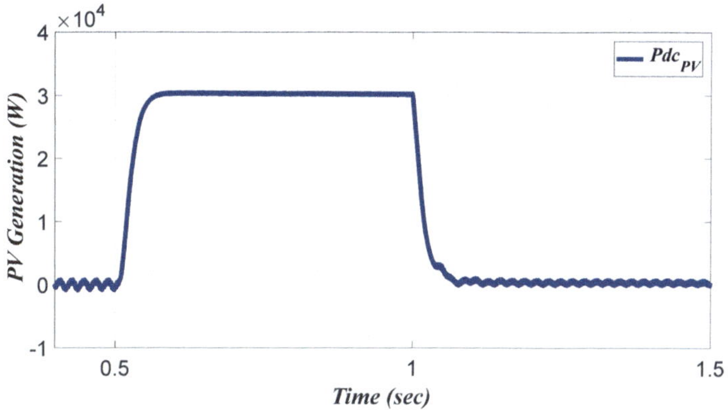

Fig. 13 PV system power generation profile in the DC model

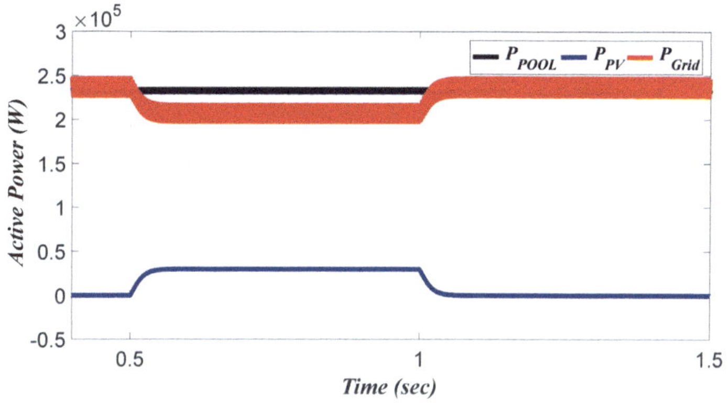

Fig. 14 Active power exchange in the AC model

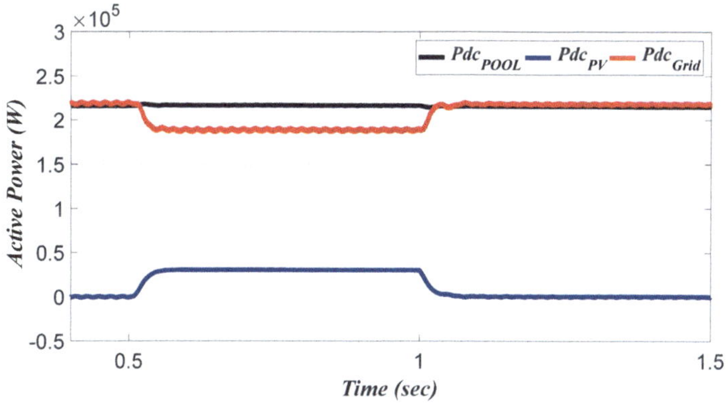

Fig. 15 Active power exchange in the DC model

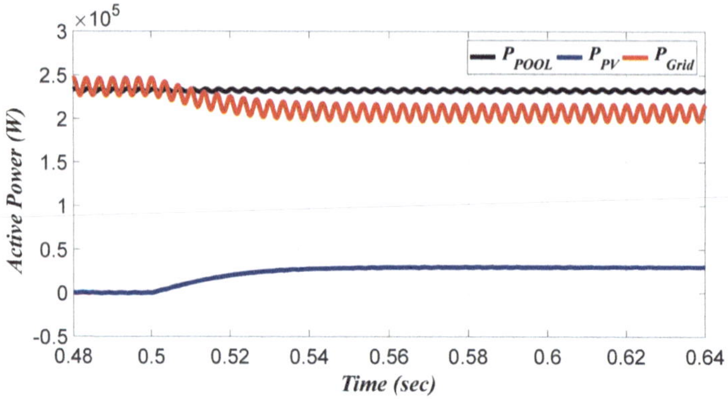

Fig. 16 Zoomed view of the active power exchange in the AC model

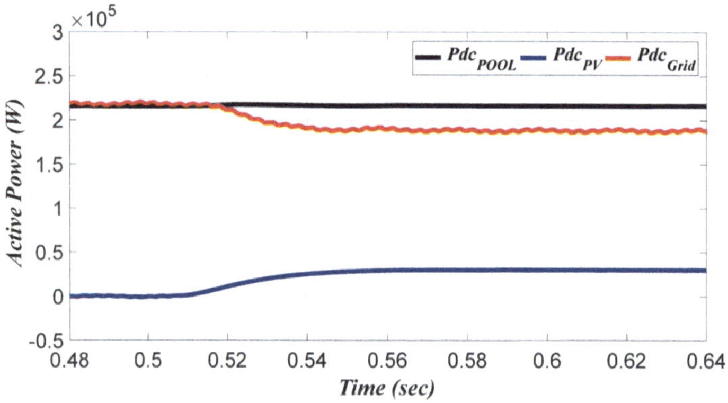

Fig. 17 Zoomed view of the active power exchange in the DC model

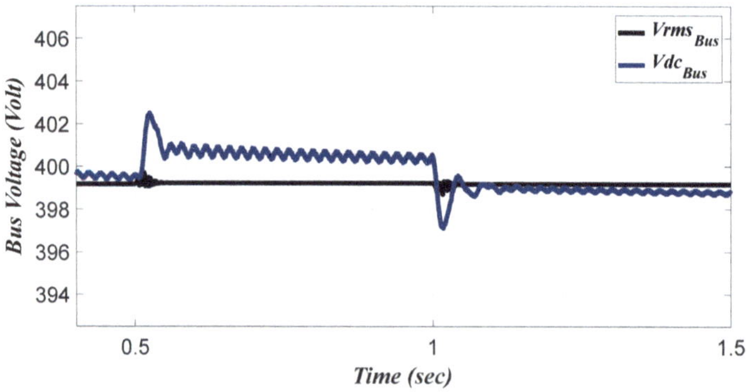

Fig. 18 Voltage across the main bus. AC and DC plant

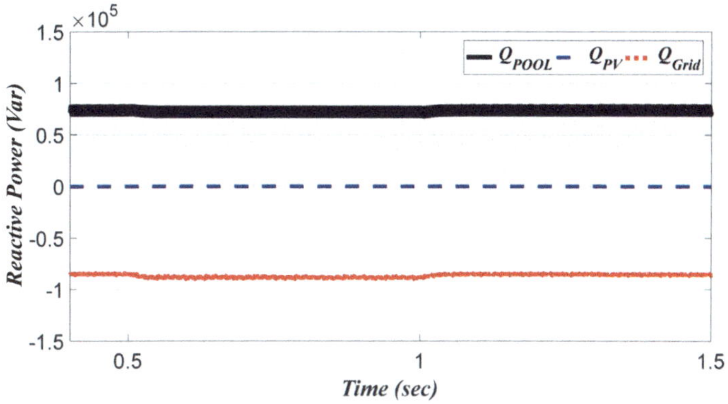

Fig. 19 Reactive power exchange in the AC system

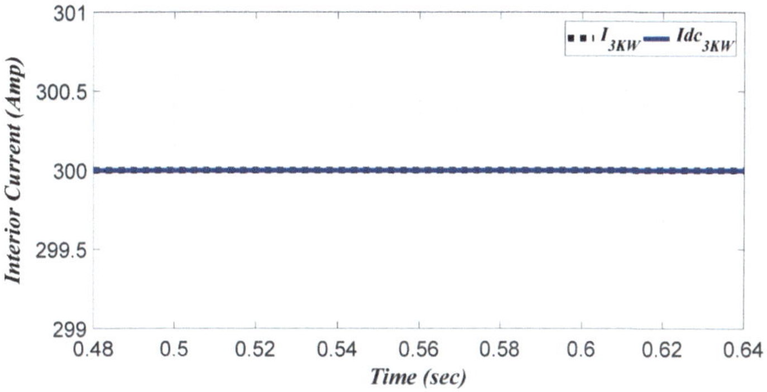

Fig. 20 Current control of the 3 kW subunits

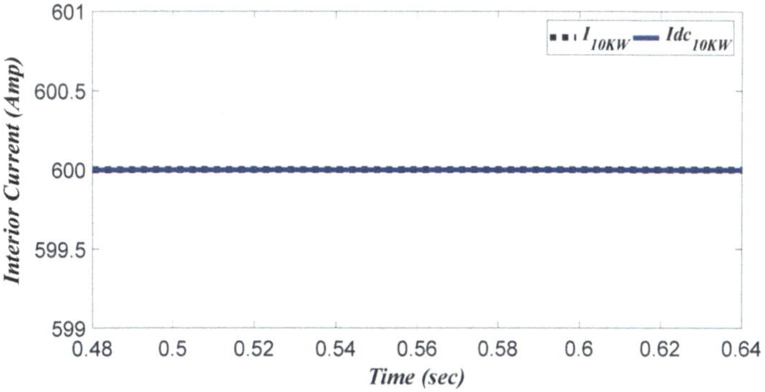

Fig. 21 Current control of the 10 kW subunits

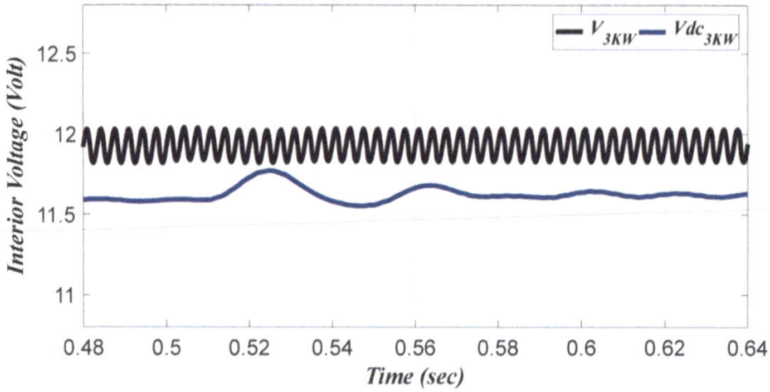

Fig. 22 Switching effect in voltage of 3 kW subunits

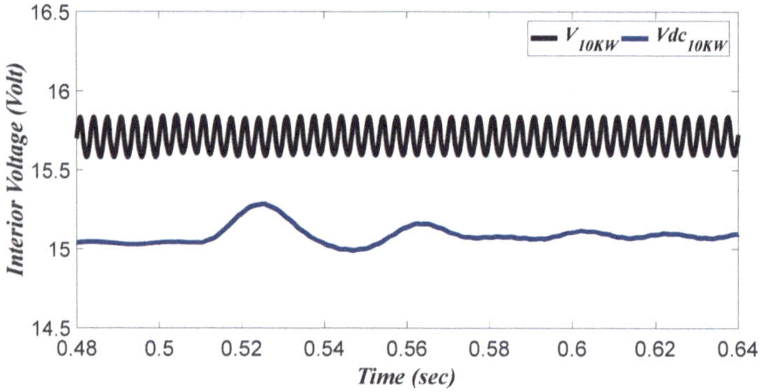

Fig. 23 Switching effect in voltage of 10 kW subunits

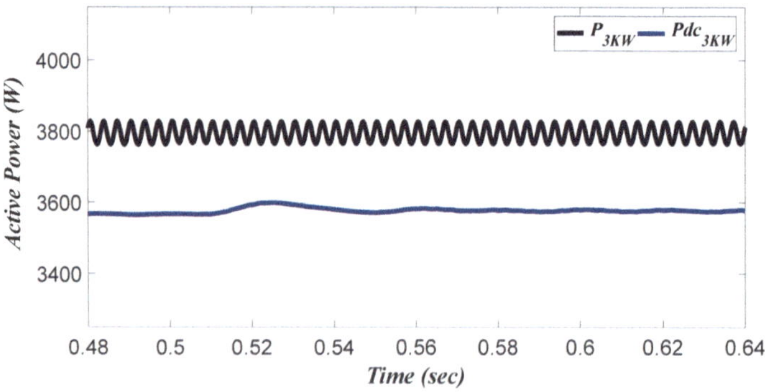

Fig. 24 Demand power of 3 kW subunits

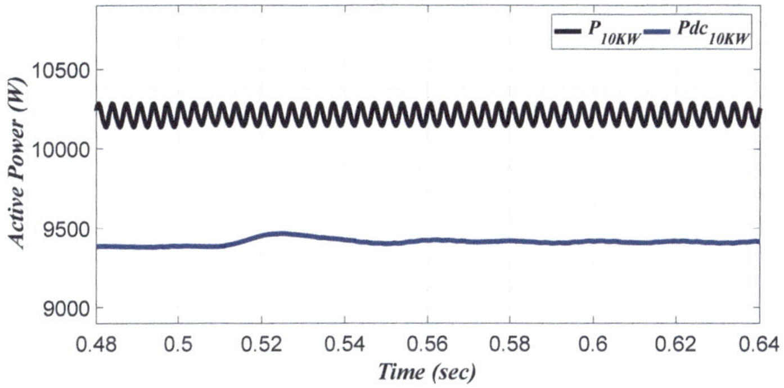

Fig. 25 Demand power of 10 kW subunits

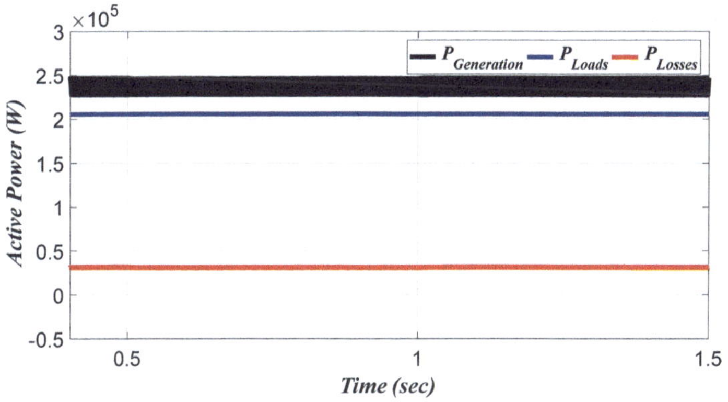

Fig. 26 Total conduction losses in AC model

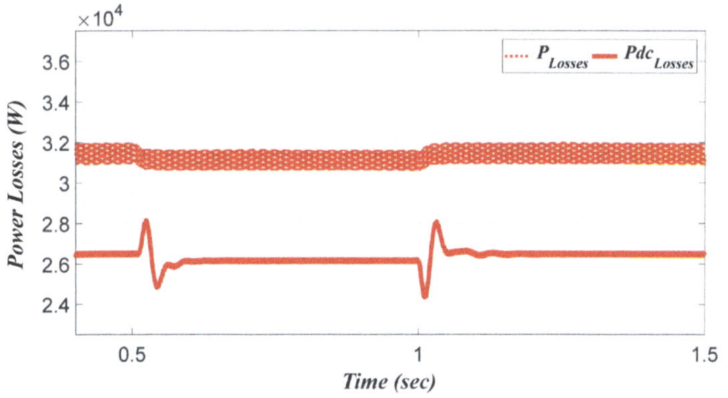

Fig. 27 Comparison of power losses in both models

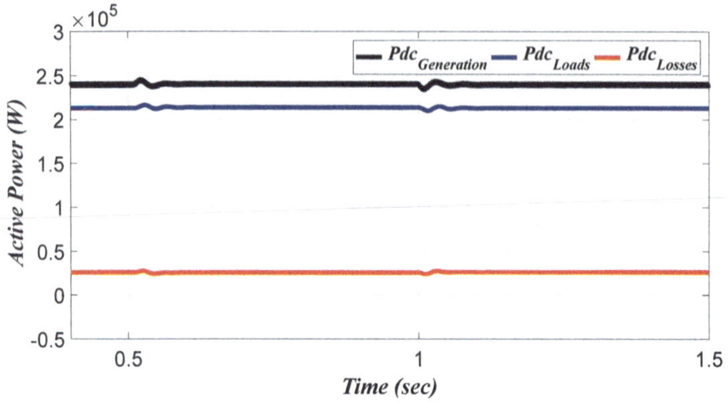

Fig. 28 Total conduction losses in DC model

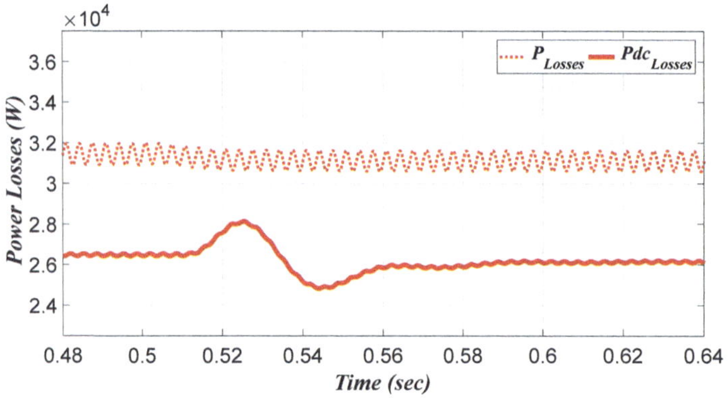

Fig. 29 Reduction of power losses in DC model

The behavior of pool and grid power signals (P_{Pool}, P_{Grid}) as PV power (P_{PV}) is injected in the system is shown in Figs. 14 and 15 in the AC and DC model, respectively. It can be noticed that the absorbed power from the grid side is reduced as a result of the contribution of the PV system.

Moreover, the behavior of the signal of the total power supplied to the pools (P_{pool}) is approximately flat, thanks to the current control mode in all of the subunits (pools) and the voltage control on 400 V bus (Fig. 18). This is verified for both the AC and DC models.

In Figs. 16 and 17, it is possible to notice that, by replacing the traditional AC structure with the proposed DC one, it results in reduced switching effects in the whole system. This implies smoother load supplying as well as reduced harmonics injection in the grid.

The exchanged reactive power in the AC system is another important issue concerning efficient power management [15]. As shown in Fig. 19, the AC system has almost 80kVAR exchanged reactive power between the pool and the grid. This reactive exchange occupies the capacity of the system and increases losses on all parts of the AC system. Reactive power costs, as well as its bad impact on the operation efficiency of the utilities, are really high in industrial applications [16]. Indeed, the reactive power reduction in AC plants requires solutions that are expensive both concerning hardware costs and energy consumption (capacitor banks and active power filters). Otherwise, the effects of reactive power are instead negligible in the DC system.

The different behavior of the power signals supplied to the AC and DC plant can be investigated by analyzing the subunit signals. From Figs. 20 to 25, the compared behavior of the 3 kW and 10 kW subunits is shown. As shown in Figs. 20 and 21, the subunits are controlled in constant current mode ($I_{3\,\mathrm{kW}}$, $I^{\mathrm{dc}}_{3\,\mathrm{kW}}$, $I_{10\mathrm{kW}}$, $I^{\mathrm{dc}}_{10\,\mathrm{kW}}$). As result of the switching effects in AC voltages ($V_{3\,\mathrm{kW}}$, $V_{10\,\mathrm{kW}}$) (see Figs. 22 and 23), the injected powers ($P_{3\,\mathrm{kW}}$, $P_{10\mathrm{kW}}$) are distorted by the switching harmonics in the AC system (see Figs. 24 and 25). The comparison of these signals confirms that, by supplying the subunits with DC voltage ($V^{\mathrm{dc}}_{3\,\mathrm{kW}}$, $P^{\mathrm{dc}}_{3\,\mathrm{kW}}$, $V^{\mathrm{dc}}_{10\mathrm{kW}}$, $P^{\mathrm{dc}}_{10\mathrm{kW}}$), the efficiency in each subunit is increased, thus enhancing the efficiency of the whole system.

The comparison of the total absorbed power in the subunits (P_{Loads}, $P^{\mathrm{dc}}_{\mathrm{Loads}}$) with the total generated power in the system ($P_{\mathrm{Generation}}$, $P^{\mathrm{dc}}_{\mathrm{Generation}}$) defines total conduction losses in the system (P_{Losses}, $P^{\mathrm{dc}}_{\mathrm{Losses}}$). In Figs. 26 and 28, a comparison of these losses in both systems is shown. It is possible to note that power losses (P_{Losses}) in the AC system (around the 32 kW) are reduced significantly in the DC system. As a result of decreased equivalent impedance of the cables and of the transformers in the DC model, power losses are reduced by about 26 kW. Hence, power loss ($\Delta P_{\mathrm{Losses}}$) reduction in the DC system is comparable with the PV system capacity (30 kW). As shown in Figs. 27 and 29, around 20% of loss reduction is achieved by using a DC voltage bus.

5 Conclusions

In this paper, an alternative DC power distribution system for industrial plants is presented. As a test case, a galvanizing plant with PV solar roof has been proposed. The proposed solution aims to improve the overall efficiency of the plant by reducing conduction losses and disturbances on the power distribution circuit.

The investigation has been performed based on the data of an existing galvanization plant. The study confirms that the replacement of the traditional AC structure with a well-designed DC system in addition with a supplementary renewable energy provider such as PV system can improve the overall efficiency of such plant with benefits both in terms of functionality and reduced power losses. To do this, a DC bus is routed along the plant and high-energy-demanding appliances are supplied with DC voltage. A comparative simulation of the classical AC approach and the proposed one highlighted the following main advantages: reduction of the switching effects and frequency distortion in the pool subunits, as well as negligible reactive power on circuit loads (transformer of the AC-DC rectifiers). These are the first major advantages of the DC model with respect to the AC one. Furthermore, the resistance of the interconnected cables and isolated transformer play a major role in the conduction losses of the system. So, the different performance of the cables, isolation transformers, and the filters on AC and DC systems is another important factor in the reduction of the losses and improvement of energy efficiency. All of these enhancements result in loss reduction and enhanced power quality in the system.

Acknowledgments This study is a part of a two-year project "AACPG" co-funded by the Regione Toscana under the Call FAR FAS 2014-PAR FAS 2007–2013 Linea A. The authors acknowledge financial support received for the work on this project.

References

1. F. Grasso, A. Luchetta, S. Manetti, et al., Measurement of electric power quantities and efficiency in nonsinusoidal conditions, in *2018 AEIT Int. Annu. Conf.*, (2018). https://doi.org/10.23919/aeit.2018.8577454
2. J. Driesen, H. De Keulenaer, M. Machiels, et al., The Leonardo power quality initiative (LPQI), an integrated approach towards E-learning in electrical engineering, in May 2003, 17th International Conference on Electricity Distribution Barcelon, 12–15 (2018)
3. E. Mamarelis, G. Petrone, B. Sahan, et al., What is the best DC/DC converter for an AC module?: Experimental analysis of two interesting solutions, in *2011 IEEE Int. Symp. Ind. Electron.*, (2011). https://doi.org/10.1109/isie.2011.5984422
4. L. Hadjidemetriou, Y. Yan, E. Kyriakides, A synchronization scheme for single-phase grid-tied inverters under harmonic distortion and grid disturbances. IEEE Trans. Power Electron. **32**(4), 2784–2793 (2017)
5. Y. Xue, K.C. Divya, G. Griepentrog, et al., Towards next generation photovoltaic inverters, in *2011 IEEE Energy Convers. Congr. Expo.*, (2011). https://doi.org/10.1109/ecce.2011.6064096
6. H. Wang, F. Blaabjerg, Reliability of capacitors for DC-link applications in power electronic converters—an overview. IEEE Trans. Ind. Appl. **50**, 3569–3578 (2014). https://doi.org/10.1109/tia.2014.2308357
7. T. Delaforge, S. Mariethoz, Optimal design of a low cost 20kW 99.1% efficiency active ZCS isolated DC-DC converter, in *2018 Int. Power Electron. Conf.*, (2018). https://doi.org/10.23919/ipec.2018.8507410
8. O. Aldosari, L.A.G. Rodriguez, J.C. Balda, S.K. Mazumder, Design trade-offs for medium- and high-frequency transformers for isolated power converters in distribution system applications, in *2018 9th IEEE Int. Symp. Power Electron. Distrib. Gener. Syst*, (2018). https://doi.org/10.1109/pedg.2018.8447788
9. D. Ronanki, S.S. Williamson, Evolution of power converter topologies and technical considerations of power electronic transformer-based rolling stock architectures. IEEE Trans. Transp. Electrif. **4**, 211–219 (2018). https://doi.org/10.1109/tte.2017.2765518
10. L. Albertoni, F. Grasso, J. Matteucci, et al., Analysis and design of full-bridge Class-DE inverter at fixed duty cycle, in *IECON 2016 - 42nd Annu. Conf. IEEE Ind. Electron. Soc*, (2016). https://doi.org/10.1109/iecon.2016.7793240
11. F. Corti, F. Grasso, A. Reatti, et al., Design of class-E ZVS inverter with loosely-coupled transformer at fixed coupling coefficient, in *IECON 2016 - 42nd Annu. Conf. IEEE Ind. Electron. Soc*, (2016). https://doi.org/10.1109/iecon.2016.7793285

12. J. Rocabert, A. Luna, F. Blaabjerg, P. Rodríguez, Control of power converters in AC microgrids. IEEE Trans. Power Electron. **27**(11), 4734–4749 (2012)
13. N.F. Roslan, J.A. Suul, J. Rocabert, P. Rodriguez, A comparative study of methods for estimating virtual flux at the point of common coupling in grid-connected voltage source converters with LCL filter. IEEE Trans. Ind. Appl. **53**(6), 5795–5809 (2017)
14. J. Grainger, W. Stevenson, *Power System Analysis* (McGraw-Hill, New York, 1994)
15. W. Qin, P. Wang, X. Han, X. Du, Reactive power aspects in reliability assessment of power systems. IEEE Trans. Power Syst. **26**(1), 85–92 (2011)
16. D. Feng, J. Zhong, D. Gan, Reactive market power analysis using must-run indices. IEEE Trans. Power Syst. **23**(2), 755–765 (2008)

Droop Control Strategy for Voltage Source Converters Containing Renewable Power Sources

Iván Andrade, Rubén Peña, Ramón Blasco-Gimenez, Javier Riedemann, and Cristian Pesce

Abstract This paper presents a strategy to control the active and reactive power in the point of common connection (PCC) of a wind generation system operating in islanded mode. A full back-to-back voltage source converter (VSC) is connected between each wind generator and the PCC. The control scheme considers voltage and frequency regulation for each VSC. The voltage and frequency references are obtained from *P-V* and *Q-f* droop characteristics of the generators, where *Q* and *P* are the reactive and active power supplied by each VSC to the PCC. Proportional-integral (PI) controllers process the voltage and frequency errors and set reference currents (in d-q frame) to be imposed by the converter. The strategy has been validated by mean of simulations, and results are presented showing the performance of the control strategy proposed.

I. Andrade
Department of Electrical Engineering, Universidad de Magallanes, Punta Arenas, Chile
e-mail: ivan.andrade@umag.cl

R. Peña
Department of Electrical Engineering, Universidad de Concepción, Concepción, Chile
e-mail: rupena@udec.cl

R. Blasco-Gimenez
Institute of Automatics and Industrial Informatics, Universitat Politécnica de Valencia, Valencia, Spain
e-mail: rblasco@upv.es

J. Riedemann (✉)
School of Electrical Engineering, Pontificia Universidad Católica de Valparaíso, Valparaíso, Chile
e-mail: javier.riedemann@pucv.cl

C. Pesce
Department of Electrical Engineering, Universidad de La Frontera, Temuco, Chile
e-mail: cristian.pesce@ufrontera.cl

W. Zamboni, G. Petrone (eds.), *ELECTRIMACS 2019*, Lecture Notes in Electrical Engineering 615, https://doi.org/10.1007/978-3-030-37161-6_22

1 Introduction

The increase in the world population and environmental pollution, along with the decrease in the fossil-based fuels, has made the renewable energy sources a very attractive alternative for electrical power generation. It is expected that the energy generation from renewable sources will play an important role in the world of 2050 [1]. Among the different renewable energy sources, wind energy takes attention due to its high availability and compatibility with the advanced technologies. As described in [2], in last years, there have been a steep increase in the installation of the wind plants with favorable future projections. In this scenario, offshore wind farms are interesting and suitable for large-scale generation, because a higher energy harvest can be obtained in the middle of the sea than in the coast (due to more constant and stronger winds). However, as this type of systems are located far away from the final consumers, it is necessary to transport the energy generated to the distribution centers located in the mainland. The typical configuration consists of connecting every wind generator to a full back-to-back VSC; then, the VSCs are connected to a point of common connection (PCC), forming an islanded AC grid [3, 4], and from this point, the energy is supplied to the load (see Fig. 1). On the other hand, to control electrical generation systems containing power electronic converters, different strategies have been suggested in the literature, mainly based on classic statism curves *P*-*f* and *Q*-*V* (*P*, *Q*, *V*, and *f* are the active and reactive power, voltage, and frequency at the PCC) aiming to obtain a correct integration with traditional power systems based on synchronous generators [5, 6]. In general, droop control (statism) is suitable to transfer and distribute the active/reactive power between the generator units and the PCC.

In this paper, the topology shown in Fig. 1 is considered, and the statism curves are implemented in every individual generator. The output of the statism curves will set the references for voltage and frequency controllers. The objective is to allow the generation units to share the load of the system. Since, in the present case, the

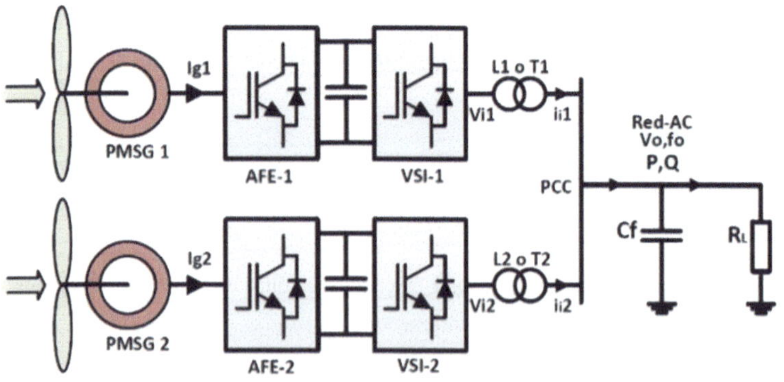

Fig. 1 Topology proposed

model results in a relation between the active power and the PCC voltage and a relation between reactive power and the PCC frequency, then the system will have inverse statism with characteristics *P*-*V* and *Q*-*f*.

2 System Model

In the proposed scheme, shown in Fig. 1, the VSCs are connected to the PCC through a step-up transformer. To model the system, it is considered that the DC-link voltages are controlled, and there is a LC filter in the output of every power inverter formed by the coupling transformer inductance and a capacitor installed in the PCC side [7].

The following expressions, in abc frame, are obtained:

$$v_i^{\mathrm{abc}}(t) = R_{\mathrm{s}} i_i^{\mathrm{abc}}(t) + L_{\mathrm{s}} \frac{d}{\mathrm{d}t} i_i^{\mathrm{abc}}(t) + v_{\mathrm{o}}^{\mathrm{abc}}(t) \tag{1}$$

$$i_i^{\mathrm{abc}}(t) = C_{\mathrm{f}} \frac{d}{dt} v_{\mathrm{o}}^{\mathrm{abc}}(t) + i_{\mathrm{o}}^{\mathrm{abc}}(t) \tag{2}$$

The above equations are transformed into a synchronous rotating frame dq. This frame is synchronized with the grid voltage v_{o}, with frequency ω_{o}, and then, the q-axis voltage will be 0 ($v_{q\mathrm{o}} = 0$), hence, as follows:

$$\frac{\mathrm{d}i_i^d(t)}{\mathrm{d}t} = -\frac{R_{\mathrm{s}}}{L_{\mathrm{s}}} i_i^d(t) + \omega_{\mathrm{e}} i_i^q(t) + \frac{1}{L_{\mathrm{s}}} v_i^d(t) - \frac{1}{L_{\mathrm{s}}} v_o^d(t) \tag{3}$$

$$\frac{\mathrm{d}i_i^q(t)}{\mathrm{d}t} = -\frac{R_{\mathrm{s}}}{L_{\mathrm{s}}} i_i^q(t) - \omega_{\mathrm{e}} i_i^d(t) + \frac{1}{L_{\mathrm{s}}} v_i^q(t) \tag{4}$$

$$\frac{\mathrm{d}v_{\mathrm{o}}^d(t)}{\mathrm{d}t} = \frac{1}{C_{\mathrm{f}}} i_i^d(t) - \frac{1}{C_{\mathrm{f}}} i_{\mathrm{o}}^d(t) \tag{5}$$

$$\omega_{\mathrm{e}} v_{\mathrm{o}}^d(t) = \frac{1}{C_{\mathrm{f}}} i_i^q(t) - \frac{1}{C_{\mathrm{f}}} i_{\mathrm{o}}^q(t) \tag{6}$$

where R_{s} and L_{s} are the equivalent resistance and inductance of the coupling transformer in the PCC side, and C_{f} is the filter capacitance. The currents i_i^d and i_i^q are controlled by the VSC by imposing the voltages v_i^d and v_i^q. Moreover, it can be noted in Eq. (5) that the PCC voltage v_{o}^d can be regulated by adjusting i_i^d, and from Eq. (6), it is noted that the frequency of the system can be controlled by means of i_i^q.

3 Control Scheme

In this section, the control strategy is thoroughly explained considering the different loops to regulate: output currents, PCC voltage, system frequency, and PCC active/reactive power.

3.1 Control of Currents i_i^d and i_i^q

To design the inverter current controller, Eqs. (3–4) are transformed into Laplace domain:

$$V_i^d(s) = U_d(s) - \omega_e L_s I_i^q(s) + V_o^d(s) \quad (7)$$

$$V_i^q(s) = U_q(s) + \omega_e L_s I_i^d(s) \quad (8)$$

where

$$U_i^d(s) = sL_s I_i^d(s) + R_s I_i^d(s) \quad (9)$$

$$U_i^q(s) = sL_s I_i^q(s) + R_s I_i^q(s) \quad (10)$$

are the transfer functions of the dq-axis currents, and Eqs. (7–8) are compensation terms used to improve the dynamic behavior of the control system. Figure 2 shows a block diagram of the current control loop.

3.2 Control of Voltage v_o^d

As mentioned above, the output voltage v_o^d can be controlled by means of regulating the current i_i^d. Equation (5) in Laplace domain is given by the following:

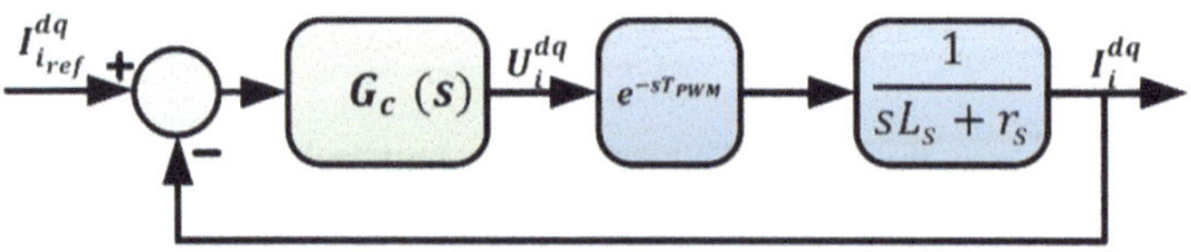

Fig. 2 Current control loop

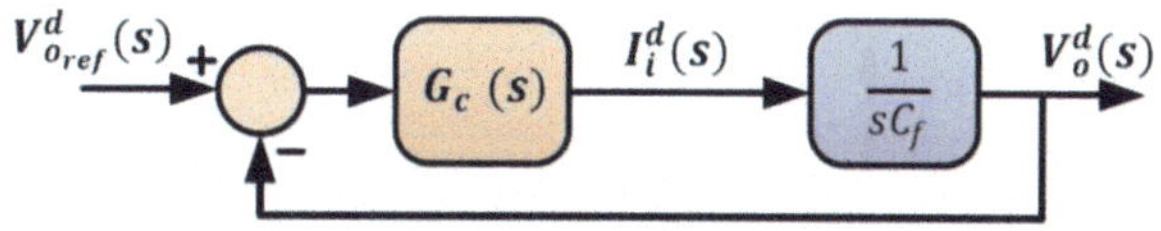

Fig. 3 Voltage control loop

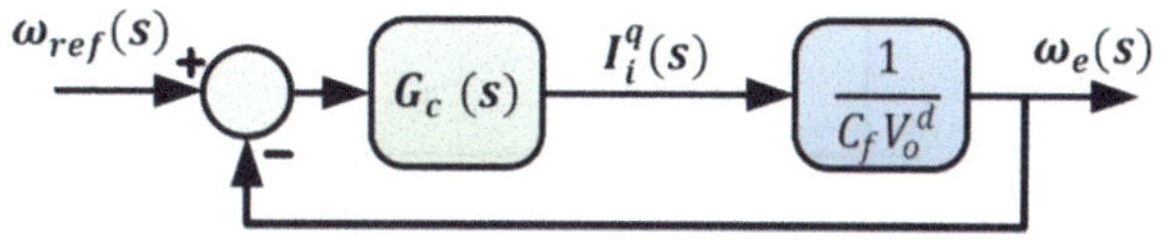

Fig. 4 Frequency control loop

$$I_i^d(s) = sC_\mathrm{f}V_\mathrm{o}^d(s) + I_\mathrm{o}^d(s) \tag{11}$$

In this equation, I_o^d is the load current that is considered a perturbation from the control point of view; therefore, it is not considered in the block diagram for the controller design (Fig. 3).

3.3 *Frequency Control*

The frequency of the system must be controlled accurately aiming to keep it inside the ranges established by electrical normatives. As stated in Sect. 2, the frequency can be controlled by means of i_i^q current. Equation (6) expressed in Laplace domain is as follows:

$$I_i^q(s) = \omega_\mathrm{e}C_\mathrm{f}V_\mathrm{o}^d(s) + I_\mathrm{o}^q(s) \tag{12}$$

Similar to Eq. (11), in this equation, the output current I_o^q is considered a perturbation and is not considered for controller design purposes. A block diagram of the control loop is shown in Fig. 4.

3.4 *Control of Active/Reactive Power*

The output power inverters of the back-to-back VSCs are connected in parallel to the PCC and contribute to control the voltage $v_\mathrm{o}^\mathrm{abc}$ and the frequency ω_e of the AC grid. Aiming to share the control of the system, increasing or decreasing the contribution of the inverters to the total active and reactive power delivered, a droop control *P*-*V* and *Q*-*f* is proposed. Figure 5 shows the type of curves used.

For tracking the reference power, the droop control is modified with voltage step ΔV that is obtained from a PI controller [7].

$$\Delta v_i = k_{\mathrm{p}P}\left(P_\mathrm{ref} - P_{\mathrm{f}i}\right) + k_{iP}\int\left(P_\mathrm{ref} - P_{\mathrm{f}i}\right)\mathrm{d}t \tag{13}$$

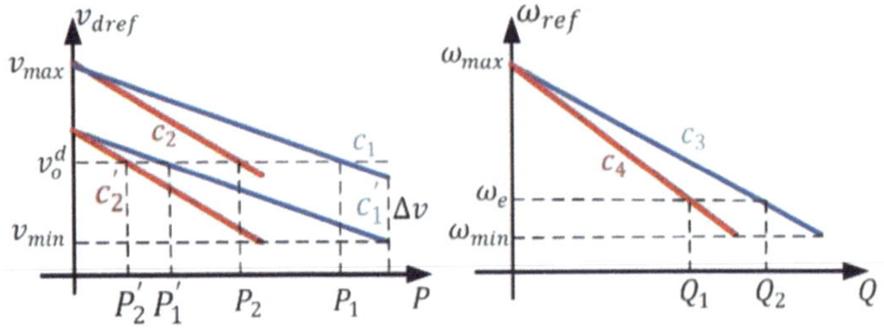

Fig. 5 Statism curves *P*-*V* (left) and *Q*-*f* (right)

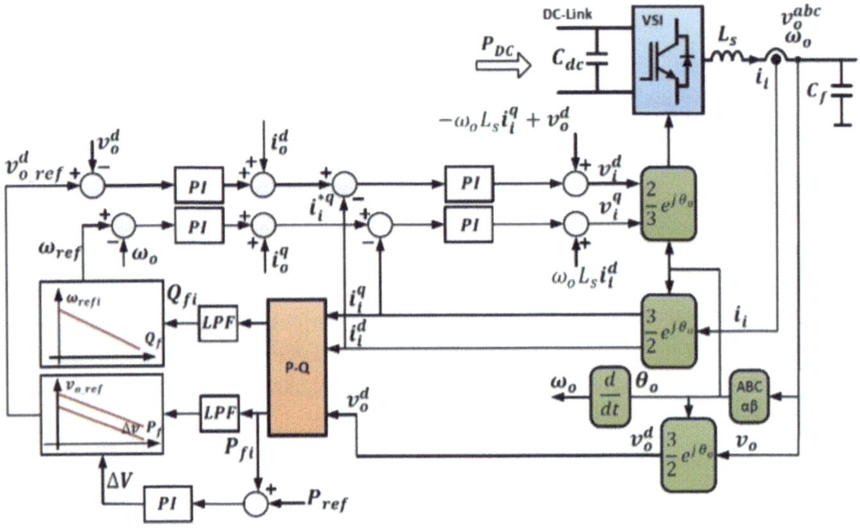

Fig. 6 Control system implemented

Then, the modified droop curve used in each voltage source inverter (VSI) is as follows:

$$V_{\text{ref}i} = V_{\text{omax}} - m_i P_{\text{max}} + \Delta V_i \tag{14}$$

Finally, the balance of reactive power Q considers a curve of constant slope (m_i) in each inverter:

$$\omega_{\text{ref}i} = \omega_{\text{max}} - m_i Q_i \tag{15}$$

The overall control system implemented is shown in Fig. 6.

4 Results

The entire system has been modeled using Matlab/Simulink platform. Figure 7 shows the schematic implemented and Table 1 shows the simulation parameters.

In Fig. 7, the diagram implemented in Simulink for the simulation of the system is shown, where the two blocks that contain the C code for the control of the converters can be depicted.

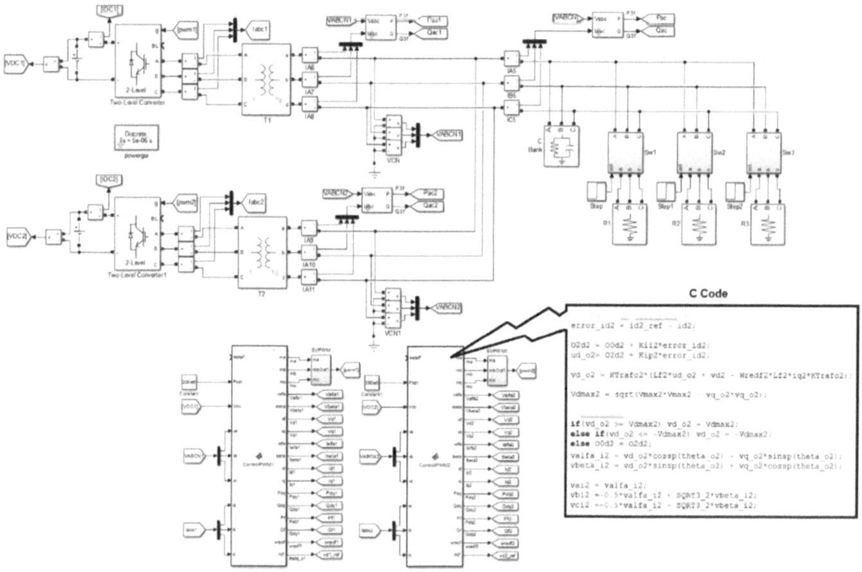

Fig. 7 Simulation scheme

Table 1 Simulation parameters

Description	Value
Rated power T_1, T_2	300 MVA
Transformer inductance	0.1 pu
Transformer resistance	0.01 pu
PCC line-to-line voltage	30 kV
PCC frequency	50 Hz
Switching frequency	2 kHz
Inverters DC-link voltage	5 kV
Capacitor bank	50 MVAr

4.1 Load Impact at PCC

In this test, each of the VSIs has a droop curve type with a fixed slope and with $\Delta v = 0$. VSI-1 has the curve (P-V) C_1 and the curve (Q-f) C_3, while VSI-2 has the curve (P-V) C_2 and the curve (Q-f) C_4. In Fig. 8a, the line voltage present in the PCC is shown and in Fig. 8b, c; the line currents of the VSIs are shown during 100 MW load impacts. The load impacts are carried out every 1 s. As the load connected to the PCC increases, it can be seen that the voltage decreases; this is because in the curve (P-V) the active power connected to the PCC increases and the output voltage of the VSIs decreases. Initially, the PCC load is 100 MW and the line voltage has a fundamental value of 35 kV. When the first 100 MW load is applied, the voltage at the PCC decreases to 33.5 kV. When the second 100 MW load is connected, the voltage decreases to 32 kV. When the last 100 MW is connected, the fundamental voltage reaches 30 kV, which corresponds to the nominal voltage for the nominal power of the system. It is also observed from Fig. 8b, c that the VSI line currents increase due to the load impacts. At rated power (400 MW), the VSI-1 line current is 70 kA and the VSI-2 line current is 50 kA.

Figure 9a shows the active power delivered by each of the VSIs during the load impacts. The VSI-1 initially has 40 MW and the VSI-2 supplies 30 MW. When the last load impact is applied to the PCC, the power supplied by the VSIs is 220 MW and 180 MW, respectively. In Fig. 9b, the d-axis line currents are shown for each VSI. It can be seen that the d-axis currents have the same shape as the active power supplied by each of the VSIs.

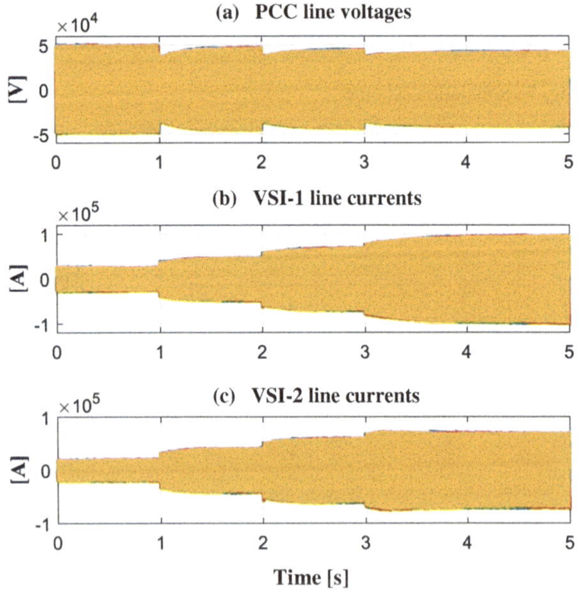

Fig. 8 Load impact in the PCC: (**a**) PCC line voltages, (**b**) VSI-1 line currents, (**c**) VSI-2 line currents

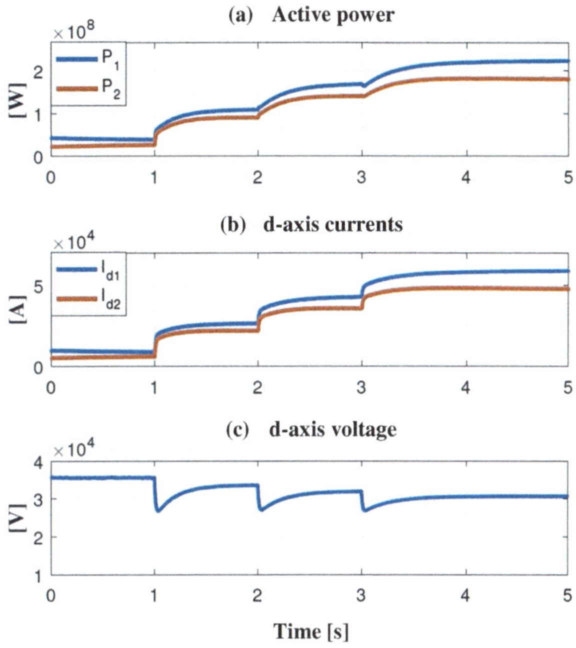

Fig. 9 Resistive load impact: (**a**) active powers P_1–P_2, (**b**) d-axis currents I_{d1}–I_{d2}, (c) d-axis voltage V_d

In Fig. 9c, the PCC d-axis voltage V_d is shown. The voltage decreases as the load connected to the PCC increases. The d-axis voltage corresponds to the line voltage of the PCC. Figure 10a shows the distribution of the reactive power of the system among the VSIs. This power is initially capacitive due to the capacitor bank that is connected to the PCC. As the active power increases and the PCC voltage decreases, the reactive power decreases. In Fig. 10b, the VSI q-axis currents are shown, and in Fig. 10c, the PCC frequency, set by the grid-forming converters, is shown. The PCC frequency is regulated at about 50 Hz.

4.2 *Changes in the Power References of the VSIs*

Another test performed was the power tracking of each VSI to different power references. In order to do that, the droop curves are moved, for example, from C_1 to C'_1. The reference power tracking is performed by the PI power controller output Δv at each VSI. In Fig. 11a, the line voltage increases as the power references of each VSI increases. In Fig. 11b, c, the line currents of the VSIs are shown.

In Fig. 12a, the active power tracking performance to step active power reference changes are shown. In VSI-1, the initial power is 20 MW; at the $t = 1$ s, the active power rises to 60 MW; at $t = 2$ s, the active power increases to 100 MW; at $t = 3$ s,

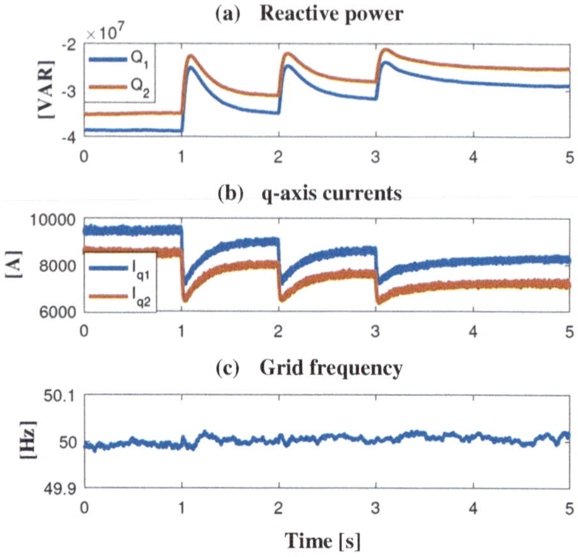

Fig. 10 Resistive load impact: (a) reactive powers Q_1-Q_2, (b) q-axis currents I_{q1}-I_{q2}, (c) grid frequency

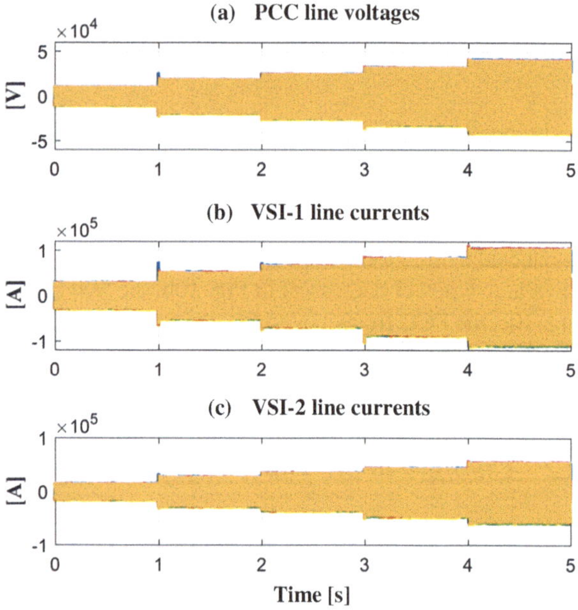

Fig. 11 Change in reference power: (**a**) PCC line voltages, (**b**) VSI-1 line currents, (**c**) VSI-2 line currents

the active power increases to 160 MW; and finally, at $t = 4$ s, the active power increases to 260 MW. In VSI-2, initially, the active power is 10 MW; at $t = 1$ s, the active power rises to 30 MW; at $t = 2$s, the power creases to 50 MW; at $t = 3$s, the power increases to 80 MW; and finally, at $t = 4$ s, the active power increases to 140 MW. In Fig. 12b, the d-axis currents are shown; these currents regulate the

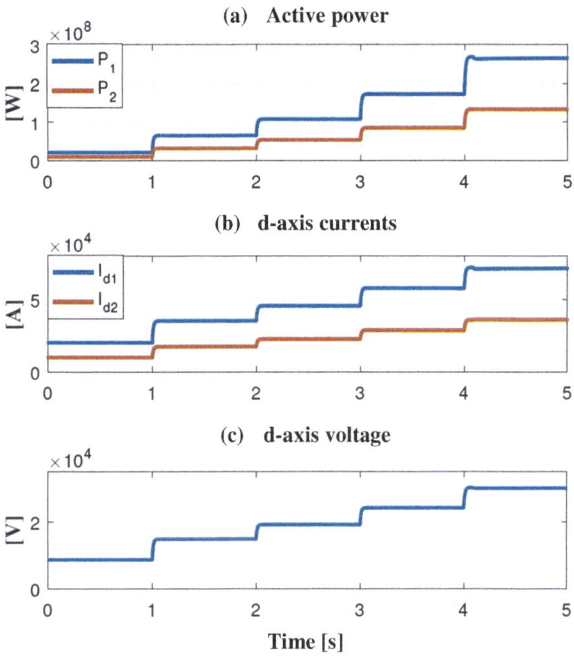

Fig. 12 Change in reference power: (**a**) active powers P_1–P_2, (**b**) d-axis currents I_{d1}–I_{d2}, (**c**) d-axis voltage V_d

output voltage of the VSIs. In Fig. 12c, the PCC d-axis voltage is shown. The PCC voltage increases as the power at that PCC increases. In Fig. 13a, the VSI reactive powers are shown. The reactive power is capacitive due to the capacitor bank connected at the PCC. The reactive power increases as the PCC voltage increases. In Fig. 13b, the VSI q-axis currents are shown, and in Fig. 13c, the PCC frequency is depicted as the active power step changes are carried out.

4.3 Impact of a Capacitor Bank at the PCC

Another test performed was to apply capacitive ten MVAR impacts at the PCC. This is done at $t = 2.5$ s. Figure 14a shows the reactive power which increases when the capacitive bank is connected. In Fig. 14b, the VSI q-axis currents are shown. In Fig. 14c, the PCC frequency is depicted. The frequency decreases to 49.4 Hz when connecting the capacitors. The 50 Hz frequency is after regained in about 300 ms. In Fig. 14d, the PCC voltage is shown.

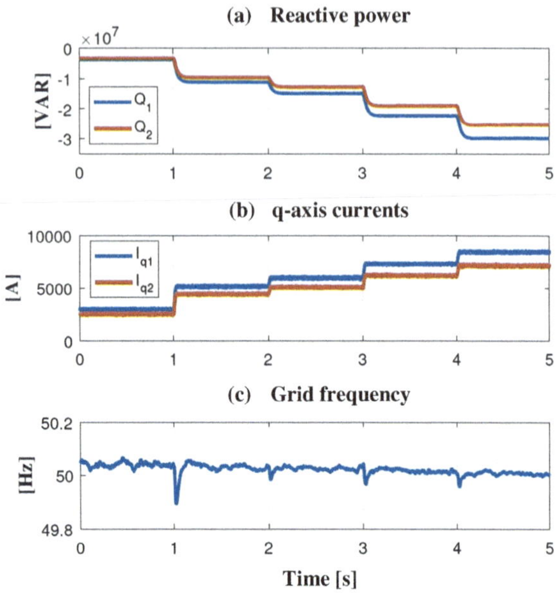

Fig. 13 Change in reference power: (**a**) reactive powers Q_1–Q_2, (**b**) q-axis currents I_{q1}–I_{q2}, (**c**) grid frequency

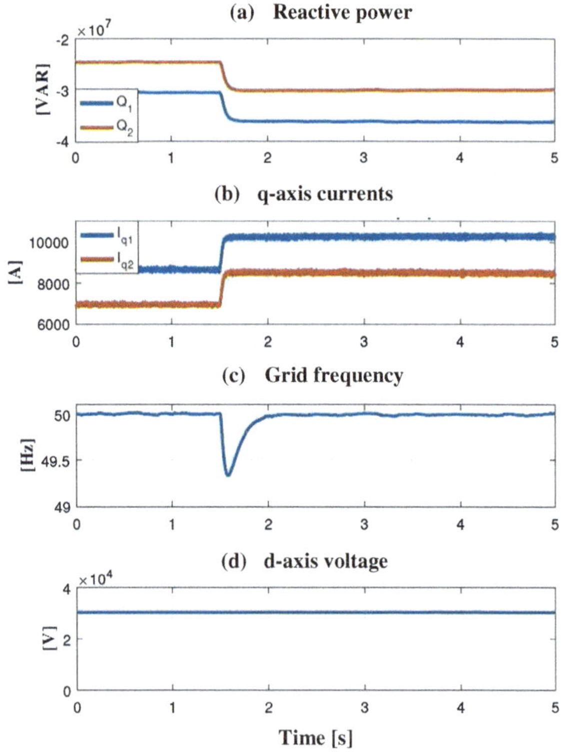

Fig. 14 Impact of a capacitive bank: (**a**) reactive powers Q_1–Q_2, (**b**) q-axis currents I_{q1}–I_{q2}, (**c**) grid frequency, and (**d**) d-axis voltage

5 Conclusions

This work has validated, through simulation, an integrated voltage, frequency, and active and reactive control system of voltage for an AC isolated. The topology and control strategy is applicable to electrical systems with generation sources located far from the electrical network. The model of the system has naturally led to the development of a control strategy of the droop type (P-V) and (Q-f), instead of the more traditional methods of the (Q-V) and (P-f) type. The droop scheme allows the distribution of active and reactive powers in each generating source. The proposed approach leads to an independent design of voltage and frequency controllers. The proposed control is capable of transferring power from the DC links of the VSIs to the AC network by shifting the (P-V) droop curves. The reactive power is distributed within the energy sources using (Q-f) droop curves. The control shows good performance to load impacts, changes in power references for each of the inverters, and the impact of a bank of capacitors connected to the PCC.

Acknowledgments The financial contribution provided by Fondecyt, Chile, under contract 11180092 and CONICYT/FONDAP/15110019 is kindly acknowledged as well as the support of the University of Concepci6n, the University of Magallanes, Chile, and the Universitat Politécnica de Valencia, Spain. Additionally, the authors would like to thank the support of the Spanish Ministry of Science, Innovation and Universities and EU-FEDER funds under grant DPI2017-84503-R.

References

1. E. E. P. I. Association et al., Global market outlook for photovoltaics 2014–2018, EPIA report. Disponível em: http://www. epia. org/home (2014)
2. G. W. E. Council, Global wind report 2015, www.gwec.net/indexphp (2016)
3. A. Chauhan, R. Saini, A review on integrated renewable energy system based power generation for stand-alone applications: configurations, storage options, sizing methodologies and control. Renew. Sustain. Energy Rev. **38**, 99–120 (2014)
4. M. Liserre, R. Cardenas, M. Molinas, J. Rodriguez, Overview of multi-mw wind turbines and wind parks. IEEE Trans. Industr. Electron **58**(4), 1081–1095 (2011)
5. D. Wu, F. Tang, J.C. Vasquez, J. M. Guerrero, Control and analysis of droop and reverse droop controllers for distributed generations, in Multi-Conference on Systems, Signals & Devices (SSD), 2014 11th International (2014), pp. 1–5
6. S.-H. Hu, C.-Y. Kuo, T.-L. Lee, J.M. Guerrero, Droop-controlled inverters with seamless transition between islanding and grid-connected operations, in Energy Conversion Congress and Exposition (ECCE), 2011 IEEE (2011), pp. 2196–2201
7. R. Blasco-Gimenez, S. Añó-Villalba, J. Rodríguez-D'derléeF. Morant, S. Bernal, Voltage and frequency control of sg based wind farms with uncontrolled hvdc rectifier," in 2010 IEEE International Symposium on Industrial Electronics. IEEE, 2010 (2010), pp. 2499–2504

Part III
Design, Optimisation, Identification and Diagnosis

ISO26262-Compliant Development of a High Dependable Automotive Powertrain Item

Jacopo Sini, Massimo Violante, and Riccardo Dessi

Abstract In the automotive domain, the electrical and electronic items are playing day after day a more central role. Since most of these units are in charge of safety-relevant functionalities, a strict development process is required. The ISO26262 automotive functional safety standard describes a mandatory process to design, validate and verify item designs. The aim of this work is to describe a suitable way to overcome some safety life cycle issues. The description starts from the concept phase, with the Hazard Analysis and Risk Assessment, in where the safety goals are defined, and an Automotive Safety Integrated Level is assigned to each of them. After that preliminary phase, it will be shown how it is possible to check the reliability of the obtained hardware design keeping into account the failure detection and mitigation capabilities of both hardware and software. To achieve this goal, a simulation-based Failure Mode and Effect Analysis assessment technique is applied to assess the hardware design's possible sources of failures and to analyse detection, isolation and mitigation capabilities. To achieve this result, the hardware model and the embedded software have been implemented using the Model-Based Software Design approach. This approach has been demonstrated on an electrical vehicle powertrain design.

1 Introduction

Thanks to electrical and electronic (E/E) items, nowadays, vehicles are embedding day by day more complex functionalities. Hence, these devices are becoming more important from the functional safety point of view. The development of Advanced Driver-Assistance Systems (ADAS) and, in future, fully autonomous

---

J. Sini (✉) · M. Violante
DAUIN – Politecnico di Torino, Turin, Italy
e-mail: jacopo.sini@polito.it; massimo.violante@polito.it

R. Dessi
General Motors – Global Propulsion Systems, Turin, Italy
e-mail: riccardo.dessi@gm.com

W. Zamboni, G. Petrone (eds.), *ELECTRIMACS 2019*, Lecture Notes in Electrical Engineering 615, https://doi.org/10.1007/978-3-030-37161-6_23

driving vehicles requires drive-by-wire capabilities. To achieve the needed safety level, it is necessary to design, implement and verify these devices in a very structured way.

There exists an international standard, ISO26262, about the functional safety of E/E components of lightweight vehicles. This standard, published in 2011, is now waiting to be updated in 2019 to keep into account its application on semiconductors and an adaption for motorcycles.

This paper is about a methodology suitable to develop high dependable automotive items. As a benchmark application, an inverter for electrical vehicle powertrains is considered. Of course, the authors do not want to present a novelty about six-step inverters, but to propose a way, compliant with mandatory standards and current industry best practices, to obtain more dependable items.

The rest of the paper is organised as follows: Sect. 2 reports background material. Sections 3 and 4 present the proposed methodology and the experimental results obtained on the benchmark application. Finally, Sect. 5 draws some conclusions.

2 Background

The ISO26262 [1] standard imposes the requirements to be followed during the development of safety-relevant items (safety life cycle). The standard is divided into ten parts. Of these, the most important related to our work are the third, fifth and sixth. Part III, "concept phase", imposes in 3–7 section to perform a Hazard Analysis and Risk Assessment activities, in which the safety goals of the item have to be determined. An Automotive Safety Integrated Level (ASIL), that is, an indication of how much the safety-related effects of a miss of the considered goal are severe, has to be assigned. Each safety goal has its own ASIL. Part V is about the hardware-level product development. In this part, the most interesting sections for our purposes are the 5–6 "hardware architectural metrics" and 5–9 "evaluation of violation of the safety goal due to random HW failures". For what regards the embedded software, described in Part VI, the most interesting section for the sake of our work is the 6–10 "software integration and testing".

The design process starts from a high level of abstraction description of the item functionalities. From this document, it is possible to start the Hazard Analysis and Risk Assessment (HARA) phase [1, 2]. During this stage, the safety-relevant effects of the item behaviour have to be considered. As requested from the standard, a list of safety goals has to be provided for the item functionalities. As said before, at the end of this process, an Automotive Safety Integrated Level (ASIL) has to be assigned to each of the goals. It is determined by summarising three different parameters: severity, controllability and exposure. Severity is a metric to represent the worst-case consequences, due to the safety goal violation, on people inside or in the surroundings of the vehicle. Controllability is a parameter to describe the capability of the average human driver to mitigate the effect of the misbehaviour. Severity and controllability depend on the driving situation. Since designers have to summarise these parameters by only considering the worst-case ones, another parameter is required from the analysis: the exposure. This parameter is related to the amount of time over the total operational one when the people are subject to the hazardous situation.

The second stage presented in this work is the Failure Mode and Effect Analysis (FMEA). This technique can be used to compute the mandatory "hardware architectural metrics" and the "violation of safety goals due to random HW failures". The main idea is to list all the possible failures of each component in the hardware designs and to determine, for each of them, if the misbehaviour is safe or dangerous and if the failure detection system of the item is able to detect and mitigate the failure. Usually, this activity is performed by experts that classify the effect starting from their knowledge of the schematics. Anyway, in this work, we used a simulation-based classification approach. More information about this methodology can be found in [3] and in Sect. 4.

Another main point of the approach is Model-Based Software Design (MBSD). It is a software development methodology that relies on two development instruments:

- Mathematical modelling tool
- An automatic software generator, able to "translate" the model into C/C++ source code

In the industrial practice, the most used modelling tool is [4] while, for the software generator, there are two competitors: MathWorks Embedded Coder™ and dSPACE TargetLink™. Thanks to MBSD, it is possible to speed up the development process, to use the model itself as code documentation and to simplify the verification activities. Therefore, it is possible to compare the behaviour of the model run by the modelling tool, to generate the code and run it on the development machine (software-in-the-loop, SIL) or on the target microprocessor (processor-in-the-loop, PIL) or to run the software integrated with the firmware and physical I/O (hardware-in-the-loop, HIL) [5].

Hardware-in-the-loop [6] is a software verification technique. It is mainly used to perform real-time verification during the software integration phase, e.g. when the application software (in this case, the controller) is integrated with the firmware level. This methodology relies on a hardware simulator, equipped with transceivers and capable to run the plant model in real time. In this way, since the signals are equivalent to the real one from both the electrical and timing point of view, it is possible to trick the system under test (SUT) that it is dealing with the actual plant.

If the application software is developed by Model-Based Software Design (MBSD), it will accelerate the setup of the simulation environment.

For our case study, we decide to consider a trapezoidal brushless direct current (BLDC) motor type [7]. The choice to consider this kind of motor is justified by the fact that the main focus of this article is to describe a suitable validation and verification process for automotive safety-critical items. The alternative solutions (sinusoidal type) are more complex from an electrical and control point of view, making it more difficult to properly design models for the benchmark system [8].

Differently to the usual FMEA [9] approach, based on a handmade analysis from experts of the effects of each possible failure in the adopted failure model, we want to determine the level of reliability, hence, the safety of a given device, by analysing simulation results, so we need a suitable mathematical model. This choice is evident and is justified by the following statements:

• Easy setup, high repeatability and resource saving. This feature allows to perform repetitive tests, in multiple different conditions, saving time of the operations. Thus, it can be implemented through specific automated software test suites, and the presence of an operator is not needed.

• Cost-benefit. In FMEA environment, fault injections are performed, and the effects on the system are studied. Therefore, the same mathematical model is used multiple times instead of several real physical components with noticeable cost reduction.

The complexity of the model is another crucial aspect that needs to be clarified. In general, we would like to have models that perfectly fit with the physical processes. But then we should deal with the limitations which come from the hardware target devices that are used to perform the simulation. This aspect is crucial to make it possible to use these models also for HIL validation phase, when simulations are performed in real time. Therefore, a reasonable approach could set the level of complexity according to what is really required, and relaxation techniques are welcome to simplify models.

One more effort is required to implement FMEA-compatible models. They have to accomplish a fault injection pattern system. Failures could be injected with external sources or directly implemented through specific models. Therefore, parallel to the physical model, the needed fault models need to be designed, and in general, this is not an easy task. Thus, nowadays, fault modelling techniques are taking relevance from an academic point of view [10, 11].

From all the previous considerations, it emerges that an important contribution of this work is related to the modelling phase either for the physical or failure part which allow us to perform all the FMEA analysis reported in Sect. 4. In the first part, we describe the BLDC motor model. In a second, we focus on the inverter side together with the implemented control logic. Finally, an overall overview of the simulation environment is given.

3 Hazard Analysis and Risk Assessment

The HARA consists to analyse the safety-related effects of a device. This phase is preliminary to the actual design of the item, and it is based only on its functionalities.

The item, as shown in Fig. 1, is composed of the following:

- Gas pedal equipped with three independent variable resistors
- Conditioning state (3×) able to amplify the three voltages independently obtained by the gas pedal
- Microprocessor (μP), to generate the needed Pulse-Width Modulation (PWM) signals

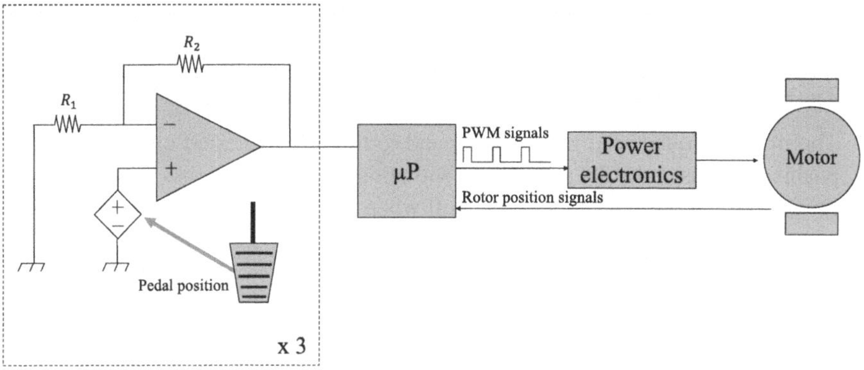

Fig. 1 Architecture of the system

- Power electronic stage composed of the six insulated-gate bipolar transistors (IGBTs) needed to drive the motor
- BLDC motor
- Hall-effect sensors (3×) to determine the rotor position

The supply unit needed to power the microprocessor is not considered in the analysis. Anyway, we determined the maximum failure rate this unit can have in order to achieve ASIL D (the strictest of ISO26262) level metrics. In any case, as described later, since power unit failures affect only the Safety Goal 2 that has been determined to have an ASIL B classification, it is possible to adopt less restrictive metrics.

As stated by the ISO26262, "the HARA shall be conducted in three stages:

- Situation analysis and hazard identification (SA/HI)
- Hazard classification (HC)
- ASIL determination"

These three phases will now be described in detail.

3.1 *Situation Analysis and Hazard Identification*

The device acts on the torque applied on the wheels. In the situation analysis, we can find these two worst conditions:

1. *Situation*: Lack of torque (or negative torque) while overtaking another car in a highway.

   Hazard identification: The car stops in the fast lane, and it can be hit by a car incoming at high speed.
2. *Situation*: Unintended acceleration at a crossing with red lights on while pedestrians and cars are crossing the street.

Hazard identification: The car is unable to stop at the red light and hits pedestrians and other cars crossing the street.

The safety goals for the analysed items are as follows:

1. The motor torque/speed must correspond to the one requested by the driver by pushing the gas pedal. Its most severe violation can be summarised as a condition in where the motor torque/speed is null when a non-null action is requested (no motor rotation when requested).
2. The motor torque when the gas pedal is completely released has to be zero (or correspond only in a regenerative braking action). The most severe violation of this safety goal can be described as an unintended vehicle acceleration.

3.2 Hazard Classification

Now, we have to classify the worst-case hazard for those two safety goals.

3.2.1 Safety Goal 1

Severity: Death of one or more people → (S3—life-threatening or fatal injuries).

Controllability: The driver can move to the safety lane before the vehicle stops, but it requires high driving capabilities → (C3—difficult to control or incontrollable).

Exposure: Overtaking → it is classified as <1% of operating time in Table B.2 [1] part 3 (E2—low probability).

3.2.2 Safety Goal 2

Severity: Death of one or more people → (S3—life-threatening or fatal injuries).

Controllability: The driver can only press the brake pedal, but it is very difficult to stop the vehicle on time → (C3—difficult to control or incontrollable).

Exposure: Stop at traffic lights → it is classified as >10% of operating time in Table B.2 [1] part 3 (E4—high probability).

3.3 ASIL Classification

Since there are no mutually exclusive situations for the safety goals and the item has to provide both of them, the ASILs will be determined by the worst obtained conditions.

By combining the S, C and E parameters, thanks to the ASIL determination table of the Standard, we find those level as follows:

Safety Goal 1) $S = 3$ $C = 3$ $E = 2 \rightarrow$ ASIL B
Safety Goal 2) $S = 3$ $C = 3$ $E = 4 \rightarrow$ ASIL D

4 Failure Mode and Effect Analysis

After we have designed the complete schematics of the item and we have developed the application layer of the embedded software, we have to perform the FMEA. Since this phase is based on the design, it is better to first describe the actual architecture of the device.

As said in Sect. 2, for our case study, we decide to consider a trapezoidal brushless direct current (BLDC) motor type.

Parallel to physical system design, a proper failure model has been chosen. A good approach is to reproduce the fault in a real scenario and then, from the analysis of data acquisition, try to build a coherent mathematical model. Thus, it is not always obvious since, apart from the effects, causes, which lead to a possible fault condition, also need to be known from the beginning, and this not always true. Therefore, a complete and exhaustive failure model is difficult to be met and cannot cover all possible scenarios. To implement the fault injection on IGBTs, we used the MOSFET block provided by the SimScape Toolbox, that is, the built-in SPICE-level simulator of MathWorks Simulink™.

But all this effort is not required for our specific application since our objective is to reproduce the damage of the IGBTs without investigating the boundary conditions which cause the break (i.e. overcurrent condition, junctions' breakdown). This aspect is out of our topic and is well-treated in other works [12]. Here, we are more interest to evaluate the possible effects at the item level. Thus, to reduce drastically the complexity of the failure model and to simplify its implementation, we represented the failure modes affecting the IGBTs as simple manual switches that can be set at any time during the simulation. As described in [3], these switches can be configured by saboteurs, in order to reduce possibilities of errors and to speed up the simulations. Doing so, we can reproduce the fault case where the single IGBT is not more drivable and stays permanently in an open- or a closed-circuit configuration. Since we are using only a simplified fault model for IGBTs, with respect to the state of the art [13], we sum the failure probabilities of both a failure of a microcontroller pin and the IGBT. This leads to a failure rate of 86% that the failure leads to a permanent open circuit and 14% that the failure leads to a permanent closed circuit.

Since we are adopting a simulation-based approach to perform FMEA, we are able to simulate also the embedded software. This fits particularly our problem since disparity and cross-checks on the incoming signals are performed at software level. It has to check that all the sensors' readouts are coherent, in a "data fusion"-like

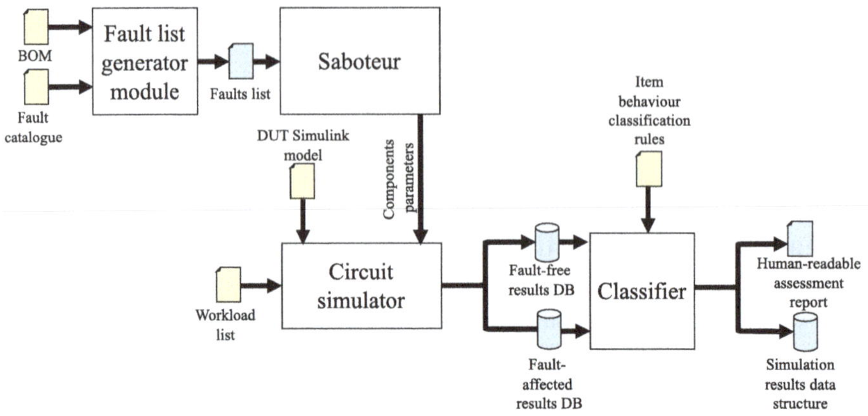

Fig. 2 Simulation-based FMEA tool architecture (Adapted from [3])

Table 1 Failure effect classification rules

Safety Goal 1	*Safe*: the motor is able to rotate *Dangerous*: the motor is unable to rotate with an angular speed greater or equal to 3 rad(s)	Detection mechanism: rotor position encoders + embedded software
Safety Goal 2	*Safe*: the motor does not turn at a faster speed with respect to the requested one *Dangerous*: the motor turns at a faster speed with respect to the requested one. In case of a stopped vehicle, if there is an output torque >5 Nm	Detection mechanism: comparison between the requested speed and the current speed at steady state

fashion. If not, it is able to determine which is the failed hardware item. In particular, the software implements disparity checks on the gas pedal triple redundancy module readouts (see Fig. 1) and cross-checks between the signals coming from the hall effects encoders and the requested speed.

The simulation environment architecture is shown in Fig. 2.

The fault list generator module takes the BOM and the fault catalogue and generates the fault list. This list is provided to the saboteur, which changes the component parameter in order to simulate the considered failure. The circuit simulator performs the SPICE-level simulation of the PCB components and run the embedded software. The item behaviour (firing signals for the invert switches) at different workload (in this case gas pedal position) is stored in databases and compared against some item behaviour classification rules.

The item behaviour classification rules with relative detection mechanism have been reported in Table 1. Pedal-level failures are not reported since, as shown in the results, there are no single-point failures of this unit able to affect the motor torque or speed.

The classification results have been reported in Table 2.

Table 2 FMEA results

Component	Failure rate [FIT]	Failure mode	Failure mode rate of occurrence, %	Failure mode effect	Fault coverage, %
R1 (unit 1)	2.25	Open	84	SD[a]	100
		Increase	8	SD[a]	100
		Decrease	8	SD[a]	100
R2 (unit 1)	2.25	Open	84	SD[a]	100
		Increase	8	SD[a]	100
		Decrease	8	SD[a]	100
OP-AMP (unit 1)	0.75	Interruption of any pin	50	SD	100
		Short of adjacent pins	50	SD	100
R1 (unit 2)	2.25	Open	84	SD[a]	100
		Increase	8	SD[a]	100
		Decrease	8	SD[a]	100
R2 (unit 2)	2.25	Open	84	SD[a]	100
		Increase	8	SD[a]	100
		Decrease	8	SD[a]	100
OP-AMP (unit 2)	0.75	Interruption of any pin	50	SD	100
		Short of adjacent pins	50	SD	100
R1 (unit 3)	2.25	Open	84	SD[a]	100
		Increase	8	SD[a]	100
		Decrease	8	SD[a]	100
R2 (unit 3)	2.25	Open	84	SD[a]	100
		Increase	8	SD[a]	100
		Decrease	8	SD[a]	100
OP-AMP (unit 3)	0.75	Interruption of any pin	50	SD	100
		Short of adjacent pins	50	SD	100
AnalogGasPedalCh1	0.75	StuckAtVCC	50	SD[a]	100
		StuckAtGND	50	SD[a]	100
AnalogGasPedalCh2	0.75	StuckAtVCC	50	SD[a]	100
		StuckAtGND	50	SD[a]	100
AnalogGasPedalCh3	0.75	StuckAtVCC	50	SD[a]	100
		StuckAtGND	50	SD[a]	100
MotorEncoderCh1	0.75	StuckAtVCC	50	SD	100
		StuckAtGND	50	SD	100
MotorEncoderCh2	0.75	StuckAtVCC	50	SD	100
		StuckAtGND	50	SD	100
MotorEncoderCh3	0.75	StuckAtVCC	50	SD	100
		StuckAtGND	50	SD	100

(continued)

Table 2 (continued)

Component	Failure rate [FIT]	Failure mode	Failure mode rate of occurrence, %	Failure mode effect	Fault coverage, %
FaultInjSel PowerSupply	<0.45 to obtain ASIL D-compliant metrics	StuckAtVCC	50	DU	100
		StuckAtGND	50	DU	100
Switch1[b]	1.05	Stuck at open condition	86	SD[a]	100
		Stuck at closed condition	14	DD[a]	100
Switch[b]	1.05	Stuck at open condition	86	SD[a]	100
		Stuck at closed condition	14	DD[a]	100
Switch3[b]	1.05	Stuck at open condition	86	SD[a]	100
		Stuck at closed condition	14	DD[a]	100
Switch4[b]	1.05	Stuck at open condition	86	SD[a]	100
		Stuck at closed condition	14	DD[a]	100
Switch5[b]	1.05	Stuck at open condition	86	SD[a]	100
		Stuck at closed condition	14	DD[a]	100
Switch6[b]	1.05	Stuck at open condition	86	SD[a]	100
		Stuck at closed condition	14	DD[a]	100

[a]The Failure mode effect classifications have been obtained by simulation-based approach
[b]This failure mode comprises both IGBT and microprocessor pin-level failures

The random HW failure metrics obtained for the proposed design are as follows:

- Random hardware fault metric: 1.21 FIT < 10 → ASIL D)
- Single-point fault metric: >99% → ASIL D (with random HW failure rate of the power supply unit <0.45 FIT)
- Latent fault metric: 100.00% > 90% → ASIL D

5 Conclusions

In this paper, a methodology to develop a high dependable automotive drive-by-wire item has been presented.

The main objective had been to improve the reliability and repeatability of the FMEA assessment, by keeping into account the effect of the embedded software on the fulfilment of safety goals. As shown, the most problematic phases of safety life cycle are the HARA, FMEA and software integration validation. The model has been designed to accomplish a simulation-based FMEA, either to be simulated in real-time, during the software integration test phase by HIL.

As benchmark application, a powertrain for electrical vehicles has been considered, focusing on the failures of the inverter side and analysing their effects on the motor side. The model has demonstrated itself to be sufficiently accurate to allow an ISO26262-compliant FMEA.

A further extension of this work can be to demonstrate usage of this model, inside an HIL environment with an instrumented controller to perform fault injections and a real motor. In this way, it will be possible to simulate controller, IGBT and sensor failures, hence improving the FMEA classification quality by comparing the effects of such failures on a simulated and a real motor.

References

1. ISO 26262, Road vehicles – Functional safety (2011)
2. K. Beckers, D. Holling, Coté I.M., Hatebur D., A structured hazard analysis and risk assessment method for automotive systems – A descriptive study In: Reliabilty Engineering and System Safety (2017) pg. 185–195
3. J. Sini, M. Violante, An automatic approach to perform FMEDA safety assessment on hardware designs, in *24th IEEE International Symposium on on-Line Testing and Robust System Design (IOLTS)*, (Platja d'Aro, Costa Brava, Spain,. 2–4 July, 2018)
4. MathWorks Simulink. https://it.mathworks.com/products/simulink.html. Retrieved 22 August 2018
5. S. Narayanamurthy, et al. MIL/SIL/PIL Approach - A new paradigm in Model Based Development (2014) In: https://www.mathworks.com/content/dam/mathworks/mathworks-dot-com/solutions/automotive/files/in-expo-2014/mil-sil-pil-a-new-paradigm-in-model-based-development.pdf. Retrieved 20 September 2018
6. J. Ledin, Hardware-in-the-loop simulation. Embed. Syst. Program. **12**(2), 42–53 (1999)
7. P. Yedamale, *Brushless DC (BLDC) motor fundamentals*, Vol. 20 (Microchip Technology Inc, 2003, pp. 3–15)
8. B.K. Lee, M. Ehsani, Advanced BLDC motor drive for low cost and high performance propulsion system in electric and hybrid vehicle, in *Electric Machines and Drives Conference, 2001. IEMDC 2001. IEEE International. IEEE*, (2001), pp. 246–251
9. D.H. Stamatis, *Failure Mode and Effect Analysis: FMEA from Theory to Execution* (ASQ Quality Press, 2003)
10. M. Ben-Daya, A. Raouf, A revised failure mode and effects analysis model. Int. J. Qual. Reliabil. Manag. **13.1**, 43–47 (1996)

11. K. Chin, A. Chan, J. Yang, Development of a fuzzy FMEA based product design system. Int. J. Adv. Manuf. Technol. **36.7-8**, 633–649 (2008)
12. G. Cassanelli et al., Reliability predictions in electronic industrial applications. Microelectron. Reliab. **45.9-11**, 1321–1326 (2005)
13. D. Piumatti, M. Sonza Reorda, Assessing test procedure effectiveness for power devices, in *IEEE 33rd Conference on Design of Circuits and Integrated Circuits*, (2018)

A Fault Localization Method for Single-Phase to Ground Faults in LV Smart Distribution Grids

Nikolaos Sapountzoglou, Bertrand Raison, and Nuno Silva

Abstract A fault localization method for single-phase to ground short-circuit (SC) faults in low voltage (LV) smart distribution grids is presented in this paper. Both the use of rms voltage phase measurements and an analysis of symmetrical components of the voltage were investigated and compared in this study. Phase measurements were found to be more suitable for single-phase to ground faults. The described method is a three-step process beginning with the identification of the faulty branch, followed by the localization of the sector in which the fault occurred and concluding with the estimation of the fault distance from the beginning of the feeder. Fault resistance values of 0.1, 1, 5, 10, 50, 100, 500 and 1000 Ω were tested. An heterogeneity analysis was performed to test the effect of the use of various conductors on the method. Faults in all three phases were implemented and simulated on a real case of a semi-rural LV distribution network of Portugal, provided by Efacec. Finally, the method presented an average estimation accuracy of 89.33% and an increased accuracy of 93.11% for low impedance faults (up to 10 Ω of fault resistance).

1 Introduction

Electricity interruptions have huge economic and social impact. Production loss, restart costs, equipment damage and raw materials spoilage can be very costly. At the same time, uncomfortable temperatures at work or home, loss of leisure time and risk to health and safety (e.g. interrupting hospital service or industrial operations) are some of the aspects of electricity interruption's societal impact [1].

N. Sapountzoglou (✉) · B. Raison
Univ. Grenoble Alpes, CNRS, Grenoble INP, Institute of Engineering, G2Elab, Grenoble, France
e-mail: nikolaos.sapountzoglou@g2elab.grenoble-inp.fr; bertrand.raison@g2elab.grenoble-inp.fr

N. Silva
Efacec, Maia, Portugal
e-mail: nuno.silva@efacec.com

W. Zamboni, G. Petrone (eds.), *ELECTRIMACS 2019*, Lecture Notes in Electrical Engineering 615, https://doi.org/10.1007/978-3-030-37161-6_24

Faults that appear in distribution networks are responsible for the majority of customer interruptions [2]. Weather conditions (hurricane, lightning, etc.), component wearing and accidents are some of the causes of such faults. Two indices are used: (a) the system average interruption duration index (SAIDI) and (b) the Value of Lost Load (VoLL), to measure the impact of electricity interruptions in terms of time and cost, respectively. A maximum SAIDI of 371 min was noticed in Europe in 2016 [3]. Additionally, according to [4], the annual average of VoLL in Europe in 2013 was 8.37 €/kWh.

For the aforementioned reasons, a fault should be detected, located and isolated as quickly as possible. All three of them are key elements of a self-healing smart grid and they include the following steps:

(a) detecting the fault usually through threshold crossing detection techniques,
(b) localizing the fault with any of the five most popular methods [5]: impedance based which are the most widely used [6–8], knowledge based [9–11], traveling wave [12, 13], methods based on sparse measurements [14–16] and hybrid methods [17, 18] and
(c) isolating the fault by sending a crew to verify the localization of the fault and fix the problem.

Due to their increased complexity and the lack of available sensors, the LV distribution grids have not been on the centre of researchers' attention, with few exceptions [7, 15, 16, 19]. The vast majority of the studies focuses on the Medium Voltage (MV) distribution network. Furthermore, some of the most widely used methods such as the impedance based method present some drawbacks; a typical example is the identification of multiple locations of the same distance from the feeder as possible fault locations. Finally, since high impedance faults are very difficult to detect as the fault current magnitude is very close to the one under normal operation, the vast majority of the fault localization studies examined only low impedance faults.

In order to address the above problems, a fault localization method for LV distribution grids based on an idea initially developed for MV networks [20] is proposed in this paper. Single-phase to ground SC faults were studied since they are the most frequent type of faults in a distribution grid [2]. Finally, high impedance faults with fault resistance values of up to 1000 Ω were included in the study.

The paper is divided into five sections. In the following section the characteristics of the grid under study are presented. The fault localization method is presented in Sect. 3. The method accuracy is analysed in Sect. 4. The paper concludes with a synopsis of the achievements accomplished during this study.

2 Grid Characteristics

For the purpose of this study, a real case semi-rural radial LV distribution grid of Portugal was used. The connection to the MV is achieved through a distribution transformer. The grid is a three-phase-four-wire one with a solidly grounded neutral.

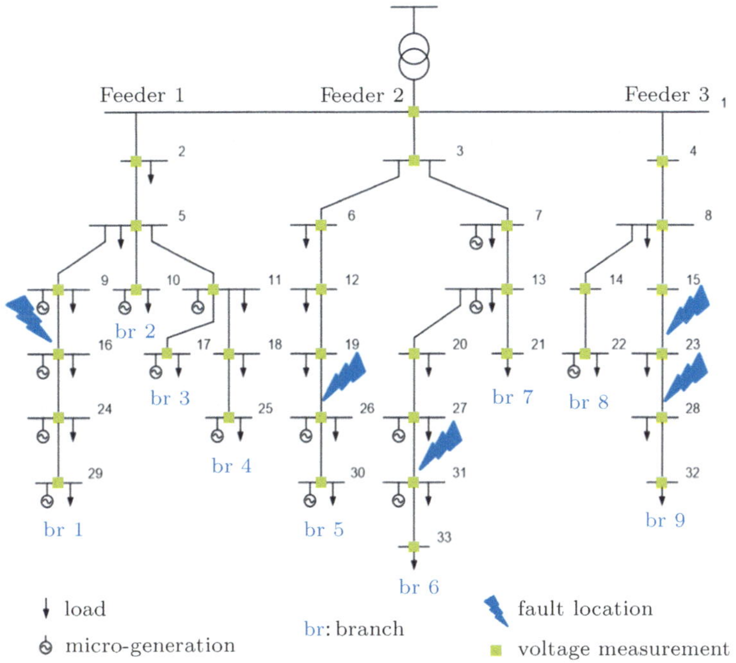

Fig. 1 Single line diagram of the LV grid with the SC fault cases. In some nodes multiple single-phase loads are connected

It consists of three main feeders and a total of thirty-three nodes with eighteen microgenerators (single-phase PV installations) and forty-eight single-phase loads. The single line diagram of the LV distribution grid is presented in Fig. 1 along with the considered as available measurements and the fault locations under study. Two basic assumptions were made at this point: (a) phase rms voltage measurements were available at each node of the grid and (b) the measurements were synchronous and non-erroneous.

Two are the basic characteristics of this grid that complexify its analysis: imbalance and heterogeneity. The imbalance term was used to describe the fact that load and microgenerators of different sizes are spread asymmetrically throughout the grid. The per feeder and per phase connection distribution of load contracted power (kVA) and microgenerator installed power (kW) is presented in Table 1. At the same time, 11 different types of conductors in terms of resistance and reactance connect the nodes with each other with lengths ranging from 35 to a maximum of 210 m, thus attributing to the grid an heterogeneous nature.

As it will be explained later, the localization method was based only on the rms voltage profile across the faulty line. Hence, the phasor mode was selected to perform the simulations in order to reduce the computational time. Furthermore, the faults were studied on their steady state, 150 ms after their occurrence. This time frame secured that the fault would appear on a steady state and before any

Table 1 Grid loads and μgens

	Contracted power (kVA)			Installed power (kW)		
Feeder	S_a	S_b	S_c	P_a	P_b	P_c
1	31.05	41.40	48.30	3.68	8.83	14.49
2	34.50	41.40	24.15	10.58	7.13	7.08
3	10.35	13.80	17.25	3.68	0.00	1.70
Total	75.90	96.60	89.70	17.94	15.96	23.27

protective device had isolated the installed microgeneration units; an action from the protective elements is expected around 200 ms from the fault occurrence. This way the influence of the microgeneration participation on the voltage profile could be studied.

At this point and before proceeding to the following section, it is necessary to introduce two notions: the branches and sectors of the grid. The branch of a grid is defined as a unique line path from the beginning of the feeder to each terminal node (e.g. first branch: from node one to node twenty-nine and fifth branch: from node one to node thirty). The nine different branches are presented in Fig. 1. Moreover, the sector is defined as the section of the grid between two consecutive available measurements. In this case, since voltage measurements are available at each node, the sector is the line connecting two adjacent nodes (e.g. first sector: from node one to node two and second sector: from node two to node five).

3 Fault Localization Method

Once an alarm signal indicates the occurrence of a fault in one of the grid feeders, the fault localization process is initiated. This process is divided into three distinct steps: (a) faulty branch identification, (b) faulty sector localization and (c) fault distance estimation. All these steps were solely based on rms voltage measurements.

3.1 Faulty Branch Identification

Identifying the faulty branch within a feeder with multiple branches is the first and most important step of the fault localization process. It is expected that the branch under fault would present the highest voltage drop within the faulty feeder. The branch to which the node with the lowest voltage belonged was identified and this branch was considered to be the one under fault.

3.2 *Faulty Sector Localization*

Following the faulty branch identification process, the faulty sector localization method is where the core idea of this study lies. The basic principle is based on the fact that across the line of a faulty branch, voltage is expected to drop linearly up until the faulty sector and stabilize to a certain value after the faulty sector. In Fig. 2, the theoretical voltage profile and the form of its curve is presented for a fault located at 405 m from the beginning of the feeder. For 0 Ω of fault resistance, the voltage will drop to zero after the faulty sector since there will be no current flowing through this part of the grid; all the current will flow from the line to the ground through their established connection during the fault leaving no circulating current in the part of the line after the fault's location. However, for any value "x" of fault resistance, the voltage will stabilize to a higher value.

As mentioned in Sect. 1, this idea was conceived in the context of the MV grid which is less complex than the LV grid. Although in reality the unbalanced and/or heterogeneous nature of the LV grid can affect the form of the voltage profile, the basic principle of the voltage stabilizing to a value after the faulty sector, still applies. Such an example is provided in Fig. 3 where the available measurements

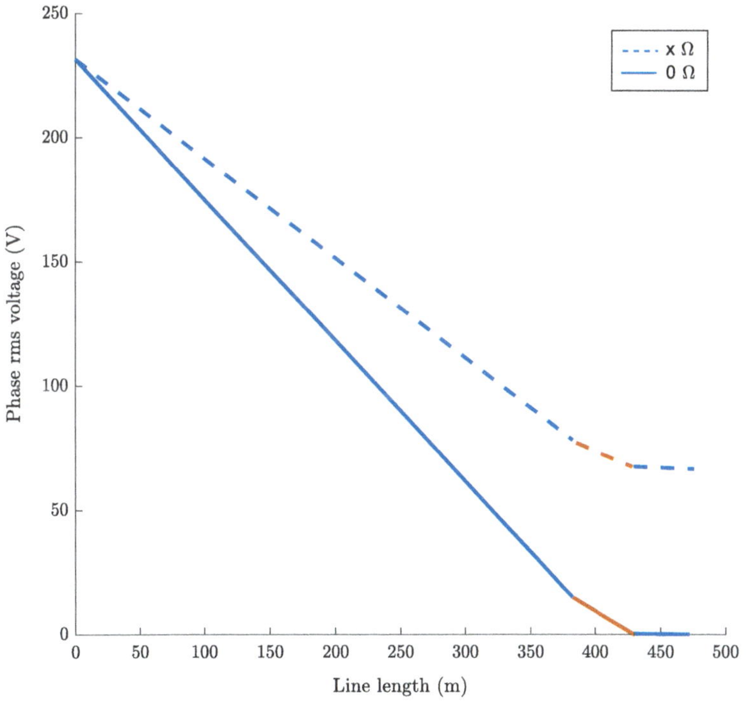

Fig. 2 Theoretical voltage profile across a faulty line for a single-phase to ground SC fault for two cases of fault resistance values: zero ("0") and non-zero ("x")

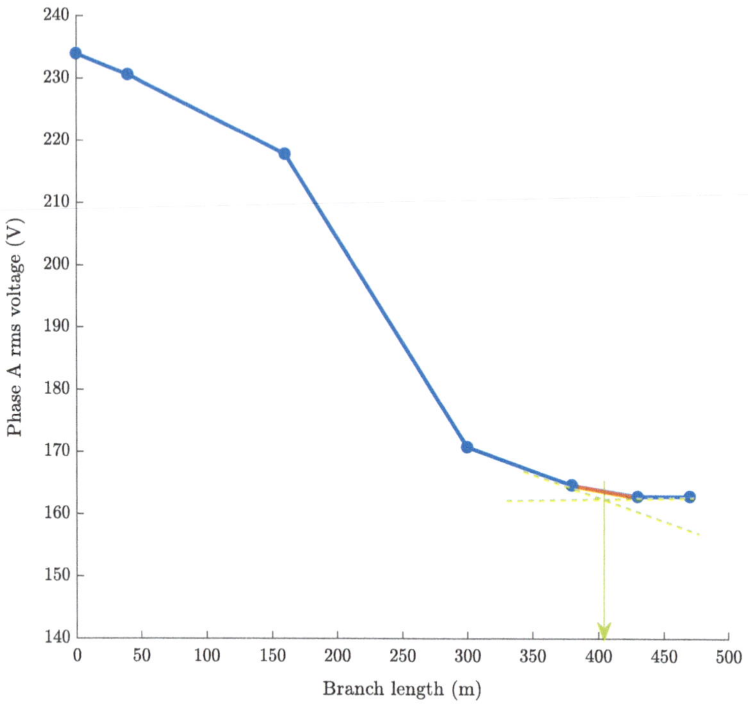

Fig. 3 Snapshot of the voltage profile along the branch length for a single-phase to ground fault (AG) between nodes 19 and 26 of a fault resistance of 1 Ω at 01 h01 min0.15 s

across the line are depicted with circles. The linear interpolation method was used to create the lines connecting these points. The change in the slope of the voltage curve is obvious as it decreases to almost zero after the faulty sector; the faulty sector is indicated by an orange line in Fig. 3.

In order to transform the above critical observation regarding the slope into an algorithm and identify the faulty sector, the two following criteria were developed:

1. if the difference between two consecutive voltage measurements was positive, signifying a change in the sign of the slope, then the previous sector was the one under fault and
2. if the absolute value of the difference between two adjacent voltage measurements was the lowest within the branch, signifying a stabilization of the curve, then the previous sector was the one under fault.

The slope of the curve is almost always negative since voltage is decreasing. However, a positive value, of the slope after the faulty sector, is possible under the presence of microgeneration units in this part of the grid thus explaining the choice of the first criterion; the effect of microgeneration units is thoroughly explained in Sect. 4.

3.3 Fault Distance Estimation

The last step in localizing the fault, after identifying the faulty branch and sector, is to estimate its location within the faulty sector. To achieve that purpose, a graphic method was implemented. From the linearly interpolated curve of Fig. 3 the lines of the sectors adjacent to the one under fault were linearly extrapolated (green dashed lines) and their intersection point was used to estimate the location of the fault inside the sector.

To measure the method accuracy the following formula described in [21] was used:

$$\text{error } (\%) = \frac{|d_{\text{estimated}} - d_{\text{actual}}|}{l_{\text{total}}} \cdot 100 \quad (1)$$

where the estimated distance is compared to the actual one and the result is normalized over the total length of the line. In Fig. 3, the fault occurs at 01 h 01 min with a fault resistance of 1 Ω, and it is located in the fifth sector of the fifth branch and at a distance of 405 m from the beginning of the feeder. The estimated distance for this case is 404.4 m presenting an accuracy of 99.38% or a deviation of 0.6 m.

As a last step, a threshold check was applied to ensure that the estimated location falls within the limits of the identified faulty sector. This corrective process helped minimize false estimations. The results of the entire fault localization method are presented in the following section.

4 Results

In order to verify the robustness of the method, the effects of the following parameters on the method were investigated. First of all, the positive sequence component of the voltage was considered as an alternative to the phase rms voltage. Secondly, the effect of using conductors of different types in terms of resistance, reactance and length was examined. Furthermore, based on the load and microgeneration profiles during a day as shown in Fig. 4, five scenarios of different load and microgeneration combinations were selected:

(a) at 04 h (0% load, 0% μgen)
(b) at 01 h (28% load, 0% μgen)
(c) at 20 h (90% load, 0% μgen)
(d) at 14 h (60% load, 49% μgen)
(e) at 12 h (50% load, 100% μgen)

Scenario (a) was used for an heterogeneity analysis and (b,c) and (d,e) to monitor how the developed method is affected by an increase of load or microgeneration, respectively. Moreover, it was assumed that a maximum of 30% of the loads would operate simultaneously. Finally, the effect of increasing the fault resistance was

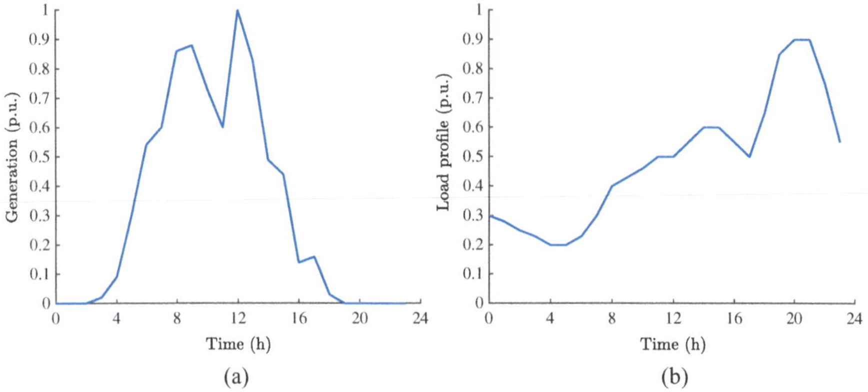

Fig. 4 Grid characteristics. (**a**) Microgeneration profile. (**b**) Load profile

Table 2 Method accuracy (%)

	Overall		Less than 50 Ω	
	v_{phase}	v_{pos}	v_{phase}	v_{pos}
Branch	75.00	66.88	91.67	89.58
Sector	58.61	82.87	81.82	99.53
Distance	89.33	86.73	93.11	88.68

studied by simulating faults with fault resistance values of: 0.1, 1, 5, 10, 50, 100, 500 and 1000 Ω. The results are discussed analytically in the following subsections.

4.1 Phase Voltage vs. Positive Sequence Component

Both the phase voltage measurements and the positive sequence component were used for all the different single-phase to ground fault cases and scenarios. In Table 2, the results for both methods are presented. It is demonstrated that the phase rms voltage measurements were more suitable for single-phase to ground faults since they presented a higher precision rate in identifying the faulty branch which is the most important step of the method. The rest of the results are presented on the basis of the phase rms voltage measurements.

4.2 High Impedance vs. Low Impedance Faults

Microgeneration units are expected to contribute to the grid voltage by increasing it while loads will cause a voltage drop. Higher impedance faults (>50 Ω) strongly affect the voltage profile bringing it very close to the normal operating conditions.

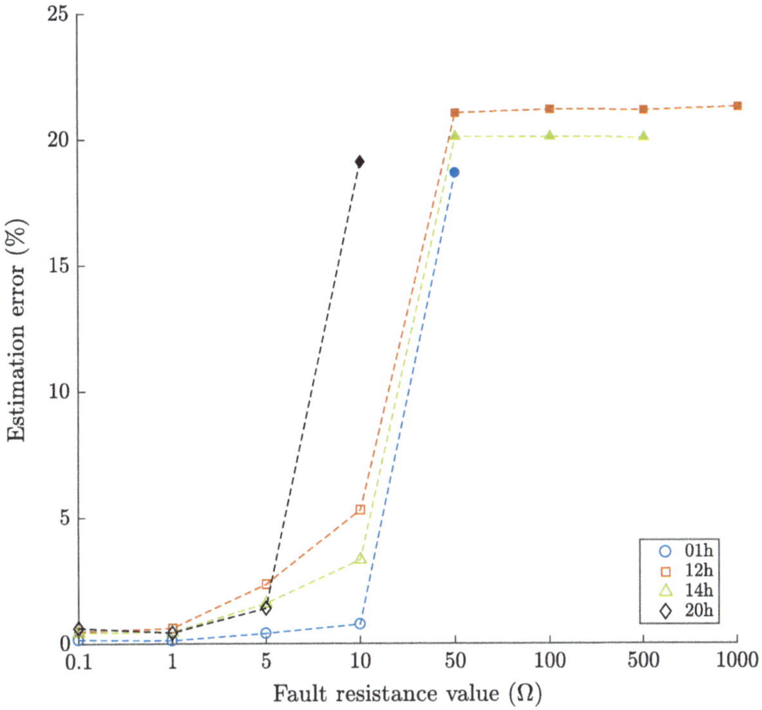

Fig. 5 Distance estimation error for a single-phase to ground fault (AG) between nodes 19 and 26. Filled markers indicate a mislocalized sector. The figure only shows results for a successful faulty branch identification

In that way the effect of the microgeneration units or the loads on the voltage profiles was amplified thus increasing the error in localizing the correct sector and estimating the fault's location.

From Table 2, it is clear that for faults of a fault resistance of less than 50 Ω the precision of the method in identifying the correct branch and sector is significantly increased by 16.67% and 23.21%, respectively. This means that by increasing the fault resistance, misidentification of a faulty branch and/or mislocalization of a faulty sector are to be expected. An illustrative example is provided in Fig. 5.

4.3 Imbalance Analysis

Load and microgeneration participation, as discussed in Sect. 2, can affect the method since they affect the form of the voltage curve. The less elements present in the faulted branch, the more accurate the method will be. Since microgeneration units increase the voltage they can, at some extent, mitigate the effect of increased load participation which tend to lower the voltage of the grid. Such an example is presented in Fig. 5. It is shown that increased load effects are more severe than the ones caused by microgeneration presence; in Fig. 5 the difference between 12 and 1 h when microgeneration increases from 49% to 100% is almost negligible. Especially for high impedance faults, increased penetration of either load or microgeneration units is likely to lead to a misidentification of the faulty branch or a mislocalization of the faulty sector. The latter will cause a significant increase of the distance estimation error.

4.4 *Heterogeneity Analysis*

A study was done without any load and microgeneration present in the grid in order to investigate the effect of the heterogeneity of the grid on the method. One parameter was set: the distance of the fault from the beginning of each feeder. Two fault cases were studied: single-phase to ground SC faults at a distance of (a) 270 m and (b) 400 m from the beginning of the feeder. The comparative results are presented in Table 3. Different distances were chosen since an increase of the fault distance would include a bigger variety of conductors per branch. An analysis was made for one branch per feeder; for the case of 400 m, two branches were selected from the second feeder since the maximum length of a branch within the first feeder is 340 m. Branch five belonging at the second feeder (f2, br5) was chosen because the differences in resistance and reactance between the conductors composing it were not that important. For the exact opposite reason, branch nine of feeder three was selected (f3, br9). Additionally, branch nine is also the longest of the grid.

Two basic conclusions can be drawn from Table 3. First of all, depending on the composition of a branch and how heterogeneous it is, a cost of a maximum of almost 5% in the accuracy of the method is possible. Increasing the number of different conductors in a line also increases the risk of losing in precision if the conductors composing the branch are of quite different type. For example, an

Table 3 Distance estimation error (%) due to heterogeneity of the grid

	270 m			400 m		
R_f (Ω)	f1, br1	f2, br6	f3, br9	f2, br5	f2, br 6	f3, br9
0.1	4.52	0.20	0.01	0.03	4.40	4.79
1	4.54	0.20	0.01	0.01	4.40	4.79
5	4.54	0.20	0.00	0.00	4.40	4.79
10	4.55	0.20	0.00	0.00	4.40	4.79
50	4.55	0.20	0.00	0.00	4.40	4.79
100	4.55	0.20	0.00	0.00	4.40	4.79
500	4.55	0.20	0.00	0.00	4.40	4.79
1000	4.55	0.20	0.00	0.00	4.40	4.79

increase of the fault distance from 270 to 400 m within branch nine which is quite heterogeneous would cost about 4.8% in accuracy of the estimation. Secondly, it should be underlined that since there were no loads or microgeneration units participating in this case, the increase of the fault resistance did not affect the localization method in any way. This is due to the fact that the stabilization of the curve to an exactly zero slope after the faulty sector was guaranteed.

5 Conclusions

A localization method for single-phase to ground SC faults in a LV distribution grid was developed in this study. Two approaches regarding the voltage measurements to be used were examined: the rms phase measurements and the positive sequence component of voltage. Different combinations of load and microgeneration participation were taken into account. Additionally, both low and high impedance faults were implemented (R_f from 0.1 to 1000 Ω). Finally, the effect of the heterogeneity of the grid was analysed as well. The basic conclusions can be summarized in the following points:

1. Phase rms voltage measurements are more suitable for single-phase to ground faults than any of the symmetrical components of the voltage.
2. The distance estimation error increases by the increase of the fault resistance. For low impedance faults distance estimation accuracy reaches 93.11%.
3. Load and microgeneration presence might lead to misidentification of the faulty branch or mislocalization of the faulty sector decreasing the precision of the method with load effects being more severe than the ones caused by microgeneration.
4. The use of different type of conductors in the studied LV grid may increase the distance estimation error up to 5%.

Acknowledgements This project has received funding from the European Union's Horizon 2020 research and innovation programme under the Marie Skłodowska-Curie grant agreement No. 675318 (INCITE).

References

1. P. Linares, L. Rey, The costs of electricity interruptions in Spain. Are we sending the right signals? Energy Policy **61**, 751–760 (2013)
2. T. Gönen, *Electric Power Distribution Engineering* (CRC Press, Boca Raton, FL, 2014). oCLC: 879683602
3. CEER, 6th Benchmarking Report on the Continuity of Electricity and Gas Supply, Brussels, Tech. Rep. C 18-EQS-86-03 (2016)
4. A. Shivakumar, M. Welsch, C. Taliotis, D. Jakšić, T. Baričević, M. Howells, S. Gupta, H. Rogner, Valuing blackouts and lost leisure: estimating electricity interruption costs for households across the European Union. Energy Res. Soc. Sci. **34**, 39–48 (2017)

5. A. Bahmanyar, S. Jamali, A. Estebsari, E. Bompard, A comparison framework for distribution system outage and fault location methods. Electr. Power Syst. Res. **145**, 19–34 (2017)
6. R. Dashti, M. Ghasemi, M. Daisy, Fault location in power distribution network with presence of distributed generation resources using impedance based method and applying π line model. Energy **159**, 344–360 (2018)
7. F.C.L. Trindade, W. Freitas, Low voltage zones to support fault location in distribution systems with smart meters. IEEE Trans. Smart Grid **8**(6), 2765–2774 (2017)
8. C. Orozco-Henao, A.S. Bretas, A.R. Herrera-Orozco, J.D. Pulgarín-Rivera, S. Dhulipala, S. Wang, Towards active distribution networks fault location: contributions considering DER analytical models and local measurements. Int. J. Electr. Power Energy Syst. **99**, 454–464 (2018)
9. M. Majidi, M. Etezadi-Amoli, M.S. Fadali, A novel method for single and simultaneous fault location in distribution networks. IEEE Trans. Power Syst. **30**(6), 3368–3376 (2015)
10. Y. Dong, C. Zheng, M. Kezunovic, Enhancing accuracy while reducing computation complexity for voltage-sag-based distribution fault location. IEEE Trans. Power Deliv. **28**(2), 1202–1212 (2013)
11. M. Sarvi, Determination of fault location and type in distribution systems using Clark transformation and neural network. Int. J. Appl. Power Eng. **1**(2), 75–86 (2012)
12. A. Borghetti, M. Bosetti, C.A. Nucci, M. Paolone, A. Abur, Integrated use of time-frequency wavelet decompositions for fault location in distribution networks: theory and experimental validation. IEEE Trans. Power Deliv. **25**(4), 3139–3146 (2010)
13. H. Nouri, C. Wang, T. Davies, An accurate fault location technique for distribution lines with tapped loads using wavelet transform, in *2001 IEEE Porto Power Tech Proceedings* (Cat. No.01EX502), vol. 3, September 2001, p. 4
14. M. Majidi, A. Arabali, M. Etezadi-Amoli, Fault location in distribution networks by compressive sensing. IEEE Trans. Power Deliv. **30**(4), 1761–1769 (2015)
15. N. Silva, F. Basadre, P. Rodrigues, M.S. Nunes, A. Grilo, A. Casaca, F. Melo, L. Gaspar, Fault detection and location in low voltage grids based on distributed monitoring, in *2016 IEEE International Energy Conference (ENERGYCON)*, April 2016, pp. 1–6
16. L. Marques, N. Silva, I. Miranda, E. Rodriges, H. Leite, Detection and localisation of non-technical losses in low voltage distribution networks, in *Mediterranean Conference on Power Generation, Transmission, Distribution and Energy Conversion (MedPower 2016)*, November 2016, pp. 1–8
17. A.C. Adewole, R. Tzoneva, S. Behardien, Distribution network fault section identification and fault location using wavelet entropy and neural networks. Appl. Soft Comput. **46**, 296–306 (2016)
18. T. Gush, S.B.A. Bukhari, R. Haider, S. Admasie, Y.-S. Oh, G.-J. Cho, C.-H. Kim, Fault detection and location in a microgrid using mathematical morphology and recursive least square methods. Int. J. Electr. Power Energy Syst. **102**, 324–331 (2018)
19. G. Niu, L. Zhou, W. Pei, Z. Qi, A novel fault location and recognition method for low voltage active distribution network, in *2015 5th International Conference on Electric Utility Deregulation and Restructuring and Power Technologies (DRPT)*, November 2015, pp. 876–881
20. A. Teninge, C. Pajot, B. Raison, D. Picault, Voltage profile analysis for fault distance estimation in distribution network, in *2015 IEEE Eindhoven PowerTech*, June 2015, pp. 1–5
21. M.M. Saha, J. Izykowski, E. Rosołowski, *Fault Location on Power Networks*. Power Systems (Springer, London, 2010). oCLC: ocn495597133

Smart Sensors in Smart Cities Collaborate for Indoor Air Quality

Paolo Baronti, Paolo Barsocchi, Erina Ferro, Fabio Mavilia, Massimo Piotto, and Lucanos Strambini

Abstract This paper presents an example of collaboration between two different air quality monitoring systems, one developed for indoor usage, the other one used in some regions of Italy as an example of citizens' collaborative work for monitoring the air quality in smart cities. The exchange of information between the two systems (the inner one and the external one) allows making a weighted decision for improving the inner air quality. By evaluating both indoor and outdoor air quality levels, a reasoner decides the best policy to be automatically adopted to improve, or at least not worsen, the indoor air quality.

1 Introduction

The deployment of wireless sensor networks (WSNs) in Smart City infrastructures has driven large amounts of information being generated each day across a variety of domains by a plethora of applications, including environmental monitoring applications. In this context, the usage of embedded devices and existing internet infrastructures is essential for the Internet of Things (IoT) vision [1]. A Smart City is just based on the usage of large amounts of IoT devices, some of them used in monitoring systems that enable the management and the remote control of wide ranges of sensor networks and actuators, used for both indoor [2] and outdoor applications [3, 4].

P. Baronti · P. Barsocchi (✉) · E. Ferro (✉) · F. Mavilia
ISTI-CNR, Pisa, Italy
e-mail: Paolo.Baronti@isti.cnr.it; Paolo.Barsocchi@isti.cnr.it; Erina.Ferro@isti.cnr.it; Fabio.Mavilia@isti.cnr.it

M. Piotto
DII - University of Pisa, Pisa, Italy
e-mail: massimo.piotto@unipi.it

L. Strambini
IEIIT-CNR, Pisa, Italy
e-mail: lucanos.strambini@ieiit.cnr.it

W. Zamboni, G. Petrone (eds.), *ELECTRIMACS 2019*, Lecture Notes in Electrical Engineering 615, https://doi.org/10.1007/978-3-030-37161-6_25

Persons spend an estimated 90% of their time indoors, so that poor indoor air quality (IAQ) both decreases the comfort level and poses a substantial risk to public health [5]. As a matter of fact, poor air quality may induce short-term health problems, such as fatigue and nausea as well as chronic respiratory diseases [6]. In this paper, we present an example of collaboration between a prototype of indoor air quality monitoring system and an outdoor air quality system, available in some regions of Italy, in order to evaluate when and if an automatic action, such as opening a window, is required for improving the inner air quality. The proposed IAQ monitoring system is based on a multi-component platform integrated in an extremely versatile monitoring system consisting of heterogeneous environmental sensors installed in a closed environment (apartment, office, some areas of an industry, etc.) for various monitoring purposes. An ad-hoc developed middleware allows a reasoner (installed on the server where all data are collected) to get the data from the sensors and to take decisions according to policies established by the user. It is not goal of this paper to present the whole monitoring system; this paper only focuses on the decision of automatically opening or not a window according to the external air quality. As far as the authors know, it is the first time in Italy that an indoor monitoring system communicates with an outdoor system for an optimal policy of maintenance of the indoor air quality.

2 The Indoor Air Quality

The term indoor air quality (IAQ) means different things to different people and there is no single accepted definition for it. There are no specific legislated standards for IAQ, although there are exposure standards set for a range of chemicals in industrial environments. IAQ is a measure/analysis of the air condition in closed spaces and it includes the physical, chemical and microbiological characterization of the air within and around buildings and structures, especially as it relates to the health and comfort of the building occupants. In the past, carbon dioxide (CO_2) was accepted as the main indicator for indoor air quality because the human beings were considered the main pollutant sources. Today, many other sources that emit pollutants different from human bioeffluents have been taken into account. In fact, many new materials and products used in indoor environments are responsible for the increase in indoor pollution concentration, especially volatile organic compounds (VOCs), which contain carbon with a vapor pressure high enough to vaporize from materials and surfaces into the indoor air at room temperatures, a process known as "off-gassing". They include alkanes, aromatics, aldehydes, ketones, alcohols and ethers, which can be usually present in indoor air (Total VOC, TVOC). High levels of indoor particulates can also be an indicator of poor air quality. Due to the lack of fresh air, air circulation and air filtration, the level of contaminants can be higher in some parts of a semi-enclosed space. In pursuit of higher energy efficiency, mechanical ventilation and circulation are necessary to manage the indoor air quality. CO_2 in the air is measured in percentage or

Table 1 Typical indoor air quality definitions based on CO_2 level picked up by most industry standards

Air quality	Recommendation	CO_2 [ppm]
Excellent	Target value	0–600
Good	Optional ventilation/airing	600–800
Fair	Ventilation/airing recommended	800–1000
Mediocre	Contaminated indoor air intensified ventilation/airing recommended	1000–1500
Bad	Heavily contaminated indoor air intensified ventilation/airing necessary	1500–2100

ppm (parts per million) and is, however, the most important indicator of indoor air quality [7]. Typical indoor CO_2 concentrations range between 700 and 2000 ppm (approximately 3657 mg/m^3) but can exceed 3000 ppm (5486 mg/m^3) during the use of unventilated appliances [8]. To keep the CO_2 concentration at a maximum of 0.15% (1500 ppm), an average of 25 m^3/h of new air per person must be added (for example, even more if you are practising a sport). According to DIN EN ISO 15011-2 standards, the ventilation rate per person is about 30 m^3/h for a house, about 20 m^3/h for a nursery and 35 m^3/h for the master bedroom. The oxygen supply necessary for our metabolism is only 10% of the new air flow needed for the exchange of air containing CO_2 and harmful substances. Tables 1 and 2 represent the typical CO_2[1] and TVOC[2] levels, respectively [9]. Outdoor environmental monitoring allows to know the status of the air quality, the water quality, the level of pollutants and so on. At the moment, there are only a couple of significant maps in real time for the quality of the air in the world: (1) the World Air Quality Index[3], which has the merit of standardizing the air quality indexes of the various countries (which are almost all different), making them comparable to each other, and (2) the map named "Che aria tira (CAT[4])?" (What air does it blow?) for Italy. CAT is the only real-time map of air quality in our country; it is made with fixed stations not belonging to the Regional Agency for Environmental Protection (ARPA—Agenzia Regionale per la Protezione Ambientale). CAT is an open source project of Active Citizenship and Citizen Science that aims at building a self-monitoring network of air quality, where citizens, associations/organizations or other institutions can build their own environmental monitoring unit and share data online on the platform created by the project. At the moment, CAT provides the values of PM2.5 and PM10 derived from 72 fixed monitoring stations in 6 provinces and 33 municipalities in Italy.

[1] https://ams.com/environmental-sensors.

[2] https://www.repcomsrl.com/wp-content/uploads/2017/06/Environmental_Sensing_VOC_Product_Brochure_EN.pdf.

[3] https://airnow.gov/index.cfm?action=aqibasics.aqi/.

[4] http://www.chearíatira.it/.

Table 2 TVOC guidelines issued by the German Federal Environmental Agency

Air quality	Recommendation	TVOC [ppb]
Excellent	Target value	0–65
Good	Ventilation/airing recommended	65–220
Moderate	Intensified ventilation/airing recommended	220–660
Poor	Intensified ventilation/airing necessary	660–2200
Unhealthy	Use only if unavoidable/intense ventilation necessary	2200–5500

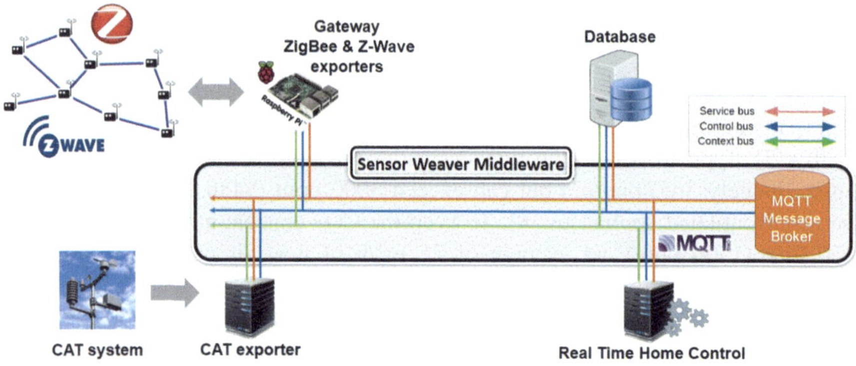

Fig. 1 Monitoring system architecture

3 The Monitoring System

The monitoring system we developed, shown in Fig. 1, is logically divided into three parts: (1) a complex wireless sensor network (WSN), constituted by a variety of heterogeneous sensor subnetworks, each one with its own monitoring peculiarities and communication protocol, (2) a secure communication platform for the exchange of sensors data and for the control of the actuators in a distributed sensor network environment (middleware); it offers a simple and technology independent abstraction layer for representing both sensors and actuators, and (3) a set of real-time applications. In this work, a real-time application related to the indoor air quality has been considered. As already stated, one element of our global monitoring network is the prototype of IAQ platform, which provides the values of air temperature, humidity and air quality in terms of ppm CO_2 and ppb TVOC. The configuration of this platform is very versatile both in terms of hardware and software. In fact, in addition to replacing the commercial sensors currently present, for example, to improve the performance of the platform itself, it is also possible to increase the functionality of the platform by introducing additional transducers for monitoring the concentration of other pollutants or particulates. This platform belongs to the ZigBee subnetwork [10], while the window open/close actuator (WA) belongs to the Z-Wave subnetwork. The middleware communication platform we developed is called *Sensor Weaver* [11]; it uses the MQTT protocol and the publish/subscribe communication paradigm, building three types of communication buses for three different types of messages: (1) Service Bus: messages related to the availability of sensors/actuators and their relative characteristics; (2) Context Bus: messages related to communications generated by sensors/actuators; (3) Control Bus: messages that control functions offered by sensors/actuators. Figure 1 shows the monitoring system architecture.

3.1 The Multi-Component Platform for IAQ

The programmable components installed give the user the ability to customize measurement protocols and to process the data collected. As a matter of fact, our IAQ platform can be considered as a smart sensor node. The transducers used are a temperature and relative humidity transducer (ENS210, AMS) and an air quality transducer (iAQ-Core, AMS); moreover, a NUCLEO-L073RZ development board (STM32 Core Board, STMicroelectronics), which installs the STMicroelectronics STM32L073RZ ARM microprocessor, and a module (WaveShare Electronics) for the implementation of the ZigBee wireless communication protocol based on the CC2530 System-on-Chip (SoC) (Texas Instruments) are also part of the platform. The EN210 transducer provides air temperature values in the operative range [−40, 100] °C with a resolution of 0.2 °C, while the relative humidity values are provided in the range [0, 100]% with a resolution of 3.5%. The iAQ-Core transducer detects a broad range of reducing gases associated with bad air quality providing a signal that is translated into CO_2 concentration in the range [450, 2000] ppm or TVOC concentration in the range [125, 600] ppb. Figure 2a shows a block diagram of the platform structure highlighting its main components. The communication between the STM32L073RZ ARM microprocessor and the transducers EN210 and iAQ-Core has been implemented using the I2C (inter integrated circuit) protocol, while the UART (universal asynchronous receiver-transmitter) protocol has been used to exchange data between the microprocessor and the CC2530 SoC.

The integration between the various components was obtained by developing an ad-hoc electronic integration board on which the components EN210, iAQ-Core and the ZigBee module were positioned. The board also supports the I2C and UART communication buses in addition to the 3.3 V power supply line for the transducers and the ZigBee module. Figure 2b shows the prototype of the IAQ platform, highlighting its individual components and the fully assembled integration board. The firmware allows the autonomous and automatic management of the various phases of the acquisition/processing/transmission protocol of the data supplied by the platform. In particular, the defined protocol includes: (1) the measurement of temperature, relative humidity and the evaluation of CO_2 and TVOC concentrations every 10 s; (2) the processing of raw data from the transducers in order to determine the levels of the physical quantities monitored; (3) the transmission to the ZigBee module of the data packets. A data packet is generated and sent when the reporting timer expires and, at the same time, the

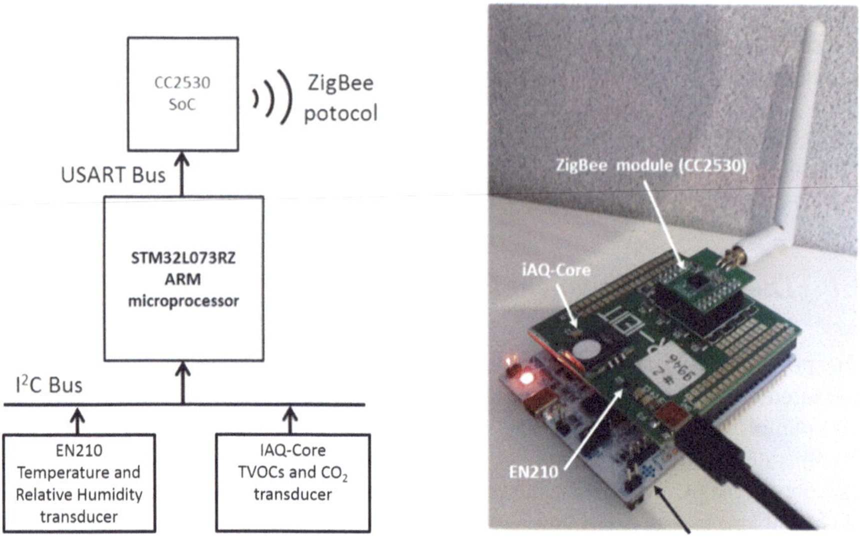

(a) (b)

Fig. 2 The platform for indoor air quality monitoring. (**a**) Block diagram highlighting components and communication protocol. (**b**) The realized prototype

variation of at least one of the monitored parameters undergoes a threshold level, defined by the user. This allows the coordinator to simply receive the messages from the nodes, without pulling them continuously. Moreover, the network coordinator provides connectivity to the middleware through the gateway.

4 The Interaction with the CAT Outdoor System

It is possible to interact with the outer CAT system via an HTTP-based API in order to retrieve the latest readings from the fixed stations, as well as the historical data. To retrieve the latest readings, a POST request is sent to the CAT server with a JSON body containing an SQL query. The response contains a JSON object with the requested information. The information that can be retrieved are the latitude/longitude of the monitoring station, the PM2.5 and PM10 readings, the timestamp of those readings and the device-id (other, more descriptive, information is also available). For instance, a body like:

```
{"stmt":"select d.geo, d.pm25, d.pm10, d.last_update, d.deviceId
    from device d where active = true"}
```

requests the most recent readings about all active devices. It is also possible to request data from a specific device with a query like:

```
{"stmt":"select d.geo, d.pm25, d.pm10, d.last_update, d.deviceId
    from device d where d.deviceId='<device-id>' and active =
    true"}
```

In this case, the response contains a reading from the specified single device:


```
{ "cols": [ "geo", "pm25", "pm10", "last_update", "deviceid" ],
  "rows": [
      [[ 11.0789829, 43.8927226 ],
       1.31, 1.41,
       1534303056465, "<device-id>"]
  ]
}
```


Note how attribute `cols` describes the meaning of the fields contained in attribute `rows`. The middleware provides a high-level interface through which its users can

- publish streams of information as *data feeds* (e.g., readings from a given sensor);
- subscribe to (e.g., request to receive) data feeds;
- offer to perform a service (e.g., perform some actuation on a device);
- request to perform a service.

In order to integrate the CAT system with our middleware, we developed a "CAT exporter" software module that is responsible for interacting with the CAT system by periodically performing HTTP requests and publishing the obtained data as middleware data feeds. Any interested client module is then able to receive readings from a specific outdoor air monitoring station by subscribing to the corresponding data feed. We performed HTTP requests to the CAT system and hence we obtained the related data feed every 240 s. A "ZigBee exporter" module performs a similar function with respect to the devices on the ZigBee network. The exporter collects indoor sensor readings and publishes them as data feeds on the middleware. It also offers a service to operate the window opener device. This actuator is a Z-Wave flush shutter device that is ideal for controlling the motors of blinds, rollers and windows. Interaction with the ZigBee network and coordinator node is hidden inside the implementation of this module, which interacts with the other modules via the middleware interface. A third "Client" module acts as a data consumer and reasoner. This module subscribes to the data feeds for indoor air quality sensing, and to the relevant data feeds for outdoor air quality sensing (the latter is chosen based on user location and closest outdoor sensing devices). Based on indoor air quality data, it determines when ventilation is appropriate and, based on outdoor air quality data, it decides whether or not to actually open the window by requesting the corresponding service. The reasoner follows several simple rules shown in Algorithm 1. Algorithm relies on four thresholds: two for the indoor CO_2 concentration estimated by our IAQ prototype, one for the PM10 and one for the PM2.5, the last two thresholds being imported from the CAT system. According to CO_2 values reported in Table 1, we defined 750 ppm and 600 ppm as thresholds

Algorithm 1 Decision rules ($th_o, th_c, limit_{PM10}, limit_{PM2.5}$)

```
while 1 do
    if state == CLOSED then
        if CO2 >= th_o && PM10 < limit_PM10 && PM2.5 < limit_PM2.5 then
            state = OPEN
    else if state == OPEN then
        if CO2 < th_c || PM10 >= limit_PM10 || PM2.5 >= limit_PM2.5 then
            state = CLOSED
```

for opening (th_o) and closing (th_c) windows, respectively. For what concern PM10 ($limit_{PM10}$) and PM2.5 ($limit_{PM2.5}$) thresholds, we consider the Italian legislative limit, set to 50 µg/m^3 and 25 µg/m^3, respectively. The windows will be opened if the indoor CO_2 concentration exceeds the opening condition threshold and, at the same time, the outer PM10 and PM2.5 values are under the respective limits. On the contrary, the windows will be closed whether CO_2 concentration falls below the closing condition threshold or PM10 and PM2.5 values exceed the respective limits.

5 Experimental Setup

In order to evaluate the potentialities of the proposed monitoring system, we conducted preliminary experiments in some indoor environments at the Research Area of the National Research Council (CNR) located in Pisa (Italy). In particular, we installed the IAQ platform prototype in an office that measures about 26 m^2 and hosts three workstations (Room A). Figure 4 shows the map of the indoor environments; the IAQ platform has been positioned 1 m high, in the middle of the Room A, while the network coordinator is in Room B. We collected sensing data for 3 days (September 25–27, 2018) and the resulting TVOC, CO_2, temperature and relative humidity time series are shown in Fig. 3. A perfect time correlation between the TVOC and CO_2 signals can be observed. This behaviour is due to the transducing method used by the iAQ-Core sensor: the CO_2 or TVOC concentrations are estimated by applying a specific algorithm to the same signal, that is the resistance variation of a single MOX (metal oxide) integrated sensor. PM10 and PM2.5 concentrations collected by CAT system in the same days are also reported in Fig. 3. The indoor CO_2 concentration exceeded the threshold for opening the windows (th_o) three times, as highlighted in the inset of Fig. 3, where an enlarged view of the data is shown. Therefore, after verifying the outdoor air quality data, the reasoner decided to actuate ventilation by opening the window of the room A, as shown in the inset of Fig. 3, where the grey regions indicate the window opening times (Fig. 4).

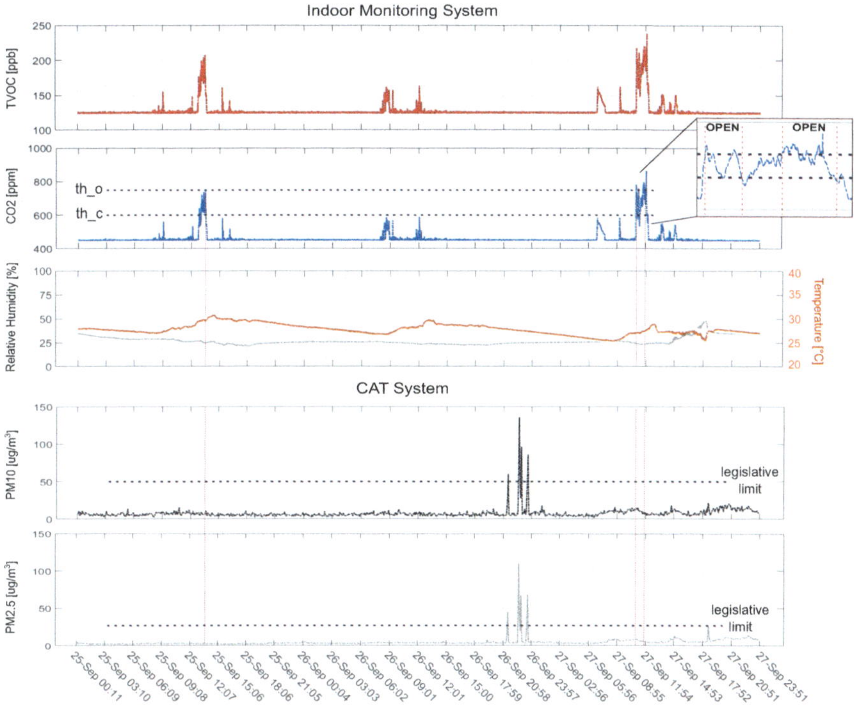

Fig. 3 Time series reporting three workdays of measurements. From top to bottom: TVOC, CO_2, temperature and relative humidity registered by IAQ platform, PM10, PM2.5 measured by CAT system

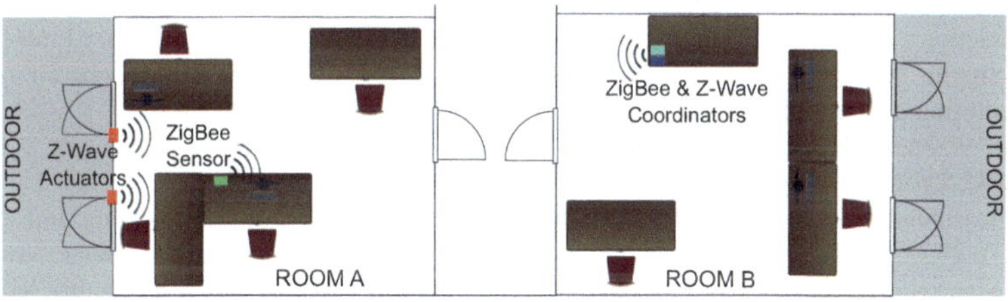

Fig. 4 Map of indoor environments where the ZigBee subnetwork has been deployed

6 Conclusions

The integration of two different air quality monitoring systems is proposed as an example of IoT application in a Smart City. One system is a platform designed for indoor environments and it is included in a wide monitoring network; the other one is CAT, an open source network for monitoring the outdoor air quality in a collaborative way, implemented in some regions of Italy. Data communication among the various sensors of the inner monitoring wireless network takes place via the developed middleware, which also allows a reasoner (installed on a server where all data are collected) to access the sensing data and to take decisions about opening or closing a window, according to the external air quality. The fused information between the two systems allows taking a weighted decision for improving the inner air quality and, as a consequence, the wellness of the occupants. As a future work, we plan to include other parameters, such as temperature and humidity, in the reasoning decisions relevant to the air quality of the inner environment, in order to implement a personalization of the environment. Readers interested in learning about the air quality in Europe are invited to read [12].

Acknowledgements This work was carried out in the framework of the SIGS project, co-funded by the Tuscany Region (Italy) under the Regional Implementation Programme for Underutilized Areas Fund (FAR FAS 2007–2013).

References

1. P. Barsocchi, P. Cassara, F. Mavilia, D. Pellegrini, Sensing a city's state of health: structural monitoring system by Internet-of-Things wireless sensing devices. IEEE Consum. Electron. Mag. **7**(2), 22–31 (2018)
2. P. Barsocchi, Position recognition to support bedsores prevention. IEEE J. Biomed. Health Inform. **17**(1), 53–59 (2013)
3. D. Bacciu, S. Chessa, C. Gallicchio, A. Micheli, E. Ferro et al., Smart environments and context-awareness for lifestyle management in a healthy active ageing framework, in *Portuguese Conference on Artificial Intelligence* (Springer, Cham, 2015), pp. 54–66
4. P. Barsocchi, A. Calabro, E. Ferro, C. Gennaro et al., Boosting a low-cost smart home environment with usage and access control rules. Sensors **18**(6), 1886 (2018)
5. A. Crivello, F. Mavilia, P. Barsocchi, E. Ferro, F. Palumbo, Detecting occupancy and social interaction via energy and environmental monitoring. Int. J. Sensor Netw. **27**(1), 61–69 (2018)
6. A. Sherin, X. Li , A cost-effective wireless sensor network system for indoor air quality monitoring applications. Proc. Comput. Sci. **34**, 165–171 (2014)
7. H.U. Wanner, *Sources of pollutants in indoor air*. IARC Scientific Publications, vol. 109 (IARC, Lyon, 1993), pp. 19–30
8. K. Arashidani, M. Yoshikawa, T. Kawamoto et al., Indoor pollution from heating. Ind. Health **34**(3), 205–215 (1996)
9. ANSI/ASHRAE Standard 62.1/2 - 2016 Ventilation for acceptable indoor air quality set
10. M. Girolami et al. The integration of ZigBee with the GiraffPlus robotic framework, in *International Joint Conference on Ambient Intelligence* (Springer, Cham, 2013)
11. P. Barsocchi, E. Ferro, L. Fortunati, F. Mavilia, F. Palumbo, EMS@ CNR: an energy monitoring sensor network infrastructure for in-building location-based services, in *High Performance Computing & Simulation (HPCS)*, pp. 857–862 (IEEE, New York, 2014)
12. Air Quality in Europe - 2018 Report, EEA Report N. 12/20118 published on October 29, 2018

A Low-Cost Monitoring and Fault Detection System for Stand-Alone Photovoltaic Systems Using IoT Technique

Adel Mellit, Amor Hamied, Vanni Lughi, and Alessandro Massi Pavan

Abstract In this paper we developed a prototype device for smart monitoring and fault detection of a stand-alone photovoltaic system (SAPVS), using an Internet of Things approach. An electronic sensing board has been designed and a web-based application has been developed in order to monitor the data (current and voltages delivered by the SAPVS, as well as air temperature and solar irradiance) in real time. The prototype has been tested experimentally at the Renewable Energy Laboratory of Jijel University, Algeria. The experimental results show the capability of the prototype to monitor data, detecting and signalling out possible faults based on the output PV power, and inform users via website about the state of the system. The faults that have been investigated are: open circuit, shading effect and dust accumulation on PV modules. The prototype is cost-effective and very easy to be implemented without any additional circuits and efforts.

1 Introduction

The photovoltaic (PV) market has seen a remarkable growth over the past years, as a result of various stimulating factors: the significant cost reduction of the PV modules on the market and more effective subsidy policies. Thanks to these factors, the return on investment of a photovoltaic system is now extremely interesting in most cases [1]. The overall capacity of photovoltaic installations around the world is about 400 GW [2]. Monitoring systems (MSs) play a very important role in supervising, controlling and performing fault detection (FD) [3] in PV systems, hence many MSs have been recently designed aiming at performing real-time monitoring of PV systems [4].

A. Mellit (✉) · A. Hamied
Renewable Energy Laboratory, Jijel University, Jijel, Algeria
e-mail: amellit@ictp.it; adel_mellit@univ-jijel.dz

V. Lughi · A. Massi Pavan (✉)
Department of Engineering and Architecture, University of Trieste, Trieste, Italy
e-mail: vlughi@units.it; apavan@units.it

W. Zamboni, G. Petrone (eds.), *ELECTRIMACS 2019*, Lecture Notes
in Electrical Engineering 615, https://doi.org/10.1007/978-3-030-37161-6_26

The Internet of Things (IoT) enables communication and data sharing among a wide variety of devices, systems and services, using the internet or others networks, without human interaction. The first clear vision of IoT was first introduced in 1999 [5]. Objects can be sensorized and/or controlled remotely through the existing network infrastructure, creating opportunities for integration of the physical world into computer-based systems. The benefits are multifold: accuracy and efficiency improvement, reduction of human intervention, reduction of costs. The IoT can create opportunities for new business models, enables the development of smart homes and smart cities, and can find use in all industries, including the health care industry. Moreover, the IoT can help users in checking and identifying malfunctioning of the systems in real time [6]. For example, the source of a fault in an SAPVS can be detected and diagnosed.

Recently, a few applications of IoT in solar energy systems have been demonstrated. For example, a solar energy MS using IoT has been designed [7]; in this application the authors used a Raspberry PI and ThinkSpeak of MathWorks to monitor the system's data. An IoT-based remote MS and control unit for solar PV systems was developed in [8]. The system facilitates preventive maintenance, historical analysis of the plant, as well as real-time monitoring and fault diagnosis. In [9] an effective implementation of an intelligent remote MS for solar PV power conditioning units (PCU) is presented. The system can be installed in solar PV-PCU to provide support for management and maintenance operations. The authors show that the system can monitor, store and manipulate data from solar PV-PCU, and that the remote monitoring functions can be realized in real time. In [10] the authors discuss a low-cost IoT-based embedded MS solar PV systems, where a GPRS module and low-cost microcontroller were used. Other applications of IoT in monitoring and supervising of PV plants can be found in [11–15].

In our previous work [16] we have designed an IoT-based prototype for monitoring current and voltage data of a PV array, as well as environmental data (such as air temperature and solar irradiance). As a further step, in this work we design a low-cost prototype for monitoring data and fault detection in a stand-alone PV System (SAPVS). For this purpose, we collect in real time the current and voltage of the system, as well the air temperature and solar irradiance. A simple algorithm is developed and integrated to detect possible faults of the PV system. The investigated faults are: one PV module in open circuit, shading and dust accumulation on a PV module. A sensing electronic circuit based on Arduino Mega2560 is also developed. IoT technology is used to transmit the monitored data on the internet using a low-cost Wi-Fi module (ESP-01). A website to store and display the monitored data online is also developed and described.

The paper is organized as follows: Sect. 2 presents a description of the investigated SAPVS, the experimental design is also given in this section. Results and discussion are reported in Sect. 3. The last section reports some concluding remarks.

2 Experimental Design and Setup

2.1 System Description

The investigated SAPVS is installed at the Renewable Energy Laboratory in Jijel University (Algeria) (see Fig. 1). It consists of 18 PV modules, 12 storage batteries, two regulators and one inverter.

The PV module characteristics are reported in Table 1.

The block diagram of the whole system is shown in Fig. 2. It consists of three PV modules connected in parallel, two batteries in series, one charge regulator with MPPT, a load (Bulb of 21 W and a resistance of 5 Ω).

Fig. 1 A stand-alone PV system installed on the rooftop of the Renewable Energy Lab at Jijel University (PV modules, batteries, charge regulator with MPPT and inverter)

Table 1 PV module specifications (PV MSX-120)

Parameter	Value
Maximum power	121 W
Maximum voltage	17.1 V
Maximum current	7.11 A
Short circuit current	7.98 A
Open circuit voltage	21.12 V

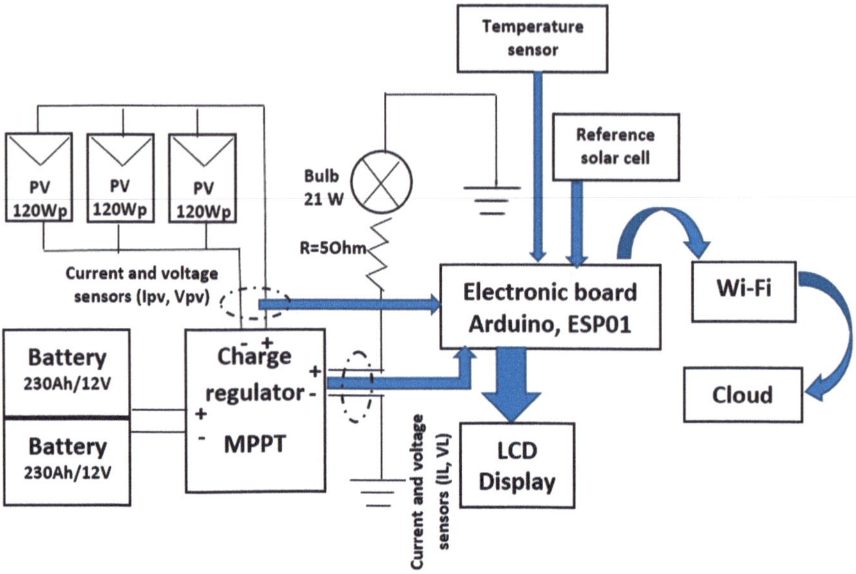

Fig. 2 Diagram of the whole system

Table 2 Employed components

Equipment	Specification	Price (Euro)
Current sensor	ACS-712/Max Current = 30 A	8
Voltage sensor	Max Voltage = 25 V	4
Arduino Mega	ATmega2560	20
Wi-Fi module	ESP-01	8
Temperature sensor	LM135	2
Reference solar cell	–	45
Voltage regulator	Diatone's BEC-MINI-3.3V	4
LCD display	LCD-16x4	8

2.2 *Experimental Step*

The system's components are presented in Table 2. With reference to Table 2 the estimated cost of the prototype is about 90 Euro.

Figure 3 shows the prototype. The acquisition board consists of an Arduino Mega 2560 microcontroller used to detect analog data from the detection circuits, a 16x4 LCD display to show the system status and monitored data in real time, an ESP-01 Wi-Fi module for data transmission to a host PC. As the ESP-01 operates at 3.3 V, a Diatone's BEC-MINI-3.3 V voltage regulator is used to reduce the supply voltage to 3.3 V.

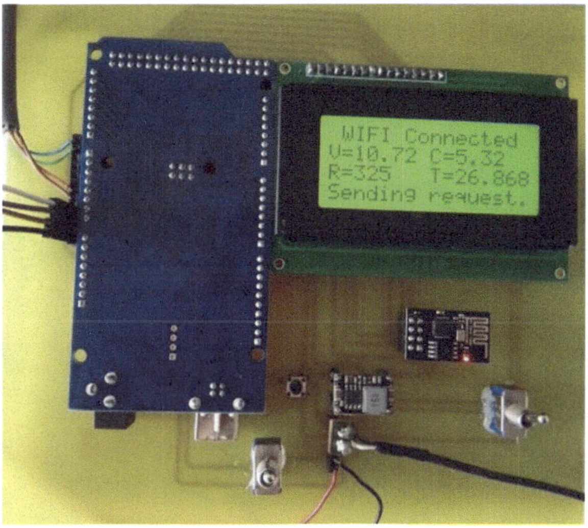

Fig. 3 The designed prototype

In the literature, numerous transmission techniques are demonstrated for monitoring purpose, most typically the Ethernet or ZigBee wireless technology. In this paper an ESP-01 Wi-Fi module is used in order to overcome some drawbacks of these technologies (i.e., the fact that Ethernet uses a network cable, a higher power consumption).

The code implemented into the Arduino Mega board is developed using Arduino IDE. The procedure of the process is shown in Fig. 4.

As sketched in Fig. 4, the process starts with the initialization of the LCD display and the Wi-Fi module ESP-01. Then the microcontroller waits for the ESP-01 to be ready. If the ESP-01 is ready, it automatically connects to the router. The monitored data is read, saved as strings and displayed on the LCD. For data transmission, ESP-01 creates a TCP connection to the server.

3 Results and Discussion

The experiments have been carried out at the Jijel University. Jijel is characterized by moderate weather. After connecting to the server, the data from the SAPVS were sent to the cloud in real time. JavaScript was used to develop the webpage. After login, the webpage shown in Fig. 5 appears, which presents in real time all the monitored data: PV array current, load current, PV array voltage, load voltage, solar irradiance and air temperature. The page can also notify users if any fault has happened, by sending an error message as well as displaying it on the LCD display.

Figure 6 shows the measured I–V curves for the different faults investigated in this work: dust accumulation, open circuit and shading. The corresponding photo of these faults is given in Fig. 7. With reference to Fig. 6 it can be observed that

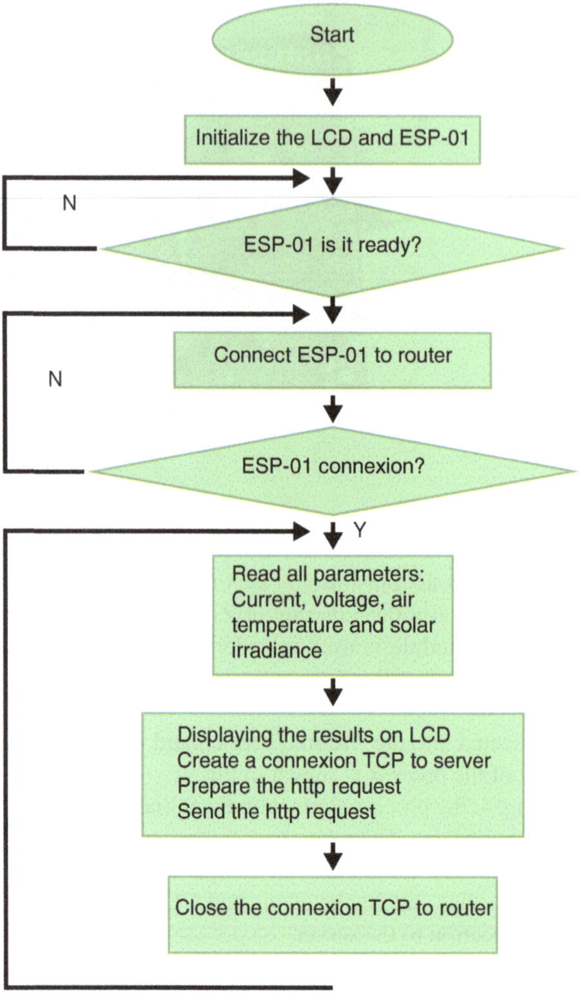

Fig. 4 Flowchart of the communication process Retour and the ESP01-Wi-Fi module

all investigated faults have a clear impact on the current (a remarkable decrease of the PV array current). On the other hand, only a slight variation in the voltage is observed for the three faults. It can also be seen that open circuit and dust accumulation faults have almost the same I–V curve.

The recorded data are posted on the website. As an example, Fig. 8 displays the evolution of the monitored data via IoT for a period of 25 min at 2 s time scale (13 h 30 min–13 h 55 min) on November 13th, 2018. As can be seen from the figure, the load current is decreased for all periods (shading effect: 13 h 35 min 29 s–13 h 38 min 3 s, open circuit: 13 h 41 min 5 s–13 h 43 min 45 s, dust accumulation: 13 h 50 min 41 s–13 h 52 min 23 s), while the voltage is relatively constant for all cases.

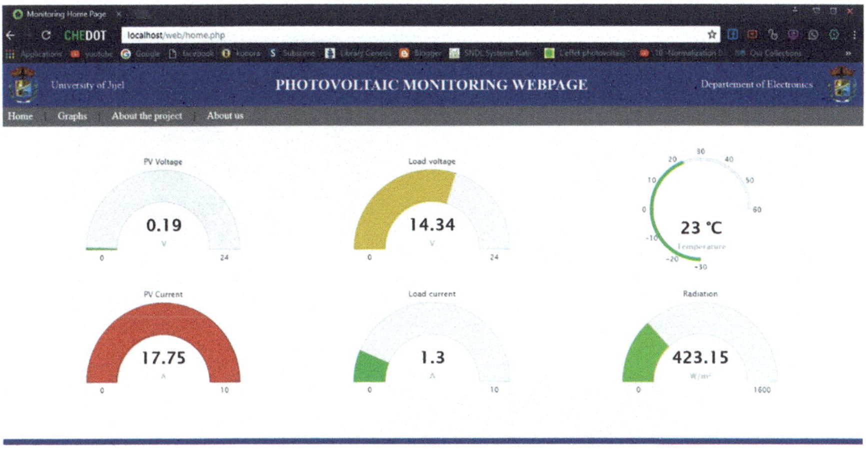

Fig. 5 Home page of the designed monitoring system

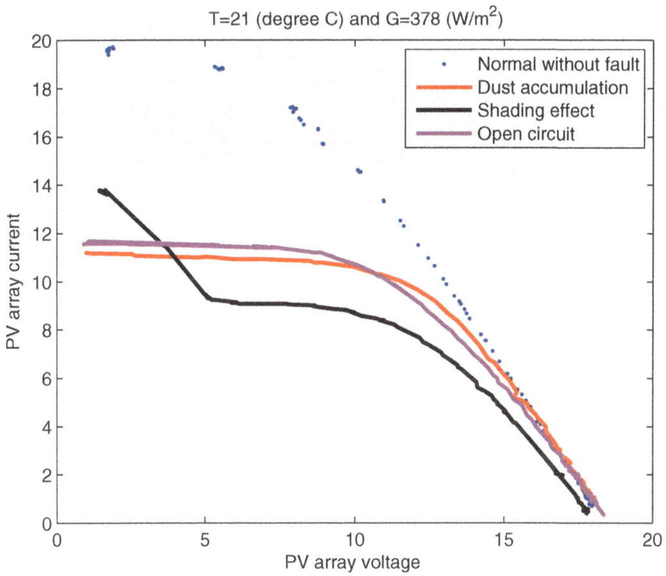

Fig. 6 Measured I–V curves (normal and faulty)

The fault can be detected based on a simple comparison between the measured output power (P_m) and the calculated one(P_c) based on an explicit model developed in [17]. The comparison formula is given as $|(P_m - P_c)| \leq \Delta$, where Δ is the error which is defined as the sum of error uncertainty of each component and the model error, which is about ± 5. The explicit model is developed in MatLab environment and implemented into the microcontroller, then the model calculate the power for a given measured solar irradiance and air temperature, after that measured power is

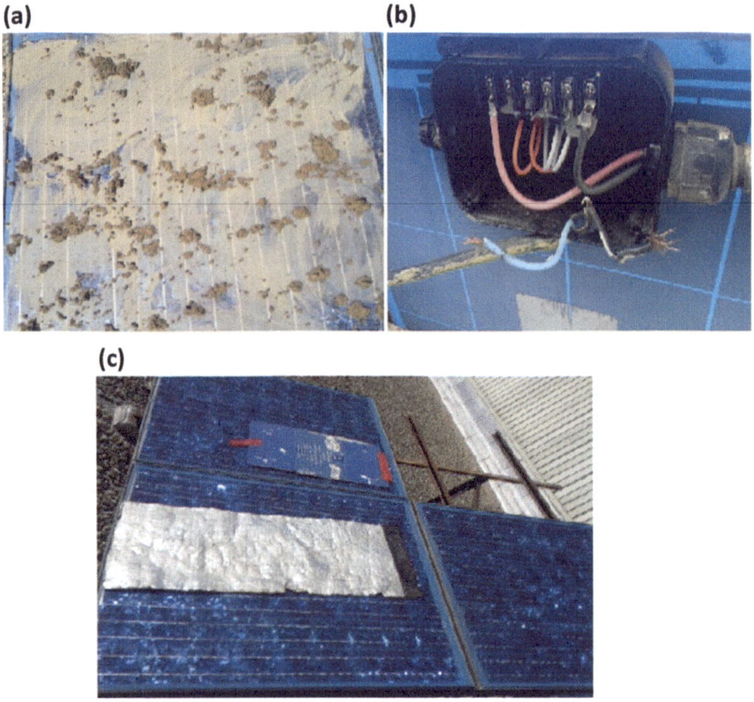

Fig. 7 The investigated faults: (**a**) dust accumulation, (**b**) open circuit, (**c**) shading effect

compared with the one expected by the model. At the moment the method cannot identify the type or the origin of the fault just a detection fault. This can help us to check online the state of the PV system.

4 Conclusions

In this paper a low-cost prototype for monitoring SAPVS data is demonstrated. Its potential for fault detection has also been shown. The recorded data have been sent to the cloud via IoT technology and are made available to the users with web-based interface, in which users can browse all recorded data in real time.

This work demonstrates that a cost-effective IoT-based system can be easily implemented for real-time monitoring of an SAPVS, thus demonstrating the potential for significantly reduced human intervention, the potential for fault detection, the possibility of historical data analysis.

This work can be further improved and extended for fault diagnosis of photovoltaic systems by using more sophisticated algorithms. The designed website also could be enhanced, e.g., by displaying more information about the faults.

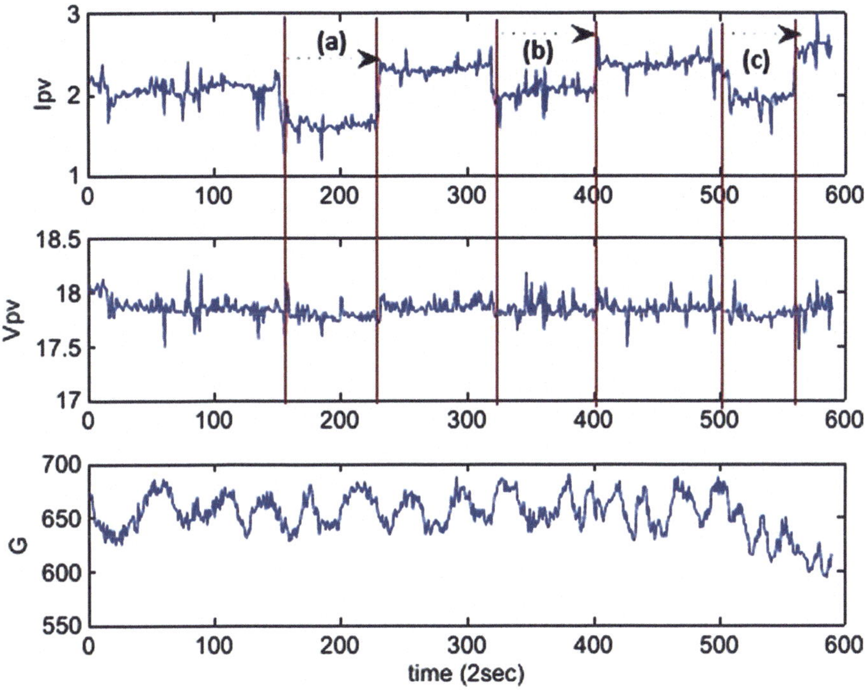

Fig. 8 The posted curves: current, voltage and solar irradiance: (**a**) period 1 (shading effect on two PV modules), (**b**) period 2 (open circuit of one PV module) and (**c**) period 3 (dust accumulation on a PV module)

Acknowledgements The second author would like to thank the ICTP for providing support to achieve the present work. The third author acknowledges the financial support provided by "MUSE – Cross-border" collaboration for a sustainable and energetically efficient university mobility, a project co-financed by the European Regional Development Fund via the cross-border cooperation program Interreg Italy-Slovenia.

References

1. S. Daliento, A., Chouder, P. Guerriero, A.M. Pavan, A. Mellit, R. Moeini, P. Tricoli, Monitoring, diagnosis, and power forecasting for photovoltaic fields: a review. Int. J. Photoenergy **2017**, 120–134 (2017)
2. IEA, Snapshot of global photovoltaic markets . Accessed March 2018
3. W. Chine, A. Mellit, V. Lughi, A. Malek, G. Sulligoi, A. Massi Pavan, A novel fault diagnosis technique for photovoltaic systems based on artificial neural networks. Renew. Energy **90**, 501–512 (2016)
4. A. Mellit, G.M. Tina, S.A. Kalogirou, Fault detection and diagnosis methods for photovoltaic systems: a review. Renew. Sustain. Energy Rev. **91**, 1–17 (2018)
5. N. Gershenfeld, *When Things Start to Think* (Macmillan, New York, 1999).

6. P. Barsocchi, P. Cassara, F. Mavilia, D. Pellegrini, Sensing a city's state of health: structural monitoring system by Internet-of-Things wireless sensing devices. IEEE Consum. Electron. Mag. **7**, 22–31 (2018)
7. S. Patil, M. Vijayalashmi, R. Tapaskar, Solar energy monitoring system using IoT. Indian J. Sci. Res. **15**, 149–155 (2017)
8. Y. Li, P. Lin, H. Zhou, Z. Chen, L. Wu, S. Cheng, F. Su, On-line monitoring system of PV array based on internet of things technology, in *IOP Conference Series: Earth and Environmental Science*, vol. 93, 2017
9. S. Adhya, D. Saha, A. Das, J. Jana, H. Saha, An IoT based smart solar photovoltaic remote monitoring and control unit, in *2nd International Conference on Control, Instrumentation, Energy and Communication (CIEC)* (2016), pp. 432–436
10. B. Shrihariprasath, V. Rathinasabapathy, A smart IoT system for monitoring solar PV power conditioning unit, in *World Conference on Futuristic Trends in Research and Innovation for Social Welfare (Startup Conclave)* (2016) pp. 1–5
11. A. Kekre, S. Gawre, Solar photovoltaic remote monitoring system using IoT, in *2017 International Conference on Recent Innovations in Signal processing and Embedded Systems (RISE)* (2017), pp. 619–623
12. M. Vestenicky, S. Matuska, R. Hudec, P. Kamencay, Sensor network proposal based on IoT for a prediction system of the power output from photovoltaic panels, in *28th International Conference IEEE Radioelektronika (RADIOELEKTRONIKA)* (2018), pp. 1–4
13. P.M. Badave, B. Karthikeyan, S.M. Badave, S.B. Mahajan, P. Sanjeevikumar, G.S. Gill, Health monitoring system of solar photovoltaic panel: an internet of things application, in *Advances in Smart Grid and Renewable Energy* (Springer, Singapore, 2018), pp. 347–355
14. H.S. Nalamwar, M.A. Ivanov, S.A Baidali, Automated intelligent monitoring and the controlling software system for solar panels, in *Journal of Physics: Conference Series*, vol. 803 (2017)
15. S.R. Madeti, S.N. Singh, Monitoring system for photovoltaic plants: a review. Renew. Sustain. Energy Rev. **67**, 1180–1207 (2017)
16. A. Hamied, A. Mellit, M.A. Zoulid, IoT-based experimental prototype for monitoring of photovoltaic arrays, in *2018 International Conference on Applied Smart Systems ICASS2018*, Médéa, 24–25 November 2018
17. A. Massi Pavan , S. Vergura , A. Mellit, V. Lughi , Explicit empirical model for photovoltaic devices. Solar Energy **155**, 647–659 (2017)

A New Approach for Effective Position/Speed Sensor Fault Detection in PMSM Drives

Saïda Ben Slimen, Manef Bourogaoui, and Houda Ben Attia Sethom

Abstract This study presents a new fault detection approach for Permanent Magnet Synchronous Motors (PMSMs), based on the Park Current Vector (PCV), which addresses the total loss position/speed information fault in electric motor drives. Two PMSMs were considered: a 53 kW PMSM was adopted for simulations, whereas experimental validation was performed using a 1 kW PMSM. Effective position/speed sensor fault detection was achieved using the proposed PCV-based approach, in which validity and feasibility are proved by simulation and experiment results.

1 Introduction

Nowadays, Permanent Magnet Synchronous Motor (PMSM)-based Adjustable Speed Drives (ASDs) are becoming increasingly attractive for several applications such as aerospace, aeronautic, military, medical, robotic, and transport applications, thanks to their many advantages, namely, high efficiency, high-speed operation, reliability, robustness, and compactness. Both low- and high-power PMSMs are used depending on the application requirements.

S. B. Slimen (✉)
LR11ES15 Laboratoire des Systèmes Electriques, Ecole Nationale d'Ingénieurs de Tunis, Université de Tunis El Manar, Tunis, Tunisia

M. Bourogaoui
LR11ES15 Laboratoire des Systèmes Electriques, Ecole Nationale d'Ingénieurs de Tunis, Université de Tunis El Manar, Tunis, Tunisia

Institut Supérieur des Technologies de l'Information et de la Communication, Université de Carthage, Tunis, Tunisia

H. B. A. Sethom
LR11ES15 Laboratoire des Systèmes Electriques, Ecole Nationale d'Ingénieurs de Tunis, Université de Tunis El Manar, Tunis, Tunisia

Ecole Nationale d'Ingénieurs de Carthage, Université de Carthage, Tunis, Tunisia
e-mail: houda.benattia@enicarthage.rnu.tn

W. Zamboni, G. Petrone (eds.), *ELECTRIMACS 2019*, Lecture Notes in Electrical Engineering 615, https://doi.org/10.1007/978-3-030-37161-6_27

However, mechanical and electric faults may affect PMSM–ASDs leading to very dangerous operating conditions and hazardous consequences that seriously disturb the normal operation of a given application [1–5].

One of the concerned PMSM–ASD components is the position/speed sensors, such as resolvers and incremental encoders. Indeed, this device is crucial for the PMSM control, especially when a closed loop operation with a given control strategy is considered in order to improve the motor operation performance. Nevertheless, these sensors are very sensitive to faults. In this context, a previous work established a deep and thorough review regarding types, causes, and consequences of position/speed sensors faults [4, 5]. Thus, several studies have been carried out to propose solutions to the problem of speed/position sensor faults detection and isolation (FDI) [5–12].

On the other hand, many fault detection techniques were performed and applied in order to detect different faults that may affect AC motor drives. Among the proposed techniques and according to the existing scientific literature, the Park Vector pattern is one of the most relevant techniques for FDI purposes in AC motor drives [13, 18].

In fact, the Park Vector approach has proved its capability to detect changes caused by the occurrence or the existence of different faults in a given PMSM–ASD. Thus, the Park Current Vector (PCV) pattern has been largely used for the diagnosis of different fault types such as broken rotor bars and short circuits in stator windings using only the measured magnitude and the phase angle, which are extracted from (d,q) currents as proven in [13–16]. Thereby, Reference [17] deals with detection and localization of faults which may affect power converter switches. Indeed, the proposed technique is based on the PCV approach in order to distinguish between open and short circuit fault in a power converter. Moreover, the authors in [18] used the PCV technique for the detection and the diagnosis of open circuit fault in a three-phase two-level inverter supplying an induction squirrel-cage motor. Simulation and experimental results highlighted the contribution of the PCV technique in the case of a faulty inverter.

In the majority of these works, the proposed approach for fault feature extraction is based on a circular representation in the (d,q) reference frame in order to extract information about different fault occurrences in AC motor drives.

In this paper, the relevant contribution consists of using a new defined angle in addition to the magnitude of (d,q) frame PMSM currents for the detection of a total loss information position sensor fault that may occur in a PMSM–ASD. On the other hand, it aims to prove the effectiveness of the fault detection technique independently of the motor power rating; thereby, the proposed approach can cover all industrial applications that include PMSM–ASDs. Within this frame, the present paper is organized as follows: First, the PMSM drive operation under position/speed sensor faulty conditions is analyzed. For this study, two PMSMs were considered, namely, a 1 kW and a 53 kW PMSM; thus, the PCV is used as a fault detection technique. Second, simulation results are presented and discussed in order to prove the effectiveness of the PCV-based fault detection technique regarding the 53 kW PMSM. Finally, experimental results, concerning the 1 kW PMSM, are presented and analyzed to validate the performance of the proposed technique.

2 PMSM Drive Operating Under Rotor Position Sensor Fault

2.1 Mathematical PMSM Modeling

The PMSM–ASD includes a PMSM, a pulse width modulation (PWM) voltage inverter, two current sensors, and a position sensor in order to ensure the closed loop operation according to a classic field-oriented control (FOC) scheme (Fig. 1). The PMSM mathematical model has been developed in the synchronous reference frame (d,q). It is described by Eqs. (1–5). In the case of salient pole PMSM, Eq. (3) becomes simplified, thanks to the equality between the axis *d* and *q* inductances.

$$v_{\mathrm{sd}} = R_{\mathrm{s}} i_{\mathrm{sd}} + l_{\mathrm{sd}} \frac{\mathrm{d}i_{\mathrm{sd}}}{\mathrm{d}t} - \dot{\theta}_{\mathrm{e}} l_{\mathrm{sq}} i_{\mathrm{sq}} \tag{1}$$

$$v_{\mathrm{sd}} = R_{\mathrm{s}} i_{\mathrm{sd}} + l_{\mathrm{sd}} \frac{\mathrm{d}i_{\mathrm{sd}}}{\mathrm{d}t} - \dot{\theta}_{\mathrm{e}} l_{\mathrm{sd}} i_{\mathrm{sd}} + \dot{\theta}_{\mathrm{e}} \Psi l_{\mathrm{sq}} i_{\mathrm{sq}} \tag{2}$$

The electromagnetic torque is given by Eq. (3):

$$C_{\mathrm{em}} = \frac{3}{2} p \left(\left(l_{\mathrm{sd}} - l_{\mathrm{sq}} \right) i_{\mathrm{sd}} i_{\mathrm{sq}} \right) + \Psi_{\mathrm{PM}} i_{\mathrm{sq}} \tag{3}$$

The mechanical equation is given by Eq. (4):

$$\frac{J}{p} = \frac{\mathrm{d}w_{\mathrm{e}}}{\mathrm{d}t} = C_{\mathrm{em}} - C_{\mathrm{r}} - f_{\mathrm{v}} \frac{w_{\mathrm{e}}}{p} \tag{4}$$

where

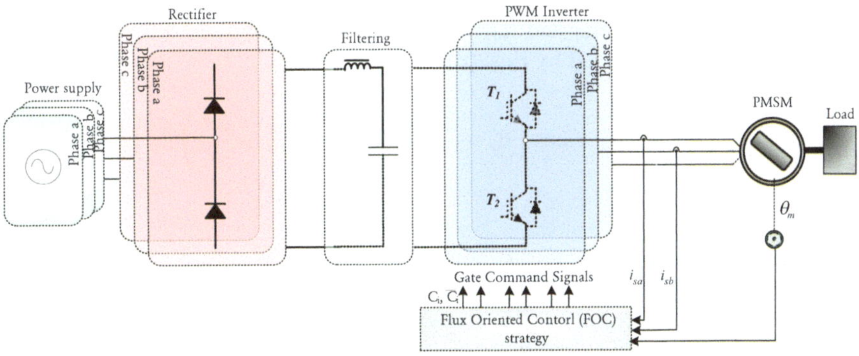

Fig. 1 PMSM-based adjustable speed drive under study

Table 1 1 kW PMSM parameters

Parameter	Notation	Value
Rated power	P	1 kW
Rated current	I_n	2.5 A
Rated speed	N_n	1000 rpm
Stator resistance	R_s	1.145 Ω
Pole pair number	p	3

Table 2 53 kW PMSM parameters

Parameter	Notation	Value
Rated power	P	53 kW
Rated current	I_n	168 A
Rated speed	N_n	600 rpm
Stator resistance	R_s	0.087 Ω
Pole pair number	p	5

$$\frac{d\theta_e}{dt} = w_e \tag{5}$$

As it is mentioned previously, two PMSMs were considered, a 53 kW PMSM for simulation results and a 1 kW PMSM for experimental validation. The main objective of this section is to analyze the effect of a position sensor fault occurrence on the system behavior and particularly on the PMSM stator currents, for both low- and high-power PMSMs. The 1 kW and the 53 kW PMSM parameters are given in Tables 1 and 2, respectively.

2.2 *The Proposed PCV-Based Sensor Fault Detection*

In this study, the proposed sensor fault detection technique is based on the Park Current Vector (PCV) (Fig. 2).

In healthy conditions, the stator current i_{sabc} Park transformation results in the Park Vector components i_{sd} and i_{sq}, which are used to apply the PCV technique. These currents are expressed by Eqs. (6) and (7).

$$i_{sd}{}^{2}\!/_{3}\left(\cos(\theta)\, i_{sa} + \cos\left(\theta_e - \frac{2\pi}{3}\right) i_{sb} + \cos\left(\theta_e - \frac{2\pi}{3}\right) i_{sc}\right) \tag{6}$$

$$i_{sd}{}^{2}\!/_{3}\left(-\sin(\theta)\, i_{sa} - \sin\left(\theta_e - \frac{2\pi}{3}\right) i_{sb} - \sin\left(\theta_e - \frac{2\pi}{3}\right) i_{sc}\right) \tag{7}$$

The proposed technique uses these currents to compute the new defined diagnostic variable, which is a new angle θ_s.

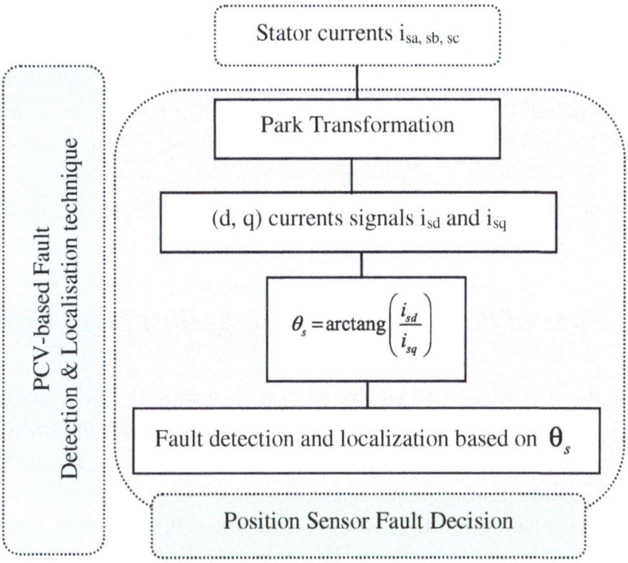

Fig. 2 The proposed PCV technique-based position sensor fault detection principle

Indeed, this new variable is given by Eq. (8).

$$\theta_\mathrm{s} = \text{arctang}\left(\frac{i_\mathrm{sd}}{i_\mathrm{sq}}\right) \tag{8}$$

Under healthy conditions, the new defined angle θ_s is null according to the FOC principle. However, when position sensor fault occurs, θ_s changes significantly. Consequently, its amplitude variation can be effectively and rapidly detected.

In the following section, the proposed fault index θ_s is used in order to detect the total loss of the position sensor information in the considered PMSM drive system. To ensure an effective detection, a threshold must be fixed independently of load conditions.

It should be noted that the contribution of this work is the use of the new defined angle, which is completely different from the classical angle used in the existing literature, basing on (d,q) PMSM currents.

3 Simulation Results

The modeling and the simulation of the 53 kW PMSM–ASD and the PCV-based fault detection technique algorithm were performed using the Matlab/Simulink® environment. A FOC was applied in order to control the PMSM speed. The simulation carried out aims to assess the proposed PCV-based position/speed sensor fault detection performance. First, load variation, which can lead to false alarms, is applied to verify its robustness against such perturbation. Next, the technique effectiveness to detect the total loss of position sensor information is analyzed. The randomness of the instant at which the position sensor fault occurs is also discussed.

The following operating conditions were considered: the motor speed is set at 600 rpm under a load torque $C_r = 0\%$, 20%, 50%, and 100% of the rated torque C_n.

3.1 Robustness of the PCV Technique Against Load Variations

In order to prove the robustness of the proposed technique under position sensor fault, the PCV was tested for different load conditions. In fact, the PMSM operates at 20 rpm under 20% of C_r. Both at $t = 0.4$ s and $t = 0.8$ s, 50% and 100% of C_r are applied to the PMSM–ASD.

The simulation results highlighted by Fig. 3 show that the new angle θ_s is not sensitive to the load torque variations. This proves definitely the robustness of this technique against load variations.

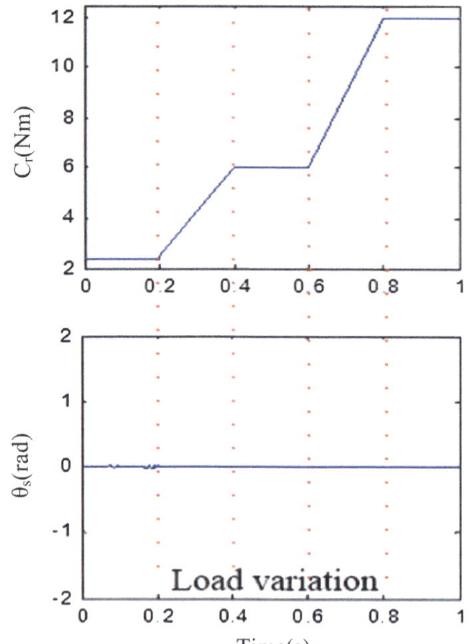

Fig. 3 Robustness of the PCV technique against load variations

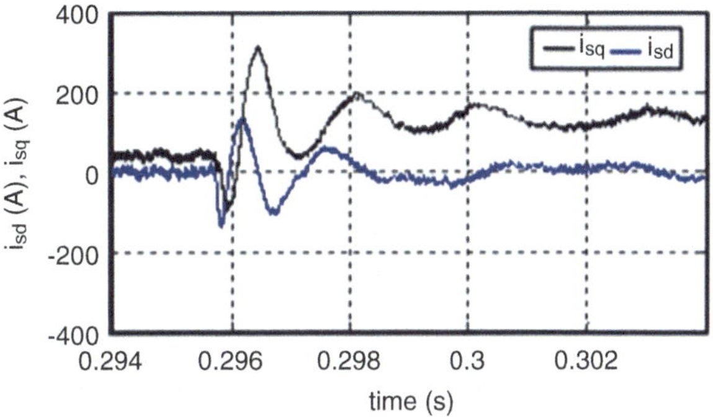

Fig. 4 The direct and the quadratic currents in the case of position sensor fault

3.2 *Impact of the Position Sensor Fault on the High-Power Rating PMSM Behavior*

At $t = 0.296$ s, $t = 0.796$ s, an abrupt total loss of position sensor information fault is applied to the 53 kW PMSM–ASD. Figure 4 depicts the direct and quadratic currents before, during, and after the total loss of position sensor information occurrence for the high-power PMSM. Regarding the direct and quadratic currents, the position sensor fault does not have a significant impact on the direct current; this current is close to zero before and after the fault occurrence, according to the FOC control strategy. However, the quadratic current increases considerably, comparing to the remaining values related to the healthy drive operating.

3.3 *Position Sensor Information Total Loss Fault Detection Using the New Defined Angle*

In this subsection, the fault detection effectiveness of the proposed PCV technique is analyzed and discussed. For this purpose, simulations were carried out for healthy and faulty PMSM drive under 0%, 20%, 50%, and 100% of the rated load torque considering two different fault occurrence instants $t = 0.296$ s and $t = 0.336$ s.

Indeed, the position sensor fault detection can be performed starting only from the new diagnostic variable θ_S. In the case of normal operating conditions, this variable is close to zero, whereas during fault occurrence it is quietly different from zero.

Noticeable results are shown in Figs. 5 and 6. The detection of the position sensor fault is possible, and it is highlighted by an important change on the diagnostic variable value when the fault occurs. This conclusion is available for different load levels and fault occurrence instants.

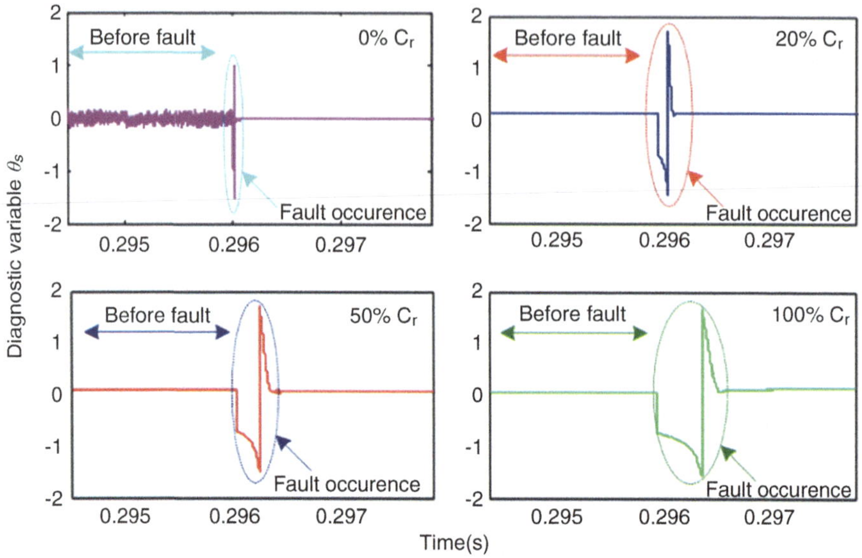

Fig. 5 Position sensor fault detection results in the case of 53 kW PMSM, fault occurrence at $t = 0.296$ s

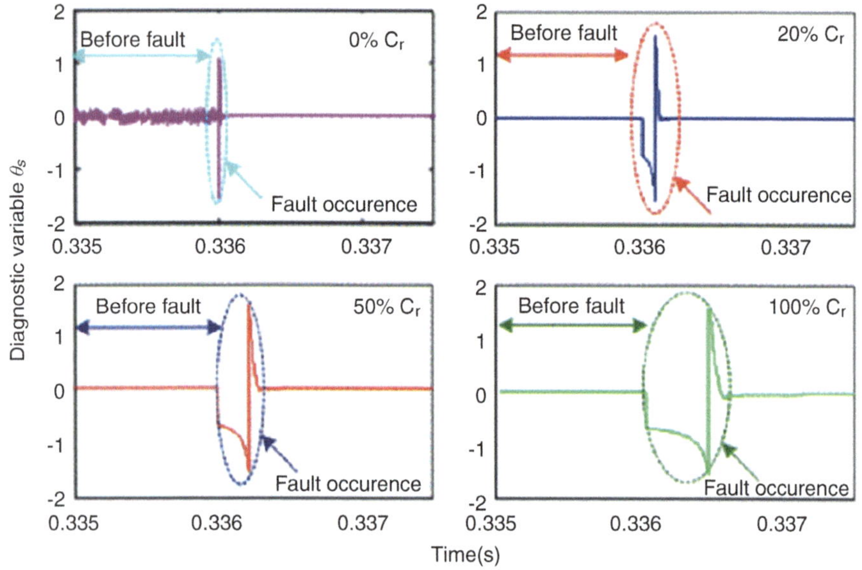

Fig. 6 Position sensor fault detection results in the case of 53 kW PMSM, fault occurrence at $t = 0.336$ s

Therefore, it can be concluded that it is possible to fix a detection threshold related to this variable in order to ensure an effective detection of the total loss of speed/position information.

The highest threshold is fixed independently of the load variation and to the fault instant occurrence. According to Figs. 5 and 6, the average value of θ_s can be fixed to 0.16 in normal conditions. When the position sensor fault occurs at $t = 0.296$ s or $t = 0.336$ s, an abrupt change in the diagnostic variable is detected and far exceeds the fixed threshold value.

4 Experimental Results

Experimental tests were carried out in real time in order to highlight the dynamic behavior of the PMSM–ASD during a total loss of the position sensor measure and to assess the effectiveness of the proposed PCV-based position sensor fault detection technique.

The experimental setup consists of a 1 kW PMSM coupled to a DC load machine, a three-phase microchip inverter, an incremental encoder, a development board "dsPICDEM" including the programmable dsPIC30F60A device, a debugger MPLAB ICD2, and a computer which is equipped with MPLAB IDE development environment used for program compilation and implementation (Fig. 7).

The incremental encoder, which is an angular position sensor, has been intended for positioning, movement control, or speed measurement applications by counting the number of peaks related to the initial position. Although the incremental encoder is less expensive than the other position sensors, it is sensitive to electric power failures which cause faults such as position sensor information total loss.

The experimental tests were carried out using the 1 kW PMSM operating in open loop and without load conditions. This step is considered as very important to validate the proposed position sensor fault detection algorithm.

First, the 1 kW PMSM operates under healthy conditions; the PMSM–ASD is well-balanced.

Second, at $t = 0.5$ s, a total loss of the encoder information is induced.

4.1 Impact of the Position Sensor Fault on the PMSM Behavior

The experimental results obtained in these conditions are shown in Fig. 8 where the impact of the position sensor fault on the direct and quadrature currents is highlighted.

Indeed, under faulty operating conditions, the position sensor fault produces important oscillations that cause abnormal rotor vibrations and degrade the drive operation performance. Figure 8a displays the experimental direct current i_{sd}. The fault occurrence produces an abrupt variation in the amplitude of this current at the time of fault and after few milliseconds; i_{sd} returns to its normal value when

Fig. 7 Experimental setup

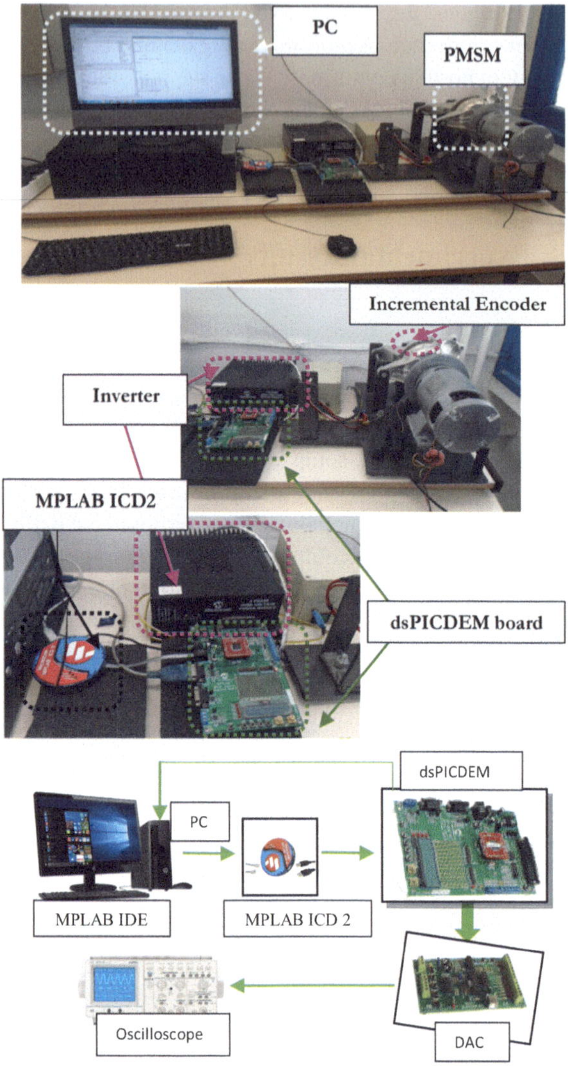

the changeover takes place. On the other hand, Fig. 8b displays the experimental quadratic current i_{sq}. Thus, the position sensor fault occurrence causes both abrupt variation in the amplitude and important oscillations of this current.

Based on the obtained results, the impact of the total loss position sensor information on i_{sd} and i_{sq} requires a rapid detection in order to avoid dangerous consequences and the system shut down.

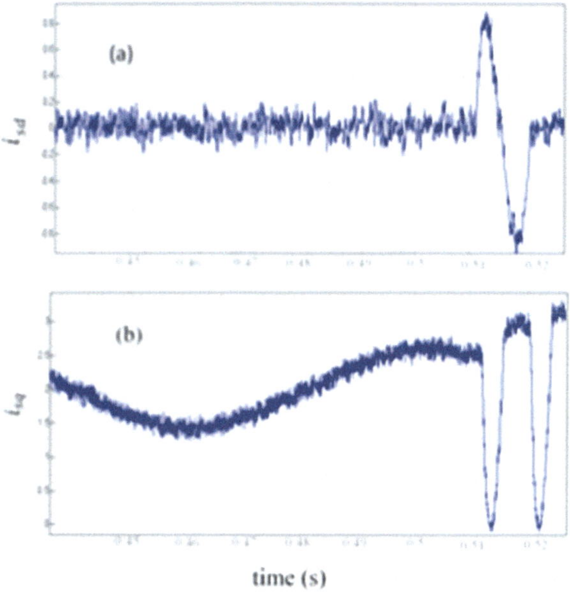

Fig. 8 Experimental direct and quadratic currents under position sensor fault conditions

4.2 *Experimental Validation of the Proposed PCV-Based Sensor Fault Detection Technique Effectiveness*

In this subsection, the main objective is to prove the effectiveness of the proposed PCV-based position sensor fault detection technique. As it was mentioned previously, this technique is based only on a new angle different from the classical one as defined in Eq. (8).

The experiments have been performed under motor speed of 200 and 500 rpm. Figures 9 and 10 show the possible real-time fault detection using the new diagnostic variable θ_s under the two considered motor speeds.

Indeed, under healthy position sensor operating and no load conditions, the direct current varies between ± 0.2 A. The θ_s average value is of the order of 0.16; thus, this value is verified by Eq. (8). For that, it is possible to fix a threshold value to be considered as unique threshold known beforehand to allow detecting if a position abrupt change occurs. In this case, this threshold should be available for all the PMSM operating ranges under no load conditions. It is chosen as the highest value above which a fault occurrence is judged to be possibly detected. Hence, this unique threshold was fixed to the highest value of θ_s relating to the variation range of i_{sd}. When this threshold is exceeded, then it necessarily means that a position sensor fault has occurred.

Indeed, as it is shown in Figs. 9 and 10, when the fault occurs at $t = 0.51$ s, an abrupt change is clearly highlighted in θ_s amplitude which far exceeds the fixed threshold value defined for healthy conditions. Consequently, the experimental

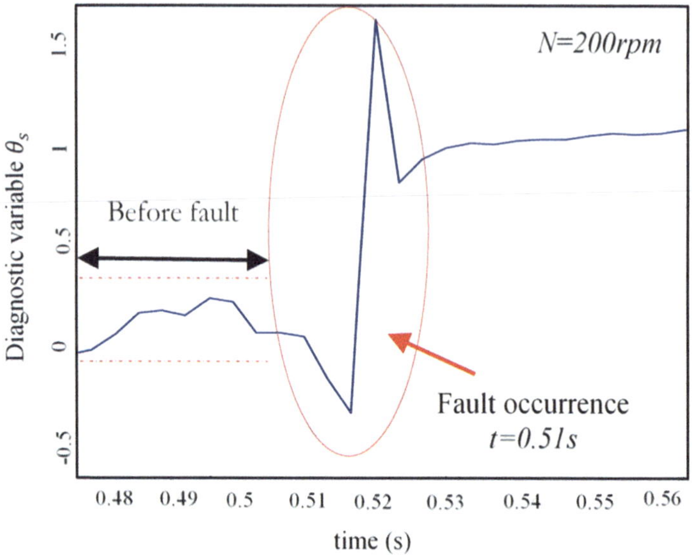

Fig. 9 The PCV-based detection and localization technique under a total loss of position sensor and a speed of 200 rpm

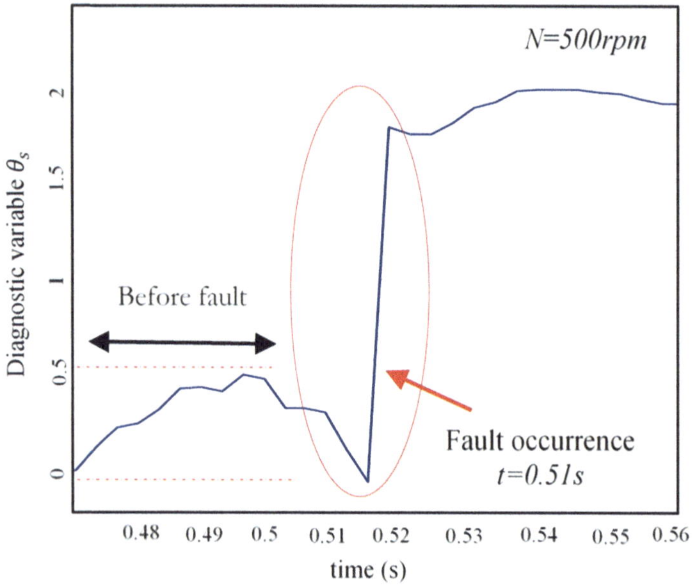

Fig. 10 The PCV-based detection and localization technique under a total loss of position sensor and a speed of 500 rpm

results agree well with the simulations in case of no load conditions. Thus, the proposed PCV-based fault detection technique shows a great ability for relevant position/speed sensor fault detection.

5 Conclusion

New aspects have been discussed and analyzed in this paper regarding the detection of an abrupt breakdown of rotor position/speed sensors in applications based on PMSM–ASDs. The Park Current Vector (PCV) has been used as a fault detection technique, thanks to its strong capabilities to provide effective detection of changes due to fault occurrences. A new angle has been defined basing on the PCV components and has been used as a new variable-based position sensor fault detection, which consists of an original contribution to improve the effectiveness of position sensor fault detection in PMSM–ASDs. Indeed, regarding the studied fault, it has been proved that the ability of the proposed fault detection variable to highlight the abrupt change in the rotor position information is strongly independent of motor power and to torque and speed operating conditions. Simulations, carried out for 53 kW salient pole PMSM, proved the effectiveness of the proposed fault detection technique. Thus, experimental results, obtained from a 1 kW PMSM-based setup, validated the simulation results and highlighted the relevant contribution of this research work.

Moreover, the impact of load variation on the PMSM behavior in closed loop will be the purpose of next investigations.

Acknowledgments This research project was supported by the Tunisian Ministry of High Education and Scientific Research, under Grant LSE-ENIT-LR11ES15.

References

1. H. Benazza, M. Moujahed, M. Jemli, M. Boussak, Implementation of improved sliding mode observer and fault tolerant control for a PMSM drive. J. Circuit. Syst. Comput. **26**, 1750032_1–1750032_16 (2017)
2. I. Jlassi, J.O. Estima, S.K. El Khil, N.M. Bellaaj, A.J.M. Cardoso, A robust observer-based method for IGBTs and current sensors fault diagnosis in voltage-source inverters of PMSM drives. IEEE Trans. Ind. Appl. **53**, 2894–2905 (2017)
3. W. Wang, J.H. Zhang, M. Cheng, Common model predictive control for permanent-magnet synchronous machine drives considering single-phase open-circuit fault. IEEE Trans. Power Electron **32**, 5862–5872 (2017)
4. S.K. El Khil, I. Jlassi, J.J.O. Estima, N.M. Bellaaj, A.J.M. Cardoso, Current sensor fault detection and isolation method for PMSM drives, using average normalised currents. Electron Lett. **52**, 1434–1436 (2016)
5. M. Bourogaoui, H.B.A. Sethom, I.S. Belkhodja, Speed/position sensor fault tolerant control in adjustable speed drives – a review. ISA Trans. **64**, 269–284 (2016)

6. M. Bourogaoui, Jlassi I., Khojet S., Khil EL, Sethom H.BA, An effective encoder fault detection in PMSM drives at different speed ranges, in IEEE International Symposium on Diagnostics for Electrical Machines, Power Electronics and Drives (2015), pp. 90–96
7. Y. Da, X. Shi, M. Krishnamurthy, A novel universal sensor concept for survivable PMSM drives. IEEE Trans. Power Electron **28**, 5630–5638 (2013)
8. G.H.B. Foo, X. Zhang, D.M. Vilathgamuwa, A sensor fault detection and isolation method in interior permanent-magnet synchronous motor drives based on an extended Kalman filter. IEEE Trans. Ind. Electron **60**, 3485–3495 (2013)
9. K. Rothenhagen, F.W. Fuchs, Doubly fed induction generator model-based sensor fault detection and control loop reconfiguration. IEEE Trans. Ind. Electron **56**, 4229–4238 (2009)
10. F. Mwasilu, J.W. Jung, Enhanced fault-tolerant control of interior PMSMs based on an adaptive EKF for EV traction applications. IEEE Trans. Power Electron **31**, 5746–5758 (2016)
11. C. Choi, K. Lee, W. Lee, Observer-based phase-shift fault detection using adaptive threshold for rotor position sensor of permanent-magnet synchronous machine drives in electromechanical brake. IEEE Trans. Ind. Electron **62**, 1964–1974 (2015)
12. I. Jlassi, S. Khojet EL Khil, A MRAS-Luenberger observer based fault tolerant control of PMSM drive. J Electr. Syst. **10-1**, 48–56 (2014)
13. O. Bilgin, M. Öğüt, H. Arabacı, Detection of induction motors rotor faults by using negative selection algorithm based on Park's vector approach, in IEEE International Conference on Electrical Machines (2014), pp. 1536–1541
14. J. Burriel-Valencia, A. Sapena-Baño, M. Pineda-Sanchez, J. Martinez-Roman, Multilayer Park's vector approach, a method for fault detection on induction motors, in IEEE International Conference on Industrial Technology (2015), pp. 775–780
15. M.W. Abitha, V. Rajini, Park's vector approach for online fault diagnosis of induction motor, in International Conference on Information Communication and Embedded Systems, pp. 1123–1129 (2013)
16. N.M.A. Freire, J.O. Estima, A.J.M. Cardoso, Multiple open-circuit fault diagnosis in voltage-fed PWM motor drives using the current Park's vector phase and the currents polarity, in IEEE Symposium on Diagnostics for Electrical Machines, Power Electronics Drives (2011), pp. 397–404
17. H.U. Shin, K.B. Lee, Fault-tolerant switched reluctance machine drives using a current Park's vector, in IEEE Conference on Energy Conversion (2015)
18. C.B. DjamalEddine, A. Bendiabdellah, M.A. Khelif, M. Bendjebbar, N. Benouzza, The enhancement of park current vectors technique for inverter fault detection, in International Conference on Systems and Control (2017)

Fractional Order System Identification Method for Online Monitoring of Humidity of Electrochemical Hydrogen Pumps

Gjorgji Nusev, Pavle Boškoski, and Gregor Dolanc

Abstract Electrochemical hydrogen pump (EHP) is a promising technology capable of extracting hydrogen from miscellaneous gas mixture and compresses it to very high pressures. The basic working principles are similar to that of a proton exchange membrane (PEM) fuel cell. Consequently, its performance is heavily dependent on humidity level of the membrane. Unlike PEM fuel cells where water is generated as a by-product, in the case of EHP the humidity has to be delivered via external humidifier. Therefore, it is paramount to have accurate information regarding the humidity in order to achieve optimal exploitation. Inaccessibility of the membranes makes it almost impossible to perform direct humidity measurements. Addressing this issue, this paper presents a method for online estimation of humidity levels based on the parameters of an equivalent circuit model (ECM). The parameter estimation is performed through a combination of evolutionary algorithm and simplex optimization. The method is evaluated on a market ready EHP device with capacity of pumping 1.4118 stl/min of H_2.

1 Introduction

Numerous studies on future energy development predict hydrogen as a main source for energy storage. In that manner it is expected that the hydrogen production will continue to increase in the years that follow [1, 2]. According to Fraile [2] the global demand for hydrogen in 2010 was around 43 Mtons and is foreseen to reach 50 Mtons by 2025. In the complete production chain one of the most

G. Nusev (✉)
Department of System and Control, Jožef Stefan Institute, Ljubljana, Slovenia

Jožef Stefan International Postgraduate School, Ljubljana, Slovenia
e-mail: gjorgji.nusev@ijs.si

P. Boškoski · G. Dolanc
Department of System and Control, Jožef Stefan Institute, Ljubljana, Slovenia
e-mail: pavle.boskoski@ijs.si; gregor.dolanc@ijs.si

W. Zamboni, G. Petrone (eds.), *ELECTRIMACS 2019*, Lecture Notes in Electrical Engineering 615, https://doi.org/10.1007/978-3-030-37161-6_28

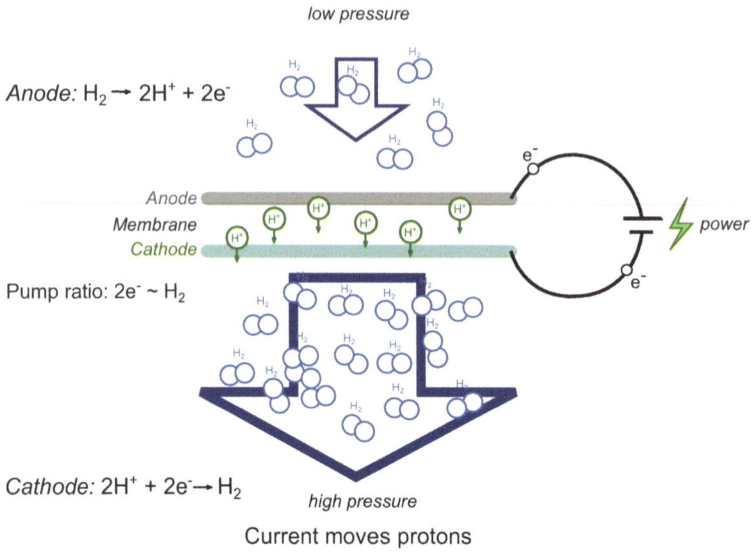

Fig. 1 Basic concept of how EHP works. It uses direct current to pull hydrogen through an impermeable membrane, very efficiently and highly selective for hydrogen [4]

energy consuming steps is the compression. There are two categories of hydrogen compressors: *mechanical* which uses adiabatic process and *non-mechanical* which mainly uses isothermal process for compression.

Non-mechanical compressors such as EHP use electricity to extract hydrogen from miscellaneous gas mixtures and compress it in order to meet different application requirements. Also the absence of moving parts makes them more efficient when comparing with the conventional mechanical compressors which are noisy and less effective than the non-mechanical [3]. The structure of EHP is similar to that of a proton exchange membrane fuel cells (PEMFC). The basic principles about how EHP works are shown in Fig. 1.

The hydrogen molecules enter the anode side (the low pressure side) of the cell where they are oxidized to protons and electrons. Due to the external voltage potential, hydrogen protons are driven through the membrane to the cathode side while the electrons are forced through an external circuit also to the cathode side, where they are recombined to form molecular hydrogen. As a result of the applied external voltage through time, the pressure at the cathode side increases with the increased number of transported hydrogen molecules. Inside the EHP, the total pressure at the cathode is larger than that at the anode, with $P_A < P_C$.

Although EHP technology has many advantages and possible applications, it is mainly used as a diagnostic test for measuring the crossover in fuel cells. Strobel et al. [5] was the first one to recognize the possibilities of the PEM-based structure for EHP. Hao et al. [6] performed ectrochemical impedance spectroscopy (EIS) on EHP with included internal humidifier with dead-end anode. Nguyen et al. [7] has shown how the temperatures of the humidifier and the temperature of the stack are influencing the performance of the stack. Common issue in the reported analysis is the control of the membrane's humidification levels.

Humidification of the membrane plays an important role in proton conduction and is tightly connected to the overall performance of the EHP. Shortage of water reduces the proton conductance, whereas flooding decreases the active surface area. Unlike PEM fuel cells, EHP does not produce water and therefore it has to be delivered via an external humidifier.

Due to physical limitations it is impractical to perform direct humidity measurements. Therefore, EIS is the only available option to perform online humidification estimation. For that purpose, this study employs an approach for fast impedance analysis that is based on discrete random binary sequence (DRBS) excitation [8]. The main advantage of DRBS excitation over the classical single-sine excitation is the time needed to perform the measurement. The impedance characteristic in the frequency band from 0.1 to 500 Hz is identified in 60 s, approximately five times faster compared to the conventional single-sine approach. Based on this impedance data, parameters of ECM are estimated. The humidity level is inferred based on the variation of parameters of the ECM parameters. Parameters estimation is performed using a hybrid optimization method comprising of genetic algorithm and simplex method.

2 Methodology

ECMs represent one approach for analysis of impedance data. Typically, the analysis is based on the link between the parameters of an ECM and the underlying chemical reactions occurring inside the electrochemical energy devices [9–11]. The accuracy of the analysis is directly related to the accuracy of the method used for the parameter estimation of the selected ECM.

There are many different methods for parameter estimation. On the one hand, there are methods which are based on complex nonlinear least squares (CNLS) deterministic algorithms such as Levenberg–Marquardt and Gauss Newton algorithm [12, 13]. The main issue with this method is the selection of good initial conditions which are crucial to provide good convergence towards the best solution. On the other hand there are different methods which are based on evolutionary algorithms (genetic algorithm). Evolutionary algorithms when compared to the CNLS are more suitable for solving general-purpose problems where prior knowledge of the parameters is not required or limited. The biggest challenge with these approaches is designing a cost function and optimization method that will converge to the global minimum of the cost function.

For the presented approach the following cost function is selected:

$$C(\mathbf{x}) = \sum_{i=0}^{N} \left[\Re(Z_M(f_i) - \hat{Z}_S(\mathbf{x}, f_i))\right]^2 + \left[\Im(Z_M(f_i) - \hat{Z}_S(\mathbf{x}, f_i))\right]^2, \quad (1)$$

where $Z_M(f_i)$ represents the measured impedance value at frequency f_i and $\hat{Z}_S(\mathbf{x}, f_i)$ represents the estimated impedance using parameters $\mathbf{x}$ at frequency f_i. Equation (1) represents a Manhattan distance and it is mainly used as an optimization function for any complex functions [14]. In our case the complex function represents the impedance.

In order to reach the global minimum in a least number of steps, the proposed approach relies on a hybrid optimization that comprises:

- Genetic algorithm;
- Bounded simplex optimization.

A block scheme of the method is shown in Fig. 2. Through the genetic algorithm multiple starting points are randomly distributed in previously defined search area. This is done in order to roughly estimate the location of the global minimum. Before it continues with the next iteration (generation), the best solutions from the current iteration are then used as an initial condition for the simplex optimization method for fine tuning. The obtained parameters from the simplex method are then sorted according to its corresponding cost function value in ascending order. Afterwards, the best individuals are propagated to the next generation. Each subsequent generation is mutated by adding a small amount of noise [13]. The process terminates when all of the following conditions are met:

- the best value of the criterion function is less than previously defined one;
- absolute change of the criterion function value during the last couple of iterations is less than 5%;
- change of the estimated parameters through each iteration is less than some predefined percent.

3 Experimental Setup

The approach was evaluated on a 5-cell EHP stack. A schematic diagram of the test bed is shown in Fig. 3.

Humidification level is controlled by altering the hydrogen supply via three solenoid valves. Solenoid valve V3 is used to deliver the hydrogen mixture to the stack directly, without humidification, while valves V4 and V10 deliver the hydrogen mixture to the stack through a humidifier. On the high pressure side there is a buffer to accumulate the pumped hydrogen and backpressure regulator V9 to control the pressure in the tank and at the high pressure side of the stack at constant value.

Fig. 2 Block scheme of the proposed optimization algorithm

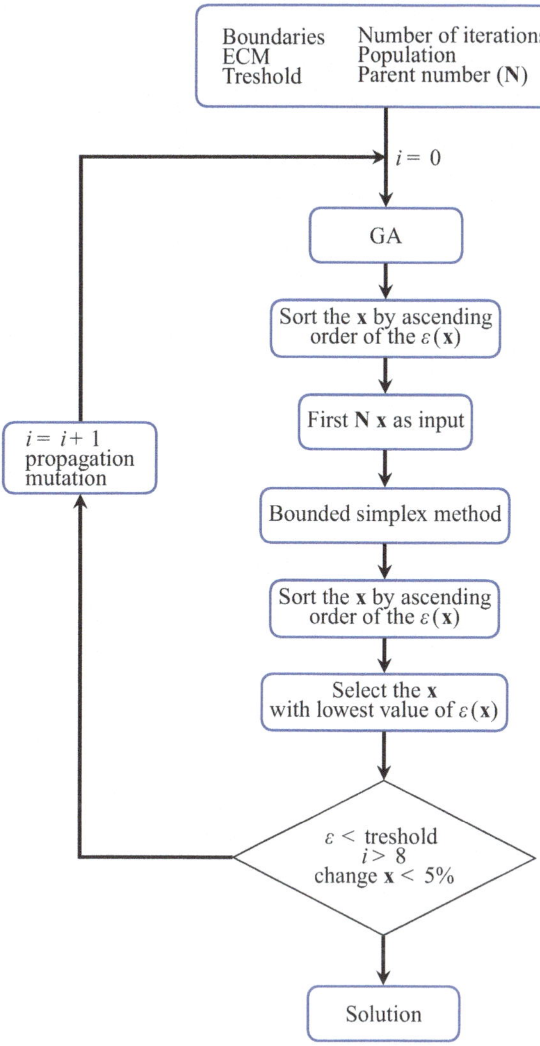

Electrical connections on the EHP are shown in Fig. 4. Characterization of the stack is performed at DC currents $I_{DC} = 20\,\text{A}$ with peak-to-peak amplitude $I_{AC} = 1\,\text{A}$ by using DRBS excitation [8]. The perturbed current is induced using programmable digital electric load (Rigol DL3031A) with excitation frequencies starting from 0.1 Hz up to 1000 Hz. The voltage response of the cells together with the current is sampled with sampling frequency $fs = 50$ kHz using 16-bit data acquisition card (NI USB 6215). Before DAQ signals are low pass filtered at $f_c = 10.8$ kHz.

When performing EIS measurements all other controllable variables are kept constant. Due to the fact that we were not able to directly measure the humidification

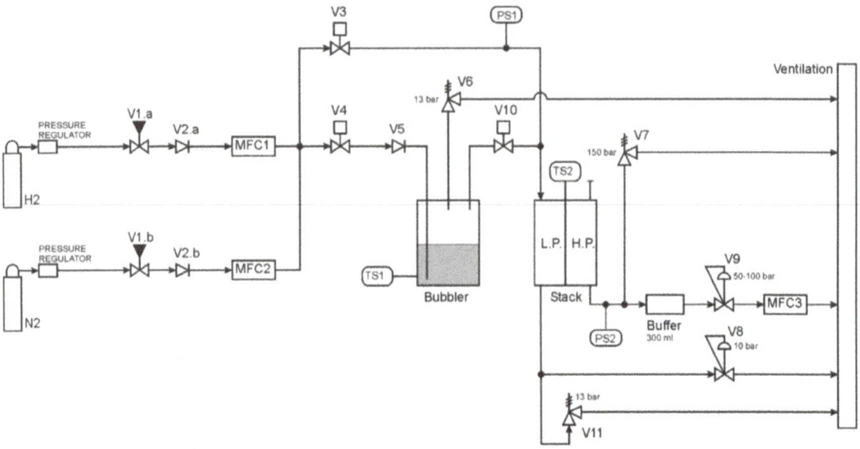

Fig. 3 Gas connections of the experimental setup for performing diagnostic experiments on EHP

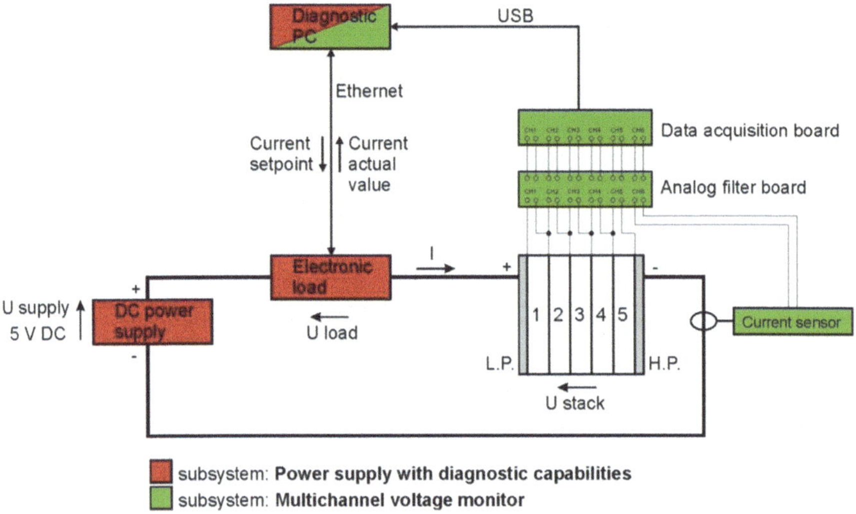

Fig. 4 Electrical connection of the EHP for performing electrochemical impedance spectroscopy

level, the first experiment was performed with the humidifier being turned on and the second experiment was performed when the humidifier turned off. The main goal was to detect the effects of membrane's humidification levels on the measured impedance.

4 Results

Since the EIS is a linear technique, the operating condition of the EHP must fulfill the linearity conditions. These conditions are associated to the value of the amplitude of the excitation signal. It states that the amplitude must be small enough in order to operate within the linear region; however, it has to be big enough to measure its response. For that reason before performing the EIS, an IV curve of the EHP was first measured (Fig. 5) in order to determine the value of the AC amplitude at desired DC point.

From the IV curve (Fig. 5) it was decided to perform EIS measurements at $I_{DC} = 20$ A with peak-to-peak amplitude $I_{AC} = 1$ A at two different operating conditions:

1. Nominal mode—with turned on humidification;
2. Drying mode—with turned off humidification.

Calculated impedances of all the 5 cells are given in Figs. 6 and 7. Solid lines are impedances at nominal conditions; dotted lines are the impedances when the cells are in drying mode. There are several effects that have to be analyzed separately.

First, the impedance of the 5th cell differs from the others. There are two different reasons. The 5th cell is located the furthest from the input and receives the least amount of humidity. Additionally, due to the connectors for measuring the cell voltage introduce additional resistance, especially the negative wire which is connected directly to the cathode where wire of the power supply is connected. The effect of this additional resistance is seen in the impedance measurement.

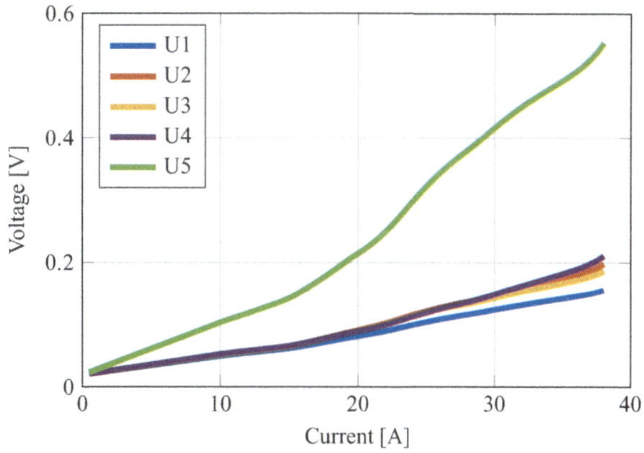

Fig. 5 IV curve of the HyET electrochemical hydrogen pump

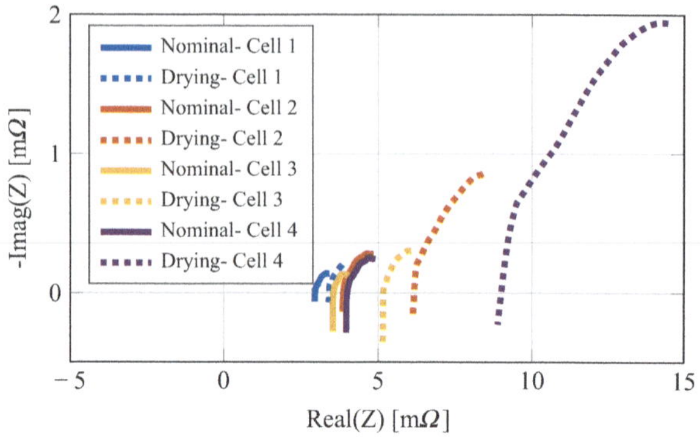

Fig. 6 Measured impedance of the first four cells in the stack at different conditions (normal and drying)

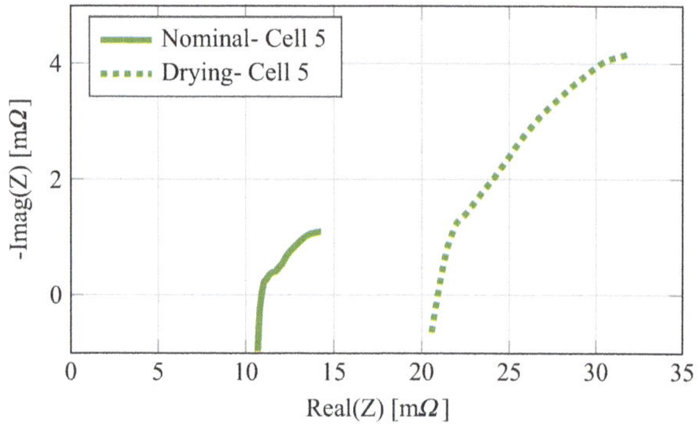

Fig. 7 Measured impedance of the fifth cells in the stack at different conditions (normal and drying)

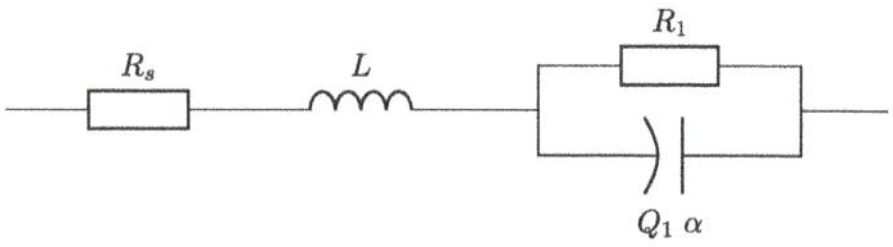

Fig. 8 Equivalent circuit model of EHP

4.1 Equivalent Circuit Model Selection

For the process of parameter estimation, the ECM topology shown in Fig. 8 has been selected for detailed analysis. The selected ECM is consisted of three elements connected in series. Resistor R_s represents the losses inside the membrane, parallel connection between resistor R_1 and constant phase element Q_1 represents the losses at the electrode, and inductance L appears due to internal wire connection between the cells inside the stack. Theoretical impedance of the selected ECM is given with the following equation:

$$Z_{SIM}(j\omega) = j\omega L + R_s + \frac{R_1}{1 + (j\omega\tau)^{\alpha}}, \qquad \tau^{\alpha} = R_1 Q_1 \tag{2}$$

4.2 Parameter Estimation

After evaluation of the hybrid optimization algorithm on the measured impedances, the parameters of the selected ECM were estimated. The values for all 5 cells in both working modes (nominal and drying) are given in Table 1. The comparison of the measured impedances in both modes, nominal and drying, and the impedance of its fitted model are shown in Figs. 9 and 10. Figure 9 shows the measured and fitted impedance of the 2nd cell in nominal mode, whereas Fig. 10 shows the impedances of the same cell in drying mode.

Drying of the cells did not have big impact on the fractional order coefficient α. Also the changes of the inductance L are insignificant. On the other hand, resistances R_S and R_1 have increased significantly, 56% and 247%, respectively. For the 4th cell the time constant τ has increased significantly.

The reason for the conductivity decrease of the membrane when in drying mode is connected to the size of the water clusters within the polymer microstructure. Dehydration of the EHP membranes leads to narrowing of the interconnecting

Table 1 ECM parameters at different operation conditions

	Parameters	Nominal mode	Drying mode
Cell 1	L [nH]	12.7	12.4
	R_S [mΩ]	2.9179	3.4068
	R_1 [mΩ]	0.94147	1.417
	τ [s]	1.317	2.7919
	α	0.38	0.40
Cell 2	L [nH]	29.5	46.8
	R_S [mΩ]	3.8012	5.9447
	R_1 [mΩ]	2.0094	6.9781
	τ [s]	1.8084	2.8614
	α	0.36	0.34
Cell 3	L [nH]	52.9	64.5
	R_S [mΩ]	3.5231	5.151
	R_1 [mΩ]	0.8457	1.853
	τ [s]	0.5	0.574
	α	0.41	0.47
Cell 4	L [nH]	58.3	90.7
	R_S [mΩ]	3.8897	7.9873
	R_1 [mΩ]	2.029	30.558
	τ [s]	2.3775	80.498
	α	0.32	0.25
Cell 5	L [nH]	198	192
	R_S [mΩ]	10.27	19.019
	R_1 [mΩ]	14.53	65.108
	τ [s]	100	100
	α	0.24	0.26

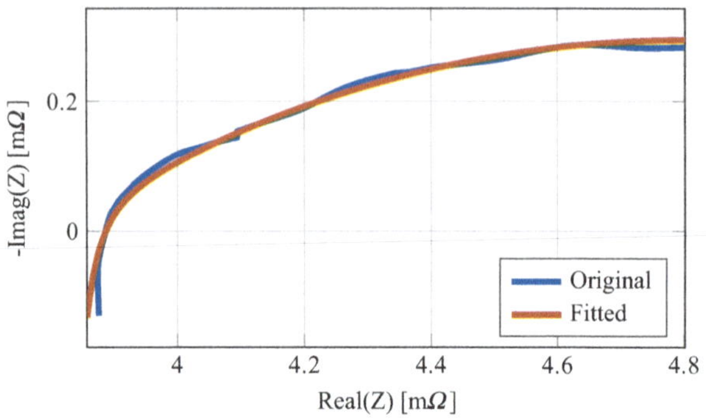

Fig. 9 Measured impedance of the cell 2 (blue line) and its fit (red line) in nominal mode

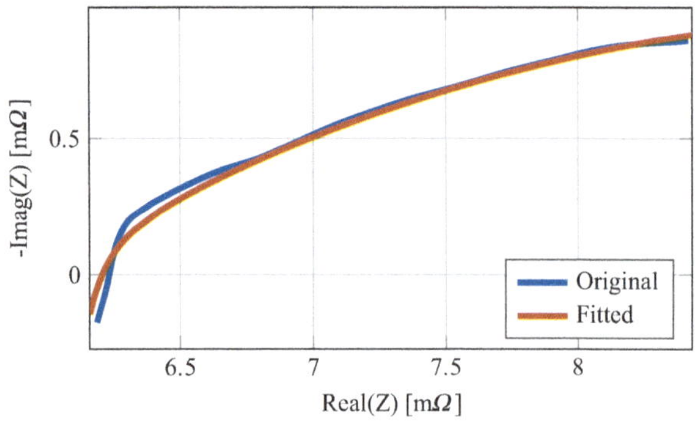

Fig. 10 Measured impedance of the cell 2 (blue line) and its fit (red line) in drying mode

channels which decreases the mobility of the protons and increases the electrical resistance [15]. On the other hand, when the EHP membranes are well humidified the microporous structure swells which leads to increase of the ionic conductance.

5 Conclusions

Performances of the EHP are highly dependent on the level of humidity presented inside the membrane. For that reason two experiments were performed, one at nominal conditions when humidification was turned on and the second at drying conditions when the humidification was turned off. By using electrochemical impedance spectroscopy, based on DRBS excitation signals, we were able to detect changes in the parameters of the selected ECM after the experiments were performed. Impedance of the EHP was measured in the frequency interval between 0.1 and 1000 Hz. Parameters of the ECM were estimated using hybrid optimization method.

From the obtained results it can be seen that drying increased the serial resistance R_S and the resistance R_1 of each of the cells. From this assumption can be concluded that by monitoring the changes of these two variables it is possible to measure the level of humidification inside the membrane while it is in operating mode. By measuring the level of humidification it is possible to prevent the drying or flooding of the EHP.

Acknowledgements The authors acknowledge the project L2-7663, and research core funding No. P2-0001 that were financially supported by the Slovenian Research Agency. The authors acknowledge financial support through the project MEMPHYS (grant agreement No 735533) within the framework of the Fuel Cells and Hydrogen 2 Joint Undertaking under the European Union's Horizon 2020 research and innovation programme, Hydrogen Europe and Hydrogen Europe research.

References

1. D. S. Satyapal, Hydrogen and fuel cells overview, U.S. Department of Energy Fuel Cell Technologies office, April 2017. [Online]. Available: https://www.energy.gov/sites/prod/les/2017/06/f34/fcto-h2-fc-overview-dla-worldwide-energy-conf-2017-satyapal.pdf
2. D. Fraile, J. Lanoix, P. Maio, A. Rangel, A. Torres, Overview of the market segmentation for hydrogen across potential customer groups, based on key application areas. CertifHy (2015)
3. K. Onda, K. Ichihara, M. Nagahama, Y. Minamoto, T. Araki, Separation and compression characteristics of hydrogen by use of proton exchange membrane. J. Power Sources **164**(1), 1–8 (2007)
4. P. Bouwman, Electrochemical hydrogen compression (EHC) solutions for hydrogen infrastructure. Fuel Cells Bullet. **2014**(5), 12–16 (2014)
5. R. Ströbel, M. Oszcipok, M. Fasil, B. Rohland, L. Jörissen, J. Garche, The compression of hydrogen in an electrochemical cell based on a PE fuel cell design. J. Power Sources **105**(2), 208–215 (2002)
6. Y. Hao, H. Nakajima, H. Yoshizumi, A. Inada, K. Sasaki, K. Ito, Characterization of an electrochemical hydrogen pump with internal humidifier and dead-end anode channel. Int. J. Hydrog. Energy **41**(32), 13879–13887 (2016)
7. M.-T. Nguyen, S. Grigoriev, A. Kalinnikov, A. Filippov, P. Millet, V. Fateev, Characterisation of a electrochemical hydrogen pump using electrochemical impedance spectroscopy. J. Appl. Electrochem. **41**(9), 1033 (2011)
8. A. Debenjak, P. Boškoski, B. Musizza, J. Petrovčič, Đ. Juričić, Fast measurement of proton exchange membrane fuel cell impedance based on pseudo-random binary sequence perturbation signals and continuous wavelet transform. J. Power Sources **254**, 112–118 (2014)
9. A. Leonide, Y. Apel, E. Ivers-Tiffee, SOFC modeling and parameter identification by means of impedance spectroscopy. ECS Trans. **19**(20), 81–109 (2009)
10. M. Hinaje, S. Raël, P. Noiying, D.A. Nguyen, B. Davat, An equivalent electrical circuit model of proton exchange membrane fuel cells based on mathematical modelling. Energies **5**(8), 2724–2744 (2012)
11. Z. Ilhan, A. Ansar, and N. Wagner, Using equivalent circuit models in understanding theperformance change of metal supported sofcs in static and dynamic conditions, in *In 8th European SOFC Forum*, (2008)
12. M. Shamsipur, S.H. Kazemi, M.F. Mousavi, Impedance studies of a nano-structured conducting polymer and its application to the design of reliable scaffolds for impedimetric biosensors. Biosens. Bioelectron. **24**(1), 104–110 (2008)

13. O. Kanoun, U. Tröltzsch, H.-R. Tränkler, Benefits of evolutionary strategy in modeling of impedance spectra. Electrochim. Acta **51**(8–9), 1453–1461 (2006)
14. A. Lasia, Modeling of experimental data, ch. 14, in *Electrochemical Impedance Spectroscopy and its Applications* (Springer, New York, 2014), pp. 301–321
15. P.D. Beattie, F.P. Orfino, V.I. Basura, K. Zychowska, J. Ding, C. Chuy, J. Schmeisser, S. Holdcroft, Ionic conductivity of proton exchange membranes. J. Electroanal. Chem. **503**(1–2), 45–56 (2001)

Report on Lithium-Ion Battery Ageing Tests

Rocco Morello, Roberto Di Rienzo, Roberto Roncella, Roberto Saletti, and Federico Baronti

Abstract Lithium-ion battery ageing modelling and prediction is one of the most relevant topics in the energy storage research field. The development and assessment of reliable solutions are not straightforward, because of the necessity to acquire information on the cell ageing processes by employing very time-consuming tests. During these tests the cells are subjected to different profiles, usually based on the repetition of several charge/discharge cycles, in order to reproduce the ageing effects in laboratory. This paper aims at accelerating the advancement in this research field by discussing a dataset containing three different ageing tests and making it available to be used by other research groups. The tests are accurately described and a preliminary analysis of the obtained results is carried out.

1 Introduction

Lithium-ion (Li-ion) batteries are used in many applications thanks to their high energy and power densities. However, their performance degrades over time and use. To optimize the battery use and enable early maintenance actions, battery management systems (BMSs) must be able to quantify this phenomenon by estimating the state-of-health (SOH) variable, which is an indicator of the Li-ion cell ageing with respect to its fresh state. Literature papers do not give a unique and universal definition of SOH. The two most used definitions consider the degradation of the cell capacity and internal resistance [1]. Several algorithms for SOH estimation have been presented. They are based on different techniques, such as the least squares method [2], Kalman filtering [3], and the incremental capacity (IC) analysis [4–8].

R. Morello (✉) · R. Di Rienzo · R. Roncella · R. Saletti · F. Baronti
Dip. di Ingegneria dell'Informazione, Università di Pisa, Pisa, Italy
e-mail: rocco.morello@ing.unipi.it; federico.baronti@unipi.it

W. Zamboni, G. Petrone (eds.), *ELECTRIMACS 2019*, Lecture Notes in Electrical Engineering 615, https://doi.org/10.1007/978-3-030-37161-6_29

The study of the ageing effects and the verification of the SOH estimation algorithms of Li-ion cells require difficult and very time-consuming tests. Cells are usually subjected to several charge and discharge cycles, with currents and operating temperatures higher than those found in the actual applications, in order to reproduce and to accelerate the ageing effects in laboratory. However, the number of cycles to apply to the cells is still very high and tests can last several months [9].

The research activities in this field can significantly benefit from sharing the ageing datasets obtained by various research groups worldwide. Examples are the CALCE battery dataset [10] distributed by the University of Maryland and the dataset provided by the NASA [11]. These datasets turned out to be useful for developing and verifying algorithms for the SOH estimation [12] and the remaining useful life (RUL) prediction [13, 14]. The NASA dataset includes the data measured on four different Li-ion cells, and in particular their voltage, current, and temperature, acquired during iterated charging and discharging cycles, and the results obtained from various electrochemical impedance spectroscopy (EIS) measurements.

The aim of this paper is to describe and make available the data obtained from ageing tests performed on three LGDBHE21865 cells, manufactured by LG Chem. The tests are carried out in a very similar way to those provided by the NASA and the University of Maryland to have consistent and uniform data, thus facilitating their use by other research groups. Compared to the already available ageing data, the tests described here can be very useful also to examine ageing effects due to different charging current amplitudes and temperatures. The tests start with a training phase performed to check the similarity of the three cell behaviours at the beginning of their life. Then, the usual ageing phases are executed at different operating conditions. Like in the NASA dataset, the ageing phases are alternated with characterization phases that in our case are based on pulsed current tests (PCTs), adopted to extract the cells parameters, like the internal resistance and the open-circuit voltage (OCV) [15–17].

The remainder of the paper is organized as follows. The next section describes the experimental setup and the Li-ion cells. This setup is employed to perform the tests reported in Sect. 3. The obtained data and some preliminary analysis results are shown in Sect. 4, while some conclusions are drawn in Sect. 5.

2 Experimental Setup

The experimental testbed used to perform the ageing tests is derived from that used in [18] (see Fig. 1). It consists of a source meter unit (SMU) Keithley 2420 controlled by a LabVIEW application running on a host PC. The SMU is used to charge and discharge the Li-ion cells and to measure the voltage and the current during the tests. The temperature of the cells under test is kept constant by using the Binder MK53 thermal chamber. The cells employed in this test campaign are the LGDBHE21865 cells by LG Chem. Their nominal capacity is 2.5 A h, which

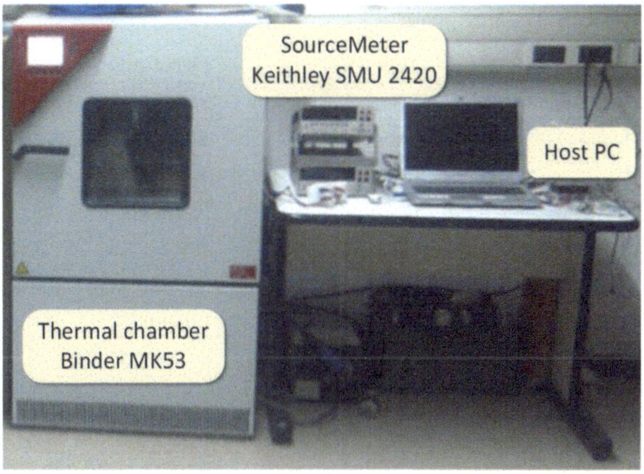

Fig. 1 Photo of the experimental testbed

is defined as the charge extracted during a constant current (CC) discharge at 0.2 C (i.e., 0.5 A) ended at 2.5 V, after a CC charge at 0.5 C (i.e., 1.25 A) with a cut-off voltage of 4.2 V followed by a constant voltage (CV) charge at 4.2 V with a cut-off current of 50 mA.

3 Experimental Tests

The experimental setup described in Sect. 2 has been used to execute various ageing experiments on the three cells under test. All the experiments start with a training phase executed at a temperature of 25 °C inside the thermal chamber. As shown in Table 1, the training phase is a repetition of charge/discharge cycles that aim at improving the cell performance and to verify that the cells have the same initial conditions before starting the real tests. The charging and the discharging processes are based on a two-stage profile: a first one with CC until the cell reaches the cut-off voltages ($V_{CV} = 4.2$ V in charge and $V_{CV} = 2.75$ V in discharge), followed by a CV phase until the cell current decreases below $|I_{cut}| = 0.25$ A.

The ageing effects are investigated by means of two different ageing profiles. Ageing A profile consists of the repetition of 50 charge/discharge cycles with a constant current $I_{CC} = 0.7$ C and $I_{CC} = 1$ C, respectively, separated by a 30 min pause, as described in Table 2. Ageing B profile is similar to the previous one, but the charge current is set to 5 A in order to investigate the dependence between ageing and charging current rate (Table 3).

A characterization procedure based on a PCT test has also been carried out in order to extract the cell parameters (Table 4). The PCT uses a train of current pulses with a given duration to change the state-of-charge (SOC) value of the cell of a well-determined percentage. The pulses have an amplitude of 1 C and a duration of 3 min in our tests, resulting in an SOC variation of 5 %. Each pulse is followed by a pause of one hour that allows us to analyse the relaxation effects and to determine the

Table 1 Training phase

Rep.	Type	Description
1	Charge	$I_{CC} = 0.7\,C$, $V_{CV} = 4.2\,V$, $\lvert I_{cut} \rvert = 0.25\,A$
1	Pause	30 min
10	Discharge	$I_{CC} = 1\,C$, $V_{CV} = 2.75\,V$, $\lvert I_{cut} \rvert = 0.25\,A$
	Pause	30 min
	Charge	$I_{CC} = 0.7\,C$, $V_{CV} = 4.2\,V$, $\lvert I_{cut} \rvert = 0.25\,A$
	Pause	30 min
1	Pause	60 min

Table 2 Ageing A profile

Rep.	Type	Description
50	Discharge	$I_{CC} = 1\,C$, $V_{CV} = 2.75\,V$, $\lvert I_{cut} \rvert = 0.25\,A$
	Pause	30 min
	Charge	$I_{CC} = 0.7\,C$, $V_{CV} = 4.2\,V$, $\lvert I_{cut} \rvert = 0.25\,A$
	Pause	30 min

Table 3 Ageing B profile

Rep.	Type	Description
50	Discharge	$I_{CC} = 1\,C$, $V_{CV} = 2.75\,V$, $\lvert I_{cut} \rvert = 0.25\,A$
	Pause	30 min
	Charge	$I_{CC} = 2\,C$, $V_{CV} = 4.2\,V$, $\lvert I_{cut} \rvert = 0.25\,A$
	Pause	30 min

OCV values, with the procedure described in [19], in which an equivalent electrical model with two RC branches is used to model the Li-ion cells. In particular, for each SOC value, the OCV is the cell voltage at the end of each pause, while the series resistance is computed as the ratio of the voltage drop at the end of the 3 min pulse to the corresponding current amplitude.

The tests performed on the three cells consist of different combinations of the phases described above, as reported in Tables 5, 6, and 7, which show also the temperature imposed by the thermal chamber during each specific phase. In fact, the same ageing profile A is applied to both cell 1 and 3, but at different temperatures, i.e., 25 °C and 40 °C, respectively.

The measurement of the cell voltage and current has been acquired and organized in Matlab structures [20], as in the NASA datasets described in Sect. 1. These structures are divided according to the step typology, i.e., charge, discharge, and pause. We also add the value of the computed extracted charge, obtained by integrating the cell current during the CC stage in the discharge phase. Furthermore, there is also a structure containing the parameter extracted from the PCT procedure for each characterization phase.

Table 4 Characterization phase

Rep.	Type	Description
1	Charge	$I_{CC} = 0.7\,C$, $V_{CV} = 4.2\,V$, $\|I_{cut}\| = 0.25\,A$
1	Pause	60 min
1	Discharge	$I_{CC} = 1\,C$, $V_{CV} = 2.75\,V$, $\|I_{cut}\| = 0.25\,A$
1	Pause	30 min
1	Charge	$I_{CC} = 0.7\,C$, $V_{CV} = 4.2\,V$, $\|I_{cut}\| = 0.25\,A$
1	Pause	30 min
20	Discharge	$I = 1\,C$, 3 min
	Pause	60 min
1	Discharge	$I_{CC} = 1\,C$, $V_{CV} = 2.75\,V$, $\|I_{cut}\| = 0.25\,A$
1	Pause	60 min
20	Charge	$I = 1\,C$, 3 min
	Pause	60 min
1	Charge	$I_{CC} = 0.7\,C$, $V_{CV} = 4.2\,V$, $\|I_{cut}\| = 0.25\,A$
1	Pause	60 min
1	Discharge	$I_{CC} = 1\,C$, $V_{CV} = 2.75\,V$, $\|I_{cut}\| = 0.25\,A$
1	Pause	30 min
1	Charge	$I_{CC} = 0.7\,C$, $V_{CV} = 4.2\,V$, $\|I_{cut}\| = 0.25\,A$
1	Pause	30 min

Table 5 Cell 1 test description

Rep.	Phase	Temperature
1	Training	25 °C
6	Characterization	25 °C
	Ageing A	25 °C
1	Characterization	25 °C

Table 6 Cell 2 test description

Rep.	Phase	Temperature
1	Training	25 °C
6	Characterization	25 °C
	Ageing B	25 °C
1	Characterization	25 °C

Table 7 Cell 3 test description

Rep.	Phase	Temperature
1	Training	25 °C
1	Characterization	25 °C
2	Ageing A	40 °C
	Characterization	25 °C

4 Tests Discussion

First, we verify the homogeneity of the three cells by comparing the parameters extracted from the first characterization phase after the training procedure. The obtained results reported in Table 8 show that the three cells behave similarly at

Table 8 Cell parameter extracted after the initial training phase

Cell	Series resistance (mΩ)	Capacity (A h)
1	21.5	2.4
2	22.3	2.4
3	21.2	2.4

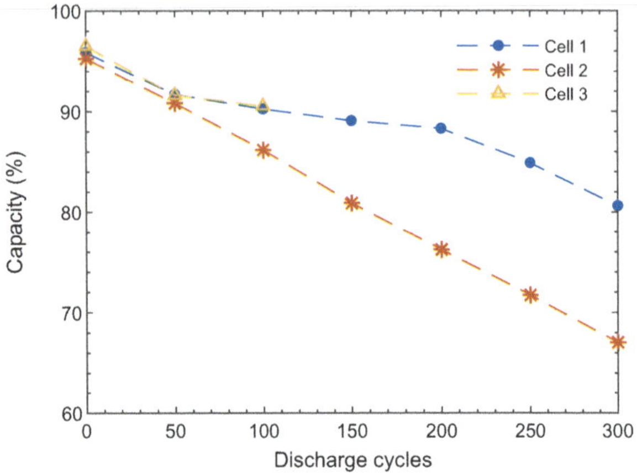

Fig. 2 Comparison of the capacity fading in percentage of the nominal capacity for the three cells under test

the beginning of their life both in terms of the internal series resistance and cell capacity.

Then, we analyse the fading of the cell capacities during the three different tests, which is reported in Fig. 2 as percentage of the cell nominal capacity. These capacity values refer to the charge extracted during the complete discharge executed at 25 °C before the PCT tests. The capacity fading in the three tests can be fairly compared, as the capacity values have been obtained in the same conditions. We can note that cell 2 is subjected to a more pronounced capacity reduction due to the different ageing procedure, which uses a higher charging current. Indeed, after 300 discharge cycles, corresponding to about 50 days of test, cell 2 has lost around 35 % of capacity with respect to its fresh state.

Another effect of the cell ageing is the increment of the series resistance. Figure 3 shows, at SOC 50 %, the increase of the series resistance for the three cells. Cell 2 is the cell in which we observe a higher variation, because of its accelerated ageing. The resistance value reaches approximately 35 mΩ after the ageing tests, starting from a value of about 22 mΩ.

The analysis of the OCV behaviour as a function of the extracted charge from the full charge condition can also provide useful information for SOH estimation. Figure 4 shows how the OCV shape changes after different ageing cycles. These modifications, showing up in the OCV ranges 3.4–3.5 V and 3.6–3.7 V, can be highlighted better by computing the IC defined as the derivative of the extracted charge Q with respect to the OCV [4–6]. As shown in Fig. 5, IC presents two peaks,

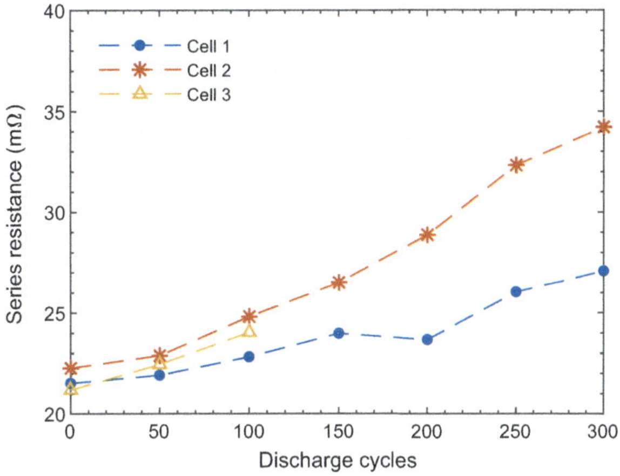

Fig. 3 Series resistance increase for the three cells during the ageing procedure, reported for an SOC 50 %

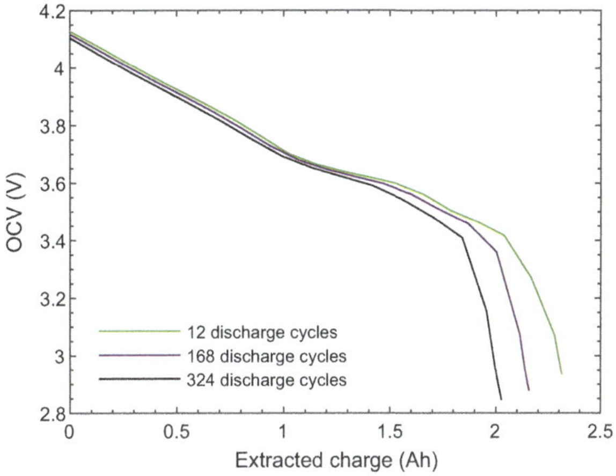

Fig. 4 OCV vs extracted charge curves obtained from the characterization tests performed during the ageing procedure on cell 2

whose position and amplitude change with ageing, as highlighted by the two arrows depicted in the figure. In particular, the peak amplitude clearly decreases when the cell ages, as observed in previous studies, such as [6, 21].

5 Conclusion

The data coming from three different ageing tests applied to three LGDBHE21865 Li-ion cells are described and discussed in this paper, in order to enable the analysis of the effects of ageing on the cell behaviour. This dataset is available and will be constantly updated with further data acquired from tests that will be executed in future. All the measured data have been organized, described, and made available to be downloaded by other research groups to accelerate the advancement in this

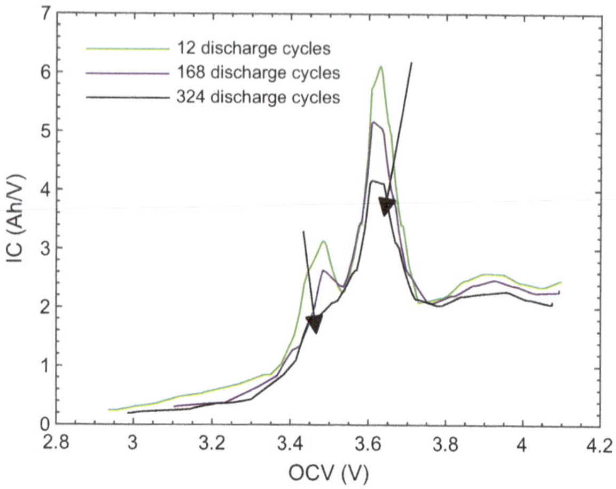

Fig. 5 Incremental capacity evolution of cell 2 during the ageing procedure. The two arrows show the evolution of the two peaks during the tests

research field. In fact, a preliminary analysis of the collected data showed that they may provide useful information for a better understanding of the effects of the charging current on the battery performance degradation. Moreover, they can encourage further developments of the IC analysis, as this work confirmed its good potentialities in SOH estimation and RUL prediction.

Acknowledgements This work has partially been supported by the Italian Ministry of Education and Research (MIUR) in the framework of the CrossLab project (Departments of Excellence) and by the University of Pisa under Grant PRA-2017-33 (Urban districts with zero environmental and energetic impact).

References

1. S. Nejad, D. Gladwin, D. Stone, A systematic review of lumped-parameter equivalent circuit models for real-time estimation of lithium-ion battery states. J. Power Sources **316**, 183–196 (2016)
2. B. Balasingam, G.V. Avvari, B. Pattipati, K.R. Pattipati, Y. Bar-Shalom, A robust approach to battery fuel gauging, part II: real time capacity estimation. J. Power Sources **269**, 949–961 (2014)
3. G.L. Plett, Dual and joint EKF for simultaneous SOC and SOH estimation, in *Proceedings of the 21st Electric Vehicle Symposium (EVS21)*, Monaco (2005), pp. 1–12
4. E. Locorotondo, L. Pugi, L. Berzi, M. Pierini, A. Pretto, Online state of health estimation of lithium-ion batteries based on improved ampere-count method, in *2018 IEEE International Conference on Environment and Electrical Engineering and 2018 IEEE Industrial and Commercial Power Systems Europe (EEEIC / I&CPS Europe)*, June (IEEE, New York, 2018), pp. 1–6
5. T. Liang, L. Song, K. Shi, On-board incremental capacity/differential voltage curves acquisition for state of health monitoring of lithium-ion batteries, in *2018 IEEE International Conference on Applied System Invention (ICASI)*, April (IEEE, New York, 2018), pp. 976–979

6. J. Tian, R. Xiong, Q. Yu, Fractional-order model-based incremental capacity analysis for degradation state recognition of lithium-ion batteries. IEEE Trans. Ind. Electron. **66**(2), 1576–1584 (2019)
7. D.-I. Stroe, E. Schaltz, SOH estimation of LMO/NMC-based electric vehicle lithium-ion batteries using the incremental capacity analysis technique, in *2018 IEEE Energy Conversion Congress and Exposition (ECCE)*, September (IEEE, New York, 2018), pp. 2720–2725
8. D. Ansean, M. Gonzalez, C. Blanco, J.C. Viera, Y. Fernandez, V.M. Garcia, Lithium-ion battery degradation indicators via incremental capacity analysis, in *2017 IEEE International Conference on Environment and Electrical Engineering and 2017 IEEE Industrial and Commercial Power Systems Europe (EEEIC / I&CPS Europe)*, June (IEEE, New York, 2017), pp. 1–6
9. D. Stroe, M. Swierczynski, A.-I. Stan, R. Teodorescu, S.J. Andreasen, Experimental investigation on the internal resistance of Lithium iron phosphate battery cells during calendar ageing, in *IECON 2013 - 39th Annual Conference of the IEEE Industrial Electronics Society*, November (IEEE, New York, 2013), pp. 6734–6739
10. CALCE Battery Research Group Datasets. Available: https://web.calce.umd.edu/batteries/data.htm
11. NASA Li-ion Battery Aging Datasets. Available: https://data.nasa.gov/dataset/Li-ion-Battery-Aging-Datasets/uj5r-zjdb
12. N. Yang, J. Feng, Q. Sun, T. Liu, D. Zhong, Online estimation of state-of-health for lithium ion batteries based on charge curves, in *2016 11th International Conference on Reliability, Maintainability and Safety (ICRMS)*, October (IEEE, New York, 2016), pp. 1–8
13. L. Li, Y. Peng, Y. Song, D. Liu, Lithium-ion battery remaining useful life prognostics using data-driven deep learning algorithm, in *2018 Prognostics and System Health Management Conference (PHM-Chongqing)*, October (IEEE, New York, 2018), pp. 1094–1100
14. S. Wang, D. Liu, J. Zhou, B. Zhang, Y. Peng, A run-time dynamic reconfigurable computing system for lithium-ion battery prognosis. Energies **9**(8), 572 (2016)
15. G.L. Plett, Extended Kalman filtering for battery management systems of LiPB-based HEV battery packs: Part 3. State and parameter estimation. J. Power Sources **134**(2), 277–292 (2004)
16. H. Dai, X. Zhang, X. Wei, Z. Sun, J. Wang, F. Hu, Cell-BMS validation with a hardware-in-the-loop simulation of lithium-ion battery cells for electric vehicles. Int. J. Electr. Power Energy Syst. **52**, 174–184 (2013)
17. R. Morello, R. Di Rienzo, R. Roncella, R. Saletti, F. Baronti, Hardware-in-the-loop platform for assessing battery state estimators in electric vehicles, in *IEEE Access* (2018), p. 1-1
18. F. Baronti, W. Zamboni, R. Roncella, R. Saletti, G. Spagnuolo, Open-circuit voltage measurement of Lithium–Iron–Phosphate batteries, in *2015 IEEE Int. Instrum. Meas. Technol. Conf. Proc.* (IEEE, New York, 2015), pp. 1711–1716
19. F. Baronti, G. Fantechi, E. Leonardi, R. Roncella, R. Saletti, Enhanced model for Lithium-Polymer cells including temperature effects, in *IECON 2010 - 36th Annu. Conf. IEEE Ind. Electron. Soc.*, November (IEEE, New York, 2010), pp. 2329–2333
20. Ageing dataset, University of Pisa. Available: https://github.com/batterylabunipi/ageing_dataset
21. L. Zhang, J. Jiang, W. Zhang, Capacity decay mechanism of the LCO + NMC532/graphite cells combined with post-mortem technique. Energies **10**(8), 1147 (2017)

Electrochemical Impedance Parameter Extraction for Online Control of Reformed Methanol High Temperature PEM Fuel Cells

Sobi Thomas, Samuel Simon Araya, Simon Lennart Sahlin, and Søren Knudsen Kær

Abstract This work focuses towards online control strategy for detecting fuel and oxidant starvation and predicting an optimal stoichiometry for operation under different fuel compositions using the electrochemical impedance spectroscopy (EIS) parameter extraction method. The tests involve three fuel compositions, namely dry hydrogen, dry reformate (H_2, CO, CO_2 and N_2) and wet reformate (H_2, CO, CO_2 and H_2O). The characterization of anode and cathode stoichiometry (both low and high) is carried out with each fuel composition by measuring electrochemical impedance spectroscopy (EIS) and current–voltage (IV) curves. The results suggest positive effects of humidified gas on the fuel cell stack performance. The changes in the mass transport resistance due to excess gas or gas starvation both on the anode and cathode could only be deduced using the EIS method. Online EIS measurement seems useful in deducing the optimal stoichiometry as the IV curves are unable to show the changes in the mass transport. Thus, to operate the fuel cell stack under an optimal fuel and oxidant utilization, an online EIS with parameter extraction algorithm can be helpful. This would ensure a better fuel and oxidant utilization and improve the system efficiency.

1 Introduction

High temperature polymer electrolyte membrane fuel cells (HT-PEMFC) are evolving as a promising technology with advantages in terms of possibility to operate with reformed fuel and reduced water management issues due to higher operating temperatures of $160-180\,°C$. These advantages reduce the fuel reforming cost which in turn reduces the overall system cost [1–3].

S. Thomas (✉) · S. S. Araya · S. L. Sahlin · S. K. Kær
Department of Energy Technology, Aalborg University, Aalborg, Denmark
e-mail: sot@et.aau.dk; ssa@et.aau.dk; sls@et.aau.dk; skk@et.aau.dk

W. Zamboni, G. Petrone (eds.), *ELECTRIMACS 2019*, Lecture Notes in Electrical Engineering 615, https://doi.org/10.1007/978-3-030-37161-6_30

To harness the various advantages of HT-PEMFC technology the operational parameters need to be continuously monitored online and controlled. Thus, the requirement of a smart adaptive control is inevitable for longer durability and reduced operational cost. Some known issues with HT-PEMFC are gas diffusion layer degradation, acid loss and membrane damage during operation [4]. A starvation condition on the anode and cathode strongly influences these degradations. Zhou et al. [5] showed that H_2 starvation resulted in accelerated degradation of catalyst layer due to carbon corrosion. Galbaiti et al. [6] observed a relationship between water transport from cathode to anode and gas stoichiometry. A higher flow resulted in more water removal from the anode outlet and a lower performance was reported. A decrease in electrical efficiency and an increase in thermal energy generation was calculated when the anode stoichiometry was changed from 1.2 to 1.6. This was attributed to higher amount of unreacted hydrogen available for the burner used for heating the reformer [7].

Sahlin [8] investigated the effect of stoichiometry on both the anode and cathode with pure hydrogen and wet reformate gas. The anode stoichiometry was varied between 1.25 and 1.4, while the cathode stoichiometry was changed between 1.3 and 1.6 since the stack was a used one, and a strong relationship between low frequency resistance (LFR) and stoichiometry was reported. While other resistances did not show significant changes. The presence of water vapour in HT-PEMFC has been reported to be beneficial especially in the presence of CO as the reaction shown in Eq. (1) is favoured [9, 10]:

$$\begin{aligned} H_2O &\rightarrow H^+ + OH^- \\ CO + OH^- &\rightarrow CO_2 + H^+ + e^- \end{aligned} \qquad (1)$$

In this manuscript, the stoichiometry effect on a 2.5 kW HT-PEMFC stack is analysed based on which a control strategy could be developed which would reduce the starvation condition as well as ensure optimal fuel utilization. For this purpose electrochemical impedance spectroscopy (EIS) was used to characterize the fuel cell stack. EIS is a useful tool used for online diagnostic purposes as it is able to provide some critical information of an operating stack which is difficult to be deduced with other techniques like IV curve or cell voltage measurement (CVM) [11–13]. The stoichiometry on the anode and cathode is changed between high and low as specified by the manufacturer.

2 Experimental Procedure

The experimental characterization was carried out using a Greenlight Innovation fuel cell test station. Galvanostatic EIS sweeps were performed using an in-house frequency analyzer, at a median AC amplitude of 2.5 A for all set points which correspond to 7.5 % of the operating current. The frequency was sweeped between

Table 1 Test matrix with the fault intensity levels for the selected parameters

Test No.	Anode composition	Anode Stoich	Cathode composition	Cathode Stoich
Change in cathode stoichiometry				
1	H_2	1.4	Air	2.0
2	H_2	1.4	Air	3.5
3	68.3% H_2 + 0.9% CO + 21.3% CO_2 + 9.5% N_2	1.4	Air	2.0
4	68.3% H_2 + 0.9% CO + 21.3% CO_2 + 9.5% N_2	1.4	Air	3.5
5	68.3% H_2 + 0.9% CO + 21.3% CO_2 + 9.5% H_2O	1.4	Air	2.0
6	68.3% H_2 + 0.9% CO + 21.3% CO_2 + 9.5% H_2O	1.4	Air	3.5
Change in anode stoichiometry				
7	H_2	1.15	Air	2.5
8	H_2	1.5	Air	2.5
9	68.3% H_2 + 0.9% CO + 21.3% CO_2 + 9.5% N_2	1.15	Air	2.5
10	68.3% H_2 + 0.9% CO + 21.3% CO_2 + 9.5% N_2	1.5	Air	2.5
11	68.3% H_2 + 0.9% CO + 21.3% CO_2 + 9.5% H_2O	1.15	Air	2.5
12	68.3% H_2 + 0.9% CO + 21.3% CO_2 + 9.5% H_2O	1.5	Air	2.5

EIS measurements were also recorded at nominal condition before the first and after the last fault intensity levels for each parameter

4000 Hz and 0.1 Hz. To minimize measurement errors, two impedance spectra were recorded at each condition. The IV curves were recorded by sweeping the current density up to 0.45 A cm^{-2} in steps of 0.03 A cm^{-2} with a rest time of 1 min.

The stack used for the test is a 2.5 kW HT-PEMFC with phosphoric acid based PBI membrane and an active area of 165 cm^2. The anode and cathode flow plates are composed of graphite plates and the number of cells are 65. The operation temperature was 160 °C for all the experiments.

The test matrix with the fuel composition and stoichiometry on the anode and cathode used for the experiments is shown in Table 1. The first six experiments (1–6) focused on changing the cathode stoichiometry, While the last six experiments (7–12) focused on the anode stoichiometry. Two EIS measurements at each set point were carried out after one hour of settling time.

3 Discussion

In Fig. 1 a comparison of performances at low and high stoichiometric ratios on the cathode and different fuel compositions on the anode is shown. At low cathode stoichiometry of 2.0 the performance curve with pure hydrogen and wet reformate gas overlaps, while the dry reformate results in lower performance, especially at higher current densities. A similar trend is observed with a stoichiometry of 3.5 with the only difference being lower performance when operated with wet reformate compared to pure hydrogen mainly due to dilution effect.

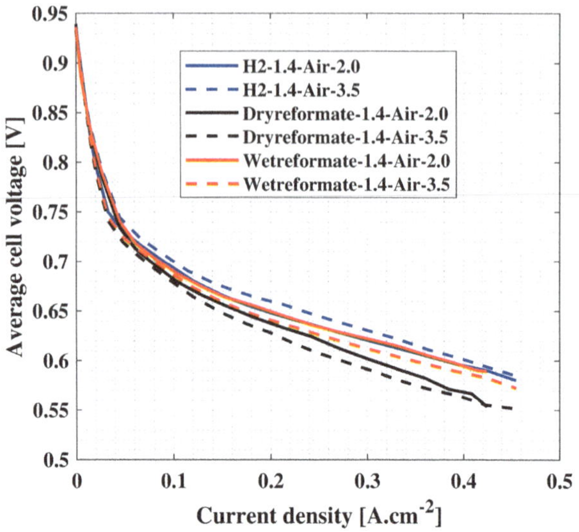

Fig. 1 Performance comparison with different cathode stoichiometry

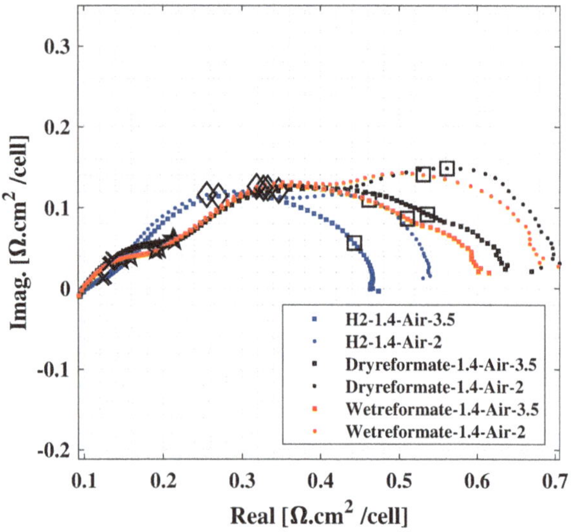

Fig. 2 EIS comparison with different cathode stoichiometry

While comparing the EIS spectra as shown in Fig. 2 we see that the changes in performance are contributed mainly by the low frequency resistance, which changes as the stoichiometric ratio is increased from 2.0 to 3.5. An improvement in the low frequency resistance in the presence of 9.5% water is also recorded. This suggests that the mass transfer is assisted in the presence of water. This is similar to what was reported in our earlier work [14]. The comparison of performance with different anode fuel composition and stoichiometry is shown in Fig. 3. At low stoichiometry of hydrogen again the wet reformed fuel and pure H_2 performances are similar, while dry reformate operation results in a lower performance. When

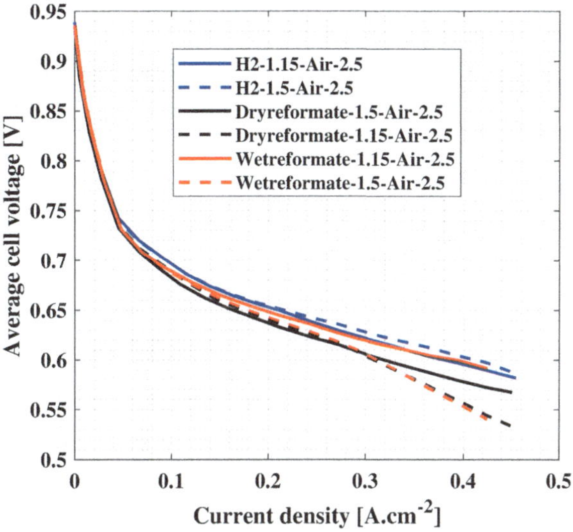

Fig. 3 Performance comparison with different anode stoichiometry

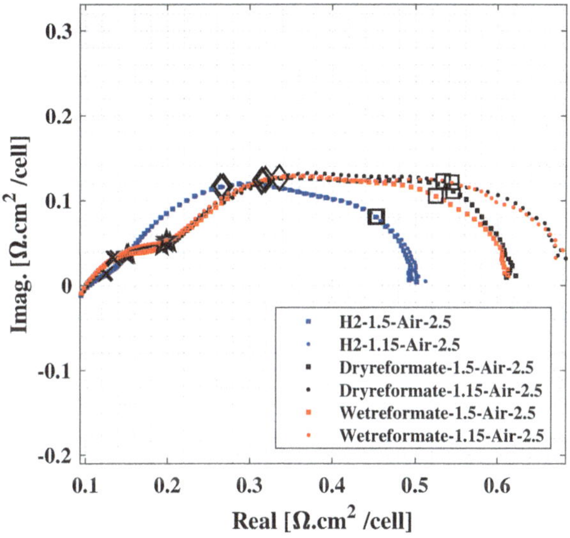

Fig. 4 EIS comparison with different anode stoichiometry

operated with higher H_2 stoichiometry the performance of wet reformate and dry reformate operation results in a lower performance compared to pure H_2 operation. This is attributed to dilution of fuel causing mass transport issues or removal of water from the cell at high flow rates is also a possibility.

The comparison of EIS spectra with different fuel composition and anode stoichiometry change from 1.5 to 1.15 is shown in Fig. 4. The EIS spectra get extended with reformed fuel both dry and wet. The wet reformate results in a lower overall resistance compared to dry reformate operation. There is only minimal difference in the resistance on changing the stoichiometry from 1.5 to 1.15 on the

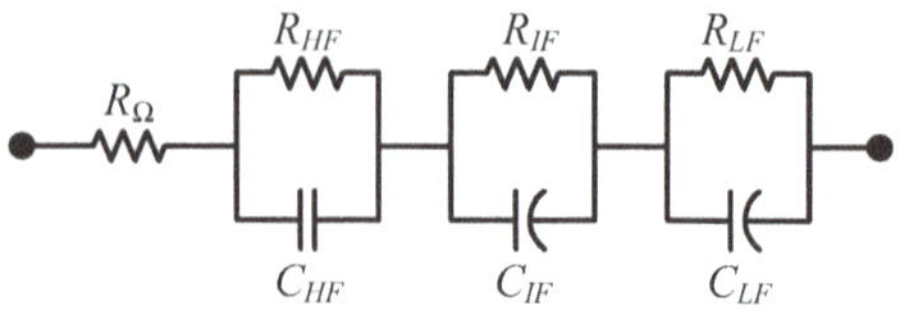

Fig. 5 Equivalent circuit used for fitting

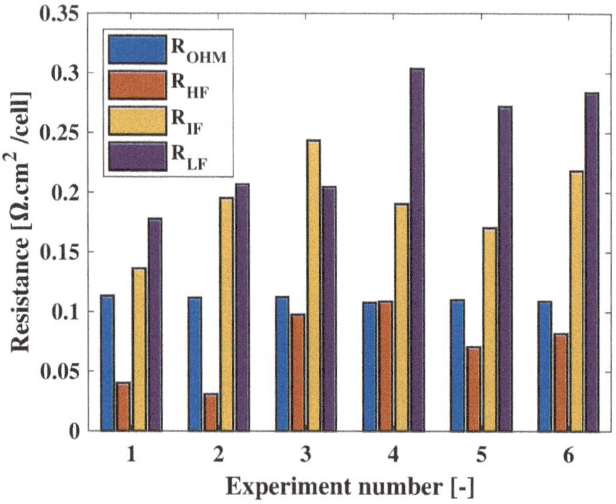

Fig. 6 Fitted resistance with different cathode stoichiometry and varying fuel composition on the anode

anode. This suggests no significant mass transport issues arise while operating at lower stoichiometries.

To better understand the EIS spectra, they were fitted with an equivalent circuit as shown in Fig. 5. The circuit consists of a series ohmic resistance (R_{OHM}) followed by three RC circuits, namely high frequency resistance (R_{HF}), intermediate frequency resistance (R_{IF}) and low frequency resistance (R_{LF}). The relationship between the resistance and physical phenomena is explained in [14].

The selected circuit was kept as simple as possible while at the same time capturing all the prominent phenomena in the EIS spectra. The intermediate and low frequency loops consist of a constant phase element with an impedance as shown in Eq. (2). The value of c is chosen as 0.9.

$$Z_{CPE} = \frac{1}{Q(j\omega)^c} \tag{2}$$

The fitted resistance is shown in Figs. 6 and 7. The resistance change with cathode stoichiometry is shown in Fig. 6. The ohmic resistance which is attributed to the cell ohmic losses shows no or minimal changes. This is in line with the experiments where there is no change in the contact or membrane resistance change over the short period of test.

The high frequency resistance is contributed by the losses caused by catalyst layer potential gradient, including in case of fuel change. While a change in the

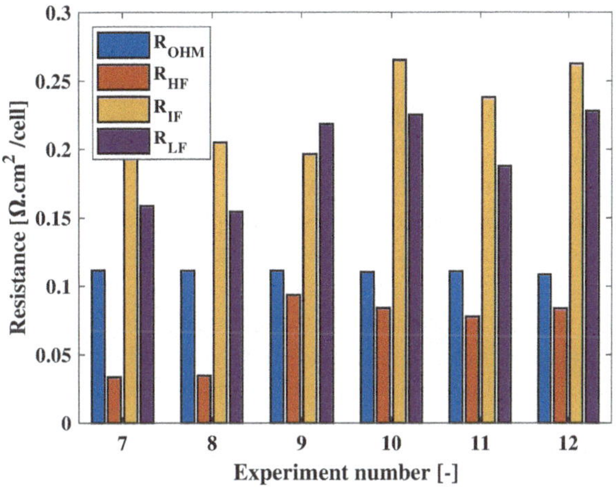

Fig. 7 Fitted resistance with different anode stoichiometry and varying fuel composition on the anode

high frequency resistance is seen on changing the fuel composition on the anode. It shows a lower resistance for wet reformed fuel operation compared to dry reformed fuel operation. The intermediate frequency which is mainly contributed by the cathode fuel composition shows a change with stoichiometry on the cathode for all the fuel compositions. The low frequency resistance which arises due to mass transport problems shows a big change with dry reformate while the changes are low with wet reformate and pure hydrogen operation. This clearly shows that the stoichiometric ratio becomes quite important when dry gas is used. The fitted resistance with stoichiometric ratio change on the anode is shown in Fig. 7. A similar trend with ohmic and high frequency resistance is seen in this case as well. However, the notable changes are more prominent with the intermediate and low frequency resistances. The low frequency resistance results in not much difference between dry reformate and wet reformate gas operation. This suggests that stoichiometry on the anode can be low for safe operation of the fuel cell. The ratio should only be adjusted in accordance with the degradation state of the fuel cell. Simon [8] showed a higher anode stoichiometry for a degraded stack.

4 Conclusions

In this work a 2.5 kW HT-PEMFC stack was characterized with three different anode fuel compositions, namely, pure dry hydrogen, dry reformate (H_2, CO, CO_2, N_2) and wet reformate (H_2, CO, CO_2 and H_2O) using electrochemical impedance spectroscopy and current–voltage sweep. The test was designed to understand the effect of high and low stoichiometry of anode fuel and air as recommended by the fuel cell manufacturer. The focus was to develop an online control strategy for adjusting the stoichiometry based on the fuel composition on the anode. This would increase the fuel utilization and thereby increase the system efficiency.

The IV curve results showed a change in performance by using dry reformate and wet reformate fuel on the anode. However, the change in mass transport is not distinguishable because the performance is seen to improve in the presence of water. Thus, using EIS online is useful to deduce the change occurring more in detail and would help to control the fuel cell stoichiometric ratio on the anode and cathode as the fuel composition changes.

The use of online EIS is also helpful for correcting the stoichiometric ratio when the stack undergoes a degraded state after several hundred hours of operation. This is concluded based on the comparison of the current study and the study reported in [8] where a degraded stack was used with dry reformate and pure hydrogen and the reported stoichiometric ratios are higher. This study could be extended to carrying out a mapping of the different stoichiometric ratios and degradation state.

Acknowledgements The authors would like to thank EUDP for funding the work through the ADDPower project. The authors would also like to acknowledge Serenergy A/S for supplying the stack for the experimental work.

References

1. T.J. Schmidt, Durability and degradation in high-temperature polymer electrolyte fuel cells. ECS Trans. **1**(8), 19–31 (2006). https://doi.org/10.1149/1.2214541. http://ecst.ecsdl.org/content/1/8/19.abstract
2. A.R. Korsgaard, M.P. Nielsen, S.R.K. Kæ r, Part two: control of a novel HTPEM-based micro combined heat and power fuel cell system. Int. J. Hydrogen Energy **33**(7), 1921–1931 (2008). http://www.sciencedirect.com/science/article/pii/S036031990800058X
3. F. Seland, T. Berning, B. Børresen, R. Tunold, Improving the performance of high-temperature PEM fuel cells based on PBI electrolyte. J. Power Sources **160**(1), 27–36 (2006). http://dx.doi.org/10.1016/j.jpowsour.2006.01.047. http://www.sciencedirect.com/science/article/pii/S0378775306001248
4. S.S. Araya, F. Zhou, V. Liso, S.L. Sahlin, J.R. Vang, S. Thomas, X. Gao, C. Jeppesen, S.K. Kær, A comprehensive review of PBI-based high temperature PEM fuel cells (2016). https://doi.org/10.1016/j.ijhydene.2016.09.024
5. F. Zhou, S.J. Andreasen, S.K. Kær, D. Yu, Analysis of accelerated degradation of a HT-PEM fuel cell caused by cell reversal in fuel starvation condition. Int. J. Hydrogen Energy **40**(6), 2833–2839 (2015). https://doi.org/10.1016/j.ijhydene.2014.12.082. http://www.sciencedirect.com/science/article/pii/S036031991403479X
6. S. Galbiati, A. Baricci, A. Casalegno, G. Carcassola, R. Marchesi, On the activation of polybenzimidazole-based membrane electrode assemblies doped with phosphoric acid. Int. J. Hydrogen Energy **37**(19) (2012). https://doi.org/10.1016/j.ijhydene.2012.07.032. http://linkinghub.elsevier.com/retrieve/pii/S0360319912016096
7. B. Najafi, A. Haghighat Mamaghani, A. Baricci, F. Rinaldi, A. Casalegno, Mathematical modelling and parametric study on a 30 kWel high temperature PEM fuel cell based residential micro cogeneration plant. Int. J. Hydrogen Energy **40**(3), 1569–1583 (2015). https://doi.org/10.1016/j.ijhydene.2014.11.093. http://www.sciencedirect.com/science/article/pii/S0360319914032273
8. S.L. Sahlin., Characterization and Modeling of a Methanol Reforming Fuel Cell System, Aalborg Universitetsforlag (Ph.D.-serien for Det Teknisk-Naturvidenskabelige Fakultet, Aalborg Universitet), 2016. https://doi.org/10.5278/vbn.phd.engsci.00059

9. F. Zhou, S.J. Andreasen, S.K. Kær, J.O. Park, Experimental investigation of carbon monoxide poisoning effect on a PBI/H_3PO_4 high temperature polymer electrolyte membrane fuel cell: Influence of anode humidification and carbon dioxide. Int. J. Hydrogen Energy. https://doi.org/10.1016/j.ijhydene.2015.09.056. http://www.sciencedirect.com/science/article/pii/S0360319915023460
10. A.D. Modestov, M.R. Tarasevich, V. Filimonov, E.S. Davydova, CO tolerance and CO oxidation at Pt and Pt–Ru anode catalysts in fuel cell with polybenzimidazole–H_3PO_4 membrane. Electrochim. Acta **55**(20), 6073–6080 (2010). https://doi.org/10.1016/j.electacta.2010.05.068. http://www.sciencedirect.com/science/article/pii/S001346861000770X, http://linkinghub.elsevier.com/retrieve/pii/S001346861000770X
11. X. Yuan, H. Wang, J. Colin Sun, J. Zhang, AC impedance technique in PEM fuel cell diagnosis—a review. Int. J. Hydrogen Energy **32**(17), 4365–4380 (2007). https://doi.org/http://dx.doi.org/10.1016/j.ijhydene.2007.05.036. http://www.sciencedirect.com/science/article/pii/S036031990700328X
12. C. Jeppesen, S.S. Araya, S.L. Sahlin, S. Thomas, S.J. Andreasen, S.K. Kær, Fault detection and isolation of high temperature proton exchange membrane fuel cell stack under the influence of degradation. J. Power Sources **359**, 37–47 (2017). https://doi.org/10.1016/j.jpowsour.2017.05.021. http://dx.doi.org/10.1016/j.jpowsour.2017.05.021
13. A. Weiß, S. Schindler, S. Galbiati, M.A. Danzer, R. Zeis, Distribution of relaxation times analysis of high temperature PEM fuel cell impedance spectra. Electrochim. Acta **230**, 391–398 (2017). https://doi.org/10.1016/j.electacta.2017.02.011. http://linkinghub.elsevier.com/retrieve/pii/S001346861730261X
14. S. Thomas, J.R. Vang, S.S. Araya, S.K. Kær, Experimental study to distinguish the effects of methanol slip and water vapour on a high temperature PEM fuel cell at different operating conditions. Appl. Energy **192**, 422–436 (2016). http://dx.doi.org/10.1016/j.apenergy.2016.11.063. http://linkinghub.elsevier.com/retrieve/pii/S0306261916316488, http://www.sciencedirect.com/science/article/pii/S0306261916316488

Multi-objective Optimization of a DC-DC Boost Converter Including Control Strategy and Stability Constraint

Davide Dell'Isola, Matthieu Urbain, Serge Pierfederici, and Farid Meibody-Tabar

Abstract In this paper, a procedure is proposed to optimally design a boost converter (including its input filter and its control) for embedded applications, where power density and high efficiency are crucial criteria. The load consists of a constant power load (CPL), which represents the most stringent case in term of stability issues. This optimization problem can be treated as a multi-objective one, which aims to maximize compactness and efficiency of the studied system. In order to find a trade-off between compactness and efficiency, the two objective functions (volume and power losses), a genetic algorithm was implemented to generate the most convenient design solutions (Pareto front). The operating conditions include both steady and fast output power load transients (step-up and step-down). Since CPL transients may lead to large deviations of the output voltage from its nominal value, the control strategy is discussed as well and included in the design procedure. The presence of a differential input filter is likely to interact with the constant power load and cause some instability issues. A constraint function excludes solutions that do not respect the feasibility, the magnetic saturation or the stability constraints.

1 Introduction

In the field of power conversion for embedded applications for more electric aircraft, as well as electric or hybrid vehicles, power density constraint, efficiency and autonomy are of prime importance. In this context, the proposed paper will focus on two criteria, namely, the efficiency and the power density.

D. Dell'Isola (✉)
GREEN – University of Lorraine, Vandoeuvre-lès-Nancy, France
e-mail: davide.dell-isola@univ-lorraine.fr

M. Urbain · S. Pierfederici · F. Meibody-Tabar
LEMTA – University of Lorraine, Vandoeuvre-lès-Nancy, France
e-mail: matthieu.urbain@univ-lorraine.fr; serge.pierfederici@univ-lorraine.fr; farid.meibody-tabar@univ-lorraine.fr

W. Zamboni, G. Petrone (eds.), *ELECTRIMACS 2019*, Lecture Notes in Electrical Engineering 615, https://doi.org/10.1007/978-3-030-37161-6_31

The optimal design of a converter is a challenging and widely discussed issue. Beside the necessity to design a compact and efficient converter, the operating performances should respect several constraints. Indeed, any application must comply with standards and electrical requirement such as input current ripple and output voltage ripple. Many parameters come into play, for instance, the ones related to the active switches, the passive components or the control stage. All of them affect, beside the converter size and the efficiency, the transient behaviour of the converter as well. The most difficult is that each parameter influences the choice of the others. Since it does not exist a unique optimum solution, which simultaneously satisfies the minimal power losses and the minimal volume value, a reasonable trade-off is necessary. A multi-objective optimization procedure was used to isolate the most favourable solutions by means of evolutionary algorithms.

In this paper, we describe a design procedure of a DC-DC boost converter, which includes a differential input LC filter and the control strategy. The differential input filter, which is used to limit the current ripple at the input terminals, may interact with the power stage and provoke instability issues, especially when it comes to dealing with a constant power load [1, 2]. If the resonance frequency of the input filter is smaller enough than one hundredth of the switching frequency, the dynamic behaviour of the converter can be modelled and studied classically by an average model. This model is no longer trustworthy for resonance frequencies close to the switching frequency. For this reason, the system stability is studied by means of limit cycles [3, 4]. This method is particularly suited for the modelling of the high-frequency behaviour of non-linear systems.

A genetic multi-objective optimization algorithm yields a set of non-dominated solutions, also known as Pareto front, i.e. a series of solutions which describe the best trade-off between the converter's size and losses. In literature [5, 6], the design procedures refer usually to the steady-state operating mode. The contribution of this paper is to take into account not only the system's behaviour in steady state but also during the transients of the load power. Moreover, the whole system small-signal stability is considered. Since the transition between two operating points is closely dependent on the control strategy, the control parameters are included in the evaluation of the Pareto front. In our case, the control structure composed of two nested feedback loops. The inner one is based on peak current mode [7], since it allows a fast response of the controlled current, which helps the reduction of the size of the converter passive components.

Using the proposed optimization approach, the boost converter has been designed in order to attain the best trade-off between volume and power losses. The simulation results show that the stability is achieved for different optimized design solutions. In addition, the input current and the output voltage ripple rates in steady state, as well as the output voltage overshoot and undershoot during the transients, stay within the admissible limits.

2 System Description

The converter under study is a DC-DC boost with a LC differential input filter which supplies a constant power load as shown in Fig. 1; u_1 and u_2 values represent the logic states of the low and high switches ($u = 1$ on-state and $u = 0$ off-state), S1 and S2, respectively. Table 1 describes the considered operating point.

The active devices are silicon carbide-based metal-oxide-semiconductor field-effect transistors (MOSFETs). Wide band-gap technology, indeed, allows to obtain converters with better efficiency and compactness than the Si technology. The input current ripple in steady state is standardized to be employed in modern aircraft environment as described in DO-160. The heatsink design has been included in the volume evaluation function. Its size depends on the semiconductor power losses, as it will be described in Sect. 6. A usual way to characterize the thermal resistance of a specific heatsink is to solve the thermodynamical equations relative to the specific geometry and material. The computational complexity of this method does not fit well with the proposed optimization routine. A more suited approach is using the information provided by the heatsink manufacturer. In this way, we can simply relate the required thermal resistance with the heatsink size. In our case, an aluminium heatsink of the ABL 146 series has been used.

3 Optimization Algorithm and Independent Variables

The optimization was performed by a genetic multi-objective optimization tool based on an elitist genetic algorithm (a variant of NSGA-II [8]), available on the MATLAB Toolbox. The design of a converter involves the choice of several

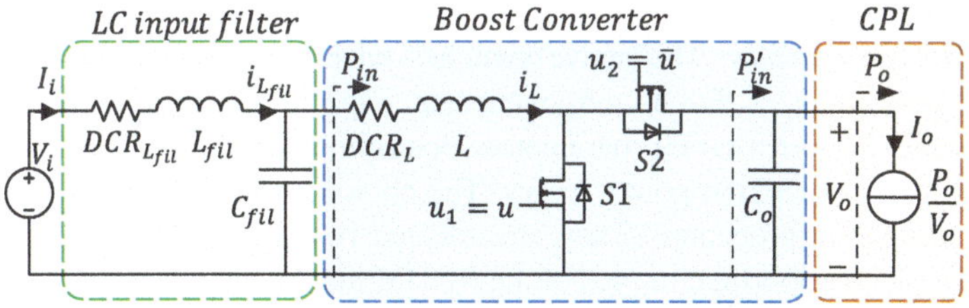

Fig. 1 Boost converter with differential input

Table 1 Operating point parameters

Symbol	Description	Value
V_i	Input voltage	270 V
V_o	Output voltage	350 V
P_o	Load power	[0–3 kW]
$\Delta v_{o_{lim}}$	Output voltage ripple limit	2% V_o
$\Delta i_{i_{lim}}$	Input current ripple limit	14% I_i

parameters. Depending on the degrees of freedom disposed, a set of independent parameters should be arbitrary chosen. The selected parameters are the independent design variables, used in the genetic algorithm, which modifies the value of these independent variables in order to minimize the objective functions (volume and power losses). Certain variables are not continuous, for instance, manufacturers only provide particular magnetic core shapes and capacitor values. In the optimization approach, the inductor cores, the capacitors and the MOSFET parameters come from the manufacturer datasheets; they are saved in a spreadsheet, and the respective indexes are used as variables in the optimization process. The creation, selection and mutation functions of the MATLAB Toolbox algorithm have been opportunely modified to operate on integer values. The chosen variables are discussed below.

Switching frequency f_s: The switching frequency affects directly the current ripple through the inductors and the voltage ripple across the capacitors, hence their inductance and capacitance values and their volume. It has a remarkable impact on the core losses and the switching losses of the active switches.

Inductor parameters: The design of the main inductor L and the filter inductor L_{fil} involves several parameters (Table 2). The inductance value is defined by the following equation:

$$L = \frac{n^2}{\Re} k_{\text{DC}}, \quad \Re = \frac{l_{\text{m}}}{A_{\text{core}} \mu_0 \mu_{\text{r}}} \tag{1}$$

where $\Re$ is the reluctance of the inductor and k_{DC} is the bias factor, which takes in account the reduction of the core permeability with the increase of the DC current trough the inductor. The selected material for the inductors core is a ferrous alloy powder with distributed air gap from Magnetics, Inc. For the main inductor, which requires greater values of inductance, we opted for the High Flux series cores. For the DM filter inductor, we choose the molypermalloy powder (MPP)cores for the better performances at high frequencies. These are un-gapped toroid cores with several available permeability values.

MOSFET parameters: The active switches have a remarkable impact on the converter performances. Each component has a specific behaviour in conduction and switching. The MOSFET parameters determine the switching times (power losses) and the conduction power losses. The choice of the active switches is part of the optimization procedure. They are selected from an assortment of available components with opportune ratings of the ROHM and Wolfspeed series.

Table 2 Inductor parameters

Symbol	Description
L or L_{fil}	Inductance value
n_{L} or $n_{\text{L}_{\text{fil}}}$	Turns number
A_{core} MLT	Magnetic core section Mean length per turn
l_{m}	Magnetic mean path length
μ_{r}	Relative permeability of the magnetic core

The parameters set $[n_{\mathrm{L}}, j_{\mathrm{L}}, i_{\mu_{\mathrm{L}}}, n_{\mathrm{Lfil}}, j_{\mathrm{Lfil}}, i_{\mu_{\mathrm{Lfil}}}, f_{\mathrm{s}}, i_{\mathrm{MOS}}]$ stands for the independent variables. The j_{L} and j_{Lfil} discrete values correspond to the inductor core indexes. Each index is associated to the geometrical dimensions of an available magnetic core in the manufacturer spreadsheet. The $i_{\mu_{\mathrm{L}}}$ and $i_{\mu_{\mathrm{Lfil}}}$ values are the indexes of the available permeability values of the cores. The index i_{MOS} leads to a specific SiC MOSFET and its respective parameters.

4 Control Strategy

The proposed control algorithm consists of a double loop control, as shown in Fig. 2. The inner loop is a peak current mode (PCM) control which regulates the main inductor current (i_{L}). The peak current controller ensures many key features, such as an inherent current limitation, a rapid dynamic response and a fixed switching frequency. The outer loop controls the output capacitor stored energy $y_{C_{\mathrm{o}}}$ and hence the output voltage v_{o}. The energy stored in the capacitor is as follows:

$$y_{C_{\mathrm{o}}} = \frac{1}{2} C_{\mathrm{o}} v_{\mathrm{o}}^2 \tag{2}$$

The output capacitor energy variation (Fig. 1) is as follows:

$$\dot{y}_{C_{\mathrm{o}}} = -P_{\mathrm{o}} + P'_{\mathrm{in}} \cong P_{\mathrm{o}} - P_{\mathrm{in}} \tag{3}$$

In Eq. (3), we suppose that the inner current loop has a fast response, and the boost inductor current i_{L} can be supposed almost equal to its reference value. Moreover, we suppose that the converter power losses are negligible. The second order control law in Eq. (4) sets the dynamic of the energy $y_{C_{\mathrm{o}}}$:

$$\ddot{e}_{y_{C_{\mathrm{o}}}} + k_{\mathrm{p}} \dot{e}_{y_{C_{\mathrm{o}}}} + k_{\mathrm{i}} e_{y_{C_{\mathrm{o}}}} = 0 \tag{4}$$

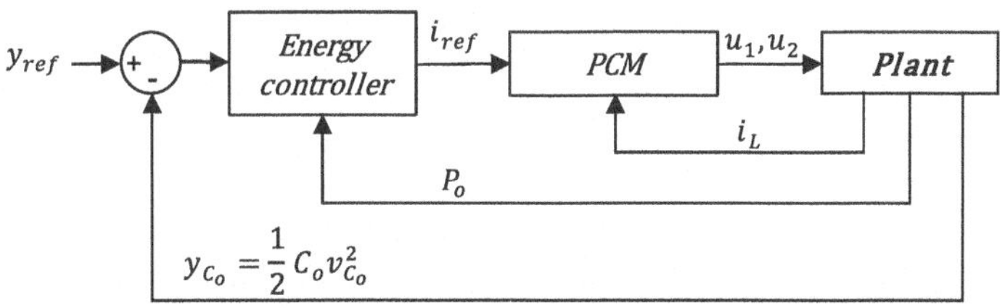

Fig. 2 Proposed control scheme

$$e_{y_{C_o}} = y_{ref} - y_{C_o},\ y_{ref} = \frac{1}{2} C_o V_o^2 \tag{5}$$

$$k_p = 2\zeta\omega_e,\ k_i = \omega_e^2, \omega_e = \frac{2\pi f_s}{50}, \tag{6}$$

where ζ and ω_e are, respectively, the damping factor and the cut-off frequency of the controlled system. The value $\omega_e/2\pi$ is chosen sufficiently smaller than the switching frequency f_s in order to decouple the dynamics of the inner and outer loops, according to Eq.(6). The reference value for the inner current loop is derived from Eqs. (3–5):

$$i_{ref} = \frac{P_o + k_p \cdot e_{y_{C_o}} + k_i \int e_{y_{C_o}} \cdot dt}{V_{in}} \tag{7}$$

5 Capacitor Design

Once the inductance values are known, then the current waveforms, the minimal capacitance values can be computed in order to keep the output voltage and input current ripples below the admissible limits. Then, the obtained capacitance value is replaced by the closest available values.

Input filter capacitor design: The differential input filter purpose is to reduce the boost inductance current ripple Δi_L, in order to keep the input current ripple within the required constraints ($\Delta i_{L_{fil}} < \Delta i_{lim}$). The input filter capacitance C_{fil}, along with L_{fil}, defines the value of the cut-off frequency and the attenuation of the filter. The required attenuation factor is given by the following equation:

$$Att_{DM} = \frac{\Delta i_{lim}}{b_I \Delta i_L},\quad C_{fil} \cong \frac{\left|\frac{1}{Att_{DM}} + 1\right|}{4\pi^2 f s^2 L_{fil}} \tag{8}$$

where b_I is the Fourier first harmonic amplitude of the boost inductor current i_L. The capacitance value of C_{fil} is determined from the input filter impedance seen by the boost inductor current at the switching frequency.

Output capacitor design: The output capacitance should ensure a maximum voltage ripple during the steady state. Moreover, far more binding, it should satisfy the required overshoot and undershoot voltage limitations during both step-up (output power rise) and step-down (output power fall) load transients [9]. By knowing the C_o energy control law and the analytical expressions of the boost inductor and capacitor currents, the minimum required value of C_o can be determined. The method is developed in the following sections for both steady state and CPL transient.

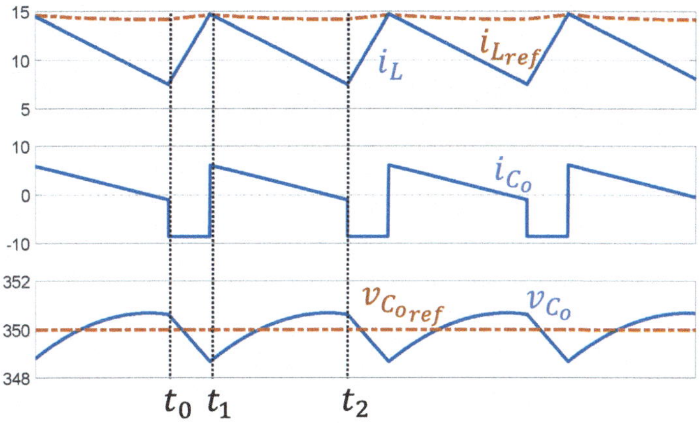

Fig. 3 Obtained results in steady-state operating mode ($L = 470\ \mu\text{H}, C_o = 100\ \mu\text{F}, f_s = 18$ kHz)

In a switching period T_s, the boost converter has two configurations ($u = 0$ or $u = 1$). The analytical expressions of the inductor and capacitor current waveforms (shown in Fig. 3) are given in Eqs. (9) and (10):

$$\begin{cases} i_L = I_i + \Delta i_L \left(-\frac{1}{2} + (t - t_0) \cdot \frac{f_S}{D}\right) \\ i_{C_o} = -I_o \qquad t \in [t_0, t_1], u = 1 \end{cases} \tag{9}$$

$$\begin{cases} i_L = I_i + \Delta i_L \left(\frac{1}{2} - (t - t_1) \cdot \frac{f_S}{1-D}\right) \\ i_{C_o} = -I_o + i_L \qquad t \in [t_1, t_2], u = 0 \end{cases} \tag{10}$$

In the first period $[t_0, t_1]$, the capacitor supplies a constant current to the load. The voltage v_{C_o} expression depends on the integral of the current through C_o:

$$v_{C_o}(t) = v_{C_o}(t_0) + \frac{1}{C_o}\int_{t_0}^{t} i_{C_o}(\tau)\, d\tau \tag{11}$$

$$v_{C_o \min} = v_{C_o}(t_1) \tag{12}$$

It obviously exhibits its minimum value at the end of this sequence (at $t_1 = D\,T_s$). As shown in [9], to determine the maximum value of v_{C_o} during the interval $[t_1, t_2]$, two cases should be considered. If the capacitor current becomes zero before $t = t_2$ (when $L < L_{\text{lim}}$), v_{C_o} reaches its maximum value at t_2' within the interval $[t_1, t_2]$; the value of t_2' can be found from Eq. (10) by setting $i_{C_o} = 0$. Otherwise, (when $L > L_{\text{lim}}$), v_{C_o} reaches its maximum value at t_2. Then, it becomes the following:

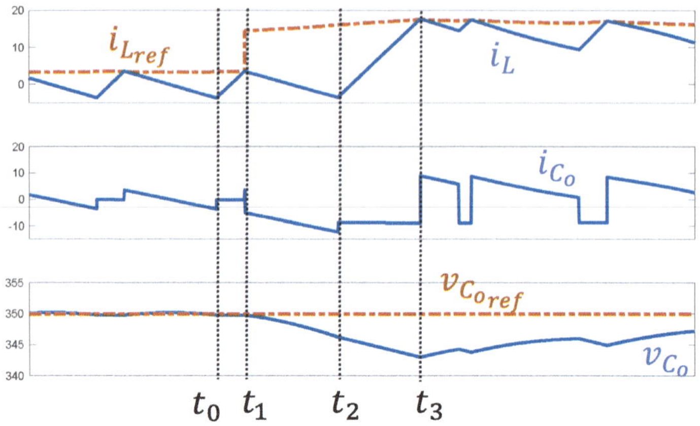

Fig. 4 Operating mode in step-up load transient state ($L = 470\ \mu$H, $C_o = 100\ \mu$F, $f_s = 18$ kHz)

$$v_{C\text{omax}} = \begin{cases} v_{C_o}(t_2') & L < L_{\text{lim}} \\ v_{C_o}(t_2) & L > L_{\text{lim}} \end{cases} \quad L_{\text{lim}} = \frac{V_{\text{in}}^2}{2 f_s P_o} \tag{13}$$

$$\Delta v_{C_o} = \left| v_{C_{\text{omax}}} - v_{C_o}(t_1) \right| < \Delta v_{o_{\text{lim}}} \tag{14}$$

The minimal C_o value is found for $\Delta v_{C_o} = \Delta v_{o_{\text{lim}}}$ in Eq. (14), and it ensures that the output voltage ripple stays within its specification limit $\Delta v_{o_{\text{lim}}}$ in steady state.

During a sudden increase of the load power, the reference current generated by the outer loop abruptly increases because of the P_o term in Eq. (7). The switch S1 stays in on-state ($u_1 = 1$) until the inductor current i_L reaches its reference value, as shown in Fig. 4. During this sequence, $i_{C_o} = -I_o$, and then, the output voltage decreases. The worst case occurs when the load transient occurs when the switch S1 has just turned off at $t = t_1$. In this case, the inductor current does not start to increase until the end of the period, then after $(1 - D)T_s$. The inductor current reaches its reference at t_3 shown on Fig. 4, and then, the switch S2 turns on and the inductor current flows towards C_o.

$$\begin{cases} i_L = \Delta i_L \left(-\frac{1}{2} + (t - t_0) . \frac{f_s}{D} \right) \\ i_{C_o} = 0 \qquad t \in [t_0, t_1], u = 1 \end{cases} \tag{15}$$

$$\begin{cases} i_L = \Delta i_L \left(\frac{1}{2} - (t - t_1) \cdot \frac{f_s}{(1-D)} \right) \\ i_{C_o} = -I_o + i_L \quad t \in [t_1, t_2], u = 0 \end{cases} \tag{16}$$

$$\begin{cases} i_L = \Delta i_L \left(-\frac{1}{2} + (t - t_2) \cdot \frac{f_s}{(1-D)}\right) \\ i_{C_o} = -I_o \qquad\qquad t \in [t_2, t_3], u = 1 \end{cases} \tag{17}$$

The minimum value of v_{C_o} is reached exactly at t_3. The instant t_3 is computed as a function of the C_o capacitor by solving the fourth degree equation $i_L = i_{ref}$ from Eqs. (17) and (7).

$$\Delta v_{C_o} = \left|v_{C_{omax}} - v_{C_o}(t_3)\right| < \Delta v_{O_{lim}} \tag{18}$$

where $v_{C_{omax}}$ is the maximal v_{C_o} value in steady state, calculated in Eq. (14). The minimum value of v_{C_o} is given by integrating the capacitor current as in Eq. (12). The minimal value of C_o to keep the undershoot under the admissible deviation is obtained when $\Delta v_{C_o} = \Delta v_{O_{lim}}$.

Similar considerations can be developed when the load power quickly steps down. In this case, the v_{C_o} exhibits an overshoot. The chosen value of C_o is the one which satisfies all of the above-mentioned constraints on ripple rate in steady state, undershoot and overshoot of output voltage.

6 Objective Functions

Volume: The volume of each passive element has been deduced from the manufacturer datasheet by the corresponding index. The design of capacitors for each independent parameter set, presented in Sect. 5, leads to the determination of the index values relative to the closest available capacitors (Vishay MKP1848C DC link series).

The heatsink volume is obtained from the required thermal resistance R_{th}. The manufacturer datasheet permits to define a relation between thermal resistance and size of the heatsink as mentioned hereafter:

$$V_{heatsink} = z\, R_{th}^{p}, \quad R_{th} = \frac{T_j - T_a}{P_{mos}} - R_{th_{j-h}} \tag{19}$$

where $R_{th_{j-h}}$ is the thermal resistance from the semiconductor junctions to the heatsink, T_j is the maximal desired temperature of the semiconductor junction and T_a is the ambient temperature.

Power losses: The power loss objective function includes the MOSFET switches and the inductor power losses.

The switch power loss model includes switching losses, conduction losses and diode losses during the dead time (see Table 3):

$$P_S = r_{DS}\, {i_{MOS}^{rms}}^2 + \left(E_{C_{oss}} + E_{on} + E_{off}\right) f_s \tag{20}$$

Table 3 MOSFET parameters

Symbol	Description
r_{DS}	MOSFET channel resistance in on-state
i_D	Diode current
i_{MOS}	MOSFET drain-source current
$E_{C_{oss}}$	MOSFET output capacitance energy loss
E_{on}, E_{off}	Turn on/off switching energy
t_{dt}	Dead time
E_{rr}	Diode recovery energy loss
V_D	Diode conduction voltage

$$P_{diode} = (2\ V_D\ I_i\ t_{dt} + E_{rr})\ f_s \tag{21}$$

The switching energy losses depend on the MOSFET parasitic capacitances and the amount of charge stored in the gate capacitance. They have been evaluated for each MOSFET as in [10] for a fixed gate resistance value.

The inductor losses are obtained by (refer to Table 2) the following equations:

$$P_{copper} = \text{DCR}\ i_{rms}^2,\ \text{DCR} = n\ \text{MLT}\frac{\rho}{A_{wire}} \tag{22}$$

$$P_{core} = V_L \frac{1}{T_s}\int_0^{T_s} k_i \left|\frac{dB_L}{dt}\right|^{\alpha} \Delta B_{pp}^{\beta-\alpha}\,dt \tag{23}$$

The parameters ρ and A_{wire} are, respectively, the copper resistivity and the copper wire section. Equation (23) is based on an improved model of the Steinmetz core losses iGSE [11], where α and β are the Steinmetz parameters and k_I is a constant coefficient related to the Steinmetz proportional parameter defined in [10].

7 Constraint Functions

The converter should respect technological and behavioural constraints. These requirements are checked by the constraint functions in each iteration for each solution. The algorithm discards unfitting solutions. The constraint functions include the inductor design constraints and the stability criterion. The constraints on input current and output voltage (in steady state or transient state) do not appear between the constraint functions, since they have already been included in the design of the passive elements.

Inductor constraints: At the maximum power operating point, the saturation of the inductor magnetic cores must be avoided. Moreover, the number of coils should not exceed the available winding area. The following conditions have to be respected:

$$B_{\max} = n \cdot \mu_0 \cdot \frac{i_{\max}}{l_{g_{eq}}} \leq B_{sat} \tag{24}$$

$$n \cdot A_{wire} \leq k_F \cdot W_A \tag{25}$$

The term W_A is the core winding area available for the winding conductors; k_F is the filling factor.

Stability constraint: In order to study the system stability for frequencies close to the switching frequency, the Poincaré mapping method is applied. It consists of a discrete modelling of the system with a sampling period equal to the switching one (T_s):

$$\boldsymbol{x}\left[(n+1)\,T_s\right] = \boldsymbol{f}\left[\boldsymbol{x}\,(nT_s)\,,d\right] \tag{26}$$

where $\boldsymbol{x} = \left[i_L, i_{L_{fil}}, v_{C_{fil}}, v_o, Y_o\right]$ is the state variable vector, with $Y_o = \int e_{y_{C_o}} dt$ the integral term used in Eq. (7). The function $\boldsymbol{f}$ is the analytic expression which allows to obtain the state vector at the time $(n+1)T_s$ from the state vector at the time $n\ T_s$ (Eq. 26). Following the same approach used in [12], we aim to find the eigenvalues λ_i of the controlled system in order to evaluate its stability ($|\lambda_i| < 1$). The duty cycle d, which represents the system input in Eq. (26), takes part in the non-explicit expression of the control surface $s(t)$ at the time $t = d \cdot T_s$:

$$s(t) = i_{ref} - i_L + m_c\left(\frac{T_s}{2} - t\right), s\,(dT_s) = 0 \tag{27}$$

Nevertheless, as shown in [12], it is possible to obtain the Jacobian matrix of the closed loop system as follows from the first-order Taylor expansion of the system in Eq. (26) and the control surface in Eq. (27).

$$J = \left[\frac{\partial f}{\partial x_i} - \frac{\partial f}{\partial d}\left(\frac{\partial s}{\partial d}\right)^{-1}\frac{\partial s}{\partial x_i}\right] \tag{28}$$

The eigenvalues λ_i of the matrix J are determined, and the stability condition for the considered discrete system is verified if $|\lambda_i| < 1$.

8 Optimization and Simulation Results

The multi-objective optimization results are depicted on Fig. 5. Each element of the Pareto fronts represents a non-dominated design solution. The genetic algorithm stopped after 115 iterations (2 h 32′) after the weighted average relative change in the spread of the Pareto solutions reached the tolerance value (set to 10^{-4}). A particular solution of the discrete variable was selected. The selected components are listed in Table 4. The expected losses and the volume are showed in Table 5.

A transient simulation was performed on Simulink/MATLAB environment for both steady state and load transient. Figure 6 shows that i_L and v_{C_0} follow the respective references. The output voltage stays within the required limits. We observe that the step-up load transient represents the most constraining case for the output capacitor design.

9 Conclusions

This paper proposes a design procedure for the minimization of the volume and the power losses for a 3 kW DC-DC boost converter for aeronautical applications. The considered converter supplies a constant power load, which represents the

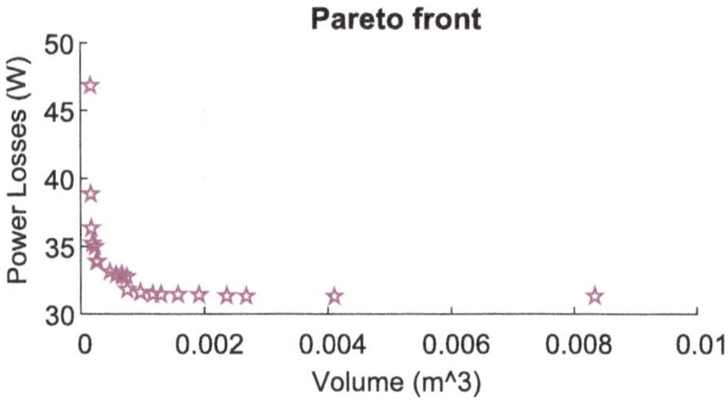

Fig. 5 Pareto front

Table 4 Selected solution

Reference	Description	Value
ROHM-SCT3030KL	MOSFET	
	Switching freq.	$f_s = 21$ kHz
MKP1848C67560JY	Output capacitor	$C_o = 75$ μF
C058714A2	Main inductor	$L = 173$ μH
MKP1848C61560JP	Filter capacitor	$C_{fil} = 15$ μF
C055586A2	Filter inductor	$L_{fil} = 45$ μH
	Gate resistance	$R_g = 7.5\ \Omega$
	Wire section	$A_{wire} = 0.79\ mm^2$
	Filling factor	$k_f = 0.3$

Table 5 Evaluated performances of the selected design solution

Objective function	S1	S2	L	L_{fil}	C_o	C_{fil}	Heatsink
Power losses (W)	12	10	10.2	3.9	–	–	–
Volume (cm^3)	–	–	1.5832	5.5090	100.63	22.40	55

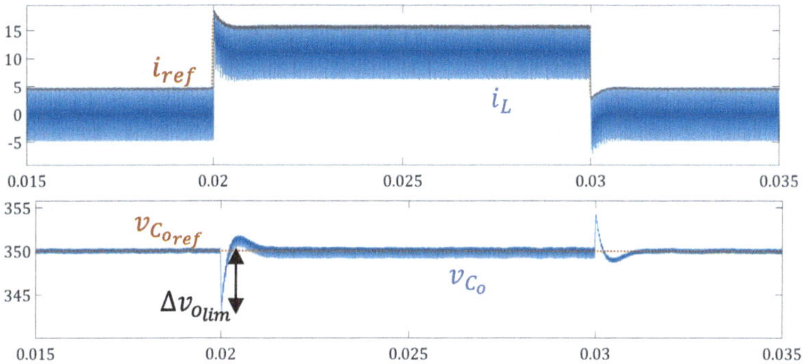

Fig. 6 Power load transient response of the inductor current and output capacitor voltage (optimization solution of Table 4)

most unfavourable case, in terms of the system stability. In order to meet the DO-160 requirements on the input current in steady state, the optimization procedure includes the design of a differential input filter. The optimization is performed by a multi-objective algorithm based on evolutionary theory. The algorithm finds the design solutions which represent the most convenient trade-off between volume and power losses. Usually, in literature, the proposed converter optimizations aim to develop design procedures only for the steady-state operating mode. In the proposed method, beyond the steady-state mode, the converter is designed to withstand fast load transients, so that the output voltage do not exceed the ripple requirements. For this reason, the control strategy and its parameters have been considered in the design procedure. Moreover, due to the presence of the input filter, the optimization algorithm includes the study of the controlled system stability and discards the unstable design solutions. The simulation results show, for several design solutions, how the performance of the converter respects the given constraints, and the passive components have not been oversized.

References

1. A. Emadi, A. Khaligh, C.H. Rivetta, G.A. Williamson, Constant power loads and negative impedance instability in automotive systems: definition, modeling, stability, and control of power electronic converters and motor drives. IEEE T. Veh. Technol **55**(4), 1112–1125 (2006)
2. P. Magne, D. Marx, B. Nahid-Mobarakeh, S. Pierfederici, Large-signal stabilization of a DC-link supplying a constant power load using a virtual capacitor: impact on the domain of attraction. IEEE T. Ind. Appl. **48**(3), 878–887 (2012)

3. J.H. Chen, K.T. Chau, Analysis of chaos in current-mode-controlled DC drive systems. IEEE T. Ind. Electron **47**(1), 67–76 (2000)
4. A. Lachichi, S. Pierfederici, J.-P. Martin, B. Davat, Study of a hybrid fixed frequency current controller suitable for DC–DC applications. IEEE T. Power Electron **23**(3), 1437–1448 (2008)
5. M. Kasper, D. Bortis, G. Deboy, J.W. Kolar, Design of a highly efficient (97.7%) and very compact (2.2 kW/dm ^{3}) isolated AC–DC telecom power supply module based on the multicell ISOP converter approach. IEEE T. Power Electron **32**(10), 7750–7769 (2017)
6. R.M. Burkart, J.W. Kolar, Comparative η–ρ–σ pareto optimization of Si and SiC multilevel dual-active-bridge topologies with wide input voltage range. IEEE T. Power Electron **32**(7), 5258–5270 (2017)
7. R.W. Erickson, D. Maksimovic, *Fundamentals of Power Electronics*, 2nd edn. (Kluwer Academic Publishers, Dordrecht, 2001)
8. K. Deb, *Multi-objective Optimization Using Evolutionary Algorithms* (John Wiley & Sons, Hoboken, NJ, 2001)
9. A. De Nardo, N. Femia, G. Petrone, G. Spagnuolo, Optimal buck converter output filter design for point-of-load applications. IEEE T. Ind. Electron **57**(4), 1330–1341 (Apr. 2010)
10. M. H. Rashid (ed.), *Power Electronics Handbook: Devices, Circuits, and Applications Handbook*, 3rd edn. (Elsevier, Burlington, MA, 2011)
11. K. Venkatachalam, C. R. Sullivan, T. Abdallah, and H. Tacca, Accurate prediction of ferrite core loss with nonsinusoidal waveforms using only Steinmetz parameters, in 2002 IEEE Workshop on Computers in Power Electronics, 2002. Proceedings, Mayaguez, Puerto Rico, 2002, pp. 36–41
12. E. Monmasson (ed.), *Power Electronic Converters: PWM Strategies and Current Control Techniques* (ISTE, Wiley, Hoboken, NJ, 2011)

Performance Testing of a Piezoelectric Device for Extracting Energy from Vibrations

Giorgia Leonardi, Fabio Passacantilli, Carmen Galassi, and Daniele Dessi

Abstract Energy harvesting from ambient sources is an interesting opportunity for wireless and self-powered electronics, increasing research efforts toward the development of new devices. Among all energy sources, vibrations seem particularly convenient for this kind of application. Piezoelectric resonant systems, though offering configurations well suited to recover energy from vibrations, suffer from narrow operational bands, and for this reason new solution to enhance performances at off-design excitation conditions is sought. In this paper a piezoelectric resonant energy harvester is developed, focusing the attention on both ceramics production method and support material choice in order to maximize the oscillation amplitudes, and consequently the energy output. The device produced at ISTEC laboratories is then compared with a commercial product under harmonic excitations. Results relative to power output show that the in-house assembled device has better performance than the commercial one in the considered tested conditions, both in absolute terms and with respect to the active piezoelectric volume of the two devices.

1 Introduction

In recent years, the growing importance of wireless and off-grid devices has increased the interest on distributed power production even equipping the devices itself. In particular, ambient and anthropic vibrations are promising energy sources worth to be explored, driving the development and optimization of many new different devices. In literature, the usefulness of vibrations in powering wireless

G. Leonardi · D. Dessi (✉)
CNR-INM – National Research Council of Italy, Institute of Engineering for Sea, Roma (RM), Italy
e-mail: giorgia.leonardi@uniroma1.it; daniele.dessi@cnr.it

F. Passacantilli · C. Galassi
CNR-ISTEC – National Research Council of Italy, Institute of Science and Technology for Ceramics, Faenza (RA), Italy
e-mail: fabio.passacantilli@istec.cnr.it; carmen.galassi@istec.cnr.it

W. Zamboni, G. Petrone (eds.), *ELECTRIMACS 2019*, Lecture Notes in Electrical Engineering 615, https://doi.org/10.1007/978-3-030-37161-6_32

and low-consumption sensors for off-grid exercise has been shown in different applications, such as monitoring of civil structures [1], buildings [2], machineries [3], and human health [4, 5], where no from-grid power supply is possible or cabling would be too expensive and complicated.

In this kind of applications, one of the most effective systems is the piezoelectric resonant energy harvester in cantilevered configuration [6, 7]. However, even if the conversion efficiency at resonant frequency is high, the performance drops as soon as the devices is working out of its narrow resonance bandwidth. This behaviour makes the piezoelectric resonators difficult to be optimized for on-field applications, since ambient vibrations usually are not confined to a single frequency, and in reverse are stochastic with most of the energy spread over a large frequency range.

Many research groups worldwide are studying different technological solutions, trying to widen the acceptable frequency range and the overall energy conversion efficiency. A critical component for increasing global performances is provided by the electrical conversion system, that rectifies and stores the energy produced by the harvester to make it usable for electronic devices. Indeed, this element of the conversion chain can optimize the power output both searching for RC resonance conditions and using active circuit for conversion (MPPT, SSHI, SECE), as pointed out by Du et al. [8–10]. Furthermore, there are many techniques to improve the mechanical performance based on tuning with excitation frequencies and on increasing the oscillation amplitude with the same excitation input. The most widespread solution is inserting a tip mass to the cantilever end [11–13], changing the resonant frequency of the system and enhancing the amplitude of oscillations. Another technique consists in making the structure impact with an external element [14], such as a pendulum; in this way, the low-frequency vibrations can be transformed into higher frequency forces, more suitable to the typical cantilever mechanical configuration [15–17]. Moreover, in order to enhance amplitude oscillations, non-linearities in the resonant system can be introduced, for example, using a magnetic tip mass interacting with another magnet fixed out of the oscillator [12, 18–20], connecting several linear resonators to each other with springs [21, 22], using bistable or tristable configurations [22–25], or adopting support elements to piezoelectric materials with a snap-through behaviour [24, 26–28].

In this paper a vibration energy conversion system is presented and tested using a dedicated experimental setup, which simulates the vibrating support of the harvester. The piezoelectric production process, carried out at the ISTEC laboratory, is described and a new support material for the resonator is adopted. A comparison of the device with a commercial piezoelectric under sinusoidal signals is made, analysing performances in terms of power production.

2 Mechanical Configuration

The energy harvester adopted in this work exploits the direct piezoelectric effect via the use of a piezoelectric cantilever vibrating because of the basement excitation, as shown in Fig. 1

The piezoelectric material is known to couple the material stress $\boldsymbol{\sigma}$ and deformation $\boldsymbol{S}$ with the electric displacement $\boldsymbol{D}$ and field $\boldsymbol{E}$, according to the following state equations:

$$\begin{bmatrix} S \\ D \end{bmatrix} = \begin{bmatrix} c^E & d^t \\ d & \varepsilon^\sigma \end{bmatrix} \begin{bmatrix} \sigma \\ E \end{bmatrix}, \tag{1}$$

where $\boldsymbol{\varepsilon}$, $\boldsymbol{d}$, and $\boldsymbol{c}$ represent the electrical, piezoelectric, and elastic properties of the material, respectively. Equations (1) model the linear relation between deformation and charge, and describes both direct and converse effect. The direct effect (or generator behaviour) manifests when there is a charge production in the material after an imposed mechanical stress. On the contrary, the converse effect (or motor behaviour) occurs when a strain in the material is induced by an external electric field (or a voltage applied).

The mechanical configuration chosen for the device is the cantilevered one, with a clamped edge and the other free (Fig. 1); this appears the best layout to exploit ambient vibrations through piezoelectricity in terms of efficiency and simplicity. It is a common practice to glue the piezoelectric lamina over a supporting plate for several reasons: mechanical stability, structural integrity, ease of installation, and no charge compensation. In the present case, the support material is harmonic steel, which provides low-damping features. The support layer thickness has been chosen to contain the neutral axis of the overall structure (Fig. 2), since this allows the piezoelectric material to be more uniformly stretched or shorten through its thickness. Indeed, if the neutral axis be placed inside the piezoelectric sheet, the material would have a partial or even total charge cancellation, and so a decreased power output.

Fig. 1 Mechanical configuration of the piezoelectric harvester

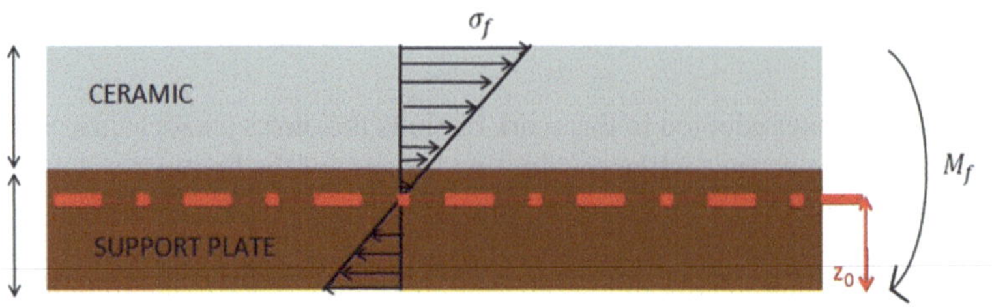

Fig. 2 Sketch of the optimal mechanical configuration

3 Piezoelectric Patch Production

In order to create a resonator well performing at low frequencies, the piezoelectric lamina was produced ad hoc at the CNR-ISTEC laboratory. The piezoelectric is produced with a tape casting process, particularly useful when a thin thickness (220 μm) needs to be reached. To create the ceramic material, a suspension based on the ceramic powder dispersed in an organic liquid with deflocculant, binder, and plasticizer is prepared [29]. The ceramic powder is lead-zirconate-titanate doped with niobium and lanthanum (PLZTN):

$$(Pb_{0.93}\ La_{0.07})\ [(Zr_{0.60}\ Ti_{0.40})_{0.9825}\ Nb_{0.0175}]\ O_3. \tag{2}$$

The additives are constituted by a dispersant, stabilizing colloidally the suspension and so avoiding agglomerates formation, a plasticizer, making tapes flexible and so possible to be bended, and a binder, binding together the particles. The solvent is an azeotropic mixture of methylethylketone (MEK) (or 2-butanone), 67% by weight, and ethanol (EtOH).

To level the thickness of the layers to 278 μm, the suspension moves on a belt and passes under a blade (Fig. 3). Then, after the drying process, the flexible solid tape can be punched or cut to obtain different shapes. Thus, the material obtained undergoes elimination of organic components (debonding) and sintering, and finally electrodes deposition by screen printing. At the end, the poling phase induces piezoelectric properties in the ferroelectric material, by orienting the electrical dipoles under a 3 kV/mm electrical field for 40 min in a silicon oil bath at 120 °C.

4 Conversion System

As the voltage output from the piezoelectric device is not constant, but varies in time due to the unsteady excitation, an electric conversion system is required to provide an output signal suitable to power an electronic device or to store temporarily the available energy. The system is schematically made of a voltage multiplier,

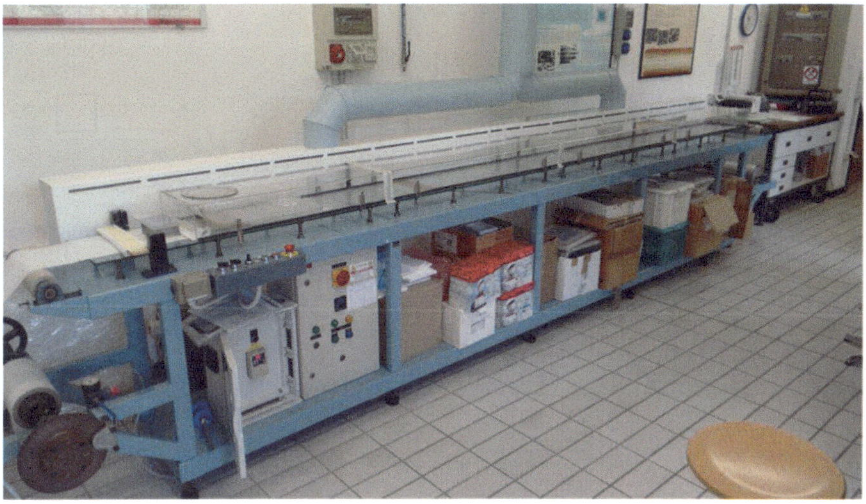

Fig. 3 Casting bench

acting also as rectifier, a charge regulator, and a capacity storage, which the load to be powered is attached to (Fig. 4). In the present case, the load is a temperature sensor. The voltage multiplier has six steps, each one made by Schottky BAT54CL diodes and tantalum capacitors of 4.7 μF, and it is able to multiply the input signal amplitude up to about six times. This circuital element makes possible for the piezoelectric harvester to reach the minimum threshold needed to activate the charge regulator BOB 09946. The output DC voltage from the BOB 09946 can take the values 1.8, 2.5, 3.3, and 3.6 V. Taking a closer look to the device, the LTC3588-1 integrated circuit (Fig. 5), optimized for high impedance devices as in piezoelectric applications, changes the input voltage into one of the aforementioned values thanks to a high efficiency buck converter. The UVLO component keeps the voltage off the buck converter until it reaches a threshold value, thanks to a capacitor optimized in order to lower the time needed to be charged. When the threshold is reached, the buck converter is activated, but it supplies voltage only when the load is connected at the output of the PLC and demands power.

Out of the BOB 09946, a capacitor (1500 μF) is used as storage to ensure a constant power flow for several seconds, enough in this case to perform a temperature measurement. The sensor indeed is activated thanks to an enable signal from BOB 09946 that controls a MOSFET: the circuit is closed only when the capacitor reaches 3.3 V and the sensor goes off when the voltage drops under 3 V. Figure 6 is a picture of piezoelectric harvester, conversion system, and temperature sensor.

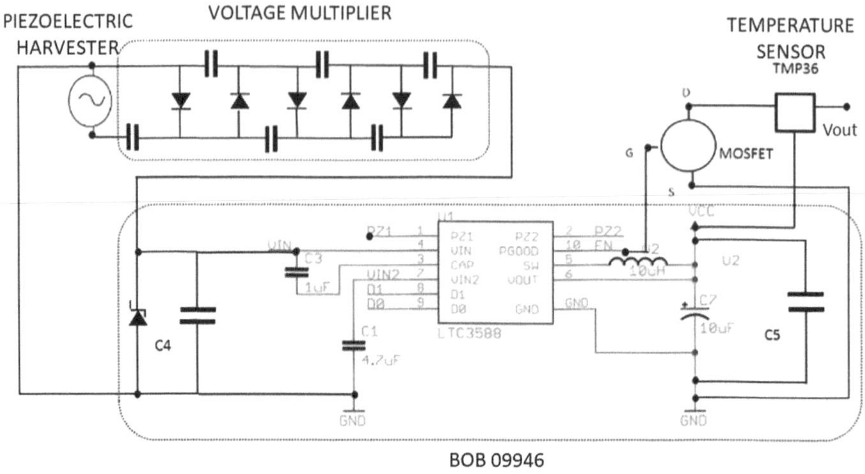

Fig. 4 Schematic representation of piezoelectric generator, conversion system, and temperature sensor

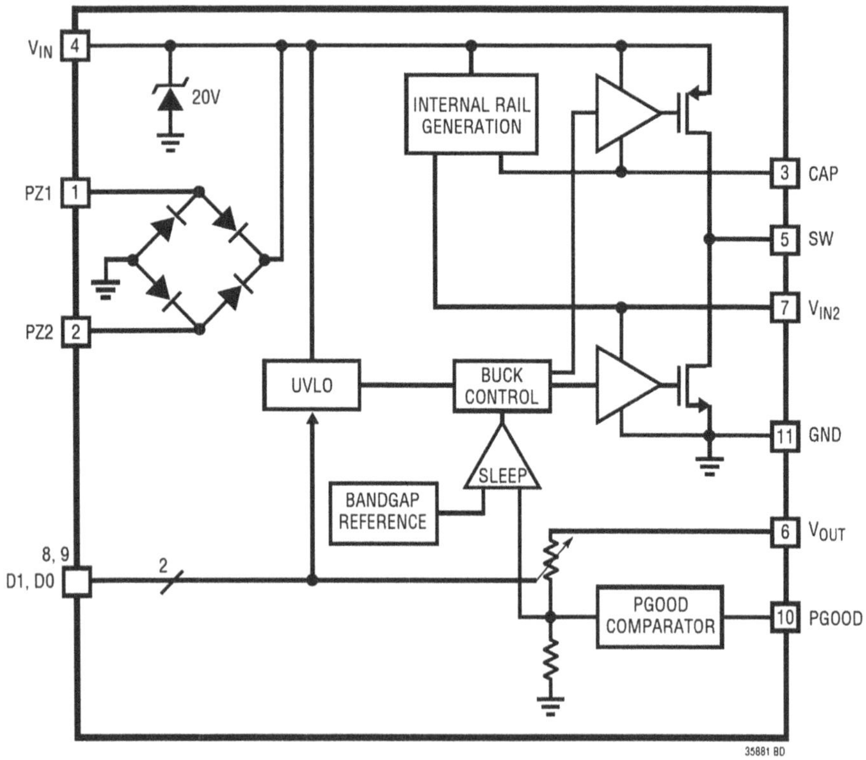

Fig. 5 Schematic representation of LTC3588-1

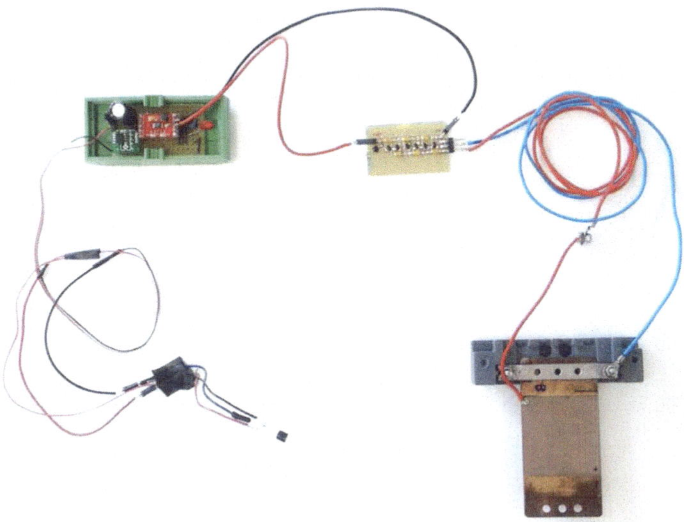

Fig. 6 Photo of piezoelectric harvester, conversion system, and temperature sensor

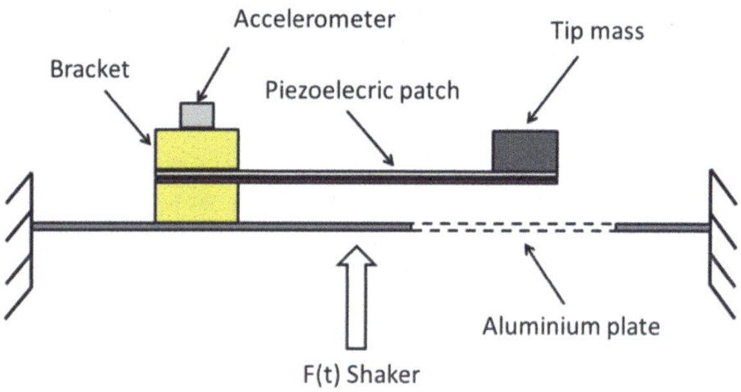

Fig. 7 Schematic configuration of the experimental setup

5 Experimental Setup

Assessing the performances of the harvester under variable conditions required to build an experimental setup made of several components, comprising a test-rig to host the vibrating plate, an excitation system via a shaker, and some sensors.

The energy harvester lies on a clamped metallic plate, simulating a real-life installation (e.g. wall or floor), excited by a force applied with the shaker (Fig. 7). The LMS acquisition system collects the signals provided by an accelerometer, placed at the clamping of the piezoelectric device, and a voltage meter between the piezoelectric electrodes. To ensure that the devices are tested under the same excitation, the accelerometer measurement is taken as reference, because though applying the same force the different mass of the devices may significantly alter the overall response. For this purpose, a reverse engineering process allows us to find which force signal must be set to obtain the desired acceleration. This can be done by

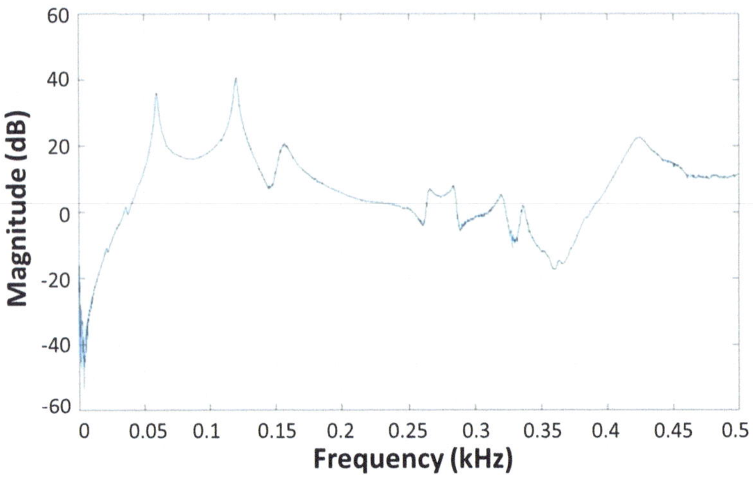

Fig. 8 Frequency transfer function of the plate

Table 1 Voltage output (rms) at $f_e = 56.6\,\text{Hz}$ harmonic excitation

Device	Voltage output	Output/input ratio
In-house assembled	9.23 V	3.69 $V \cdot s^2/m$
Commercial	7.30 V	2.92 $V \cdot s^2/m$

a trial-and-error procedure or with a general procedure that requires the calculation of the frequency transfer function (Fig. 8) between applied force and acceleration.

6 Results and Discussion

In this section the comparison between the in-house assembled device, also denoted in the following as ISTEC-S, and the commercial product (MIDE PPA 1012) will be carried out to underline the main differences in performance between the two solutions.

The first test has been performed with a harmonic seismic excitation at an rms acceleration amplitude and frequency equal to 0.1 m/s^2 and 56.6 Hz, respectively, the latter being the resonance frequency of the ISTEC-S. To bring the resonance frequency of the MIDE PPA 1012 to the same value of 56.6 Hz, a tip mass has been added. Though this feature should favour the commercial solution, the rms voltage output, in open circuit conditions, of the ISTEC-S is 9.23 V, higher than 7.3 V of PPA 1012, as shown in Table 1.

The sensitivity of the two devices to perturbation of the optimal excitation frequency has been tested too. Both ISTEC-S and PPA 1012 have been equipped with a tip mass to lower their natural frequency f_n to 20 Hz (Fig. 9). Thus, five tests with sinusoidal excitation have been carried out at the same amplitude (0.6 m/s^2

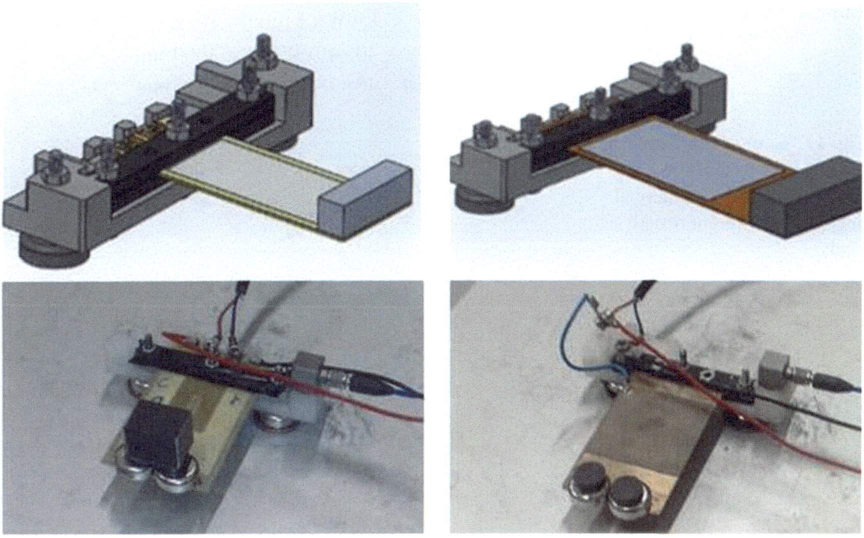

Fig. 9 Schematic and real configuration of the commercial (on the left) and the in-house assembled devices (on the right)

Fig. 10 Comparison between the commercial and in-house assembled devices output voltage for harmonic excitation at different frequencies

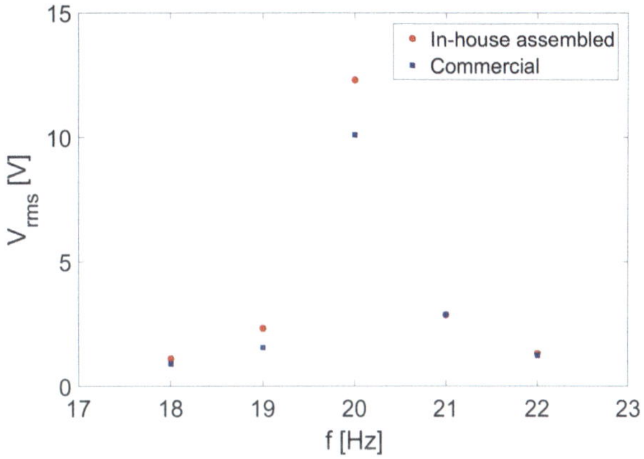

of rms), varying the excitation frequency f_e of $\pm 5\%$ and $\pm 10\%$ with respect to the initial value f_n. Results in Fig. 10 show the expected drop of performance for both devices out of the resonant frequency $f_e \neq f_n$. Anyway, ISTEC-S still provides a better performance at the resonant excitation and for $f_e > f_n$, through this difference becomes negligible for $f_e < f_n$.

In order to operate an absolute and a relative comparison between the two devices, the dimensions of the active piezoelectric materials have been reported in Table 2. Both devices have been tested for different resistive loads, varying from 0.82 kΩ till 90 kΩ (Figs. 11 and 12). The curves show an optimal load condition

Table 2 Dimensions of the active piezoelectric material

Device	L×W×T [mm]	Area	Volume
In-house	40×40×0.2	1600 mm^2	320 mm^3
Commercial	46×38.4×0.25	1766 mm^2	442 mm^3

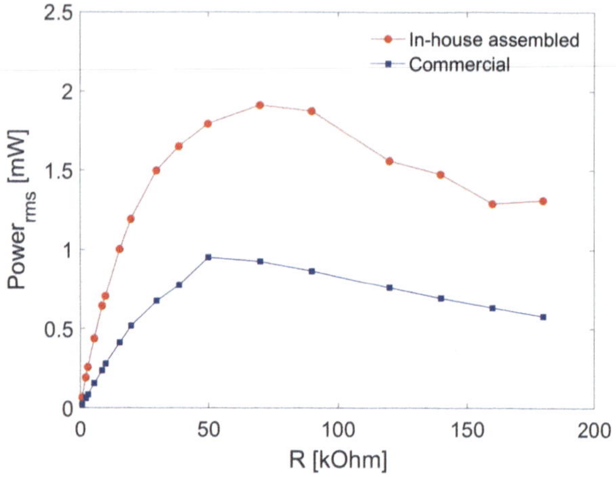

Fig. 11 Comparison between the power output of the in-house and the commercial devices for different resistive loads

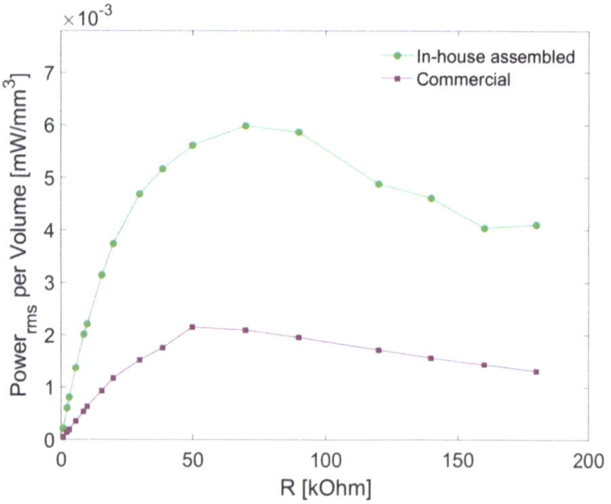

Fig. 12 Comparison between the power output per volume of the in-house and the commercial devices for different resistive loads

at 70 kΩ for the ISTEC-S device and at 50 kΩ for the PPA 1012, as suggested by the analysis of the equivalent RC circuit describing the system. Both power output and power per volume of active piezoelectric material results point out a better performance of the ISTEC-S with respect to the commercial solution.

At the end, the performances of the two devices are experimentally compared considering their response in a realistic working case, where the load is given by a resistive temperature sensor (TMP36 by Analog Devices). The test has been carried out with a harmonic seismic excitation at an rms acceleration amplitude and frequency equal to 0.1 m/s^2 and 20 Hz, respectively. The initial charge of the storage

Table 3 Recharge times of the storage capacitor

Device	First charge time	Temperature data sample time
In-house	5 min	27 s
Commercial	9 min	48 s

capacity takes a longer time than intermediate charging between sensor recordings (see Table 3), as the 3.3 V threshold has to be reached from scratch. Indeed, the intermediate charges are operated in a shorter period, since the MOSFET of the sensor opens only when the voltage drops from 3.3 to 3 V, and the capacity voltage has to rise of just 0.3 V. Both devices power successfully the load, but recharge time is 27 s for ISTEC-S against 48 s of PPA 1012, as shown in Table 3.

7 Conclusions

In this work, a piezoelectric device for recovering energy from vibrations has been tested up to verify its capability to provide enough power to measure ambient temperature within acceptable vibration levels; under harmonic excitation of 20 Hz, the whole-body 1-h exposure limit is about 10 m/s^2 according to ISO 2631, far higher than test excitation. Keeping the same cantilever configuration, the performance assessment of two different layouts in terms of materials and geometries (one built at CNR-ISTEC laboratory, the other fabricated by MIDE) has been carried out showing a larger voltage and power output for the in-house assembled device in the considered testing conditions both in absolute and relative to piezoelectric volume terms. Since one of the critical points is the tuning between the excitation and natural frequency of the cantilever, off-design excitation conditions have been tested highlighting the expected decrease of performances. On this aspect, there is the need to address more systematically off-design conditions by identifying both the acceptable operational envelope above the minimum power requirement of the sensor and the countermeasures to expand the operational margins.

Acknowledgements We wish to thank Carlo Baldisserri, Claudio Capiani, and Davide Gardini for their contribution in fabrication of ISTEC-S piezoelectric device and testing and Massimo Adriano for his support in setting the electronic components.

References

1. H. Wang, A. Jasim, X. Chen, Energy harvesting technologies in roadway and bridge for different applications—a comprehensive review. Appl. Energy. **212**, 1083–1094 (2018)
2. X.D. Xie, Q. Wang, Design of a piezoelectric harvester fixed under the roof of a high-rise building. Eng. Struct. **117**, 1–9 (2016)

3. B.C. Essink, R.B. Owen, D.J. Inman, Optimization of a zigzag shaped energy harvester for wireless sensing applications. Spec. Top. Struct. Dyn. **6**, 85–89 (2017)
4. M.T. Todaro, F. Guido, L. Algieri, V.M. Mastronardi, D. Desmaële, G. Epifani, M. De Vittorio, Biocompatible, flexible, and compliant energy harvesters based on piezoelectric thin films. IEEE Trans. Nanotech. **17**(2), 220–230 (2018)
5. N. Jackson, O.Z. Olszewski, C. O'Murchu, A. Mathewson, Shock-induced aluminium nitride based MEMS energy harvester to power a leadless pacemaker. Sensors Actuators A: Phys. **264**, 212–218 (2017)
6. Y. Kuang, M. Zhu, Characterisation of a knee-joint energy harvester powering a wireless communication sensing node. Smart Mater. Struct. **25**(5), 055013 (2016)
7. A.E. Kubba, K. Jiang, Efficiency enhancement of a cantilever-based vibration energy harvester. Sensors 2014 **14**, 188–211 (2013)
8. S. Du, Y. Jia, C.D. Do, A.A. Seshia, An efficient SSHI interface with increased input range for piezoelectric energy harvesting under variable conditions. IEEE J. Solid-State Circuits **51**(11), 2729–2742 (2016)
9. S. Du, A.A. Seshia, An inductorless bias-flip rectifier for piezoelectric energy harvesting. IEEE J. Solid-State Circuits **52**(10), 2746–2757 (2017)
10. S. Du, G.A. Amaratunga, A.A. Seshia, A cold-startup SSHI rectifier for piezoelectric energy harvesters with increased open-circuit voltage. IEEE Trans. Power Electron. **34**, 263–274 (2018)
11. N. Jackson, F. Stam, O.Z. Olszewski, R. Houlihan, A. Mathewson, Broadening the bandwidth of piezoelectric energy harvesters using liquid filled mass. Proc. Eng. **120**, 328–332 (2015)
12. Z. Hu, J. Qiu, X. Wang, Y. Gao, X. Liu, Q. Chang, Y. Long, X. He, An integrated multi-source energy harvester on vibration and magnetic field energy. AIP Adv. **8**(5), 2158–3226 (2018)
13. Y. Jia, A.A. Seshia, Power optimization by mass tuning for MEMS piezoelectric cantilever vibration energy harvesting. J. Microelectromechanical Syst. **25**(1), 108–117 (2016)
14. A. Abedini, S. Onsorynezhad, F. Wang, Study of an impact driven frequency up-conversion piezoelectric harvester, in *ASME 2017 Dynamic Systems and Control Conference, American Society of Mechanical Engineers*, pp. V003T41A005–V003T41A005 (2017)
15. D.M. Toma, M. Carbonell Ventura, D. Pujol Bresco, A. Manuel Lázaro, J. Miquel Masalles, An impacting energy harvester through piezoelectric device for oscillating water flow, in *MARTECH: 5th International Workshop on Marine Technology: 19th–20th of November*, Vilanova i la Geltrú, pp. 39–42 (2013)
16. N. Okada, S. Yabe, H. Fujimoto, M. Murai, Experiments on floating wave-power generation using piezoelectric elements and pendulums in the water tank, in *2012 Oceans-Yeosu, IEEE*, pp. 1–8 (2012)
17. C. Viñolo, D. Toma, A. Mànuel, J. del Rio, An ocean kinetic energy converter for low-power applications using piezoelectric disk elements. Eur. Phys. J. Spec. Top. **222**(7), 1685–1698 (2013)
18. A.M. Wickenheiser, Broadband and low frequency vibration-based energy harvesting improvement through magnetically induced frequency up-conversion, in *ASME 2010 Conference on Smart Materials, Adaptive Structures and Intelligent Systems, American Society of Mechanical Engineers*, pp. 611–618 (2010)
19. F. Cottone, H. Vocca, L. Gammaitoni, Nonlinear energy harvesting. Phys. Rev. Lett. **102**(8), 080601 (2009)
20. S. Zhou, J. Cao, D.J. Inman, J. Lin, S. Liu, Z. Wang, Broadband tristable energy harvester: modelling and experiment verification. Appl. Energy **133**, 33–39 (2014)
21. S. Zhou, D.J. Inman, J. Cao, A linear-element coupled nonlinear energy harvesting system, in *ASME 2015 Conference on Smart Materials, Adaptive Structures and Intelligent Systems, American Society of Mechanical Engineers*, pp. V002T07A007–V002T07A007 (2015)
22. S. Zhou, J. Cao, D.J. Inman, S. Liu, W. Wang, J. Lin, Impact-induced high-energy orbits of nonlinear energy harvesters. Appl. Phys. Lett. **106**(9), 093901 (2015)
23. J. Cao, A. Syta, G. Litak, S. Zhou, D.J. Inman, Y. Chen, Regular and chaotic vibration in a piezoelectric energy harvester with fractional damping. Eur. Phys. J. Plus **130**(6), 103 (2015)

24. S.A. Emam, J. Hobeck, D.J. Inman, Experimental study of nonlinear vibration energy harvesting of a bistable composite laminate, in *Proceedings of the ASME 2017 Conference on Smart Materials, Adaptive Structures and Intelligent Systems* (2017)
25. W. Wang, J. Cao, C.R. Bowen, D.J. Inman, J. Lin, Performance enhancement of nonlinear asymmetric bistable energy harvesting from harmonic, random and human motion excitations. Appl. Phys. Lett. **112**(21), 213903 (2018)
26. D.N. Betts, C.R. Bowen, H.A. Kim, N. Gathercole, C.T. Clarke, D.J. Inman, Nonlinear dynamics of a bistable piezoelectric-composite energy harvester for broadband application. Eur. Phys. J. Spec. Top. **222**(7), 1553–1562 (2013)
27. T.S. Mehdi, An analytical study on piezoelectric-bistable laminates with arbitrary shapes for energy harvesting, in *7th ECCOMAS Thematic Conference on Smart Structures and Materials* (2015)
28. A.J. Lee, D.J. Inman, A multifunctional bistable laminate: Snap-through morphing enabled by broadband energy harvesting. J. Intell. Mater. Syst. Struct. (2018). https://doi.org/10.1177/1045389X18770895
29. D. Gardini, M. Deluca, M. Nagliati, C. Galassi, Flow properties of PLZTN aqueous suspensions for tape casting. Ceram. Int. **36**(5), 1687–1696 (2010)

Hybrid DC–DC Converters with Topology Morphing Control and Post-fault Operation Capability

Dmitri Vinnikov, Andrii Chub, Oleksandr Korkh, Andrei Blinov, and Elizaveta Liivik

Abstract This paper discusses an alternative realization approach of the reconfigurable full-bridge/voltage doubler rectifier for implementation in hybrid DC–DC converters, i.e., the power electronic converters that can adaptively change the power circuit topology for optimization of performance or enabling the post-fault operation. The proposed rectifier is characterized by the reduced number of semiconductor components and the ability of independent use of its capacitors for forming the series resonant tank and output voltage filtering. The paper explains the derivation and operation principle of the semi-active reconfigurable rectifier and discusses the experimental results obtained by the help of a 350 W laboratory prototype. Finally, the fully controlled version of the proposed rectifier is presented, which features superior control and reconfiguration possibilities in both directions of the power flow and, therefore, could be used as a versatile power electronic building block for hybrid DC–DC converters.

1 Introduction

In the last decade, the topic of hybrid DC–DC converters has attracted pronounced interest of industry and academia. It can open new opportunities for power electronic systems such as the extension of a DC voltage gain range, flattening of power conversion efficiency over a wide input voltage and load regulation range, and improvement of reliability without any significant modifications in the hardware [1–9].

The core idea of hybrid DC–DC converters lies mainly in the inherent hardware redundancy of the full-bridge topology and possible reconfiguration into a half-bridge [1] or even a flyback [2] converter. Figure 1 shows the most popular approach to the hybrid DC–DC converter, which is based on the full-bridge inverter, an

D. Vinnikov (✉) · A. Chub (✉) · O. Korkh · A. Blinov · E. Liivik (✉)
Power Electronics Group, Department of Electrical Power Engineering and Mechatronics, Tallinn University of Technology, Tallinn, Estonia
e-mail: dmitri.vinnikov@taltech.ee; andrii.chub@taltech.ee; liisa.liivik@taltech.ee

W. Zamboni, G. Petrone (eds.), *ELECTRIMACS 2019*, Lecture Notes in Electrical Engineering 615, https://doi.org/10.1007/978-3-030-37161-6_33

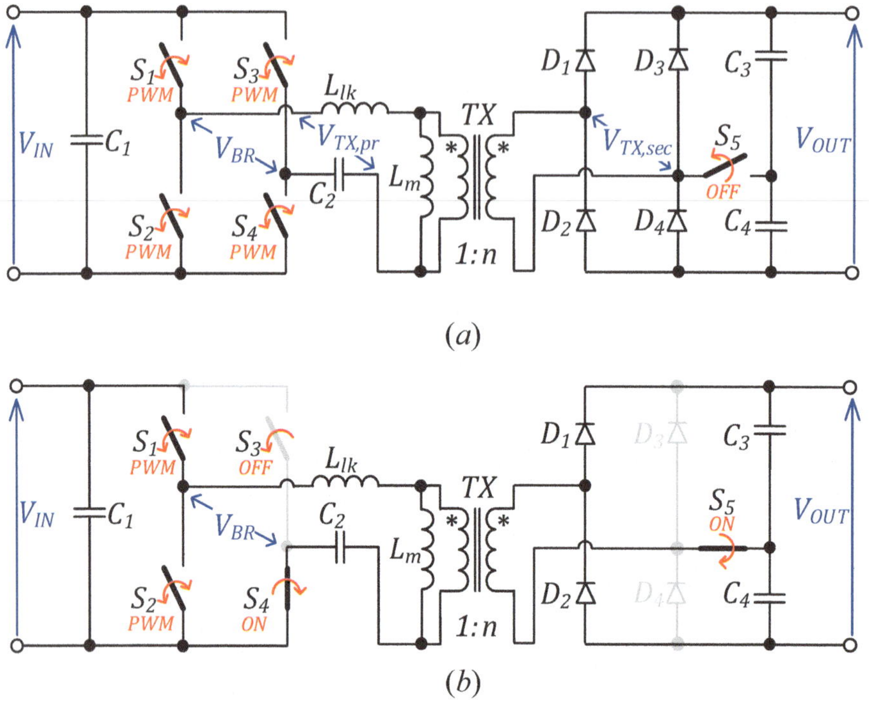

Fig. 1 Hybrid DC–DC converter and its possible topological reconfigurations: full-bridge front-end inverter and full-bridge rectifier (**a**) and asymmetric half-bridge inverter and symmetric voltage doubler rectifier (**b**)

isolation transformer, and a reconfigurable rectifier. Depending on the operation conditions, the front-end inverter could be reconfigured from a full-bridge (Fig. 1a) into a half-bridge topology, typically done "on the fly" by modifying the modulation sequence of the inverter switches [3].

One of four reconfiguration possibilities of the full-bridge inverter into the asymmetric half-fridge is illustrated in Fig. 1b. The upper switch S_3 of the second leg of the inverter is continuously turned OFF, and the opposite switch in the same leg (S_4) is continuously turned ON, thus connecting the primary winding of the transformer to the DC rail of the inverter bridge. The top and bottom switches S_1 and S_2 of the first leg are controlled alternately with short dead time. As the asymmetric half-bridge inverter generates the unipolar voltage pulses, the DC-blocking capacitor C_2 is connected in series with the primary winding of the isolation transformer, thus preventing the transformer core from saturation. In the majority of cases, this capacitor is additionally used for forming the series resonant tank with the leakage inductance of the isolation transformer L_{lk}, thus offering an opportunity of resonant switching implementation without adding extra components and increasing complexity of the converter [4].

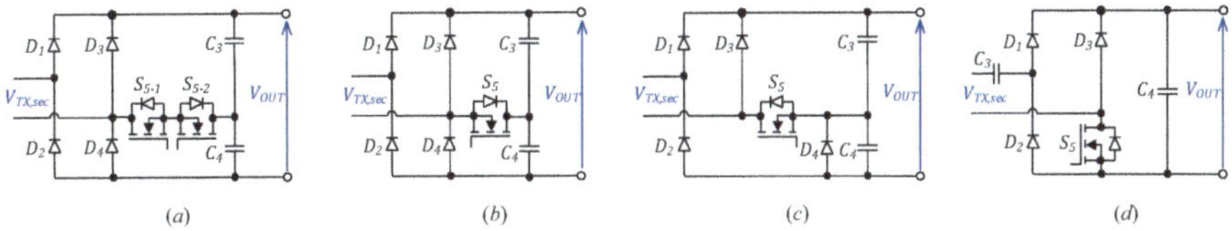

Fig. 2 Reconfigurable full-bridge/voltage doubler rectifiers: symmetric structure with reverse current blocking dual MOSFETs [5] (**a**), simplified version with single MOSFET and PWM + PFM hybrid control of front-end inverter [6] (**b**), alternative single-switch version with position changing of diode D_4 [7] (**c**), and proposed further simplification possibility, where the switch S_5 is placed instead of diode D_4 and the capacitor C_3 is moved from the output to the input side of the rectifier (**d**)

The output side of the hybrid DC–DC converter is based on the multimode rectifier, which can be reconfigured from the full-bridge to the symmetric voltage doubler topology and back by closing or opening the mode-changing switch S_5 (Fig. 1). The switch S_5 operates in the ON-OFF mode as a static switch and could be realized by the use of a mechanical relay or with two back-to-back MOSFETs in a common source configuration, as shown in Fig. 2a [5]. Using a special modulation technique that combines the pulse width modulation (PWM) and the pulse frequency modulation (PFM), the number of switches could be reduced to one (Fig. 2b) [6]. Alternatively, the single-switch version could be obtained by changing the position of the diode D_4, as shown in Fig. 2c [7].

This paper discusses further development of the hybrid DC–DC converters by the utilization of the simplest structure of the full-bridge/voltage doubler reconfigurable rectifier (Fig. 2d). In this approach, the mode-changing switch S_5 is placed instead of diode D_4, and the capacitor C_3 is moved from the output to the input side of the rectifier. The resulting advantages are in the minimal number of semiconductor components and the possibility of independent use of the capacitor C_3 as the resonant capacitor, which allows keeping C_4 large enough for smoothing the output voltage. The fully controlled version of the proposed hybrid rectifier features excellent control and reconfiguration possibilities in both directions of the power flow and, therefore, could be used as a power electronic building block (PEBB) for hybrid DC–DC converters.

2 Generalized Operation Principle of the Semi-active Hybrid Rectifier

Depending on the operation conditions, the discussed rectifier can operate in either a full-bridge or an asymmetric voltage doubler rectifier (VDR) mode, thus featuring different voltage gains and stresses of semiconductor components. The mode of the rectifier is changed through the shorting of the switch S_5.

2.1 Full-Bridge Rectifier (FBR) Mode

In the FBR configuration, the switch S_5 operates in the switching mode. The equivalent circuit of the reconfigurable rectifier is shown in Fig. 2d. Idealized steady-state waveforms in the FBR mode are presented in Fig. 3. It is assumed that the diagonal transistors of the front-end inverter bridge are switched alternately in pairs with the duty cycle close to 0.5, and the isolation transformer is supplied with the symmetric balanced pulsed voltage. For simplicity of the analysis, it was assumed that the series resonant tank is formed by the DC-blocking capacitor C_2 and the leakage inductance L_{lk}, and the converter is switched at the resonant frequency (as shown in Fig. 1). As a result, almost pure sinusoidal currents through the isolation transformer and all semiconductors were obtained. The dead time (t_1–t_2 and t_3–t_4 in Fig. 3) should be carefully dimensioned to ensure the zero

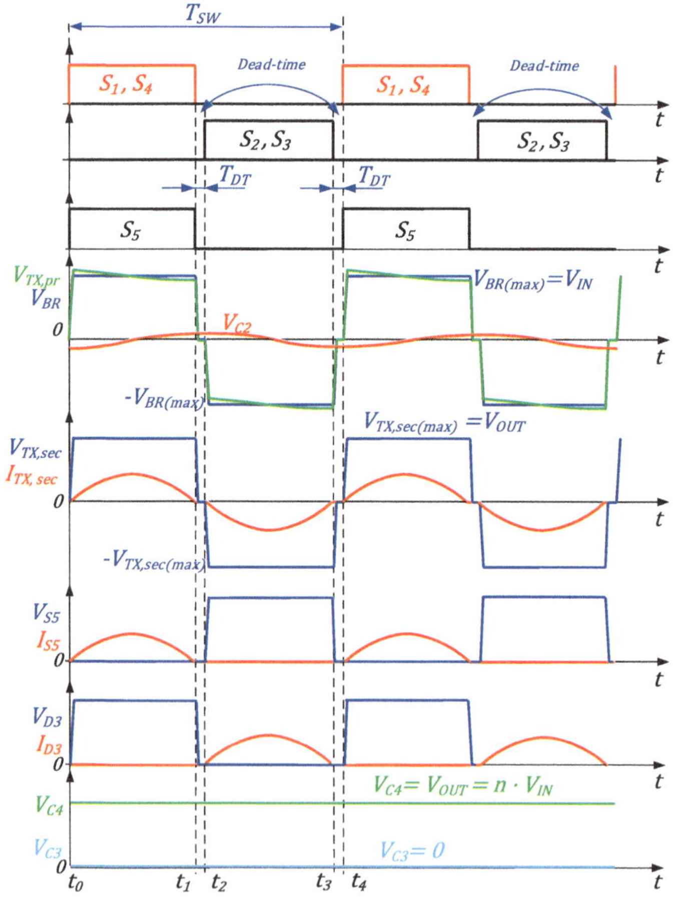

Fig. 3 Idealized steady-state waveforms of the hybrid DC–DC converter in the FBR mode

voltage switching (ZVS) and to avoid the undesired cross-conduction of the top and bottom inverter switches in any operating point of the converter. It is also assumed that capacitors C_3 and C_4 are selected large enough to avoid the influence on the resonance and ensure the ripple-free output voltage. As the converter in FBR mode features fully symmetric operation, the average voltages of the capacitors C_2 and C_3 remain zero.

As shown in Fig. 3, the switch S_5 is synchronized with the diagonal switches S_1 and S_4 of the front-end inverter and conducts current during the positive half-cycle of the switching period. In previous studies of a similar topology, the switch S_5 in the FBR mode was kept continuously open and its body diode operated as part of an uncontrolled rectifier bridge, which required the use of costly SiC MOSFET with co-packed SiC Schottky barrier diode (SBD) [8] to balance performance in both modes.

2.2 *Voltage Doubler Rectifier (VDR) Mode*

In the VDR mode, the switch S_5 is permanently turned ON, thus connecting the secondary winding of the transformer to the negative output terminal of the converter. The upper diode D_3 in the same leg of the rectifier is reverse biased by V_{OUT} permanently, and the equivalent circuit of the rectifier changes to that shown in Fig. 4. As a result, the gain of the rectifier is increased twofold.

The idealized steady-state waveforms of the hybrid DC–DC converter with a rectifier reconfigured to the VDR mode are presented in Fig. 5. It is assumed that the front-end inverter is reconfigured to the asymmetric half-bridge according to Fig. 1b, i.e., S_1 and S_2 are being switched alternately with the duty cycle close to 0.5, S_3 is permanently turned OFF, and S_4 is continuously conducting. As compared to the FBR mode, the output voltage of the inverter bridge V_{BR} becomes unipolar, and the capacitor C_2 starts acting as the DC-blocking capacitor, resulting in the average offset V_{OFS} of the capacitor C_2 voltage, which is equal to half of the input voltage (Fig. 5). Hence, the capacitance C_2 cannot be realized with multilevel ceramic capacitors due to inevitable drift of the resonant frequency with the change of the mode. Hence, foil capacitors are recommended.

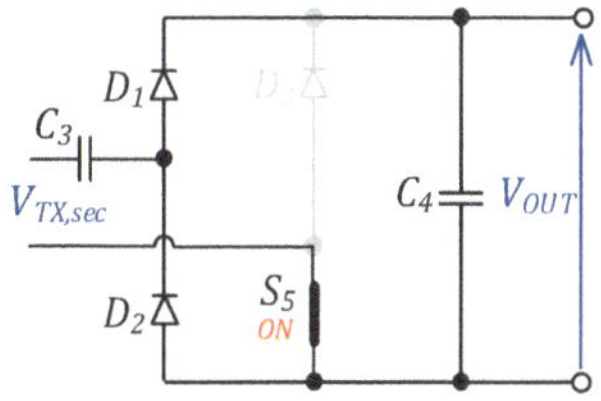

Fig. 4 Equivalent circuit of the reconfigurable rectifier in the VDR mode when the switch S_5 is permanently turned ON

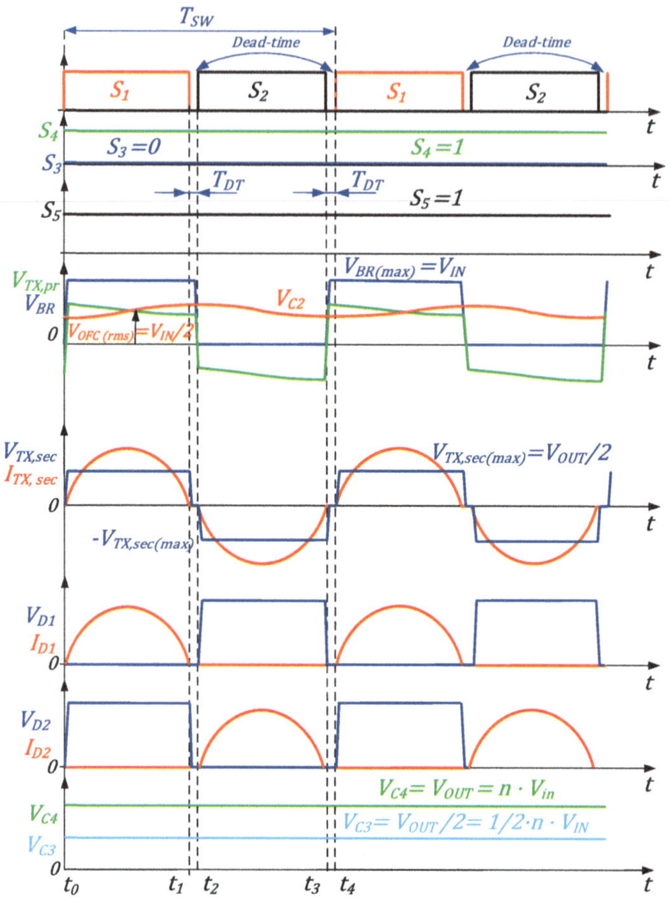

Fig. 5 Idealized steady-state waveforms of the hybrid DC–DC converter in the VDR mode

3 Experimental Results

To evaluate the performance of the hybrid DC–DC converter with the proposed reconfigurable rectifier, a 350 W laboratory prototype was assembled and tested. The types and specifications of the components used in the prototype are listed in Table 1. In all the tests, the converter was operating with the duty cycle close to 0.5 after the dead-time deduction. The series resonant tank was formed using the DC-blocking capacitor C_2 and the leakage inductance L_{lk} of the isolation transformer (Fig. 1). The front-end inverter was switched at the resonant frequency of 110 kHz, and depending on the test, it was reconfigured from the full-bridge to the half-bridge in the way shown in Fig. 1b. To keep the DC voltage gain unchanged and ensure the

Table 1 Specifications of the experimental prototype

Component	Value/type
Switches $S_1 \ldots S_4$	Infineon BSC035N10NS5100 V, 100 A, 3.5 mΩ
Diodes $D_1 \ldots D_3$	Wolfspeed C3D02060E600 V, 4 A
Switch S_5	Wolfspeed C3M0120090900 V, 23 A, 120 mΩ
Input capacitor C_1	Panasonic EEV-FK2A151M150 μF, 100 VDC
Capacitors C_2 and C_3	Panasonic ECW-FG2J305Q13 μF, 630 VDC
Output capacitor C_4	EPCOS/TDK B43501150 μF, 400 VDC
TX turns ratio n	6
TX leakage inductance L_{lk}	24 μH
TX magn. Inductance L_{m}	1 mH
Dead time t_{DT}	200 ns

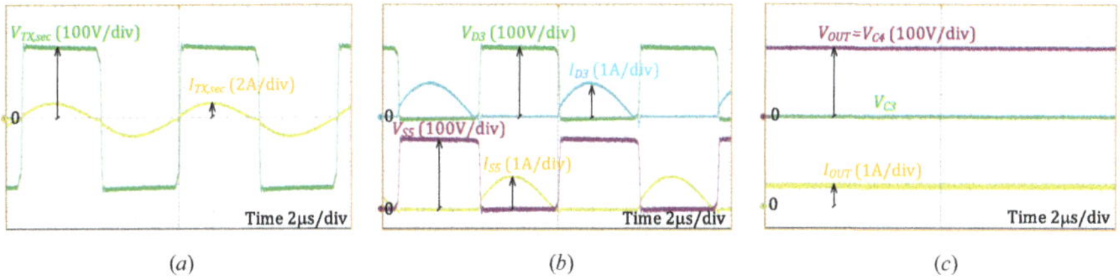

Fig. 6 Experimental steady-state waveforms in the FBR mode at $V_{\mathrm{IN}} = 50$ V and $P = 300$ W: secondary winding voltage and current of the isolation transformer (**a**), operating voltages and currents of the diode D_3 and switch S_5 (**b**), and operating voltages of capacitors C_3 and C_4 and output current of the converter (**c**)

same level of the output voltage, the rectifier was reconfigured from the full-bridge to the VDR mode, correspondingly. During the tests, the input voltage was fixed at 50 V, and the output power was gradually increased from 10% to 100%.

3.1 Steady-State Waveforms

Experimental steady-state waveforms of the secondary side of the hybrid DC–DC converter are presented in Figs. 6 and 7. For the waveform acquisition, the digital oscilloscope Tektronix DPO7254 equipped with the current probes Tektronix TCP0030A and high-voltage differential voltage probes Tektronix P5205A was used. The converter was supplied by the programmable power supply EA-PSI 9080-60 and loaded by the DC electronic load Chroma 63204.

First, the converter was tested in the full-bridge mode at $V_{\mathrm{IN}} = 50$ V and $P = 300$ W (Fig. 6). As the switching frequency is equal to the resonant frequency, the current through the isolation transformer has a pure sinusoidal waveform. As a result, the rectifier diodes and switch S_5 are operating under ZCS (Fig. 6b). As can be seen from Fig. 6c, the voltage of the capacitor C_3 is 0, and the voltage of

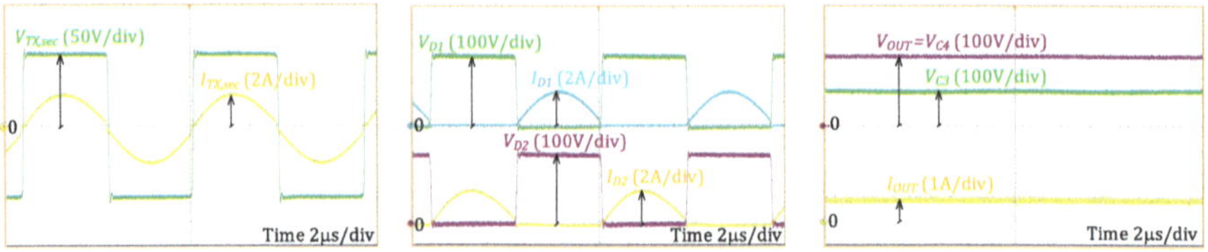

(*a*) (*b*) (*c*)

Fig. 7 Experimental steady-state waveforms in the VDR mode at $V_{\text{IN}} = 50$ V and $P = 300$ W: secondary winding voltage and current of the isolation transformer (**a**), operating voltages and currents of the diodes D_1 and D_2 (**b**), and operating voltages of capacitors C_3 and C_4 and output current of the converter (**c**)

the capacitor C_4 equals the output voltage. After the reconfiguration of the front-end inverter to the asymmetric half-fridge, the rectifier was also reconfigured into the VDR for keeping the DC voltage gain of the converter unchanged. Accordingly, the output voltage remained at the level of 300 V, similar to the full-bridge mode (Fig. 7c). However, due to the reconfiguration of the front-end side, the current through the isolation transformer and the semiconductors was doubled for the same amount of processing power as in the full-bridge mode (Fig. 7a). The rectifier diodes are operating under ZCS (Fig. 7b), and the voltages of C_4 and C_3 equal the output voltage and half of the output voltage, correspondingly.

3.2 Power Conversion Efficiency

Figure 8 shows the power conversion efficiency of the experimental hybrid DC–DC converter with the proposed reconfigurable rectifier, which was measured by the precision power analyzer Yokogawa WT1800. The efficiency was measured for both the full-bridge and half-bridge modes in the power range from 35 to 350 W. Although SiC MOSFET C3M0120090 from Wolfspeed was used as the baseline solution for switch S_5, Si MOSFET STF42N60M2 from ST and SiC MOSFET ROHM SCH2080KE with co-packed SiC SBD were also evaluated as alternative options (Table 2).

It is seen from Fig. 8a that in the FBR mode and power range from the half to full power, the compared MOSFETs feature similar performance with the resulting peak efficiency close to 97.5%. Due to the better switching dynamics, the use of SiC MOSFETs is more beneficial at power levels below 150 W, with a resulting efficiency rise of up to 2% against the Si counterpart.

In the VDR mode, the switch S_5 operates in a static mode; therefore, only the conduction losses defined by the drain-source on-state resistance $R_{\text{DS(on)}}$ of the switch directly affect the power conversion efficiency. It is seen from Fig. 8b that

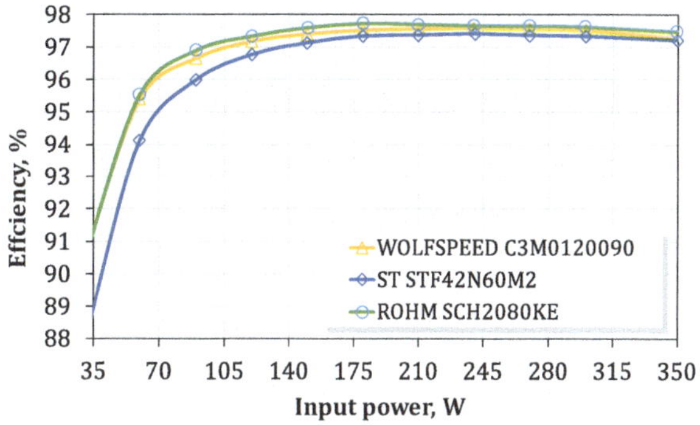

(a)

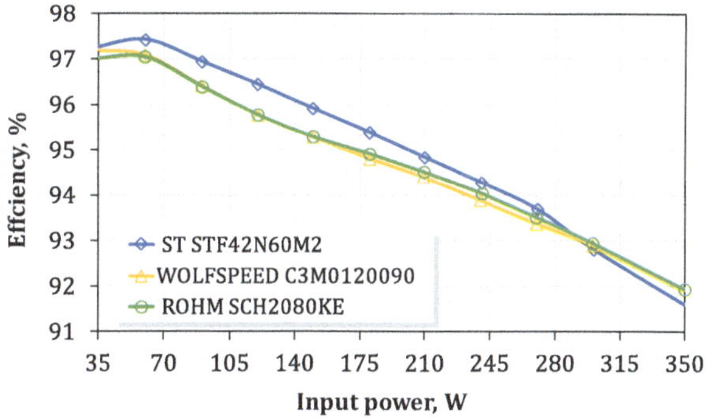

(b)

Fig. 8 Comparison of power conversion efficiency in full-bridge (**a**) and half-bridge (**b**) mode measured at $V_{IN} = 50$ V with different types of MOSFETs used for S_5

Table 2 Datasheet specifications of MOSFETs used for S_5

	Wolfspeed C3M0120090	ROHM SCH2080KE	ST STF42N60M2
V_{DS}, V	900	1200	650
$R_{DS(on)}$, mΩ	120	100	76
I_D, A	23	40	34
V_{SD}, V	4.8	1.3	1.6
t_{rr}, ns	24	37	438
Package	TO-247	TO-247	TO-220

the rectifier with Si MOSFET in the load range from 10 to 80% demonstrates up to 0.8% higher efficiency with the peak value of 97.5%. At higher power levels, the SiC MOSFET-based rectifiers demonstrated better efficiency, which is mostly caused by the better thermal management of the TO-247 casing of the given SiC transistor. The bottom line of this study was to demonstrate that the proposed concept of the reconfigurable rectifier could be realized with the generic components, as the Si and SiC MOSFETs have demonstrated rather similar performance.

3.3 Efficiency Optimization by Topology Morphing Control (TMC)

The given converter suffers from greater stresses in semiconductors when operating in the VDR mode and, consequently, features considerably lower efficiency at full power. Therefore, the VDR mode should be used at the full power only at a semiconductor fault in the input side. However, the converter shows high efficiency at light load in the VDR mode due to dominance of switching losses in semiconductors. Hence, fewer switching semiconductors result in efficiency improvement. To harness this advantage, the topology morphing control (TMC) with the transition between FBR and VDR modes dependent on the operating power was implemented. It enables flat efficiency curve across a wide input power range with efficiency swing reduction from 6% down to nearly 1%, as shown in Fig. 9.

3.4 Post-fault Reconfiguration Test

In this study, fault detection was not considered. Therefore, a short circuit in the switch S_1 was emulated by keeping it turned ON along with the switch S_5, as shown in Fig. 10. The voltage V_{BR} changes from bipolar to unipolar, which results in recharging of series capacitors C_2 and C_3. Mode transition transients last for roughly 320 μs, which means that the converter is capable of withstanding fault and performing transition to the post-fault stable operation, taking only 35 switching periods. However, the considered conditions are ideal, while a realistic scenario will include some delay between the fault and its detection by a control system, which will deteriorate the dynamics of the transients.

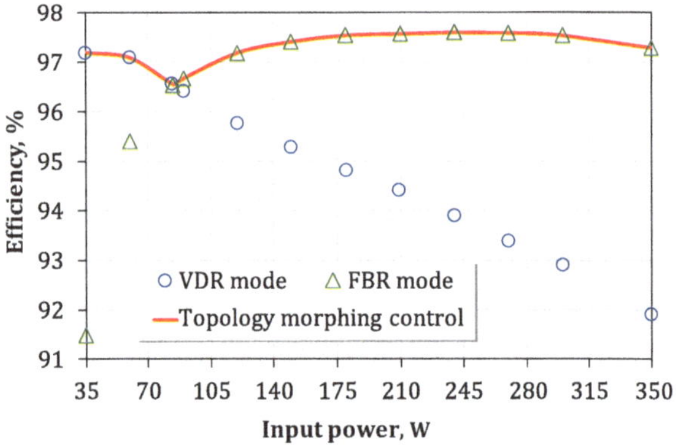

Fig. 9 Light power efficiency optimization of the hybrid DC–DC converter by topology morphing control (TMC)

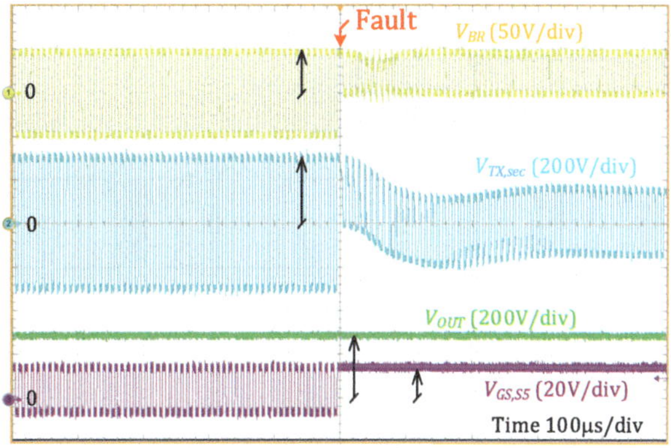

Fig. 10 Voltage waveforms before and after the post-fault reconfiguration of the experimental hybrid DC–DC converter

4 Active Reconfigurable Rectifier as a Power Electronic Building Block for Bidirectional Hybrid DC–DC Converters

The proposed concept of a semi-active reconfigurable rectifier could be further extended to a fully active version by replacing the diodes with controlled switches (Fig. 11). In this case, the reconfigurable front-end inverter and active rectifier will have an entirely identical structure that could be envisioned as a versatile power electronic building block (PEBB) for hybrid DC–DC converters. This approach will allow for unique reconfiguration flexibility and ensure the continuity of operation at the failure of any switch of the rectifier. For example, if one of the top or bottom switches is short- or open-circuited, the post-fault control sequence will be applied, thus forcing the opposite switch in the same leg to remain open or closed, correspondingly. By that, the rectifier will change its topology to the VDR, and the healthy leg will continue synchronous rectification. To maintain the same DC voltage gain, the front-end inverter will be reconfigured to the half-bridge, as described in the previous sections of the paper.

By using two reconfigurable PEBBs, a novel topology of the bidirectional hybrid DC–DC converter could be derived, which will support the bidirectional power flow control and full fault tolerance in both directions. As compared to similar concepts [9], the proposed converter (Fig. 11) features significantly reduced component count and hybridization of functions of internal capacitors C_2 and C_3, which, depending on the power flow direction and operation mode, can be used as resonant capacitors, DC-blocking capacitors, or VDR capacitors. Figure 12 shows the experimental efficiency of the proposed converter, including the light load efficiency optimization by TMC. As can be seen, the converter features almost identical performance in both directions of the power flow with the peak efficiency close to 98%.

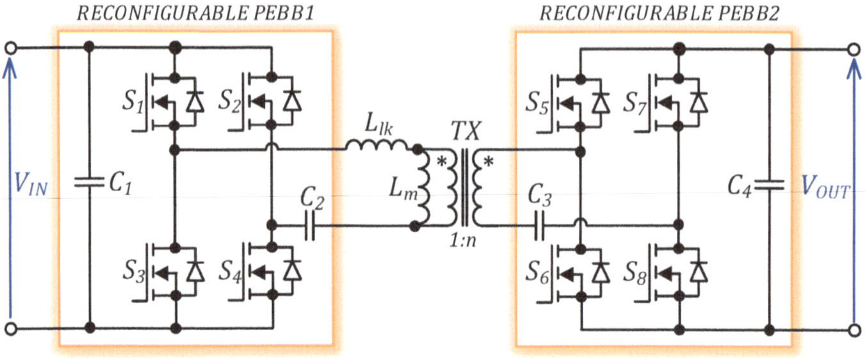

Fig. 11 Proposed concept of the fully fault-tolerant bidirectional DC–DC converter based on the reconfigurable PEBB

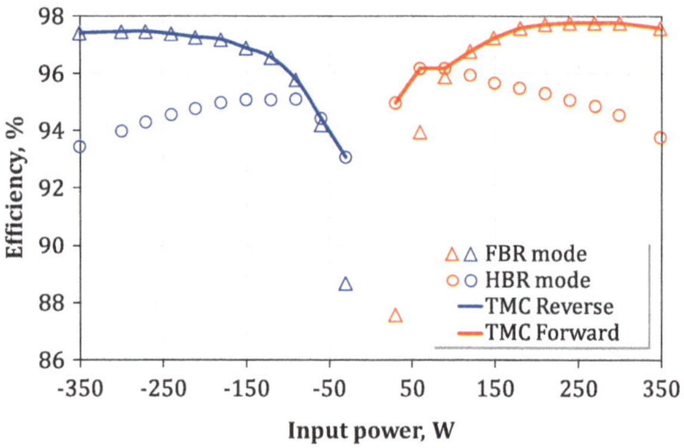

Fig. 12 Experimental efficiency and light power efficiency optimization of the proposed bidirectional hybrid DC–DC converter by using the TMC. MOSFET Infineon BSC035N10NS5 and Wolfspeed C3M0120090 were used for the low- and high-voltage bridges, correspondingly

5 Conclusions and Future Work

Application of a semi-active rectifier to the full-bridge voltage-fed converters enables a number of features beyond the reach of conventional counterparts. It can be reconfigured from the full-bridge rectifier into the voltage doubler rectifier. This results in the possibility to maintain the converter DC voltage gain in case the input side is reconfigured into half-bridge intentionally or due to a fault of an input-side semiconductor component. In the first case, the converter can utilize the topology morphing control to achieve flat efficiency across the wide input power range. In the latter case, the converter is capable of restoring its operation within short time needed to recharge the resonant capacitance. Experimental results show that the given converter achieves up to 6% efficiency improvement at light load in normal operating conditions, while its efficiency drops by no more than 5% at the full power due to increased current stress in remaining active semiconductor components after a fault.

The obtained results suggest that bidirectional series resonant dual active bridge converter with full tolerance to any possible fault of a semiconductor component can be implemented keeping all the advantages of the semi-active converter, which was verified experimentally along with the peak efficiency close to 98%. This derived converter and its wide input voltage range operation will be the focus of the future research.

Acknowledgments This research was supported by the Estonian Centre of Excellence in Zero Energy and Resource Efficient Smart Buildings and Districts (ZEBE), grant 2014-2020.4.01.15-0016, funded by the European Regional Development Fund, and by the Estonian Research Council grant PSG206.

References

1. Z. Liang, R. Guo, G. Wang, A. Huang, *A New Wide Input Range High Efficiency Photovoltaic Inverter* (2010 IEEE Energy Conversion Congress and Exposition, Atlanta, GA, 2010), pp. 2937–2943
2. S. Poshtkouhi, O. Trescases, Flyback mode for improved low-power efficiency in the dual-active-bridge converter for bidirectional PV microinverters with integrated storage. IEEE T. Ind. Appl. **51**(4), 3316–3324 (2015)
3. L.F. Costa, G. Buticchi, M. Liserre, A family of series-resonant DC–DC converter with fault-tolerance capability. IEEE T. Ind. Appl. **54**(1), 335–344 (2018)
4. A. Chub, D. Vinnikov, R. Kosenko, E. Liivik, Wide input voltage range photovoltaic microconverter with reconfigurable buck–boost switching stage. IEEE T. Ind. Electron **64**(7), 5974–5983 (2017)
5. Y. Shen, H. Wang, Z. Shen, Y. Yang, F. Blaabjerg, A 1-MHz series resonant DC-DC converter with a dual-mode rectifier for PV microinverters. IEEE T. Power Electron **34**(7), 6544–6564
6. H. Wang, Z. Li, A PWM LLC type resonant converter adapted to wide output range in PEV charging applications. IEEE T. Power Electron **33**(5), 3791–3801 (2018)
7. Y. Chen, H. Wang, Y. Liu, Improved hybrid rectifier for 1-MHz LLC-based universal AC-DC adapter, in 2017 IEEE Applied Power Electronics Conference and Exposition (APEC), Tampa, FL, 2017, pp. 23–30
8. D. Vinnikov, A. Chub, E. Liivik, F. Blaabjerg, Y. Siwakoti, Boost half-bridge DC-DC converter with reconfigurable rectifier for ultra-wide input voltage range applications, in 2018 IEEE Applied Power Electronics Conference and Exposition (APEC), San Antonio, TX, 2018, pp. 1528–1532
9. L. Costa, G. Buticchi, M. Liserre, *Bidirectional Series-Resonant DC-DC Converter with Fault-Tolerance Capability for Smart Transformer, 2016 IEEE Energy Conversion Congress and Exposition* (ECCE, Milwaukee, WI, 2016), p. 1

Generator Topologies for Horizontal Axis Tidal Turbine

Mohammad Rafiei, Francesco Salvatore, and Fabio Giulii Capponi

Abstract Over the last decade, research on technologies to exploit tidal current kinetic energy for renewable electricity generation has had a significant growth. However, as to date, there is not a consensus worldwide on standard Power Take-Off (PTO) systems, due to the current immaturity of tidal energy converter technologies. In most cases, mechanical/electrical power conversion follows well-proven technologies derived by the mature wind-energy sector. However, the peculiarities of tidal energy resource impose ad hoc technology solutions. In this paper, different generator topologies and recent developments for marine tidal energy systems are reviewed and compared. The aim is to provide an overall perspective and identify areas for further development. Among considered technologies, the direct-drive permanent magnet synchronous generator by the full-rated frequency converter (FFC) represents an appealing solution, for reduced system complexity and maintenance requirements and possibility to develop smart Maximum Power Point Tracking (MPPT) strategies.

1 Introduction

The constant growth of world population and economy, with the highest rates in developing countries, drives a continuous increase in global energy demand as it has never been experienced in the past. Most of this generalized growth is still based on the consumption of fossil fuels, although it is nowadays fully understood that this is no longer sustainable. The environmental impact of greenhouse gas emissions in terms of climate change and global warming is apparent, and time left

M. Rafiei · F. Giulii Capponi (✉)
Department of Astronautical, Electrical and Energy Engineering, University of Rome 'La Sapienza', Roma, Italy
e-mail: mohammad.rafiei@uniroma1.it; fabio.giuliicapponi@uniroma1.it

F. Salvatore
CNR-INM, Roma, Italy
e-mail: francesco.salvatore@cnr.it

W. Zamboni, G. Petrone (eds.), *ELECTRIMACS 2019*, Lecture Notes in Electrical Engineering 615, https://doi.org/10.1007/978-3-030-37161-6_34

to invert the established trend is quickly reducing. The conversion to a new system based on the exploitation of renewable energy sources like hydropower, solar, wind and biomasses is then becoming an imperative rather than an option [1]. In order to further increase the impact of renewables, it is essential to complement well-established sources with new technologies capable to diversify the energy mix and to enhance clean energy security [2].

In this scenario, ocean energy is attracting an increasing interest, due to the amount of energy potential stored in different forms, such as waves, currents, thermal and salinity gradients, which could meet more than the total electricity demand globally. While research efforts are focused on developing efficient approaches to harvest ocean energy at competitive costs, technology challenges are being faced in order to demonstrate systems with the adequate level of reliability and robustness in the harsh marine environment over a full life cycle [3, 4].

Among alternative ocean energy types, systems to convert the kinetic energy of marine currents present the highest technology readiness level (TRL). The terminology marine currents includes both tidal currents, generated by gravitational forces exchanged among marine water masses, moon and sun, and ocean currents, driven by thermal and salinity gradients of water masses in oceans at a different latitude. In both cases, the kinetic energy in the currents is captured by turbines, while a Power Take-Off (PTO) system converts mechanical power at the turbine shaft into electricity.

While the cost of energy from submerged turbines is still very high as compared to well-established green energy technologies, tidal energy is already considered as a promising short-term solution to contribute to the renewable energy mix in coastal areas and remote islands [5].

The present work's aim is to present an overview of state-of-the-art technologies with a specific focus on the components in converting mechanical energy into electrical energy.

Considering various types of existing PTO system and also new concepts at the development stage, the problem of design and selection of cost-effective architectures for tidal turbines is an interesting issue, with a number of multidisciplinary implications [6]. The generator, being the main part that converts mechanical power into electrical energy, plays a vital role in choosing the PTO system topology. Hence, in this paper, the tidal current conversion systems and different generator topologies which are suitable for horizontal axis tidal turbine application are reviewed.

In the following pages, distinguishing aspects characterizing tidal energy in terms of resource and conversion technologies are reviewed. Landmark projects are described and PTO implementations analysed and compared. Results of this review provide the basis for the development of mechanical/electrical conversion strategies as the subject of ongoing research in the Underpinning Laboratory for Studies on Sea Energy Systems (ULYSSES) Project by the Marine Engineering Institute of the Italian National Research Council (CNR-INM).

2 Tidal Current Source

The combined effects of gravitational forces by the moon and sun and centrifugal forces associated with the Earth rotation determine the periodic variation of the water level (tidal range) in seas and oceans with a period of 12 h and 25 min. Within this period, tidal currents associated with tidal range oscillations invert the direction of water mass motions (flood and ebb cycle).

At a given location, the relative positions between the moon and the sun cause a variation of tidal range and tidal current intensity over a period of 14 days. Specifically, the highest tidal range occurs with the so-called spring tides, when moon and sun are in phase, while the lowest tidal range (neap tides) occurs when the moon and sun are nearest to 90° to each other [7].

Regarding energy harvesting techniques, tidal currents present a close similarity with wind. In both cases, the kinetic energy of the fluid can be extracted by a suitable type of turbine rotor.

The power P (W) available from an onset flow can be written as follows:

$$P = \frac{1}{2}\rho C_{\mathrm{P}} A V^3. \tag{1}$$

where ρ is the fluid density, V is the mean velocity of flow and A is the rotor cross-sectional area. The nondimensional quantity C_P (power coefficient) denotes fraction of the power which can be extracted from the flow incoming to the rotor disc. For wind generators, C_P has a theoretical upper limit of 0.593 (the Betz limit) and typical values up to 0.45–0.5 [8]. For tidal turbines, characterized by relatively small span/chord-length ratio blades, C_P is in the range 0.35–0.45 [9, 10]. These ranges are typical for horizontal axis turbines, in which the rotor axis is ideally aligned with the direction of the onset flow. Smaller C_P values characterize cross-flow (e.g. vertical axis) turbines.

Compared to the largest existing wind turbines, both rated power and the size of tidal turbines deployed at sea are still small, since unit upscaling still needs further technology assessment. Nevertheless, key advantages of tidal energy as compared to wind energy are the extremely high predictable resource (see above) and the energy density, as the sea water is about 800 times denser than air. For the same rotor swept area, water moving at 1.5 m/s (three knots) has about the same power extraction potential as a wind blowing at 50 km/h [11].

3 Tidal Current Conversion Systems

The main tidal turbine component arrangements are shown in Fig. 1, including a turbine rotor and the conversion system from mechanical energy into electrical energy or PTO system that incorporates a mechanical drive train, a generator and

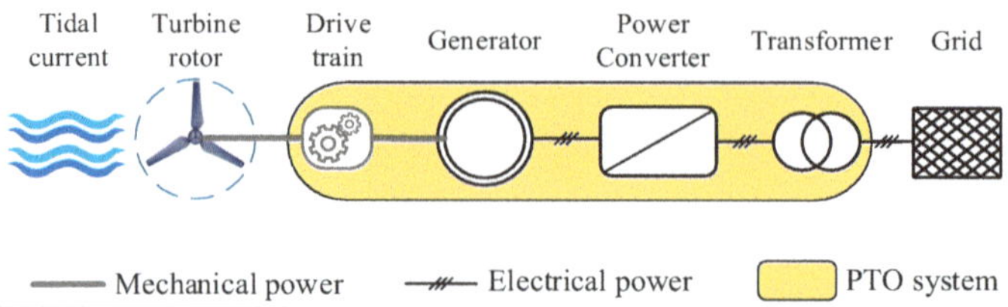

Fig. 1 Main tidal turbine component arrangements

power electronic interface (PEI), and a transformer in order to send the electrical power production to the main grid.

The turbine rotor, which captures the kinetic power from tidal current and converts it to the mechanical power, usually has three blades as a trade-off between hydrodynamic performance and manufacturing costs.

The challenge with the tidal current conversion systems is to convert a quite low and variable input, that is, the tidal flow at typically 2–3 m/s with up to 15% fluctuations, into a much faster and stable output, in order to deliver 50–60 Hz alternative current suitable for connection to the network.

Usually, converting the high-torque and low-speed mechanical power into high-frequency electrical power is done by utilization of a standard speed induction generator (IG) with a gearbox. Gearboxes are used when the rotor rotational speed is significantly different from the rotational speed required by the generator [12]. But they have some disadvantages, such as low efficiency, need of lubrication and frequent maintenance, resulting into lower reliability of the whole system. Hence, in the last decades, there have been considerable research and development activities on gear technologies, and both research institutes and industry have become increasingly attracted to the magnetic gears (MGs). In [13], the progress and development of MG technologies has been reviewed. MGs, because of their contact-less operation and low maintenance requirement, potentially have high reliability as well as efficiency; however, these points are just in theoretical point of view and still not became industrialized. The advantages that this technology could bring are expected to be significant.

An alternative way to convert mechanical power into the electrical power is to use a direct-drive synchronous generator (SG), where the turbine directly drives the generator, hence the denomination 'direct drive'. In this topology, a PEI as a full-rated frequency converter (FFC) has been used between generator and main grid; hence, the generator rotational speed is well-regulated over a wide range, including very low speeds that are typical for slow-turning marine rotors. Since the generator operates in the low speed to produce sufficient power, it requires the high torque. To achieve to these qualifications, minimize winding loss value and enhance the efficiency, the direct-drive generators with multipole large diameter are preferably designed.

Both gearbox and direct-drive topologies are being used in the energy market, but recently, a significant interest for the direct-drive generator in comparison to the geared generator has been created due to high reliability and low maintenance of drive train by avoiding gearbox and increasing the efficiency. However, large diameter, huge weight and high cost are the drawbacks for direct-drive generators [14].

The PEI is an essential part of the PTO system that is used to match the turbine characteristics (frequency, voltage, active and reactive power) by the main grid. Usually, a transformer is located inside the nacelle of turbines to avoid the transmission cable losses due to the high current flowing.

4 Review of Electrical Generator Technologies

Nowadays, there are few companies which develop and build generators for tidal energy applications. In terms of performance, tidal turbines are broadly similar to usual wind turbines; hence, some of the generator topologies which have been developed for wind turbines have been extended in tidal devices as well [15].

The first type of typical generator systems is a fixed speed with gearbox and a standard IG, which is connected directly to the main grid. The second topology, which can be considered as a limited variable speed, is a doubly fed induction generator (DFIG) with gearbox. The last one is a full variable speed that is decoupled from the main grid by a FFC, and it can be classified in two configurations: a standard generator with gearbox and a direct-drive synchronous generator (DDSG) [16].

The advantages and disadvantages of three generator topologies which are suitable for horizontal axis turbine are presented here below and compared in Table 1.

Table 1 Generator topology comparison

Generator type	Advantages	Disadvantages
DFIG	• Speed control for optimal energy use • Active and reactive power control ability • Low cost	• Limited speed range • Complex control system • Need slip rings • Need gearbox (low reliability)
SCIG	• Full-speed range • Simple structure • Pulse reduction in turbine torque	• Full-rated conversion system (higher cost) • Need slip rings • Need gearbox (low reliability)
PMSG	• Full-speed range • High efficiency and energy yield • Brushless • No need gearbox (high reliability)	• Full-rated conversion system (higher cost) • High cost of permanent magnet • Demagnetization probability of the magnet • High weight and dimension

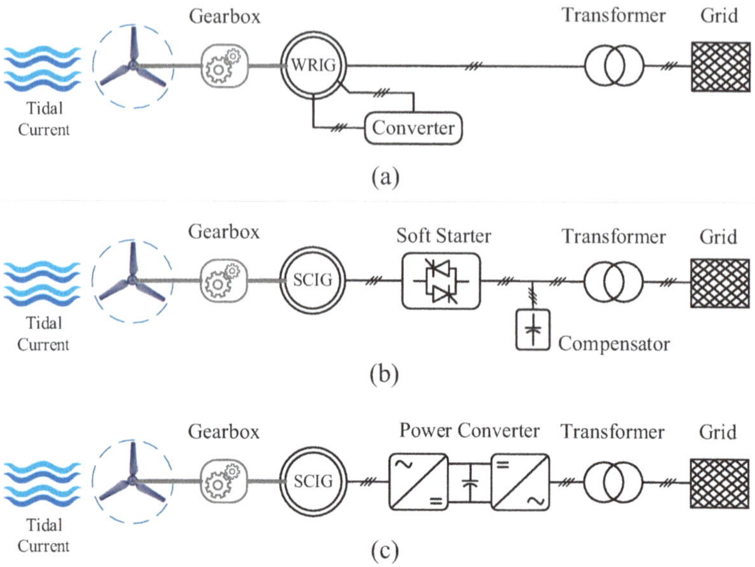

Fig. 2 Induction generator topology. (**a**) WRIG topology. (**b**) SCIG fixed speed topology. (**c**) SCIG variable speed topology

4.1 Induction Generator (IG)

In a general classification, induction generators (IG) set in two categories according to the type of rotor: the wound rotor induction generator (WRIG) and the squirrel cage induction generator (SCIG).

As shown in Fig. 2a, WRIG are equipped with an induction generator with winding rotors which gives the possibility to change the machine slip by changing the variable rotor resistance. The change in rotor resistance is accomplished by the power electronic devices installed on the rotor. In this type, depending on the rotor's resistance, the generator speed can vary up to 10%.

Figure 2b shows a SCIG which is directly connected to the main grid through a soft starter and a transformer, where a soft starter is used to limit the high starting currents of SCIG. In addition, a capacitor bank is being used to provide reactive power for IG.

In order to solve the problems of compensator and soft starter in SCIG, it has been proposed to replace these two parts by a FFC (Fig. 2c). In this configuration which is able to operate in variable speed, the generator output frequency does not need to be the same as the grid frequency. However, in comparison to the conventional SCIG, since the power is being processed via the power converter, the losses are higher, and also, it is more expensive because of FFC [17].

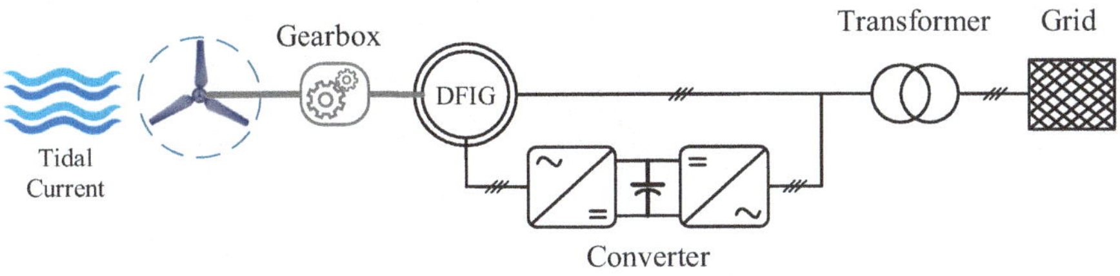

Fig. 3 DFIG topology

4.2 Doubly Fed Induction Generator (DFIG)

The DFIG is the most common type of generator used for large wind integration. Turbines which are equipped by a DFIG with multiple-stage gearbox are able to operate in a wider range of their maximum efficiency due to their variable speed. The stator is connected directly to the main grid and constructed in the same way as a SCIG, and the rotor windings, as shown in Fig. 3, is connected to the main grid via the PEI. The advantage of this configuration is to absorb flow fluctuations by varying the rotation speed of the rotor and not moving them to the grid. The control system of a DFIG turbine, in addition to providing the necessary conditions for the grid, can provide the maximum energy transmission with the best efficiency. However, the DFIG system requires components for dissipation of friction-induced heat in the gearbox and external synchronization circuit needed between the grid and stator to avoid the high starting current. Moreover, both gearbox and the brush-slip ring set bringing power to the rotor require dedicated maintenance [18, 19].

4.3 Synchronous Generator (SG)

SGs are usually used as variable speed concept, and they can be classified in two categories; the first is with an electrically excited rotor and the other with a permanent magnet rotor [20]. The grid connection scheme of these generators is shown in Fig. 4.

The electrically excited synchronous generator (EESG) is made by a rotor equipped with a winding fed with a DC excitation. The EESG gives more possibilities to control the flux compared to the Permanent Magnet Synchronous Generator (PMSG), to minimize losses in various power ranges [21]. However, the PMSG, due to enhanced energy yield, high reliability and reduced failure, has a high potential for the marine tidal conversion systems. In [18] and [22–27], the benefits of direct-drive PMSG have been discussed.

The high cost of the magnets and its demagnetization on high temperatures have always been disadvantages of PMSG, but over the last years, by improving performance and decreasing the cost of PMs and also the cost of PEI, the use of PM

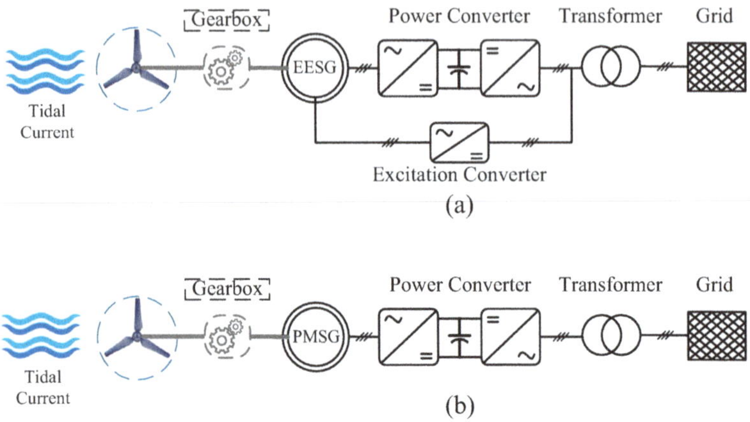

Fig. 4 Synchronous generator topology. (**a**) EESG topology. (**b**) PMSG topology

machines has become more attractive. Therefore, these trends make the direct-drive PMSG with a FFC more considerable particularly for horizontal ducted turbines [12, 28].

Variable speed concept, with a medium-speed SG and a single-stage gearbox shown by a dotted line in Fig. 4, is considered as an appealing solution with a lower number of components as compared to a DFIG with a multiple-stage gearbox and higher speed than DDSG concept. The interesting possibility in this category is using the same frame both for the generator and for the gearbox, that is, all the active parts of the generator are mounted around the secondary transmission stage, so the generator and gearbox can use the same bearings, while gearbox lubrication oil can be used to cool the generator that leads to having the lowest weight and compact size [29].

5 Marine Current Turbine Technologies

The last decade has been characterized by an impressive development of technologies for tidal energy harvesting. In this section, landmark projects with megawatt (MW)-scale horizontal axis devices deployed at sea and grid-connected are described. Main features of energy conversion mechanisms are presented and compared in Table 2 [33–35].

SeaGen-S was the first megawatt-scale project demonstrated at sea. The design consists of a twin horizontal axis two-bladed rotor. The rotors utilize an active blade pitching system and are self-starting when the tide reaches an average speed of about 1 m/s. Rated power of 2×600 kW is reached at design speed of 2.5 m/s. It was installed in 2008, and it has produced 10 GWh electricity up to 2017 (Fig. 5a) [30].

Table 2 Landmark tidal turbine technology characteristics

Device	Company	Pitch control	Gearbox	Generator	Yawing system	Rated power (kW)
SeaGen-S	Marine Current Turbines	Yes	Planetary	SCIG	Yes	2000
Alstom TGL	General Electric (GE)	Yes	Epicyclical	IG	Yes	1000
AR1500	Atlantis Resources	Yes	Epicyclical	PMSG	Yes	1500
HS1000	Andritz Hydro Hammerfest	Yes	Yes	IG	No	1000
Sabella D10	Turbine Technology	No	No	PMSG	No	1000
SIT	Schottel Hydro	No	Planetary	IG	No	54–74

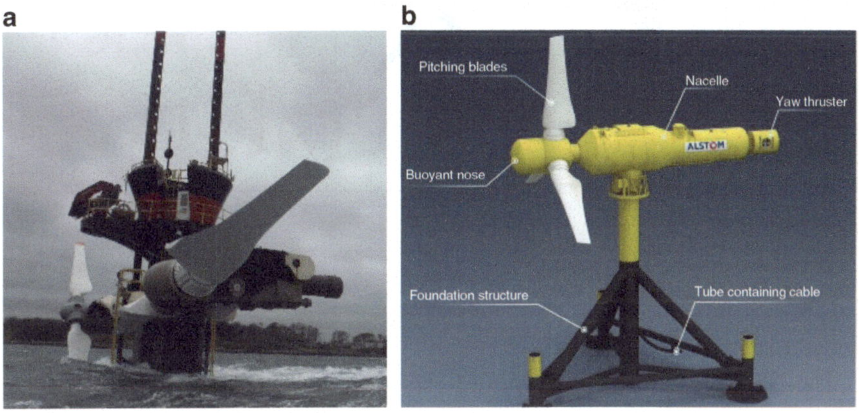

Fig. 5 (**a**) SeaGen-S [30]; (**b**) Alstom TGL 1MW [31]

Alstom (General Electric) developed the TGL 1MW, a three-bladed rotor with 18 m diameter and pitching system, an epicyclical gearbox, induction generator, frequency converter and transformer. The nacelle weighs about 150 tons and is installed on a separate seabed-mounted foundation. The nacelle features a yawing system to minimize the rotor wake effect and the structural loads. Turbine operation at sea was the subject of the ReDAPT project concluded in 2017 (Fig. 5b) [31, 32]

The MeyGen project, owned and operated by Atlantis Resources, is one of the first examples of tidal array globally. In the first phase of the project (completed), two different turbines have been deployed: one Atlantis Resources AR1500 (designed by Lockheed Martin) and three Andritz Hydro Hammerfest HS1000 (Fig. 6).

The AR1500 is an active pitch and full yaw turbine with 1.5 MW rated power at flow speed of 3 m/s. The turbine is designed for a 25-year life, with three services at 6¼ year intervals. Total mass of the unit is 150 tons in air. The generator is an

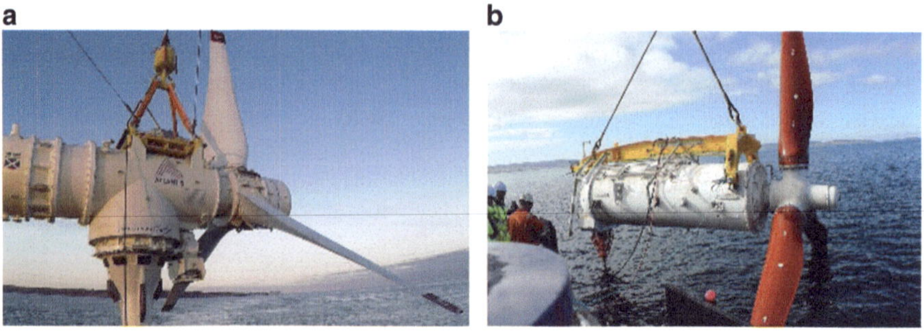

Fig. 6 (a) AR1500 [30]; (b) HS1000 [36]

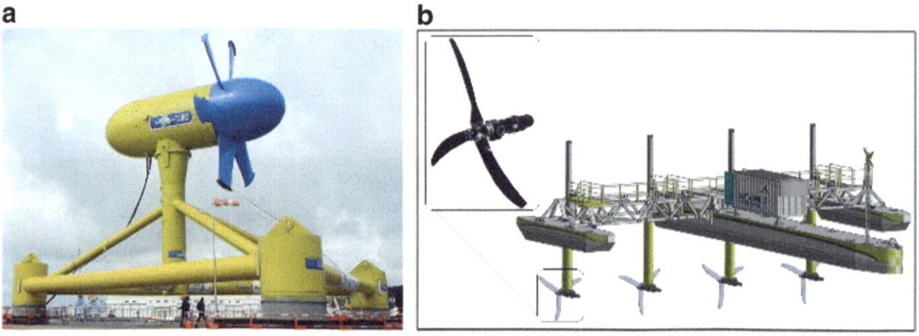

Fig. 7 (a) Sabella D10 [37]; (b) SIT [38]

integrated medium-speed PMG from The Switch Company with a two-stage and epicyclical gearbox (Fig. 6a) [30].

The HS1000 is designed to produce electrical power when the speed of tidal current arrived to 1 m/s. The pitching system allows to power generation during both ebb and flood tides. Turbine is designed for a 25-year life, with four services at 5-year intervals. The total mass of the nacelle is approximately 130 tons, and substructure is approximately 150 tons in air. The nacelle is optimized to minimize the rotor wake effect (Fig. 6b) [36].

The French company Sabella has developed the D10, a megawatt-scale six-bladed turbine with 10 m diameter deployed in 2015 in Fromveur Passage (Iroise Sea, Brittany) and first tidal turbine connected ever to the grid in France. The device is 17 m high, 450 tons of weight, and it stands on a gravity-based foundation and operates at about 50 m depth. Innovative aspects of this technology are bidirectional blade sections to avoid pitching systems, a direct-drive PMSG and a ruggedized design for offshore operations in harsh conditions (Fig. 7a) [37].

The Schottel Instream Turbine (SIT) by Schottel Hydro (Fig. 7b) is a lightweight generator with rotor diameters between 4 and 6 m and rated power between 54 and 120 kW. SIT units are designed to operate in clusters in floating platforms to reach MW-scale power output. Reducing turbine size yields a better power/weight ratio.

In contrast to large-scale tidal turbines with nacelle weights more than 130 tons, a single SIT weighs only one ton, and hence, a megawatt-scale installation can be obtained by clustering 20 SIT turbines at a total weight of only 20 tons [38].

6 Conclusions

State-of-art technologies for the conversion of mechanical into electrical energy in marine current (tidal) turbines have been reviewed. Reviewing current tidal energy demonstration projects around the world, a requirement is that devices may be free of maintenance service for long periods (about 5 year), because access to turbine is extremely expensive at sea. This yields a significant difference between a wind turbine and a tidal turbine that leads to design a simple, robust and well-proven Power Take-Off (PTO) system which is asked to have good efficiency, high reliability, low cost and long service life with affordable maintenance requirements. Therefore, despite the doubly fed induction generator (DFIG) with multiple-stage gearbox is widely used in wind turbines; this solution is not adequate for tidal turbine applications. Moreover, the variable speed induction generator (IG) with gearbox is not a suitable solution because of maintenance requirement, except for some floating turbines where regular maintenance is possible.

Therefore, the variable speed direct-drive permanent magnet synchronous generator (DDPM) topology with fixed blades and without yaw system represents an appealing solution, in that variable pitch and yawing mechanism, gearbox and brushes are not required. This yields reduced system complexity and maintenance requirements and also enhanced reliability. Furthermore, this solution has the advantages of a full-rated frequency converter (FFC) suitable for developing a Maximum Power Point Tracking (MPPT) control strategy for turbines.

Acknowledgements The work has been partially supported by the H2020 MaRINET-2 Project (GA 731084) and by CNR under the ULYSSES Project.

References

1. OECD/IEA, *OECD Green Growth Studies: Energy* (OECD/IEA, Paris, 2012)
2. IEA, *Renewables 2018: Analysis and Forecasts to 2023* (IEA, Paris, 2018)
3. R. Pelc, R.M. Fujita, Renewable energy from the ocean. Marine Policy **26**(6), 471–479 (2002)
4. World Energy Council, *Marine Energy/World Energy Resources 2016, Technical Report* (World Energy Council, London, 2016)
5. B. Multon (ed.), *Marine Renewable Energy Handbook* (ISTE: John Wiley & Sons, London, 2012)
6. M. Beam, B. Kline, B. Elbing, W. Straka, A. Fontaine, M. Lawson, Y. Li, R. Thresher, M. Previsic, Marine hydrokinetic turbine power-take-off design for optimal performance and low impact on cost-of-energy, in 32nd International Conference on Ocean, Offshore and Arctic Engineering (OMAE 2013), France (2013)

7. http://modelling.seacore.online
8. T. Burton, N. Jenkins, D. Sharpe, E. Bossanyi, *Wind Energy Handbook* (John Wiley & Sons, Ltd, Hoboken, NJ, 2011)
9. C. Garrett, P. Cummins, The efficiency of a turbine in a tidal channel. J. Fluid Mech. **588**, 243–251 (2007)
10. R. Vennell, Exceeding the Betz limit with tidal turbines. Renew. Energy **55**, 277–285 (2013)
11. H. Chen; N. Aït-Ahmed; E. H. Zaïm; M. Machmoum, Marine tidal current systems: state of the art, in 2012 IEEE International Symposium on Industrial Electronics, May (2012)
12. G. Marsh, D. Robb, Patently innovative: Imagination in wind turbine technology continues to flourish. Refocus **8**(2), 30–35 (2007)
13. P.M. Tlali, R.-J. Wang, S. Gerber, Magnetic gear technologies: a review, in International Conference on Electrical Machines (ICEM) (2014)
14. J. Zhang, L. Moreau, M. Machmoum, P. Guillerm, State of the art in tidal current energy extracting technologies, 2014 First International Conference on Green Energy ICGE (2014)
15. D. Magagna, R. Monfardini, A. Uihlein, *JRC Ocean Energy Status Report: 2016 Edition. EUR 28407 EN. JRC104799* (Publications Office of the European Union, Luxembourg, 2016)
16. H. Polinder, D.J. Bang, H. Li, Z. Chen, Concept Report on Generator Topologies, Mechanical & Electromagnetic Optimization, Project UpWind (2007)
17. Z. Chen, Characteristics of induction generators and power system stability, 2005 International Conference on Electrical Machines and Systems, September 2005
18. S. Benelghali, M.E.H. Benbouzid, J.F. Charpentier, Generator systems for marine current turbine applications: a comparative study. IEEE J. Ocean Eng. **37**(9), 554–563 (July 2012)
19. Z. Chen, J.M. Guerrero, F. Blaabjerg, A review of the state of the art of power electronics for wind turbines. IEEE Trans. Power Electron. **24**(8), 1859–1875 (2009)
20. D. Bang, H. Polinder, G. Shrestha, J. Ferreria, Review of generator systems for direct-drive wind turbines, in European Wind Energy Conference & Exhibition, Belgium (2008)
21. A. Grauers, *Design of direct-driven permanent-magnet generators for wind turbines, Ph.D. dissertation* (Chalmers University of Technology, Göteborg, 1996)
22. H. Polinder, F.F.A. van der Pijl, G.J. de Vilder, P. Tavner, Comparison of direct-drive and geared generator concepts for wind turbines. IEEE Trans. Energy Convers. **21**, 725–733 (2006)
23. R. Poore, T. Lettenmaier, *Alternative Design Study Report: WindPACT Advanced Wind Turbine Drive Train Designs Study. Report No. NREL/SR-500-33196* (NREL, Golden, CO, 2003)
24. H. Weh, H. Hoffmann, J. Landrath, H. Mosebach, J. Poschadel, Directly-driven permanent-magnet excited synchronous generator for variable speed operation, in Proceedings of the 1988 European Wind Energy Conference, pp. 566–572
25. G. Bywaters, V. John, J. Lynch, P. Mattila, G. Nortor, J. Stowell, M. Salata, O. Labath, A. Chertok, D. Hablanian, *Northern Power Systems windPACT Drive Train Alternative Design Study Report. Report no. NREL/SR-500-35524* (NREL, Golden, CO, 2004)
26. P. Lampola, *Directly driven, low-speed permanent-magnet generators for wind power applications, Ph.D. dissertation* (Helsinki University of Technology, Helsinki, 2000)
27. M. Dubois, *Optimized permanent magnet generator topologies for direct drive wind turbines, Ph.D. dissertation* (Delft University of Technology, Delft, 2004)
28. Y. Duan; R.G. Harley, Present and future trends in wind turbine generator designs, IEEE Power Electronics and Machines in Wind Applications, June 2009.
29. https://theswitch.com/wave-and-tidal/permanent-magnet-generators/
30. https://simecatlantis.com/services/turbines/
31. http://redapt.eng.ed.ac.uk/index.php?p=library_redapt_reports
32. https://www.alstom.com
33. S. Benelghali, M.E.H. Benbouzid, J.F. Charpentier, Marine tidal current electric power generation technology: state of the art and current status, in Electric Machines & Drives Conference, IEMDC'07. IEEE Int. **2**, 1407–1412 (2007)
34. Z. Zhou; F. Scuiller; J. F. Charpentier; M. Benbouzid; T. Tang, An up-to-date review of large marine tidal current turbine technologies, in IEEE International Power Electronics and Application Conference and Exposition, November 2014

35. http://www.emec.org.uk/about-us/our-tidal-clients/
36. https://www.andritz.com/products-en/hydro/products/tidal-current-turbines
37. http://www.sabella.fr
38. https://www.schottel.de

Part IV
Advances in Photovoltaic Applications

Real-Time Procedure to Detect Losses in Photovoltaic Generators Using the Instantaneous and the Translated Performance Ratio

Francisco J. Sánchez-Pacheco, Llanos Mora-López, Michel Piliougine, Juan Ramón Heredia-Larrubia, and Mariano Sidrach-de-Cardona

Abstract The energy yield of a photovoltaic plant depends on the performance of each of its single photovoltaic modules. In this paper, a methodology is presented that allows calculating the losses produced in a module and identifying if it is operating properly according to given climatic conditions. The proposed model is based on the comparison between the theoretical performance ratio of a module and the resulting experimental one under given climatic conditions. The results obtained allow estimating the losses that affect a module nonoperating properly. By means of real-time monitoring at module level, the herein described experimental system together with the proposed methodology allows to quantify the module losses. The immediate corrective actions will avoid further losses in the generated energy.

1 Introduction

Stand-alone or grid-connected PV plants generate an energy that depends on the photovoltaic cell technology efficiency for given climatic conditions. This efficiency is essentially a consequence of the cell manufacturing process and materials used, as well as of the module assembly technology [1].

When these modules are working under real operating conditions (ROC), the values of the climatic conditions may vary significantly in comparison with those defined by the manufacturers at the standard test conditions (STC), which

F. J. Sánchez-Pacheco (✉) · J. R. Heredia-Larrubia
Departamento de Tecnología Electrónica, University of Málaga, Málaga, Spain
e-mail: fsanchezp@uma.es

L. Mora-López · M. Piliougine
Departamento de Lenguajes y Ciencias de la Computación, University of Málaga, Málaga, Spain
e-mail: llanos@uma.es

M. Sidrach-de-Cardona
Departamento de Física Aplicada II, University of Málaga, Málaga, Spain
e-mail: msidrach@uma.es

W. Zamboni, G. Petrone (eds.), *ELECTRIMACS 2019*, Lecture Notes in Electrical Engineering 615, https://doi.org/10.1007/978-3-030-37161-6_35

correspond to an irradiance of 1000 W/m^2, 25 °C of cell temperature, and air mass (AM) index AM 1.5. Furthermore, the energy generated can be significantly reduced due to predictable and unpredictable losses.

The performance ratio (PR) is a metric commonly used for measuring solar photovoltaic plant efficiency. The PR is defined in IEC61724 [2] in terms of energy yield in a given period of time. Actually, the PR defines an energy balance between two points of a system. It allows quantifying the overall losses of the PV system, in comparison with the maximum energy it might deliver under the STC [3–5]. It can be calculated either for the whole system (considering both the AC and the DC sides of the generator) or for a part of it. It will then report us the losses that have been produced.

Likewise, this balance can also be instantaneous. In this case, as we propose in this paper, an instantaneous PR can be defined (PR_i) in terms of power. In our case, a methodology is proposed where two different values of performance ratio [6] in terms of power are calculated, in which the timescale has been reduced to the instantaneous value at a given moment in time and the deployment boundary has been reduced to the DC side of the PV generator up to a PV module granularity level.

The losses of a PV module can be due on the one hand to the fact that the PV module is not working under the referred STC. These climatic condition-related losses are non-avoidable. On the other hand, there are losses due to partial shadowing of the PV modules, presence of dust or soiling, defective ones, or severe maximum power point tracking (MPPT) controller misfunction or failure. These losses can be considered as avoidable and can be consequently minimized [7].

This topic has been treated in the literature under different scopes. The convenience of detecting faulty photovoltaic modules is highlighted in [8–10]. In [11], the authors describe a wireless-based monitoring circuit, aimed at monitoring the electrical parameters of a PV module, in order to quantify in real time their performance. Furthermore, in [12], an extensive analysis is made of actual procedures and methods for PV module fault detection. Generally, the performance ratio is related to a 1-year period. In [13], the authors describe the results of experimental measurements of photovoltaic modules under outdoor conditions. In most of the cases, the performance estimation is related to the amount of energy generated in a given time frame period in comparison with the maximum energy that theoretically might have been generated. In [14], the authors emphasize the importance of the module temperature in order to estimate its performance.

In this work, we propose a methodology to estimate in real time the losses produced in a photovoltaic generator. The power generated in a given moment in time is compared with the maximum power that might be extracted according to applying ROC. For this purpose, PV module operating parameters, namely, voltage (V), current (I), and module backplane temperature (T_m), are required to be monitored. This allows to estimate the losses that might apply [7, 15, 16].

Although the procedure herein described has been validated by means of a laboratory prototype demonstrator, the method can be scaled up to a complex power PV generator as well.

In the literature, several theoretical models are described to determine the characteristic *I–V* curve of the PV module. In [17], different considerations are made on the one-diode and two-diode models.

In order to determine the performance of the PV modules, it is mandatory to have real-time information of their operating parameters and their evolution as well. This task is done by the monitoring system prototype proposed in [18]. In this paper, a real-time smart monitoring and communications module is described, which is able to perform the monitoring of both PV module voltage and current issued in real time, as well as the backplane temperature. The obtained parameters are then routed by means of power line communications technology, to the central computer system for further processing. This topology, without an extensive extra-cost (no additional wiring is required), can be deployed up to a large-scale PV generator.

This paper is organized as follows: after this introduction, in Sect. 2, the proposed methodology is described and further validated with experimental data in Sect. 3; finally, in Sect. 4, resulting conclusions obtained in this work are reported.

2 Proposed Methodology

Once the module operating electrical parameters under ROC have been monitored by means of described experimental system, instantaneous generated power (P_i) can be calculated. This value will then be compared with the ROC conditions translated generated power (P_t), from the maximum power at STC (according to data sheets) (P_mSTC). In this case, the Osterwald model is applied [19]:

$$P_\mathrm{t} = P_\mathrm{mSTC} \cdot \frac{G}{G_0} \cdot [1 + \gamma\,(T_\mathrm{m} - T_0)] \tag{1}$$

where P_t is the ROC translated power from the STC conditions (P_mSTC); P_mSTC is the PV module maximum power at STC, taken from the manufacturer's data sheets; G is the incident irradiance on the module surface; G_0 is the global irradiance at STC (1000 W/m^2); T_m is the PV module backplane measured temperature; and γ is the module power temperature coefficient.

In order to quantify the losses, two different PR are defined. The instantaneous ($_i$ superscript) performance ratio is calculated from the following equation:

$$\mathrm{PR}_\mathrm{DC}^{i} = \frac{\frac{P_\mathrm{i}}{P_\mathrm{mSTC}}}{\frac{G}{G_0}} \tag{2}$$

where P_i is the output power of the PV module; P_{mSTC} is the module power at STC, taken from the manufacturer data sheet; G is the incident irradiance on the module surface; and G_0 is the global irradiance at STC (1000 W/m^2). This PR^i_{DC} gives the overall losses of the module [7].

The obtained result of PR^i_{DC} does not always give consistent information in order to determine if the PV module is operating optimally. In case of not, the issue is to know how far it is operating from the optimal performance.

For this purpose, a new value of PR, based on ROC translated parameters ($_t$ superscript) called PR^t_{DC}, is calculated that allows us to obtain additional information related to the system losses. It can be obtained by the following equation:

$$PR^t_{DC} = \frac{\frac{P_t}{P_{mSTC}}}{\frac{G}{G_0}} \quad (3)$$

where P_t is the theoretical maximum power that the PV module might generate under real operating conditions (ROC) as calculated in Eq. (1).

This is the maximum performance ratio that theoretically might be obtained from the module under existing ROC. This calculation can be done in an embedded system with a simple and effective formulation.

In both cases, PR^i_{DC} and PR^t_{DC} give values corrected in temperature and in effective irradiance (compensating the angle of incidence (AOI) and the AM index), since they are calculated considering the calibrated cell measured incident irradiance (G). They can be compared, and the differences between both of them allow estimating the possible origin of the losses, which might be due to shadowing (partial, total, or temporary), soil, MPPT tracker misfunction, or other system failure.

If the modules are operating properly, the resulting PR^t_{DC} should be similar to the experimentally obtained value of PR^i_{DC}. A significant difference between both of them means that the module is affected of losses that are not due to ROC. That is, the difference indicates the losses are avoidable and are due to upgradable adverse conditions.

In order to evaluate significant deviations in the module performance, instantaneous losses are calculated, which represent the overall losses that the PV module are submitted to and are given by the following equation:

$$L^i_{DC} = 1 - PR^i_{DC} \quad (4)$$

On the other hand, theoretical losses are quantified to be those that the module might be subject to at ROC, cannot be, avoided and are given by the following equation:

$$L^t_{DC} = 1 - PR^t_{DC} \quad (5)$$

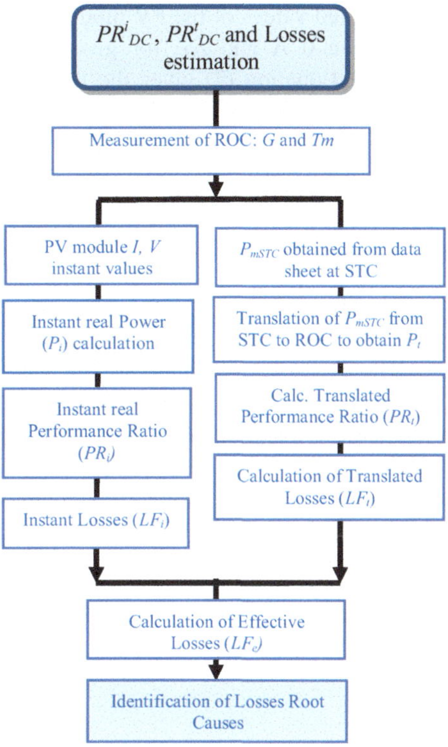

Fig. 1 Proposed methodology flow diagram

The effective losses are subsequently given by the following equation:

$$L^e_{\mathrm{DC}} = \left| L^t_{\mathrm{DC}} - L^i_{\mathrm{DC}} \right| \tag{6}$$

This parameter identifies the losses that might be avoided, by applying corresponding corrective actions [7]. The process is summarized in flow diagram of Fig. 1.

3 Model Validation and Experimental Results

In order to demonstrate the viability of the proposed procedure, it has been applied to experimental I–V curve measurements made on Yocasol PCB-195 54 cell polycrystalline 195 W PV modules, located at the Renewable Energies Laboratory of the University of Málaga and recorded during a clear sky daylong. This validation has been made by the comparison between estimated peak power values, using proposed models, with consolidated ones obtained by measurement of field-installed PV modules.

Translated values of the PV module output voltage (V), for the range of current (I) between 0 A and I_{SC} for given ROC, can be quantified by the following equations. Equation (7) gives the ROC translated value of I_{SC}, (I_{SCt}), while Eq. (8) is applied to PV module translated V_{OC} (V_{OCt}):

$$I_{SCt} = I_{SC0} \cdot \frac{G}{G_0} \cdot [1 + \alpha (T_m - T_0)] \tag{7}$$

$$V_{OCt} = V_{OC0} \cdot [1 + \beta (T_m - T_0)] \cdot \left[1 + \delta \cdot \ln\left(\frac{G}{G_0}\right)\right] \tag{8}$$

where 0 subscript variables are the manufacturer's data sheet STC referred ones, G is the incident irradiance in the plane of the module (in W/m^2), G_0 is the reference irradiance at STC (1000 W/m^2), α stands for the current temperature coefficient (°C^{-1}), T_m is the module backplane temperature, T_0 is the module temperature at STC, β is the voltage temperature coefficient (°C^{-1}), and δ is the irradiance temperature coefficient (°C^{-1}). In our case, the module temperature and the effective irradiance parameters are measured by our experimental system [7].

The relative error (RE) has been calculated in % using the following equation:

$$\text{RE}\,(\%) = \left|\frac{P_i - P_t}{P_i}\right| \times 100 \tag{9}$$

In Fig. 2, obtained results are presented. The proposed model is capable of predicting well the peak power of the module with a relative error smaller than 3% along the day (mainly due to experimental errors in the measurement of the module

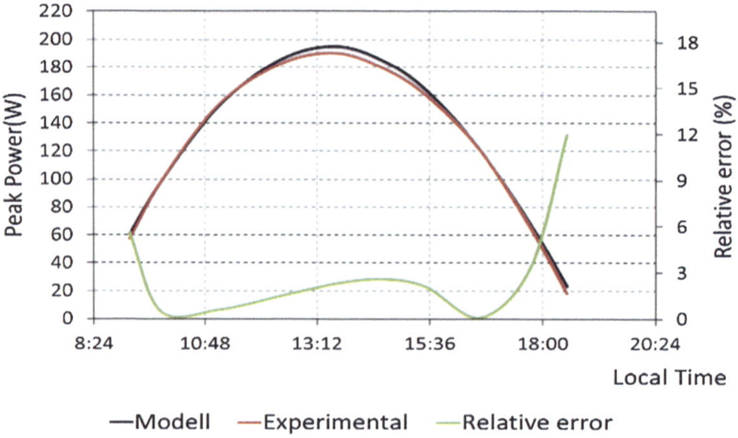

Fig. 2 One daylong module generated peak power, predicted one, and resulting relative error

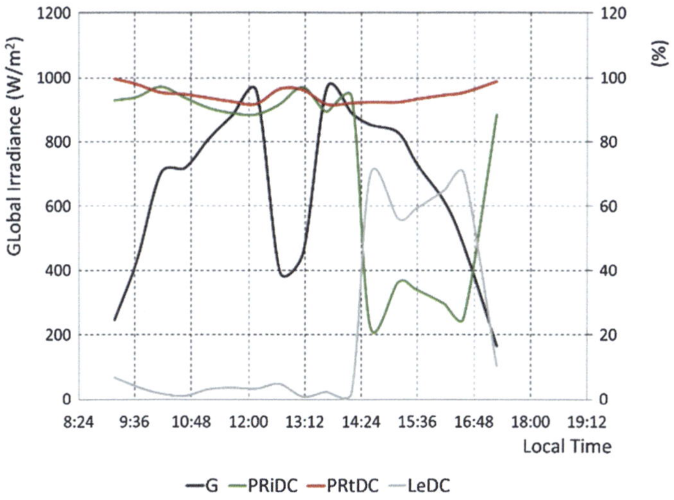

Fig. 3 Incident irradiance, G (W/m^2), instantaneous $\mathrm{PR}^i_{\mathrm{DC}}$, translated $\mathrm{PR}^t_{\mathrm{DC}}$, and resulting effective losses L^e_{DC} expressed in %

temperature) except in the hours close to sunrise and sunset where the irradiation is very small.

Once the validity of the model has been verified, it has been applied to predict the effective losses according to the proposed methodology. In Fig. 3, global irradiance measured values are depicted (black line), as well as calculated values of PR, namely, $\mathrm{PR}^i_{\mathrm{DC}}$ (green line) and $\mathrm{PR}^t_{\mathrm{DC}}$ (red line), and resulting effective losses L^e_{DC} (gray line). As it can be seen, obtained results of PR are very similar during all the day, independently of the received irradiance. This signifies that the modules are operating properly. Nevertheless, starting by 14:31 h, a significant variation of the real PR in comparison with the theoretical one appears, according to the incident irradiance on the module. This is due to an intentionally generated partial shadowing to simulate a situation of severe losses. Consequently, the effective losses that so far during the day were very small now suffer a significant increase, reaching values bigger than 60%.

A more detailed study can be made of the *I*–*V* curve of a partially shadowed module as depicted in Fig. 4.

In this figure, the theoretical module *I*–*V* curve under STC (black line) is represented; the theoretical expected curve, when an incident irradiance G of 878 W/m^2 is applied on the module, which reaches a temperature of 52 °C (red line), is obtained by applying Eqs. (7) and (8) and the experimentally measured curve (green line). The shape of the experimentally measured *I*–*V* curve (green line) explains the fact that the power at the maximum power point is significantly decreased, which reduces the module $\mathrm{PR}^i_{\mathrm{DC}}$ at that moment and the fact that the effective losses L^e_{DC} reach values bigger than 60%, as it is depicted in Fig. 3.

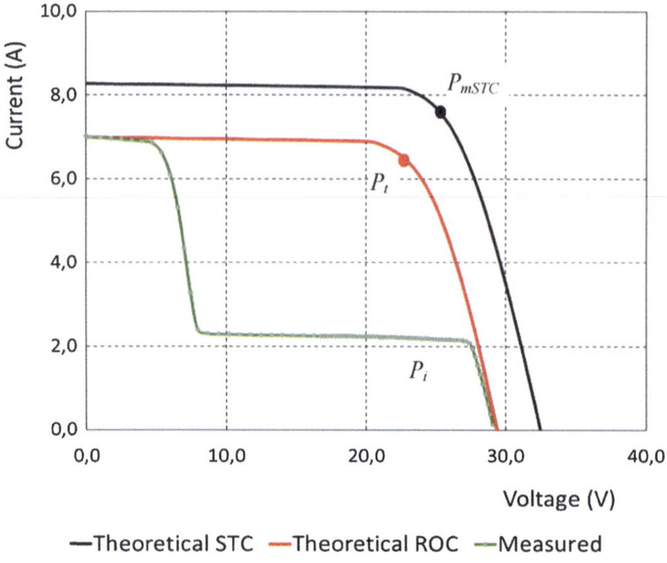

Fig. 4 Theoretical STC, ROC translated and experimentally measured *I–V* curves of a shadowed PV module

In the curves depicted in Fig. 4, obtained by means of Eqs. (7) and (8), the translated value of the maximum power point (P_t = 147.5 W) obtained from the one at STC (P_{mSTC} = 192.7 W) is equivalent to the one resulting by applying the Osterwald model of Eq. (1). In referred equations, the following experimentally obtained temperature coefficients have been used: $\alpha = 0.08\%/°C$, $\beta = -0.32\%/°C$, $\gamma = -0.38\%/°C$, and $\delta = -0.35\%/°C$. This confirms the fact that it is not necessary to obtain the *I–V* curve, in the identification of P_i, which would require the disconnection of the PV module and further connection of it to an *I–V* curve tracer. The P_i value obtained by the SMCM can be directly used in Eq. (2) to obtain the PR^i_{DC}.

These losses have been quantified by the proposed methodology, and resulting values of L^e_{DC} have been obtained without the need to measure the *I–V* curve of each module.

This also happens when the incident irradiance suffers a sudden change, which implies resulting values of L^e_{DC} very close to zero and, in any case, into the estimated error range of the proposed methodology. In these cases, it can be affirmed that the losses of the module are due to the applying real operating conditions, and no additional problems can be evidenced. Nevertheless, when the modules are submitted to shadowing, the resulting values of PR^i_{DC} and PR^t_{DC} diverge which produces a significant increase of the L^e_{DC} value. This indicates the % of losses in the module due to this problem.

4 Conclusions

In this paper, a methodology has been presented that allows real-time detection of defective PV modules. By means of a real-time monitoring system, it is possible to detect if a single PV module that a generator is configured of is behaving abnormally.

Although, in this case, the experimental results have been obtained from the laboratory prototype setup, the solution can be scaled up to a string or even plant level, so that each single PV module might be monitored in real time. Hence, the losses due to its faulty operation can be quantified.

This procedure is based on the analysis of the two PR proposed, which quantify instantaneously the power generation performance, namely, PR^{i}_{DC} and PR^{t}_{DC}. For this purpose, the SMCM delivers the instantaneous values of the PV module operational parameters V, I, and T_{m}. The theoretical translated performance ratio (PR^{t}_{DC}) requires the ROC monitored data T_{m} and G, which allows to compensate the PR^{t}_{DC} in terms of effective irradiance and ambient temperature. The comparison of both terms allows to reliably estimate the losses at PV module level. The calculation of these parameters allows to estimate the instant and translated losses L^{i}_{DC} and L^{t}_{DC}, which result in corresponding effective losses L^{e}_{DC}. It has also been demonstrated that the resulting effective losses (L^{e}_{DC}) can be clearly identified in real time in case of PV modules subject to severe adverse conditions. This procedure allows to significantly improve the PV plant yield and maintenance operations [7].

References

1. M.A. Green, K. Emery, Y. Hishikawa, W. Warta, E.D. Dunlop, Solar cell efficiency tables (version 43). Prog. Photovoltaics **22**(1), 1–9 (2014)
2. IEC 61724, Photovoltaic system performance monitoring. Guidelines for measurement, data exchange and analysis. CENELEC (2000)
3. B. Marion, J. Adelstein, K. Boyle, H. Hayden, B. Hammond, T. Fletcher, B. Canada, D. Narang, A. Kimber, L. Mitchell, G. Rich, T. Townsend, Performance parameters for grid-connected pv systems, in 31st IEEE Photovoltaics Specialists Conference, Florida (2005)
4. A. Dolara, F. Grimaccia, S. Leva, M. Mussetta, R. Faranda, M. Gualdoni, Performance analysis of a single-axis tracking PV system. IEEE J Photovoltaics **2**(4), 524–531 (2012)
5. A. Dolara, S. Leva, G. Manzolini, E. Ogliari, Investigation on performance decay on photovoltaic modules: snail trails and cell microcracks. IEEE J. Photovoltaics **4**, 5 (2014)
6. R. Moreno-Sáez, M. Sidrach-de-Cardona, L. Mora-López, Data mining and statistical techniques for characterizing the performance of thin-film photovoltaic modules. Exp. Syst. Appl. **40**(17), 7141–7150 (2013)
7. F.J. Sánchez Pacheco, Doctoral Thesis, Photovoltaic Systems Distributed Monitoring for Performance Optimization, Málaga (2015). https://riuma.uma.es/xmlui/bitstream/handle/10630/9922/TD_Sanchez_Pacheco.pdf?isAllowed=y&sequence=1
8. N. Gokmen, E. Karatepe, B. Celik, S. Silvestre, *Simple Diagnostic Approach for Determining of Faulted PV Modules in String Based PV Arrays, Solar Energy* (Elsevier, Amsterdam, 2012)

9. J.D. Bastidas-Rodriguez, G. Petrone, C.A. Ramos-Paja, G. Spagnuolo, Photovoltaic modules diagnostic: an overview, in 39th Industrial Conference of the Industrial Electronics society, IECON (2013)
10. J.D. Bastidas-Rodriguez, E. Franco, G. Petrone, C.A. Ramos-Paja, G. Spagnuolo, Quantification of photovoltaic module degradation using model based indicators. J. Math. Comput. Simulat. **131**, 101–113 (2017)
11. P. Guerriero, V. d'Alessandro, L. Petrazzuoli, G. Vallone S. Daliento Effective real-time performance monitoring and diagnostics of individual panels in PV plants, in International Conference on Clean Electrical Power (ICCEP) (2013)
12. A. Mellit, G.M. Tina, S.A. Kalogirou, Fault detection and diagnosis methods for photovoltaic systems: a review. Renew. Sustain. Energy Rev. **91**, 1–17 (2018)
13. M. Piliougine, J. Carretero, L. Mora López, M. Sidrach de Cardona, Experimental system for current-voltage curve measurement of photovoltaic modules under outdoor conditions. Prog. Photovoltaics Res. Appl. **19**, 591–602 (2011)
14. P. Mora-Segado, J. Carretero, M. Sidrach-de-Cardona, Models to predict the operating temperature of different photovoltaic modules in outdoor conditions. Prog. Photovoltaics Res. Appl. **23**(10), 1267–1282 (2014). Published on-line in Wiley Online Library (www.wileyonlinelibrary.com)
15. P.L. Carotenuto, P. Manganiello, G. Petrone, G. Spagnuolo, Online recording a PV module fingerprint. IEEE J Photovoltaics **4**(2), 659–668 (2014)
16. D.L. King, W.E. Boyson, J.A. Kratochvill, *Photovoltaic Array Performance Model, Sandia Report SAND2004-3535* (Sandia National Laboratories, Livermore, CA, 2004)
17. N. Femia, G. Petrone, G. Spagnuolo, M. Vitelli, *Power Electronics and Control Techniques for Maximum Energy Harvesting in Photovoltaic systems* (CRC Press, Boca Raton, FL, 2013). ISBN 978-1-4665-0690-9
18. F.J. Sánchez-Pacheco, P.J. Sotorrío-Ruiz, J.R. Heredia-Larrubia, F. Pérez-Hidalgo, M. Sidrach-de-Cardona, PLC-based PV plants smart monitoring system: field measurements and uncertainty estimation. IEEE Trans. Instrum. Meas. **43**(9), 2215–2222 (2014)
19. C.R. Osterwald, Translation of device performance measurements to reference conditions. Sol. Cells **18**(3–4), 269–279 (1986)

Matlab/Simulink Interface Design and Implementation for PV Arrays Reconfiguration

Mariana Durango Florez, Luz Adriana Trejos Grisales, and Daniel Gonzalez Montoya

Abstract One of the major problems in photovoltaic (PV) arrays operation is the mismatching effect due to partial shading; it produces significant drops in the power delivered by the PV system. The power losses due to partial shading for a given operating condition can be mitigated in part by changing the scheme connection of the PV array. This paper introduces a solution to reconfigure a 3×3 PV array by means of a graphic interface using Matlab and Simulink. A mathematical model is used to calculate the configuration which provides the maximum power under a given operating condition. Then, a connection matrix is identified and the connections on the PV array are implemented by using an ARDUINO DUE and a relay based reconfiguration board. The proposed solution is validated by comparing the power vs voltage ($P - V$) characteristics of different configuration schemes obtained from experimental tests.

1 Introduction

The interest of the scientific community in the development and implementation of renewable energies, especially, solar energy, has increased gradually through the last few years: in 2017, the power generation capacity increased an estimated 178 GW globally, with photovoltaic (PV) capacity as the main addition [1]. To improve the efficiency in PV systems, the study of the losses which affect the extraction of the maximum power is imperative. One of the main causes of power drops is mismatching: different modules parameters or partial shading [2]. Partial shading occurs when part of a PV array is shaded (due to passing clouds, parts of other

M. Durango Florez (✉) · D. Gonzalez Montoya
Department of Electronic and Telecommunications, Instituto Tecnologico Metropolitano, Medellin, Colombia
e-mail: marianadurango134392@correo.itm.edu.co; danielgonzalez@itm.edu.co

L. A. Trejos Grisales
Department of Electromechanics and Mechatronics, Instituto Tecnologico Metropolitano, Medellin, Colombia
e-mail: adrianatrejos@itm.edu.co

W. Zamboni, G. Petrone (eds.), *ELECTRIMACS 2019*, Lecture Notes in Electrical Engineering 615, https://doi.org/10.1007/978-3-030-37161-6_36

buildings, trees, adjacent PV panels, etc.) while the rest of the array remains fully illuminated [3], this producing the activation of the bypass diodes connected in antiparallel to the PV modules. When that happens, the power vs voltage ($P - V$) characteristic exhibits multiple power peaks which can introduce errors in the operation of the algorithms used to track the global maximum power point (*GMPP*).

1.1 Reconfiguration of Photovoltaic Arrays

The process of rearranging solar PV panels either physically or electrically is known as reconfiguration. Such a process is carried out to mitigate the effects of partial shading on the power provided by the PV system [4, 5]. However, a large number of switches and sensors are required to implement different techniques reported in literature [4] and the high complexity of some algorithms increases the cost of such applications [6]. On the other hand, another form of reconfiguring a PV array can be considered, this by performing changes in the connection scheme of the array, i.e., changing the connection of the array using conventional configurations also called *regular configurations* which follows a defined pattern as illustrated in Fig. 1 where

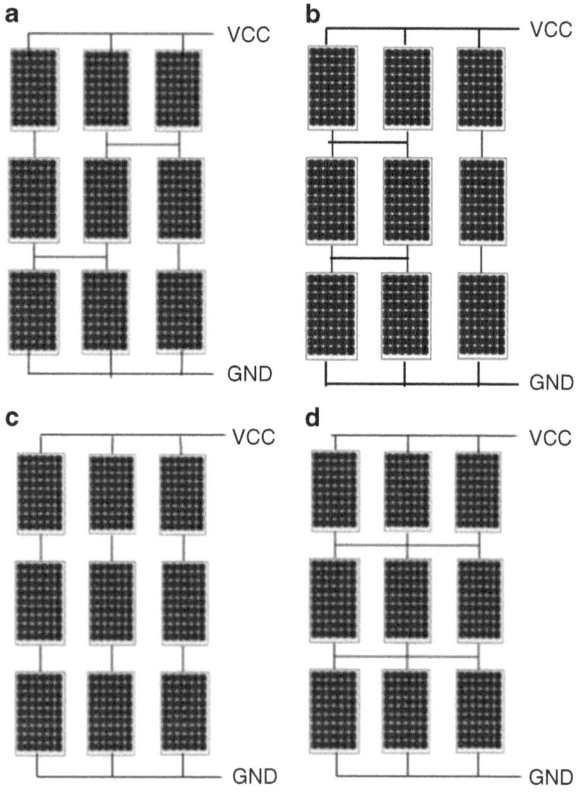

Fig. 1 PV configurations example. (**a**) HC (**b**) Irregular (**c**) SP (**d**) TCT

item (a) represents a Honey-comb configuration (HC), item (c) a serial-parallel configuration (SP), and item (d) a total-cross-tied configuration (TCT). Moreover, other configurations known as irregular or hybrids [7, 8], illustrated in Fig. 1 item (b), which are connected randomly or without a pattern, can also be considered. This kind of reconfiguration requires less number of switches reducing the cost and complexity of its implementation. One of the requirements for performing the reconfiguration between conventional or irregular structures is a technique to calculate the best configuration (in terms of produced power) for a given operating conditions.

For this purpose, procedures as the one introduced in [7] can be used. Such a procedure is able to calculate the currents and voltages and then the power of a PV system with any configuration operating under uniform or partial shading conditions. In this way, it is possible to know which connection is more suitable for mitigating the effects of partial shading; this information can be used in a switching board to perform the physical connections in the PV array. This paper presents the application of the modeling procedure introduced in [7] implemented in Matlab environment, adapting it in order to evaluate every possible configuration, conventional and irregular, for a 3×3 PV array, finding the best connection that delivers the highest MPP. The different connections in the PV array are performed through a switching matrix which is managed by means of a graphic interface designed in Simulink, using an Arduino DUE. This paper is organized as follows: In Sect. 2 the main algorithm of the modeling procedure and the application proposed will be described. Section 3 is dedicated to define the elements of the experimental setup: Sect. 3.1 elements that are used to implement the reconfiguration process and in Sect. 3.2 the experimental results are presented. Finally, in Sect. 4 conclusions close the paper.

2 Proposed Application of Modeling Procedure

2.1 Modeling Procedure

The algorithm described in [7] calculates the current of any conventional or irregular PV array, with N rows and M columns of panels ($N \times M$), operating under uniform or partial shading conditions, to obtain the *I-V* and *P-V* curves of the PV array. The model is based on the single diode model (*SDM*) including bypass diode; such a model requires seven parameters: photovoltaic current I_{ph}, saturation current I_{sat}, ideality factor n, series resistance R_s, parallel resistance R_{sh}, bypass diode saturation current $I_{sat,db}$, and bypass diode ideality factor n_{db}. In order to represent the ties (connections between the array strings), the procedure makes use of a $(N-1) \times (M-1)$ matrix known as *connection matrix*; such matrix is filled with "0" or "1": a "0" indicates no connection between strings and a "1" indicates connection. As an example, connection matrices (M_{conn}) corresponding to HC and the irregular configuration described in Fig. 1 are shown in (1). For TCT and SP, the connection matrices $M_{conn-TCT}$ and $M_{conn-SP}$ are a matrix filled with ones and a matrix filled with zeros, respectively.

$$M_{conn-Irregular} = \begin{bmatrix} 1\,0 \\ 1\,0 \end{bmatrix} \quad M_{conn-HC} = \begin{bmatrix} 0\,1 \\ 1\,0 \end{bmatrix} \tag{1}$$

In this work, the modeling procedure reported in [7] will be used to evaluate the *I-V* characteristics of all the possible connections, that means, the switches states corresponding with the regular or irregular configurations for a 3×3 PV array under a given irradiance profile in order to acquire the scheme which mitigates best the partial shading effect. In this way, the connection matrix concept will be used for obtaining the signals which will be sent to the switching matrix to perform the suitable connection in the PV array.

2.2 *Identification of Connection Matrices*

The automation of the connection matrices calculation and the recognition of all possible configuration schemes regular and irregular have been added to the procedure in [7] by changing the manual entry of connection matrices for an automated process developed in this application and represented in Algorithm 1. In the next sub-sections these contributions will be explained.

Algorithm 1 Identification and creation of connection matrices

INPUT: N, M, array dimensions
OUTPUT: *Nt*, *Tco*
Set gu=1
Calculate *Nm* with (2)
For j=1 **to** *Nm* **do**
Calculate *F(j)* with (3)
end for
Calculate *Tco* with (6)
For i=1 **to** *Nm* **do**
Set *Array*(1,i)=1
Save in *Fr* all possible permutations of the elements of the vector *Array*
end for
For ij=1 **to** length of the largest array dimension in F **do**
Identify $nF_{(ij,1)}$ and a= rows of $nF_{(ij,1)}$
For mm=1 **to** a **do**
Set Nodo(gu)=reshape $nF_{(ij,1)}$ into (N-1)x(M-1)=*Nt* and set gu=gu+1
end for
end for
Return *Nt* and *Tco*

To identify the possible configurations in an $N \times M$ PV array, first the amount *Nm* of elements in the connection matrix needs to be settled [9], for this purpose, (2) is applied and referenced in line 2 of the pseudocode:

$$Nm = (N - 1)(M - 1) \tag{2}$$

The connection matrix depends on the number of ties or ("1") in it, then, it is necessary to place one tie ("1") in all the position in the matrix while the rest of the positions are ("0") then with two until *Nm* ties, and calculate the possible combinations *F* that can be obtained with every configuration; *for* loop in lines 3–5 of the pseudocode and (3) where *j* is the variable of iteration inside the *for* loop, the number of these possible combinations can be known. An example of the position change inside the matrix is shown in (4) and (5).

$$F(j) = \frac{Nm!}{(Nm - j)!j!} \tag{3}$$

$$M_{conn-(1,1)} = \begin{bmatrix} 1\,0 \\ 0\,0 \end{bmatrix} - - > M_{conn-(1,2)} = \begin{bmatrix} 0\,1 \\ 0\,0 \end{bmatrix} \tag{4}$$

$$M_{conn-(1,1;1,2)} = \begin{bmatrix} 1\,1 \\ 0\,0 \end{bmatrix} - - > M_{conn-(1,1;2,1)} = \begin{bmatrix} 1\,0 \\ 1\,0 \end{bmatrix} \tag{5}$$

The total number of possible configurations *Toc* is the result of adding together the *F* of every iteration as is shown in (6); the plus 1 is the SP configuration which has no ties between strings.

$$Tco = \sum_{j=1}^{Nm} F(j) + 1 \tag{6}$$

The connection matrix has to be created for each possible configuration; the *for* loop in lines 7–10 of the pseudocode go over an initial vector *Array* that will set the elements and a vector *Fr* that will process them, both of dimensions $(1 \times Nm)$, making permutations in a similar process as described above, the first element will be the first tie ("1"), i.e., (3). Each resulting permutation is a different configuration of the connection matrix; therefore, these vectors will be saved in a cell *nF* of dimensions $(Nm \times 1)$, as indicated in line 9.

To get the final connection matrices associated with the PV array, the vectors *Fr* inside *nF* must be reformed into a $(N-1)$ rows and $(M-1)$ columns matrices $((N-1) \times (M-1))$, based on the dimensions of the photovoltaic array. The process is carried out by means of *for* loops in lines 11–16 and 13–15 of the pseudocode, in charge of re-arrange and save the outcoming matrices in an auxiliary cell *Nt*; the auxiliary variables *a*, *gu* and the cell *Nodos* are introduced to simplify the code.

2.3 Obtainment of the Optimum Configuration Scheme

The recognition of the connection matrices makes possible to evaluate regular and irregular configurations schemes: *For* loops traverse the *Nt* cell within the modeling procedure in [7], where it is possible to calculate the voltage *V* and current *I* values for every configuration found. Now, by making use of the Ohm's law (7), the power can be calculated.

$$P = VI \tag{7}$$

The proper configuration scheme is the one with the highest power, then, the maximum values of each connection matrix are saved in a vector *Pconf* of dimensions (1×Toc), by using a simple comparison between the vector data to find the maximum, the GMPP can be found, and because of the indexed of the *Nt* cell, also the optimum configuration. It is simple to identify whether it is a regular or an irregular configuration, with the connection matrices defined above for regular configurations (SP, TCT, and HC) and *if* conditionals, the comparisons between the connection matrices in *Nt* cell and the known configuration schemes will define which regular configuration matches or in case of irregular configuration will only display such connection matrix.

The corresponding connection matrix is deployed through a specific code that communicate Matlab and Simulink environment where the electrical reconfiguration will be made by sending the proper digital signals through the Arduino DUE to the reconfiguration board.

3 Performance of the Proposed Application

3.1 Experimental Setup

An illustrative diagram of the experimental setup is shown in Fig. 2. The optimal configuration scheme is deployed in a reconfiguration board based on a switches matrix, also, connected in series with the solar panel there is a blocking diode whose purpose is to avoid wrong connections inside the photovoltaic array. The graphic interface allows to communicate the algorithm with the experimental platform through the Arduino and has a fast identification of the type of configuration scheme, regular or irregular, that reached the GMPP.

3.1.1 Reconfiguration Board

In order to implement the application proposed in this paper, a proper system has been designed at Universidad Nacional of Colombia and produced at the

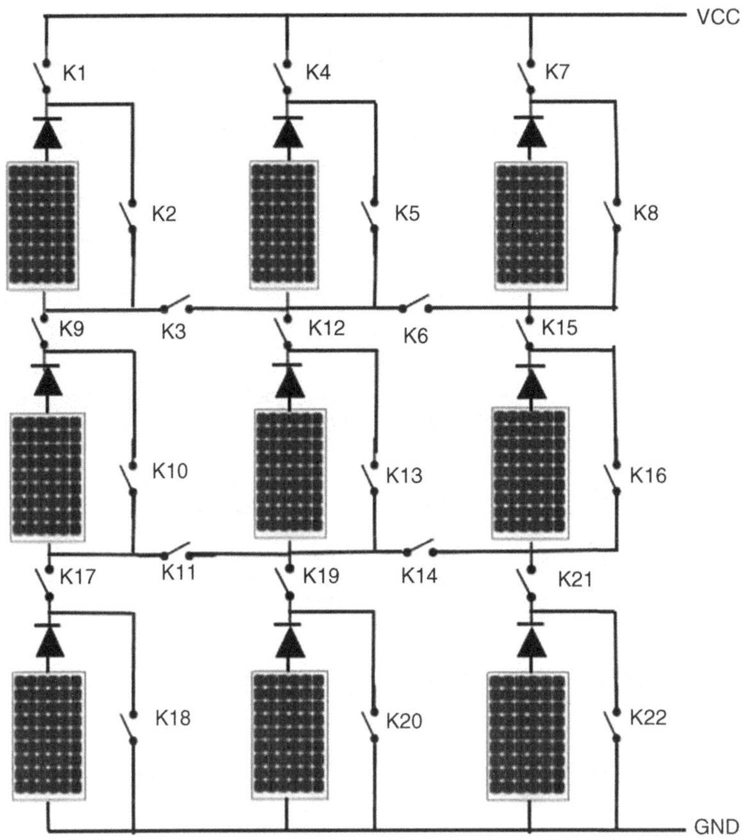

Fig. 2 Experimental setup

Instituto Tecnologico Metropolitano to reconfigure nine solar panels (3×3 matrix). The system described is an electronic board shown in the photograph of Fig. 3 that contains IN and OUT elements, switches (relays), and protection elements:

- Twenty two relay NO (normally open) with a control stage,
- Terminal blocks that provide direct connection with the photovoltaic array and the power supply for the relays,
- Contact pins for the Arduino board,
- Blocking diodes and transistors,
- LED diodes

The relay control stage consists of a transistor whose base is connected to a digital output from the Arduino, this energizes the coil inside the relay and acts as a current amplifier. A LED has been incorporated to verify visually the relay state, and a freewheeling diode which absorbs the current peak from the coil when

Fig. 3 Reconfiguration board

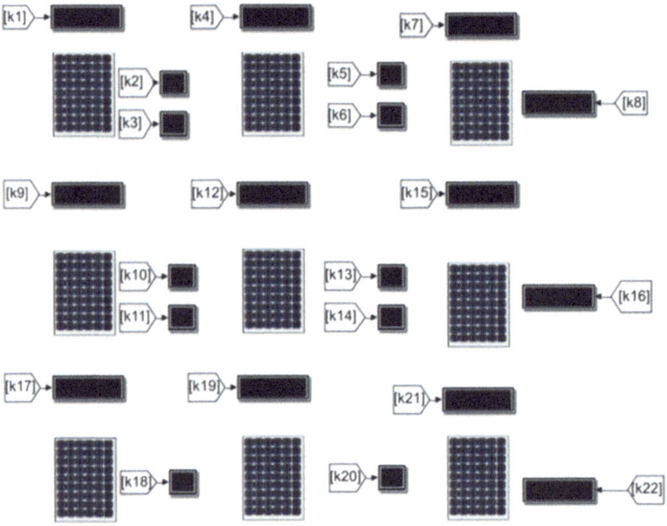

Fig. 4 Graphic interface in Simulink

is connected or disconnected, due without it, the transistor will be exposed to overvoltages that eventually will damage it.

3.1.2 Interface Design with Arduino

A graphic interface in Simulink has been designed to visualize which configuration scheme has been implemented on the board, showing the state of every single relay as is illustrated in Fig. 4. Using the Simulink libraries that allow to communicate Matlab and Simulink with Arduino via serial port. Each relay has attached a digital output of the Arduino, enumerated from K1 until K22, corresponding also with the diagram exposed in Fig. 2. K1, K6, K11, and K14 are the equivalent of the elements in the connection matrix. The graphic interface has 22 displays that show the state of each relay (one for ON and zero for OFF) and 9 solar panels, with the aim of facilitating the view of the actual configuration scheme.

3.2 Experimental Results

In order to evaluate the system described above, a 3×3 PV array has been employed, simulation and experimental results were founded using nine ERDM 10 solar panels, whose main electrical characteristics are: $P_{max} = 10\,\text{W}$, $I_{mp} = 0.58\,\text{A}$, $V_{mp} = 17.40\,\text{V}$, $V_{oc} = 21.8\,\text{V}$, and $I_{sc} = 0.63\,\text{A}$. The effectiveness of the implementation of the graphic interface on the reconfiguration target was tested considering one irradiance profile (i.e., Fig. 5) under solar irradiance conditions without significant changes.

For the 3×3 PV array, 16 possible configurations have been found; after calculating the I and V values for each one of them, a comparison has been made for 3 regular configurations, as is shown in Fig. 6 the $P - V$ characteristic curves of the HC, SP, and TCT configurations show an inflection point due to the bypass diode enabling. This result shows TCT as the optimum configuration with 28.9387 W for this irradiance profile with a 34.54% of improvement over HC configuration (18.9424 W) and 38.54% over SP configuration (17.7856 W). The connection matrices for the regular configurations worked are shown in Figs. 7, 8 and 9; where items (a) of each figure represent the digital signals ("0" for OFF and "1" for ON) for the corresponding outputs to the Arduino and the state of the relays K3, K6, K11, and K14 in the reconfiguration board that allows the connection between strings, items (b) display the interface once the correct ties between strings are deployed.

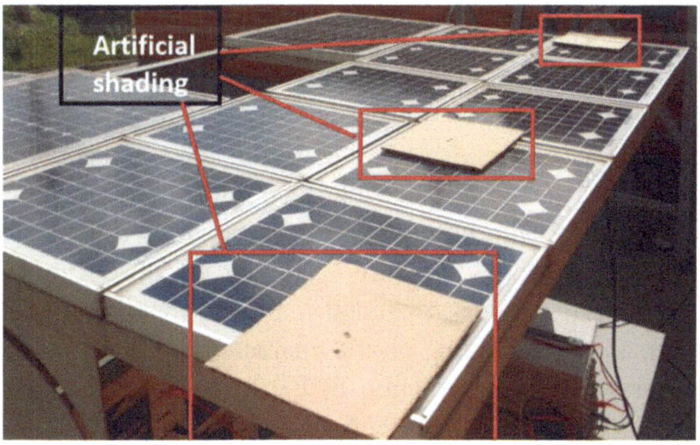

Fig. 5 Experimental test

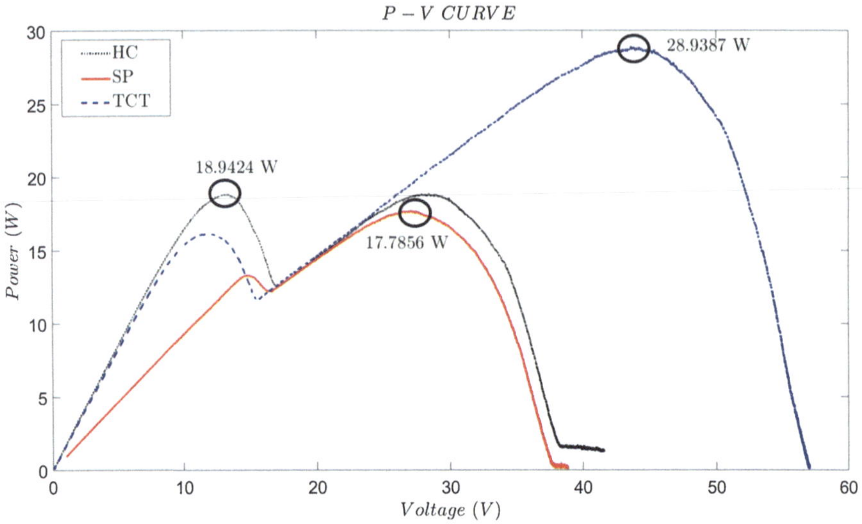

Fig. 6 HC, SP, TCT experimental $P - V$ curves

4 Conclusion

An application of the modeling procedure reported in [7] has been designed and implemented. The identification of all configuration schemes and the automation of the connection matrices calculation for a $(N \times M)$ PV array introduced in this work make possible to consider them for an appropriate evaluation of the I and V characteristics. In this way, a suitable configuration scheme that reaches the GMPP to ensure the maximum power extraction can be found. This application shows high compatibility between Simulink, Matlab, and Arduino, also, is an affordable and more suitable setup for PV systems efficiency modeling. The proposed solution has been validated by experimental tests consisting in the evaluation and comparison of three conventional configurations (HC, SP, and TCT) exposed to the same irradiance profile. The results showed that TCT configuration was the suitable connection providing higher power at the GMPP of 34.54% and 38.54% in comparison with HC and SP configurations, respectively. The evidence of the deployment of the connection matrices related with the regular configurations worked has been illustrated in the graphic interface with the correct ties between strings, also, the designed reconfiguration board can be utilized for PV arrays of a maximum power of 296.055 W. As a future work, the proposed application can be developed in a neuronal network to make real-time identifications with less processing time, also, a proper shading profile characterization is under investigation to improve the accuracy of the prediction of the power in PV arrays. This could be a previous stage that can be added to the solution presented in this paper.

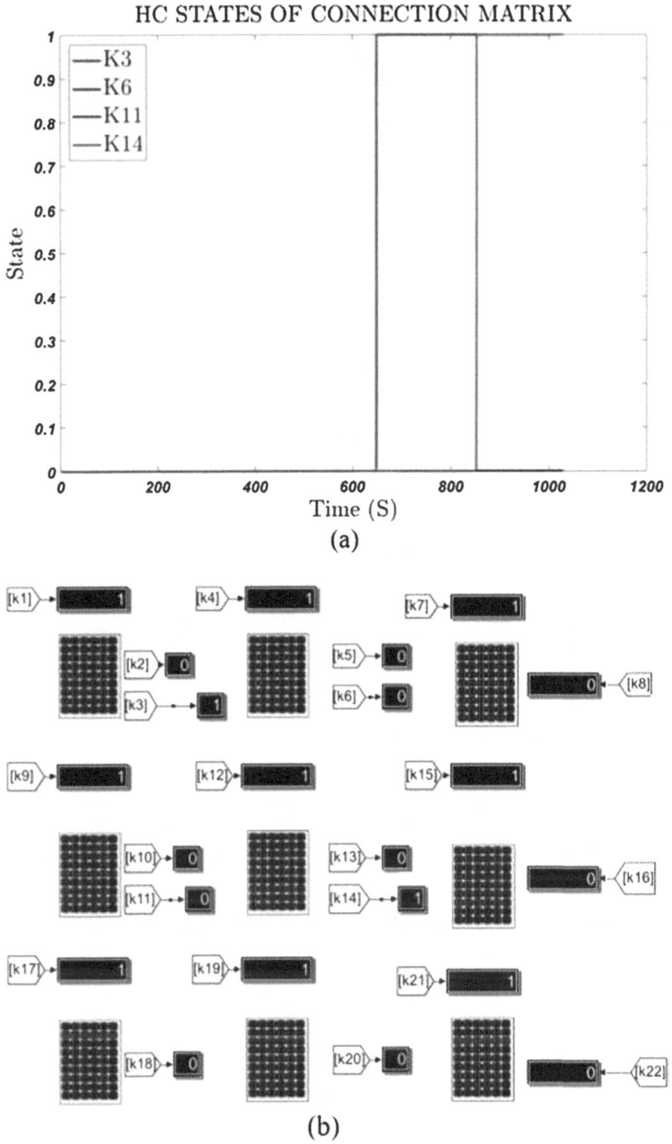

Fig. 7 Result from HC configuration. (**a**) Representation of the digital signals in HC configuration. (**b**) HC configuration in graphic interface

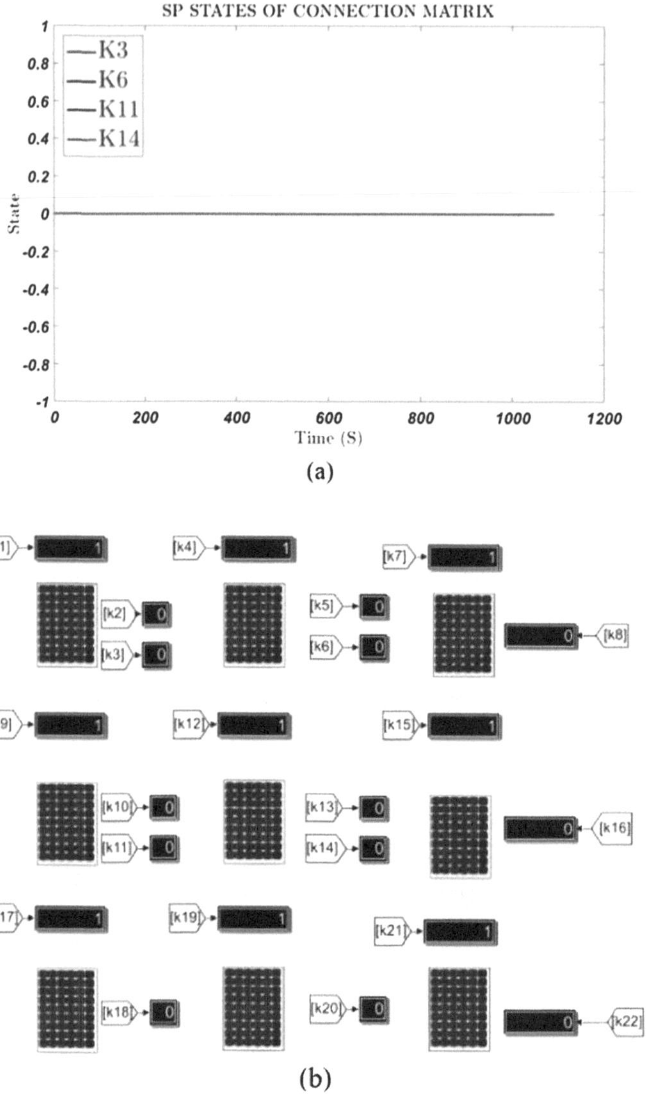

Fig. 8 Result from SP configuration. (**a**) Representation of the digital signals in SP configuration. (**b**) SP configuration in graphic interface

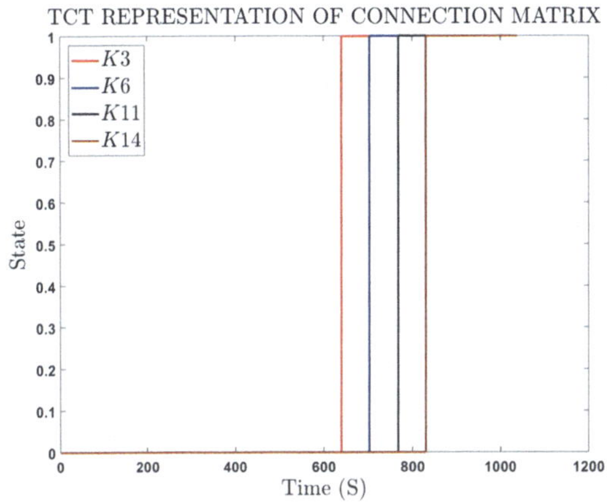

(a)

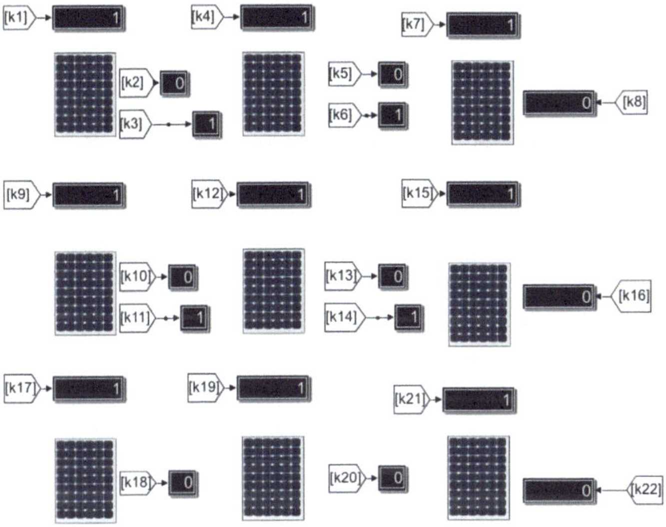

(b)

Fig. 9 Result from TCT configuration. (**a**) Representation of the digital signals in TCT configuration. (**b**) TCT configuration in graphic interface

Acknowledgements This work was supported by the research groups of Automática, Electrónica y Ciencias Computacionales (AECC), and Materiales Avanzados y Energia (MATyER) of the Instituto Tecnológico Metropolitano and the "Programa Jovenes Investigadores e Innovadores ITM."

References

1. REN21, Renewables 2018 Global Status Report. Renewable Energy Policy Network For The 21st Century, pp. 29–47 (2018)
2. G. Cipriani, V. Di Dio, N. Madonia, R. Miceli, F. Pellitteri, F.R. Galluzzo, Reconfiguration strategies to reduce mismatch effects on PV array: an Arduino-based prototype. Paper presented at the 2014 international symposium on power electronics, electrical drives, automation and motion, SPEEDAM 2014, pp. 1003–1008 (2014). https://doi.org/10.1109/SPEEDAM.2014.6872115. Retrieved from www.scopus.com
3. Y. Mahmoud, E. El-Saadany,Fast reconfiguration algorithm for improving the efficiency of PV systems, in *8th International Renewable Energy Congress*, vol. IREC 2017 (2017). https://doi.org/10.1109/IREC.2017.7925993. Retrieved from www.scopus.com
4. B.I. Rani, G.S. Ilango, C. Nagamani, Enhanced power generation from PV array under partial shading conditions by shade dispersion using Su Do Ku configuration.IEEE Trans. Sustainable Energy **4**, 594–601 (2013). https://doi.org/10.1109/TSTE.2012.2230033
5. D.S. Pillai, N. Rajasekar, J.P. Ram, V.K. Chinnaiyan,Design and testing of two phase array reconfiguration procedure for maximizing power in solar PV systems under partial shade conditions (PSC).Energy Convers. Manag. **178**, 92–110 (2018). https://doi.org/10.1016/j.enconman.2018.10.020
6. A. Fathy,Recent meta-heuristic grasshopper optimization algorithm for optimal reconfiguration of partially shaded PV array. Sol. Energy **171**, 638–651 (2018). https://doi.org/10.1016/j.solener.2018.07.014
7. J.D. Bastidas Rodriguez, L.A. Trejos Grisales, D. Gonzalez Montoya, C.A. Ramos Paja, G. Petrone, G. Spagnuolo,General modeling procedure for photovoltaic arrays. Electr. Power Syst. Res. J. Renewable Sustainable Energy **155**, 67–79 (2018). https://doi.org/10.1016/j.epsr.2017.09.023
8. N. Mishra, A. Singh-Yadav, R. Pachauri, K. Yogesh, K. Vinod, Performance enhancement of PV system using proposed array topologies under various shadow patterns. Sol. Energy **157**, 641–656 (2017). https://doi.org/10.1016/j.solener.2017.08.021
9. S. Malathy, R. Ramaprabha, Reconfiguration strategies to extract maximum power from photovoltaic array under partially shaded conditions. Renew. Sust. Energ. Rev. **81**, 2922–2934 (2018). https://doi.org/10.1016/j.rser.2017.06.100

Experimental Verification of a Method to Model the Operation of PV Modules During Irradiance Transitions

Kari Lappalainen and Seppo Valkealahti

Abstract Photovoltaic (PV) systems are prone to deep, steep and frequent irradiance fluctuations, mainly originated from overpassing cloud shadows, which cause fluctuations in PV power production. These irradiance transitions have been modelled by using a mathematical function to study their behaviour in a systematic way. Although the used methods and obtained results seem to be reliable, the simulation model has not been verified in detail. In this paper, the accuracy of the used theoretical model for irradiance transitions has been verified experimentally. The results show that the simulation model is accurate enough to study the irradiance transitions caused by moving clouds and their effects on the operation of PV systems.

1 Introduction

Photovoltaic (PV) systems are prone to irradiance fluctuations, mainly originated from overpassing cloud shadows, which cause fluctuations in PV power production. These fluctuations can be deep, steep and frequent [1]. Partial shading resulting from cloud shadows may also lead to mismatch power losses and to the occurrence of multiple maximum power points (MPPs) leading to failures in MPP tracking, thereby causing extra losses.

During the past few years, several studies have been presented regarding the operation of partially shaded PV systems, e.g. [1–5]. The PV modules are usually modelled through the one-diode model, which provides a good trade-off between accuracy and complexity. However, the actual correctness and accuracy of the used simulation approaches and tools have obtained little attention. A comparison between the one- and two-diode models of a PV cell under partial shading conditions has been presented in [6] using MATLAB Simulink. An LTspice simulation model,

K. Lappalainen (✉) · S. Valkealahti
Electrical Engineering, Tampere University, Tampere, Finland
e-mail: kari.lappalainen@tuni.fi; seppo.valkealahti@tuni.fi

W. Zamboni, G. Petrone (eds.), *ELECTRIMACS 2019*, Lecture Notes in Electrical Engineering 615, https://doi.org/10.1007/978-3-030-37161-6_37

based on the one-diode model, has been presented and validated in [7]. In [8], a one-diode model-based simulation model of PV arrays has been presented and compared with two earlier published approaches.

In [9, 10], studies on the effects of the length of a simulation time step, the size of a basic simulation unit, and the resolutions of *I–U* curves and irradiance data on the accuracy of the simulations of PV system operation under partial shading conditions have been presented. It has been found that the irradiance accuracy of 1.0 W/m^2 and the number of *I–U* curve points of 1000 are sufficient for PV string and array-level partial shading studies. The effect of the simulation unit size up to the size of a PV module has been found to be moderately small in long-term PV analysis, and a time step of 1 s has been found to be short enough for most studies. However, sharp irradiance transitions may reveal an exception.

This paper presents an experimental verification of a method to model the operation of PV modules during irradiance transitions resulting from cloud shadows. The modelling method utilises the one-diode model and a mathematical irradiance transition model presented in [11]. Irradiance and temperature measurements and *I–U* curve measurements on a single PV module with sampling frequencies of 10 and 1 Hz, respectively, of 4 days have been analysed. The analyses show that the model of change in irradiance based on the modified sigmoid function gives accurate simulation results compared to simulations based directly on the measured irradiance and temperature of the PV module. Thus, the model of the irradiance transitions caused by moving clouds provides a realistic and usable way for PV power fluctuation studies.

2 Methods

2.1 Simulation Model for the PV Modules

The one-diode model provides the following relationship between the current I and voltage U of a PV module:

$$I = I_{\mathrm{ph}} - I_{\mathrm{o}}\left(e^{\frac{U+R_{\mathrm{S}}I}{AN_{\mathrm{S}}kT/q}} - 1\right) - \frac{U + R_{\mathrm{s}}I}{R_{\mathrm{sh}}}. \tag{1}$$

The parameters of this model are the light-generated current I_{ph}, the dark saturation current I_{o}, the series resistance R_{s}, the ideality factor A, the temperature T and the shunt resistance R_{sh}. N_{s} is the number of PV cells in the module, k the Boltzmann constant and q the elementary charge. In this study, the bypass diodes of the module were modelled using Eq. (1) with the assumptions that I_{ph} is zero, R_{sh} is infinite and the bypass diode temperature is equal to the temperature of the PV cells.

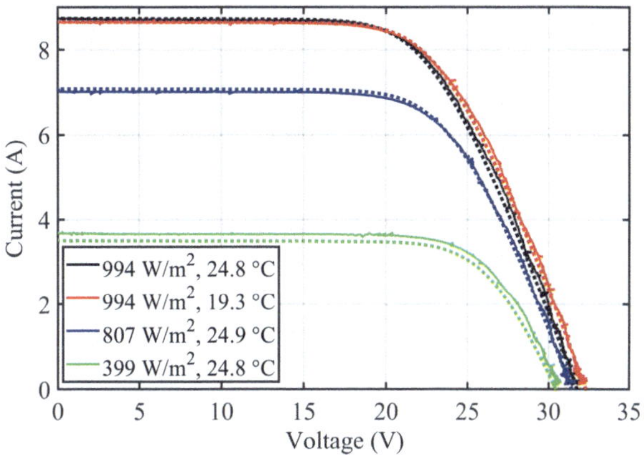

Fig. 1 Measured (solid lines) and simulated (dashed lines) *I–U* curves under various operating conditions

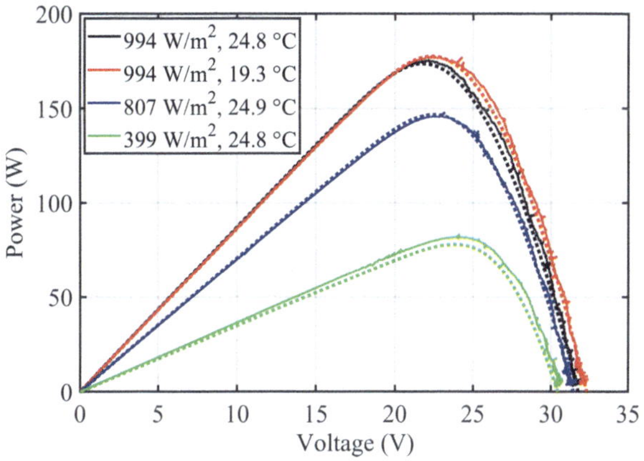

Fig. 2 Measured (solid lines) and simulated (dashed lines) *P–U* curves under various operating conditions

In Figs. 1 and 2, the measured and simulated *I–U* and *P–U* curves are presented under various operating conditions. The one-diode model was fitted to the measured *I–U* curves of a NAPS NP190GKg PV module. The simulated curves were obtained using the one-diode model and measured module irradiance and temperature at the time of curve measurement as input values. The black curves represent closely standard test conditions (STC) (1000 W/m^2 and 25 °C), the red curves lower temperature with the STC irradiance and the blue and green curves lower irradiances with the STC temperature. The differences in the short-circuit (SC) current and open-circuit (OC) voltage and in the MPP power, voltage and current between the measured and simulated curves of Figs. 1 and 2 are compiled into Table 1.

Although the simulation model is quite accurate under good irradiance conditions, the accuracy is lower on low irradiance levels. This is a known feature of the one-diode model. The inaccuracy on low irradiance levels appears especially as a difference in current and, thereby, in power.

Table 1 Accuracy of the simulation model in the cases of Figs. 1 and 2 with respect to the measured I–U curves

Case	1	2	3	4
Irradiance (W/m^2)	994.4	994.5	806.7	399.4
Temperature (°C)	24.83	19.34	24.88	24.80
Difference in U_{OC} (%)	0.156	0.188	0.266	0.840
Difference in I_{SC} (%)	0.100	0.002	0.853	4.632
Difference in P_{MPP} (%)	0.915	0.252	0.519	5.176
Difference in U_{MPP} (%)	1.420	0.612	1.405	0.551
Difference in I_{MPP} (%)	0.496	0.228	2.305	4.415

It is worth noting that the experimental measurements of the PV module were not performed at the terminals of the module. Thus, the measurements include the resistance of quite long cables (0.363 Ω), which is taken into account in the adjusting of the simulation model. Measurements also reveal that the actual module SC current is larger, and the OC voltage is smaller in STC than declared by the manufacturer.

2.2 *Irradiance Transitions*

Measurements done at the solar PV power station research plant of Tampere University have been used in this study. By utilising the method described in [11], 201 decreasing and 199 increasing irradiance transitions were identified in 4 days (June 22nd–25th 2018) of data measured with a single irradiance sensor attached to the studied PV module. The median duration of these transitions was 12.4 s.

The mathematical model based on the modified sigmoid function was used to model irradiance transitions caused by cloud edges [11]. The model gives irradiance G during a transition as follows:

$$G(t) = \frac{G_{us} - G_{s}}{1 + e^{(t-t_0)/b}} + G_{s}, \tag{2}$$

where G_{us} and G_s are the irradiances of an unshaded and a fully shaded situation, respectively, and t is the time. Parameter t_0 defines the midpoint of the transition. Parameter b is related to transition steepness, and its sign defines whether the transition is decreasing or increasing. The operation of the mathematical model has been validated with around 43,000 measured irradiance transitions in [12]. The average of the normalised root-mean-square deviation (NRMSD) between the curve fits and the measured irradiance transitions, normalised to the irradiance change during the transition, was 2.7%.

By using the mathematical model, irradiance transitions can be defined by four independent variables: b; shading strength, i.e. the irradiance attenuation due to shading; apparent speed; and apparent direction of movement [12]. The combined use of the mathematical model and the simulation model of the PV module enables realistic simulations of the electrical behaviour of PV generators during partial shadings caused by clouds using measurements of only three irradiance sensors [12].

The irradiance given by the curve fit of the modified sigmoid function (Eq. 2) to the measured irradiance data was used as an input for the simulation model. The temperature of the PV module was assumed to stay constant during an irradiance transition, and the temperature at the beginning of the transition was used as an input for the simulation model. As a theoretically ideal reference case, the same one-diode model of the PV module was applied by using the measured irradiance and temperature during the transition as inputs.

3 Results

First, the accuracy of the used one-diode model was studied utilising the measurements of 4 days. Daily measurement time was from 9 a.m. to 6 p.m. (UTC + 2). The simulation results, obtained using the measured irradiance and temperature, were compared to the measured *I*–*U* curves. The results of this analysis are presented in Sect. 3.1.

In Sect. 3.2, the operation of the PV module during the identified irradiance transitions was studied. The simulation results obtained by utilising the measured irradiance and temperature values and by using the mathematical irradiance transition model were compared to the measured *I*–*U* curves. The results are further discussed in Sect. 3.3.

3.1 Accuracy of the One-Diode Model

The measured irradiance and the MPP power of the PV module for a 24 min period on June 23rd 2018 are presented in Fig. 3. The chosen period contained many irradiance transitions caused by moving clouds and is therefore a good example of partly clouded conditions. As can be seen in the figure, the irradiance transitions can be very deep and steep. The measured power follows the irradiance closely, but the amplitude of the changes is smaller. The reason for this is that the efficiency of PV cells decreases with increasing cell temperature, which is caused by increasing irradiance.

The measured and simulated MPP power, voltage and current during a 5 min period on June 23rd 2018 are presented in Fig. 4. The differences between the simulation results and measurements were reasonably small. It can be seen in Fig. 4 that the simulation model somewhat smoothens the fastest fluctuation spikes in the measurements. By comparing Figs. 3 and 4, it can be noticed that the MPP power and current increase with the increasing irradiance, whereas the MPP voltage

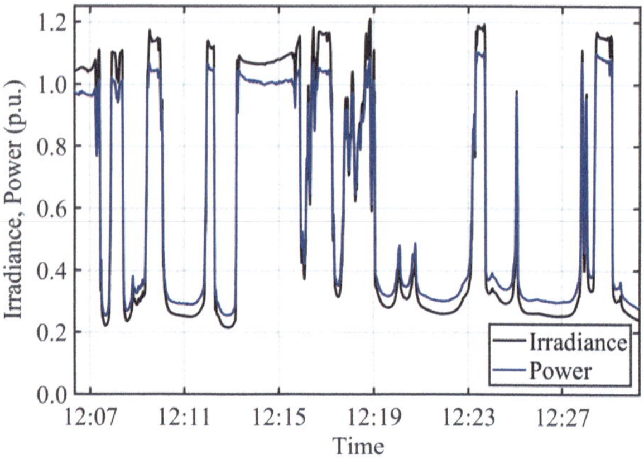

Fig. 3 Measured irradiance and PV module power during a 24 min period on June 23rd 2018. The irradiance is shown relative to the STC irradiance and the power relative to the nominal MPP power in STC

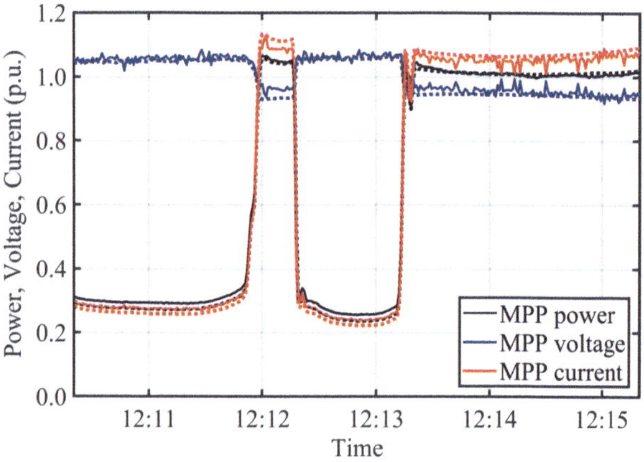

Fig. 4 Measured (solid lines) and simulated (dashed lines) MPP power, voltage and current during a 5 min period on June 23rd 2018

behaves in reverse. The decrease of the MPP voltage with the increasing irradiance is caused by the increase of the operating temperature of the PV cells.

The NRMSDs between the simulated and measured MPP power, voltage and current during the period of 4 days are compiled into Table 2. The NRMSD value of each quantity was normalised to the average of the measured values of the quantity. The NRMSD of the MPP voltage was clearly smaller than the NRMSD of the MPP power and current. The reason for this is that, as illustrated in Fig. 4, the changes in the MPP power and current during the studied period were much larger than in the MPP voltage, which is almost constant with irradiance values higher than 400 W/m^2 [12].

As stated earlier, the accuracy of the simulation model is worse with low irradiances than with high irradiances. Thus, the NRMSDs of the MPP power, voltage and current have been presented separately for low and high irradiances in Table 2. The NRMSD values of the MPP power and current were about three and two times, respectively, as large as at low irradiances than at high irradiances. On the contrary, the NRMSD of the MPP voltage actually decreased with the decreasing

Table 2 NRMSD values of the MPP power, voltage and current between the measurements and simulations during the period of 4 days

Level of irradiance	P_{MPP} (%)	U_{MPP} (%)	I_{MPP} (%)
All irradiances	4.26	1.26	4.54
Below 400 W/m^2	5.93	1.25	5.83
Above 700 W/m^2	2.02	1.54	2.91

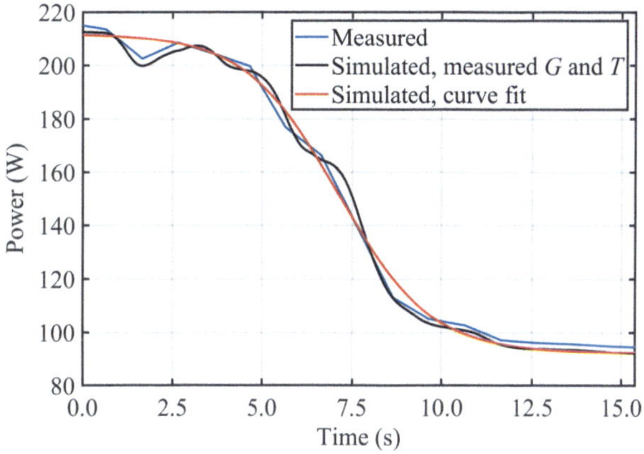

Fig. 5 Measured and simulated MPP power of the PV module during an identified irradiance transition

irradiance. During the studied period, the low irradiance values below 400 W/m^2 appeared 80% of the time leading to large overall NRMSD values of the MPP power and current. Only 10% of the time irradiance was more than 700 W/m^2.

3.2 Accuracy of the Irradiance Transition Model

An example of the behaviour of the MPP power of the PV module during a typical irradiance transition is presented in Fig. 5. As can be seen, the simulated power, obtained by using the measured irradiance and temperature as inputs, follows the measured power very well. Also, the simulated power, obtained with the curve fit, matches well to the measured power with an NRMSD of 3.3%. In general, the largest differences between the simulated MPP power using the curve fit and the experimental MPP power appear during fluctuation spikes occurring during the transitions (such as around 1.5 s in Fig. 5) and during clearly asymmetrical transitions since the fitted function is symmetrical.

The NRMSDs between the two simulated and the measured MPP power, voltage and current during the identified irradiance transitions are presented in Table 3. The NRMSD values between the simulated MPP power and current using the measured G and T and the measured ones were only somewhat higher than during the whole period of 4 days (Table 2), while the NRMSD of the MPP voltage was even a bit lower. This demonstrates that the one-diode simulation model works well also

Table 3 NRMSD values between the two simulated and the measured MPP power, voltage and current during the identified irradiance transitions

Simulation method	P_{MPP} (%)	U_{MPP} (%)	I_{MPP} (%)
Measured G and T	4.68	1.25	5.19
Curve fit	7.18	1.31	7.72

during the transitions. The NRMSD values between the simulated MPP power and current using the curve fit and the measured ones were only around 50% larger that when using the measured G and T, whereas there was only negligible difference in the NRMSD of the MPP voltage between the two simulation methods. The small difference between the NRMSD values of the two simulation approaches confirms that the mathematical model based on the modified sigmoid function describes the irradiance transitions caused by cloud edges very well.

3.3 *Discussion*

The differences between the simulation results based on the one-diode model and the measured I–U curves are caused partly by the simplifications made in the used PV cell model but result also from other sources. The temperature used in the simulations was measured from the backside of the module. This temperature, naturally, can be up to 1 °C lower than the actual operating temperature of the PV cells in the module. Moreover, the operating conditions of all the cells were assumed to be identical with each other. In practice, there are always some differences in the operating conditions of the cells. Especially during sharp irradiance transitions, irradiance and temperature differences can be notable. Also, when the mathematical irradiance transition model was applied, the temperature of the PV module was assumed to stay constant during an irradiance transition which causes some error to the results. Anyway, the differences between the simulation results and the measured I–U curves were relatively small.

4 Conclusions

An experimental verification of a method to model the operation of PV modules during irradiance transitions caused by cloud shadows was presented in this paper. The method uses the one-diode model for PV modules and a mathematical model based on the modified sigmoid function for irradiance transition. The one-diode model was shown to be accurate enough for most of the PV module and system analyses under highly varying climatic conditions. The modified sigmoid function has been used earlier to study irradiance transitions caused by moving clouds in a systematic way. It was verified experimentally in this study that irradiance transitions can be modelled accurately enough with the modified sigmoid function to study the irradiance transitions caused by moving clouds and their impact on the operation of PV systems.

References

1. K. Lappalainen, S. Valkealahti, Output power variation of different PV array configurations during irradiance transitions caused by moving clouds. Appl. Energy **190**, 902–910 (2017)
2. S.R. Potnuru, D. Pattabiraman, S.I. Ganesan, N. Chilakapati, Positioning of PV panels for reduction in line losses and mismatch losses in PV array. Renew. Energy **78**, 264–275 (2015)
3. N. Rakesh, T.V. Madhavaram, Performance enhancement of partially shaded solar PV array using novel shade dispersion technique. Front Energy **10**, 227–239 (2016)
4. S. Vijayalekshmy, G.R. Bindu, S. Rama Iyer, A novel Zig-Zag scheme for power enhancement of partially shaded solar arrays. Sol. Energy **135**, 92–102 (2016)
5. R. Yan, T.K. Saha, P. Meredith, A. Ananth, M.I. Hossain, Megawatt-scale solar variability study: an experience from a 1.2 MWp photovoltaic system in Australia over three years. IET Renew. Power Gen. **10**, 1229–1236 (2016)
6. O.W. Yin, B.C. Babu, Simple and easy approach for mathematical analysis of photovoltaic (PV) module under normal and partial shading conditions. Optik **169**, 48–61 (2018)
7. S. Gallardo-Saavedra, B. Karlsson, Simulation, validation and analysis of shading effects on a PV system. Sol. Energy **170**, 828–839 (2018)
8. J.D. Bastidas, E. Franco, G. Petrone, C.A. Ramos-Paja, G. Spagnuolo, A model of photovoltaic fields in mismatching conditions featuring an improved calculation speed. Electr. Power Syst. Res. **96**, 81–90 (2013)
9. K. Lappalainen, S. Valkealahti, Simulation resolution of PV system partial shading studies, in 33rd European Photovoltaic Solar Energy Conference Amsterdam, pp. 1973–1977, (2017)
10. K. Lappalainen, S. Valkealahti, Size of a basic simulation unit in PV system partial shading studies, in 35th European Photovoltaic Solar Energy Conference Brussels, pp. 1647–1651, (2018)
11. K. Lappalainen, S. Valkealahti, Recognition and modelling of irradiance transitions caused by moving clouds. Sol. Energy **112**, 55–67 (2015)
12. K. Lappalainen, Output power variation and mismatch losses of photovoltaic power generators caused by moving clouds, Doctoral Thesis, Tampere University of Technology (2017)

Optimal Power Dispatching in the DC Microgrid with Clear Sky Irradiance Model

Hongwei Wu, Wenshuai Bai, Fabrice Locment, and Manuela Sechilariu

Abstract The optimization of power dispatching has been proved to be useful for reducing the operation energy cost of a microgrid based on photovoltaic source. However, the formulation of the optimization problem needs the weather forecast to predict photovoltaic generation. The current hourly forecast is always available and often lacks accuracy. Thus, this work proposes the optimization based on a clear sky model to predict the solar irradiance. This model has the advantage of simplicity, since it depends only on the geographical coordinates. The analyses have been done to compare the weather data during 5 months, and the validation of the proposed model is carried out by simulation. The results show the optimization results of the proposed model are slightly better than a common hourly forecast weather provided by a meteorological website.

1 Introduction

The DC microgrid has been the focus of research recently, because it can integrate effectively renewable resources, such as photovoltaic (PV) panels [1]. Particularly, the microgrid can be integrated into a building with rooftop PV panels to make a zero-energy or positive-energy building [2].

Similar to the traditional power grid, optimal power dispatching can reduce the power losses and the operation cost of the microgrid [3]. Till now, a lot of optimization algorithms have been developed, and among them, the mixed integer linear programming (MILP) method is widely used for being fast and effective [4, 5]. However, their effects depend highly on the accuracy of the load and production predictions [6]. For traditional grid, it involves only the load consumption prediction, which has been thoroughly studied [7] and can be adapted directly to the microgrid problems. However, in the context of microgrids based on

H. Wu (✉) · W. Bai · F. Locment · M. Sechilariu
Sorbonne University – Université de Technologie de Compiègne, Compiègne, France
e-mail: hongwei.wu@utc.fr; wenshuai.bai@utc.fr; fabrice.locment@utc.fr; manuela.sechilariu@utc.fr

W. Zamboni, G. Petrone (eds.), *ELECTRIMACS 2019*, Lecture Notes in Electrical Engineering 615, https://doi.org/10.1007/978-3-030-37161-6_38

renewable energy, the generation prediction is essential too [7]. Regarding the PV panels, the prediction concerns mainly on the solar irradiance for a given site [6]. Indeed, this is still a challenge. Even though the weather satellites can give precise information on a large scale, the high-resolution solar irradiance forecast is hardly accessible or costing at a high expense.

Large amount of precedent research concentrates on the local irradiance forecast problem. The solutions include using the numerical weather prediction model as stated in [8], analysing the satellite images [9] and using all-sky imager for cloud tracking [10]. Also, the statistical methods can be useful, such as the switching Markov model [11]. Recently, it is also popular to predict by the machine learning algorithms. In [12], the tree-based ensemble method predictions are presented and compared with a support vector regression predictor. In [13], the prediction is done by combining wrapper mutual information and extreme learning machine. Besides, the probabilistic forecast is another way to deal with the high variability of the renewable sources, such as presented in [14–16]. Unlike deterministic forecast, which outputs a series of fixed values, the probabilistic forecast produces most probable values with a certain interval. However, these methods need either the extra data, such as satellite images and all-sky image, or the long-term historical irradiance data to train the artificial intelligence model or establish the statistical model. Thus, the application is limited by the availability of these conditions.

This paper proposes to use the simple clear sky (CS) model to realize irradiance prediction. The advantage is that only the geographical coordinates of the site and the PV panel orientation are needed, which are accessible easily. The prediction results are compared with the real recorded data in Compiègne in Northern France, as well as the free accessible forecast data made by *Météo-France* (MF), during a period of 5 months in 2018. Moreover, the two sets of prediction are inputted into a MILP-based optimization solver to do the power dispatching, and the real data based simulations are done, in order to study the effectiveness of the prediction.

This paper is organized as follows: The clear sky model and MF forecast are presented and compared in Sect. 2, and the optimization problem of a DC microgrid is given in Sect. 3, followed by the simulation and the results presented in Sect. 4. The final conclusion is given in Sect. 5.

2 Weather Forecast

In the context of microgrid power dispatching, the weather forecast is often expected to give the solar irradiance prediction for the day-ahead and intraday optimization. Thus, the prediction time horizon should cover the whole PV operation period. In this paper, the daily period is set from 9:00 to 18:00.

2.1 Clear Sky Irradiance Model

The CS model is a method based on geometric calculation to predict solar irradiance on a specific surface while no cloud influence is considered. Obviously, it is accurate only in sunny days, but its accuracy is acceptable when the sky is not heavily covered. This is particularly useful for PV panels, since they are mostly installed in sunny regions.

Since the CS model is classical, one common method is adopted in this work, and more details can be found in [17].

The beam radiation on earth g_{b} can be simply expressed as follows:

$$g_{\mathrm{b}} = g_{\mathrm{sc}} \cdot T^{m}. \tag{1}$$

g_{sc} is the extraterrestrial solar irradiation and can be set as 1367 W/m^2. T is the atmospheric transmittance for short wave solar irradiance and can be seen as 0.7 for simplicity. m is the air mass coefficient, and it is approximately expressed by the local zenith angle θ_{z} as given in Eq. (2):

$$m = \frac{1}{\cos\left(\theta_{\mathrm{z}}\right)}. \tag{2}$$

The beam radiation on a given sloped surface can be then determined as the following:

$$g_{\mathrm{bg}} = g_{\mathrm{b}} \cdot \cos\theta. \tag{3}$$

θ is the angle of incidence between the beam irradiance on the surface and the normal to the surface. An approximate expression of θ is given in Eq. (4).

$$\begin{aligned} \cos\theta = {} & \sin\delta \cdot \sin\varphi \cdot \cos\beta - \\ & \sin\delta \cdot \cos\varphi \cdot \cos\beta \cdot \cos\gamma \\ & + \cos\delta \cdot \cos\varphi \cdot \cos\beta \cdot \cos\omega \\ & + \cos\delta \cdot \sin\varphi \cdot \sin\beta \cdot \cos\gamma \cdot \cos\omega \\ & + \cos\delta \cdot \sin\varphi \cdot \sin\gamma \cdot \sin\omega \end{aligned} \tag{4}$$

where δ is the declination angle of the sun at the solar noon, φ is the latitude, β is the slope angle of the given surface, γ is the surface azimuth angle which is 0 if due south and ω is the hour angle. It is evident that these data can be derived from the geographical coordinates and the slope of the given surface, i.e. the installed PV panels. All of them are easily accessible for any PV installation.

Besides, the diffuse radiation on the surface g_{dg} is obtained by Eq. (5).

$$g_{\mathrm{dg}} = 0.3 \cdot \left(1 - T^{m}\right) \cdot \frac{\left(1 + \cos\beta\right) g_{\mathrm{sc}}}{2} \cdot \cos\theta. \tag{5}$$

The reflect radiation may exist if there exists any building near the PV panels. However, it needs a complex geometric modelling and differs from one site to another. Thus, it is not considered in this work. As a result, the total radiation g_{tg} obtained by the clear sky model is the sum of the beam radiation and the diffuse radiation.

$$g_{tg} = g_{bg} + g_{dg}. \tag{6}$$

2.2 *Météo-France Forecast*

MF is a public establishment that offers meteorological information in France. It offers high-resolution metrological forecast covering all the French territory under open licence [18]. The forecast is generated by the AROME model, and the geographical resolution is 0.025°, which corresponds to an area of 6.25 km^2. The hourly forecast is made several times per day including the accumulated ground solar radiation in J/m^2. Aiming at the day-ahead optimization, this work focuses on the forecast made at midnight. In order to obtain the instant solar irradiance, a hypothesis is made assuming that the irradiance is homogeneous in the area and uniformly distributed in the 1-h interval.

2.3 *Case Study*

A group of PV panels are installed in the parking area of Université de Technologie de Compiègne as depicted in Fig. 1, whose coordinates are 49.401°N, 2.796°E. The PV panels are with a slope of 4° and an azimuth angle of −97°, almost in the direction of east. In addition, a sensor system is installed to record the instant solar irradiance on the panels at every 10 s.

In order to validate the CS model, the instant irradiance prediction given by CS model is compared with the recorded data for sunny days as depicted in Fig. 2. The blue and red curves are the beam radiation and total radiation obtained by the CS model, and the yellow curve is the recorded real data.

In Fig. 2a, the real irradiance is close to the beam radiation except for the early hours in the morning. On the contrary, the real irradiance is close to the total radiation in Fig. 2b. This is because the buildings nearby are sheltering the PV panels from the diffuse radiation, when the sun position is high. This is the case of Fig. 2a. When the sun position is relatively low, the same buildings reflect the solar irradiance on the PV panels.

Since the hourly forecast from MF is of ladder form and the visual comparison to the continuous curve in Fig. 2 is not significant, the hourly average irradiances are compared in Fig. 3, in which the yellow bar represents the MF forecast. It can be seen that this forecast gives a higher prediction than the total radiation from

Fig. 1 Photovoltaic panels in the campus in Compiègne, France

the CS model but close to the real irradiance on the 26th June 2018. However, the forecast on the 10 October is far from the reality. The real irradiance of this day is even higher than the total radiation of the CS model, meaning the atmospheric transparency T is exceptionally higher than 0.7. Besides the graphs, some usual statistical indicators can help to compare the CS prediction and the MF forecast with the recorded data, such as mean absolute error (MAE), mean bias error (MBE), root mean square error (RMSE), mean absolute percentage error (MAPE) and Pearson correlation coefficient (ρ). They are defined as follows:

$$\mathrm{MAE} = \frac{1}{n}\sum_{k=1}^{n}\left|g_{\mathrm{pred,k}} - g_{\mathrm{real},k}\right|. \tag{7}$$

$$\mathrm{MBE} = \frac{1}{n}\sum_{k=1}^{n}\left(g_{\mathrm{pred,k}} - g_{\mathrm{real},k}\right). \tag{8}$$

$$\mathrm{RMSE} = \sqrt{\frac{1}{n}\sum_{k=1}^{n}\left(g_{\mathrm{pred},k} - g_{\mathrm{real},k}\right)^{2}}. \tag{9}$$

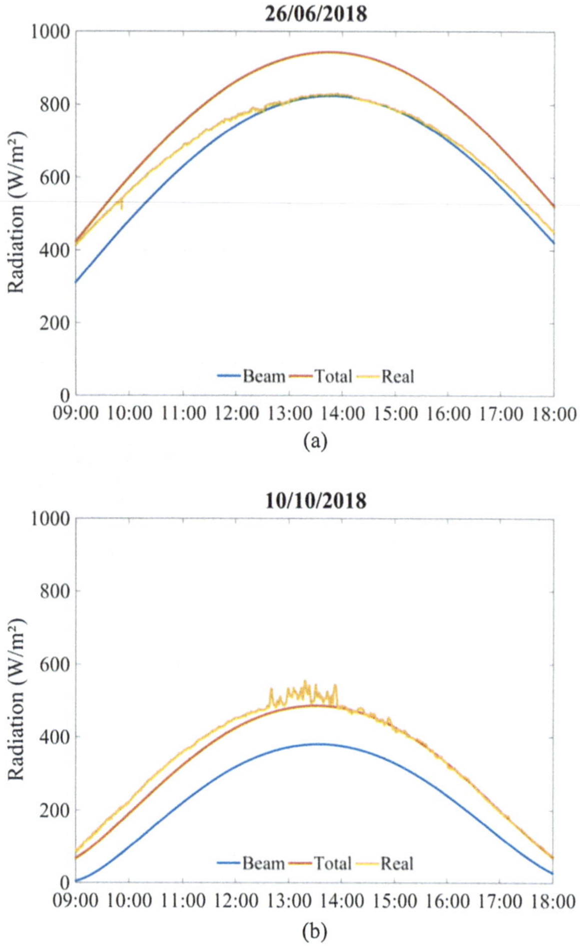

Fig. 2 Instant irradiance comparison between the prediction given by CS model and recorded data: (**a**) on the 26th June 2018; (**b**) on the tenth October 2018

$$\mathrm{MAPE} = \frac{1}{n}\sum_{k=1}^{n}\left|\frac{g_{\mathrm{pred},k} - g_{\mathrm{real},k}}{g_{\mathrm{real}k}}\right| \cdot 100\%. \tag{10}$$

$$\rho = \frac{\mathrm{cov}\left(g_{\mathrm{pred}}, g_{\mathrm{real}}\right)}{\sigma_{\mathrm{g}}\mathrm{pred} \cdot \sigma_{\mathrm{g}}\mathrm{real}}. \tag{11}$$

$g_{\mathrm{pred},k}$ is kth predicted irradiance and $g_{\mathrm{real},k}$ is kth real irradiance. $\mathrm{cov}(g_{\mathrm{pred}}, g_{\mathrm{real}})$ indicates the covariance between the prediction and the reality; $\sigma_{\mathrm{g\_pred}}$ and $\sigma_{\mathrm{g\_real}}$ are the standard deviation of the predicted irradiance and that of the real irradiance, respectively.

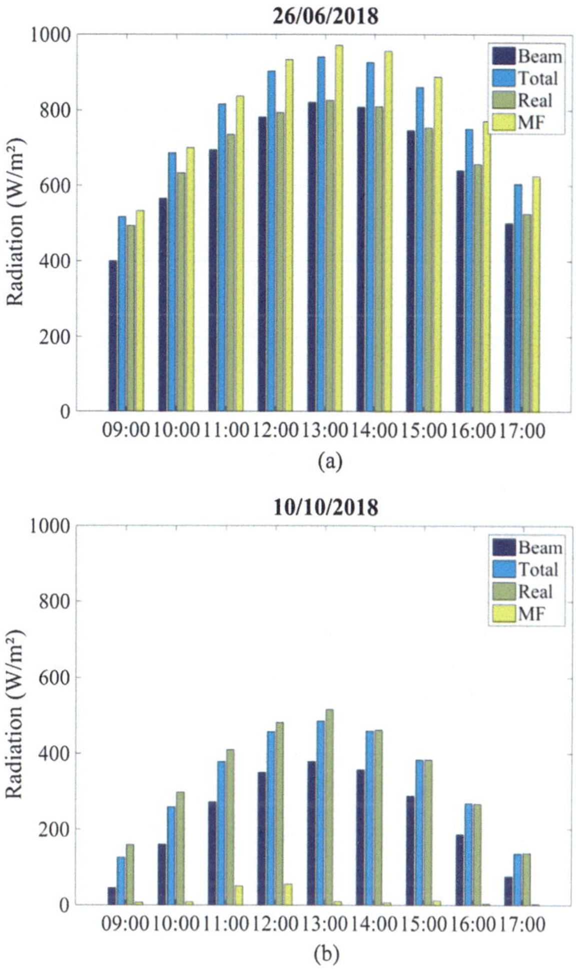

Fig. 3 Hourly average irradiance of the predictions and the recorded data: (a) on the 26th June 2018; (b) on the tenth October 2018

MAE and MBE tell the mean accuracy of the prediction without and with the bias. A positive bias means the over-prediction, whereas negative means the under-prediction. RMSE value is more impacted by the large errors than by the small ones. MAPE reflects the average relative error. ρ describes the linear similarity between the prediction and the reality. From Table 1, it can be seen that the CS prediction is superior to the MF forecast, since the values of MAE, MBE, RMSE and MAPE are smaller, meaning less error, and the Pearson coefficient is higher. It validates the CS model in case of sunny days.

Further validation must be done for longer period. The studied site is located in Northern France, and in winter day, little PV generation can be made. Thus, a 5-month study, dated from the third June to the 31st October 2018, is carried out (except the eighth October, data unavailable due to system maintenance). The

Table 1 Statistical comparison of 2 sunny days

Statistics	26th June 2018		10th October 2018	
	Clear sky	MF	Clear sky	MF
MAE	86.69	110.02	18.96	329.00
MBE	86.69	110.02	−17.75	−329.00
RMSE	91.85	115.48	24.86	350.97
MAPE	10.80%	15.59%	7.49%	95.64%
ρ	99.39%	99.26%	99.16%	48.28%

Table 2 Statistical comparison for 5 months

Statistics	Clear sky	MF
Average MAE	197.35	208.60
Average MBE	184.62	141.22
Maximal RMSE	627.44	549.56
Average MAPE	33.69%	97.85%
Average ρ	62.32%	62.75%

statistics are done each day between 9:00 and 18:00, and the results are given in Table 2.

The mean MAE of the CS model is close to the mean MBE, since in cloudy days the model always tends to over-predict. On the contrary, the average MBE for MF forecast is obviously less, since it can both over- and under-predict the irradiance. The maximal RMSE of CS model is quite large, and it is surely due to the bad weather, such as the heavy rain. But the MAPE of 33% is acceptable for such a simple model, especially when compared to the nearly 100% of MF forecast. In the end, both predictions present a Pearson coefficient of 62%, meaning in general the predictions are synchronized with the reality.

3 Prediction in the Optimization Problem for DC Microgrid Power Dispatching

The power dispatching problem in this work is formulated according to a building-integrated microgrid as depicted in Fig. 4. It is composed of a group of PV panels, an energy storage system, a connection to the public grid and the loads in the building. All these components are connected to a common capacitive DC bus via power converters. The PV panels are driven by a maximal power point tracking (MPPT) algorithm in most of time, but it is possible to shed some PV power p_{PV_S} to limit the power injected into the bus p_{PV} if needed. The energy storage system is classical lead acid batteries and can be discharged or charged with the power p_{S_D} or p_{S_C} to supply the load or absorb the excessive PV generation. Similarly, the grid can supply power to the microgrid with the power p_{G_S} as well as absorb the power p_{G_I} from the microgrid. The loads in the building are seen as a single power demand

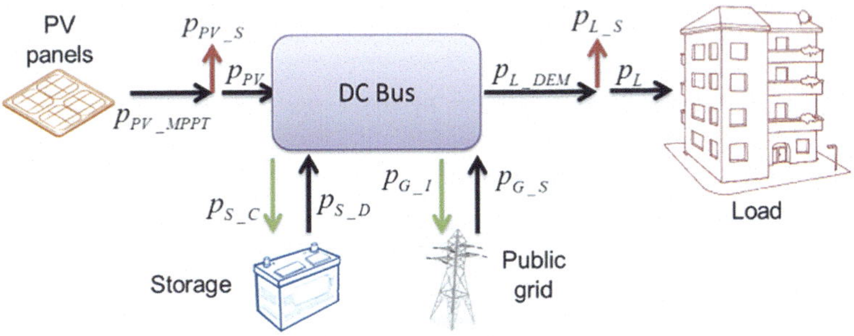

Fig. 4 Topology of the studied DC microgrid

$P_{\mathrm{L\_DEM}}$. If the available power is insufficient, the real load power P_{L} can be shed to keep the power balance.

Some constraints, physical and regulatory, must be respected in the microgrid operation. For example, the state of charge (SOC) of the storage, as defined in Eq. (12), must be kept in a certain range to avoid over-discharge and overcharge. In Eq. (12), v_{S} is the storage voltage and C_{REF} is the battery capacity in Ah, while SOC_0 is the initial value.

$$\mathrm{soc}\,(t_i) = \mathrm{SOC}_0 + \frac{\sum_{t=t_0}^{t_i} \left[p_{\mathrm{S}}\mathrm{C}\,(t_i) - p_{\mathrm{S}}\mathrm{D}\,(t_i)\right] \Delta t}{3600 \cdot v_{\mathrm{S}} \cdot C_{\mathrm{REF}}}. \tag{12}$$

The power supplied or absorbed by the storage and the grid must be constrained owing to the component capacity. Furthermore, the microgrid is neither allowed to charge the battery by the grid power nor inject the power into the grid while discharging the battery.

The goal of the optimal power dispatching is to maximize the PV generation and respond to the load power demand. In this objective, the operation energy cost for each element is defined in Eq. (13):

$$\begin{aligned}
C_{\mathrm{PV}}\mathrm{S} &= \sum_{t_i=t_0}^{t_F} T_{\mathrm{PVS}} \cdot \Delta t \cdot p_{\mathrm{PV}}\mathrm{S}\,(t_i) \\
C_{\mathrm{S}} &= \sum_{t_i=t_0}^{t_F} T_{\mathrm{S}} \cdot \Delta t \cdot \left[p_{\mathrm{S}}\mathrm{C}\,(t_i) + p_{\mathrm{S}}\mathrm{D}\,(t_i)\right] \\
C_{\mathrm{G}} &= \sum_{t_i=t_0}^{t_F} T_{\mathrm{G}} \cdot \Delta t \cdot \left[p_{\mathrm{G}}\mathrm{S}\,(t_i) - p_{\mathrm{G}}\mathrm{I}\,(t_i)\right] \\
C_{\mathrm{L}}\mathrm{S} &= \sum_{t_i=t_0}^{t_F} T_{\mathrm{LS}} \cdot \Delta t \cdot p_{\mathrm{L}}\mathrm{S}\,(t_i)
\end{aligned} \tag{13}$$

Based on the above definitions and constraints, the power dispatching optimization problem can be formulated as following:

$$
\begin{aligned}
&\text{Minimize } C_{\text{TOTAL}} = C_{\text{G}} + C_{\text{S}} + C_{\text{PVS}} + C_{\text{LS}} \\
&\quad \text{with respect to} \\
&p_{\text{PV}}(t_i) + p_{\text{GS}}(t_i) + p_{\text{SD}}(t_i) = p_{\text{L}}(t_i) + p_{\text{SC}}(t_i) + p_{\text{GI}}(t_i) \\
&\text{SOC}_{\text{MIN}} \leq \text{soc}(t_i) \leq \text{SOC}_{\text{MAX}} \\
&0 \leq p_{\text{PVS}}(t_i) \leq p_{\text{PVMPPT}}(t_i) \\
&0 \leq p_{\text{SD}}(t_i) \leq P_{\text{S MAX}}, \quad 0 \leq p_{\text{SC}}(t_i) \leq P_{\text{S MAX}} \\
&0 \leq p_{\text{GI}}(t_i) \leq P_{\text{GI MAX}}, 0 \leq p_{\text{GS}}(t_i) \leq P_{\text{GS MAX}} \\
&\text{if } p_{\text{PV}}(t_i) - p_{\text{L}}(t_i) \geq 0, \text{ then } p_{\text{SD}} = 0, p_{\text{GS}} = 0 \\
&\text{else}, \quad p_{\text{SC}} = 0, p_{\text{GI}} = 0
\end{aligned}
\tag{14}
$$

MILP solver is suitable for this problem, but the accurate PV generation prediction is essential. For a given PV panel, its generation depends mostly on the solar irradiance. If the predicted irradiance is not accurate enough, the optimization results can be invalidated. Thus, the accuracy of the irradiance prediction has a direct impact on the operation energy cost of the microgrid.

4 Simulation Validation

In order to test the CS prediction in the realistic scenarios, the simulation of the microgrid is realized with the 5-month data. The simulation includes a two-layer structure. The upper layer inputs the predictions into IBM CPLEX optimization solver and outputs the power dispatching reference. The lower layer, consisting of a MATLAB/Simulink model of the microgrid, computes the real operation cost while taking the recorded irradiance data and the power dispatching reference into account. The values of the parameters are given in Table 3.

Aiming at simplifying the comparison, a same load profile is used in the optimization and simulation for all the 150 days. The optimization is operated twice for each day, using, respectively, the CS prediction and the MF forecast. A typical profile is shown in Fig. 5, in which the difference between the prediction and the real production can be seen.

As a result, the CS prediction has a total cost of 12.4 € for the 150 days, and the MF forecast has 14.7€. For a more detailed analysis, a daily relative cost difference Δ is defined as in Eq. (15)

Table 3 Microgrid parameters

Parameters	Values	Parameters	Values
PV peak power	4140 Wc	T_{PVS}	0.7 €/kWh
Storage capacity	185 Ah	T_{LS}	1.5 €/kWh
Storage voltage	96 V	T_{S}	0.01 €/kWh
Storage power limit	±3000 W	T_{G}	0.1 €/kWh
Grid power limit	±3000 W	SOC limit	[20%, 80%]

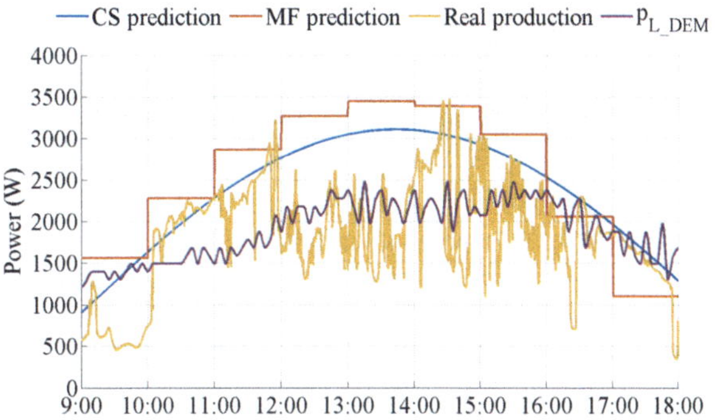

Fig. 5 PV generation and load profiles on the 31st August 2018

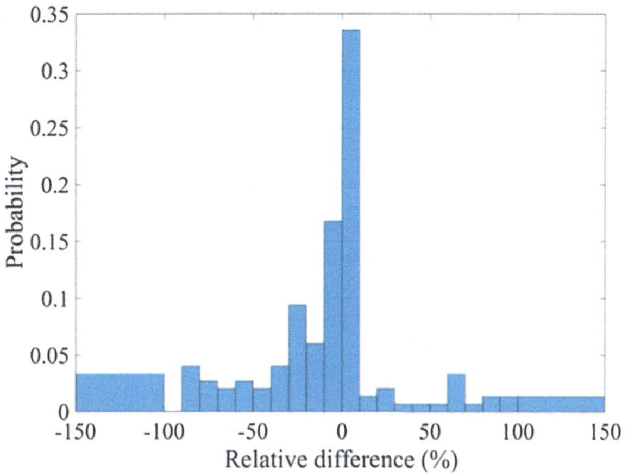

Fig. 6 Distribution of relative difference of the operation cost

$$\Delta = \frac{C_{CS} - C_{MF}}{|C_{MF}|} \cdot 100\%. \tag{15}$$

C_{CS} and C_{MF} are, respectively, the daily cost based on CS prediction and on MF forecast. Hence, Δ is positive if the CS prediction leads to a higher cost. The distribution of Δ during the 150 tested days is shown in Fig. 6. It can be figured out that the relative difference is mostly in the range of 0–10%, meaning in most days the operation cost for CS prediction is slightly higher than that of MF forecast. Though, the probability of Δ being negative is higher than that of being positive. That explains why the total cost of CS prediction is lower than the other.

The results show that the open licence MF forecast is not accurate enough for the microgrid power dispatching, since the geographical resolution is not high enough and the hourly forecast cannot cover the high variability of the solar irradiance. Hence, the CS prediction can be useful for sites for which more precise forecast is unavailable.

5 Conclusions

The power dispatching of a PV-based microgrid can help in improving the performance, but the prediction of the solar irradiance is necessary. In this paper, a CS model is presented to predict the solar irradiance on the PV panels. This model involves only geometric calculation and thus can be widely applied and requires no extra instrument or historical data. A case study is carried out to compare the CS prediction with the free available MF forecast during 5 successive months in 2018. Moreover, the optimal power dispatching is done with the same data. The simulation results show that the proposed model can lead to less operation cost than the MF forecast, even though some prediction errors persist. In conclusion, the clear sky irradiance model can play a key role for low-cost optimal power dispatching.

References

1. M. Sechilariu, F. Locment, D.C. Urban, *Microgrid: Intelligent Control and Power Flow Optimization* (Elsevier, Amsterdam, 2016). ISBN: 978-012803736-2
2. E. Rodriguez-Diaz, J.C. Vasquez, J.M. Guerrero, Intelligent DC homes in future sustainable energy systems: when efficiency and intelligence work together. IEEE Consum. Electron Mag. **5**(1), 74–80 (2016)
3. A. Anvari-Moghaddam, J.M. Guerrero, J.C. Vasquez, H. Monsef, A. Rahimi-Kian, Efficient energy management for a grid-tied residential microgrid. IET Gener. Transm. Distrib. **11**(11), 2752–2761 (2017)
4. L. Bartolucci, S. Cordiner, V. Mulone, V. Rocco, J.L. Rossi, Renewable source penetration and microgrids: Effects of MILP-based control strategies. Energy **152**, 416–426 (2018)
5. M. Sechilariu, B.C. Wang, F. Locment, Supervision control for optimal energy cost management in DC microgrid: design and simulation. Int. J. Electr. Power Energy Syst. **58**, 140–149 (2014)
6. T. Beck, H. Kondziella, G. Huard, T. Bruckner, Assessing the influence of the temporal resolution of electrical load and PV generation profiles on self-consumption and sizing of PV-battery systems. Appl. Energy **173**, 331–342 (2016)
7. Y. Wang, S. Mao, R.M. Nelms, *Online Algorithms for Optimal Energy Distribution in Microgrids* (Springer International Publishing, Berlin, 2015)
8. T. Kato, Y. Manabe, T. Funabashi, K. Yoshiura, M. Kurimoto, Y. Suzuoki, A study on several hours ahead forecasting of spatial average irradiance using NWP model and satellite infrared image, 2016 International Conference on Probabilistic Methods Applied to Power Systems (PMAPS), pp. 1–8 (2016)
9. J.M. Bright, S. Killinger, D. Lingfors, N.A. Engerer, Improved satellite-derived PV power nowcasting using real-time power data from reference PV systems. Sol. Energy **168**, 118–139 (2018)
10. C. David, R. Walter, G.-D. Benjamín, E.S. Les, L. Ricardo Guerrero, First results of a low cost all-sky imager for cloud tracking and intra-hour irradiance forecasting serving a pv-based smart grid in La Graciosa Island, in 2017 IEEE 44th Photovoltaic Specialist Conference (PVSC), pp. 1–5 (2017)
11. A. Shakya, S. Michael, C. Saunders, D. Armstrong, P. Pandey, S. Chalise, R. Tonkoski, Solar irradiance forecasting in remote microgrids using markov switching model. IEEE Trans. Sustain. Energy **8**(3), 895–905 (2017)

12. M.W. Ahmad, M. Mourshed, Y. Rezgui, Tree-based ensemble methods for predicting PV power generation and their comparison with support vector regression. Energy **164**, 465–474 (2018)
13. H. Bouzgou, C.A. Gueymard, Fast short-term global solar irradiance forecasting with wrapper mutual information. Renew. Energy **133**, 1055–1065 (2019)
14. W. El-Baz, P. Tzscheutschler, U. Wagner, Day-ahead probabilistic PV generation forecast for buildings energy management systems. Sol. Energy **171**, 478–490 (2018)
15. H. Verbois, A. Rusydi, A. Thiery, Probabilistic forecasting of day-ahead solar irradiance using quantile gradient boosting. Sol. Energy **173**, 313–327 (2018)
16. P. Mathiesen, J.M. Brown, J. Kleissl, Geostrophic wind dependent probabilistic irradiance forecasts for coastal California. IEEE Trans. Sustain. Energy **4**(2), 510–518 (2013)
17. J.A. Duffie, W.A. Beckman, *Solar Engineering of Thermal Processes*, 4th edn. (John Wiley & Sons, Hoboken, NJ, 2013)
18. Météo-France. https://donneespubliques.meteofrance.fr

Centralized Control in Photovoltaic Distributed Maximum Power Point Tracking Systems

Ramón López-Erauskin, Ander González, Johan Gyselinck, Giovanni Petrone, and Giovanni Spagnuolo

Abstract Photovoltaic energy harvest in distributed maximum power point tracking systems has demonstrated to be superior to the traditional photovoltaic systems under mismatch conditions. The distributed architecture usually consists of series-connected DC/DC converters forming a string, dedicated to process the power of individual photovoltaic panels. However, the classical approach assumes an independent control of the DC/DC converters preventing them from knowing the operating condition of the other converters in the string. The adoption of centralized algorithms allows full control of the variables in distributed maximum power point tracking systems and hence further increases the energy harvest. This paper proposes a novel centralized control that matches distributed and central maximum power point tracking functions, as well as an innovative functionality that improves the dynamic performance in photovoltaic applications.

1 Introduction

In the last decades, the interest in settling the negative effects of mismatching phenomena in photovoltaic (PV) systems, e.g., shading or uneven aging of the PV panels, has led to a broad spectrum of techniques and solutions in solar energy applications. Mismatch conditions in a string composed of several series-connected PV panels adversely affect the energy generation of the entire string. The use of bypass diodes helps alleviating the mismatch situation, but may lead to a power vs. voltage (P-V) curve of the panels with several local maximum power points (MPP), making the string MPP tracking (MPPT) operation inefficient. Online techniques in traditional central MPPT (CMPPT) systems generally search for the global

R. López-Erauskin (✉) · A. González · J. Gyselinck
BEAMS – Université Libre de Bruxelles, Brussels, Belgium
e-mail: johan.gyselinck@ulb.ac.be

G. Petrone · G. Spagnuolo
DIEM Università degli Studi di Salerno, Fisciano, Italy
e-mail: gpetrone@unisa.it; gspagnuolo@unisa.it

W. Zamboni, G. Petrone (eds.), *ELECTRIMACS 2019*, Lecture Notes in Electrical Engineering 615, https://doi.org/10.1007/978-3-030-37161-6_39

MPP, e.g., making a periodical voltage sweep in the P-V curve. Unfortunately, these techniques only process part of the available power in the PV string and not the sum of the power the PV panels are able to deliver by considering them independently. Distributed MPPT (DMPPT) architectures are thought to constitute an alternative to the traditional approach. A DMPPT system consists of DC/DC converters dedicated to track the MPP of few or single PV panels. The output terminals of these converters are connected in series creating a string that substitutes the string of PV panels. The increased PV power generation in DMPPT systems compared to the traditional PV systems has been demonstrated in [1–6]. These approaches employ autonomous DC/DC converters with their own MPPT control working independently from the rest of the system. Such control strategy makes it impossible for each converter to know the operating condition of the other converters in the string, worsening the power processed to the DC bus. Authors in [7, 8] introduce the concept of the multi-variable (MV) control that evaluates the power at the DC bus. In both papers, they consider a constant DC bus voltage, which may not be optimal under mismatch conditions between PV panels. Indeed, the DC bus voltage control can extend the MPP operating range of DMPPT systems as demonstrated in [9, 10]. The study in [11] claims for a proper coordination of the DMPPT and CMPPT functions. A hybrid-MV (H-MV) control has been presented in [12], including the DC bus voltage control in the MV control sequence. In addition, H-MV deals with the dynamic performance of MV algorithms with a new functionality derived from the vectorial MV control (MVPOV) in [8]. Nonetheless, the dynamic performance of the H-MV algorithm is not optimized and presents PV and output voltage deviations, deteriorating the power generation. Such deviations can be avoided if the algorithm in charge of controlling the system takes into account the slope of the irradiance variation and adapts the control variables accordingly.

In this paper, an innovative centralized MV (CMV) control is presented, able to correctly track any irradiance variation speed thanks to the new adaptive strategy. Furthermore, the power is measured at the output terminals of the inverter instead of at the DC bus as in [12], so that the optimization process keeps into account the efficiency of the overall power processing system, including both the DC and the AC stages. The proposed CMV control exploits the benefits of the MV approach improving both static and dynamic performances as well as boosting the overall efficiency of the DMPPT system.

2 System Description

The DMPPT architecture is shown in Fig. 1. Each DC/DC converter performs the MPPT of the corresponding PV panel. Henceforth, the group consisting of a PV panel and its dedicated DC/DC converter will be referred to as *module*. The output terminals of these modules are connected in series in order to obtain

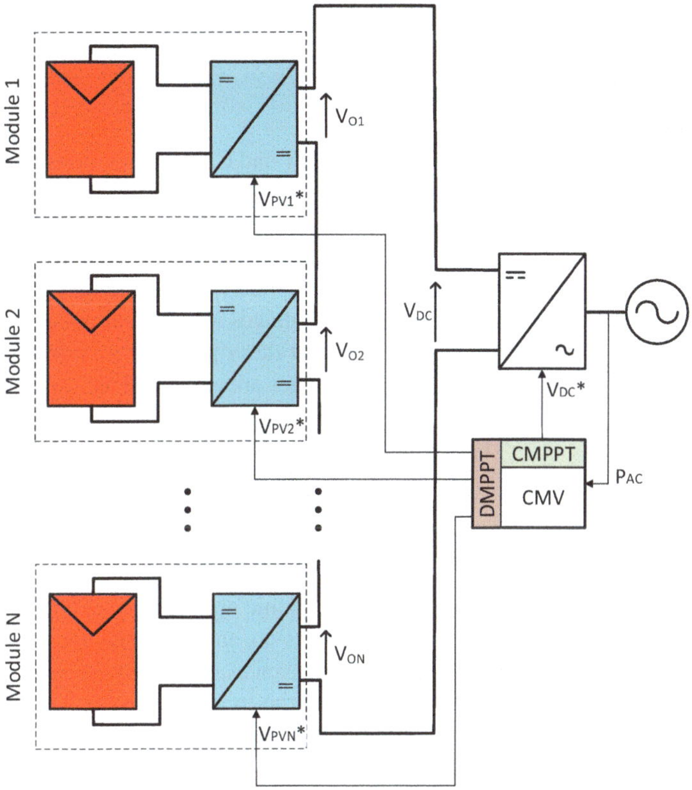

Fig. 1 DMPPT architecture of the grid-connected PV system

a high DC bus voltage, requirement for the inverter to work. The converters adopted in this paper are the interleaved boost-TPC topology [13], but the application of the proposed control approach is not limited to a specific DC/DC converter topology. For a comprehensive understanding of the topology, the reader can find exhaustive analysis, modulation and control strategies in [14] and [15]. The inverter is implemented in a 2-D look-up table that considers different efficiency curves with respect to the DC bus power and voltage.

A supervisory controller is in charge of the control of the whole system: it analyzes the AC power measured at the output of the inverter and sets the operating point for PV and DC bus voltages. The goal is to maximize the output power by finding the best combination of operating points for each conversion stage of the system.

3 Joint Operation of DMPPT and CMPPT

3.1 Background

A straightforward method to coordinate DMPPT and CMPPT functions is to include the DC bus voltage control into the MV operating sequence. Authors in [12] validate this approach proposing the hybrid MV (H-MV) perturb & observe (P&O) control. The H-MV control sequence is: $\{|V_{ec}|, \theta_1, \theta_2, \ldots, \theta_{N-1}, V_{DC}\}$, being N the number of modules, V_{DC} the DC bus voltage, and $|V_{ec}|$ and $\theta_{1\ldots N-1}$ the vector magnitude and the angles of the vector defining the PV voltages in spherical polar coordinates [8], respectively. The vectorial operating point description for a system with N=3 modules is depicted in Fig. 2 and the PV voltages are expressed as:

$$V_{PV1} = |V_{ec}| \cdot \sin\theta_1 \cdot \sin\theta_2 \tag{1}$$

$$V_{PV2} = |V_{ec}| \cdot \cos\theta_1 \cdot \sin\theta_2 \tag{2}$$

$$V_{PV3} = |V_{ec}| \cdot \cos\theta_2 \tag{3}$$

The H-MV control in [12] improves the dynamic performance of MV algorithms under changing irradiance conditions. Thanks to an additional measurement in between two consecutive MPPT sampling instances,it analyzes the output power and determines if any significant variation of the irradiance exists. Under varying irradiance conditions, H-MV triggers the blocking of the variables θ_1, θ_2, V_{DC} and implements the P&O only on $|V_{ec}|$. Consequently, all PV voltages change simultaneously avoiding undesired large deviations between output voltages of the modules and ensuring system stability [12]. For this operation, the irradiance level S and its variation speed need to be known. The irradiance level S is related to the DC power P_{DC} and it is evaluated at each sampling instance according to the following normalized expression:

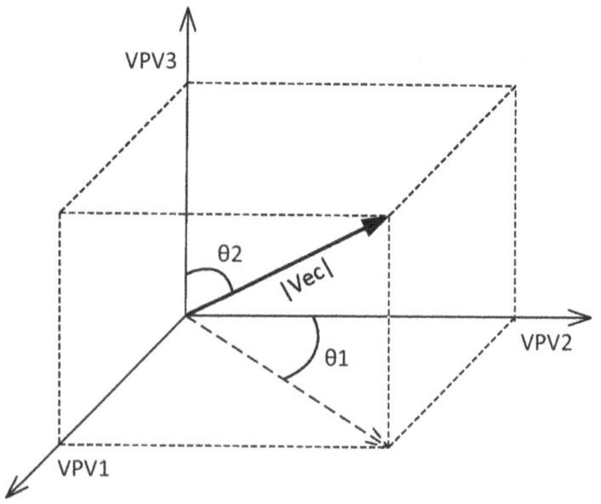

Fig. 2 Operating point description of the vectorial approach with N=3 PV modules

$$S = 1\,\mathrm{kW/m^2} \cdot \frac{P_{\mathrm{DC}}}{P_{\mathrm{DC@STC}}} \tag{4}$$

and the irradiance variation $\dot{S}$ is calculated substituting P_{DC} for the DC power variation ΔP_{DC} in (4). It is worth to note that although the H-MV is able to identify the presence of the irradiance variation, it is not able to detect whether the origin of the power variation at the DC bus is due to the change in one or more PV panels. Therefore, the H-MV dynamic performance is evaluated under the same irradiance variations for all the PV panels.

3.2 *Proposed Centralized Multi-Variable Control*

The CMV approach presented in this paper adds modifications to the H-MV algorithm presented in [12], (1) improving the dynamic response of the system for any irradiance variation speed, (2) evaluating the power at the AC-side of the inverter, thus also accounting for the inverter conversion efficiency. Therefore, the irradiance level S in (4) is now related to the AC power.

According to [7], the PV voltage perturbation step (ΔV_{PV}) is designed for a given maximum irradiance variation $\dot{S}_{\max}$:

$$\Delta V_{\mathrm{PV}} = \sqrt{\dot{S}_{\max} \cdot k} \tag{5}$$

where k groups several characteristic parameters that depend on the chosen converter and PV panel under a given initial irradiance level (S_{init}) [7].

Unfortunately, for a new irradiance variation $\dot{S}$ faster than $\dot{S}_{\max}$, ΔV_{PV} is no longer sufficient to correctly track the MPP, thus a new value has to be settled. Therefore, if $\dot{S} > \dot{S}_{\max}$, the new ΔV_{PV} becomes:

$$\Delta V_{\mathrm{PV-a}} = \sqrt{\frac{\dot{S}}{\dot{S}_{\max}}} \cdot \Delta V_{\mathrm{PV}} \tag{6}$$

where $\Delta V_{\mathrm{PV-a}}$ is the adapted value of ΔV_{PV}. This condition holds if $S \geq S_{\mathrm{init}}$. The CMV algorithm performs a good MPPT if the power variation ΔP_{PV} due to ΔV_{PV} perturbation is bigger than the power variation $\Delta P_{\dot{S}}$ caused by the irradiance variation $\dot{S}$:

$$\Delta P_{\mathrm{PV}} > \Delta P_{\dot{S}} \tag{7}$$

The P-V curve of a PV panel based on silicon technology tends to be more flat as the level of the irradiance S lowers [11]. Figure 3 shows a zoomed view of the characteristic P-V curve of a PV panel in the neighborhood of the MPP at different S (S_{low} and S_{high}, solid lines). The curves change due to a variation of the

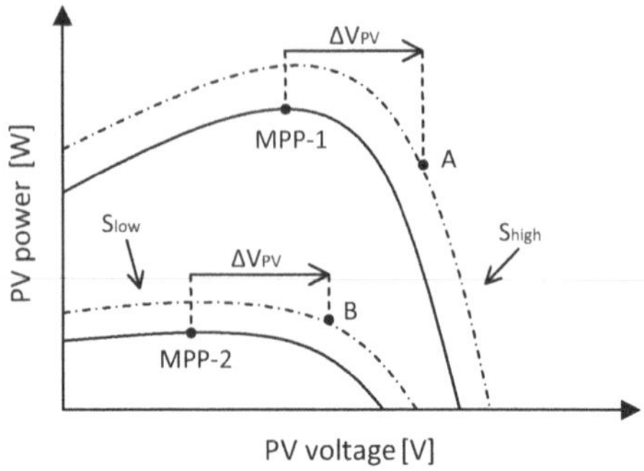

Fig. 3 Operating point of PV panel in presence of an irradiance variation at two different irradiance levels

irradiance between two consecutive sampling instances (dashed lines). The change in power depends on the derivative of the PV power ΔP_{PV} with respect to ΔV_{PV} $(\Delta P_{PV}/\Delta V_{PV})$. For the same ΔV_{PV} in Fig. 3, if the converter is working at MPP-1 the next operating point will be point A, satisfying the inequality (7). On the contrary, if the converter is working at MPP-2, the next operating point will be point B, thus detecting an increase in power and V_{PV} will further increase, moving the operating point away from the MPP.

Adding a correction factor to ΔV_{PV-a} in (6) improves the MPPT performance of the H-MV under the different irradiance conditions. As ΔP_{PV} is inversely proportional to the irradiance level S, yields:

$$\Delta V_{PV-a} = \sqrt{\frac{\dot{S}}{\dot{S}_{max}}} \cdot \frac{S_{init}}{S} \cdot \Delta V_{PV} \tag{8}$$

Translating (8) to the polar environment [8], we obtain the adapted value for the perturbation step of $|V_{ec}|$:

$$\Delta V_{ec-a} = \sqrt{3} \cdot \Delta V_{PV-a} \tag{9}$$

4 Simulation Results

The simulations are carried out in MATLAB-Simulink environment. Each converter adopts an averaged model of non-isolated boost-type DC/DC converters which takes conduction losses into account. The system consists of $N = 3$ modules composed of one BENQ-260 Wp PV panel connected to an interleaved boost-TPC and an inverter. The main electrical characteristics of the PV panel in standard test conditions ($S = 1000\,\mathrm{W/m^2}$, cell temperature = 25 °C, and air mass AM = 1.5) are: short-circuit current $I_{sc} = 8.83$ A, open-circuit voltage $V_{oc} = 37.7$ V, MPP current and voltages $I_{mpp} = 8.34$ A and $V_{mpp} = 31.2$ V, respectively. The inverter is

Table 1 Perturbation amplitudes of the variables for the different control methods

Symbol	Description	Value
$\Delta V_{\mathrm{SV,\,min}}$	min. SV voltage perturbation	0.2509 V
$\Delta V_{\mathrm{ec,\,min}}$	min. V_{ec} perturbation	0.4345 V
$\Delta\theta_{1,\,\mathrm{min}}$, $\Delta\theta_{2,\,\mathrm{min}}$	min. θ_1 & θ_2 perturbation	0.55°

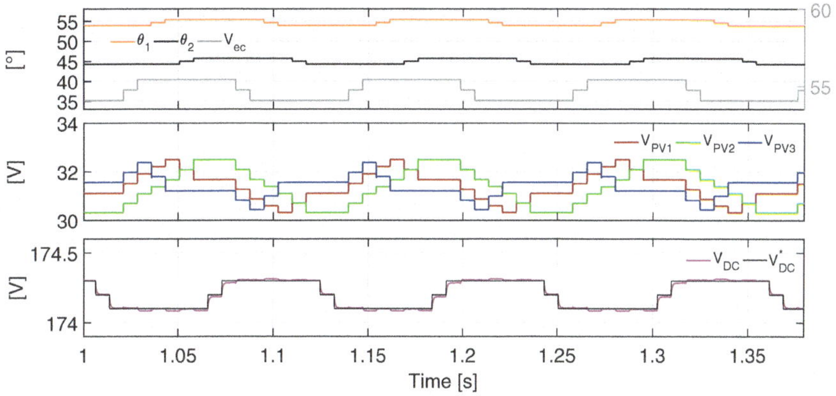

Fig. 4 Steady state operation of the CMV control technique

modeled using the efficiency curves of a scaled-down commercial solar inverter with DC bus voltage control in a range of [70, 240] V and a nominal power of 1.5 kW.

Table 1 gathers the minimum voltage steps for the different controls that can be adopted in this system. According to the guidelines in [16] the minimum time interval that ensures that the system reaches the steady state after a voltage step perturbation is $t_a = 3.7$ ms. The DC bus voltage step is chosen to be $\Delta V_{\mathrm{DC}} = 0.1$ V.

Simulation results in Fig. 4 show the steady state behavior of the system. Top plot corresponds to the variables in polar coordinates and middle plot shows the behavior of PV voltages. V_{DC} in the bottom plot is performed after setting the vectorial variables.

4.1 DC Bus Voltage Control

In Fig. 5 the performances of MVPOV and CMV are compared. When the MVPOV performs, V_{DC} is fixed to 174.2 V as the DC bus voltage providing the maximum AC power under homogeneous irradiance condition $S = 700\,\mathrm{W/m^2}$. From top to bottom, PV power, AC power, and the DC bus voltage for the adaptive case are appreciated. The simulation starts with the PV panels under nonuniform irradiance levels ($S_1 = 700\,\mathrm{W/m^2}$, $S_2 = S_3 = 1000\,\mathrm{W/m^2}$). At time $t = 2$ s, the irradiance becomes uniform for all the PV panels at $S = 700\,\mathrm{W/m^2}$. CMV seeks

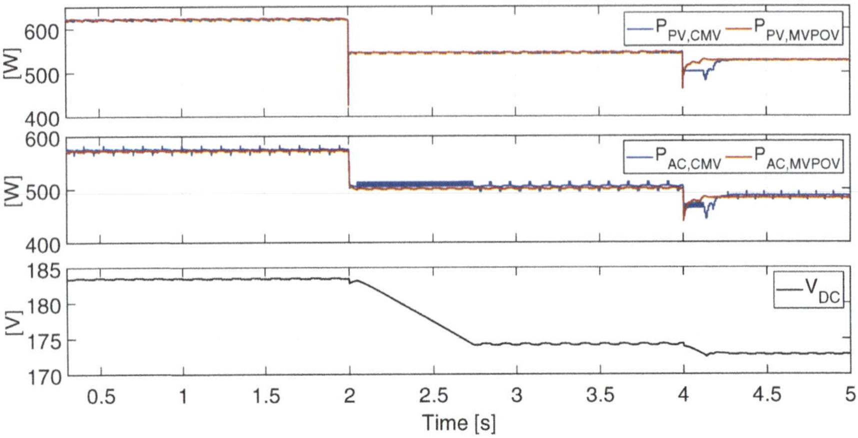

Fig. 5 System response adopting the multi-variable algorithm with (blue) and without (red) DC bus voltage control

a DC bus voltage level that ensures the highest conversion efficiency for the new generation conditions, adapting the DC bus voltage stepwise from around 183.7 V to the vicinity of 174.2 V. At time $t = 4$ s, a sudden ambient temperature change from $T_a = 25\,^{\circ}\mathrm{C}$ to $T_a = 50\,^{\circ}\mathrm{C}$ is imposed on one of the PV panels. Though this condition is not realistic, it shows the ability of the algorithm to adapt the DC voltage operating point also at different temperature conditions.

The MPPT performance is comparable for both MVPOV and CMV techniques (Fig. 5, top plot). Only when a change in power occurs MVPOV reaches its steady state faster than CMV. This is due to the simultaneity of the power transition event and the variable that is being perturbed. As long as V_{DC} control is being performed, the PV voltages will remain unchanged. From Fig. 5, it can be concluded that CMV ensures the highest overall efficiency of the PV system, by also considering the DC bus voltage accommodation in the control sequence.

The CMV algorithm is able to correct the operating point of the DC bus voltage as well as the PV voltages under non-homogeneous irradiance conditions. Although the performance of the overall system is barely improved, the simulation shown in Fig. 5 validates the proposed joint operation of the distributed and centralized control.

The performance improvement in this kind of PV systems depends on the ratio between the nominal DC power of the inverter and the PV peak power, the number of DC/DC converters, their topology, and the efficiency curve of the inverter operating in the system. These aspects are under investigation and will be the objective of authors' future research.

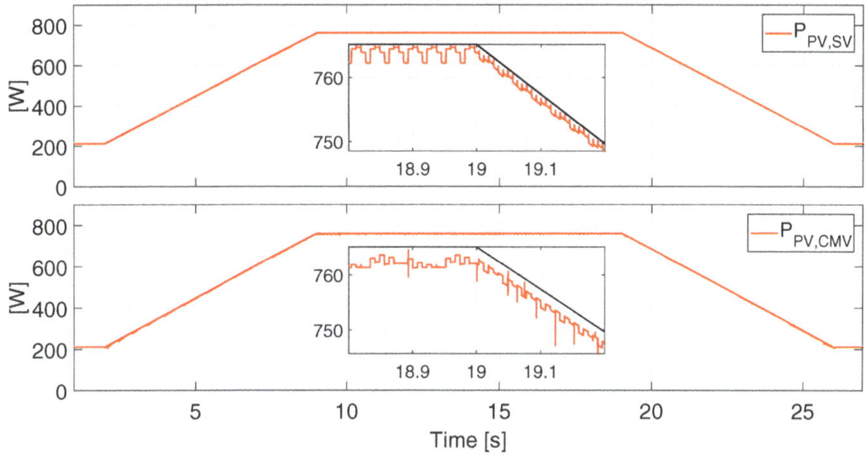

Fig. 6 EN 50530 dynamic performance test adopting both SV and H-MV control algorithms

4.2 Dynamic Performance of the Proposed Control Algorithm

The CENELEC standard EN 50530 proposes a test procedure to evaluate the dynamic performance of algorithms dedicated to control PV systems [17]. Two test sequences with different irradiance levels are proposed (sequence A fluctuates between 100 and 500 W/m^2 and sequence B between 300 and 1000 W/m^2) specifying a range of irradiance variation speeds. For the sake of simplicity, sequence B and the fastest irradiance variation of $\dot{S} = 100\,Wm^{-2}s^{-1}$ are chosen. The simulations are carried out under homogeneous irradiance variation for all the PV panels to prove the validity of the proposed algorithm. The non-homogeneous PV generation can be compensated using batteries at module level and mimic homogeneous irradiance conditions. Figure 6 shows a comparison between the conventional single-variable (SV) and the CMV controls. Both SV and CMV algorithms perform small perturbation steps in steady state, designed to track a maximum irradiance variation of $\dot{S}_{max} = 20\ Wm^{-2}s^{-1}$ as the threshold to switch between static and dynamic behavior. ΔV_{PV} and ΔV_{ec} values for SV and CMV controls, respectively, (Table 1) are calculated using the methodologies described in [7] and [8]. Once the irradiance starts ramping-up, ΔV_{PV} and ΔV_{ec} are adapted according to (8) and (9), respectively.

For the test in Fig. 6 the energy yield results in $Y_{E,\,SV} = 236.34$ Wh and $Y_{E,\,CMV} = 236.36$ Wh for SV and CMV approaches, respectively. The efficiencies of each conversion stage under the different irradiance conditions gathered in Table 2 are computed within the execution period shown in Fig. 6. The total efficiency is obtained by computing the MPPT, converters' and inverter's efficiencies: $\eta_{Total} = \eta_{MPPT} \cdot \eta_{conv} \cdot \eta_{Inv}$. The total efficiency is considerably higher for CMV compared to SV at lower irradiance levels, given by an increase in the efficiency of

Table 2 MPPT and conversion efficiencies

S (W/m^2)	300	300 → 1000	1000	1000 → 300	300
$\eta_{\text{MPPT, SV}}(\%)$	99.78	99.69	99.87	99.71	99.81
$\eta_{\text{MPPT, CMV}}(\%)$	99.58	99.52	99.59	99.62	99.56
$\eta_{\text{conv, SV}}(\%)$	92.19	95.74	95.46	95.55	92.19
$\eta_{\text{conv, CMV}}(\%)$	96.30	95.67	95.40	95.59	95.80
$\eta_{\text{Inv, SV}}(\%)$	91.14	93.37	96.78	96.19	91.64
$\eta_{\text{Inv, CMV}}(\%)$	89.75	94.31	96.96	95.57	90.77
$\eta_{\text{Total, SV}}(\%)$	83.84	89.12	92.27	91.64	84.32
$\eta_{\text{Total, CMV}}(\%)$	86.07	89.79	92.12	91.01	86.58

the converters. This is most probably affected by the DC voltage control in the CMV approach. Actually, a reduction of voltage at the output of the converters could boost η_{conv}, as the voltage difference between input and output of the converter is smaller.

Figure 7 shows the performance of the proposed CMV algorithm in detail. In steady state, CMV performs the vectorial multi-variable P&O. Under $\dot{S} = 100$ Wm^{-2}s^{-1} speed, the P&O control acts on $|V_{\text{ec}}|$, with its characteristic three-point-behavior. Thanks to the adaptive action, the amplitude of the PV voltages is larger at lower S (Fig. 7a and d) while at high S it is reduced (Fig. 7b and c).

5 Conclusion

In DMPPT systems, the dynamic response of the MV control needs to be treated differently from individual MPPT control, with a specific functionality that avoids the divergence of the voltage variables from their respective optimal operating points under changing irradiance conditions. The proposed modifications in the CMV control allow the PV voltages to be kept close to their optimal operating point, improving its dynamic performance, that is comparable to SV control. Since the design of the perturbation variable does not only depend on the variation speed of the irradiance, but also on its level, in this paper a solution that improves the MPPT performance of the algorithm by increasing the efficiency of the system is proposed, taking into account the irradiance level.

In addition, the inclusion of the DC bus voltage control in the MV control sequence has proved to be an effective and straightforward method to coordinate the control between distributed and central MPPT. By evaluating the power at the AC-side, the proposed control performs overall system optimization and individually performed optimization of MPPT, converters' and inverter's efficiency is not necessary.

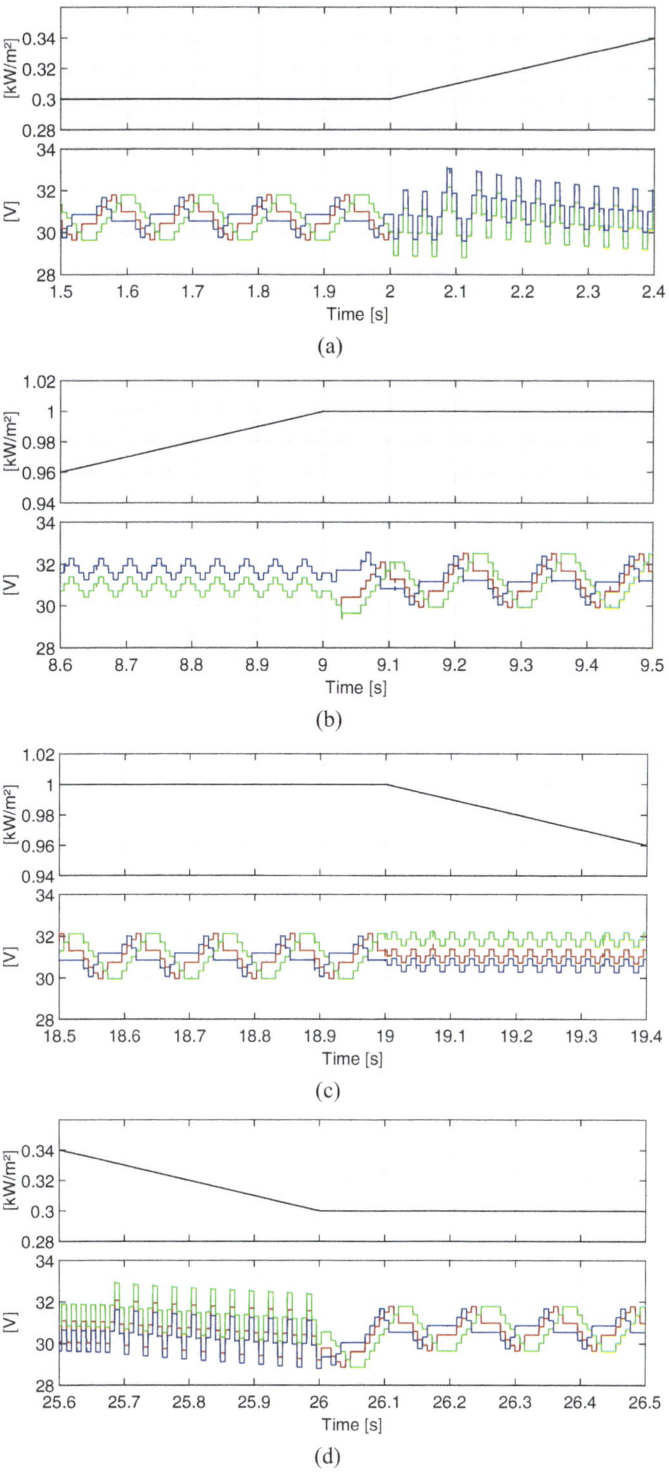

Fig. 7 Details of static and dynamic behavior of the PV voltages under EN 50530 dynamic performance test

Acknowledgements This work was supported by Belgian Walloon region under the project BATWAL. The authors would also like to thank the Università Degli Studi di Salerno for their close collaboration in the development of the work in this paper.

References

1. E. Roman, R. Alonso, P. Ibanez, S. Elorduizapatarietxe, D. Goitia, Intelligent PV module for grid-connected PV systems. IEEE Trans. Ind. Electron. **53**(4), 1066–1073 (2006)
2. G.R. Walker, J.C. Pierce, Photovoltaic DC-DC module integrated converter for novel cascaded and bypass grid connection topologies design and optimization, in *2006 37th IEEE Power Electronics Specialists Conference*, pp. 1–7, June 2006
3. G. Adinolfi, N. Femia, G. Petrone, G. Spagnuolo, M. Vitelli, Energy efficiency effective design of DC/DC converters for DMPPT PV applications, in *2009 35th Annual Conference of IEEE Industrial Electronics*, pp. 4566–4570, Nov 2009
4. L. Linares, R.W. Erickson, S. MacAlpine, M. Brandemuehl, Improved energy capture in series string photovoltaics via smart distributed power electronics, in *2009 Twenty-Fourth Annual IEEE Applied Power Electronics Conference and Exposition*, pp. 904–910, Feb 2009
5. S.M. MacAlpine, R.W. Erickson, M.J. Brandemuehl, Characterization of power optimizer potential to increase energy capture in photovoltaic systems operating under nonuniform conditions. IEEE Trans. Power Electron. **28**(6), 2936–2945 (2013)
6. A.J. Hanson, C.A. Deline, S.M. MacAlpine, J.T. Stauth, C.R. Sullivan, Partial-shading assessment of photovoltaic installations via module-level monitoring. IEEE J. Photovoltaics **4**(6), 1618–1624 (2014)
7. G. Petrone, C.A. RamosPaja, G. Spagnuolo, M. Vitelli, Granular control of photovoltaic arrays by means of a multioutput maximum power point tracking algorithm. Prog. Photovoltaics: Res. Appl. **21**(5), 918–932 (2012), https://onlinelibrary.wiley.com/doi/abs/10.1002/pip.2179
8. C.A. Ramos-Paja, G. Spagnuolo, G. Petrone, S. Serna, A. Trejos, A vectorial MPPT algorithm for distributed photovoltaic applications, in *2013 International Conference on Clean Electrical Power (ICCEP)*, pp. 48–51, June 2013
9. R. López-Erauskin, A. González, J. Gyselinck, Control strategies for modular grid-tied DMPPT systems integrating photovoltaic generation and battery storage at module level, in *2018 20th European Conference on Power Electronics and Applications (EPE'18 ECCE Europe)*, pp. P.1–P.9, Sept 2018
10. A. González, R. López-Erauskin, J. Gyselinck, Grid-tied modular and scalable photovoltaic distributed maximum power point tracking system with storage at module level using non-isolated three-port converters, in *2018 20th European Conference on Power Electronics and Applications (EPE'18 ECCE Europe)*, pp. P.1–P.9, Sept 2018
11. M. Vitelli, On the necessity of joint adoption of both distributed maximum power point tracking and central maximum power point tracking in PV systems. Prog. Photovoltaics: Res. Appl. **22**(3), 283–299 (2012). https://onlinelibrary.wiley.com/doi/abs/10.1002/pip.2256
12. R. López-Erauskin, A. González, G. Petrone, G. Spagnuolo, J. Gyselinck, Multi-variable perturb & observe algorithm for grid-tied PV systems with joint central and distributed MPPT configuration. Trans. Sustainable Energy (under review)
13. A. González, R. López-Erauskin, J. Gyselinck, Interleaved three-port boost converter for photovoltaic systems including storage, in *2017 19th European Conference on Power Electronics and Applications (EPE'17 ECCE Europe)*, pp. P.1–P.9, Sept 2017
14. A. González, R. López-Erauskin, J. Gyselinck, Analysis, modeling, control and operation of an interleaved three-port boost converter for DMPPT systems including PV and storage at module level. Heliyon **5**(3), e01402 (2019)

15. A. González, T.K. Chau, R. López-Erauskin, H.H.-C. Iu, T. Fernando, J. Gyselinck, Nonlinear MIMO control of interleaved three-port boost converter by means of state-feedback linearization, in *2018 18th International conference on power electronics and motion control*, Sept 2018
16. N. Femia, G. Petrone, G. Spagnuolo, M. Vitelli, Optimization of perturb and observe maximum power point tracking method. IEEE Trans. Power Electron. **20**(4), 963–973 (2005)
17. Overall efficiency of grid connected photovoltaic inverters. CENELEC, Standard, May 2010

Optimal PV Panel Reconfiguration Using Wireless Irradiance Distributed Sensing

Antonino Laudani, Gabriele Maria Lozito, Martina Radicioni, Francesco Riganti Fulginei, and Alessandro Salvini

Abstract The aim of this paper is to provide a dynamic reconfiguration method for partially shaded photovoltaic arrays. The implemented strategy is able to increase power production of the array with respect to the initial topology in real time and with any shading pattern. The array is supposed to be made of strings of modules interconnected in parallel and each module is constituted by series-connected photovoltaic cells. Irradiance values are calculated through a closed-form relation given the operating point of the modules, their temperatures, and their equivalent circuit model. This procedure frees the system from the necessity of costly pyranometers. The implemented method has been validated in Matlab environment simulating random shading conditions and implemented on a low-cost 32-bit microcontroller with wireless connectivity capabilities. The results prove the efficiency of the proposed solution.

1 Introduction

Mismatching conditions are due to different properties of interconnected cells and heavily influence photovoltaic (PV) power generation, leading to great efficiency reductions; the main mismatch affecting PV systems is non-uniform irradiation of the PV plant, corresponding to a partial or full shading condition [1, 2]. Shadows often occur in PV arrays, in particular when embedding PV modules in buildings. As a matter of fact, shadows coming from other buildings, dirt, dust, approaching clouds, falling leaves and atmospheric fluctuations are just some of the factors that can determine a shading condition. The most serious problem arising is determined by power losses that can be significant. Considering the structure of a PV panel, it can be seen that it is made of groups of series-connected cells, representable as

A. Laudani (✉) · G. M. Lozito · M. Radicioni · F. R. Fulginei · A. Salvini
Dipartimento di Ingegneria, Universita degli Studi Roma Tre, Roma (RM), Italy
e-mail: antonino.laudani@uniroma3.it; gabrielemaria.lozito@uniroma3.it; martina.radicioni@uniroma3.it; francesco.rigantifulginei@uniroma3.it; alessandro.salvini@uniroma3.it

W. Zamboni, G. Petrone (eds.), *ELECTRIMACS 2019*, Lecture Notes in Electrical Engineering 615, https://doi.org/10.1007/978-3-030-37161-6_40

power sources. When a cell, or a group of cells, is affected by shading, its current is reduced since the cell(s) starts acting as a load, absorbing power: this determines a reduction in the power delivered by the whole module [3]. The standard practice adopted to mitigate the effects of shading is the use of bypass diodes connected in parallel to PV cells or groups of cells; in this way the shaded portion never operates with negative voltages; however, this solution does not allow to retrieve the lost power, but just cuts-off the ill cell(s). Moreover, the insertion of diodes modifies the $P - V$ characteristic of PV arrays [4, 5], introducing further maximum power points (MPPs) and so, making the assessment of the unique global MPP (GMPP) hard. To face this problem, several maximum power point tracking (MPPT) algorithms have been developed able to detect the MPP in fast and accurate ways and not to remain trapped in the local solutions (LMPP) [6]. Some of the main problems of these methods are their complexity and costs; moreover, they often need embedded sensors. Another key issue is hot spotting; hot spots are areas of the array characterized by higher temperatures with respect to the rest of the system and this is caused by differences in the level of irradiance, [7, 8]. The resulting overheating may bring the cell(s) temperature well above the limit to which they can operate, with possible irreversible damages of the case of the cell and of the PV device itself. Hence, the need of accurate measurements of the level of irradiance on the whole array. In this work, an analytical model is used to calculate the exact value of irradiance of a cell given the operating point and the temperature [9]. As a matter of fact, measuring temperature is quite trivial, while irradiance is a quantity whose determination is often hard: irradiance sensors need a control system and are quite costly, the price of a good quality pyranometer ranges from 200 to 500 dollars; moreover, their placement is quite tricky and requires the exact knowledge of the inclination of sun rays on the surface of the system. The method here adopted provides an accurate and low-cost solar irradiance sensing for PV systems: the cost of the whole system is around 20 dollars. Other methods can be found in literature for sensing accurately and monitoring solar irradiance, such as in [10–13].

In the development of a technique aimed at mitigating shading effects the observation of the particular shading pattern is generally a key point [14]; as a matter of fact, reconfiguration methods are based on this concept. The modules that make up the array are rearranged in ways aimed at retrieving power with respect to the initial configuration, according to the shading pattern. As a consequence, each configuration is suitable for a certain shading scenario and several topologies have been proposed and studied in literature [15–17]. In this paper, a real-time reconfiguration based method is proposed able to provide an optimal configuration for each shading scenario. After having identified the optimal configuration, a network of switches is supposed to be driven by an MCU based master device to rearrange the PV modules accordingly. The switching topology of this work which allows complete reconfiguration of the devices must be considered ideal. Switch boxes seldom offer the possibility to rearrange freely the PV devices in a network. In general, PV devices can be either excluded from the series by means of a smart bypass or at most switched with devices belonging to adjacent strings.

The paper will be structured as following: in Sect. 2 the method for calculating irradiance analytically is shown; in Sect. 3 two shading compensation strategies are presented: the traditional insertion of bypass diodes and the reconfiguration technique; Sect. 4 presents the proposed dynamic reconfiguration and the algorithm developed; in Sect. 5 the implementation of the method on a 32-bit microcontroller unit is shown; Sect. 6 presents the conclusions.

2 Analytical Irradiance Extraction

2.1 One Diode Circuital Model for PV Cells and Arrays

The most common model used to characterize the I–V curve of a PV device is the one diode mode, illustrated in Fig. 1. This tool can be used to characterize either a single PV cell or an array made of series or parallel connected cells. In this way, the current generated by the cell or module can be expressed through a relation taking into account the values of five parameters: the shunt resistance R_p, the series resistance, R_s, the photocurrent, I_{irr}, the diode inverse saturation current, I_o, and the ideality factor, n. Moreover, the number of cells connected in series N_s and in parallel N_p have to be considered:

$$i_{pv} = N_p I_{irr} - N_p I_o \left[e^{\frac{q\left(v_{pv}+\frac{i_{pv} N_s R_s}{N_p}\right)}{N_s nkT}} - 1 \right] - \frac{v_{pv} + \frac{i_{pv} N_s R_s}{N_p}}{N_s R_p / N_p} \tag{1}$$

In Eq. (1) $q = 1.602 \times 10^{-19}$ C is the electronic charge, $k = 1.3806503 \times 10^{-23}$ J/K represents the Boltzmann constant, and T is temperature.

Each of the five parameters appearing in Eq. (1) is a function of environmental conditions, i.e., solar irradiance, G, and temperature, T, according to the following relations:

$$I_{irr} = \frac{G}{G_{ref}} (I_{irr,ref} + \alpha_T (T - T_{ref})) \tag{2}$$

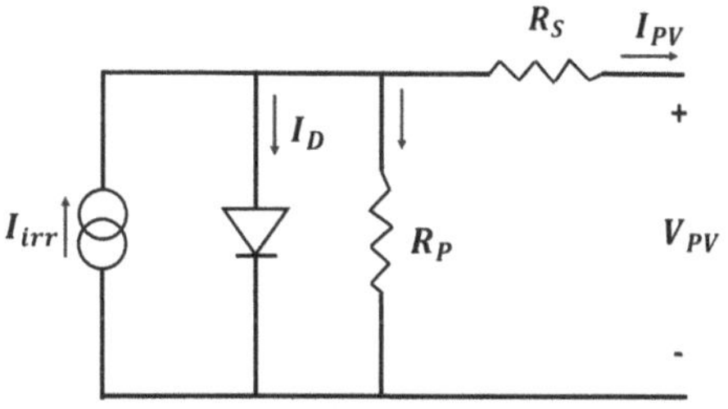

Fig. 1 One diode model

$$I_0 = I_{0,ref}\left(\frac{T}{T_{ref}}\right)^3 e^{\left[\frac{E_{g,ref}}{kT_{ref}} - \frac{E_g}{kT}\right]} \tag{3}$$

$$R_p = R_{p,ref}\frac{G_{ref}}{G} \tag{4}$$

$$R_s = R_{s,ref} \tag{5}$$

$$n = n_{ref} \tag{6}$$

In equations from (2) to (6) the subscript ref refers to quantities at standard reference conditions (SRC), i.e., $T = T_{ref} = 25\,^{\circ}\mathrm{C}$, $G = G_{ref} = 100\,\mathrm{W/m^2}$. The five parameters can be extracted both from datasheet information or from experimental I–V curves: several procedures are available for this purpose [18–20]; in this work the method proposed in [19] is implemented to provide an accurate identification of the PV device through the one diode circuital model. If more accurate identifications are needed, it is worth considering the uncertainty introduced by the error declared in the devices datasheet. A strategy to reduce as much as possible this uncertainty is described in [21].

2.2 *Closed-Form Irradiance Formulation*

As mentioned above, the I–V characteristic of the cell or module is strongly dependent on atmospheric conditions; as a consequence of this, the operating point, as well as the power produced, deeply changes according to the values of T and G; hence, the need of assessing these quantities accurately. The followed approach makes use of a closed-form expression of solar irradiance, knowing the operating point of the device, i.e., v_{pv} and i_{pv}, and its temperature. The following equation, derived from algebraic manipulations of (1), allows to derive the expression of irradiance as a function of temperature:

$$\frac{G}{G_{ref}}\left(N_p I_{irr,ref} + N_p \alpha_T\left(T - T_{ref}\right) - \frac{v_{pv} + i_{pv} N_s R_{s,ref}/N_p}{N_s R_{p,ref}/N_p}\right) = i_{pv} + N_p I_0\left[e^{\left(\frac{v_{pv}+i_{pv}N_s R_{s,ref}/N_p}{N_S nkT}\right)} - 1\right] \tag{7}$$

3 Shading Compensation Techniques

3.1 Bypass Diodes

Nowadays, PV modules are generally equipped with several bypass diodes connected in parallel to each cell or group of cells. When a cell of a string gets shaded its photocurrent decreases and becomes smaller than that of the other cells. The failing cell becomes reverse biased and starts operating like an external load, consuming power; the function of bypass diodes is to provide an alternative path to the string current in order to bypass the shaded cell. This strategy allows the power production to be unaffected by shading. However, some drawbacks still persist. The presence of bypass diodes severely alters the P–V characteristic of the PV device; as a matter of fact, instead of a unique GMPP, multiple peaks, LMPP, appear.

The risk of MPPT traditional algorithms of falling in local maxima causes their failure in finding the exact solution of the problem, Fig. 2. Computational intelligence strategies based on artificial neural networks (ANNs) [22], fuzzy logic [23], and meta-heuristic methods [24], such as particle swarm optimizers or genetic algorithms, have been successfully applied to improve the searching capability. These algorithms are able to escape local minima, but they are more computationally demanding and the continuous operating-point commutation can damage the DC-DC stage. Moreover, further power losses are involved because of voltage drops of the conducting bypass diodes.

3.2 Reconfiguration Techniques

The process of changing the physical position of the PV modules that constitute an array in order to provide a new arrangement often allows to enhance power extraction. Besides the particular configuration of the system, the shading scenario is a factor of main relevance. As a matter of fact, several studies can be found in

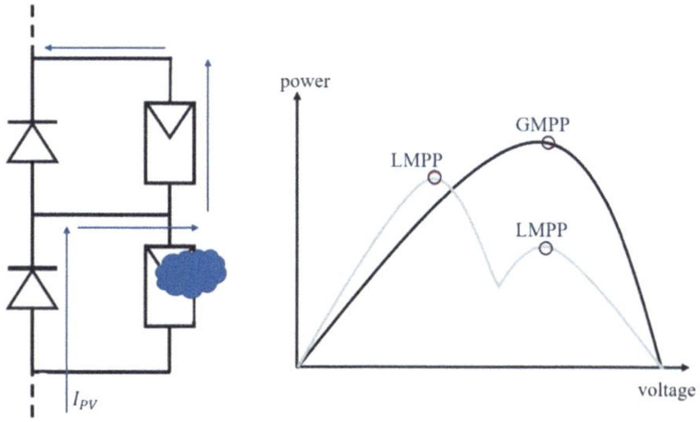

Fig. 2 Bypass diodes operation and P–V characteristic

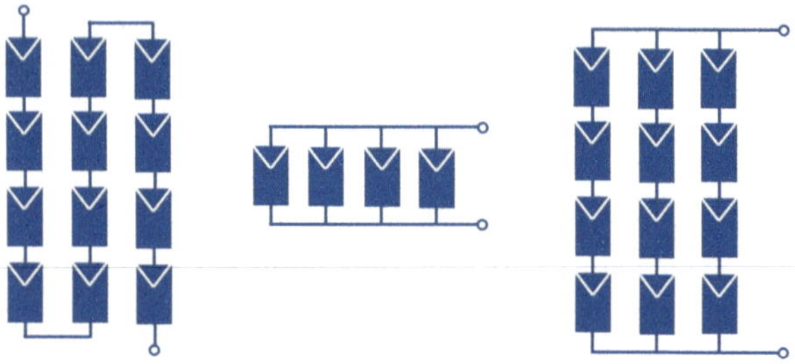

Fig. 3 Regular configurations. From left to right: S, P, and SP

literature providing rearrangement based solutions, but most of them are not able to give an absolutely optimal strategy that can be applied to any shading pattern. Among regular configurations total cross tied (TCT) one is worthy of note, since has been demonstrated to be the best solution in most cases [25]. The basic series (S) and parallel (P) configurations both present severe shortcomings: S scheme is affected by current reduction when a cell gets shaded since the overall current of the string is forced to that of the ill cell; P scheme, instead, even if guarantees a higher power production with respect to S, suffers from drawbacks concerning voltage. Hence, a series parallel (SP) solution, taking the benefits of both schemes, should be preferred, Fig. 3.

The class of the regular configurations includes, also, Bridge Linked (BL) and Honey Comb (HC) schemes that allow to diminish the number of interconnections with respect to TCT one and, thus, reducing the power losses due to cabling. However, the best performances are achieved through irregular reconfigurations, among which the Sudoku and the Zig Zag are the best known [26].

Moreover, reconfiguration methods can be classified into two main groups: dynamic and static techniques, to which the aforementioned schemes belong. Rearranging dynamically an array means varying the physical location of its panels by mean of switches, sensors, and controllers. The approach that is going to be shown is based on this latest concept; as a matter of fact, such a strategy allows not to take into account the particular shading scenario since the system will be able to reconfigure itself in real time, providing, to each situation, a suitable arrangement of the panels.

4 Smart Switching Reconfiguration: Method and Results

The proposed algorithm, implemented in MATLAB environment, takes under exam an $S - P$ scheme, in which a number of N PV modules are connected in series to provide a string, and M strings are connected in parallel to form an array. The aim of the work is that of developing a simple rearranging scheme able to suit any atmospheric condition, also in rapidly changing scenarios, and to retrieve the power that, otherwise, would be lost. At this purpose, a random irradiance pattern has been generated providing unevenly shading of the array. The $I–V$ characteristic of each

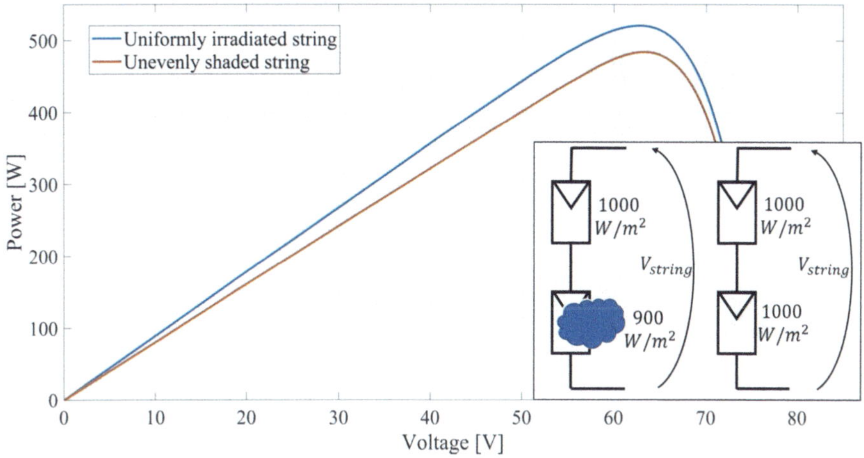

Fig. 4 Comparison between power generated by differently shaded strings

of the modules used to test the algorithm has been extracted by means of the one diode model from datasheets information by using the procedure presented in [18]. Results are reported for Mitsubishi Electric Photovoltaic Module PV-MLE260HD. Each module is constituted by 120 monocrystalline silicon cells of 78×156 mm.

Once the benchmark is generated, the P–V curve of each of the string is extracted. The power of a single string is computed by knowing the open-circuit voltage of the string, that is, $v_{oc} = \sum_{k=1}^{N} v_{oc,k}$, where $v_{oc,k}$ is the open-circuit voltage of the k-th module of the string and imposing $i_{sc} = min_k(i_{sc,k})$, i.e., the short-circuit current of the string is limited by the cell with the lowest irradiance. Thus, from the knowledge of the voltage of each, V_{pv}, it is possible to calculate its current, I_{pv}, by means of an explicit formulation based on the use of the *Lambert-W* function, such as in [19]. So, the power of the string is the product $P_{pv} = V_{pv} I_{pv}$. As can be seen from Fig. 4, a unique shaded module is enough to limit the power generation of the whole array. Once the power generated by a string is known, its modules are rearranged to provide higher power. As far as the power of the whole array is concerned, it is calculated as the product of the summation of the previously calculated current of the strings, $I_{array} = \sum_{i=1}^{M} I_{pv,i}$, and the voltage range of the string associated to the lowest open-circuit voltage.

The algorithm presented rearranges the modules in ascending order with respect to their irradiance level; in this way, modules with similar values of irradiance G are positioned close to each other. Figure 5 shows the reconfigured array. In this way, the shortcomings about power reduction can be limited. In Table 1 the results concerning three different simulated shading patterns are reported. In the table, $\Delta G(\%)$ refers to the maximum difference of irradiance between modules of the same string while the second column represents the percentage portion of the array that is shaded. The results are analyzed in terms of P_{MPP}: it can be noticed that, at the same level of $\Delta G(\%)$, the more the shaded portion, the less is the percentage

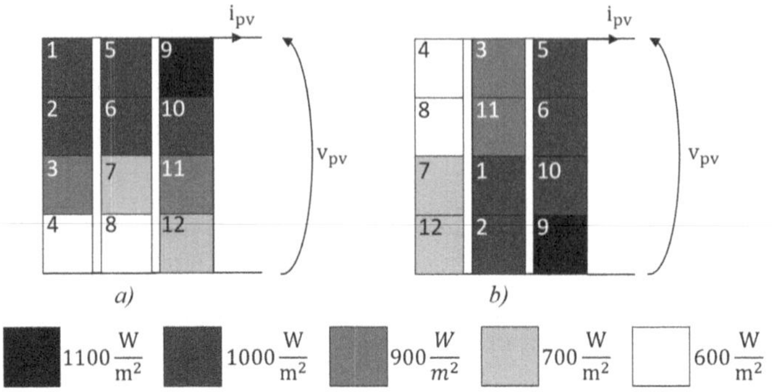

Fig. 5 Comparison between the initial configuration of the array (**a**), and the reconfigurated array (**b**)

Table 1 Percentage power gain at MPP with respect to the maximum power of the initial configuration

ΔG (%)	Shaded portion (%)	ΔP_{MPP} (%)
40	25	20.14
40	50	11.30
30	25	15.18
30	50	6.08
20	25	10.38
20	50	2.26

in power gain achieved through reconfiguration; on the other side, fixing the shaded portion, it is evident that the method is more efficient when dealing with higher values of $\Delta G(\%)$. Hence, the benefits of this approach can be highly appreciated especially when dealing with strings with large irradiance differences and lower shaded portions of the array. Lastly, it has to be highlighted that the scheme allows to avoid the use of bypass diodes, since all modules are exploited and there is no need to cut-off any of them.

5 Implementation on a Low-Cost 32-Bit MCU

5.1 Irradiance Estimation Equipment

The algorithm for the irradiance estimation must be implemented on a microcontroller (MCU) unit with two important characteristics. First, the MCU must be able to provide analog and digital interfaces for the measurement of the panel voltage, panel current, and panel temperature. Second, since the optimal reconfiguration strategy is based on the knowledge of the irradiance on every panel in the system, the MCU must be able to implement wireless communications.

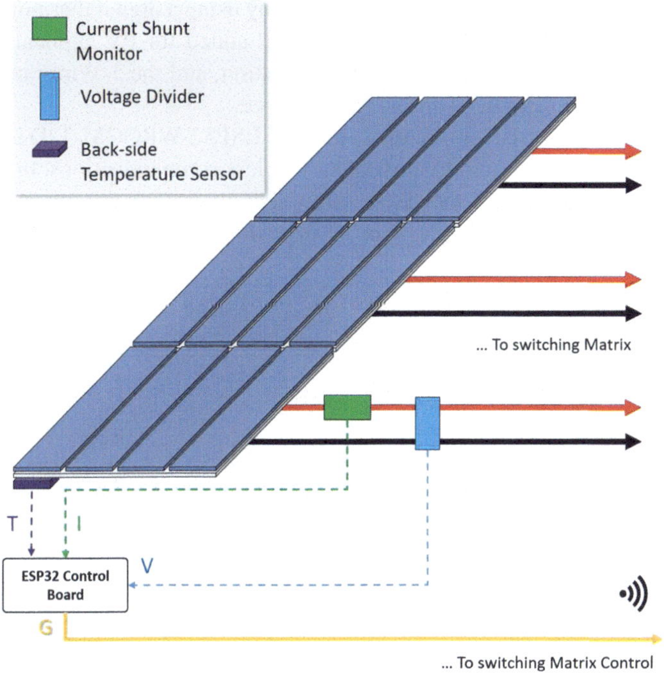

Fig. 6 MCU scheme for irradiance assessing

The MCU chosen for this application is the low-cost 32-bit device ESP32 microcontroller. This microcontroller features two accurate 12-Bit ADCs, a wide set of digital interfaces (SPI, UART, I2S, and I2C), 52 kB of SRAM, and a full stack for both Wi-Fi 802.11 b/g/n and Bluetooth 4.2. The microcontroller is usually sold on a micro-board where additional components such as FLASH memory and the RF antenna are added. The ESP32-WROOM-32D, shown in Fig. 6, features 4 MiB of external flash to be used as program memory and a printed RF antenna. The whole system can be easily programmed from the Arduino IDE after the installation of the board package for the ESP32 microcontrollers.

The proposed MCU features all the three characteristics specified for the project, assuming some servicing circuitry is provided. The measurement of the PV panel voltage can be done directly from the MCU ADC, providing a resistive divider for the voltage, which can be implemented with simple resistors (the use of a buffering circuit should not be needed considering the large input resistance of the ADC). Current measurement is more delicate and must be implemented sensing the voltage on small series resistance, properly amplified. Alternatively, integrated circuits, such as the Texas Instruments INA260, provide voltage, current, and power measurements that can be accessed via I2C interface. This solution is more costly, but reduces the burden in terms of analog circuit design.

Temperature measurement can be performed by using a digital thermometer such as the DS18B20. This thermometer is very well suited for PV application due to the temperature range, the programmable resolution, and the 1-Wire interface that allows easy communications with the MCU device.

The second characteristic is featured since the ESP32-WROOM-32D is equipped with all the hardware required to implement wireless communications. In particular, considering the application can have large spatial extension, a mesh topology (supported by the chosen device) can be appealing thanks to its robustness and redundancy.

5.2 Switching Matrix Control

The information on irradiance for single panels must be collected by a master device where the algorithm described in Sect. 4 determines the optimal reconfiguration and then arranges the switching matrix accordingly (Fig. 7). The device characteristics in this case are less stringent than the ones for the slave devices (i.e., the ones on the panels measuring the irradiance) because there is no need to perform actual measurements. On the other hand, the device must be able to support wireless communication with the slaves, and most importantly, feature enough computational capabilities to run the reconfiguration algorithm. The ESP32-WROOM-32D features a 240 MHz 32-bit microcontroller with a large FLASH memory bank that allows implementing the algorithm proposed in Sect. 4 and most of the algorithms present in literature. However, if more complex algorithm should be implemented, such as ones involving evolutionary computation, or neural networks, a more advanced platform featuring at least a floating-point unit co-processor should be considered for the master.

6 Conclusions

This work has analyzed the issue of partial shading as well as its effects on PV array. An analytical method has been used to assess the irradiance pattern on the array, avoiding the use of standard solar sensors, whose installation is often tricky. Moreover, this application is useful in the evaluation of the degradation and aging of PV devices, for which real-time irradiance sensing can be a valuable asset [27]. An algorithm has been implemented in MATLAB providing a solution for the rearrangement of the modules constituting the array in order to achieve higher power levels with respect to the initial configuration. The solution suits any shading scenario since it is based on the knowledge of the actual values of irradiance that are calculated in real time. This computation step is performed by a system of MCU slave units. Moreover, the reconfiguration algorithm is implemented on a MCU master unit that collects irradiance data from the slave network via Wi-Fi and

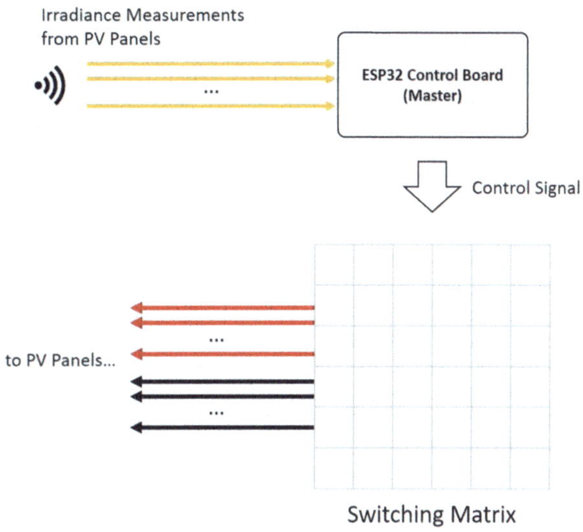

Fig. 7 Control block for the switching

drives the switches. The proposed method allows to improve the performances of a unevenly shaded array and, moreover, allows not to resort to traditional solar sensors since the pyranometer is directly implemented on MCU units. Since the proposed approach considers an $S-P$ reconfiguration strategy, sorting modules according to irradiance takes into account the main factor for current mismatch. A further step could include a comparison between the proposed method and a reconfiguration strategy based on rearrangement according to MPP values: the two approaches are computationally similar; a good discrimination between the two should come from a sensitivity analysis with respect to the $\{v, i, T\}$ triplet. Even without considering an electrical simulation of the switch box, associating a cost-function to the switching activity, thus creating a hysteretic behavior could greatly benefit the approach in terms of stability. Indeed, if such cost-function were to be implemented, the problem should be formulated as a multiobjective one. In this case, suitable optimization algorithms should be used. Once this framework is in place, computing the cost-function through the switch-box simulation should be an easy task. Moreover, to account for the actual spatial distribution of the modules, such cost-function should penalize the switching according to the distance.

References

1. S.R. Pendem, S. Mikkili, Modelling and performance assessment of PV array topologies under partial shading conditions to mitigate the mismatching power losses. Solar Energy **160**, 303–321 (2018)
2. S. Silvestre, S. Kichou, A. Chouder, G. Nofuentes, E. Karatepe, Analysis of current and voltage indicators in grid connected PV (photovoltaic) systems working in faulty and partial shading conditions. Energy **86**, 42–50 (2015)

3. T. Sebbagh, R. Kelaiaia, A. Zaatri, An experimental validation of the effect of partial shade on the IV characteristic of PV panel. Int. J. Adv. Manuf. Technol. **96**(9–12), 4165–4172 (2018)
4. J. Bai, Y. Cao, Y. Hao, Z. Zhang, S. Liu, F. Cao, Characteristic output of PV systems under partial shading or mismatch conditions. Solar Energy **112**, 41–54 (2015)
5. K.A. Kim, P.T. Krein, Reexamination of photovoltaic hot spotting to show inadequacy of the bypass diode. IEEE J. Photovoltaics **5**(5), 1435–1441 (2015)
6. J. Ahmed, Z. Salam, A critical evaluation on maximum power point tracking methods for partial shading in PV systems. Renew. Sustain. Energy Rev. **47**, 933–953 (2015)
7. T. Ghanbari, Permanent partial shading detection for protection of photovoltaic panels against hot spotting. IET Renew. Power Gener. **11**(1), 123–131 (2016)
8. D. Rossi, M. Omaña, D. Giaffreda, C. Metra, Modeling and detection of hotspot in shaded photovoltaic cells. IEEE Trans. Very Large Scale Integr. (VLSI) Syst. **23**(6), 1031–1039 (2015)
9. M. Carrasco, A. Laudani, G.M. Lozito, F. Mancilla-David, F. Riganti Fulginei, A. Salvini, Low-cost solar irradiance sensing for PV systems. Energies **10**(7), 998 (2017)
10. A. Oliveri, L. Cassottana, A. Laudani, F.R. Fulginei, G.M. Lozito, A. Salvini, M. Storace, Two FPGA-oriented high-speed irradiance virtual sensors for photovoltaic plants. IEEE Trans. Ind. Inf. **13**, 157–165 (2017)
11. E.I.O. Rivera, F.Z. Peng, Algorithms to estimate the temperature and effective irradiance level over a photovoltaic module using the fixed point theorem, in *Power Electronics Specialists Conference, 2006. PESC'06. 37th IEEE* (IEEE, New York, 2006), pp. 1–4
12. F. Mancilla-David, F. Riganti-Fulginei, A. Laudani, A. Salvini, A neural network-based low-cost solar irradiance sensor. IEEE Trans. Instrum. Meas. **63**, 583–591 (2014)
13. W.T. da Costa, J.F. Fardin, B. Lauro de Vilhena, D.S.L. Simonetti, Estimation of irradiance and temperature using photovoltaic modules. Solar Energy **110**, 132–138 (2014)
14. S. Bana, R. Saini, Experimental investigation on power output of different photovoltaic array configurations under uniform and partial shading scenarios. Energy **127**, 438–453 (2017)
15. B. Dhanalakshmi, N. Rajasekar, Dominance square based array reconfiguration scheme for power loss reduction in solar photovoltaic (PV) systems. Energy Convers. Manage. **156**, 84–102 (2018)
16. F. Belhachat, C. Larbes, Modeling, analysis and comparison of solar photovoltaic array configurations under partial shading conditions. Solar Energy **120**, 399–418 (2015)
17. B. Celik, E. Karatepe, S. Silvestre, N. Gokmen, A. Chouder, Analysis of spatial fixed PV arrays configurations to maximize energy harvesting in BIPV applications. Renew. Energy **75**, 534–540 (2015)
18. A. Laudani, F.R. Fulginei, A. Salvini, Identification of the one-diode model for photovoltaic modules from datasheet values. Solar Energy **108**, 432–446 (2014)
19. A. Laudani, F.R. Fulginei, A. Salvini, High performing extraction procedure for the one-diode model of a photovoltaic panel from experimental I–V curves by using reduced forms. Solar Energy **103**, 316–326 (2014)
20. A. Laudani, F. Riganti Fulginei, A. Salvini, G.M. Lozito, S. Coco, Very fast and accurate procedure for the characterization of photovoltaic panels from datasheet information. Int. J. Photoenergy **2014** (2014). http://dx.doi.org/10.1155/2014/946360
21. S. Coco, A. Laudani, G.M. Lozito, F. Riganti Fulginei, A. Salvini, Sensitivity analysis of the reduced forms of the one-diode model for photovoltaic devices. Int. J. Numer. Modell. Electron. Netw. Dev. Fields **32**, e2327 (2018)
22. P. Kofinas, A.I. Dounis, G. Papadakis, M. Assimakopoulos, An intelligent MPPT controller based on direct neural control for partially shaded PV system. Energy Build. **90**, 51–64 (2015)
23. C.S. Chin, M.K. Tan, P. Neelakantan, B.L. Chua, K.T.K. Teo, Optimization of partially shaded PV array using fuzzy MPPT, in *Proceedings of the Science and Engineering 2011 IEEE Colloquium Humanities*, December (2011), pp. 481–486
24. K. Sundareswaran, S. Peddapati, S. Palani, MPPT of PV systems under partial shaded conditions through a colony of flashing fireflies. IEEE Trans. Energy Convers. **29**(2), 463–472 (2014)

25. A. Laudani, G.M. Lozito, V. Lucaferri, M. Radicioni, F.R. Fulginei, On circuital topologies and reconfiguration strategies for PV systems in partial shading conditions: a review. AIMS Energy **6**(5), 735–763 (2018)
26. S. Vijayalekshmy, G. Bindu, S.R. Iyer, A novel Zig-Zag scheme for power enhancement of partially shaded solar arrays. Solar Energy **135**, 92–102 (2016)
27. A. Faba, S. Gaiotto, G.M. Lozito, A novel technique for online monitoring of photovoltaic devices degradation. Solar Energy **158**, 520–527 (2017)

Unity Efficiency and Low-Cost MPPT Method for Single-Stage Grid-Connected PV System

Yacine Triki, Ali Bechouche, Hamid Seddiki, and Djaffar Ould Abdeslam

Abstract This paper presents a new maximum power point tracking (MPPT) algorithm based on the adaptive linear neuron concept. This algorithm is designed to extract the maximum power in single-stage grid-connected photovoltaic (PV) system configuration. In the considered system, a PV panel is directly connected to a grid through a three-phase pulse-width modulation inverter. The control is achieved in the synchronous *dq* frame, and the proposed MPPT estimates directly the optimal *d*-axis duty cycle component. Furthermore, in order to achieve a unity power factor operation, the *q*-axis reference current is set to zero. In this work, only one proportional-integral controller is used to maintain the reactive power to zero value. To verify the effectiveness of the proposed algorithm, the grid-connected system is implemented and simulated under MATLAB-Simulink software. The obtained results are compared to those achieved by the conventional perturb and observe based MPPT technique under fast and slow irradiance changes. The simulation results show that the proposed method leads to achieve incomparable performances such as unity efficiency and zero oscillations in the PV panel in both transient and steady-state operations.

1 Introduction

Solar energy has become an attractive and competitive renewable energy source. However, the generated power from the photovoltaic (PV) panels depends on the temperature and irradiance conditions. In addition, the PV panels exhibit extremely nonlinear voltage-current characteristics which also vary with environmental conditions. Moreover, for each operating condition, there is a unique maximum power

Y. Triki · D. Ould Abdeslam (✉)
IRIMAS Institute, University of Haute Alsace, Mulhouse, France
e-mail: yacine.triki@uha.fr; djaffar.ould-abdeslam@uha.fr

A. Bechouche · H. Seddiki
L2CSP Laboratory, Mouloud Mammeri University, Tizi Ouzou, Algeria

W. Zamboni, G. Petrone (eds.), *ELECTRIMACS 2019*, Lecture Notes in Electrical Engineering 615, https://doi.org/10.1007/978-3-030-37161-6_41

point (MPP), where the PV system generates its maximum power. So, a maximum power point tracking (MPPT) technique is required as control strategy to track the MPP in order to maximize the produced energy [1].

Various MPPT methods have been developed and published in relevant scientific literature. The most discussed MPPT methods are perturb and observe (P&O) [2], incremental conductance [3], fractional open-circuit voltage [4], and hill climbing [5]. Recently, artificial neural network (ANN)-based methods have provided an important interest and have been successfully implemented for MPPT strategies [6]. Their main advantage is the ability to learn and improve their performances throughout previous experiences. So, ANN methods feature several advantages which significantly increase the PV system efficiencies. However, attractiveness of the ANN-based methods depends on their complexity.

Grid-connected PV system is defined as electricity-generating PV power system that is connected to the power grid. Commonly, this connection is achieved using two stages of converters [7]. The first stage is a boost converter used to increase the PV output voltage and to achieve the MPPT function. The second stage is a PWM inverter used to realize the grid-connected function. On the other hand, single-stage grid-connected systems can be also used. It consists of the panel and inverter only. The two-stage grid-connected system has the advantage that it is easy to design its control scheme, but it has some drawbacks such as complex topology, lower efficiency, and higher cost [8]. On the contrary, single-stage topology provides many advantages such as simple topology, higher power efficiency, and lower cost. However, since the system contains only one stage of power conversion, all the control objectives need to be realized simultaneously, such as MPPT control, harmonics reduction, and synchronization with the power grid, so that the complexity of control scheme is much increased [7].

In this paper, ANN-based MPPT method for single-stage grid-connected PV system is developed. The proposal is based on the adaptive linear neuron (ADALINE). The use of the ADALINE is motivated by its simple structure, convergence speed, and a possible physical interpretation of its weights [9]. In addition, the ADALINE training is performed online, which eliminates the need for repetitive offline training. The three-phase single-stage grid-connected inverter is chosen to interface PV panel system with the power grid. This topology is used, especially to reduce the power loses and to simplify the system configuration. The main contributions of the proposed MPPT and the use of the single-stage inverter can be summarized as reduction of the transient and steady-state oscillations and improvement of the efficiency. In order to prove the effectiveness of the proposed MPPT method, comparative study with the P&O algorithm is performed through numerical simulations. Obtained simulation results show perfect performances of the proposal in transient and steady states.

2 Grid-Connected PV System Modeling

2.1 Description of the Controlled Grid-Connected PV System

The proposed single-stage grid-connect PV system consists of three parts: a set of PV panels, a three-phase single-stage grid-connected inverter, and the grid. The block diagram of the proposed system is depicted in Fig. 1. In the control unit, two controllers operate simultaneously; the first one is used to perform the MPPT, and the second one is used to ensure unity power factor (UPF) operation. The main contribution of this work is in the first controller where a new MPPT control strategy based on the ADALINE is proposed. The block diagram of this strategy is presented in Fig. 5.

2.2 PV Panel Model

Ten PV panels of IFRI260-60 type are connected in series and used to experiment the proposed MPPT method. The electrical specifications of a PV panel at standard test conditions (STC) are given in Table 1 [10]. Among the many available equivalent circuits for PV panel modeling, the single-diode model is adopted in this paper.

The output current (I) of the PV panel can be expressed as [11] follows:

$$I = I_{\mathrm{pv}} - I_{\mathrm{o}}\left[\exp\left(\frac{V + R_{\mathrm{s}}\, I}{a V_{\mathrm{t}}}\right) - 1\right] - \frac{V + R_{\mathrm{s}}\, I}{R_{\mathrm{p}}} \tag{1}$$

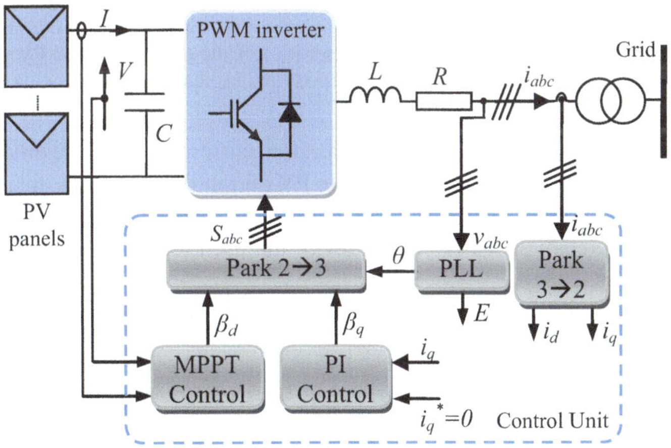

Fig. 1 Block diagram of the grid-connected PV system

Table 1 Overall system parameters

IFRI260-60 PV panels at STC		
P_{max}	Maximal power	260 × 10 W
I_{sc}	Short-circuit current	8.65 A
V_{oc}	Open-circuit voltage	38.1 × 10 V
N_s	Number of cells in series	60 × 10
PWM inverter		
R	Resistance	0.1 Ω
L	Inductance	5 mH
C	Capacitor	100 μF
E	Grid voltage	100 V
MPPT algorithms		
μ	ADALINE learning rate	adaptive
$\Delta\alpha$	P&O step size	0.00075
T_e	Sampling period	0.0001 s

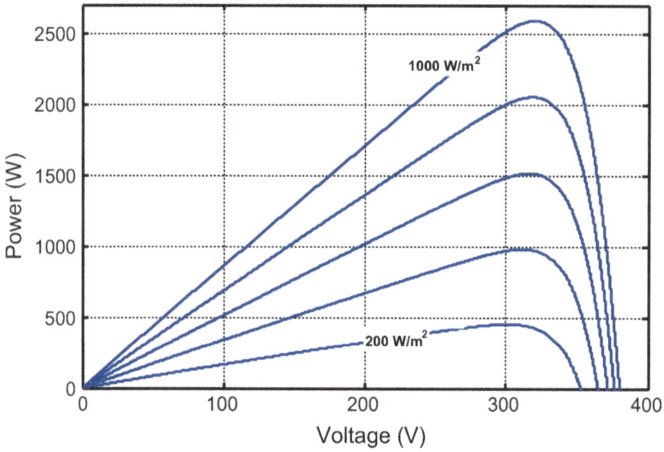

Fig. 2 Power-versus-voltage (*P*-*V*) curve characteristics of the ten IFRI260-60 PV panels connected in series at 25 °C for different irradiance levels (graphs plotted with step of 200 W/m^2)

where I_{pv} is the PV current, I_o is the PV saturation current, and a is the diode ideality factor. V_t is the thermal voltage and V is the PV output voltage. R_p is the equivalent parallel resistance and R_s is the equivalent series resistance.

The power-versus-voltage (*P*-*V*) curves of the used ten PV panels in this work are plotted in Fig. 2.

2.3 *Grid-Connected PWM Inverter Modeling*

The power circuit of the grid-connected three-phase PWM inverter is presented in Fig. 3. This inverter is connected to the grid through a three-phase AC filter

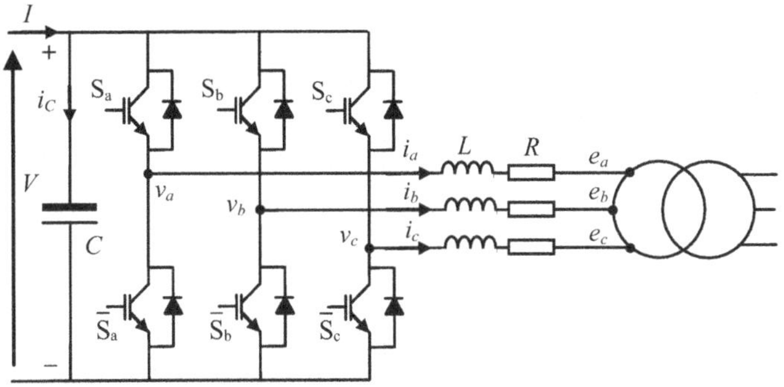

Fig. 3 Topological structure of the grid-connected three-phase PWM inverter

containing inductances (L) and series with resistances (R). e_a, e_b, and e_c are the grid voltages; i_a, i_b, and i_c are the inverter output currents; v_a, v_b, and v_c are the control voltages; V is the input DC voltage (PV panel output voltage).

The PWM inverter model can be described in d-q frame as follows:

$$\begin{bmatrix} e_d \\ e_q \end{bmatrix} = -L\frac{\mathrm{d}}{\mathrm{d}t}\begin{bmatrix} i_d \\ i_q \end{bmatrix} - \begin{bmatrix} R & -\omega L \\ \omega L & R \end{bmatrix}\begin{bmatrix} i_d \\ i_q \end{bmatrix} + \frac{V}{2}\begin{bmatrix} \beta_d \\ \beta_q \end{bmatrix} \tag{2}$$

where e_d, e_q, i_d, and i_q are, respectively, the d- and q-axis components of the grid voltages and the inverter's output currents; β_d and β_q are the d- and q-axis components of the modulating signals; ω is the angular frequency of the grid voltages. The DC bus voltage dynamics can be written in the following way:

$$I - C\frac{\mathrm{d}V}{\mathrm{d}t} = \frac{3}{4}\left(\beta_d i_d + \beta_q i_q\right) \tag{3}$$

The active P and reactive Q powers of the PWM inverter are finally calculated as follows:

$$P = \frac{3}{2}\left(e_d i_d + e_q i_q\right) \tag{4a}$$

$$Q = \frac{3}{2}\left(e_q i_d - e_d i_q\right). \tag{4b}$$

2.4 Unity Power Factor Operation

By setting the grid voltage vector according to the d-axis ($e_d = E$, $e_q = 0$) and under unity power factor (UPF) operation ($i_q = 0$), the PWM inverter model becomes the following equations:

$$-L\frac{\mathrm{d}i_d}{\mathrm{d}t} - Ri_d + \frac{1}{2}V\beta_d = E \tag{5a}$$

$$-\omega Li_d + \frac{1}{2}V\beta_q = 0 \tag{5b}$$

$$I = C\frac{\mathrm{d}V}{\mathrm{d}t} + \frac{3}{4}i_d\beta_d \tag{5c}$$

3 ADALINE-Based MPPT

3.1 The ADALINE Theory

ADALINE is a powerful ANN technique used in many applications in power systems. It's a multi-input single-output topology which is equivalent to one neuron. It's composed by an input vector $X(k) = [x_1(k) \ \ldots \ x_m(k)]$, an adjustable weight vector $W(k) = [w_1(k) \ \ldots \ w_m(k)]^T$, a linear activation function, and an estimated output $y_{\mathrm{est}}(k)$. Architecture of an ADALINE is shown in Fig. 4.

The estimated output can be calculated for any input $X(k)$ at sample time k as follows:

$$y_{\mathrm{est}}(k) = \sum_{i=1}^{m} w_i(k).x_i(k). \tag{6}$$

ADALINE is an online learning process. Its weights are adjusted to minimize the error $e(k)$ between the estimated output $y_{\mathrm{est}}(k)$ and the desired response $y_{\mathrm{d}}(k)$. The estimation error $e(k)$ is then defined as follows:

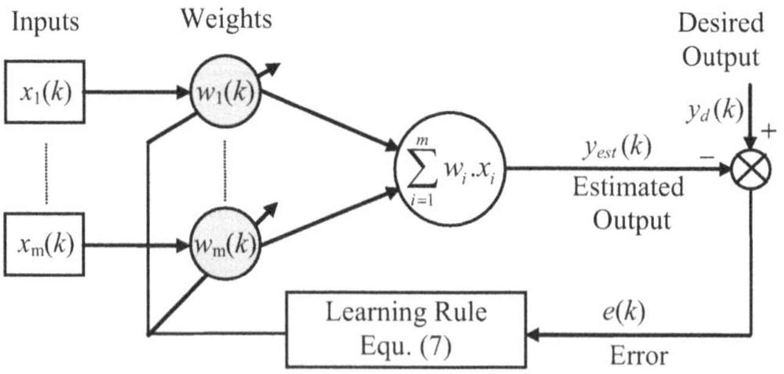

Fig. 4 Architecture of the ADALINE

$$e(k) = y_{\text{d}}(k) - y_{\text{est}}(k) = y_{\text{d}}(k) - \sum_{i=1}^{m} w_i(k).x_i(k). \tag{7}$$

When inputs are applied to the network, its output $y_{\text{est}}(k)$ is compared to a target $y_{\text{d}}(k)$. Based on the generated error, a learning rule is used to adjust the weights in order to move the ADALINE output closer to the target. The most known learning rule is that called α-LMS algorithm given as follows:

$$W(k+1) = W(k) + 2\mu\, e(k)X(k) \tag{8}$$

where μ is the learning rate. The choice of μ controls stability and speed of convergence of the ADALINE [9]. For input pattern vectors, stability is ensured for most practical purposes if it is as follows:

$$0 < \mu < 2 \tag{9}$$

3.2 Proposed MPPT Method

The task of the proposed control is to extract the maximum power from the PV source by generating an appropriate reference d-axis duty cycle component β_d. To set the control, we suppose that the inductor drop voltage is neglected since the sun irradiance doesn't change rapidly. Moreover, the resistance is generally low and can be neglected. Then, Eq. (5a) is written as follows:

$$\frac{1}{2} V \beta_d = E \tag{10}$$

The MPPT methods are based on the fact that the derivative of the output power with respect to the panel voltage is equal to zero at the MPP. Therefore, the following equation leads to the MPP condition in terms of panel voltage V, i.e.:

$$V = -I\frac{\mathrm{d}V}{\mathrm{d}I} \tag{11}$$

The substitution of Eq. (11) in Eq. (10) leads to the following equation:

$$-\frac{I\mathrm{d}V}{2\mathrm{d}I}\beta_d = E \tag{12}$$

From Eq. (12) and in order to operate at the MPP, the d-axis duty cycle component must be tuned continuously to satisfy the following equality:

$$\beta_d(k) = -\frac{2E\Delta I(k)}{I(k)\Delta V(k)} \tag{13}$$

At this stage, the ADALINE strategy is introduced to solve Eq. (13) in the aim to find the optimal duty cycle. Considering the ADALINE structure defined in Sect. 3.1, the following equalities are posed:

$$-\frac{2E\Delta I(k)}{I(k)\Delta V(k)} = y_{\mathrm{d}}(k) \tag{14a}$$

$$\beta_d(k) = w(k) = y_{\mathrm{est}}(k) \tag{14b}$$

$$x(k) = 1 \tag{14c}$$

The implementation of the proposed control is easy; $y_{\mathrm{d}}(k)$ is recurrently compared to $\beta_d(k)$, and according the generated error e, the duty cycle $\beta_d(k+1)$ is increased or decreased up to the error becomes zero. Then, by means of Eq. (8), the ADALINE-based MPPT algorithm is expressed by the following equation:

$$\beta_d\,(k+1) = \beta_d(k) + 2\mu\,(y_{\mathrm{d}}(k) - \beta_d(k)) \tag{15}$$

As seen, the duty cycle is iteratively adapted with constant learning rate μ. However, during the simulation tests, it's found that the better results, at transient state, are found by using variable learning rate. Consequently, to adapt the duty cycle, tanh(x) function is used in this paper. The MPPT algorithm becomes the following equation:

$$\beta_d\,(k+1) = \beta_d(k) + \tanh\,(e(k)) \tag{16}$$

with

$$\tanh\,(e(k)) = \frac{1 - e^{-2e(k)}}{1 + e^{-2e(k)}} = \frac{1 - e^{-2(y_{\mathrm{d}}(k)-\beta_d(k))}}{1 + e^{-2(y_{\mathrm{d}}(k)-\beta_d(k))}} \tag{17}$$

Considering Eq. (16), the block diagram of the proposed MPPT algorithm is given by Fig. 5.

4 Simulation Results

To verify the effectiveness of the proposed method, dynamic and steady-state performances are examined according to fast (±400 W/m^2/s) and slow (±40 W/m^2/s) sun irradiation slop change [12]. In addition, the sun irradiance G changes between low irradiance level (200 W/m^2) and high irradiance level (1000 W/m^2). The PWM

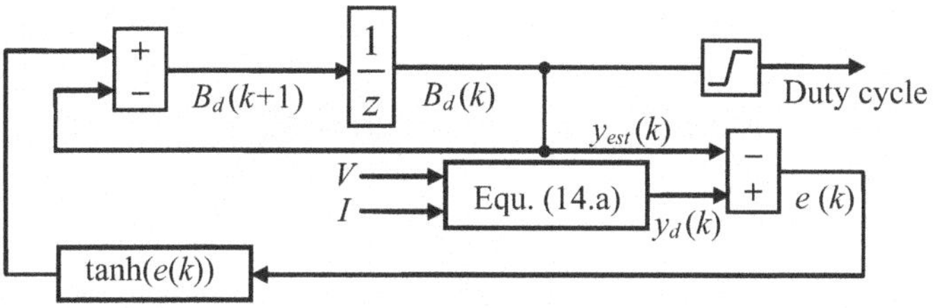

Fig. 5 Block diagram of the proposed MPPT algorithm

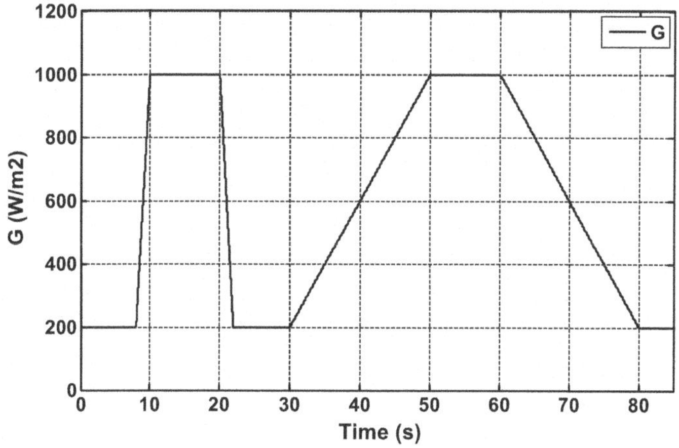

Fig. 6 Irradiance test sequence

inverter operates at UPF operation and is connected to 100 V–50 Hz line to neural three-phase grid voltages. The irradiance test sequence is illustrated in Fig. 6.

The overall system parameters including the used IFRI260-60 PV panels at STC are given in Table 1.

Figure 7 depicts the shape of the extracted power when the test sequence of irradiance is applied. As can be seen from Fig. 7a, the two methods are able totrack the maximal power under a constant and changing irradiance conditions. However, the zoom illustrated in Fig. 7b demonstrates the superiority of the proposed method whereabouts 100% of efficiency is provided in steady state. Furthermore, the PV power doesn't present any oscillations by means of the presented method unlike with the P&O technique where large oscillations and losses are observed. Figure 7c shows the dynamic response of the proposed MPPT compared to P&O. This test considers a variable solar irradiance profile with fast and slow rate of change. It can be seen that 100% of efficiency is still reached with the proposed technique like the steady-state operation. As for the P&O method, losses and oscillations in the PV power areobserved.

The PV current under constant and variable irradiance level is shown in Fig. 8. In Fig. 8a, it can be seen that the optimal current is clearer and more stable by using the

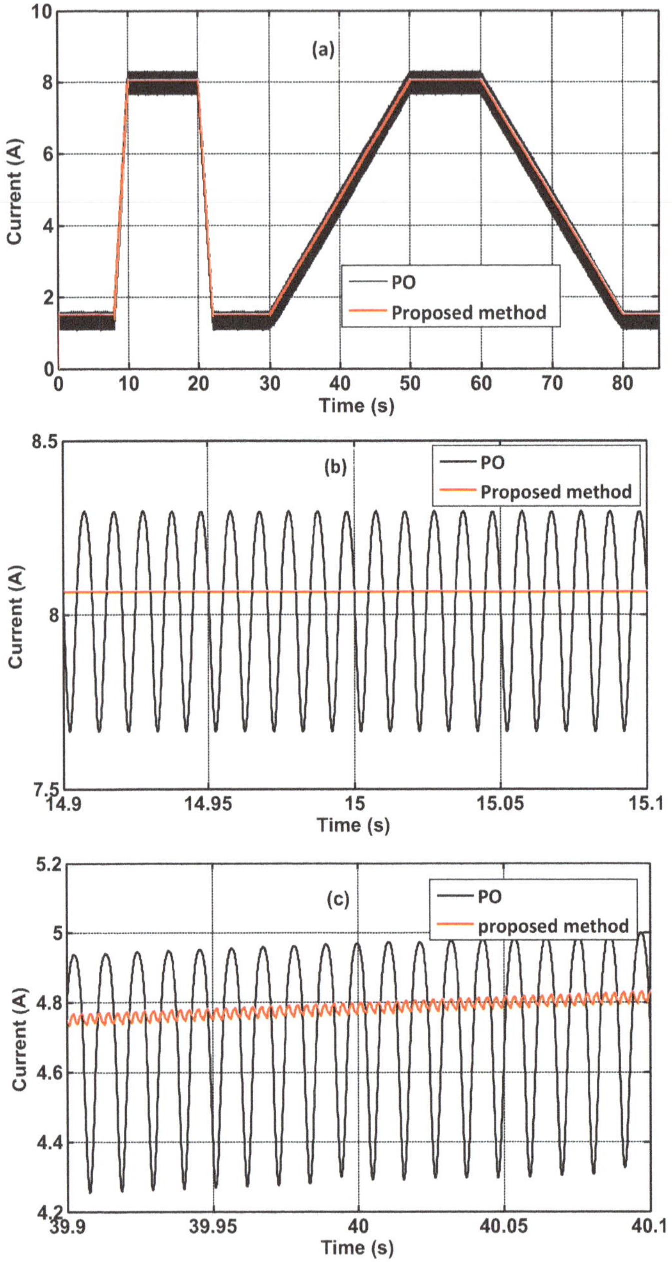

Fig. 7 PV current under constant and variable irradiance profile: (**a**) overall current response, (**b**) zoom of the current in steady state, and (**c**) zoom of the current in transient state

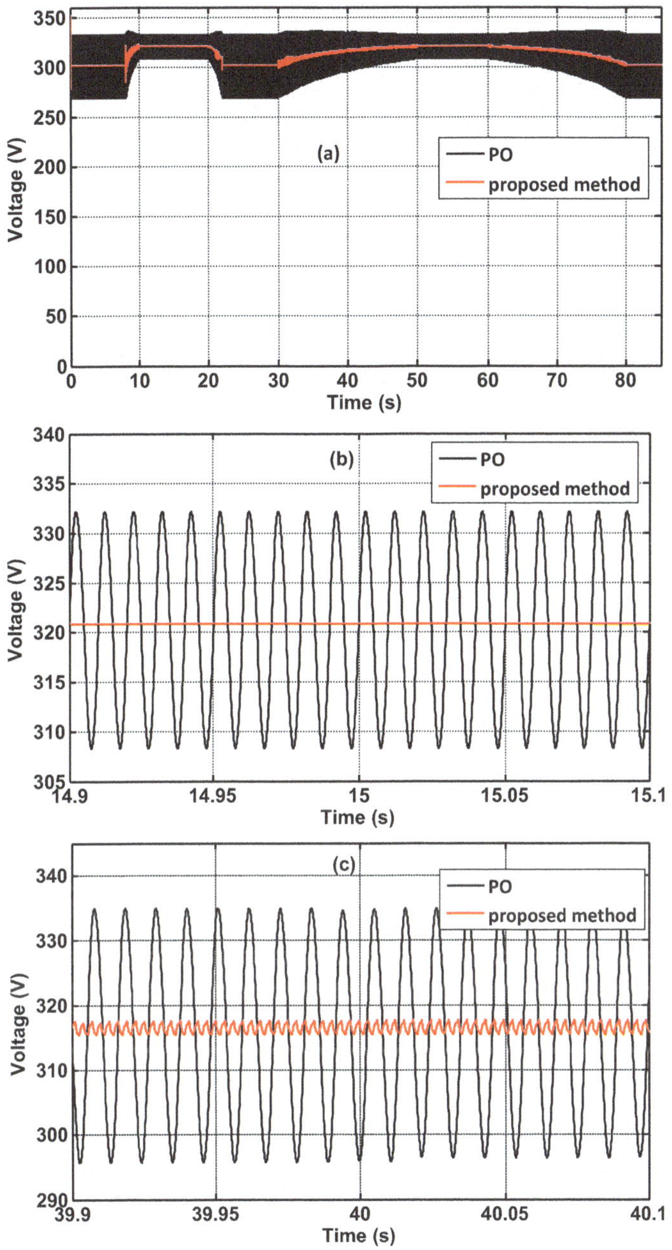

Fig. 8 PV panel voltage under constant and variable irradiance profile: (**a**) overall response of the methods, (**b**) zoom of the voltage in steady state, (**c**) zoom of the voltage in transient state

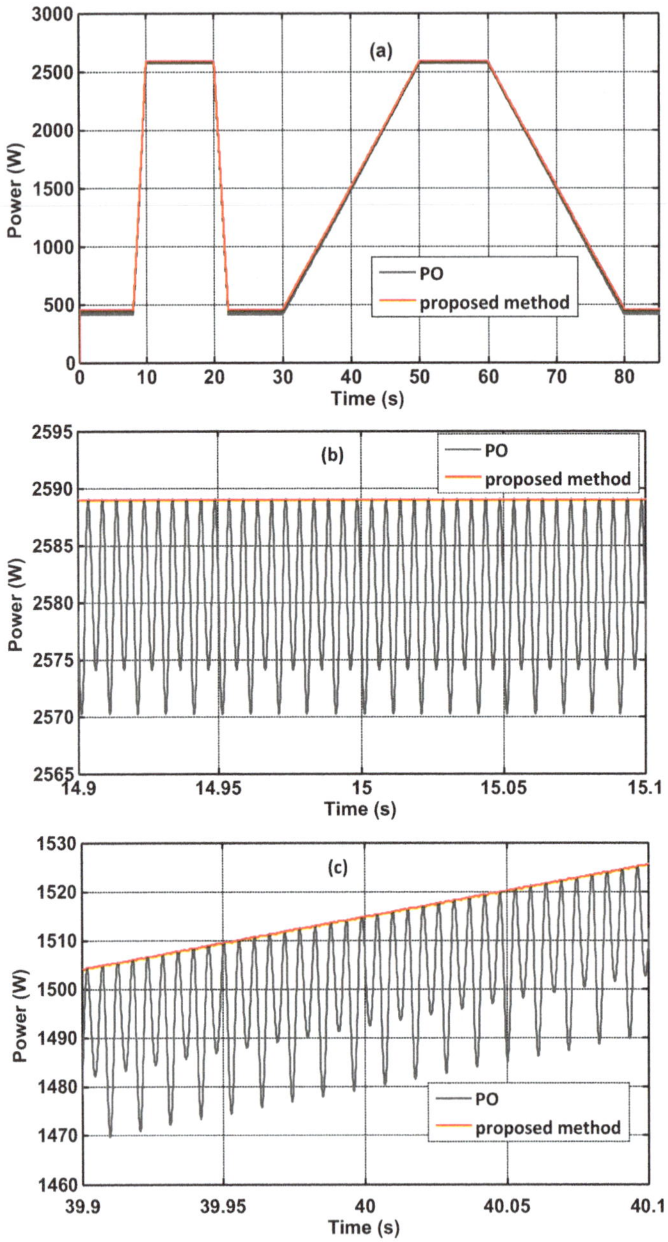

Fig. 9 PV power under constant and variable irradiance profile: (**a**) overall power response, (**b**) zoom of the power in steady state, and (**c**) zoom of the power in transient state

proposed method. Indeed, in steady state, the current shape is a straight line without oscillations by means of the proposed technique (see Fig. 8b). This involves the absence of stress in the panel and certainly leads to increase the lifetime. Likewise, during transient state and referring to Fig. 8c, minor oscillations are observed in the current but remain insignificant compared to those generated by the P&O method.

The PV panels' voltage waveform is shown in Fig. 9. It is clear that the shape of the voltage is better by means of the proposed MPPT algorithm as shown in Fig. 9a. In steady state, the voltage is a pure line without any oscillation unlike the P&O technique where high oscillations are observed (see Fig. 9b). Therefore, this MPPT method based on the ADALINE strategy leads to increase the lifetime of the used panel. In variable atmosphere conditions, the voltage curve of the PV panel has very little oscillations compared to the conventional P&O method as displayed in the zoomed Fig. 9b.

5 Conclusions

In this paper, a new MPPT method based on the ADALINE strategy is presented. This MPPT is designed to extract the maximum power from PV panel in single-stage grid-connected system. The use of the proposed method reduces considerably the system complexity since the *d*-axis duty cycle reference is directly fed from the MPPT to the active power control. Numerical simulations have been carried out under variable atmospheric conditions. The obtained results are compared to those achieved by the conventional P&O algorithm. Supreme and incomparable results are found with the proposed method such as unity efficiency in transient and steady-state operations. Furthermore, the oscillations in the PV panels' voltage and current are fully removed and lead to stress less of the PV panel. Finally, this method solves the drawbacks of the conventional methods such as the trade-off between speed convergence and oscillations in steady state and masterfully improves the poor efficiency in variable conditions, especially in low irradiance level.

Acknowledgments This work was supported by the Franco-Algerian cooperation program PHC-TASSILI (Project No. 17MDU995).

References

1. E.I. Batzelis, Simple PV performance equations theoretically well founded on the single-diode model. IEEE J. Photovolt. **7**(5), 1400–1409 (2017)
2. V.R. Kota, M.N. Bhukya, A novel linear tangents based P&O scheme for MPPT of a PV system. Renew. Sustain. Energy Rev. **71**(Suppl C), 257–267 (2017)
3. M.A. Elgendy, D.J. Atkinson, B. Zahawi, Experimental investigation of the incremental conductance maximum power point tracking algorithm at high perturbation rates. IET Renew. Power. Gener. **10**(2), 133–139 (2016)

4. A. Thangavelu, S. Vairakannu, D. Parvathyshankar, Linear open circuit voltage-variable step size-incremental conductance strategy based hybrid MPPT controller for remote power applications. IET Power Electron. **10**(11), 1363–1376 (2017)
5. N. Kumar, I. Hussain, B. Singh, B.K. Panigrahi, Framework of maximum power extraction from solar PV panel using self-predictive perturb and observe algorithm. IEEE Trans. Sustain. Energy. **9**(2), 895–903 (2018)
6. S. Messalti, A. Harrag, A. Loukriz, A new variable step size neural networks MPPT controller: review, simulation and hardware implementation. Renew. Sustain. Energy Rev. **68**(Part 1), 221–233 (2017)
7. Y. Singh, I. Hussain, B. Singh, S. Mishra, Grid integration of single-phase single-stage SPV system using an Active Noise Isolation based control algorithm, in *IEEE Int. Conf. on Power. Electron., Drives and Energy Systems (PEDES), Trivandrum*, (IEEE, Piscataway, NJ, 2016), pp. 1–5
8. C. Jain, B. Singh, Single-phase single-stage multifunctional grid interfaced solar photo-voltaic system under abnormal grid conditions. IET Gener. Transm. Dis. **9**(10), 886–894 (2015)
9. B. Widrow, M.A. Lehr, 30 years of adaptive neural networks: perceptron, madaline, and backpropagation. Proc. IEEE **78**(9), 1415–1442 (1990)
10. *IFRI260-60 PV Module Datasheet*. Available at: http://www.ifrisol.com
11. M.G. Villalva, J.R. Gazoli, E.R. Filho, Comprehensive approach to modeling and simulation of photovoltaic arrays. IEEE Trans. Power Electron. **24**(5), 1198–1208 (2009)
12. R. Bründlinger, N. Henze, H. Häberlin, B. Burger, A. Bergmann, F. Baumgartner, prEN 50530–the new European standard for performance characterisation of PV inverters, in *24th EU PV Conf., Hamburg, Germany*, (WIP-Renewable Energies, Munich, 2009)

Distributed Converters in Large PV Plants: Performance Analysis Supported by Behavioral Models

Giovanni Nobile, Mario Cacciato, Giuseppe Scarcella, Giacomo Scelba, Ester Vasta, Agnese Giuseppa Federica Di Stefano, Giuseppe Leotta, Paola Maria Pugliatti, and Fabrizio Bizzarri

Abstract This paper is aimed to assess the performance of distributed converters in large PV plants through the analysis of a case study represented by a 2 MW PV plant in Central Italy. The electrical layout of a 500 kW subfield has been modified performing the installation of DC/DC converters at string level in order to create an independent MPPT control for every string. This kind of performance analysis is usually carried out using data acquired by the plant data logger. Unfortunately, the presence of partial unavailability, monitoring system faults, shutdown for maintenance activities, etc. can create several issues in data processing. To support data elaboration, a novel behavioral modeling approach has been developed and exploited in this work. This novel approach, based on an integrated state-space average model, can improve the performance analysis ensuring a satisfactory accuracy but keeping a low computation effort. Validation is performed considering real operating scenarios in case study.

1 Introduction

In large PV plants, the conversion system is usually realized using central inverters whose rated power is in the range from hundreds of kW to some MW. Such converters are usually installed into electrical cabins with fan cooling. The control system of central inverters often implements a single maximum power point tracking (MPPT) for the PV subfield connected to the DC bus of the inverter. In the literature, several papers deal with central inverters, and the research activities are focused mainly on converter topology [1], control system [2, 3], and modeling

---

G. Nobile (✉) · M. Cacciato · G. Scarcella · G. Scelba · E. Vasta
Department of Electrical, Electronics Engineering and Computer Science DIEEI, University of Catania, Catania, CT, Italy
e-mail: gnobile@dieei.unict.it

A. G. F. Di Stefano · G. Leotta · P. M. Pugliatti · F. Bizzarri
Enel Green Power SpA, Catania, Italy

W. Zamboni, G. Petrone (eds.), *ELECTRIMACS 2019*, Lecture Notes in Electrical Engineering 615, https://doi.org/10.1007/978-3-030-37161-6_42

[4]. On the contrary, there are no comprehensive analyses that investigate on distributed converters for large PV plants although their installation can provide significant benefits in terms of yield and reliability. Such benefits are related to the compensation of mismatch effects caused by nonuniform shadowing, faults in some PV modules, misaligned trackers, etc. [5]. In case of PV plants in operation from several years, distributed converters can limit the losses caused by the nonuniform aging of PV modules [6].

This work investigates on a specific distributed conversion typology represented by DC/DC converters in SiC technology installed at string level, close to the parallel string boxes. The case study is represented by a large PV plant in Central Italy. The layout of a subfield has been modified in August 2018 in order to allow the installation of such DC/DC converters. This paper reports the results of the experimental tests under different evaluation criteria. A novel integrated state-space average model has been developed and exploited with the purpose to replace wrong data occurring in case of faults in monitoring system, unavailability, missing data due to maintenance shutdown, etc.

The results shown in this paper can be effectively exploited in design process and in business planning of new PV plants as well as in retrofitting of existing fields.

2 Case Study

The case study of this work is represented by a ground-mounted 2 MW PV field in Central Italy, shown in Fig. 1. The plant is in operation since 2011. Technical data of the main components of the PV field are listed in Table 1. The rated power of PV modules is 235 W except for some strings in which 230 or 240 W PV modules have been used depending on their availability on the market during construction in 2011. The plant is divided into four subfields having similar rated power (about 500 kW). Each subfield is connected to a central inverter. There are two 1250 kVA transformers with double low-voltage windings each one fed by a central inverter.

This plant is chosen for the experimental tests because it is affected by nonuniform aging degradation in PV modules leading to relevant mismatch phenomena between strings. These phenomena, combined with local shadowing effects depending on the morphology of the terrain, have caused during the years significant differences between the subfields in terms of energy production.

With the purpose of reducing the gap between the yearly energy amount theoretically available and that really produced by the less performing 500 kW subfield (subfield 4), the string power optimizers in SiC technology have been installed in August 2018. Figure 2 shows some of the string optimizers during the assembly works on the rear side of the string boxes. As a consequence, the control system of the central inverter connected to the experimental subfield has been modified disabling the original MPPT control while regulating the DC link voltage value in a narrow range around 730 V. The obtained configuration for the experimental subfield is shown in Fig. 3, while Fig. 4 shows the standard power

Fig. 1 Pictures of the 2 MW PV plant in Central Italy that represents the case study of this work

Table 1 Technical specifications of the main power components in the 2 MW PV plant in Central Italy

PV modules (poly, 60 cells)	
P_{module}	235 W (230 or 240 in some strings)
V_{oc}	36.65 V (235 W module)
I_{sc}	8.59 A (235 W module)
V_{MPP}	28.70 V (235 W module)
I_{MPP}	8.19 A (235 W module)
P_{module}/T, V_{oc}/T, I_{sc}/T	−0.43%/°C, −0.35%/°C, +0.03%/°C
Strings and string boxes	
Modules in a string	20
Number of power inputs	16
Monitoring channels for string current	8 (the measured current is the sum of two strings)
String power optimizers	
Rated power	10 kW
Power inputs	1 PV array = 2 strings
Maximum efficiency	99.5%
Power device technology	SiC
Central inverters	
Number of inverters	4
Rated AC power	500 kVA
AC voltage, frequency	400 V, 50/60 Hz
MPPT DC voltage range	450–820 V
Maximum efficiency	98.6%

Fig. 2 Photos show the mounting of the string optimizers. Left: cabling modification at the string box. Right: string optimizers mounted at the rear side of the string box

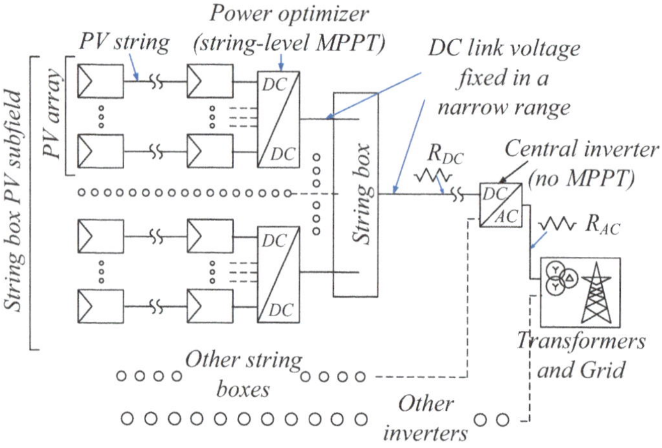

Fig. 3 Single line diagram of the experimental subfield whose original layout has been modified mounting the power optimizers at string level

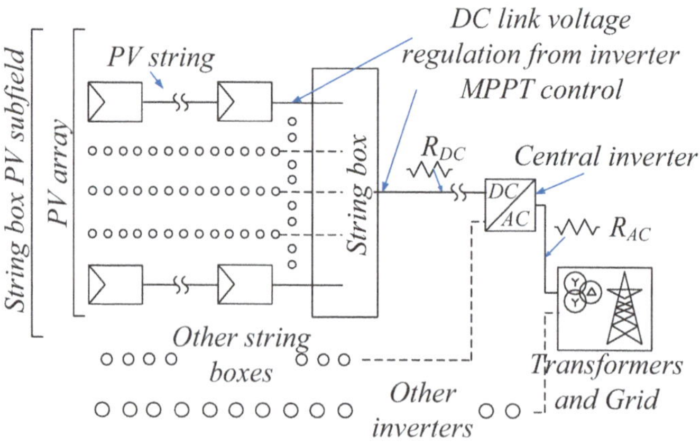

Fig. 4 Single line diagram of the subfields whose layout has maintained the original configuration

layout of the other three subfields (subfield 1, 2, 3) that have kept the original configuration. The monitoring system of the PV plant and of its subfields consists of several sensors and meters, whose accuracy can be assigned in the range from 1.5% to 2.5%.

3 Behavioral Modeling Approach

A suitable modeling approach, based on an integrated state-space average technique, has been developed to support data analysis for the performance assessment of distributed converters in large PV fields. Such behavioral model can be used to replace wrong or missing data in these cases:

- Faults in sensors and power analyzers
- Faults in monitoring system
- Partial or total unavailability due to electrical faults
- Shutdown of a subfield for maintenance activities
- Electrical quantities not provided by data loggers

Moreover, the model can compute all the electrical quantities in every section of the PV plant, thus allowing a simple calculation of losses, voltage drops, etc. On the contrary, in most real PV plant, the monitoring system does not provide all the electrical quantities for each section of the conversion system.

The basic general-purpose PV system configuration considered in this work is shown in Fig. 5. For both DC/DC converter and DC/AC conversion stage, it is possible to take into account simple topologies since the target is to build a behavioral model avoiding the complexity of the detailed models.

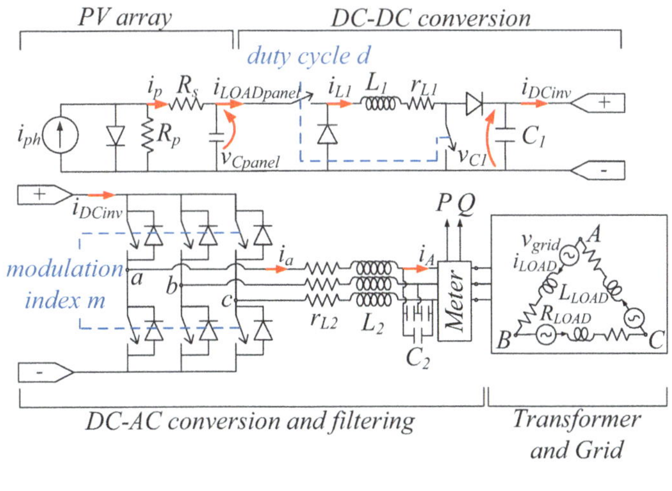

Fig. 5 Basic general-purpose configuration of a utility-scale PV field

3.1 Integrated State-Space Average Model

The integrated state-space average model used in this work is a modified version of the basic modeling approach presented by the authors in [7]. Such approach is a good trade-off between accuracy and computational complexity. The latter requirement is crucial in the performance assessment of utility-scale PV fields due to the large amount of data to be processed. Following the modeling approach described in [7], each component belonging to the conversion system of the PV plant in Fig. 5 has been modeled using its state-space average representation.

About the DC/DC converters, their model can be built as described in [8, 9] using the state-space average method. The implementation of the same method for DC/AC converters is usually performed in a different way, for example, using the symmetrical components as in [10]. In most cases, both the conversion stages are present in PV fields as for the case study. Unfortunately, the creation of an integrated model including all the conversion stages is difficult because of their different modeling approaches.

The key parameter to allow this matching is the DC link current i_{DClink}. Some authors have investigated the calculation of its average or RMS value exploiting power balances [11] or integration methods [12], in some cases with a reduced-order Fourier transform. The main drawback of these methods is the computational effort related to data storage in integrals. To overcome these issues, a novel calculation method has been developed obtaining a simple formula in which the average DC link current is a linear function of the symmetrical components of the AC currents:

$$\begin{aligned} i_{\text{DCinv}} = (i_{\text{DCinv}})_0 = \tfrac{\sqrt{3}m}{2}\Big[&x_1 \cos\left(\varphi_a - \tfrac{\pi}{6}\right) - x_2 \sin\left(\varphi_a - \tfrac{\pi}{6}\right) \\ &+ x_3 \cos\left(\varphi_b - \tfrac{\pi}{6}\right) - x_4 \sin\left(\varphi_b - \tfrac{\pi}{6}\right) \\ &+ x_5 \cos\left(\varphi_c - \tfrac{\pi}{6}\right) - x_1 \sin\left(\varphi_c - \tfrac{\pi}{6}\right)\Big] \end{aligned} \tag{1}$$

where $x_1, \ldots, x_6$ are the real and imaginary parts of the direct components of the virtual line AC currents:

$$\begin{aligned} (i_{ab})_1 &= x_1 + jx_2 \\ (i_{bc})_1 &= x_3 + jx_4 \\ (i_{ca})_1 &= x_5 + jx_6 \end{aligned} \tag{2}$$

and m is the modulation index of the inverter.

This result allows to integrate the models of converters belonging to multistage conversion systems. Moreover, a suitable procedure for obtaining the state-space models of the PV array, the filters, the transformer, and the grid and for merging all these models in a comprehensive representation of the whole PV plant is finally obtained, as it is reported in [7]. Using such approach, the PV plant is represented as a single state-space system whose inputs are the irradiance and module temperature and the outputs are the electrical quantities exploitable for loss evaluation and for performance analysis. The analytical formulation of the final integrated state-space average model is in [7].

It is worth noting that, thanks to the adopted approach, technical details about the converters topology and their control system are no longer necessary. Since the state-space representation has a general validity, the proposed model can be implemented in any simulation platform.

3.2 Parameter Identification

The integrated state-space representation of the PV system allows the exploitation of several well-known identification methods in literature. Such methods can be also used to tune the regulator parameters in control systems. With the aim to implement an identification method with general validity for any plant configuration, a constrained minimum formulation can be applied in order to minimize the deviation between the model output and the real data collected by data loggers [13, 14]. To explain this method, let $y_{\text{measured}}(k)$ be an electrical variable acquired by the data logger in the form of a time series with N time samples:

$$[Y_{\text{measured}}] = \begin{bmatrix} y_{\text{measured}}(1) \\ y_{\text{measured}}(2) \\ \cdots\cdots\cdots \\ y_{\text{measured}}(k) \\ \cdots\cdots\cdots \\ y_{\text{measured}}(N) \end{bmatrix} \tag{3}$$

The corresponding quantity provided by the model is named $y_{\text{model}}(k)$. The latter can be considered the linear combination of the p_n parameters (unknown, to be identified) and of the terms h_{ki} that fix the relations between $y_{\text{model}}(k)$ and each p_i parameter:

$$y_{\text{model}}(k) = h_{k1}p_1 + h_{k2}p_2 + \cdots + h_{kn}p_n \tag{4}$$

For N time samples, this relationship becomes the following:

$$\begin{bmatrix} y_{\text{model}}(1) \\ y_{\text{model}}(2) \\ \cdots\cdots\cdots \\ y_{\text{model}}(k) \\ \cdots\cdots\cdots \\ y_{\text{model}}(N) \end{bmatrix} = \begin{bmatrix} h_{11} & h_{12} & \ldots & h_{1n} \\ h_{21} & h_{22} & \ldots & h_{2n} \\ \cdots & \cdots & & \cdots\cdots \\ h_{k1} & h_{k2} & \ldots & h_{kn} \\ \cdots & \cdots & & \cdots\cdots \\ h_{N1} & h_{N2} & \ldots & h_{Nn} \end{bmatrix} \begin{bmatrix} p_1 \\ p_2 \\ \cdots \\ p_n \end{bmatrix} \tag{5}$$

that is, in compact form:

$$[Y_{\text{model}}] = [H][P] \tag{6}$$

Thanks to the constrained minimum formulation, the optimal set of parameters can be easily obtained as follows:

$$[P] \simeq \left([H]^T [H]\right)^{-1} [H]^T [Y_{\text{measured}}] \tag{7}$$

This formulation has been applied for each PV array in the experimental subfield, considering the rated power of the PV modules in each string (see Table 1).

4 Performance Analysis in Case Study

4.1 Implementation and Accuracy of the Behavioral Model

The usefulness of the proposed model can be pointed out with an example. Figure 6 shows the power curve of a PV array in the experimental subfield measured during a sunny day in the autumn 2018. The power waveform provided by the behavioral model is superimposed to the measured one. From 10:40 to 12:00, a maintenance shutdown occurs. Thanks to the model, it is possible to estimate with a good accuracy the theoretical energy during this shutdown. The relative error in the pink area (≃1.8%) is calculated from the following:

$$\varepsilon_{\%} = \frac{E_{\text{model}} - E_{\text{measured}}}{E_{\text{measured}}} 100 \tag{8}$$

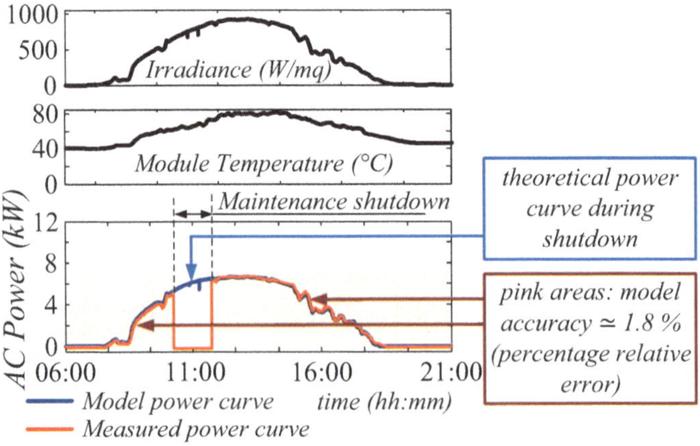

Fig. 6 Comparison between measured power and power curve provided by the behavioral model during a sunny day in autumn 2018. The behavioral model is used to trace the theoretical power curve during a maintenance shutdown

To establish an average reference value for the accuracy of the proposed modeling approach, the power curve provided by the model has been compared to the one measured in the field (at given irradiance and temperature) during several days in 2018 and 2019. The accuracy level calculated by Eq. (8) is in the range from 1.5% to 3.0% that is acceptable in comparison to the accuracy of standard power meters and sensors.

4.2 Execution Time Performance

The computational complexity of the behavioral model has been tested in comparison to the execution time of the detailed model referring to the basic PV system in Fig. 5 considering two different cases:

- Case A: The same simulation step size 1×10^{-6} s is assigned to the state-space average model and to the detailed one.
- Case B: A larger sample time equal to 2×10^{-5} s is assigned to the state-space average model. The step size of the detailed model is equal to the previous case; otherwise, such model cannot run properly.

For both cases, two different operating scenarios have been implemented and run 1000 times to get a large statistical database. The relative difference in execution times is summarized in Table 2. As expected, the behavioral model ensures a significant reduction in computational effort.

Table 2 Execution time performance evaluation, proposed vs. detailed model

	Case A, execution time	Case B, execution time
Scenario 1	Detailed model mean = 72.2 s, std = 3.0 s	Detailed model mean = 72.2 s, std = 3.0 s
	Proposed model, mean = 27.7 s, std = 4.8 s	Proposed model, mean = 3.5 s, std = 0.2 s
	Relative difference, mean = −61.6% std = 6.6%	Relative difference, mean = −95.1% std = 0.3%
Scenario 2	Detailed model mean = 72.1 s, std = 0.8 s	Detailed model mean = 72.1 s, std = 0.8 s
	Proposed model, mean = 30.0 s, std = 1.5 s	Proposed model, mean = 3.5 s, std = 0.1 s
	Relative difference, mean = −62.6% std = 2.2%	Relative difference, mean = −95.1% std = 0.1%

Simulation end time instant: 0.3 s. Scenario 1: 1000 W/m^2, 25 °C; scenario 2: 800 W/m^2, 45 °C; nonzero reactive power is forced after 0.1 s. Computer hardware: Intel Core i3-4005U CPU at 1.70 GHz (four CPUs), RAM 4096 MB, HD SSD 240 GB read speeds up to 545 MB/s, 64 bit

4.3 DC Cable Voltage Drops

In the experimental subfield, the measurement of the current in the cables that connect the optimizers and the DC link of the central inverter is not available. The cable is represented by its own resistance R_{DC} in Fig. 3. The integrated state-space average model described in this paper is able to calculate the average DC link current (flowing in DC cables) in a straightforward way as a linear function of the symmetrical components of the AC currents (see Sect. 3.1). In such a way, it is possible to evaluate in a direct way the actual voltage drop and Joule losses at each DC cable. Figure 7 shows some plots about the voltage drops in some DC lines of the PV plant. The behavioral model is used to replace data that are not provided by the monitoring system. Thanks to the higher DC voltage value, the voltage drop and the Joule losses are lower in the experimental subfield in comparison to the other subfields. However, in the PV plant under investigation, such effects can be neglected, thanks to the short lengths of DC cables and to their large cross sections.

5 Performance Analysis in Case Study

5.1 Daily PR of the PV Arrays in the Experimental Subfield

Figure 8 shows the distribution plots of the daily performance ratio (PR) for some PV arrays connected at the input of the DC/DC optimizers in the experimental subfields for the time periods:

- February 2018 to July 2018: "before" distributed converters (bDC period)
- August 2018 to February 2019: "after" distributed converters (aDC period)

The daily PR of each PV array is as follows:

$$\mathrm{PR}_{\mathrm{array}} = \frac{E_{\mathrm{array}}}{P_{\mathrm{array}}\mathrm{Rad}} \tag{9}$$

where P_{array} is the rated power of the array (sum of the rated power of two strings), Rad is the daily solar radiation, and E_{array} is the daily energy calculated from the power curve of the selected array. This power is calculated using the current value at the optimizer output and the DC link voltage; the latter also take into account the voltage drop on DC cables. Repeating the calculation in Eq. (9) for all the PV arrays in the experimental subfield and comparing bDC and aDC periods, the aggregate result in Fig. 9 shows that the average benefit obtained in terms of PR is about 2%.

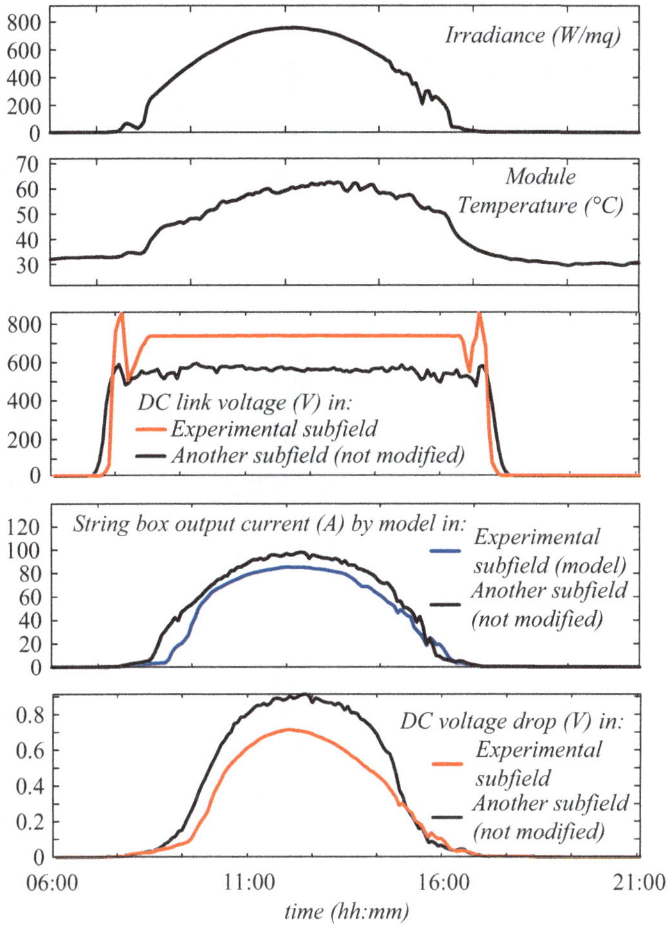

Fig. 7 Calculation of voltage drop during a sunny day in winter 2019. Comparison between the experimental subfield and another subfield with the original configuration

5.2 Central Inverter Operation

It is useful to investigate the behavior of the central inverter whose control system has been modified removing the original MPPT control and fixing the DC link voltage in a narrow range around 730 V. Figure 10 shows the main electrical quantities acquired by the data logger of the central inverter in the experimental subfield during 2 days belonging to the bDC and aDC period.

The following can be observed:

- The maximum efficiency in the aDC period is equal to the one measured in the bDC period (dashed line).

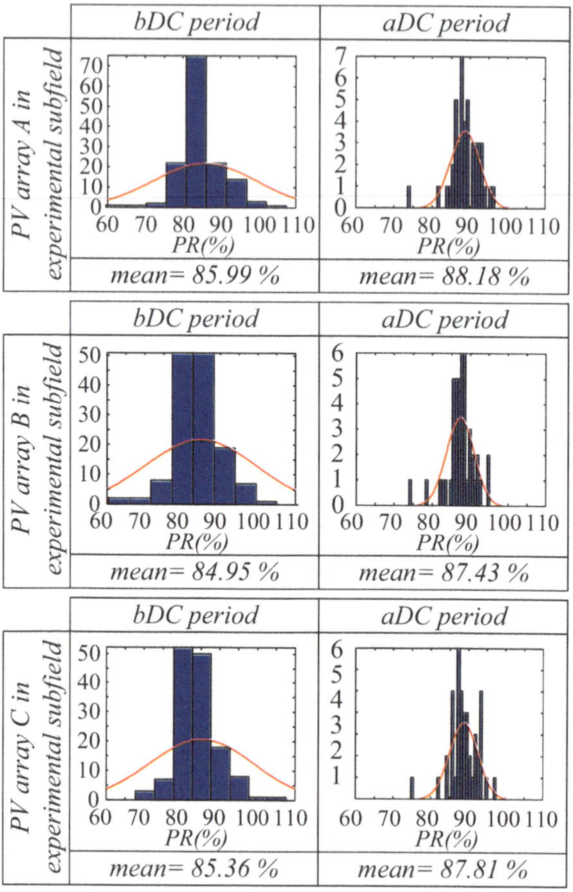

Fig. 8 (**a–c**) Distribution plots of the daily PR in some arrays of the experimental subfield for the periods bDC and aDC

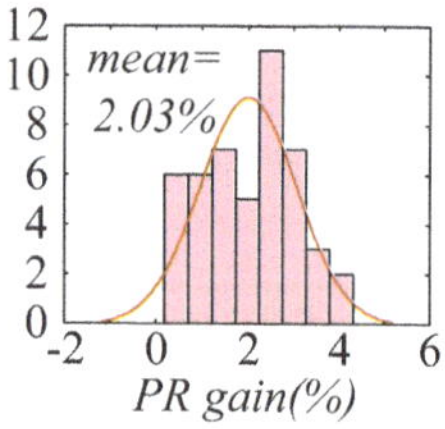

Fig. 9 Distribution plot of the PR gain in all the arrays of the experimental subfield, time period February 2018 to February 2019

- In the aDC period, during sunrise and sunset, the DC link voltage exhibits large fluctuations (yellow areas).
- The same phenomenon occurs in case of fast power variations (cyan areas) causing temporary significant decrease in efficiency level.

The modification of the control strategy in the central inverter implies bad effects on its operation in case of large power fluctuations, leading to a reduction in the energy gain measured at the AC side in variable weather conditions.

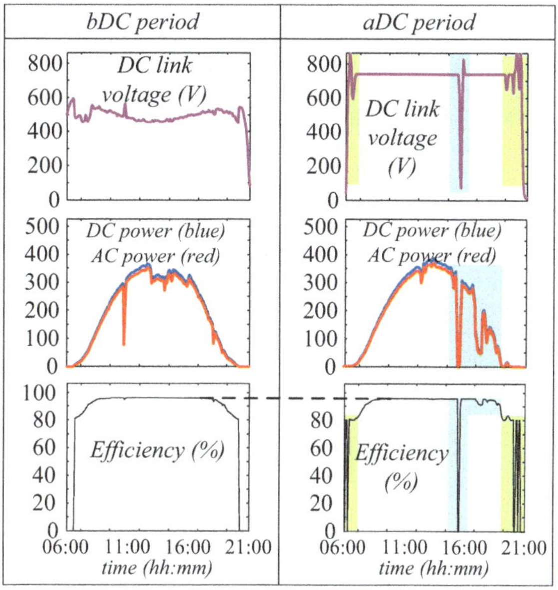

Fig. 10 Main electrical quantities acquired by the data logger of the central inverter in the experimental subfield during 2 days close to the installation date of the string optimizers

6 Conclusions

Although the performance assessment of large PV plants with distributed conversion systems is very topical, in literature, there is no a comprehensive overview on the available technical solutions. This paper contributes to fill this lack of information reporting the results coming from the performance evaluation of a large PV plant in Central Italy where DC/DC string optimizers have been mounted in a 500 kW experimental subfield. The average benefit in terms of PR at string level is about 2%. A suitable behavioral model has been developed and implemented to support the performance analysis in case of missing or wrong data. However, the operation of the inverter connected to string optimizers requires further investigations for future installations because of its impact on the energy recovery.

References

1. A. Cabrera-Tobar, E. Bullich-Massaguè, M. Aragues-Penalba, O. Gomis-Bellmunt, Topologies for large scale photovoltaic power plant. Renew. Sustain. Energy Rev. **59**, 309–319 (2016)
2. K. Zeb, W. Uddin, M.A. Khan, Z. Ali, M.U. Ali, N. Christofides, H.J. Kim, A comprehensive review on inverter topologies and control strategies for grid connected photovoltaic system. Renew. Sustain. Energy Rev. **94**, 1120–1141 (2018)
3. M. Cacciato, A. Consoli, New regenerative active snubber circuit for ZVS phase shift Full Bridge converter, in *IEEE Applied Power Electronics Conference and Exposition*, (IEEE, Piscataway, NJ, 2011), pp. 1507–1511

4. P. Han, Z. Lin, L. Wang, G. Fan, X. Zhang, A survey on equivalence modeling for large-scale photovoltaic power plants. Energies **11**, 1–14 (2018)
5. N. Agarwal, A. Agarwal, Mismatch losses in solar photovoltaic array and reduction techniques. MIT Int. J. Elect. Instr. Eng. **4**, 16–19 (2014)
6. P. Manganiello, M. Balato, M. Vitelli, A survey on mismatching and aging of PV modules: the closed loop. IEEE Trans. Indust. Elect. **62**, 7276–7286 (2015)
7. G. Nobile, M. Cacciato, G. Scarcella, G. Scelba, A.G.F. Di Stefano, G. Leotta, P.M. Pugliatti, F. Bizzarri, Comparison between central and string inverters performance for the utility-scale PV plant in Nova Olinda Brazil, in *35th European Photovoltaic Solar Energy Conference and Exhibition EU PVSEC*, (WIP, Munich, 2018), pp. 1935–1941
8. R.H.G. Tan, L.Y.H. Hoo, DC-DC converter modeling and simulation using state-space approach, in *IEEE Conference on Energy Conversion CENCON*, (IEEE, Piscataway, NJ, 2015), pp. 42–47
9. G. Kanimozhi, J. Meenakshi, V. Sreedevi, Small signal modeling of a DC-DC type double boost converter integrated with SEPIC converter using state space averaging approach. Energy Procedia **117**, 835–846 (2017)
10. Z. Lin, H. Ma, Modeling and analysis of three-phase inverter based on generalized state space averaging method, in *39th Annual Conference of the IEEE Industrial Electronics Society IECON*, (IEEE, Piscataway, NJ, 2013), pp. 1007–1012
11. J.W. Kolar, S.D. Round, Analytical calculation of the RMS current stress on the DC-link capacitor of voltage-PWM converter systems. IEEE Proc. Elect. Power Appl. **153**, 535–543 (2006)
12. X. Pei, W. Zhou, Y. Kang, Analysis and calculation of DC-link current and voltage ripples for three-phase inverter with unbalanced load. IEEE Trans. Power Electron. **30**, 5401–5412 (2015)
13. M. Galad, P. Spanik, M. Cacciato, G. Nobile, Comparison of common and combined state of charge estimation methods for VRLA batteries, in *ELEKTRO Conference*, (IEEE, Piscataway, NJ, 2016), pp. 220–225
14. G. Nobile, M. Cacciato, G. Scarcella, G. Scelba, Performance assessment of equivalent-circuit models for electrochemical energy storage systems, in *43rd Annual Conference of the IEEE Industrial Electronics Society IECON*, (IEEE, Piscataway, NJ, 2017), pp. 2799–2806. Chapter 46

Diagnose Algorithm and Fault Characterization for Photovoltaic Arrays: A Simulation Study

Luis D. Murillo-Soto and Carlos Meza

Abstract The performance of photovoltaic installation is highly affected by faults in single modules. Faults in photovoltaic arrays are difficult to detect, locate and diagnose due to the way in which modules are configured. Given that photovoltaic arrays are formed by modules in series, a fault in a single module affects the whole system. Therefore, the technology to detect and diagnose faults inside solar arrays is emerging, the present paper proposes several expressions that help to detect and diagnose failures using a proposed algorithm. The expressions were obtained by an inductive approach based on the analysis of simulation cases where different faults were tested. The array model used for the simulation was built in Spice software based on the five-parameter model of a solar module.

1 Introduction

A photovoltaic (PV) system is a carbon-free mechanism that uses an inexhaustible and free energy source to produce electricity. This form of power generation is modular and contains no moving parts but its energy density is low, which is why photovoltaic installations may be composed of thousands of PV panels organized in parallel combinations of strings. A collection of interconnected PV panels, also called PV array, injects energy to the utility grid through an inverter which is in charge of (1) extracting the maximum amount of energy from the modules and (2) converting the generated DC current into AC current [1] (Fig. 1). If an electric fault occurs in one of the panels of the array the energy production will be affected to a degree that depends on the nature, severity and location of the fault.

PV array faults cause energy losses, degradation and hazardous situations for technicians and operators, that is why in the recent years there have been an increase interest in this topic [2]. Research in PV array faults has been focused on their causes

L. D. Murillo-Soto (✉) · C. Meza
Costa Rica Institute of Technology, Cartago, Costa Rica
e-mail: lmurillo@tec.ac.cr; cmeza@tec.ac.cr

W. Zamboni, G. Petrone (eds.), *ELECTRIMACS 2019*, Lecture Notes in Electrical Engineering 615, https://doi.org/10.1007/978-3-030-37161-6_43

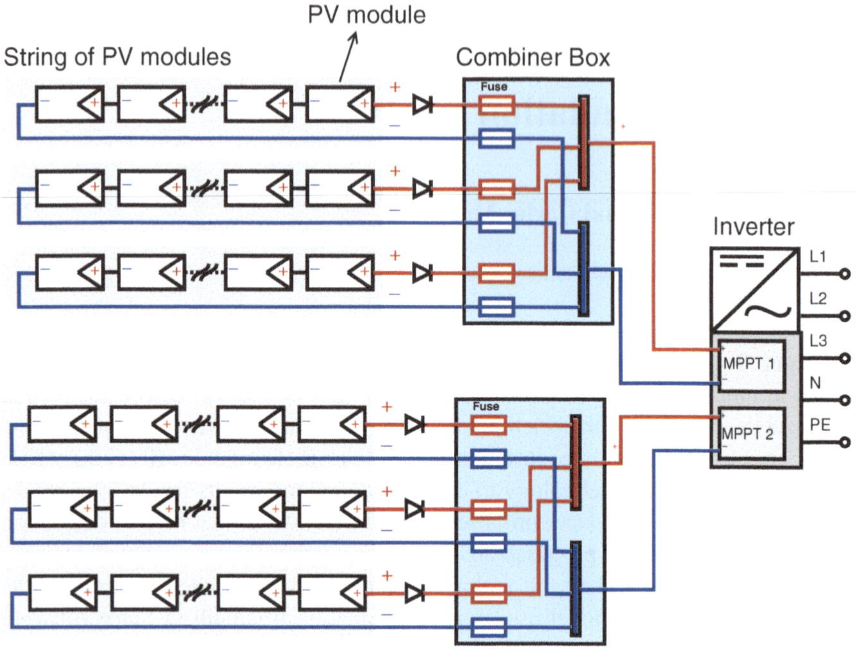

Fig. 1 Schematic of a PV installation comprising of two arrays and one inverter

(e.g., [3, 4]), their location (e.g., [5]) and their diagnosis. Several papers such as [6] and [2] have identified the most typical faults in PV arrays (see Fig. 2), i.e.,

1. Short-circuit between one node and ground.
2. Short-circuit between two nodes in the same string.
3. Short-circuit between two nodes located in different strings.
4. Low and middle impedance paths between one node and ground.
5. Low and middle impedance paths between nodes inside the array.
6. Open-circuit in the string.
7. Electric degradation of the internal PV parasitic resistance.
8. Partial shadows over the array.

Regarding PV array fault diagnosis techniques, Mellit in [7] classifies them as (1) image analysis and (2) electrical characterization. Imaging methods are currently expensive and time consuming, whereas electrical characterization is cheaper and more flexible [8]. The latter can be further divided into the following techniques:

- signal processing and statistical methods (e.g., [6, 9])
- $i - v$ curve characteristics analysis (e.g., [10, 11])
- power losses analysis (e.g., [12, 13])
- current and/or voltage measurements (e.g., [14, 15]), [5]).
- artificial intelligent methods (e.g., [16–18])

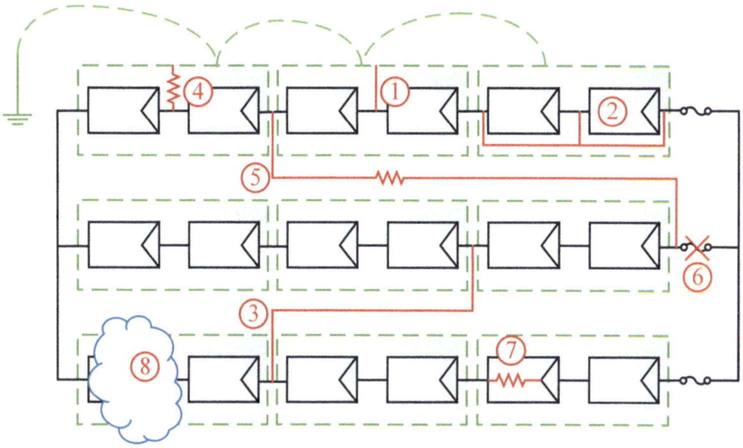

Fig. 2 Common electric faults in a PV Array. In this figure each panel has two modules

The present paper focuses on current and voltage measurements to diagnose faults in PV panels. Measuring electrical variables in the PV array allows to identify and locate faults in real time (also called on-line method as stated in [2]). A major disadvantage to this approach is the cost of the additional circuitry required to measure currents and voltages in multiple points in the array. Nevertheless, as stated in [19] it is possible to develop low cost measurement circuits. In this regard, based on PV module voltage measurements, this work derives a set of mathematical conditions that are explicitly related to open-circuit, short-circuit and ground short-circuit faults. To the authors' knowledge the derived mathematical conditions have not been presented in the PV array fault-diagnosis literature.

The next section of this paper presents the used mathematical model of the PV module. Later, mathematical expressions to characterize PV array fault conditions are presented. The next section shows and analyses results of a simulation experiment to validate the fault characterization expressions. Section 5 proposes an algorithm that uses the derived characterization to diagnose faults in PV arrays and finally the main conclusions and future work are presented.

2 Assumptions and Models

The present paper assumes that the electrical behaviour of a PV panel can be described based on the single-diode five-parameter PV model, which relates the generated current, i_{pv}, and voltage, v_{pv} according to the following expression:

Table 1 Parameters and variables of the PV module model

Parameter or variable	Description
$I_{\mathrm{ph}}(T, G)$	Photocurrent
$I_{\mathrm{s}}(T)$	Saturation current
$R_{\mathrm{p}}(G)$	Parallel internal resistance
$R_{\mathrm{s}}(T, G)$	Series internal resistance
$I_{\mathrm{ph}}^{\mathrm{stc}}$	Photo current at standard test conditions (STC), i.e., when $(G = 1000W/m^2, T = 298.15°K)$
$I_{\mathrm{s}}^{\mathrm{stc}}$	Saturation current at STC
$R_{\mathrm{s}}^{\mathrm{stc}}$	Series resistance at STC
$R_{\mathrm{p}}^{\mathrm{stc}}$	Parallel resistance at STC
N_s	Number of PV cells connected in series
N_p	Number of PV cells connected in parallel
A_q	Ideality factor
k	Boltzmann constant $k = 1.38 \times 10^{-23}$
q	Charge of the electron $q = 1.6 \times 10^{-19}$
V_T	Thermal voltage, $V_T = kT/q$
κ	$\kappa = 1/(N_s A_q V_T)$
K_i	Short circuit thermal coefficient
E_g	$E_g = 1.121(1 - 0.0002677(T - 298.15))$, energy gap

$$
\begin{aligned}
i_{\mathrm{pv}} =& I_{\mathrm{ph}}(T, G) - I_{\mathrm{s}}(T)\left(e^{(v_{\mathrm{pv}}+i_{\mathrm{pv}}R_{\mathrm{s}}(T,G))\kappa} - 1\right) + \\
& - \frac{(v_{\mathrm{pv}} + i_{\mathrm{pv}}R_{\mathrm{s}}(T, G))}{R_{\mathrm{p}}(G)}
\end{aligned} \tag{1}
$$

where some variables are not constants but function of the module's temperature, T, and the incident irradiance, G, as expressed mathematically in (1), (2), (3), (4) and (5) according to [11, 20–22] and [23]. The parameters used in the previously mentioned expressions are described in Table 1. The five-parameter model has been validated with experimental data in [10] and [24].

$$
I_{\mathrm{ph}}(T, G) = (I_{\mathrm{ph}}^{\mathrm{stc}} + K_i(T - 298.15))\frac{G}{1000} \tag{2}
$$

$$
I_{\mathrm{s}}(T) = I_{\mathrm{s}}^{\mathrm{stc}}\left(\frac{T}{298.15}\right)^3 e^{\left(\frac{qE_g}{kA_q}\left(\frac{1}{298.15} - \frac{1}{T}\right)\right)} \tag{3}
$$

$$
R_{\mathrm{s}}(T, G) = R_{\mathrm{s}}^{\mathrm{stc}}\frac{T}{298.15} \cdot \left(1 - 0.217\ln\left(\frac{G}{1000}\right)\right) \tag{4}
$$

$$
R_{\mathrm{p}}(G) = R_{\mathrm{p}}^{\mathrm{stc}}\frac{G}{1000} \tag{5}
$$

3 Characterization of Fault Conditions

A typical PV array is shown in Fig. 3. The array is made of n strings of x photovoltaic modules connected in series. It is possible to characterize fault conditions using the following variables:

- the voltage in each PV module, $V_m(i, j)$.
- the current in each string, $I_s(i)$, and,
- the PV array voltage, V_{op}.

Measuring the voltage in each module is possible to determine its state. More specifically, we define three state categories for the PV module: faulty, affected and normal. Faulty modules are those that are short- or open-circuited. Affected modules are located in the same string of the faulty ones and have voltages that differ from the normal condition. Normal modules are located in strings that do not have faults.

Next, expressions for the PV module voltage that characterizes its condition are shown for short-circuit, short-circuit to ground and open-circuit faults.

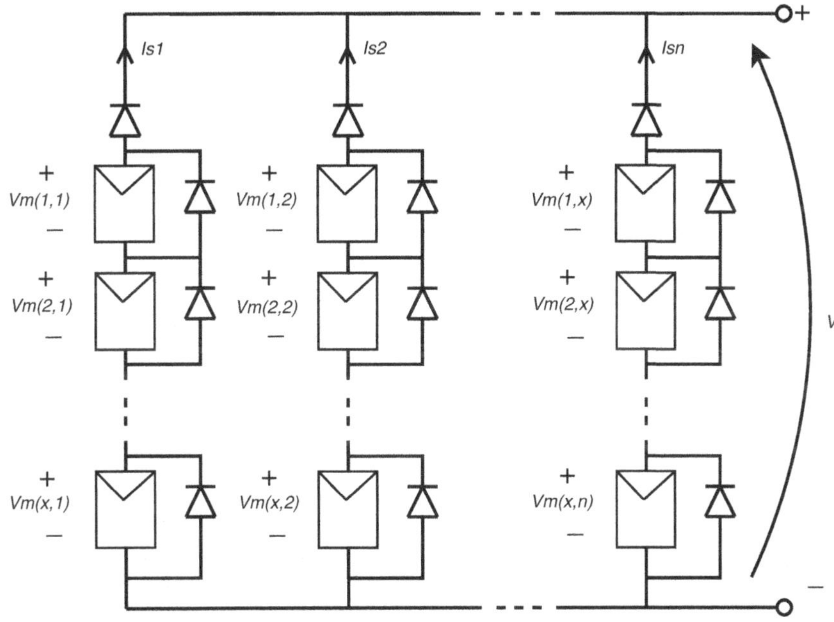

Fig. 3 Diagram of a typical photovoltaic array

3.1 Short-Circuit Faults

Short-circuited modules and ground to short-circuit faults can be characterized using the same expression. Using circuit analysis it is possible to derive the number of modules that are directly experienced short-circuit faults in the j-string, y_j, i.e.,

$$y_j = x - \left\lceil \left\lfloor \frac{V_{op}}{\max\left(V_m(:, j)\right)} \right\rceil \right\rfloor \tag{6}$$

where $\max\left(V_m(:, j)\right)$ is the maximum module voltage in the j-string, and $\lceil\lfloor\cdot\rceil\rfloor$ is the half up rounding function, i.e., $\lceil\lfloor 2.3\rceil\rfloor = \lceil\lfloor 2.5\rceil\rfloor = 2$. This function is valid even when the diodes are not consider ideal.

If the voltage is measured in the PV module, it is possible to characterize its state. When the module is faulty its voltage is

$$V_m = V_{m_f} = 0 \tag{7}$$

where sub-index f denotes the faulty condition.

On the other hand, affected modules present a voltage equals to

$$V_{m_a} = \begin{cases} \dfrac{V_{op}}{x - y} & \text{if } I_s > 0 \\ V_{m_{OC}} & \text{if } I_s = 0 \end{cases} \tag{8}$$

where sub-index a represents the affected condition. In this case the voltage across the affected module is defined using two expressions depending on the operating voltage of the array, V_{op}. If the affected string current is zero it means that all the affected modules have voltages equal to their open-circuit, $V_{m_{OC}}$. Take into account that if non-ideal diodes are considered, there will be a small variation from the values presented in (8).

Finally, voltage in non-affected modules is equal to

$$V_{m_n} = \frac{V_{op}}{x} \rightarrow I_{s_n} \geq I_{s_a}, V_{m_n} \leq V_{m_a} \tag{9}$$

Sub-index n is used for normal modules. Non-faulty string current, I_{s_n}, can be greater or equal than faulty strings, I_{s_a}. Affected modules can have voltages greater or equal to normal modules voltages.

3.2 Open-circuit

Following a similar procedure it is possible to obtain expressions that characterize open-circuit faults, i.e.,

$$V_{m_f} = V_{op} - (x - 1) \cdot V_{mOC} \rightarrow I_s = 0 \quad (10)$$

$$V_{m_a} = V_{mOC} \rightarrow I_s = 0 \quad (11)$$

$$V_{m_n} = \frac{V_{op}}{x} \rightarrow I_s \approx I_{op}/(n - 1) \quad (12)$$

where n represents the number of the strings in the array and I_{op} is the direct current generated by the PV array.

4 Simulation Experiment

In order to validate the derived mathematical expression a simulation experiment has been performed. The results obtained in this experiment can be extrapolated to other cases, i.e., the effects of the faults on a specific PV array can be extrapolated to a $x \times n$ array, where x is the number of modules in series and n is the number of strings. Such generalization is possible due to the intrinsic symmetry of PV arrays, given that they are formed by strings of the same sized using the same type and model of PV modules.

The simulated case consists of a PV array made of two strings connected in parallel. Each string has four modules connected in series, as shown in Figs. 4, 6, 8. The model was simulated with a Spice-based software [25] and the values of the parameter of the module model were obtained from the data-sheet according to [24] and [26] and are presented in Table 2. Non-ideal bypass and blocking diodes were used.

The simulation experiment considers the following faults:

- short-circuit provoked by a faulty diode or very low parasitic parallel resistance,
- short-circuit between one node in the string and the ground and
- open-circuit in any point of the string that occurs due to burned fuses, open blocking diodes, bad contact inside the connectors or broken solder joints of the cells inside the module.

The cases used to analyse each fault are detailed next.

4.1 *Fault 1: Short-Circuited PV Module*

To emulate a short-circuit in a module, resistances $\{R_{17}, R_{18}, R_{19}, R_{20}\}$ in Fig. 4 were set to two possible values $\{0.01, 100\text{M}\}\Omega$, where the former represents the faulty condition and the latter the normal one.

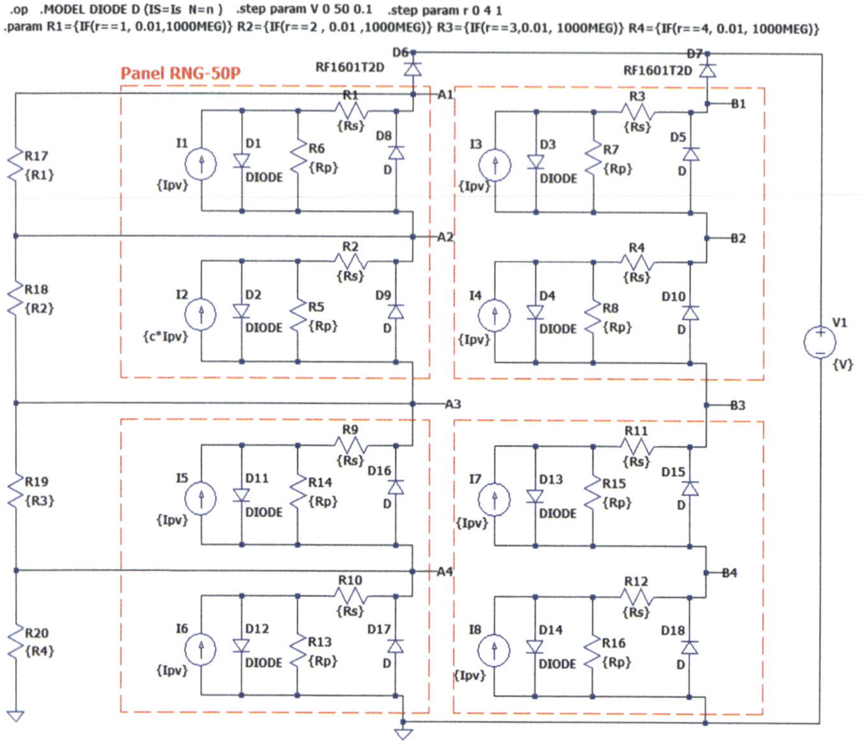

Fig. 4 Spice schematic circuit and model used for the simulation of short-circuits in each module of one string

Table 2 Value of the parameters the PV module

Parameter	Value
$I_{\text{ph}}^{\text{stc}}$	2.95
$I_{\text{s}}^{\text{stc}}$	8.09E−12
$R_{\text{s}}^{\text{stc}}$	0.37
$R_{\text{p}}^{\text{stc}}$	1500
N_s	18
N_p	1
κ	2.38
K_i	1.18E−3

Four different simulation tests were performed, in each one of them a different module was short-circuited. The faulty module is the one that is short-circuited and the affected ones are the rest of the modules that are located in the same string. The simulation results of all the tests are shown in Fig. 5. As expected the faulty module presents a zero voltage, $V_{m_f} = 0$, the affected modules have all the same voltage which is greater or equal than the voltage of the unaffected or normal modules, i.e., $V_{m_a} \geq V_{m_n}$.

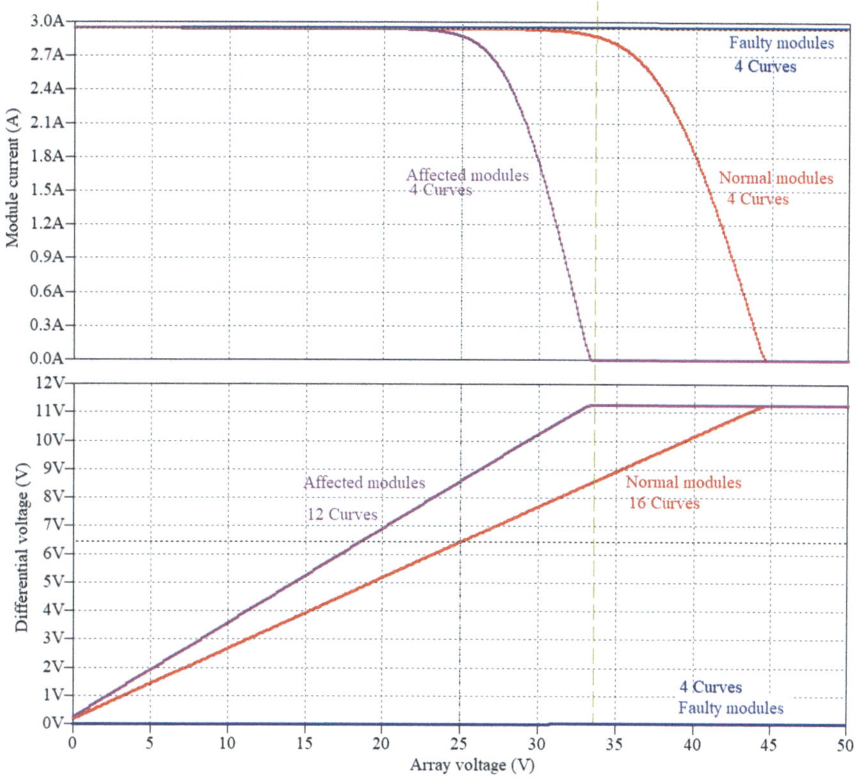

Fig. 5 Simulation results of short-circuit faults in PV array. The upper graph shows the current in the strings and the bottom graph represents the differential voltage of the modules

4.2 *Fault 2: Short-Circuit to Ground*

The short-circuit between the nodes $\{A1, A2, A3, A4\}$ and ground is simulated with the model of Fig. 6. The results generated are presented in Fig. 7, again the upper chart shows the string current while the below graph is the differential voltage of the module.

When a short-circuit to ground is present in one node, the modules located between it and ground are considered faulty, whereas the remaining modules are affected. Figure 7 depicts the simulation results for normal operation (green) and when nodes A2 (cyan), A3 (magenta), A4 (grey) are connected to ground. Notice that, no matter which node is connected to ground, all the faulty modules present the same voltage (0 voltage). For each fault condition, the affected modules present also the same voltage, not distinguished in the graph because the curves are superimposed. It was verified that the voltage between the terminals of the normal,

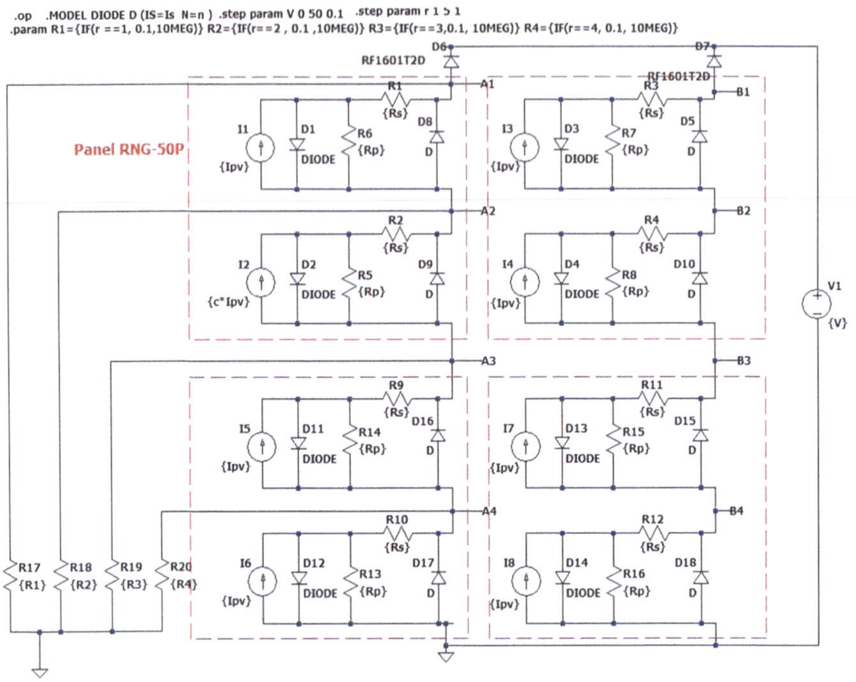

Fig. 6 Spice schematic circuit and model used for the simulation of the short-circuit to ground fault

affected and faulty models can also be defined with the expressions derived in the previous section, i.e., (7)–(9).

4.3 *Fault 3: Open-Circuit*

Figure 8 shows the schematic used to simulate open-circuit faults. Four different simulations were performed, in each one of them a different node, A1, A2, A3 or A4, was open-circuited. The simulation results are shown in Fig. 9, notice that when $I_s \neq 0$ in the string without fault then $V_{m_a} > V_{m_n} > V_{m_f}$.

5 Algorithm for the Diagnosis Based on Rules for the PV Array

Based on the obtained conditions, we propose a novel algorithm to diagnose faults in a PV array inspired in the inference method as described in [27]. The proposed

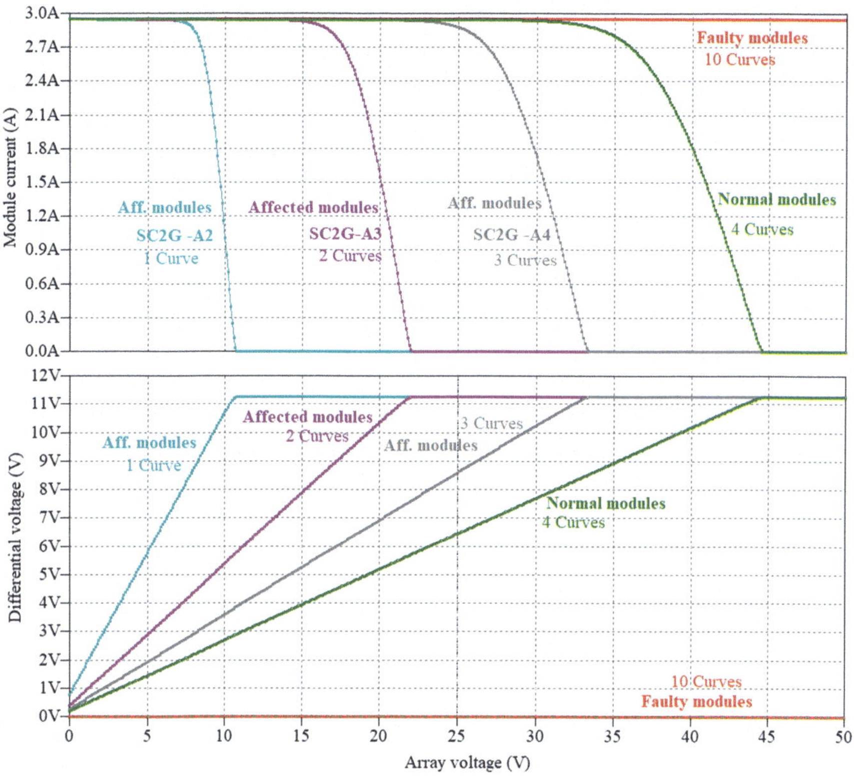

Fig. 7 Simulation results of short-circuit to ground faults. The upper graph is the current in the strings and the bottom graph represents the differential voltage of the modules

algorithm, shown in Algorithm 1 and in Fig. 10, distinguishes between the three failures characterized with the mathematical expressions presented in Sect. 3. It gives the position of the fault using an array **M** of $x \times n$ elements, where x is the number of modules per string, n is the number of strings and the i, j element in **M** represents the i module in the j string of the PV array.

The algorithm takes as an input the differential voltage, expressed as a matrix, $\boldsymbol{Vm} \in \Re^{x \times n}$, where $\boldsymbol{Vm}(i, j)$ is the voltage for the i-th module in the j-th string. The current on each string is also used as an input and it is represented as a vector $\boldsymbol{Is} \in \Re^{n}$, where $\boldsymbol{Is}(j)$ is the current in the j-th string. The algorithm also requires the output voltage of the array V_{op} and the open-circuit voltage of the module $V_{m_{OC}}$.

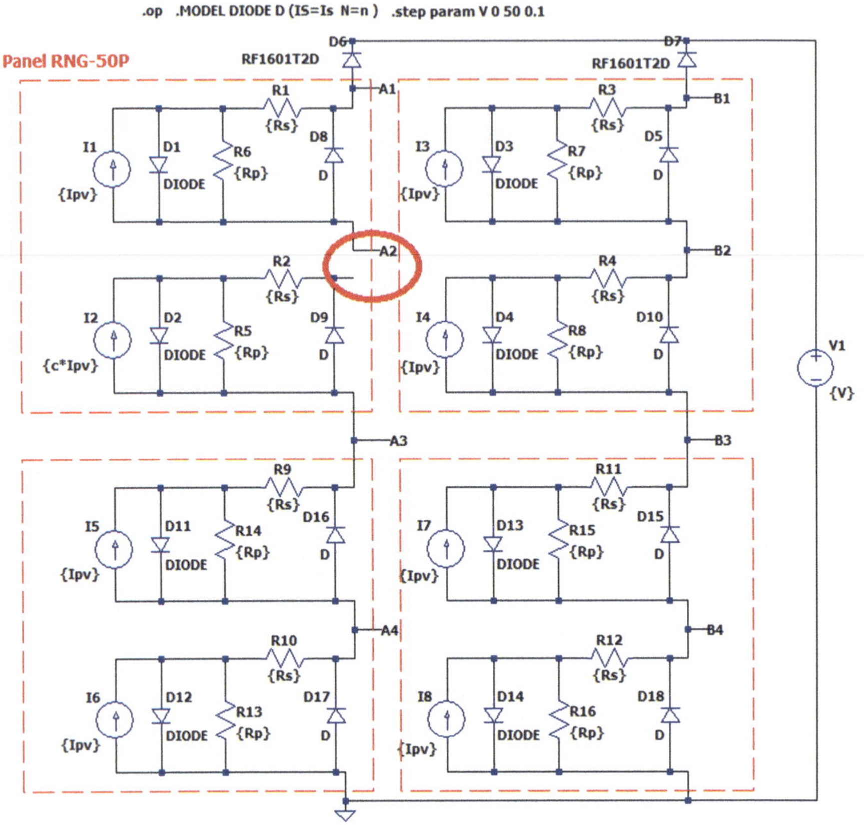

Fig. 8 Spice schematic circuit and model used for the simulation open-circuit faults

Algorithm 1: Diagnosis of faults at module level

Input data: $\boldsymbol{Vm}$, $\boldsymbol{Is}$, V_{op}, $V_{m_{OC}}$
Output data: **M**
Constants: n, x
M $\leftarrow$ "Normal operation"
for j=*1:n* **do**
  $Vm_{max}(j) \leftarrow MAX(\boldsymbol{Vm}(:,j))$
  $y \leftarrow x - \lceil \lfloor \frac{Vop}{Vm_{max}(j)} \rfloor \rceil$
  for i=*1:x* **do**
    if $\boldsymbol{Is}(j)=0 \wedge \boldsymbol{Vm}(i,j)=(V_{op} - (x-1) \cdot V_{m_{OC}})$ **then**
      M(i,j)$\leftarrow$ "Module with open circuit"
    if $\boldsymbol{Vm}(i,j)=0 \wedge y > 0$ **then**
      if $\boldsymbol{Vm}(x,j)=0$ **then**
        M(i,j)$\leftarrow$ "Module's ShortCircuit2Ground"
      else
        M(i,j)$\leftarrow$ "Module with short circuit"

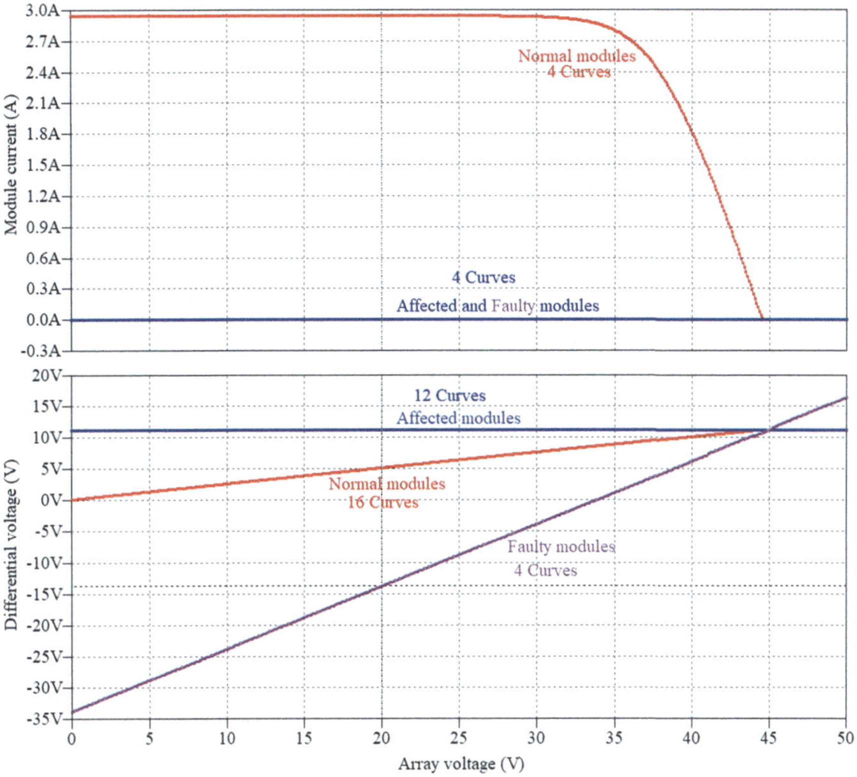

Fig. 9 Simulation results of open-circuit faults (open-circuit at A1, A2, A3 and A4). The upper graph shows the current in the strings for all the cases and the bottom graph represents the differential voltages of the modules for all the cases. Notice that some curves are superimposed

6 Conclusions

This article characterizes mathematically the most common PV array faults, i.e., open-circuit, the short-circuit and the short-circuit to ground faults. A proposed algorithm, which uses the derived mathematical expression, is able to diagnose (detect, locate and identify) these common faults. The proposed algorithm has several advantages such as it is easy to implement, it is of one order $O(n)$ and it can identify faults in different strings at the same time.

Fig. 10 Flowchart of the algorithm proposed

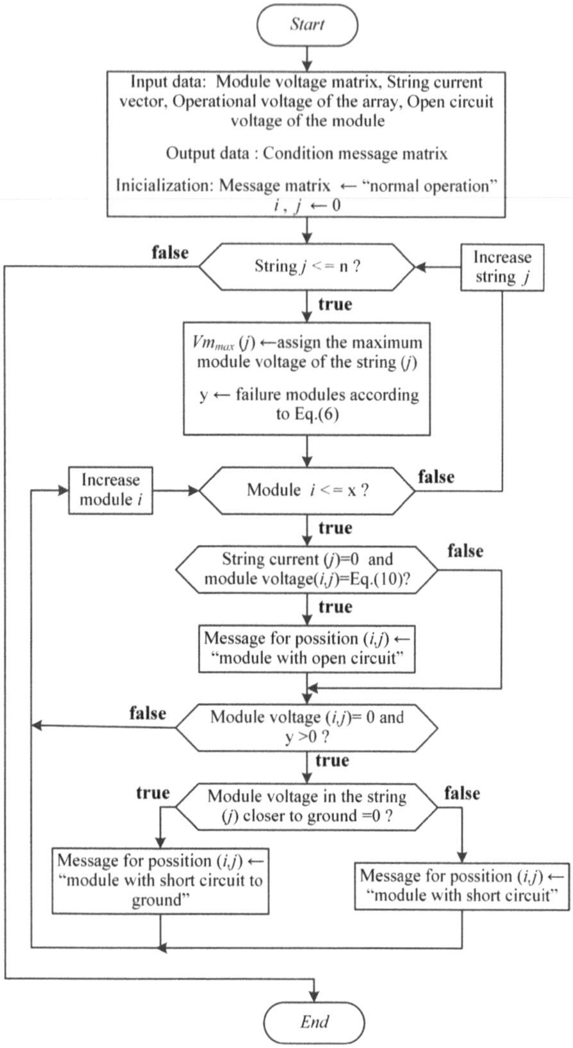

Acknowledgements This article was supported by the scholarship program of the Costa Rica Institute of Technology and the VIE project number: 5402-1341-1701.

References

1. C. Meza, J.J. Negroni, D. Biel, F. Guinjoan, Energy-balance modeling and discrete control for single-phase grid-connected PV central inverters. IEEE Trans. Ind. Electron. **55**(7), 2734–2743 (2008)

2. J. Bastidas-Rodriguez, G. Petrone, C. Ramos-Paja, G. Spagnuolo, Photovoltaic modules diagnostic: an overview, in *IECON 2013–39th Annual Conference of the IEEE Industrial Electronics Society*, November (IEEE, New York, 2013), pp. 96–101
3. M. Köntges, S. Kurtz, C. Packard, U. Jahn, K. Berger, K. Kato, T. Friesen, H. Liu, M. Van Iseghem, Review of Failures of Photovoltaic Modules. Technical report, 2014
4. A. Triki-Lahiani, A. Bennani-Ben Abdelghani, I. Slama-Belkhodja, Fault detection and monitoring systems for photovoltaic installations: a review. Renew. Sustain. Energy Rev. **82**, 2680–2692 (2018)
5. M. Banavar, H. Braun, S.T. Buddha, V. Krishnan, A. Spanias, S. Takada, T. Takehara, C. Tepedelenlioglu, T. Yeider, Signal processing for solar array monitoring, fault detection, and optimization. Synth. Lect. Power Electron. **7**(1), 1–95 (2012)
6. Y. Zhao, B. Lehman, R. Ball, J. Mosesian, J.-F. de Palma, Outlier detection rules for fault detection in solar photovoltaic arrays, in *2013 Twenty-Eighth Annual IEEE Applied Power Electronics Conference and Exposition (APEC)* (2013), pp. 2913–2920
7. A. Mellit, G. Tina, S. Kalogirou, Fault detection and diagnosis methods for photovoltaic systems: a review. Renew. Sustain. Energy Rev. **91**, 1–17 (2018)
8. H. Chen, H. Yi, B. Jiang, K. Zhang, Z. Chen, Data-driven detection of hot spots in photovoltaic energy systems. IEEE Trans. Syst. Man Cybernet. Syst. **PP**, 1–8 (2019)
9. S. Vergura, G. Acciani, V. Amoruso, G.E. Patrono, F. Vacca, Descriptive and inferential statistics for supervising and monitoring the operation of PV plants. IEEE Trans. Ind. Electron. **56**(11), 4456–4464 (2009)
10. W. Chine, A. Mellit, A.M. Pavan, V. Lughi, Fault diagnosis in photovoltaic arrays, in *2015 International Conference on Clean Electrical Power (ICCEP)*, June (IEEE, New York, 2015), pp. 67–72
11. J.D. Bastidas-Rodriguez, E. Franco, G. Petrone, C.A. Ramos-Paja, G. Spagnuolo, Model-based degradation analysis of photovoltaic modules through series resistance estimation. IEEE Trans. Ind. Electron. **62**, 7256–7265 (2015)
12. A. Chouder, S. Silvestre, Automatic supervision and fault detection of PV systems based on power losses analysis. Energy Convers. Manage. **51**, 1929–1937 (2010)
13. C. Huang, L. Wang, Simulation study on the degradation process of photovoltaic modules. Energy Convers. Manage. **165**, 236–243 (2018)
14. M. Davarifar, A. Rabhi, A. El Hajjaji, Comprehensive modulation and classification of faults and analysis their effect in DC side of photovoltaic system. Energy Power Eng. **5**(04), 230 (2013)
15. D. King, B. Hansen, J. Kratochvil, M. Quintana, Dark current-voltage measurements on photovoltaic modules as diagnostic or manufacturing tool, in *Conference Record of the 26 IEEE Photovoltaic Specialists Conference*, September (1997), pp. 1125–1128
16. Y. Huang, J. Lu, C. Liu, X. Xu, W. Wang, X. Zhou, Comparative study of power forecasting methods for PV stations, in *2010 International Conference on Power System Technology*, October (IEEE, New York, 2010), pp. 1–6
17. W. Chine, A. Mellit, V. Lughi, A. Malek, G. Sulligoi, A. Massi Pavan, A novel fault diagnosis technique for photovoltaic systems based on artificial neural networks. Renew. Energy **90**, 501–512 (2016)
18. L.D. Murillo-Soto, G. Figueroa-Mata, C. Meza, Identification of the internal resistance in solar modules under dark conditions using differential evolution algorithm, in *2018 IEEE International Work Conference on Bioinspired Intelligence (IWOBI)* (IEEE, New York, 2018), pp. 1–9
19. L.D. Murillo-Soto, C. Meza, Voltage measurement in a reconfigurable solar array with series-parallel topology, in *2017 IEEE 37th Central America and Panama Convention (CONCAPAN XXXVII)* (Managua), November (IEEE, New York, 2017), pp. 1–5

20. H. Braun, S. Buddha, V. Krishnan, C. Tepedelenlioglu, A. Spanias, M. Banavar, D. Srinivasan, Topology reconfiguration for optimization of photovoltaic array output. Sustain. Energy Grids Netw. **6**, 58–69 (2016)
21. H. Ibrahim, N. Anani, Variations of PV module parameters with irradiance and temperature. Energy Proc. **134**, 276–285 (2017)
22. H. Tian, F. Mancilla-david, K. Ellis, P. Jenkins, E. Muljadi, A detailed performance model for photovoltaic systems. Solar Energy J. (2012). Preprint
23. M.G. Villalva, J.R. Gazoli, E.F. Ruppert, *Modeling and Circuit-Based Simulation of Photovoltaic Arrays*, vol. 14(1) (2009)
24. D. Sera, R. Teodorescu, P. Rodriguez, PV panel model based on datasheet values, in *2007 IEEE International Symposium on Industrial Electronics*, June, vol. 4 (IEEE, New York, 2007), pp. 2392–2396
25. Analog-Devices, LTspice, 2018
26. M.C. Di Piazza, G. Vitale, *Photovoltaic sources: modeling and emulation*. Green Energy and Technology (Springer, London, 2013)
27. R. Isermann, Model-based fault-detection and diagnosis–status and applications. Annu. Rev. Control **29**(1), 71–85 (2005)

Computation of the Lambert W Function in Photovoltaic Modeling

Efstratios I. Batzelis, Georgios Anagnostou, Chandan Chakraborty, and Bikash C. Pal

Abstract Recently, the Lambert *W* function has emerged as a valuable mathematical tool in photovoltaic (PV) modeling and other scientific fields. This increasing interest is because it can be used to reformulate the implicit equations of the single-diode PV model into explicit form. However, the computation of the Lambert *W* function itself is still not clear in the literature; some studies use the iterative built-in functions in MATLAB or other computational platforms, while others adopt their own approximation formulae. This paper takes a deeper look at the ways the Lambert *W* function is evaluated in PV models and carries out a comparative study to assess the most commonly used methods in terms of accuracy, computational cost, and application range. These alternatives are implemented in a modern computer and a typical microcontroller to evaluate their performance in both simulations and embedded applications. The analysis concludes that some series expansions are good options for PV modeling applications, requiring less execution time than the built-in MATLAB *lambertw* function and exhibiting negligible approximation error.

1 Introduction

The field of photovoltaic (PV) modeling has been steadily attracting the research interest for more than three decades. Various models for different PV technologies have been proposed in the literature, the *single-diode PV model* still being the most widely adopted approach. One of the main computational challenges of this model is the implicit nature of the fundamental current-voltage (I-V) equation (i.e., $f(I, V) = 0$), which dictates numerical or iterative solution that hinders evaluation.

E. I. Batzelis (✉) · G. Anagnostou · B. C. Pal
Department of Electrical and Electronic Engineering, Imperial College London, London, UK
e-mail: e.batzelis@imperial.ac.uk; georgios.anagnostou11@imperial.ac.uk; b.pal@imperial.ac.uk

C. Chakraborty
Department of Electrical Engineering, Indian Institute of Technology Kharagpur, Kharagpur, India
e-mail: cc.iitkgp@gmail.com

W. Zamboni, G. Petrone (eds.), *ELECTRIMACS 2019*, Lecture Notes in Electrical Engineering 615, https://doi.org/10.1007/978-3-030-37161-6_44

An alternative was first proposed in 2005 by Jain et al. [1], further developed afterwards by Petrone et al. [2]: they used the *Lambert W function* to reformulate the *I-V* equation of the PV cell into an equivalent explicit form (i.e., $I = f(V)$ or $V = g(I)$). This allowed for easier and straightforward calculation, avoiding numerical solution difficulties such as increased computational time, initialization challenges, convergence issues, etc. [3, 4]. It also provided theoretical insight on how each variable and parameter relate to each other, and permitted derivation of analytical models for more complicated PV structures, such as the PV string and array under partial shading [2–4].

Historically, the mathematical expression we^w was first studied by Lambert in 1758 and afterwards by Euler in 1783, but it did not get much attention from their contemporaries at that time. It was revived in the 1990s when developers of the *Maple* computational platform, Corless et al., managed to give an exact solution to the *double-well Dirac delta function model* in quantum mechanics and other mathematical problems of the sciences [5]. The inverse relation of we^w was then identified as a function on its own, denoted Lambert *W* function (also found as *product logarithm* or *omega function* in the literature). In PV modeling, several published studies and a book [3] have adopted this approach since 2005; a SCOPUS search using the keywords "photovoltaic" and "lambert W" yields 75 papers and more than 1000 citations in total, while the actual number of relevant papers is hundreds. Clearly, the use of Lambert *W* function in PV modeling has created a new trend in the field.

However, when it comes to the computation of this function, the literature is not very clear; as it is an elementary function, it can be calculated either through iterative algorithms or approximation formulae. In this paper, the most common calculation approaches used in PV modeling are examined [4–9], including the MATLAB *lambertw* built-in function, and compared in terms of accuracy, computational performance, and formulation complexity; this comparison takes place in a modern computer and a typical microcontroller (MCU). Common implementation challenges are discussed and the developed code in MATLAB and C for each alternative is made available. Notably, this is the first comparative study to assess the implementation methods of the Lambert *W* function in PV modeling.

2 Single-Diode PV Model

The circuit of the single-diode PV model is shown in Fig. 1, being characterized by the so-called *five parameters*: photocurrent I_{ph}, diode saturation current I_s, diode modified ideality factor a, series resistance R_s, and shunt resistance R_{sh} [2–4, 7–11]. This model describes any PV generator (cell, module, array, etc.) under uniform operating conditions.

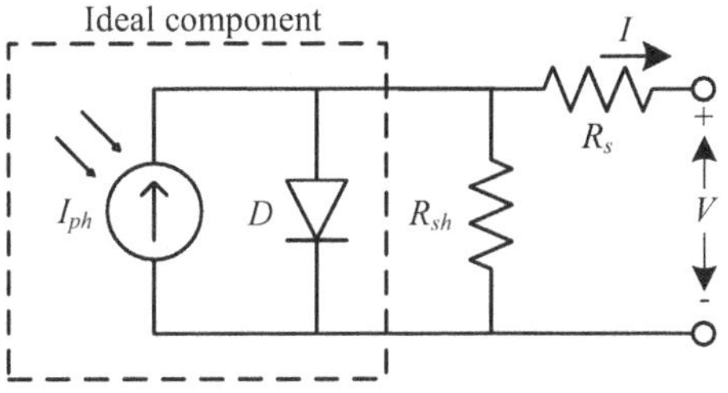

Fig. 1 Electrical equivalent circuit of the single-diode PV model [8]

The current-voltage (I-V) equation of this circuit is given conventionally in implicit form by

$$I = I_{ph} - I_s\left(e^{\frac{V+IR_s}{a}} - 1\right) - \frac{V + IR_s}{R_{sh}} \tag{1}$$

where I and V are found in both sides of the equation; to solve (1), one has to employ a numerical or iterative algorithm with all associated difficulties. The equivalent explicit forms of (1) as $I = f(V)$ or $V = g(I)$ are found by applying the Lambert W function $W\{x\}$ [1–4, 8, 9]:

$$I = \frac{R_{sh}(I_{ph} + I_s) - V}{R_s + R_{sh}} - \frac{a}{R_s}W\left\{\frac{R_s R_{sh} I_s}{a(R_s + R_{sh})}e^{\frac{R_s R_{sh}(I_{ph}+I_s)+R_{sh}V}{a(R_s+R_{sh})}}\right\} \tag{2}$$

$$V = R_{sh}(I_{ph} + I_s) - (R_s + R_{sh})I - aW\left\{\frac{R_{sh} I_s}{a}e^{\frac{R_{sh}(I_{ph}+I_s-I)}{a}}\right\} \tag{3}$$

3 Calculation of the Lambert W Function

The Lambert W function $W\{x\}$ is the inverse of the function we^w, i.e., the root of the equation $we^w = x$ (e.g., $we^w = 2 \leftrightarrow w = W\{2\} = 0.8526$). In general, this relation has several branches in the complex plane, thus not being a function in the conventional sense. In PV applications, however, the argument x takes real positive values (see (2) and (3)), which restricts $W\{x\}$ to the *principal branch* [5]. The principal branch is a function, thereafter referred to simply as the Lambert W function; an indicative plot is shown in Fig. 2 along with the logarithmic function for comparison.

The Lambert W function cannot be expressed in terms of other elementary functions and it is calculated either numerically or through approximation formulae.

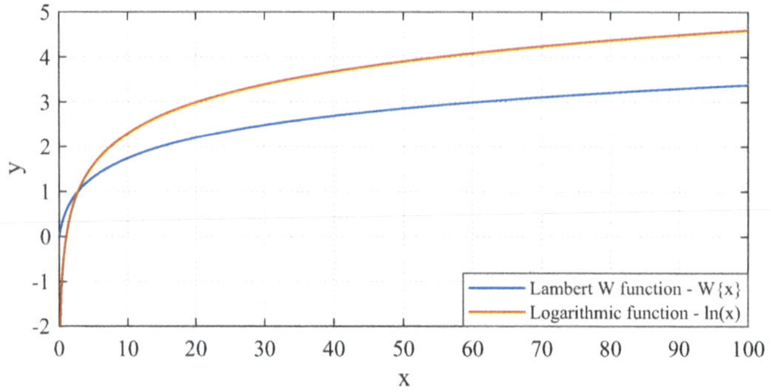

Fig. 2 The Lambert *W* and logarithmic functions

More information on the theoretical background and calculus may be found in [5] and other publications of these authors. In the following, the computational methods usually adopted in PV models in the literature are described and discussed.

3.1 The MATLAB lambertw Function

The *lambertw* function in MATLAB provides solution to the general multi-valued relation in the complex plane; to get the principal branch, one can write $lambertw(0, x)$ or simply $lambertw(x)$. The evaluation is made numerically, using the Haley's method with an iteration step of [6]:

$$W_{j+1} = W_j - \frac{W_j e^{W_j} - x}{e^{W_j}(W_j + 1) - \frac{(W_j+2)(W_j e^{W_j} - x)}{2W_j+2}} \tag{4}$$

where W is the Lambert W function of x and j indicates the iteration step. The majority of the relevant studies in PV modeling adopt this approach, as it is readily available and achieves machine accuracy. Nevertheless, the iterative nature renders this method computationally less efficient compared to the series expansions, especially when it comes to MCU implementation as shown later in the paper.

3.2 The Asymptotic Formula

The *asymptotic formula* was proposed by Corless et al. in 1996 [5]:

$$W\{x\}=L_1-L_2+\frac{L_2}{L_1}+\frac{L_2(-2+L_2)}{2L_1^2}+\frac{L_2(6-9L_2+2L_2^2)}{6L_1^3}$$
$$+\frac{L_2(-12+36L_2-22L_2^2+3L_2^3)}{12L_1^4}$$
$$+\frac{L_2(60-300L_2+350L_2^2-125L_2^3+12L_2^4)}{60L_1^5}$$
$$+\,O\left[\left(\frac{L_2}{L_1}\right)^6\right] \tag{5}$$

where $L_1 = ln(x)$ and $L_2 = ln(ln(x))$. This formula is suitable for large arguments only; the *Wolfram MathWorld* suggests $x \geq 3$ [12], while the number of terms of (5) used in the literature varies: the first 7 terms in [4, 10], the first 4 terms in [7]. As shown in Sect. 4, the maximum approximation error for 7 terms and $x \geq 3$ is 1.32%.

A common difficulty faced when applying (5) is numerical error for very large arguments, as reported in [11] (returning Inf or NaN, calculation fails, etc.). For example, if $x = 800e^{800}$ which is larger than the maximum floating-point number in IEEE double precision (64bit arithmetic), L_1 and L_2 become Inf and (5) yields NaN (not a number), although clearly $W\{800e^{800}\} = 800$ which is a perfectly finite value.

To overcome this limitation, the argument could be given in the form of $x = ae^b$, effectively making $L_1 = ln(a) + b$ and $L_2 = ln(ln(a) + b)$ in (5). In other words, instead of evaluating the large term x and then taking the logarithms, calculate directly L_1 and L_2 as functions of the finite a and b components of x. This trick is used in the implementation code of this formula and most of the following series expansions (see Appendix).

3.3 *The Hybrid Calculation Formula*

To cope with the unsuitability of the asymptotic formula at small arguments, a *hybrid formula* is proposed in [4] which combines the asymptotic expansion (5) (7 terms) with a series expansion found in [13] that is accurate for small values:

$$W\{x\} = \begin{cases} W_1, & \text{when } 0 \leq x < 9 \\ W_2, & \text{when } x \geq 9 \end{cases} \tag{6}$$

where

$$W_1 = u + \frac{u}{1+u}p + \frac{u}{2(1+u)^3}p^2 - \frac{u(2u-1)}{6(1+u)^5}p^3$$

$$+\frac{u(6u^2-8u+1)}{24(1+u)^7}p^4-\frac{u(24u^3-58u^2+22u-1)}{120(1+u)^9}p^5 \tag{7}$$

$$\begin{aligned}W_2 = L_1 - L_2 + \frac{L_2}{L_1} + \frac{L_2(-2+L_2)}{2L_1^2} + \frac{L_2(6-9L_2+2L_2^2)}{6L_1^3} \\ + \frac{L_2(-12+36L_2-22L_2^2+3L_2^3)}{12L_1^4} \\ + \frac{L_2(60-300L_2+350L_2^2-125L_2^3+12L_2^4)}{60L_1^5}\end{aligned} \tag{8}$$

having $u = x/e$, $p = 1 - x/e$ and $L_1 = ln(x)$, $L_2 = ln(ln(x))$. This formula yields a relative error lower than 0.1% for the entire real non-negative argument range [4].

3.4 The Simple Approximation Formula

In [8], a much simpler calculation series is proposed, based on a 1961 study [14]:

$$W\{x\} = ln(x)\left[1 - \frac{ln(ln(x))}{ln(x)+1}\right] \tag{9}$$

reported to yield errors less than 1.5% for $x \geq 2$ [8]. At smaller arguments, the calculation error is very large and (9) is not suitable for application.

3.5 The Analytical Approximation Formula

A similar in nature approximation was introduced by Barry et al. in 2000 [15], adopted recently in a PV model [9]:

$$W\{x\} = (1+\epsilon)ln\left(\frac{\frac{6}{5}x}{ln\left[\frac{\frac{12}{5}x}{ln\left(1+\frac{12}{5}x\right)}\right]}\right) - \epsilon ln\left[\frac{2x}{ln(1+2x)}\right] \tag{10}$$

where $\epsilon = 0.4586887$ is a constant. The approximation error of this formula is claimed to be 0.196% for $x \geq 0$. However, during our investigation we found that this level of accuracy is achieved for approximately $x \geq 3 \times 10^{-5}$ in 64bit double precision arithmetic ($x \geq 3 \times 10^{-3}$ in 32bit single precision arithmetic), since (10) does not seem to be convergent at near-zero values of the argument as explained later. The implementation code in this paper employs this consideration.

4 Simulation Results

In this section, the accuracy and computational performance of the various Lambert *W* implementations are assessed and compared in simulations environment; these are denoted here as *MATLAB*, *Asymp7* (asymptotic formula with 7 terms), *Asymp4* (asymptotic formula with 4 terms), *Hybrid*, *Simple*, and *Analyt*, respectively. First, they are applied to the general case of real positive arguments, and then to an *I*-*V* curve of a PV array. The implementation code is made publicly available (see Appendix), while the MATLAB approach (i.e., the *lambertw* function) is employed as a benchmark for assessing accuracy. All simulations are performed in MATLAB R2017b, in a computer with a 3.5-GHz CPU and 64-GB RAM at the default 64bit arithmetic.

4.1 Evaluation at Real Positive Arguments

Here the six alternative methods are evaluated at real positive values of the argument; Fig. 3a–b depicts the function outputs and the relative errors for the indicative case of $0 < x \leq 100$ (one million values uniformly distributed).

It seems that the results of all methods match quite well in Fig. 3a, the main difference being the range of the argument x: the Asym7 (red line) and Asym4 (yellow line) approaches are defined for $x \geq 3$, and the Simple formula (green line) for $x \geq 2$. A clearer picture on the accuracy is given in Fig. 3b: the errors are larger at small arguments, the Asym7, Asym4 and Simple methods exceeding 1% for some values. The Hybrid formula (purple line) seems to be the most accurate among the series expansions.

To investigate the argument range of the Analyt method, the results of the Analyt and MATLAB approaches at very small arguments up to 10^{-4} are shown in Fig. 4. It is evident that the formula (blue line) is not convergent at near-zero values; a lower bound of 3×10^{-5} has to be applied to the input range to achieve a relative error of about 0.2%, which is close to 0.196% considered by the authors [9, 15].

The complete picture on the accuracy, argument range, and execution time is given in Table 1, where bold font indicates most favorable values. The Hybrid formula is the most accurate, followed by Analyt method, while the rest yield higher max errors above 1%. As for the calculation cost, clearly the MATLAB function exhibits the highest execution time due to the iterative algorithm involved; the approximation formulae require approximately $\frac{1}{3}$ of the MATLAB time, except for the more demanding Analyt method due to the larger number of the logarithmic function calls.

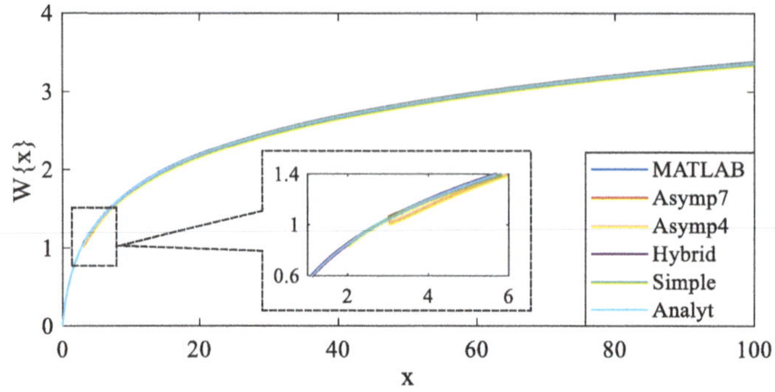

(a)

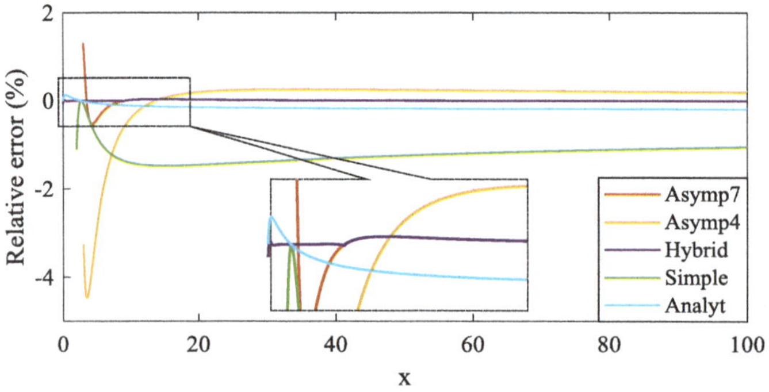

(b)

Fig. 3 (**a**) The Lambert *W* function outputs and (**b**) the respective relative errors at real positive arguments

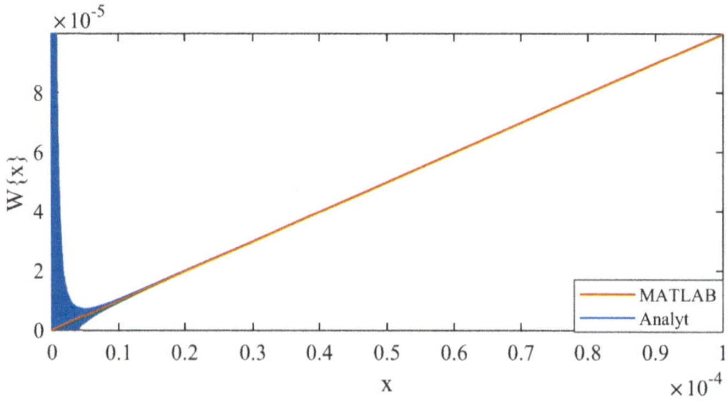

Fig. 4 The Analyt method at very small positive arguments

Table 1 Performance of the six Lambert *W* function implementations at real positive arguments

Method	Range	Max error	Time per evaluation
MATLAB	**Entire**	–	103 ns
Asymp7	$x \geq 3$	1.32%	35 ns
Asymp4	$x \geq 3$	4.46%	33 ns
Hybrid	**Entire**	**0.06%**	36 ns
Simple	$x \geq 2$	1.47%	**31** ns
Analyt	$x \geq$3e-5	0.19%	63 ns

Table 2 Five parameters of the PV array

STC parameters					
I_{ph}	15.88 A	I_s	7.44e–10 A	a	14.67 V
R_s	2.04 Ω	R_{sh}	425.2 Ω		

4.2 *Evaluation of the I-V Curve of a PV Array*

The results of the previous section are somewhat theoretical and do not indicate clearly the effectiveness of the Lambert *W* function implementations when it comes to a PV model. To assess this, the six alternatives are employed here to produce the *I-V* characteristic of a 4 kW PV array at standard test conditions (2 strings, 12 Yingli YL-165 modules per string, parameters in Table 2). The *I-V* curve is found in two ways: calculating I as a function of V based on (2) (*I-approach*) and the opposite using (3) (*V-approach*). The results in the two cases are shown in Fig. 5a–b (1000 points).

Apparently all curves coincide, but some of them are incomplete; the arrows indicate the range of application. The Asymp7, Asymp4, and Simple methods (purple, yellow, and green colors) can produce very limited part of the *I-V* characteristic in either figures, while the Analyt formula provides a larger part but still only a portion of the curve. This is due to the lower bound of the input range of these methods that leads to exclusion of some parts towards the short-circuit current region. In other words, the argument of the $W\{x\}$ terms in (2) and (3) varies within [9.3e–10, 17.7] and [2.2e–8, 1.5e+192], respectively, the lower bounds being out of the input ranges of these methods (see Table 1). Only the MATLAB and Hybrid approaches provide the entire *I-V* characteristic in both cases.

The complete picture is given in Table 3. It is quite interesting how the seemingly small lower bounds in the argument range of the Asymp7, Asymp4, Simple, and Analyt approaches restrict that much their application in calculating the *I-V* curve. If these bounds are relaxed, larger parts of the characteristic are produced, but they appear to be distorted and highly erroneous. These results correspond to the study-case 4 kW PV array; for different PV system sizes, larger or smaller portions of the *I-V* curve can be produced.

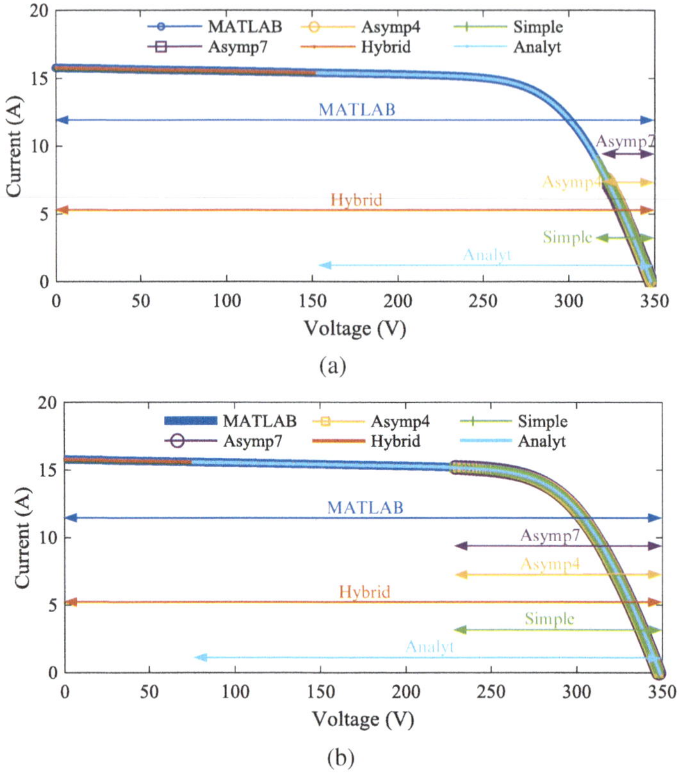

Fig. 5 *I-V* curve produced: (**a**) I-approach and (**b**) V-approach

The accuracy is measured using the normalized root mean square error (NRMSE) metric: the results seem acceptable for all methods, the Hybrid formula exhibiting the lowest errors in both I-approach and V-approach. As for the computational time per *I-V* curve point, similar results to the previous section are extracted, except that all times all slightly higher due to the additional overhead of (2)–(3). This analysis concludes that the Hybrid formula is the safest and most accurate method to produce the *I-V* curve among the series expansions.

5 Experimental Results

The Lambert *W* function has found recently embedded applications to PV systems, such as in model-based MPPTs and other control algorithms [16]. In these cases, the computational complexity is even more important due to the limited calculation capacity of the common MCUs. To investigate this aspect, this section discusses some experimental results from the implementation of the six Lambert *W* function

Table 3 Performance of the six Lambert *W* function implementations in calculating the *I-V* curve of a 4 kW PV array (times per points)

	I-approach			V-approach		
Method	**Range**	**NRMSE**	**Time**	**Range**	**NRMSE**	**Time**
MATLAB	**Entire**	–	203 ns	**Entire**	–	450 ns
Asymp7	>323 V	0.16%	39 ns	>229 V	0.001%	54 ns
Asymp4	>323 V	1.28%	**36** ns	>230 V	0.009%	52 ns
Hybrid	**Entire**	**4.7e–5**	72 ns	**Entire**	**1.3e–6**	71 ns
Simple	>317 V	0.82%	40 ns	>226 V	0.021%	**50** ns
Analyt	>153 V	0.03%	65 ns	>76 V	0.178%	96 ns

Table 4 Computational performance of the six methods when implemented in an MCU (times per evaluation)

	Positive arguments		I-approach		V-approach	
Method	**Range**	**Time**	**Range**	**Time**	**Range**	**Time**
MATLAB	**Entire**	26.2 μs	**Entire**	15.0 μs	<294 V	31.4 μs
Asymp7	$x \geq 3$	7.7 μs	>323 V	8.9 μs	>229 V	7.9 μs
Asymp4	$x \geq 3$	**7.4** μs	>323 V	8.6 μs	>230 V	7.6 μs
Hybrid	**Entire**	7.7 μs	**Entire**	**8.0** μs	**Entire**	7.9 μs
Simple	$x \geq 2$	7.3 μs	>317 V	8.3 μs	>226 V	**7.5** μs
Analyt	$x \geq$3e−3	17.9 μs	>221 V	21.2 μs	[90,294] V	21.5 μs

alternatives in a typical MCU. The Texas Instruments model TMS320F28335 is used, which is often found in power electronics and motor drive applications; this MCU has an 150 MHz clock and a 32bit hardware multiplier, but no division unit. The respective C code is properly optimized for this MCU (see Appendix).

The three case studies of the previous section are reexamined here. In terms of accuracy, the results are the same as the respective simulation ones, except from the Analyt method; the latter faced worse convergence issues at low arguments due to the 32bit arithmetic (64bit in simulations) which necessitated increase of the lower bound to 3×10^{-3} (from 3×10^{-5}). The findings on the application range and execution times are shown in Table 4.

Comparing the application ranges of Table 4 to Table 1 and Table 3, it seems that Analyt is more restricted at low voltages, and that there is now an upper limit to both MATLAB and Analyt methods in the V-approach. The former is due to the increased lower bound of Analyt, whereas the latter is because the maximum value of the argument 1.5e+192 (see Sect. 4.2) is not supported in the 32bit arithmetic of the MCU (as opposed to the 64bit MATLAB arithmetic). On the contrary, the rest of the approximation formulae do not face this difficulty, as they employ the large argument manipulation trick discussed in Sect. 3.2; unfortunately, this is not applicable to MATLAB's (4) or Analyt's (10).

As for the execution cost, the MCU times are of a few μs here, as opposed to the ns simulation values of Sect. 4. This becomes a critical computational burden to the MCU when the control algorithm requires evaluation of the Lambert *W* function at every switching cycle. For example, for a typical switching period of 50 μs (switching frequency 20 kHz), MATLAB and Analyt take up almost half of the calculation window, as opposed to the other much more cost-effective approximation formulae.

The main conclusion from this investigation is that when it comes to embedded applications, proper handling of large arguments and computational efficiency should be the main criteria in selecting the calculation method of the Lambert *W* function. The analysis concludes that the Hybrid approach exhibits the most favorable performance in this regard.

6 Conclusions

In this paper, we examine how the Lambert *W* function is calculated in PV models in the literature, including the MATLAB *lambertw* function and five approximation formulae. The accuracy, computational cost, and applicability are assessed on a modern computer and a typical microcontroller.

The results indicate that the argument of the Lambert *W* function in the PV equations takes near-zero values close to the short-circuit and very large values close to open-circuit regions; this dictates that the calculation formula should be applicable to a wide range of real positive numbers (no lower or upper bounds if possible). The MATLAB *lambertw* function exhibits the best accuracy at the cost of higher execution time, but may fail at very large arguments when implemented in an MCU. Alternatively, the Hybrid formula given in [4] proves more computationally efficient and robust, but yields a calculation error of the order of 10^{-5} or 10^{-6}.

Acknowledgements Dr. E. Batzelis' work has received funding from the European Union's Horizon 2020 research and innovation programme under the Marie Sklodowska-Curie grant agreement No 746638. The rest of the authors' work has been conducted as part of the research project "Joint UK-India Clean Energy Centre (JUICE)" which is funded by the RCUK's Energy Programme (contract no: EP/P003605/1).

Appendix

The implementation code in MATLAB and C of the six Lambert *W* function alternatives is available online in the GitHub repository: https://github.com/ebatzelis/Lambert-W-function-in-PV-modeling.git.

The C code is properly optimized for MCU implementation (minimum number of divisions and logarithmic/exponential evaluations, factoring terms, etc.). The implementation of the asymptotic, Hybrid, and Simple formulae overcomes the large argument contingency by inputting the Lambert W argument in the form of

$x = ae^b$ (see Sect. 3.2). For example in (3), instead of calculating first the possibly large $x = \frac{R_{sh} I_s}{a} e^{\frac{R_{sh}(I_{ph}+I_s-I)}{a}}$ and then the Lambert W term $W\{x\}$, one can directly find $W\{x\} = lambertW\left(\frac{R_{sh} I_s}{a}, \frac{R_{sh}(I_{ph}+I_s-I)}{a}\right)$.

References

1. A. Jain, A. Kapoor, A new approach to study organic solar cell using Lambert W-function. Sol. Energy Mater. Sol. Cells **86**(2), 197–205 (2005)
2. G. Petrone, G. Spagnuolo, M. Vitelli, Analytical model of mismatched photovoltaic fields by means of Lambert W-function. Sol. Energy Mater. Sol. Cells **91**(18), 1652–1657 (2007)
3. G. Petrone, C.A. Ramos-Paja, G. Spagnuolo, *Photovoltaic Sources Modeling*, 1st edn. (Wiley-IEEE Press, New York, 2017)
4. E.I. Batzelis, I.A. Routsolias, S.A. Papathanassiou, An explicit PV string model based on the Lambert W function and simplified MPP expressions for operation under partial shading. IEEE Trans. Sustainable Energy **5**(1), 301–312 (2014)
5. R.M. Corless, G.H. Gonnet, D.E.G. Hare, D.J. Jeffrey, D.E. Knuth, On the Lambert W function. Adv. Comput. Math. **5**(4), 329–359 (1996)
6. C. Moler, The Lambert W Function (2013), https://blogs.mathworks.com/cleve/2013/09/02/the-lambert-w-function/
7. Y.M. Roshan, M. Moallem, Maximum power point estimation and tracking using power converter input resistance control. Sol. Energy **96**, 177–186 (2013)
8. E.I. Batzelis, S.A. Papathanassiou, A method for the analytical extraction of the single-diode PV model parameters. IEEE Trans. Sustain. Energy **7**(2), 504–512 (2016)
9. E. Moshksar, T. Ghanbari, A model-based algorithm for maximum power point tracking of PV systems using exact analytical solution of single-diode equivalent model. Sol. Energy **162**, 117–131 (2018)
10. Y.M. Roshan, M. Moallem, Maximum power point tracking using boost converter input resistance control by means of Lambert W-Function, in *Proceedings of 2012 3rd IEEE International Symposium on Power Electronics for Distributed Generation Systems*, Aalborg, pp. 195–199, June 2012
11. A. Xenophontos, A.M. Bazzi, Model-based maximum power curves of solar photovoltaic panels under partial shading conditions. IEEE J. Photovoltaics **8**(1), 233–238 (2018)
12. E.W. Weisstein, Lambert W-function, http://mathworld.wolfram.com/LambertW-Function.html
13. R.M. Corless, D.J. Jeffrey, D.E. Knuth, Sequence of series for the Lambert W function, in *Proceedings of the 1997 International Symposium on Symbolic and Algebraic Computation, ISSAC*, Maui, HI, pp. 197–204, July 1997
14. W. Borsch-Supan, On the evaluation of the function $\phi(\lambda)$ for real values of λ. J. Res. **65B**(4), 245–250 (1961)
15. D. Barry, J.-Y. Parlange, L. Li, H. Prommer, C. Cunningham, F. Stagnitti, Analytical approximations for real values of the Lambert W-function. Math. Comput. Simul. **53**(1–2), 95–103 (2000)
16. E.I. Batzelis, S. Papathanassiou, B.C. Pal, PV system control to provide active power reserves under partial shading conditions. IEEE Trans. Power Electron. **31**(11), 9163–9175 (2018)

Part V
Advances in Power Electronics for Energy Storage and Hardware Emulation

Lithium-Ion Capacitors: Characterization and Modeling at Both High and Low Temperatures

Zeineb Chabrak Payet, Alexandre De Bernardinis, Pascal Venet, and Richard Lallemand

Abstract The lithium-ion capacitor is a recent energy storage component. Although it has been commercialized for several years, its hybridization still requires further investigation to characterize it. The literature has studied some of its characteristics focusing on experimentation at positive temperatures. This paper aims to enlarge the tests to include very low temperatures, showing the difference between Nyquist plots at 65 and $-30\,^{\circ}$C. It also presents the Ragone plot for several temperatures, with a comparison between three storage systems: a battery, a supercapacitor, and the lithium-ion capacitor. Finally, a model of the LIC is proposed, for low and high temperatures, with experimental validation.

1 Introduction

With the emergence of applications requiring increasing energy storage systems (ESSs), there is a growing need to study and develop new ones. ESSs can be found in almost every application, from transportation with hybrid and electrical means of transport, to renewable energy sources with photovoltaic and wind turbine firms, all requiring ESSs to store the produced energy.

The mostly known storage components are the battery, with its different compositions, and the supercapacitor or the electrochemical double layer capacitor (EDLC). These components have been deeply studied in literature and their use is widely spread in all the applications. Recently, a new component came to surface and even

Z. Chabrak Payet (✉) · A. De Bernardinis · R. Lallemand
Systèmes et Applications des Technologies de l'Information et de l'Energie, ENS Paris-Saclay, CNRS, Université Paris-Saclay, IFSTTAR, Versailles, France
e-mail: zeineb.chabrak@ifsttar.fr; alexandre.de-bernardinis@ifsttar.fr; richard.lallemand@ifsttar.fr

P. Venet
Univ Lyon, Université Claude Bernard Lyon 1, Ecole Centrale de Lyon, INSA Lyon, CNRS, Ampère, Lyon, France
e-mail: pascal.venet@univ-lyon1.fr

W. Zamboni, G. Petrone (eds.), *ELECTRIMACS 2019*, Lecture Notes in Electrical Engineering 615, https://doi.org/10.1007/978-3-030-37161-6_45

if it has been commercialized for several years now, his use is still not frequent and the studies about it are still limited. This component is the lithium-ion capacitor (LIC), a combination between a lithium-ion battery (LIB) and a supercapacitor (SC).

The lithium-ion capacitor combines a negative electrode from the battery, composed of graphite pre-doped with lithium-ions Li+, and a positive electrode from the supercapacitor, composed of activated carbon. This allows the LIC to acquire a higher energy density than the SC, while conserving a high power density and a long lifetime. The LIC has been commercialized by several manufacturers, but only few are available on the market, like JSR Micro and YUNASKO. The difference between the LICs of the two manufacturers is the capacitance and the type of electrodes.

Some applications started to take place, showing the important impact the LIC could have. Two types of applications could be found in the literature. The first one is the use of LICs in renewable energy, for example, to replace the classic capacitor in a DC bus, to ameliorate the power output of a wind turbine [1]. The second type is the use of LICs in transportation, mainly public transport whether they are on-board of an electric bus or in stationary, on the wayside of a tramway line to save the energy from braking and restore it to the vehicles during accelerations [2].

In the literature, recent papers about this component have been published. They are mainly trying to characterize and model this component in different ways. For the modeling, the first trials were about building a model of a LIC similar to that of a supercapacitor developed in [3] and [4]. But these models were not suitable to all constrains, which is normal since the role of the battery-type electrode in the model cannot be neglected. Other models were established, using electrical components (R and C) and the temperature effect, but they do not take into account the frequency [5]. As for the thermal modeling, it can be found in [6].

Other studies on characterizing this new component focused on the aging tests that are necessary to estimate its lifetime. Lately, this has been studied in [7]. The authors of the paper used a type of LIC with no specified names but with characteristics different from those of the JSR Micro, the mostly known LIC constructor. In [8] and [9], the aging of the LICs from JSR Micro is studied in floating mode and in cycling, confirming the long lifetime of the component.

In all the studies mentioned above, the LIC has always been characterized at positive high temperatures. The experiments in low temperatures are very rare and are limited to a simple Nyquist plot if existing.

This research aims at closing the gap in the characterization of the LIC. We present some new characteristics that, as far as we know, have not been considered in existing literature. The frequency and time domain characterizations will be reproduced, but, rather than focusing only on the positive temperatures, the study will go throughout all the operating temperature range of the LIC, i.e., from −30 to 65 °C. The time domain characterization will allow to plot, for the first time, the Ragone diagram of the LIC for several temperatures, including the lower ones, as far as the discharge current can go. Then, in order to compare those plots, Ragone diagrams of a Li-ion battery and a supercapacitor of nearly the same capacitance will be plotted as well. Afterwards, a model based on the frequency characterization will be introduced and tested to confirm its accuracy, also for all temperatures range.

2 Experimental Procedure

The experimental part has been performed on two types of prismatic cells of LIC from the same manufacturer, JSR Micro: ULTIMO 2300F ULR and ULTIMO 3300F LR, their characteristics are presented in Table 1. The results of the next sections are those of the LIC 3300F only, since they are almost the same for the two types.

This type of LIC is composed of a positive electrode of activated carbon similar to a supercapacitor's electrode. The negative one is composed of graphite pre-doped with Li-ions Li+, similar to a Li-ion battery's electrode. The electrolyte between the electrodes is an organic one, similar to that of a Li-ion battery, composed of a $LiPF_6$ salt. This hybrid composition given in Fig. 1 increases the energy density of the LIC while keeping a high power density.

The study in this paper concerns only this one type of lithium-ion capacitors, and can be generalized to all cells with the same composition of electrodes. Since other types of LICs exist with completely different electrodes (two electrodes, both with lithium and carbon, for example), they cannot be included in the general results of this paper.

2.1 *In Frequency Domain*

The characterization in frequency domain is mainly based on electrochemical impedance spectroscopy (EIS), performed using an IM6 Zahner workstation. This technique allows to determine the different electrochemical processes happening inside the cell while applying a small perturbation of voltage or current over a specified frequency range.

Table 1 Characteristics of JSR Micro LICs

Measurement	Ultimo 2300F ULR	Ultimo 3300F LR
Temperature range	−30...70 °C	−30...70 °C
Voltage range	2.2...3.8 V	2.2...3.8 V
Capacitance	2300 F	3300 F
Resistance at 1 kHz	0.6 mΩ	0.7 mΩ
Energy density	8 Wh/Kg	13 Wh/Kg
Weight	0.365 kg	0.350 kg

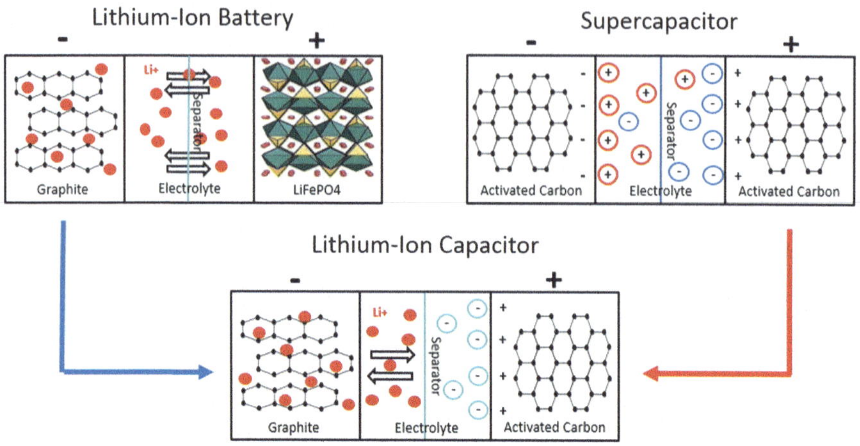

Fig. 1 Internal composition of the LIC cell

EIS measurements are applied on the LIC for temperatures going from −30 to 65 °C (temperatures of the climatic chamber). The LIC cell stays overnight in the climatic chamber ESPEC at the defined temperature and tests are completed the next day. As for the voltage, measurements are applied for five different voltages (i.e., different states of charge, SoC): 2.2, 2.6, 3, 3.4, and 3.8 V. The galvanostatic mode (fixed current value) has been used to measure the impedance, with an AC maximum current amplitude equal to 5 A. The scanning frequency varied between 10 mHz and 100 kHz.

2.2 *In Time Domain*

The characterization in time domain is based on a simple charge/discharge process. This has been done using a home-made test-bench with a maximum charge current of 18.9 A and a discharge current up to 400 A. The same temperatures as for frequency domain tests have been selected. The LIC has been charged and discharged fully, i.e., between 2.2 and 3.8 V even if the LIC is theoretically able to accept current up to 1100 A or more [10] for 25 °C, discharge current amplitudes used for tests raised from 10 to 400 A when temperature constrains allow it (since high current amplitudes were impossible to apply for very low temperatures).

3 Results and Discussion

3.1 *Ragone Plot*

The characterization in time domain has allowed to determine the gravimetric energy and power densities of the Li-Ion capacitor. These densities are needed to plot the Ragone diagram, highly used to compare the performances of different storage systems for a given application regarding their position on the plot. In literature, the existing Ragone diagrams showing the new features of the LIC, i.e., high gravimetric energy density compared to an EDLC and high gravimetric power density compared to a Li-ion battery, are global diagrams plotted at 20 °C, giving only an approximate position of each component. The aim of this section is to plot Ragone diagram of the LIC, with measured values, not only for the usual temperature, but throughout all the possible temperatures. This way, the plotted diagram is more precise and gives indication on the nature of the LIC that have not been observed yet.

The following equations have been used to calculate the peak gravimetric power and the gravimetric energy densities, only for the discharge part, for every discharge current.

- Gravimetric peak power density: $P = max(\frac{VI}{M})$
  P is the gravimetric power density in W/kg, V the voltage in V, I the current in A and M the weight in kg.
- Gravimetric energy density: $E = \frac{\int VIdt}{3600*M}$
  E is the gravimetric energy density in Wh/kg.

3.1.1 Ragone Plot for LIC 3300 F

Figure 2 represents the Ragone diagram for all tested temperatures on a LIC3300F cell. As it can be seen, for high temperatures, the energy density of the LIC is quite high (13 Wh/kg) and the difference between the plots at 25, 45, and 65 °C is almost nonexistent. When temperature decreases, the energy density drops, especially for high discharge current, starting from 10 °C, while the power density stays almost constant. For negative temperatures, the energy density of the LIC becomes very low. The cell cannot be discharged at high currents anymore (maximum current discharge 100 A at −30 °C).

The gravimetric power density does not change a lot with the temperature decrease when remaining over 0 °C. Indeed, the calculation protocol takes into account a constant discharge current, and the peak voltage, which is always near 3.8 V, since the resistance at those temperatures is still small. When evolving towards lower temperatures, the resistance increases considerably, leading to a more significant voltage drop, which results in a much smaller peak voltage at the

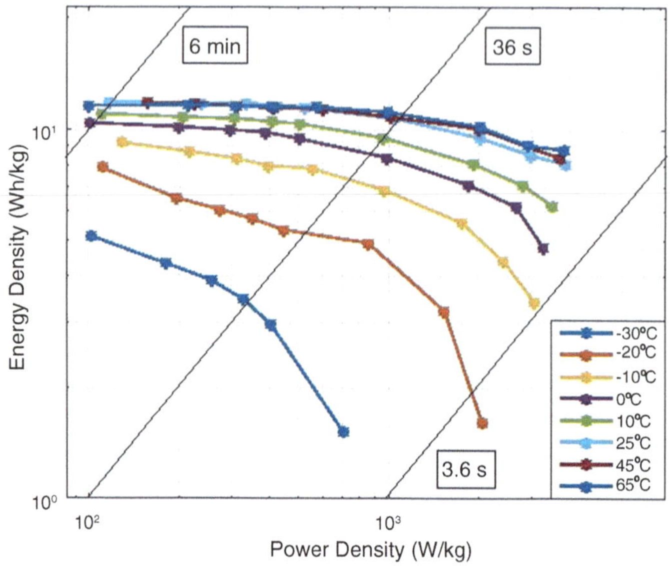

Fig. 2 Ragone diagram of a LIC 3300F at all tested temperatures

beginning of the discharge. This can explain the reason of the decrease of the power density at low temperatures.

To explain the energy density's drop at low temperatures, the LIC will be considered as a thermodynamic system. This way, the energy is equal to the ΔG, the Gibbs energy, with ΔG proportional to $V_p - V_n$. V_p is the potential of the positive electrode, it is the one responsible of the global potential of the cell. V_n is the potential of the negative electrode. For usual temperatures, it is very low and constant, it barely participates in the total cell polarization. This is explained in [11] where authors added a third lithium-reference electrode to study the potentials of the positive and the negative electrodes versus lithium one. When going toward low temperatures, it has been found that the potential of the negative electrode rises significantly, and its contribution to the cell polarization becomes quite important, while the positive electrode's potential remains constant. This results in a decrease of the difference between the two potentials, leading to a smaller ΔG and consequently, a drop in the energy density.

Those results have been confirmed from an electrochemical point of view in [12], where authors studied the electrodes separately to observe the inner processes that take place at low temperatures.

3.1.2 Comparison Between LIC, Battery, and Supercapacitor

The same tests of charge/discharge described above have also been performed for a battery of 1.1 Ah and a supercapacitor of 3000 F. Ragone diagrams of these components have been plotted the same way as the LIC, using the same equations to calculate their gravimetric power and energy densities.

- For the battery: since the maximum current allowed is 30 A, the discharge has been performed for 1, 5, 10, 20, and 30 A. Also, since the maximum temperature is 60 °C, the maximum applied temperature chosen was 45 °C to compare with the other cells.
- For the supercapacitor: the discharge currents applied are the same as those applied to the LIC with also the same temperatures.

Figure 3 shows Ragone diagrams for all three components at four different temperatures −10, 0, 25, and 45 °C, one color for each one. As it can be noticed, a significant gap exists between the energy density of the battery compared to the other cells, with 83 Wh/kg at 25 °C. On the contrary, its power density is quite insufficient, knowing that the power densities of the LIC and the supercapacitor could be much higher since the maximum allowed currents could not be attained, due to the experimental test bench design.

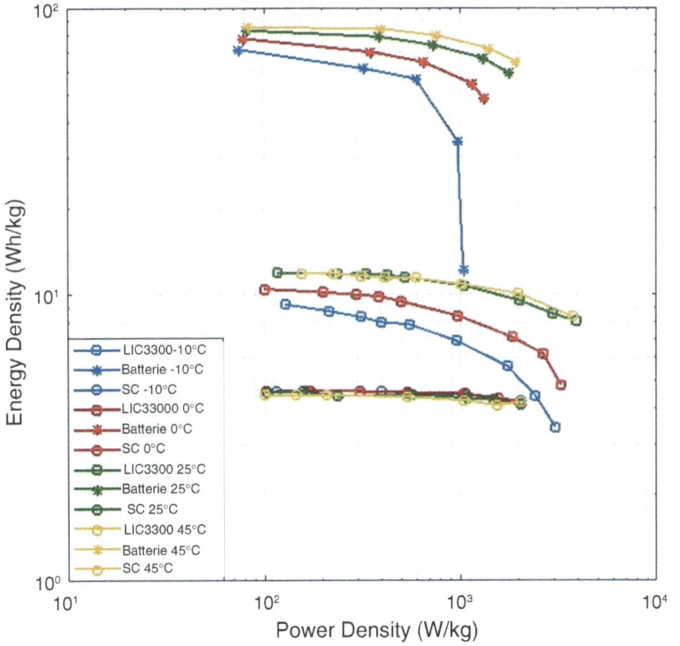

Fig. 3 Ragone diagram for the 3 cells at several temperatures

When comparing the LIC and the supercapacitor, the energy density of the LIC cell is found to be twice the EDLC's one, with also a higher power density for the same current. The difference is that the supercapacitor presents very stable diagrams that barely change with temperature. It is capable of providing the same specific energy at 65 and −30 °C without any degradation. In contrary, LIC is much more sensitive to the temperature gradient. Indeed, its specific energy decreases a lot and can even reach values smaller than those of the supercapacitor at very low temperature.

From this comparison, it is clear that, for high temperatures, the LIC has inherited the supercapacitor's characteristics, with a higher energy density. However, when the temperature drops, it behaves more like a battery, since the contribution of the negative graphite electrode increases at those temperatures.

3.2 Nyquist Plot

The Nyquist plot is obtained from EIS measurements, and it represents the opposite of the imaginary part of the impedance of the cell as a function of its real part. This plot is a good starting point to determine the inner electrochemical processes happening in the LIC, considering the results in the existing literature.

At high frequencies, the cell has an inductive behavior that will not be discussed below. Only the positive part of -Im(Z) of the plot will be studied, representing the capacitive part.

3.2.1 At Low Temperatures

The Nyquist plots represented in Fig. 4 at low temperatures exhibit at −30 °C a large semi-circle at medium frequencies, similar to batteries. It represents the charge transfer in the cell, and its radius is proportional to the charge transfer resistance. For a LIC, the charge transfer happens only at the negative electrode, where the oxydo-reduction reaction takes place. It consists of the de-solvation of the lithium ions, their transport into the solid /electrolyte interface and their reduction in charge (the opposite process happens in discharge) [13]. The radius of the semi-circle diminishes when going toward 0 °C, since the charge transfer resistance decreases a lot with high temperatures.

After the semi-circle, at low frequencies, a vertical line appears, representing the solid-state diffusion into the active material (graphite) of the electrode, which is, in general, the slowest process in charge or discharge.

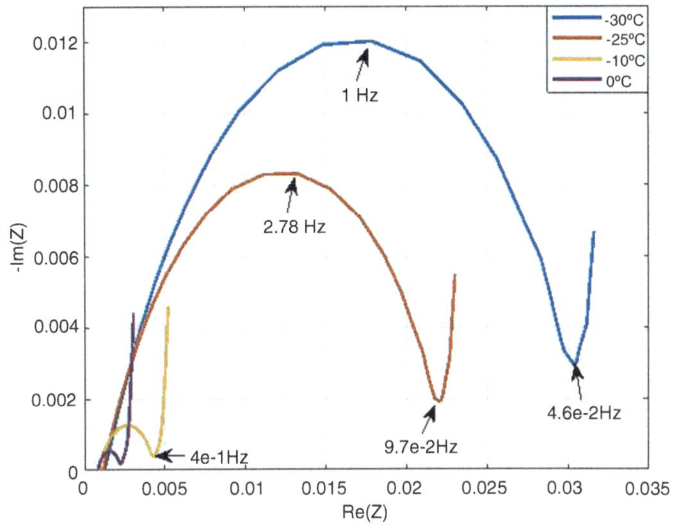

Fig. 4 Nyquist plot at low temperatures for V=3.8V

3.2.2 At High Temperatures

Temperature will be considered as high when it is positive. In the Nyquist plots at high temperatures represented in Fig. 5, it can be noticed that the semi-circles present at low temperatures have disappeared, and they have been replaced by a straight line. In [14], authors explain that the kinetics of the negative graphite electrode tend to be faster than the kinetics of the positive electrode at high temperatures which could explain the form of the Nyquist plot, alike a supercapacitor's plot. The diffusion part remains the same.

3.3 Capacitance vs Voltage

The plot of the capacitance in function of the voltage in Fig. 6 has been calculated from the EIS measurements, where the considered capacitance is taken at the lowest frequency of the Nyquist, i.e., at 10 mHz. In the literature, some papers [15] have mentioned the particularity of this plot for a lithium-ion capacitor, since it shows a minimum of the capacitance at 3 V at positive temperatures. When going towards low temperatures, this particularity begins to disappear and the capacitance tends to vary almost linearly to the voltage above $-10\,^{\circ}$C, resembling to a supercapacitor's C=f(V) plot. This shows that the participation of the negative electrode in the total capacitance of the LIC can be neglected at low temperatures, and the behavior of the total cell becomes almost identical to a supercapacitor. Knowing that the capacitance of a Li-ion battery decreases sharply with negative temperatures, the result from the plot seems adequate.

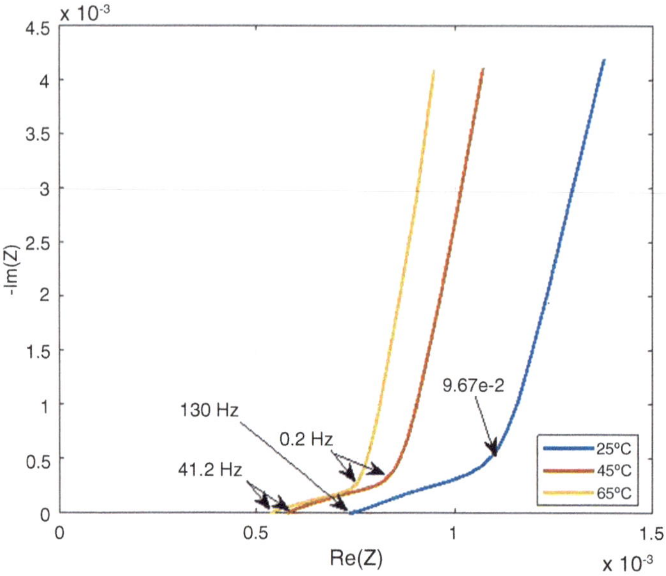

Fig. 5 Nyquist plot at high temperatures for V = 3.8 V

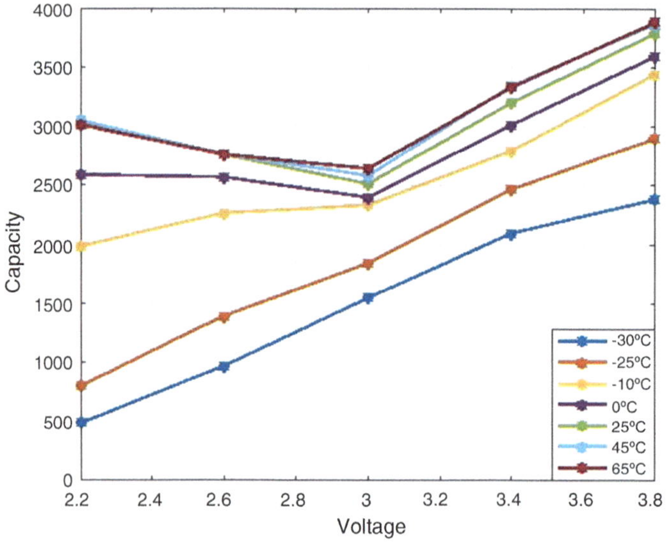

Fig. 6 Capacitance vs voltage for tested temperatures

3.4 *Modeling and Test*

3.4.1 Model

The elaborated model is mainly based on the electrochemical impedance spectroscopy (EIS) measurements detailed in a previous section. From the Nyquist plots obtained, the internal phenomena have been determined based on the plots of batteries and supercapacitors in the literature.

This model, represented in Fig. 7, is composed of:

- a resistance R_0, at the intersection of the plot with the real axis, when the imaginary part is equal to zero. It represents the ohmic resistance of the cell.
- a constant phase element CPE_1 (Q_1 and n_1) in parallel of a resistance R_1, to model the interfacial phenomenon happening in the cell at medium frequencies. It represents the charge transfer.
- a second constant phase element CPE_2 (Q_2 and n_2) representing the diffusion process happening at low frequencies.

The inductive part at high frequencies is not represented, since it is independent of the temperature and the voltage.

The analytic function of the model is equal to the total impedance Z_{tot} of the cell:

$$Z_{tot} = R_0 + \frac{R_1}{1 + R_1 Q_1 (jw)^{n_1}} + \frac{1}{Q_2 (jw)^{n_2}} \quad (1)$$

To optimize the function above, the ZfitGUI function has been used. It is an open source Matlab function specially designed for Nyquist plots. The function permits to optimize the real and the imaginary part at the same time to determine the best values for the circuit's components, so that the optimized plot fits the best the measured plot with the smallest residue.

Simulations performed as mentioned show good results and fit accurately the experimental data for almost all temperatures. Only for −10 and 0 °C, the residual error is more important at the low frequency part. The results are illustrated in Fig. 8, for both temperatures −25 and 45 °C.

Indeed, even if Nyquist plots at low and high temperatures are quite different, all processes happening inside the cell remain the same, but their contribution is more or less significant with temperature variations. In the Table 2, values of all the

Fig. 7 Equivalent model of LIC in frequency domain

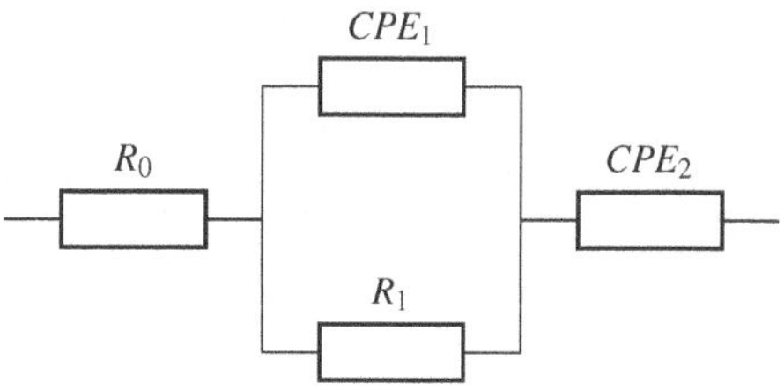

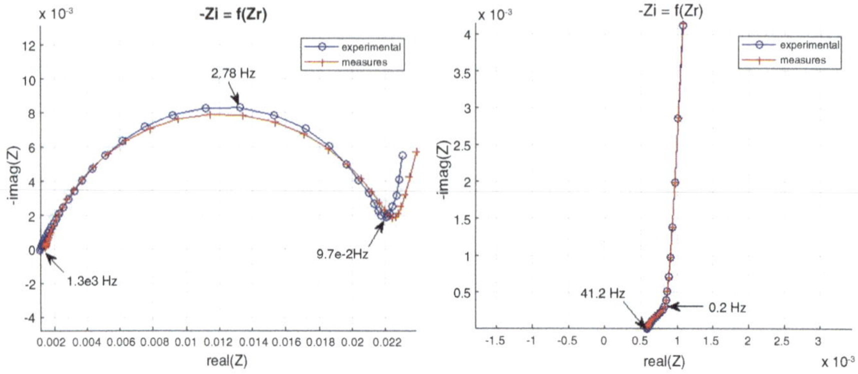

Fig. 8 Experimental and optimized plots at −25 and 45 °C

Table 2 Values of the model parameters obtained by simulation

	−25 °C	−10 °C	0 °C	25 °C	45 °C	65 °C
R0	1.41e–3	9.52e–4	8.53e–4	7.43e–4	5.77e–4	5.62e–4
Q1	3.86	5.19	5.33	1.22e3	1.12e3	1.29e3
n1	0.81	0.83	0.86	0.73	0.83	0.88
R1	2.12e–2	3.25e–3	1.29e–3	3.67e–4	2.88e–4	2.06e–4
Q2	1.87e3	1.59e3	1.58e3	3.30e3	3.51e3	3.54e3
n2	0.85	0.67	0.62	0.95	0.96	0.97

variables have been listed for all temperatures. As it can be seen, for Q1 and Q2 which represent mostly the charge transfer and the diffusion, their values decrease with the decrease of temperature, which confirm the fact that these processes go slower at low temperatures. In the contrary, resistances' values tend to be higher at low temperature, confirming results from EIS measurements.

4 Conclusions

Our results support that the lithium-ion capacitor presents interesting characteristics. For instance, this storage component has inherited some functions from the battery and the supercapacitor, while its hybridization generates new features, as explained in the above. In fact, the difference between all forms of the Nyquist plot at both high and low temperature is a very interesting feature, proving that the behavior of the cell evolves a lot with temperatures. Moreover, the EIS measurements show that the plot of the capacitance vs the voltage of the cell presents a new aspect with a minimum at 3 V that disappears at low temperatures. To our knowledge, this study is first to introduce this result. Experiments at both low and high temperatures show interesting different results that need to be deeply investigated in further publications, since they are very promising for future applications of the LIC.

References

1. G. Mandic, A. Nasiri, E. Ghotbi, E. Muljadi, Lithium-Ion capacitor energy storage integrated with variable speed wind turbines for power smoothing. IEEE J. Emerg. Select. Topics Power Electron. **1**(4), 287–295 (2013)
2. F. Ciccarelli, A. Del Pizzo, D. Iannuzzi, Improvement of energy efficiency in light railway vehicles based on power management control of wayside lithium-ion capacitor storage. IEEE Trans. Power Electron. **29**(1), 275–286 (2014)
3. S. Barcellona, F. Ciccarelli, D. Iannuzzi, L. Piegari, Modeling and parameter identification of lithium-ion capacitor modules. IEEE Trans. Sustainable Energy. **5**(3), 785–794 (2014)
4. N. El Ghossein, A. Sari, P. Venet, A lithium-ion capacitor electrical model considering pores size dispersion, in *2018 IEEE International Conference on Industrial Technology (ICIT)* (2018)
5. Y. Firouz, N. Omar, J-M. Timmermans, P. Van den Bossche, J. Van Mierlo, Lithium-ion capacitor - characterization and development of new electrical model. Elsevier Energy **83**, 597–613 (2015)
6. Y. Firouz, N. Omar, P. Van den Bossche, J. Van Mierlo, Electro-thermal modeling of new prismatic lithium-ion capacitors, in *2014 IEEE Vehicle Power and Propulsion Conference (VPPC)* (2014)
7. M. Uno, A. Kukita, Cycle life evaluation based on accelerated aging testing for lithium-ion capacitors as alternative to rechargeable batteries. IEEE Trans. Ind. Electron. **63**(3), 1607–1617 (2016)
8. N. El Ghossein, A. Sari, P. Venet, Accelerated cycle aging tests applied to lithium-ion capacitors, in *2017 IEEE Vehicle Power and Propulsion Conference (VPPC)* (2017)
9. N. El Ghossein, A. Sari, P. Venet, Effects of the hybrid composition of commercial lithium-ion capacitors on their floating aging, in *2018 IEEE Transactions on Power Electronics*(2018)
10. J. Ronsmans, B. Lalande, Combining energy with power: lithium-ion capacitors, in *2015 International Conference on Electrical Systems for Aircraft, Railway, Ship Propulsion and Road Vehicles (ESARS)* (2015)
11. P.H. Smith, T.N. Tan, T.L. Jiang, J. Chung, Lithium-ion capacitor: electrochemical performance and thermal behavior. Elsevier J. Power Sources **243**, 982–992 (2013)
12. J. Zhang, J. Wang, Z. Shi, Z. Xu, Electrochemical behavior of lithium ion capacitor under low temperature. J. Electroanalyt.l Chem. **817**, 195–201 (2018)
13. T. Richard Jow, S.A. Delp, J.L. Allen, J-P. Jones, M.C. Smart, Factors limiting Li+ charge transfer kinetics in Li-ion batteries. J. Electrochem. Soc. **165**(2), A361–A367 (2018)
14. M.C.Smart, J.F.Whitacre, B.V.Ratnakumar, K.Amine, Electrochemical performance and kinetics of Li1+x(Co1/3Ni1/3Mn1/3)1xO2 cathodes and graphite anodes in low-temperature electrolytes. J. Power Sources **168**(2), 501–508 (2007)
15. N. El Ghossein, A. Sari, P. Venet, Interpretation of the particularities of lithium-ion capacitors and development of a simple circuit model, in *2016 IEEE Vehicle Power and Propulsion Conference (VPPC)* (2016)

Theoretical Comparison Analysis of Six-Phase Interleaved Boost Converter Based on SiC Semiconductor and Inverse Coupled Inductor for Fuel Cell Electric Vehicle

Hanqing Wang, Arnaud Gaillard, and Daniel Hissel

Abstract This paper deals with the comparison analysis of the proposed six-phase interleaved boost converter (IBC) for fuel cell electric vehicle (FCEV) application. Silicon carbide (SiC) semiconductor and inverse coupled inductors have been used to improve converter's performance. According to the comparison analysis, the proposed converter's input current ripple has been reduced, while fuel cell stack's life span can be extended. Then, the total volume and weight of magnetic component has been decreased due to the inverse coupled inductor. Furthermore, benefiting from the SiC semiconductor, high switching frequency has been selected and low switching loss has been obtained. The power loss of the proposed converter has been reduced, while the reliability, efficiency, thermal performance, and power density have been increased.

1 Introduction

During the last decade, conventional internal combustion engine leads to increase of pollution emission with a high ratio. Proton exchange membrane fuel cell (PEMFC), which owns advantages as zero pollutant emission, fast start-up, low operating temperature, long cell, and stack life, is treated as the most interesting green power source [1]. In order to satisfy FCEV powertrain's DC bus voltage requirement, normally, a power conditioning unit is requested to be connected with PEMFC which is a low-voltage high-current power source. Plenty of DC/DC converters, bidirectional DC/DC converters, and DC/AC inverters are published for FCEV [2,

H. Wang (✉) · A. Gaillard
Univ. Bourgogne Franche-Comte, UTBM, FCLAB, FEMTO-ST Institute, CNRS, Belfort, France
e-mail: hanqing.wang@utbm.fr; arnaud.gaillard@utbm.fr

D. Hissel
Univ. Bourgogne Franche-Comte, FCLAB, FEMTO-ST Institute, CNRS, Belfort, France
e-mail: daniel.hissel@univ-fcomte.fr

W. Zamboni, G. Petrone (eds.), *ELECTRIMACS 2019*, Lecture Notes in Electrical Engineering 615, https://doi.org/10.1007/978-3-030-37161-6_46

3]. Although the conventional boost converter (BC) is mostly utilized, the FC life span can be decreased by high-current ripple [4]. Hence, the interleaved boost converter (IBC) has been studied in the previous work by combining with a low source current ripple, modularity, and high reliability [5, 6].

The wideband gap (WBG) technology is more and more attractive in the application of high-frequency, high-power, and high-temperature areas during recent years [7]. Compared with the classical Si semiconductors, ones based on SiC have lower switching losses, better thermal capacities, and higher reverse voltage capabilities.

As the magnetic component influences the converter's weight and volume strongly, the inductor based on coupled structure is attractive to obtain a compact structure and lighter system [8–10]. This is important in practical application to respect the application field of FCEV.

In this study, a six-phase IBC combined with inverse coupled inductors and SiC semiconductors (IC-IBC) has been proposed to obtain higher inductance density, gain good steady-state and transient performances, and reduce weight and volume of converter. Therefore, the proposed converter can obtain high compactness and high-power density.

In Sect. 2, the theoretical comparison analysis of the proposed converter is presented. The influences that are introduced by the inverse coupled inductors based on cascade structure to inductor current, fuel cell stack current, and magnetic component volume are discussed. Then, the efficiency of converter has been analyzed and compared. The paper is concluded in Sect. 3.

2 Theoretical Comparison Analysis of the Proposed Converter

The topology of the proposed six-phase IC-IBC for FCEV application is presented in Fig. 1, and the inductor's structure is illustrated in Fig. 2.

In this section, the conventional boost converter (BC), the six-phase IBC based on uncoupled inductor and SiC semiconductor (UC-IBC), and IC-IBC which is proposed in this study have been compared together. Some primary characteristics of these converters are presented in Table 1.

2.1 Comparison Analysis of Inductor Current

The specifications of converter are presented in Table 2. Clearly, high-power and high-input currents are required. Even though high switching frequency (100 kHz) has been selected for UC-IBC and IC-IBC, low switching losses are able to be obtained benefiting from the use of SiC semiconductor [11].

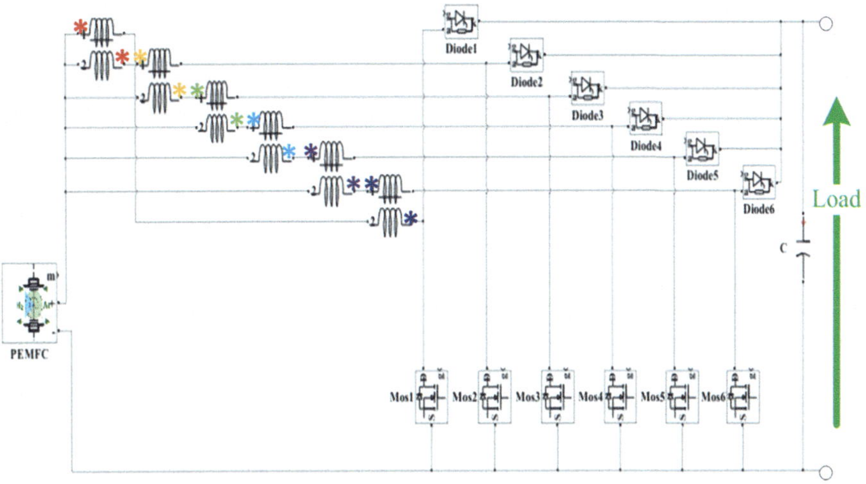

Fig. 1 The proposed six-phase IC-IBC

Fig. 2 Inductor's physical structure (left) and electric structure (right) of IC-IBC

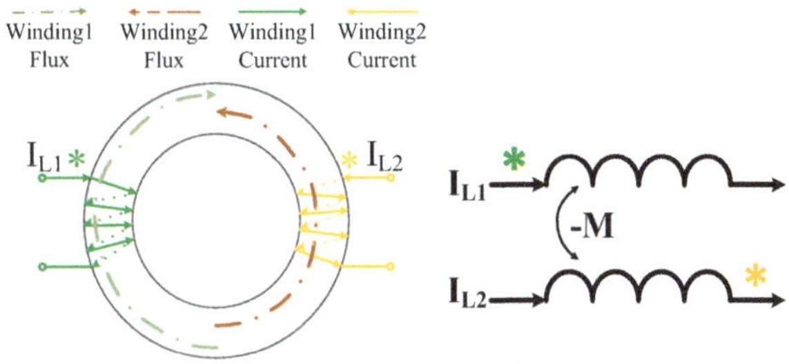

Table 1 DC/DC boost converters in our study

Abbreviation	Topology	Inverse coupled inductor	SiC semi-conductor	Number of phases
BC	Boost	×	×	1
UC-IBC	Interleaved boost	✗	✓	6
IC-IBC	Interleaved boost	✓	✓	6

Table 2 The specifications of DC/DC boost converters in our study

Parameter	Symbol		Value	Unit
Input voltage	V_{in}		70	V
Nominal power	P_N		21	kW
Output voltage	V_{out}		350	V
Inductor current ripple	Δi_L		$20\% \times I_L$	A
Switching frequency	BC	f_{s1}	20	kHz
	UC-IBC	f_{s2}	100	
	IC-IBC	f_{s3}	100	

The conventional pulse-width modulation (PWM) control is widely used for boost converter. For the boost converter based on interleaved structure, no matter coupled or uncoupled, the phase-shifted control strategy is generally employed.

For these three converters, the DC bus voltage (V_{out}) is related to duty cycle and can be obtained by Eq. (1):

$$V_{out} = V_{in} \times \frac{1}{1 - D} \tag{1}$$

To be emphasized, the inductor current ripple is also influenced by duty cycle. Meanwhile, the fuel cell stack current ripple is strongly linked to inductor current ripple for interleaved boost converter. For IC-IBC, each phase is coupled with its two neighbor phases. The average model of six-phase IC-IBC can be obtained as Eq. (2) based on the principled of volt-second balance of inductor under the steady state. Here, phase 1 of IC-IBC is treated as an example to analyze the coupling relationship. Phase 1 is inversely coupled with phase 2 and 6. For the co-owned inductor between phase 1 and 2, L_{12} stands for phase 1's self-inductance and L_{21} stands for phase 2's self-inductance. Once current flows through phase 2, it also acts to phase 1 and leads to the mutual inductance which is denoted by M_{21}. M_{12} is just the opposite. The same case occurs for the co-owned inductor between phase 1 and 6. i_{L1}–i_{L6} stands for the inductor average current of each phase. R_L and R_{on} denote the inductor resistance and MOSFET conduction resistance separately.

The inductor current waveforms of IC-IBC and UC-IBC are illustrated by Fig. 3. Here, the duty cycles of both converters are all set as 0.8, and only one cycle has been presented.

The peak-peak value of inductor current ripple is closely related to inductor average current. No doubt, the inductor current ripple of BC is the largest. The inductor current of BC and UC-IBC is easy to be analyzed. Nevertheless, due to the existence of mutual inductance, the inductor current changing process of IC-

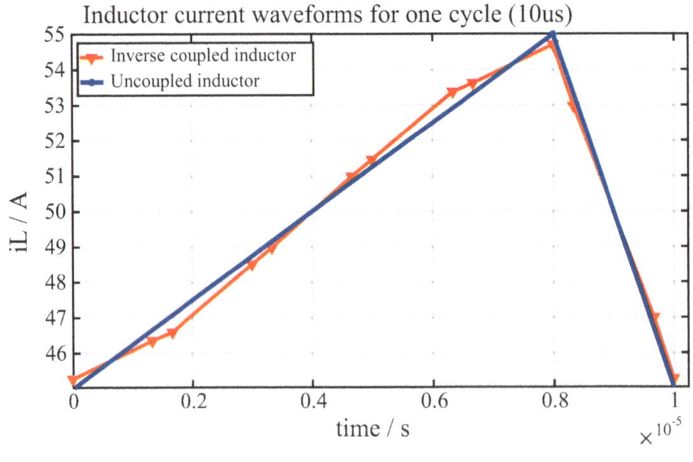

Fig. 3 Inductor current comparison (BC, UC-IBC, IC-IBC)

IBC is much more complicated. According to Fig. 3, there are 12 stages during one operating cycle.

$$
\begin{cases}
\left(L_{12} \cdot \frac{di_{L1}}{dt} - M_{21} \cdot \frac{di_{L2}}{dt}\right) + \left(L_{16} \cdot \frac{di_{L1}}{dt} - M_{61} \cdot \frac{di_{L6}}{dt}\right) \\
\quad = v_{in} - (1 - d_1) \times v_o - i_{L1} \times (R_L + R_{on}) \\
\left(L_{23} \cdot \frac{di_{L2}}{dt} - M_{32} \cdot \frac{di_{L3}}{dt}\right) + \left(L_{21} \cdot \frac{di_{L2}}{dt} - M_{12} \cdot \frac{di_{L1}}{dt}\right) \\
\quad = v_{in} - (1 - d_2) \times v_o - i_{L2} \times (R_L + R_{on}) \\
\left(L_{34} \cdot \frac{di_{L3}}{dt} - M_{43} \cdot \frac{di_{L4}}{dt}\right) + \left(L_{32} \cdot \frac{di_{L3}}{dt} - M_{23} \cdot \frac{di_{L2}}{dt}\right) \\
\quad = v_{in} - (1 - d_3) \times v_o - i_{L3} \times (R_L + R_{on}) \\
\left(L_{45} \cdot \frac{di_{L4}}{dt} - M_{54} \cdot \frac{di_{L5}}{dt}\right) + \left(L_{43} \cdot \frac{di_{L4}}{dt} - M_{34} \cdot \frac{di_{L3}}{dt}\right) \\
\quad = v_{in} - (1 - d_4) \times v_o - i_{L4} \times (R_L + R_{on}) \\
\left(L_{56} \cdot \frac{di_{L5}}{dt} - M_{65} \cdot \frac{di_{L6}}{dt}\right) + \left(L_{54} \cdot \frac{di_{L5}}{dt} - M_{45} \cdot \frac{di_{L4}}{dt}\right) \\
\quad = v_{in} - (1 - d_5) \times v_o - i_{L5} \times (R_L + R_{on}) \\
\left(L_{61} \cdot \frac{di_{L6}}{dt} - M_{16} \cdot \frac{di_{L1}}{dt}\right) + \left(L_{65} \cdot \frac{di_{L6}}{dt} - M_{56} \cdot \frac{di_{L5}}{dt}\right) \\
\quad = v_{in} - (1 - d_6) \times v_o - i_{L6} \times (R_L + R_{on})
\end{cases}
\tag{2}
$$

2.2 Comparison Analysis of Fuel Cell Stack Current

According to the architecture of conventional BC, the ratio between input current ripple and inductor current ripple equals to 1. For UC-IBC, the coupling coefficient can be treated as zero, and the ratio between input current ripple and inductor current ripple is only depending on the duty cycle. Different from UC-IBC, the coupling coefficient has to be taken into consideration during the analysis of IC-IBC's input current ripple.

$$
\frac{\Delta i_{FC}}{\Delta i_L}
= \begin{cases}
\frac{(6D-1)k^3+(1-6D)k^2+(4-24D)k+24D-4}{Dk^3+(3-D)k^2-4Dk+4D-4}, & 0 < D < \frac{1}{6} \\
\frac{(18D^2-9D+1)k^3-(18D^2-9D+1)k^2-4(18D^2-9D+1)k+72D^2-36D+4}{(3D^2-6D+1)k^3-(3D^2-9D)k^2-(12D^2-12D+2)k+12D^2-12D}, & \frac{1}{6} < D < \frac{2}{6} \\
\frac{(18D^2-15D+3)k^3-(18D^2-15D+3)k^2-(72D^2-60D+12)k+72D^2-60D+12}{(3D^2-6D+1)k^3-(3D^2-3D-2)k^2-(12D^2-12D+2)k+12D^2-12D}, & \frac{2}{6} < D < \frac{3}{6} \\
\frac{(18D^2-21D+6)k^3-(18D^2-21D+6)k^2-(72D^2-84D+24)k+72D^2-84D+24}{(3D^2-2)k^3-(3D^2-3D-2)k^2-(12D^2-12D+2)k+12D^2-12D}, & \frac{3}{6} < D < \frac{4}{6} \\
\frac{(18D^2-27D+10)k^3-(18D^2-27D+10)k^2-(72D^2-108D+40)k+72D^2-108D+40}{(3D^2-2)k^3-(3D^2+3D-6)k^2-(12D^2-12D+2)k+12D^2-12D}, & \frac{4}{6} < D < \frac{5}{6} \\
\frac{(6D-5)k^3+(5-6D)k^2+(20-24D)k+24D-20}{(D-1)k^3-(2+D)k^2+(4-4D)k+4D}, & \frac{5}{6} < D < 1
\end{cases}
\tag{3}
$$

The peak value of input current occurs at the maximum value of each phase's current. Similarly, the minimum point of input current appears at the minimum value of each phase's current. Thus, one phase's current can be selected as the reference, and other phases' current values at the chosen phase's maximum and minimum point can be achieved depending Eq. (2). As the input current is the sum of each inductor current, the input current ripple can be obtained. The "ratio" between input current ripple (Δi_{FC}) and inductor ripple (Δi_L) is achieved as in Eq. (3) and illustrated in Fig. 4. k denotes the coupling coefficient and D is the duty cycle. As the coupling coefficient can be both positive and negative, six-phase IBC, which is constructed with direct-coupled inductor with cascade structure (DC-IBC), is also taken into consideration in order to clearly explain the influence of coupling coefficient to the ratio.

Firstly, the ratio of BC is equal to 1 all the time as analyzed previously. Secondly, UC-IBC is the critical condition whose maximum ratio reaches 1 when the duty cycle equals 0 or 1. Thirdly, the ratio of IC-IBC is bigger than UC-IBC in whole range, while the ratio of DC-IBC is just in opposite. In other words, the ratio of DC-IBC is below one no matter how duty cycle changes. Similarly, it is also possible for IC-IBC to obtain a ratio below one when a suitable duty cycle is selected. The last point is that when the duty cycle equals an integral multiple of time delay ($T_s/6$, T_s is the switching period), theoretically, the ratio can reach zero which means the fuel cell stack current only contains DC component.

Depending on the study of the converters' operating processes and the relationship between fuel cell current and single-phase inductor current, the fuel cell stack current waveforms of BC, UC-IBC, and IC-IBC can be obtained as in Fig. 5.

Due to the interleaved structure and the phase-shifted control strategy, the cycle of fuel cell stack current is one-sixth of inductor current. The duty cycle of three converters is 0.8, and the coupling coefficient of IC-IBC is −0.5. The fuel cell current ripples are 0.13%, 0.06%, and 20% for IC-IBC, UC-IBC, and BC, respectively.

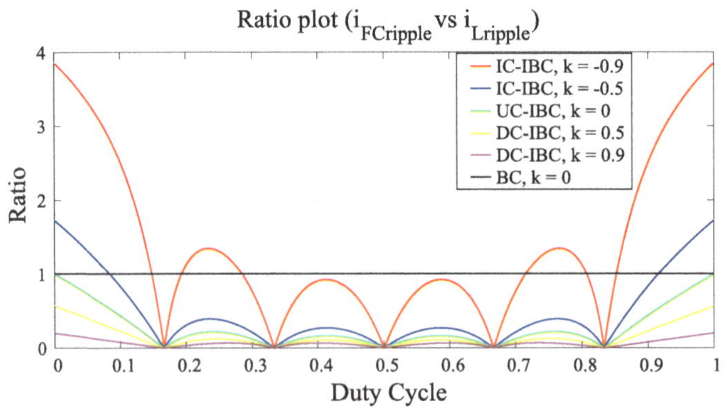

Fig. 4 The analysis of relationship between Δi_{FC} and Δi_L for BC, IC-IBC, DC-IBC, and UC-IBC

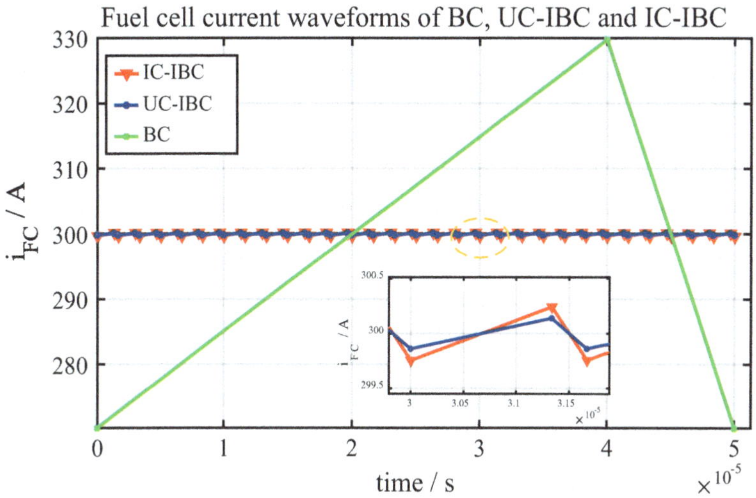

Fig. 5 Fuel cell stack current waveforms of three converters

Hence, IBC helps a lot to decrease FC current ripple, and PEMFC's life span can be extended based on the specific structure.

2.3 *Comparison Analysis of Inductor Volume*

Power inductor is the main magnetic component in power converter. Magnetic component's volume influences strongly power converter's volume. The volume of magnetic component can be estimated by inductor value, current ripple, max flux density, and current's RMS value. Depending on [12], geometricalconstant K_g of magnetic core is a figure of merit which indicates the magnetic core's effective electric size. Equations (4) and (5) are used to determine the minimum core geometrical constant of uncoupled inductor core and coupled inductor core separately.

$$K_{\text{g min\_uncoupled}} = \frac{\rho_{\text{cu}} \times L^2 \times I_{\text{Lmax}}^2}{B_{\text{sat}}^2 \times K_{\text{u}} \times R_{\text{L}}} \tag{4}$$

$$K_{\text{g min\_coupled}} = \frac{\rho_{\text{cu}} \times L_{\text{M}}^2 \times i_{\text{M,sat}}^2 \times (I_{\text{L1}} + I_{\text{L2}})^2}{B_{\text{sat}}^2 \times K_{\text{u}} \times P_{\text{cu\_tot}}} \tag{5}$$

In both equations, the wire effective resistivity ρ_{cu} is 1.724×10^{-6} Ω cm, the saturation flux density B_{sat} is 0.2 T, and the winding fill factor K_{u} is 0.4. The inductances are 47×10^{-6} H and 56×10^{-6} H for the converter BC and UC-IBC separately. For DC-IBC and IC-IBC, the magnetizing inductances are the same that equals 56×10^{-6} H. The maximum inductor currents are 330 A and 55 A of BC and UC-IBC separately.

The DC flux of IC-IBC will be partially or totally canceled which depends on the configuration of core. The magnetizing current of IC-IBC can be treated as the inductor current ripple, while the one of DC-IBC is composed by the inductor current ripple and two times of DC component of inductor current. According to the specification of converter, the maximum magnetizing current of IC-IBC is 5 A and 110 A for DC-IBC.

The relationship between quantity of interleaved phases (N) and magnetic core geometrical constant is illustrated in Fig. 6. Obviously, the core geometrical constant of BC is a constant value because there is only a single phase in this kind of converter. For the other converters, more interleaved branches can lead to smaller

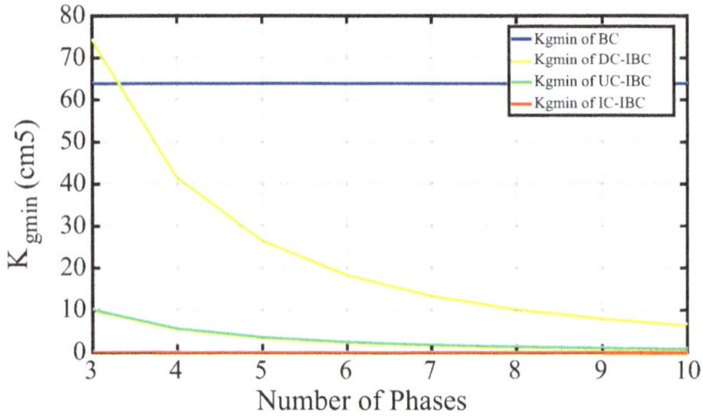

(a)

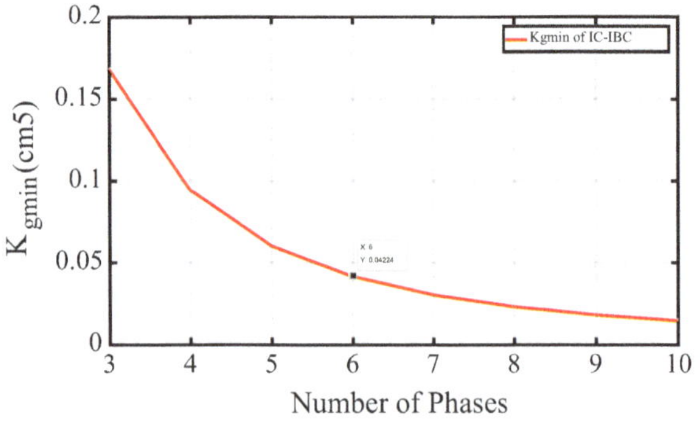

(b)

Fig. 6 Inductor volume comparison (BC, IC-IBC, UC-IBC, and DC-IBC). (**a**) Relationship between K_{gmin} and number of interleaved phases. (**b**) Zoom for K_{gmin} of IC-IBC core geometrical constant. IC-IBC possesses the minimum K_{gmin} in the whole range. For DC-IBC, it is just in opposite. When the number of phase equals 3, DC-IBC even has a bigger K_{gmin} than BC. The K_{gmin} of UC-IBC situates just between these two curves. The zoom of IC-IBC's K_{gmin} has been illustrated in Fig. 6b. Clearly, the minimum core geometrical constant is only 0.04224 cm^5 and has been decreased by almost 98% compared with UC-IBC.

Therefore, from the view of the geometrical sizing of magnetic component, IBC combined with cascade constructed inverse coupled inductors is more attractive than the one based on uncoupled or direct coupled inductor.

2.4 Comparison Analysis of Converter Efficiency

The WBG semiconductor is more and more attractive to satisfy the requirement of high-power, high-frequency, and high-temperature applications recently. Discrete SiC power MOSFETs C2M0025120D of CREE® (Wolfspeed group) have been used to gain volume and weight for UC-IBC and IC-IBC. The SiC Schottky diodes C4D40120D, which are produced by CREE® (Wolfspeed group), have been chosen due to low diode reverse recovery losses. For the converter BC, Si IGBT module FZ400R12KE4 (Infineon®) have been selected to satisfy the high current. Meanwhile, three Si Schottky diode IDW100E60 (Infineon®) have been used in parallel to share the high current.

MOSFET's power losses consist of the conduction losses (P_{on_mos}) as in Eq. (6) and the switching losses (P_{turnon_mos} and $P_{turnoff\_moss}$) as in Eq. (7). At the same time, the gate driver also leads to power losses (P_{gate_driver}) as in Eq. (8).

$$\begin{cases} P_{on\_mos} = I_{mos\_rms}{}^2 \times R_{on\_mos} \\ I_{mos\_rms} = \sqrt{D \times I_{L\_dc}{}^2 + D \times \frac{\Delta i_L{}^2}{12}} \end{cases} \tag{6}$$

$$\begin{cases} P_{turnon\_mos} = f_s \times \frac{E_{test\_on} \times V_{switch}}{V_{sw\_test\_on} \times I_{sw\_test\_on}} \times I_{switch} \\ P_{turnoff\_mos} = f_s \times \frac{E_{test\_off} \times V_{switch}}{V_{sw\_test\_off} \times I_{sw\_test\_off}} \times I_{switch} \end{cases} \tag{7}$$

$$P_{gate\_driver} = Q_{gate\_mos} \times V_{gs} \times f_s \tag{8}$$

Schottky diode's power losses consist of the conduction losses (P_{on_diode}) as in Eq. (9) and the reverse recovery losses (P_{rev_diode}) as in Eq. (10). To be emphasized, Eq. (10) can only estimate Si-based diode's reverse recovery losses. For SiC- and GaN-based Schottky diodes, almost zero reverse recovery can be obtained [13].

$$\begin{cases} P_{\text{on\_diode}} = I_{\text{L\_dc}} \times (1-D) \times V_{\text{F}} + \left(I_{\text{dio\_rms}}\right)^2 \times R_{\text{on\_diode}} \\ I_{\text{dio\_rms}} = \sqrt{(1-D) \times I_{\text{L\_dc}}{}^2 + (1-D) \times \frac{\Delta i_{\text{L}}{}^2}{12}} \end{cases} \tag{9}$$

$$P_{\text{rev\_diode}} = 0.5 \times V_{\text{rev}} \times Q_{\text{rr}} \times \left(I_{\text{dio\_dc}}/I_{\text{rev}}\right) \times f_{\text{s}} \tag{10}$$

Inductor's power losses consist of the copper losses (P_{copper}) as in Eq. (11) and the core losses (P_{core}) as in Eq. (12). The calculation of P_{core} is normally depending on the parameters provided by the manufacturer.

$$P_{\text{copper}} = I_{\text{L\_dc}}^2 \times R_{\text{L\_dc}} + \Delta i_{\text{L\_rms}}^2 \times R_{\text{L\_ac}} \tag{11}$$

$$P_{\text{core}} = K_{\text{fe}} \cdot (\Delta B)^{\beta} \cdot A_{\text{c}} \cdot l_{\text{m}} \tag{12}$$

The total theoretical power losses of these converters can be obtained as in Eq. (13) and are shown in Fig. 7.

$$P_{\text{losses}} = N \times \begin{pmatrix} P_{\text{on\_mos}} + P_{\text{turnon\_mos}} + P_{\text{turnoff\_mos}} + P_{\text{gate\_driver}} \\ + P_{\text{on\_diode}} + P_{\text{rev\_diode}} \\ + P_{\text{copper}} + P_{\text{core}} \end{pmatrix} \tag{13}$$

T_{j} equals 150 °C is the worst case. Under this condition, each converter's upper limit power losses can be achieved. In real application, the power losses will be lower than these values for each topology. According to Fig. 7, for each converter, IC-IBC obtains the lowest power losses, while BC obtains the highest power losses. Compared with BC (2612W), the total power losses have been reduced by 36.5% and 30.1% for IC-IBC (1650W) and UC-IBC (1798W), respectively, when T_{j} equals to 150 °C.

The conduction losses of SiC MOSFET are low for IC-IBC and UC-IBC because the current flow through the semiconductor is lower than the one through Si IGBT

Fig. 7 Total power losses of BC, UC-IBC, and IC-IBC at nominal power ($T_{\text{j}} = 150$ °C)

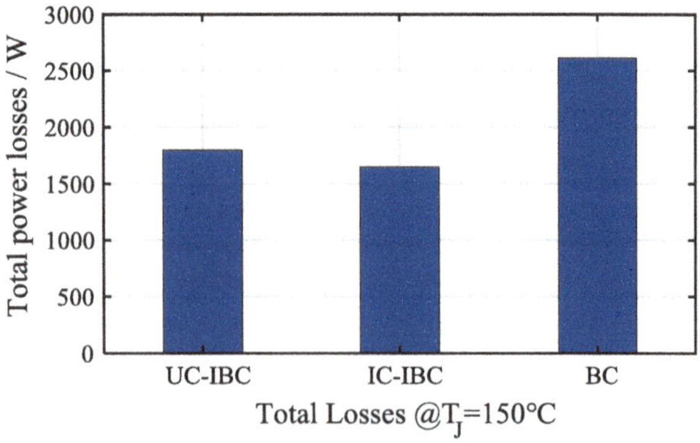

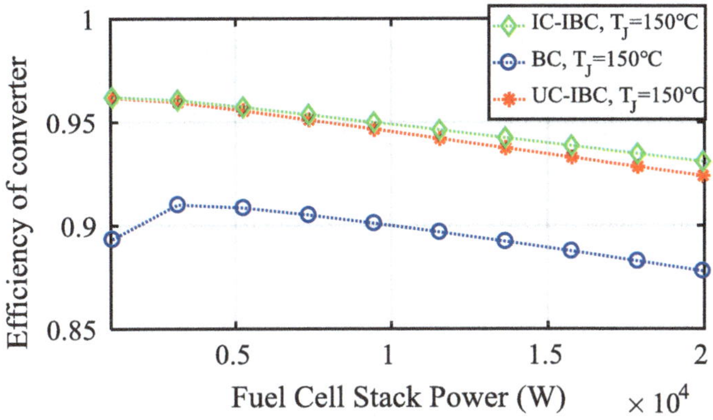

Fig. 8 The efficiency comparison of BC, UC-IBC, and IC-IBC with different semiconductor junction temperatures

for BC. The switching energy required by SiC MOSFET is also smaller than Si IGBT. This is an important advantage of SiC MOSFET compared with Si-based power switches. Benefiting from the use of SiC Schottky diode, the reverse recovery loss of IC-IBC and UC-IBC is negligible. Meanwhile, IC-IBC's core losses have been reduced significantly compared with UC-IBC and BC which benefits from the elimination of DC magnetic flux in the inverse coupled inductor.

The theoretical efficiency comparison of each converter with different semiconductor junction temperatures is presented in Fig. 8. IC-IBC achieves the highest efficiency in the whole range. Its highest efficiency is 96.2%, and the efficiency can be maintained over 93% at the worst case. On the other hand, the maximum efficiency of BC is below 90.88% which is at a low level compared with other two converters. The efficiency of UC-IBC is close to IC-IBC. The difference between these two converters is mainly caused by the magnetic components. A distinct difference can be found at the nominal power of fuel cell stack. Therefore, from the point of view of converter efficiency, the proposed IC-IBC is the best choice for FCEV application. Furthermore, an active air cooling system is possible to be utilized for IC-IBC due to the low-power losses. Oppositely, according to the power losses of BC, the liquid cooling system is required. In other word, the thermal dissipation system of BC will be more complicated and can be restricted by the limited inner space of actual FCEV.

3 Conclusion

In this study, a 21 kW six-phase IBC, which is based on SiC semiconductors and inverse coupled inductors, is proposed for FCEV application. IBC helps reduce the fuel cell stack current ripple and extend its life span. Benefiting from the use of SiC semiconductor, high switching frequency is applied in this study, and the switching loss is kept at a low level. Relatively high efficiency is obtained by the proposed converter. Meanwhile, the volume of magnetic component is decreased. Benefitting from the inverse coupled inductor, total volume and weight of the proposed converter have been decreased dramatically. The power density of the converter has also been increased which is attractive for FCEV application. The idea proposed in this study will be verified on a medium power scale fuel cell system test bench.

References

1. D. Hissel, M.-C. Péra, Diagnostic & health management of fuel cell systems: issues and solutions. Annu. Rev. Contr. **42**, 201–211 (2016)
2. A. Kolli, A. Gaillard, A. De Bernardinis, O. Bethoux, D. Hissel, Z. Khatir, A review on DC/DC converter architectures for power fuel cell applications. Energ. Conver. Manage. **105**, 716–730 (2015)
3. H. Wang, A. Gaillard, D. Hissel, Six-phase soft-switching interleaved boost converter based on SiC semiconductor and coupled inductor for fuel cell vehicles, in *2017 IEEE Vehicle Power and Propulsion Conference (VPPC)*, (IEEE, Piscataway, NJ, 2017), pp. 1–6
4. B. Wahdame, L. Girardot, D. Hissel, F. Harel, X. François, D. Candusso, M.C. Pera, L. Dumercy, Impact of power converter current ripple on the durability of a fuel cell stack, in *IEEE International Symposium on Industrial Electronics, 2008. ISIE 2008*, (IEEE, Piscataway, NJ, 2008), pp. 1495–1500
5. H. Wang, A. Gaillard, D. Hissel, 6-Phase soft-switching interleaved boost converter based on sic semiconductor for fuel cell vehicles, in *Vehicle Power and Propulsion Conference (VPPC)*, (IEEE, Piscataway, NJ, 2016), pp. 1–7
6. P. Thounthong, L. Piegari, S. Pierfederici, B. Davat, Nonlinear intelligent DC grid stabilization for fuel cell vehicle applications with a supercapacitor storage device. Int. J. Elect. Power Energy Syst. **64**, 723–733 (2015)
7. K. Hamada, M. Nagao, M. Ajioka, F. Kawai, SiC—emerging power device technology for next-generation electrically powered environmentally friendly vehicles. IEEE Trans. Elect. Dev. **62**(2), 278–285 (2015)
8. B.C. Barry, J.G. Hayes, M.S. Ryłko, CCM and DCM operation of the interleaved two-phase boost converter with discrete and coupled inductors. IEEE Trans. Power Electron. **30**(12), 6551–6567 (2015)
9. W. Martinez, C. Cortes, M. Yamamoto, J. Imaoka, K. Umetani, Total volume evaluation of high-power density non-isolated DC–DC converters with integrated magnetics for electric vehicles. IET Power Electron. **10**(14), 2010–2020 (2017)
10. Y. Yang, T. Guan, S. Zhang, W. Jiang, W. Huang, More symmetric four-phase inverse coupled inductor for low current ripples & high-efficiency interleaved bidirectional buck/boost converter. IEEE Trans. Power Electron. **33**(3), 1952–1966 (2018)
11. H. Wang, Design and Control of a 6-Phase Interleaved Boost Converter based on SiC Semiconductors with EIS Functionality for Fuel Cell Electric Vehicle (Doctoral dissertation), 2019
12. R.W. Erickson, D. Maksimovic, *Fundamentals of Power Electronics* (Springer Science & Business Media, Berlin, 2007)
13. https://www.wolfspeed.com/media/downloads/109/C4D40120D.pdf

Spatial and Temporal Temperature Homogenization in an Automotive Lithium-Ion Pouch Cell Battery Module

Markus Gepp, Vincent Lorentz, Martin März, Fanny Geffray, Elsa Guyon, and Fabrice Chopard

Abstract A battery system with a thermally optimized module design with regard to boundary conditions in automotive applications is developed. Measures for spatial and temporal temperature homogenization are realized. Highly thermal conductive pyrolytic graphite sheets as heat spreaders replace conventional metallic cooling sheets in a lightweight module design. Efficient space utilization with a novel phase change material for thermal peak-shaving enables benefits in thermal management and lifetime. Heat-conductive adhesives and elastomer-based gap filler sheets further reduce the thermal resistance and the rise in temperature. Measurements showed a maximum temperature difference between the cells of 4.3 K and a maximum thermal resistance between cells and coolant of 0.12 K/W. By integrating thermal solutions, the gravimetric and volumetric overhead was reduced by 25% and 10% compared to the state of the art.

Abbreviations

DCR	Direct current resistance
DOD	Depth of discharge
EIS	Electrochemical impedance spectroscopy
EOL	End of life
NMC	Lithium nickel manganese cobalt oxide
PCM	Phase change material
PGS	Pyrolytic graphite sheets
SEI	Solid electrolyte interface
SOC	State of charge
SOH	State of health

M. Gepp (✉) · V. Lorentz · M. März
Fraunhofer Institute for Integrated Systems and Device Technology, Erlangen, Germany
e-mail: markus.gepp@iisb.fraunhofer.de; vincent.lorentz@iisb.fraunhofer.de

F. Geffray · E. Guyon · F. Chopard
Hutchinson SA Centre de Recherche, Chalette-sur-Loing, France
e-mail: fabrice.chopard@hutchinson.com

W. Zamboni, G. Petrone (eds.), *ELECTRIMACS 2019*, Lecture Notes in Electrical Engineering 615, https://doi.org/10.1007/978-3-030-37161-6_47

1 Introduction

Specifications for battery systems in automotive applications demand cost-effective solutions with high power and energy densities. From a module designer's view, this restricts the volumetric and gravimetric overhead that is available for battery design and thermal management referred to as the cumulated cell's volume and weight.

Regarding the thermal management, automotive batteries face various and unsteady operation conditions, such as varying ambient conditions due to external heat sources or a wide range of mission profiles and fluctuations in ambient temperatures, placing high demands on thermal management. Furthermore, thermal stress and overheating caused by peak loads need to be limited by derating strategies.

Another specification is a sufficient battery service life, which is directly connected to temperature-related cycle and calendrical aging. High temperatures—according to Arrhenius law—as well as low temperatures, lead to accelerated aging mechanisms and degradation of the cell. On module level, temperature differences between parallel- and series-connected cells or local hot spots have adverse impact on the system service life.

In order to meet the requirements and challenges, a battery module with advanced thermal management is designed. Temperature-related aging and performance is considered for a pouch cell module in NMC/graphite chemistry, to derive the potential and the requirements for the thermal management such as temperature homogenization both spatial and temporal. The module design comprises a thermal separation of the cells for temperature homogenization and the use of flexible highly thermal conductive graphite sheets to dissipate heat. A gap filler material is developed and a highly thermal conductive glue is used to reduce contact resistances, while temperature peaks are compensated with a new developed phase change material with an enhanced thermal conductivity.

A variety of thermal management approaches and concepts based on PCM and graphite sheets is available, evaluated on simulation level or on test bench with simplified setups [1–3]. Aim of this investigation is to achieve application-oriented and comparable results, with a detailed constructed module including the integrated thermal management, and to meet the essential requirement of a low design space overhead. An assessment of temperature-related aging with regard to the selected cell type and chemistry enables an evaluation of the thermal management measures.

2 Temperature-Dependent Aging and Performance

2.1 Aging Phenomena

General statements concerning battery aging, caused by side reactions in the battery, are only possible to a limited extent due to a large variety of Li ion chemistries [4]. In [5] several studies on temperature-dependent aging with different chemistries are reviewed. Aging phenomena are identified in capacity decay and power fading, i.e., rise of the internal impedance, resulting from various processes and their interactions [6].

Capacity decay is analyzed in d*V*/dSOC measurements, whereas voltage peaks are representing transitions between voltage plateaus of the negative electrode. These peaks are shifted or vanished over cycling due to aging-related loss of active material or lithium [7].

The rise of impedance is measured with electrochemical impedance spectroscopy (EIS), assuming an equivalent model with capacitor-resistor elements. The electrode polarization resistance is connected to the charge transfer resistance at low frequency while high-frequency resistances are associated with SEI layer growth. The ohmic resistance results from the electrolyte, separator, and current collectors [8].

The aging process can be separated in mechanisms at the cathode and anode during cycling and storage, depending on SOC, DOD, currents, and temperature level [6]. In the following, the cycle aging is described for NMC/graphite cells.

2.2 Aging Mechanisms and Their Temperature Dependency

A good overview of aging mechanisms is given in [6, 9], and [8]. The major source for aging at the negative graphite electrode is considered to be connected to the electrode-electrolyte interface SEI [4, 8]. The anode voltage operates outside the electrochemical stability related to the organic solvent. Electrolyte decomposition, electrode corrosion, and irreversible consumption of lithium result in capacity fade and impedance rise but also lead to the formation of the protective and passivating SEI. This effect is mainly dependent on electrolyte and therefore increased for elevated temperatures, accelerating the reaction kinetics. Due to diffusing particles, the SEI growth with morphology changes and reformation is ongoing after its initial formation. SEI components are converted to resistive inorganic products with lower ionic conductivity, increasing the internal resistance of the battery.

When the battery is cycled, the intercalation of lithium ions and volume changes mechanically stress the electrode and result in particle cracking, gas evolution, and graphite exfoliation. The carbon structure seems to be destroyed by voltage-level transitions at high depth of discharge (DOD), leading to capacity fade [7]. The mechanical stability of the electrode is also worsened due to the continuous decomposition of the binder at high SOC and high temperatures.

Both electrodes are deteriorated by power fade due to current collector corrosion and SEI growth, decreasing accessible contact surface area and porosity.

At low temperatures, the intercalation potential of the carbonaceous anode is close to the potential of lithium metal [10]. Slow lithium kinetics at high currents lead to local and inhomogeneous lithium plating, where the electrolyte is decomposed by metallic lithium, resulting in loss of lithium and electrolyte (capacity and power fade) [11]. The functional capability of the battery is also affected by a false capacity reduction due to an increased internal resistance and polarization effect at low temperatures. This false reduction of usable capacity can still be discharged at low currents [12].

At the positive electrode, the major source for aging is considered to be capacity fade through structural changes such as phase transitions and disordering in the bulk material and SEI layer formation at high temperatures [6, 8]. In addition, at high temperatures chemical decomposition and dissolution of the cathode leads to metal transition to anode resulting in anode SEI layer growth.

2.3 Impact on Cycle Life for Elevated Temperatures

The state of health (SOH) is a predefined measure of aging depending on cycle life and calendar life; its indicators are capacity retention and impedance rise. According to [13], the definition of end of life (EOL) for batteries is application specific. EOL is reached at 80% of the nominal capacity rated by the manufacturer or at a doubled actual internal resistance. While a capacity reduction is generally definable, the frequency-dependent impedance needs to be regarded as application specific. For power applications, the direct current resistance (DCR) is typically used [14]. The DCR, which is the sum of ohmic and polarization resistance, provides a sufficient basis for the following estimation, but it cannot completely describe the dynamic behavior of the battery.

To estimate the impact of temperature-dependent aging and performance, a search on aging tests of commercial large format pouch cells with NMC/graphite chemistry is conducted. In [8] the necessary data for temperature-dependent cycle life, according to the above-described EOL conditions, can be extracted. The DCR doubles at cycle 1600 at room temperature, cycle 1200 at 45 °C, and cycle 300 at 65 °C while the capacity limit of 80% is reached after 2600 cycles at room temperature, 2000 cycles at 45 °C, and 800 cycles at 65 °C. It can be concluded, considering both EOL limits, that cycle life is reduced by about 25% for elevated temperatures and 75% for high temperatures.

On module level, temperature gradients between the cells are regarded. Simply said, not considering any internal or cooling structure, a temperature distribution over the module with a hot spot in the middle is expected. The temperature rise is caused by the heat generation of the cells and is in linear relationship to the current, because of a quadratic dependence of power losses and an exponentially decreasing internal resistance [15, 16]. Consequently, higher currents or internal resistances, e.g., at low ambient temperatures, result in higher temperature gradients. For example, a temperature difference of 6 K in a non-homogenized pouch module after a 3C discharge at 25 °C is given [17]. In [18] and [19] similar values were determined for modules with different cell types and chemistries.

The temperature gradients lead to different aging rates of the single cells depending on their position in the module. For example, for parallel connected cells, a linear dependency between capacity loss and temperature difference was found in [20] and [21]. By interpolation of temperature-related cycle life according to the specified EOL limits and considering a typical temperature difference of 6 K in a module, lifetime of the single cells differs roughly 5% for moderate, 8% for elevated, and 20% for high temperatures.

To quantify the impact on module lifetime, additional module-specific information regarding the temperature distribution and, due to the temperature dependency of the maximum temperature gradient, typical temperature profiles while in operation are necessary. Besides, voltage balancing and derating strategies compensate the impact of diverging SOHs on the system lifetime.

Considering the temperature-dependent aging mechanisms and their impact on service life enables to draw the following conclusions:

- The aging rate at the lower and upper limits of the temperature range is significantly increased.
- Temperature differences in the module arise during cycling depending on electrical and thermal boundary conditions.
- Temperature differences lead to diverging states of health of the cells, reducing the system lifetime.
- The impact of temperature differences on state of health is increased for higher temperatures.

These conclusions are addressed by measures of temporal and spatial homogenization with the following approach:

3 Module Concept and Construction

3.1 Description of the Concept and Construction

The module structure is described in Fig. 1, with a section view seen from the top side.

Casted PCM parts are located on both lateral sides of the cells and align them within the vacuum molded plastic frames. Soft cellular silicone foams provide initial pressure and swelling compensation. As heat spreaders, PGS graphite foils are

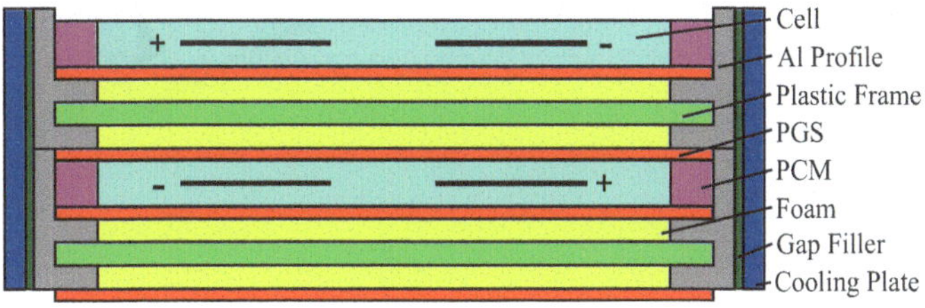

Fig. 1 Concept of the module in cross-section view

directly adhered to the cell surfaces and connect them with PCM and metal profiles by thermal conductive glue. Milled aluminum profiles are designed as mold inlays for the frames and provide mechanical stability and a thermal connection to the cooling plates. Commercial liquid cooling plates are combined with conformable gap filler sheets to compensate geometrical production-related tolerances and thus eliminate air gaps for a reduced thermal resistance. All cells are connected in series by ultrasonic welding.

As previously described, the design of the prototype module is restricted to production processes, which are economical for low quantities. For example, vacuum casting instead of injection molding demands higher wall thicknesses. Nevertheless, the focus is on a lightweight construction to reduce gravimetric and volumetric design overhead. Figure 2 shows the assembled prototype.

3.2 Integration of Advanced Thermal Material Solutions

The thermal concept of the battery module was developed in close cooperation with industrial partners, supporting the integration of thermal material solutions. The integrated components are shown in Fig. 3.

Fig. 2 Battery module prototype

Fig. 3 Frame including PGS foil and PCM inlays

3.2.1 Panasonic Pyrolytic Graphite Sheets (PGS)

The synthetic pyrolytic graphite sheets with a thickness of 70 μm have an in-plane thermal conductivity of 1000 W/(m K) and provide flexibility and a low minimal bending radius with high bending cycle stability. The material is used as a heat spreader between cells, frames, and PCM inlays. With its flexibility PGS can be directly adhered to the pouch cell and also compensates relative movements due to cell swelling with relief folds. The high thermal conductivity together with the low thermal contact resistance provides similar properties as metallic cooling sheets.

3.2.2 Hutchinson PCM PCsMart®

The PCsMart® is an elastomer-based smart phase change material, which stores latent heat during its melting and releases it during its solidification. It has a micro-encapsulated structure to maintain its solid state while melting. Two functions are aimed to be fulfilled in the battery system:

First is safety; overheating and thermal runaway have to be avoided. The PCsMart® is in thermal contact with the graphite sheets and cell edges and tabs to absorb the dissipated heat and store as latent heat. It provides a thermal buffer to the upper temperature limit. The velocity of the storage phase is improved by decreasing the thermal resistance of the PCM material with additives. The PCsMart® product has an increased thermal conductivity (1.5 W/(m K)) compared to common pure PCMs.

Its second function is to avoid accelerated aging at higher temperatures and extend lifetime. The variety of PCMs actually presents a wide range of melting temperature ranges whose width and peak depend on its chemical nature and the heat flow rate. Therefore the type of PCM material formula is adapted to this application. Regarding the battery cell NMC chemistry and its operation temperature, the optimum in order to fulfill both functions is a melting range from 35 to 55 °C.

3.2.3 Hutchinson Gap Filler Material

Hutchinson gap filler material is a thermally conductive interface pad. It compensates the geometrical production-related tolerance of the frames because of its mechanical properties. Consequently, it allows reducing the thermal contact resistance due to air gaps and thus optimizing heat transfer. The gap filler sheet with a high through-plane conductivity of 4 W/(m K) leads to a preferential thermal path. It helps to homogeneously distribute heat on the cooling plate and thus improve the efficiency of the latter. With its low density (1 g/cm^3) and its UL 94 V-0 classification, Hutchinson gap filler is suitable in automotive applications.

3.2.4 Polytec PT Thermal Conductive Adhesive

The pasty adhesive has an improved thermal conductivity of 2 W/(m K) due to nonabrasive additives. It provides mechanical flexibility and cures at room temperature. The adhesive has a good adhesion on metal and plastics and is used to thermally connect all components along the thermal path in the battery module, e.g., PGS, PCM inlays, frames, and gap filler.

3.3 Temporal and Spatial Temperature Homogenization

Temporal and spatial temperature homogenization are addressed with the following measures (beginning with temporal temperature homogenization):

- Phase change material (PCM) reduces the overall temperature rise due to its additional thermal mass. In this concept, the integrated PCM mass has one-third of the specific heat (J/K) referred to as specific heat of the cells.
- The PCM is also used for thermal peak-shaving by adjusting the phase change of the material to the upper temperature range of the cells. The enthalpy consumed for the phase change corresponds to the enthalpy for heating up the PCM and cells by 9 K.
- By thermally connecting the PCM in parallel to the heat path, an adverse impact on the thermal resistance and on the temperature rise by PCM is avoided.

Spatial temperature homogenization is achieved by the reduction of the thermal resistances by:

- Development of new conformable and high through-plane conducting gap filler sheets to reduce contact resistance
- Thin and flexible highly thermal conductive PGS for heat spreading connected with a heat-conducting adhesive

Internal constructive asymmetries are avoided by foams, presenting a thermal barrier between the cells. This prevents the formation of a spatial temperature gradient and different temperature-induced aging rates.

4 Experimental Setup and Test Procedure

The experimental setup is described in Figs. 4 and 5. The battery module within a temperature chamber is cooled by a thermostat. Inlet and outlet temperatures as well as the flow rate of the coolant are measured. A source and sink provide the electrical load cycle; voltage and temperature at the hot spot of all cells are logged.

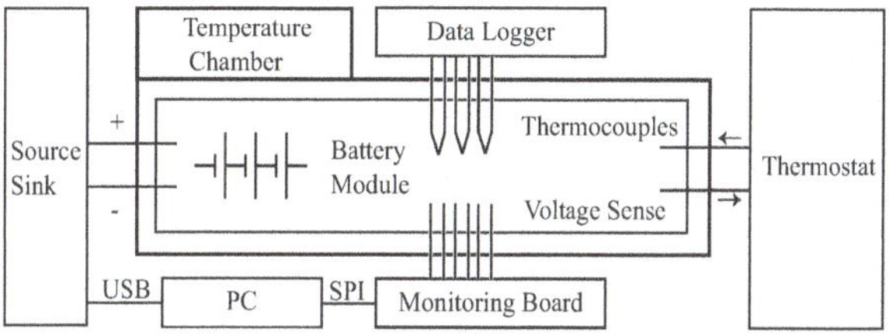

Fig. 4 Schematic test setup

Fig. 5 Experimental setup

5 Results and Discussion

The prototype is assessed with regard to three criteria. Spatial homogenization is shown by measuring the maximum temperature difference between the cells at a test cycle. Temporal homogenization is evaluated by the maximum thermal resistance that determines the rise in temperature referred to as the dissipated heat. Third, the overhead in weight and volume is specified, making the mechanical concept comparable.

5.1 Spatial Temperature Homogenization

As test cycle, 50 A (2C) charge current with 10 A cutoff at 48 V and 130 A (5C) discharge current with cutoff at 35 V are applied (Fig. 6). Initial temperature, ambient temperature, and coolant temperature are set to 20 °C. The coolant is a water-glycol mixture with a flow rate of 3.7 L/min.

The temperatures of the single cells are measured and the arithmetic average is plotted in Fig. 7. A stationary state with a maximum temperature of 45 °C is reached after the second discharge cycle at 45 min. The rise in temperature in stationary state is 12 °C.

In Fig. 8, the maximum temperature difference between the hottest and coldest cell is plotted with a peak of 4.3 °C. The distribution of temperature is described as having a hot spot at the middle cells while the outer cells are colder. Due to warming of the fluid, the distribution is slightly shifted between coolant inlet and

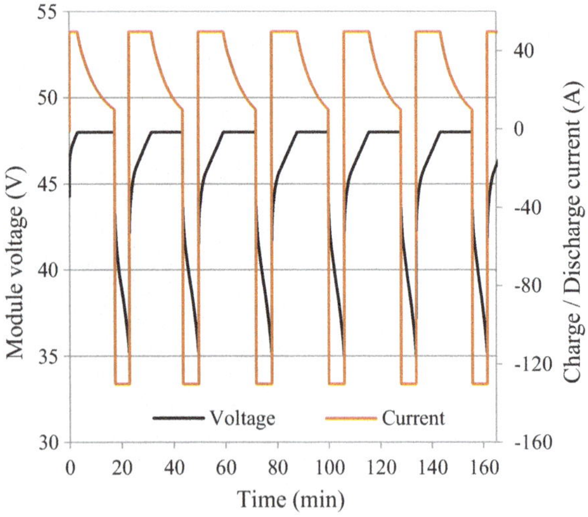

Fig. 6 Electrical load cycle

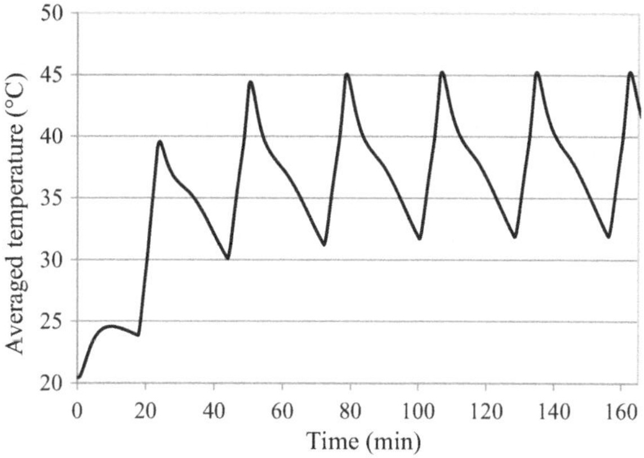

Fig. 7 Averaged temperature of the cells in module

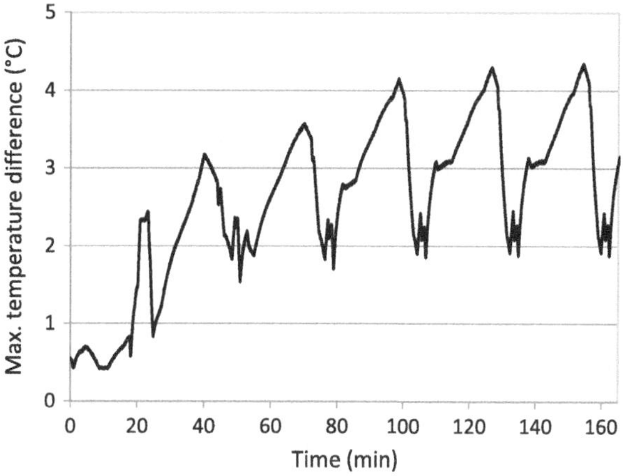

Fig. 8 Maximum temperature difference between hottest and coldest cell in module

outlet of the cold plates. A measurement with similar conditions is conducted in [17] with a maximum temperature difference of 6 °C.

5.2 *Maximum Thermal Resistance*

The liquid-cooled module within the insulated temperature chamber is continuously cycled with 2C/2C charge/discharge current rates to apply roughly constant heat dissipation until stationary state is reached. The temperatures of the cells and the cold plates are measured and averaged, resulting in a maximum temperature difference of 8 °C in stationary state (Fig. 9). The extracted heat of 69 W is calculated by the coolant flow rate, its temperature rise, and the specific heat capacity of the water-glycol mixture. A reduced flow rate leads to higher temperature rise of the coolant

Fig. 9 Average mean temperatures of cells and cold plates

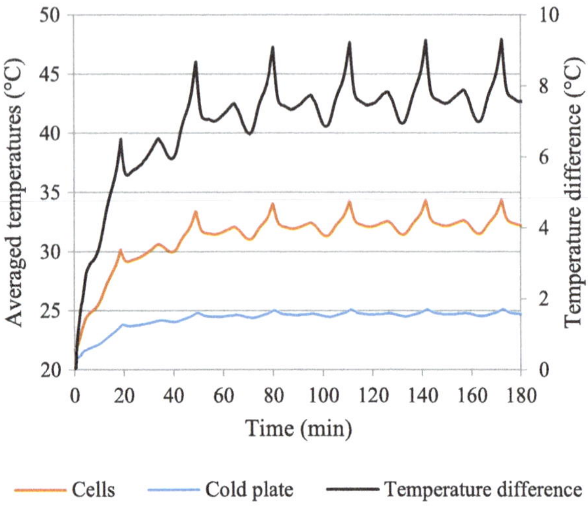

and improves the measurement accuracy. As result, a maximum thermal resistance between the averaged cell temperature and coolant of 0.12 K/W is determined.

5.3 *Overhead Evaluation and Specifications*

The gravimetric and volumetric overhead referred to as the cell's weight and volume is presented in Table 1. As reference, the Opel Ampera battery module with similar pouch cells from LG Chem is chosen. The frames concept using thin and light graphite sheets as heat spreaders reduces the overhead in weight by 25% and the overhead in volume by 10%. This also includes the PCM with an overall sensitive heat of 1.85 kJ/K and a latent heat of 70 kJ/K providing benefits in thermal management.

Finally, it should be remarked that mechanical safety measures besides non-flammability, e.g., measures against mechanical deformation in case of incidents, would require additional overhead, which was not the focus of this study.

6 Conclusion

A battery module is designed, assembled, and evaluated in order to demonstrate spatial and temporal temperature homogenization. Conventional aluminum cooling sheets are replaced by thin and light graphite foils. Novel elastomer-based gap filler pads as well as casted compressible elastomer-based PCM inlays are developed and integrated in the system. The thermal conductivity of the materials is enhanced by

Table 1 Mechanical, electrical, and thermal specifications

Specifications	Characteristics
Max. dimensions L × W × H (mm^3)	174.3 × 185.5 × 243.7
Module weight (kg)	11.15
Module volume (L)	7.88
Gravimetric overhead	40% (25% less than state of the art[a])
Volumetric overhead	65% (10% less than state of the art[a])
Configuration	12s1p
Nominal capacity (Ah)	25.9
Nominal voltage (V)	44.4
Max continuous current rate	6C
Temperature range (°C)	−30 ... +45
Conductivity PGS (W/(m K))	1000
Conductivity gap filler (W/(m K))	4 (through-plane)
Conductivity glue (W/(m K))	2 ± 0.2
Conductivity PCM (W/(m K))	1.5 (ASTM D5930-9)
Sensitive heat of incl. PCM (kJ/K)	1.85
Sensitive heat of incl. cells (kJ/K)	5.65
Latent heat of incl. PCM (kJ)	70
Melting range PCM (°C)	40 (35 ... 55)

[a]Compared to Opel Ampera battery module, LG Chem pouch cells, liquid cooled, market launch 2012

additives and conductive glue. As result, the measurements showed a maximum temperature difference between the cells of 4.3 °C in a continuous cycle with maximum discharge current, a maximum thermal resistance of 0.12 K/W on module level, and a reduced gravimetric (−25%) and volumetric overhead (−10%) compared to the state of the art.

Acknowledgments The research leading to these results has received funding from the European Union's Seventh Framework Programme for Research, Technological Development and Demonstration under grant agreement no. 608770 ("eDAS").

The research leading to these results has also received funding from European Union's Horizon2020 Programme for Research and Innovation under grant agreement no. 770019 ("GHOST").

The authors acknowledge the Panasonic Device Solution Business Division for the supply of PGS and their technical support and collaboration.

The authors also acknowledge the Polytec PT GmbH for the supply of thermal conductive adhesives and their technical support and collaboration.

References

1. P. Goli, S. Legedza, A. Dhar, R. Salgado, J. Renteria, A.A. Balandin, Graphene-enhanced hybrid phase change materials for thermal management of Li-ion batteries. J. Power Sources **248**, 37–43 (2014)
2. N. Javani, I. Dincer, G.F. Naterer, G.L. Rohrauer, Modeling of passive thermal management for electric vehicle battery packs with PCM between cells. Appl. Therm. Eng. **73**, 307–316 (2014)
3. W.Q. Li, Z.G. Qu, Y.L. He, Y.B. Tao, Experimental study of a passive thermal management system for high-powered lithium ion batteries using porous metal foam saturated with phase change materials. J. Power Sources **255**, 9–15 (2014)
4. M. Broussely, P. Biensan, F. Bonhomme, P. Blanchard, S. Herreyre, K. Nechev, R.J. Staniewicz, Main aging mechanisms in Li ion batteries. J. Power Sources **146**, 90–96 (2005)
5. T.M. Bandhauer, S. Garimella, T.F. Fuller, A critical review of thermal issues in lithium-ion batteries. J. Electrochem. Soc. **158**, R1–R25 (2011)
6. J. Vetter, P. Novák, M.R. Wagner, C. Veit, K.-C. Möller, J.O. Besenhard, M. Winter, M. Wohlfahrt-Mehrens, C. Vogler, A. Hammouche, Ageing mechanisms in lithium-ion batteries. J. Power Sources **147**, 269–281 (2005)
7. M. Ecker, N. Nieto, S. Käbitz, J. Schmalstieg, H. Blanke, A. Warnecke, D.U. Sauer, Calendar and cycle life study of Li(NiMnCo)O2-based 18650 lithium-ion batteries. J. Power Sources **248**, 839–851 (2014)
8. K. Jalkanen, J. Karppinen, L. Skogström, T. Laurial, M. Nisula, K. Vuorilehto, Cycle aging of commercial NMC/graphite pouch cells at different temperatures. J. Appl. Energy **154**, 160–172 (2015)
9. F. Leng, C.M. Tan, M. Pecht, Effect of temperature on the aging rate of li ion battery operating above room temperature. Sci. Rep. **5**, 12967 (2015)
10. J. Jaguemont, L. Boulon, P. Venet, Y. Dubé, A. Sari, Low temperature aging tests for lithium-ion batteries, in *IEEE International Symposium for Industrial Electronics*, (IEEE, Piscataway, NJ, 2015)
11. D. Burow, K. Sergeeva, S. Calles, K. Schorb, A. Börger, C. Roth, P. Heitjans, Inhomogeneous degradation of graphite anodes in automotive lithium batteries under low-temperature pulse cycling conditions. J. Power Sources **307**, 806–814 (2016)
12. X. Gong, C.C. Mi, Temperature-dependent performance of lithium ion batteries in electric vehicles, in *IEEE Applied Power Electronics Conference and Exposition, March*, (IEEE, Piscataway, NJ, 2015)
13. B. Scrosati, J. Garche, W. Tillmetz, *Advances in Battery Technologies for Electric Vehicles*, Woodhead Publishing Series in Energy Number 80 (Elsevier Ltd., Amsterdam, 2015)
14. W. Waag, S. Käbitz, D.U. Sauer, Experimental investigation of the lithium-ion battery impedance characteristic at various conditions and aging states and its influence on the application. J. Appl. Energy **102**, 885–897 (2013)
15. J. Shim, R. Kostecki, T. Richardson, X. Song, K.A. Striebel, Electrochemical analysis for cycle performance and capacity fading of a lithium-ion battery cycled at elevated temperature. J. Power Sources **112**, 222–230 (2002)
16. T. Waldmann, G. Bisle, B.I. Hogg, S. Stumpp, M.A. Danzer, M. Kasper, P. Axmann, M. Wohlfahrt-Mehrens, Influence of cell design on temperatures and temperature gradients in lihtium-ion cells: an in operando study. J. Electrochem. Soc. **162**, A921–A927 (2015)
17. C. Veth, D. Dragicevic, C. Merten, Thermal characterizations of a large-format lithium ion cell focused on high current discharges. J. Power Sources **267**, 760–769 (2014)
18. M. Gepp, R. Filimon, S. Koffel, V.R.H. Lorentz, M. März, Advanced thermal management for temperature homogenization in high-power lithium-ion battery systems based on prismatic cells, in *IEEE International Symposium for Industrial Electronics*, (IEEE, Piscataway, NJ, 2015)

19. S. Paul, C. Diegelmann, H. Kabza, W. Tillmetz, Analysis of ageing inhomogeneities in lithium-ion battery systems. J. Power Sources **239**, 642–650 (2013)
20. N. Yang, X. Zhang, B. Shang, G. Li, Unbalanced discharging an aging due to temperature differences among the cells in a lithium-ion battery pack with parallel combination. J. Power Sources **306**, 733–741 (2016)
21. T. Bruen, J. Marco, Modelling and experimental evaluation of parallel connected lithium ion cells for an electric vehicle battery system. J. Power Sources **310**, 91–101 (2016)

A Two-Level Fuzzy Logic Machine-Based Control Algorithm for Resilient Microgrids in ICT Applications

Giovanni Brunaccini, Marco Ferraro, Davide Aloisio, Giorgio Dispenza, Nico Randazzo, Francesco Sergi, and Vincenzo Antonucci

Abstract Resilient microgrids have been the subject of growing interest from information and communications technology (ICT) service providers to assure service availability (and therefore revenue) even in the case of grid fault due to bad weather conditions, as an internal storage capacity as uninterruptible power supply is used. However, such storage equipment represents an unavoidable cost in terms of initial investment, maintenance, and operational efficiency. In this work, starting from a previous development of a prototype supply system for a landline station, the control algorithm of the storage devices was investigated to optimize the cost/benefit ratio. A fuzzy logic system controller was developed to exploit the revenue opportunities offered by the energy market, converting a landline station into an active system that exchanges power through the grid. Besides this, a fuel cell generator was integrated to achieve further benefits (system resiliency and battery size reduction). The simulation results indicated a well-reactive behavior for energy price, battery state of charge, and grid fault probability variations.

1 Introduction

Nowadays, ICT equipment (e.g., radio base stations, data centers) needs stable and continuous power supply, usually assured by dedicated storage systems. However, such systems represent an additional cost to their owner and operator. From the customer's point of view, the economic losses are caused by service interruption, while from the operator's point of view, the extra costs are connected to contract penalties toward their customers [1].

G. Brunaccini (✉) · M. Ferraro · D. Aloisio · G. Dispenza · N. Randazzo · F. Sergi · V. Antonucci
Consiglio Nazionale delle Ricerche – Istituto di Tecnologie Avanzate per l'Energia "Nicola Giordano" (CNR-ITAE), Messina, Italy
e-mail: giovanni.brunaccini@itae.cnr.it; marco.ferraro@itae.cnr.it; davide.aloisio@itae.cnr.it; giorgio.dispenza@itae.cnr.it; nico.randazzo@itae.cnr.it; francesco.sergi@itae.cnr.it; vincenzo.antonucci@itae.cnr.it

W. Zamboni, G. Petrone (eds.), *ELECTRIMACS 2019*, Lecture Notes in Electrical Engineering 615, https://doi.org/10.1007/978-3-030-37161-6_48

Usually, the main causes of system downtime are ascribable to the weather conditions [2], and even in the last decade, enormous phenomena such as floods and hurricanes caused lot of system downtime events.

In a distributed generation (DG) scenario (as in [3]), microgrids can take an active part on the grid regulation, as users ("prosumers" in this case). Such an optimization process may be accomplished both by the hardware design (e.g., by energy content and power peak value for the storage system, nominal power for the internal generation system) and by the control logic to regulate the energy flows through the grid exploiting the installed resources (generation and storage). In this view, assuring both UPS functions and grid services, the design should be addressed to provide the system with resiliency features. Despite not still defined by international standards, "resiliency" (or "resilience") has operational definitions (for critical infrastructure), as in [4–6], rooting in the concurrent presence of robustness, resourcefulness, and rapid recovery (in case of disruptive events) of the assessed infrastructure. Indeed, since the unexpected occurrences are mostly grid faults, this process would be based on a predictive algorithm approach that, in literature [7–9], leads to the development of not fully deterministic control logic techniques, but is based on the probability of the event occurrence.

Among these techniques (artificial intelligence) to implement predictive algorithm, fuzzy logic (FL) has the capability of mapping blurry concepts in membership sets and, as reported in Tables 1 and 2, appears flexible enough to support the resilience implementation in systems operating in a DG scenario. This feature suggests using FL to exploit weather forecast report data (such as rain probability, expected wind and sky clarity, and so on) to guess if a grid fault has a relevant occurrence probability or not.

Table 1 Literature comparison between FL and other approaches/techniques

Other approaches in literature	Difference with fuzzy logic (FL)
Hardware redundancy: supporting the distribution system by decentralized plants and even combining the two approaches [10]	FL implies absence of cost of new hardware (since fuzzy logic control algorithm can be implemented on very cheap controller). Moreover, hardware redundancy is usually (in literature) represented by additional microgrid (i.e., nodes) to support the electric distribution system
Artificial intelligence (AI) techniques	FL belongs to artificial intelligence approach; a more detailed comparison among these techniques is presented below and in the main text
Demand response (DR) incentive (for power delivery) scheme in an energy market to simplify the load forecast [11]	Incentives to DR (to simplify load forecast) can be included in FL, but their value (per time step) is "modulated" by weather conditions and system state (SoC in our case)

Table 2 Literature comparison between FL and other artificial intelligence techniques

Other AI techniques (references in squared brackets)	Reason to prefer fuzzy logic (FL)
Artificial neural networks (ANNs)—e.g. [12]	ANNs require long training to learn system behavior. Moreover, due to the rarity of the forecasted events, the number of faults in the training data is exiguous even with a lot of input combinations
Multi-agent system (platforms) (MAS)—e.g. [13]	These platforms improve the overall grid resilience, but are not adequate to enhance the single node (as addressed by the present work). FL, operating local optimization (even thanks to the simplified dataset), has quicker response representation. Moreover, they require continuous data exchange between nodes that can be absent during the emergency states (that is the target of the proposed investigation)
System with inductive learning (IL)—e.g. [14]	IL requires sensor network and, therefore, besides the cost increment, can (successfully) only work at run-time. However, it is a good alternative once the design and set-up phase is concluded

In parallel, the exploitation of storage capacity availability (i.e., the expected idle/emergency state, according to the weather prediction) can be optimized by an algorithm based on the data coming from energy market (buying and/or selling price) in the "day-ahead" price databases. FL was used to integrate the price value in the overall control algorithm, as detailed in the following discussion. The developed algorithm was addressed to model the hybridization of a battery with a solid oxide fuel cell (SOFC) system as source of internal generation (whose prototype and load profile were already studied in a previous work [1]) and to optimize the storage capacity operation.

The block diagram of the prototype is reported in Fig. 1 and the load profile in Fig. 2. In this work, the control and optimization by artificial intelligence are presented.

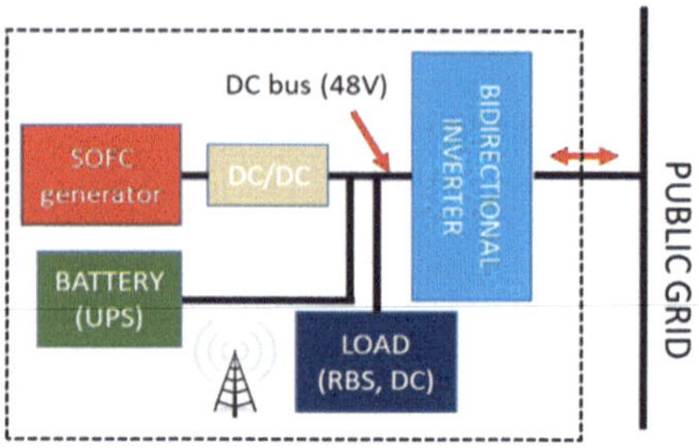

Fig. 1 Block diagram of the hybrid power supply system and connections to load (telecommunication station) and electric grid

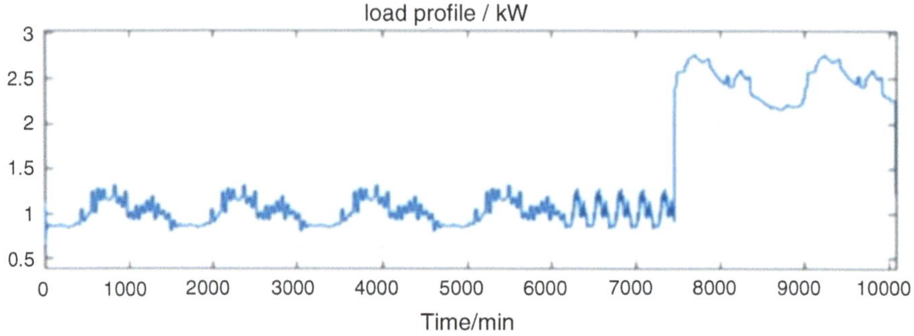

Fig. 2 Telecommunication station load profile. The peak load period occurs during the weekend; this depends on the positioning of the station on the territory. The choice of such unbalanced distribution represents an additional challenge to the hybrid system design

2 Numerical Simulations

2.1 Data Overview

To evaluate the potential benefits of the above-mentioned approach, a simulation tool was developed in MATLAB Simulink environment.

The FL algorithm was implemented by using the following four signals (Tables 3, 4, and 5 show the partition of each overall signal span into overlapping intervals):

For this reason, the FL algorithm has been modified to be fully operative only in a large range, but not 0–100%. Besides this, the thresholds would allow the system (in the perspective of more flexible energy market) to have a reservation of capacity both to store and release power from and to the grid. From the overall logic point of view, the algorithm was based on a two-level FL scheme (Fig. 3), i.e., a first layer that, based on the weather forecast signals, only determines if the grid faults have a relevant probability to occur or they should be negligible, respectively identified as "resilient" or "normal." Such states identify the operation mode of the latter FL layer. The latter FL layer has two FLMs (FL machines) to determine the charge/discharge rate of the battery. The charge or discharge operation is the algebraic result of the difference between actual load and internal power production

Table 3 Membership function assignment (first-layer FLM)

Variable	Description
Wind sp. speed	From a weather station near the installation site and distributed in five intensity intervals
Weather signal	Combination of "sky clarity," "rain probability," and "rain intensity" from weather report
Energy price	Varies according to the "day-ahead" price of the Italian energy market for electrical power production/consumption [15]
State of Charge (SoC)	The actual storage (battery) SoC that is particularly critical once the usually unused capability (UPS function) is utilized to support the grid services

Table 4 Membership function assignment (first-layer FLM)

Variable	Membership intervals (overlapping with different probability)				
Wind sp. (m/s)	Low <3	Moderate 2–7	Average 5–11	Intense 7–18	High >11
Weather signal	No rain		Moderate		Heavy rain
Mode (OUTPUT)		Normal >0.4		Resilient <0.6	

Table 5 Membership function assignment (second-layer FLMs)

Variable	Low range		Medium range	High range	
SoC% (min max)	0 30	25 40	35 60	50 80	70 100
Price/c€/kWh	4–6		5.5–8		7–12
Grid disconnection expectation	0–0.3		0.22–0.6	0.5–1.0	

(fuel cell generation). In each time step, the "net load" (negative when FC generation power is greater than the actual load power) must be satisfied by the algebraic sum of the power from battery and power from grid. The simulation was performed by using 3-min time slots. This output acts like a switch for the latter-level FLM that defines the normal or the resilient fuzzy mode algorithms. The forecast signals were collected to determine the state profile (*normal/resilient*) by a first fuzzy logic machine whose inputs are the weather (as a linear weighted combination of sig_wx and rain probability) and the wind speed; both (input) signals are reported in Fig. 4. Wind speed, weather signal, energy price, and battery SoC were converted into partially overlapping intervals named after logic labels, explained in the following single-layer FLM descriptions:

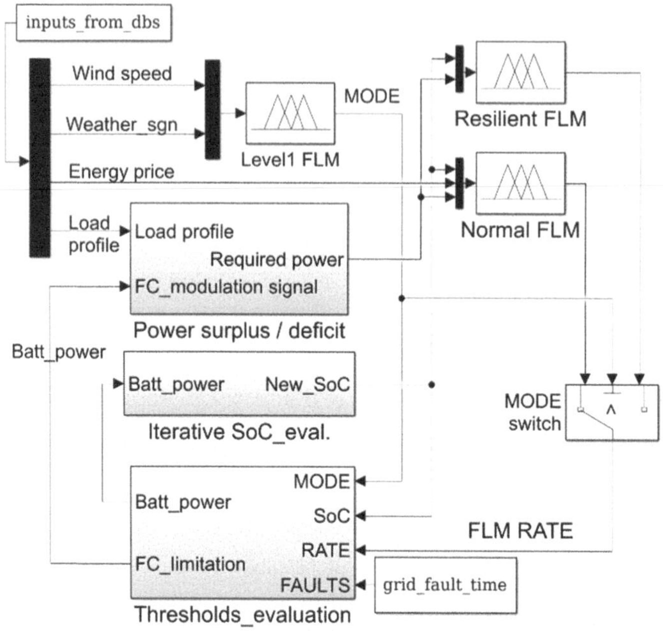

Fig. 3 Implemented two-layer fuzzy logic controller ("normal" and "resilient" modes). During "emergency" mode (i.e., grid outage), the FL is bypassed to, since the only function is the surviving of the load (telecommunication station)

2.2 *Implementation of First-Layer FLM: Raw Signals from Weather Forecast—Collection and Treatment*

The two input signals of the first-level fuzzy machine (weather and wind) were elaborated, through the tabled fuzzy rules (inference engine), to determine a single output and select the operation mode of the controller (i.e., the system state: *normal* or *resilient*). This signal was associated to a "fuzzy" threshold (with a smooth transition in the range of 0.4–0.6, as indicated in Table 4) for the defuzzification process. The resulting (output) mode profile is reported in Fig. 5.

2.3 *Second-Layer FLM: Signals from Energy System: Collection and Treatment*

In the second layer of the fuzzy controller, the overall algorithm distinguishes a *normal* and a *resilient* operation state and determines how the discrepancy (either surplus or deficit) between the actual generation and the load consumption is

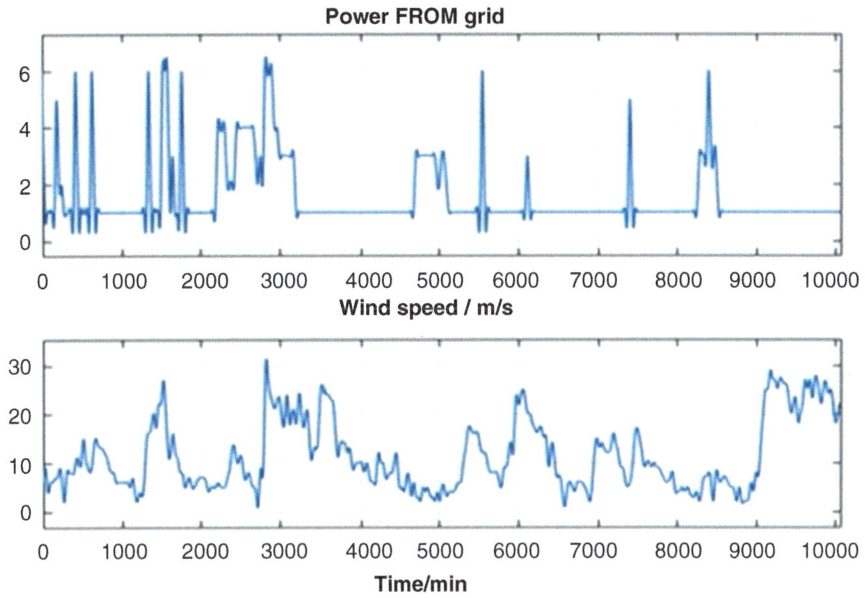

Fig. 4 Weather signals recorded (1 week) as input variables of the first layer of the FLM (the corresponding output is the operational mode "normal/resilient")

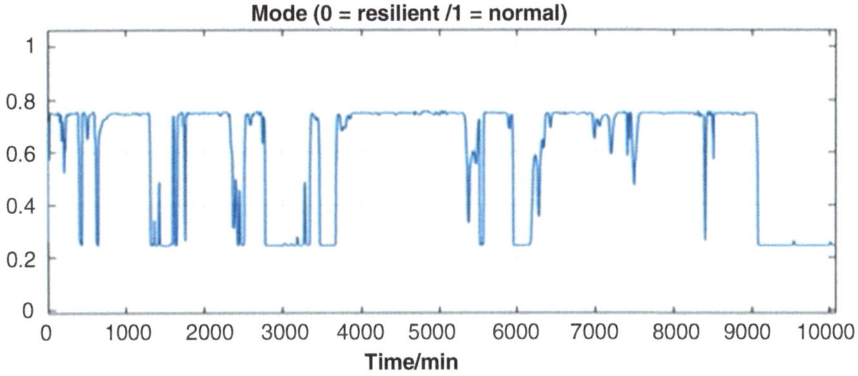

Fig. 5 Operational mode (*normal/resilient*) as output of the first-layer FLM. This signal switch (in each time step, 3 min) between the two corresponding control algorithms

distributed between the battery and the grid. Thus, it requires actual energy price (recorded with hourly base for a week, Fig. 6) and SoC as input variables.

The battery SoC values were divided into five intervals and the selected membership functions are in Table 5. To cover the whole span of each variable, the intervals are overlapping, like in Table 5.

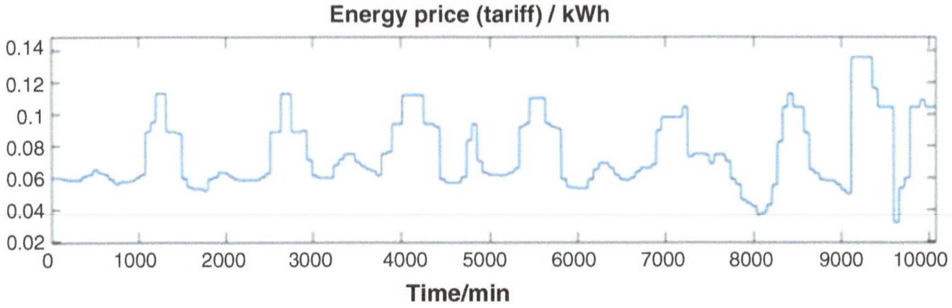

Fig. 6 Weekly profile of the energy price (c€/kWh) in the considered week (one sample per hour)

Table 6 Power surplus/deficit distribution (battery charge/discharge rate)

Interval name	Actual grid current/nominal battery current
Battery ≫ grid	0–0.4
Battery > grid	0.2–0.55
Balanced	0.45–0.67
Battery < grid	0.55–0.9
Battery ≪ grid	0.8–1.0

To define a correspondence between the input and the output variable intervals, this approach requires a table of rules (listed by a priority order) to univocally assign the combination of two or more input variables to a single output interval. Such rules work as Boolean combination of membership functions within "IF … THEN" conditional statements, as in [16]. For example if the wind speed is "high" and weather is "storm," the response will belong with high probability to "resilient"; otherwise it will belong with lower probability to it if they are, respectively, "average" and "moderate." Analogously, in the second layer, if SoC is low, price is low, and there is a surplus, the grid rate tends to charge the battery with a higher current; when the price is "high" ("low"), the grid power request decreases (increases) consequently. On the other hand, in "resilient" mode the FLM aims to bring the SoC to high values. The target is injecting the surplus preferably to increase the battery SoC instead of toward the grid, regardless of the energy price. The output variables result in a variation of the batteries' charging/discharging power, through a "defuzzification" process, which is able to translate the output states in numbers to control the actuators. Except for the extreme values of the SoC (qualitatively described above), the fuzzy layer determines the distribution of the power surplus/deficit between batteries and grid, whose "nominal" interval limits are highlighted in Table 6.

Basically, since the resilient case has relevant expectation of a forthcoming grid failure, the algorithm does not take into account, as visible in Fig. 3, the energy price. On the contrary, the "threshold and emergency controller" evaluates, in the *resilient* mode only, an extra function "fast charge" to accumulate more energy (since the probability of a grid fault is relevant) in the batteries to overcome the grid fault periods. This defuzzification threshold value (0.4) was set to bias the control logic toward a conservative mode and adapted to trigger the "fast charge output." Indeed, the "fast charge function" determines an interruption of the basic fuzzy logic to start an extra charging process. Moreover, if the grid failure occurs, it is necessary to consider an emergency state that is determined by the absence of energy exchange through grid, so that the battery SoC is a direct consequence of the power surplus/deficit.

3 Results

To analyze the behavior of the FLM in different cases, in the simulation, different artificial grid faults were forced. This was done to assess the effect of correct (*resilient* state) or wrong (*normal* state) outage prediction impact on the surviving capability of the system.

3.1 Analysis of Extreme Cases in Normal Mode

However, in case of extreme values of the considered variables (e.g., battery SoC greater than about 85% or lower than about 30%), the algorithm optimizes the storage capacity utilization for grid services by reducing the effect of the variable "energy price" in the decision algorithm. To analyze this criticism of the algorithm, the implemented simulation was performed in the normal mode only. This is visible in the plots in Fig. 7. In A, A′, and A″ sections, due to the very high SoC (those peak values reach 85%), the system injects (for just one time slot per peak) power to the grid even if the price is low to discharge the battery a little and avoid the probability of being unable to offer reserve capacity to the grid.

On the other hand, in section B, despite the high price of energy, the system prevents the battery from deep discharge to preserve both capacity for grid services and battery life, as well as the residual capacity from further displacement that would be detrimental for the necessary (even short) UPS function in case of an abrupt disruptive event.

3.2 Dynamic Behavior of the System to Abrupt Signal Variation

A close-up on a valuable price variation (Fig. 8) allows to verify the effect of an abrupt (decreasing) variation of the hourly energy price. For instance, in the time window from minute 4000 to 4500, the discrepancy between generated and consumed power has some sign variations and the energy prices move quite rapidly

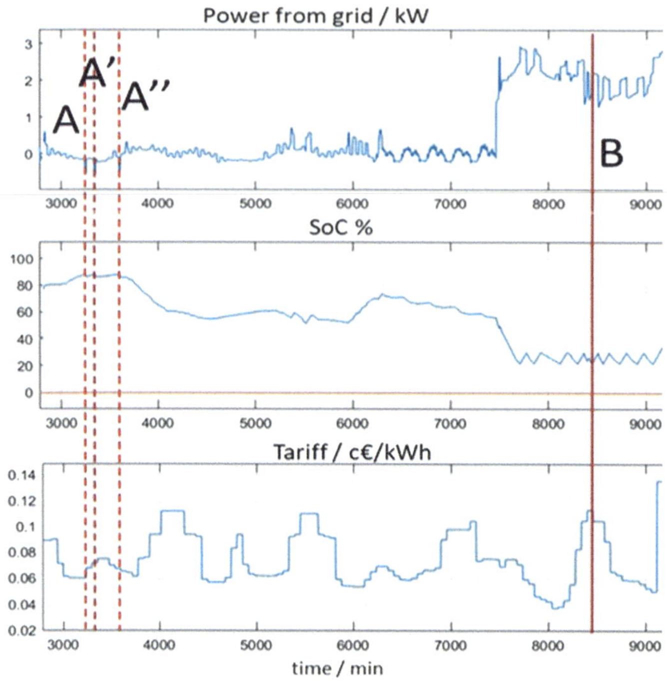

Fig. 7 "Normal" mode—effect of different SoC: in A, A′, and A″ the power is injected into the grid even at low energy price, due to the compensation effect of the high SoC. In B, extremely low SoC makes the system to inject power into grid even at low price

from the highest values to the minimum. The comparison of the plots ranging around 4100–4250 shows that while the fuel cell generation exceeds the load consumption, since the price is high, the power surplus goes entirely to the grid (i.e., SoC is constant). In particular, it is evident that, due to the rapidly diminished energy price, despite a positive discrepancy between generation and load, the power drained from the grid has an increasing trend, so that the low energy price signal is interpreted as the right time to re-charge the battery with higher power instead of injecting power to the grid. Hence, in this case, the battery is re-charged both from grid and SOFC generator. On the contrary, in the time slots in which the load exceeds the FC generation, the most part of the necessary power comes from the battery and a small part (with a decreasing trend) from the grid (this is because the SoC is in the intermediate range).

Moreover, once the price has its most relevant drop (around minute 4260), the grid contribution to load supply increases and the battery contribution decreases. These two results demonstrate that fuzzy logic controller has effective consequences both in long-time observation windows and in fast dynamics (comparable with the examined phenomena).

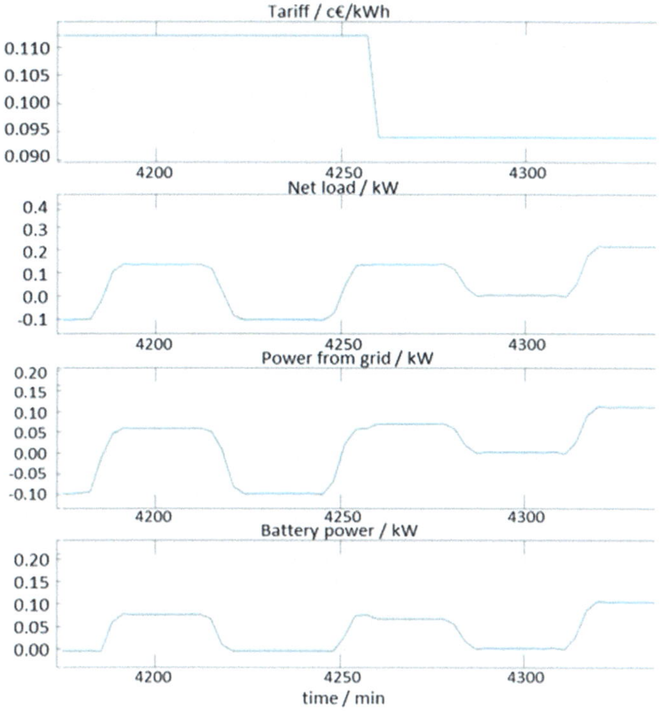

Fig. 8 At high energy price values, the battery injects power to the grid. Once the energy price abruptly drops, even if the load requires more power than the FC generation, the grid is used to contemporarily supply the load to re-charge the battery

3.3 *Analysis in Emergency Mode: Limits, Constraints, and System Corrections*

Since in *emergency* mode the fuzzy control system cannot operate, only solutions at system level were implemented. Due to the very unbalanced profile, the selected load profile can severely stress the storage regulation.

In particular, two opposite case have been simulated:

1. A long grid failure occurs while the SoC has *very high* value, in a working day.
2. A long grid failure occurs while the SoC has *intermediate* value, during the weekend with high power consumption.

In the former case, as visible in Fig. 9, the SoC may reach 100% due to the average excess of the generated power with respect to the average load in the working days. For this reason, in Fig. 9, from section A to section B, since the grid power must be zero, the battery SoC would increase over 100%. Therefore, the algorithm was refined to balance this point with a programmed (in slow steps,

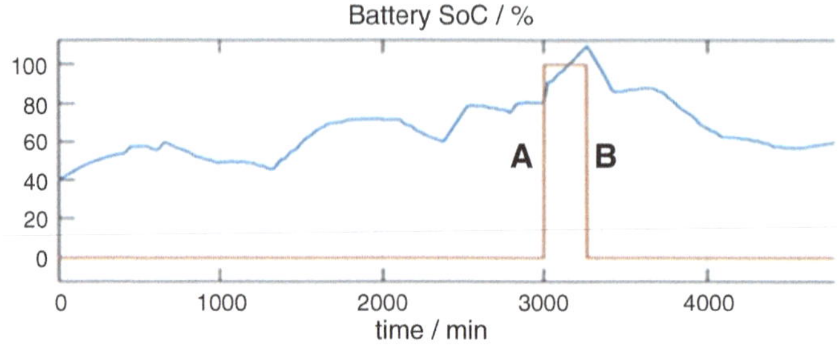

Fig. 9 Emergency mode—SoC drift due to the low load consumption over time without the possibility of injecting generation excess into the grid. The correction is operated by limiting the SOFC production in compliance with the system dynamics

compliant with manufacturer indications) reduction of fuel cell power generation down to half of the nominal power to limit the battery SoC increment. Such correction introduces a loss of efficiency in the energy conversion process, but lets the algorithm operate without risk of being shut-down to avoid malfunctioning.

The carried out analysis highlighted that, to be compliant with the power dynamics indicated by the manufacturer, it is possible to set a SoC threshold value of 87% to start the power generation reduction. However, this event occurred (in Italy) extremely rarely in the last few years.

4 Conclusions

In the presented work, a control strategy based on fuzzy logic was developed and analyzed to exploit the storage resources (UPS/battery) to provide energy services to the grid while satisfying the regular necessity of continuous operation of a landline station for telecommunications. To reach an appropriate balanced goal between revenue from energy market and service surviving, a fuzzy logic-based algorithm was tested against different conditions of the system operation. In particular, during normal operation, the controller outputs were analyzed at the most rapid variations of the energy price, to verify the adaptation of the battery charging/discharging rate regulation even according to the SoC value evolution. Weather, price, consumption, and generation were considered to calculate the energy flows through the battery and the grid. This led even to discharging battery toward load and grid in case of high price and high SoC or, in contrast, draining power to satisfy the load in case of low energy price depending on the battery SoC. On the other hand, the *resilient* working mode, acting as a pre-alert state, did not include the energy cost in the battery rate calculation, since the system downtime is considered (in those time slots of the simulation) by far disadvantageous than the purchase from grid even at high price. Concerning the *emergency* mode, it was studied by forcing grid outage periods at different combinations of weather forecast signal, power load, and energy price levels, to analyze the effect of the weather forecasting and FL, reducing the probability of service down-time.

Acknowledgments The research activity was supported by the Italian Ministry of Economical Development through the "Research for the Development of the National Power Grid Fund" in the frame of the "PT 2015–2017—Electrochemical Systems for the Energy Storage" project.

References

1. G. Brunaccini et al., Modeling of a SOFC-HT battery hybrid system for optimal design of off-grid base transceiver station. Int. J. Hydrogen Energy **42**(46), 27962–27978 (2017)
2. F.H. Jufri et al., Analysis of determinants of the impact and the grid capability to evaluate and improve grid resilience from extreme weather event. Energies **10**(11), 1779 (2017)
3. B. Brooks et al., Incorporating user utility in a smart microgrid with distributed generation and elastic demand, in *2017 IEEE International Conference on Communications Workshops (ICC Workshops)*, (IEEE, Piscataway, NJ, 2017)
4. https://www.dhs.gov/sites/default/files/publications/niac-critical-infrastructure-resilience-final-report-09-08-09-508.pdf
5. ENISA, *ENISA Report. Measurement Frameworks and Metrics for Resilient Networks and Services. Technical Report* (ENISA, Heraklion, 2011)
6. H.H. Willis, K. Loa, *Measuring the Resilience of Energy Distribution Systems* (RAND, Santa Monica, CA, 2015)
7. T. Wang et al., Neural networks for power management optimal strategy in hybrid microgrid. Neural Comput. Applic. **31**, 2635 (2019)
8. L. Hernandez et al., Short-term load forecasting for microgrids based on artificial neural networks. Energies **6**(3), 1385–1408 (2013)
9. P.-C. Chen, M. Kezunovic, Fuzzy logic approach to predictive risk analysis in distribution outage management. IEEE Trans. Smart Grid **7**(6), 2827–2836 (2016)
10. M. Mukherjee et al., Distributed generator sizing for joint optimization of resilience and voltage regulation, in *2018 North American Power Symposium (NAPS)*, (IEEE, Piscataway, NJ, 2018)
11. X. Yang et al., A novel microgrid based resilient Demand Response scheme in smart grid, in *2016 17th IEEE/ACIS International Conference on Software Engineering, Artificial Intelligence, Networking and Parallel/Distributed Computing (SNPD)*, (IEEE, Piscataway, NJ, 2016)
12. S. Zarrabian et al., Real-time smart grids control for preventing cascading failures and blackout using neural networks: experimental approach for N-1-1 contingency. Int. J. Emerg. Elect. Power Syst. **17**(6), 703 (2016)
13. S. Rivera et al., A multi-agent system transient stability platform for resilient self-healing operation of multiple microgrids, in *ISGT 2014*, (IEEE, Piscataway, NJ, 2014)
14. W. Kohn et al., A micro-grid distributed intelligent control and management system. IEEE Trans. Smart Grid **6**(6), 2964–2974 (2015)
15. http://www.mercatoelettrico.org/it/
16. D.G. Schwartz, A system for reasoning with imprecise linguistic information. Int. J. Approx. Reason. **5**, 463–488 (1991)

Luenberger Observer for Lithium Battery State-of-Charge Estimation

Stefano Barsali, Massimo Ceraolo, Jiajing Li, Giovanni Lutzemberger, and Claudio Scarpelli

Abstract One of the main concerns regarding energy storage systems during their normal operation is the possibility to perform an accurate state-of-charge estimation. This cannot be done by simple ampere-hour counting, unless drift correction means are put in place to avoid accumulation of measurement errors over time. In this paper, a state-of-charge estimation algorithm is widely analysed and tested on a nickel manganese cobalt oxide (NMC) lithium cell. The procedure consists of the utilisation of an equivalent electrical network battery model and the implementation of a Luenberger technique for a runtime correction, from the measure of battery's voltage and current. Although application of Luenberger-style estimation is not new in literature for application to batteries, new expressions of battery model parameters and more detailed simulations are shown, to imply much higher estimation accuracy than in the past. After setting the model parameters, different test cycles have been considered, to evaluate the robustness of the proposed technique.

1 Introduction

Energy storage systems based on lithium batteries are more and more considered for stationary and mobile applications. A critical aspect about batteries' design and utilisation is how to precisely estimate their state-of-charge (SOC). The easiest technique is based on evaluating SOC through capacity counting, by simply evaluating the integral of the current profile measured at a terminal. However, if particular adjustments are not introduced, the accuracy of this method is affected by self-discharge phenomena and increase over time of inevitable measuring errors.

Therefore, one of the most usual corrections to avoid this drift is the relation between OCV and SOC, which needs to maintain the lithium battery at zero current

S. Barsali · M. Ceraolo · J. Li · G. Lutzemberger (✉) · C. Scarpelli
DESTEC, Università di Pisa, Pisa, PI, Italy
e-mail: lutzemberger@dsea.unipi.it

W. Zamboni, G. Petrone (eds.), *ELECTRIMACS 2019*, Lecture Notes in Electrical Engineering 615, https://doi.org/10.1007/978-3-030-37161-6_49

for at least 1 h. After that, the battery's voltage measured corresponds to the open circuit voltage (OCV). This is defined as the voltage at the battery terminals when no current flowed for a sufficient time, thus allowing the internal dynamic transients to vanish, and the voltage stabilises. Experience shows that this interval is very long, depending on type of battery and on SOC. Typically, for a good precision, at least 1 h must be waited. The need for a long period in which the battery delivers no current represents an evident inconvenience of this method, mostly in applications characterised by a highly variable current.

This paper analyses and tests an algorithm for state-of-charge (SOC) estimation of lithium batteries, to be used also in the case of highly variable currents. There are several papers dealing with SOC estimation, typically using Kalman filters, in order to reach also measuring error filtering [1–5], or Luenberger-based state estimator [6–9]. This paper falls below this second category. In particular, the Luenberger algorithm has been already used by authors of this paper, for the SOC estimation of supercapacitors [10].

The importance of the topic makes the addition of this study not redundant in comparison with the previous ones. This paper's implementation is based on the exploitation of a testing cycle that the authors have presented in the past [11, 12] and on the accurate choice of battery mathematical models for the purpose, for which the authors have also a broad experience for different types of batteries [13, 14].

To evaluate if the proposed SOC estimation technique efficiently works, it must be mentioned that this requires a proper battery mathematical model. Therefore, this paper preliminarily focuses on the battery model's calibration and determination of its numerical parameters, then on the algorithm setup and finally on testing the quality of the proposed technique on different test cycles, under different initial SOC conditions. In all the cases analysed, the evaluated SOC has been compared to the reference one, obtained by the integration of the current profile, measured through very-high-precision lab instruments.

2 Model Formulation

2.1 Equivalent Circuit Representation

The typical model of a lithium cell that we find in literature is made by an equivalent electric circuit able to properly simulate the cell's voltage output, under a defined current profile input. The common topology shown in Fig. 1 is taken as reference.

Below, the electrical equation used in the circuit is shown:

$$
\begin{aligned}
V &= \mathrm{OCV} - \sum_{i=1}^{n} V_i - I \cdot R_0 \\
\dot{V}_1 &= \frac{I}{C_1} - \frac{V_1}{R_1 C_1} \\
\dot{V}_n &= \frac{I}{C_n} - \frac{V_n}{R_n C_n}
\end{aligned}
\tag{1}
$$

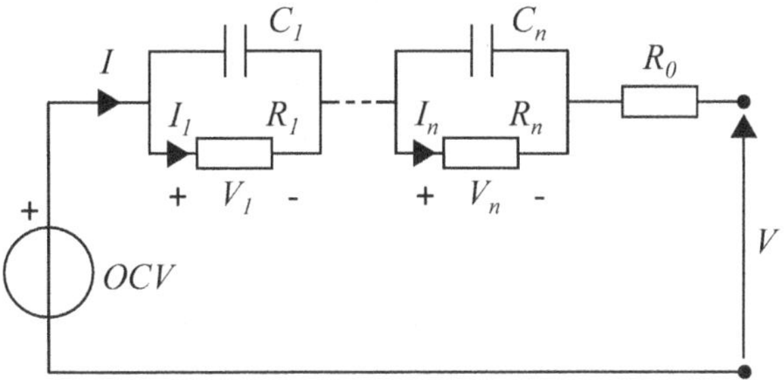

Fig. 1 Equivalent electrical circuit of the lithium cell

It is convenient to select the number n of the parallel R-C blocks as a trade-off between the desired accuracy and the ease to use, as described in [10–12]. To prevent much overcomplexity of the requested algorithm to evaluate the model parameters, n can be assumed as $n = 1$.

The state-of-charge (SOC) is traditionally used for electrochemical batteries, as the indicator of the stored level of charge. It is defined as:

$$\mathrm{SOC} = 1 - \frac{Q_e}{Q_n} \tag{2}$$

where Q_e indicates the extracted charge and C_n the nominal capacity.

2.2 *Evaluation of Parameters*

Experiment with the aim of estimating the parameter values of R_0, R_1 andC_1 in function of SOC has been performed.

Experimental setup is constituted by a charging system driven by a 60 V–250 A Ametek® programmable DC power supply (model SPS60X250-K02D). The discharging system is a 60 V–500 A Zentro-Elektrik® electronic load (model EL6000). Both the systems can be controlled remotely via a GPIB standard interface, through a software developed in LabVIEW®. Regarding the measurement system, the current is measured using a shunt, class 0.5 and full-scale 150 A; the acquisition system is a DAQ device (model NI 9219), with an accuracy of 0.3% of reading. The cell tested is in a climate chamber Binder® MK53, characterised by a temperature range of −40 °C to 180 °C. All the experimental tests have been executed with the climate chamber set at 22 °C. The main rated characteristics of the cell under test [15] are shown in Table 1.

Under the so-called *multiple step test* (MST), the cell was firstly completely charged and then subjected to ten discharging pulses with 1-h rest phase between one step and the consequent one, till the cell was fully discharged. A portion of this test is shown in Fig. 2.

Table 1 NMC lithium cell characteristics

Max voltage (V)	4.15
Nominal voltage (V)	3.7
Min voltage (V)	2.5
Nominal capacity (Ah)	20
Max continuous current (A)	100
Max peak current (A)	200
Mass (kg)	0.43

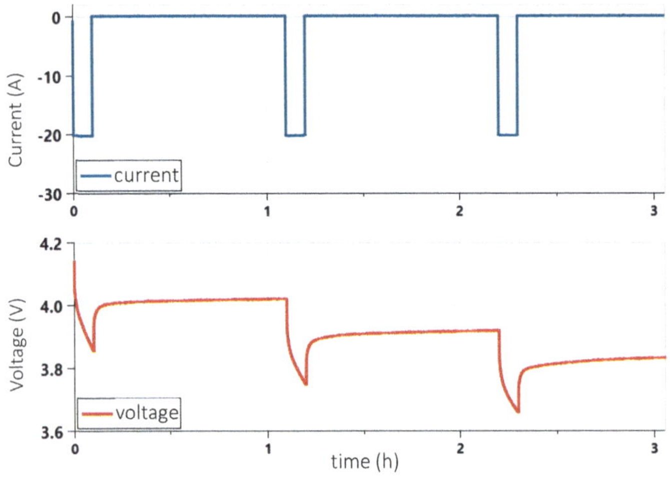

Fig. 2 Portion of MST test: current (negative if discharging) and voltage

Despite the fact that the cell's parameter values are generally functions of both the SOC and cell internal temperature, feasible results can also be obtained neglecting this last dependence, as followed by this paper.

Starting from the MST results, a numerical identification algorithm was applied, to minimise the error between actual and simulated voltage, using the following equation:

$$\varepsilon = \min\left\{\sum_{k=1}^{K}\sqrt{\left(v_{k,\text{actual}} - v_{k,\text{model}}\right)^2}\right\} \tag{3}$$

where k are the individual set of values, each one calculated at time instants t_k. Hence, we can evaluate the values of R_0, R_1 andC_1, at different SOCs. On the other hand, the OCV values have been directly taken from voltage measurements, at the end of each 1-h rest phase. The obtained set of parameters at different SOCs is summarised in Table 2.

The above collected points were directly used inside the model and introduced through look-up tables. Two adjacent values are then linearly interpolated. A

Table 2 Identification of cell model parameters as SOC function

SOC	OCV (V)	R_0(mΩ)	R_1(mΩ)	R_1C_1(s)
1	4.15	4.3	4.7	116
0.9	4.01	4.3	5.1	160
0.8	3.90	4.1	5.1	96
0.7	3.82	4.0	4.3	60
0.6	3.73	4.0	4.3	60
0.5	3.69	4.0	4.9	93
0.4	3.66	4.2	6.0	115
0.3	3.63	4.5	6.4	102
0.2	3.57	4.5	10.0	101
0.1	3.52	5.2	14.6	45
0	3.30	5.2	15.0	45

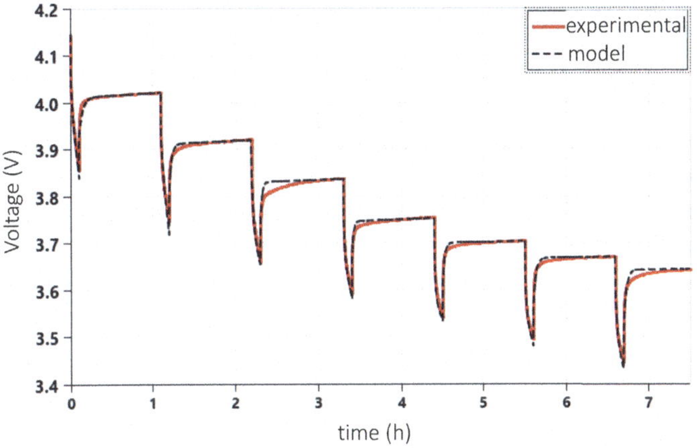

Fig. 3 Comparison between experimental and simulated voltage during seven steps of MST test

detailed discussion on possible approximation of the trends collected in Table 2 is explained in [16].

Figure 3 shows the comparison between experimental and simulated voltage, considering cell's behaviour in response to the full *multiple step test* (MST). As we can observe, the model well matches the experimental voltage in large parts of the test cycle.

3 SOC Estimation

One of the most common techniques used for SOC estimation is the OCV-SOC correlation. It consists of executing a simple capacity through ampere-hour counting, together with the inclusion of a voltage compensation. This correction is necessary to avoid the accumulation of measuring and calculation errors that over time brings large errors to SOC estimation. In fact, ampere-hour counting continuously measures the cell current $i(t)$, integrating it, to calculate the total extracted charge Q_e. In case the measuring process was unaffected by errors, according to Eq. (1), the SOC would be simply calculated through:

$$\mathrm{SOC}(t) = 1 - \frac{\int i(t)\,\mathrm{d}t}{C_n} \tag{4}$$

It is rather noticeable that pure ampere-hour counting cannot give a valid over time SOC measure, since in real applications the current measure can be altered by measuring errors to be integrated over time. In addition, further errors can be caused by integral operations. Therefore, compensation of cumulated errors can be implemented when the cell's current is null for a time adequate to damp transients, e.g. 1 h, and the cell voltage is assumed equal to the OCV.

The concept of Luenberger observer SOC estimation is still based on computing the battery's extracted charge by integrating the current and correcting the results. However, as distinct from the previous method, the Luenberger technique allows the execution of this correction continuously. The general scheme of this approach is shown in Fig. 4. The actual and simulated battery voltages are compared to each other, and the outcoming error triggers a feedback on the computed charge, requiring low computational costs.

The *Li-cell real* block represents the real device that is subjected to the experimental charging/discharging current I_{exp}, whilst V_{exp} is the voltage experimentally measured. The *Li-cell model* block contains the very same circuit shown in Fig. 1: its input is the current I_{mod} provided by the upper feedback correction; hence, it calculates V_{est} (the simulated terminal voltage). The consequent state-of-charge estimation, $\mathrm{SOC}_{\mathrm{est}}$, is lastly calculated according to Eq. (2).

If we consider an initial condition where the simulated voltage V_{est} is equal to the experimental cell voltage V_{exp}, the voltage error e_{v} is zero and so is the current error

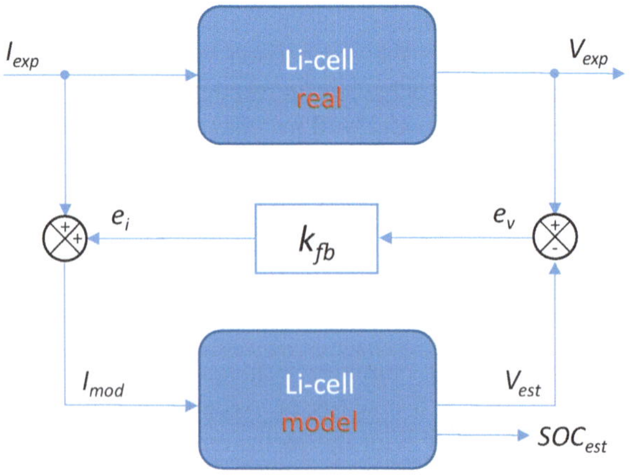

Fig. 4 Luenberger-style SOC estimation technique applied to the lithium cell under study

e_i. In this case, the *Li-cell model* block receives an input current I_{mod} identically equal to I_{exp}.

Due to measuring errors and model uncertainties, some divergences between V_{est} and V_{exp} emerge. This generates a voltage error and also a current error. In this case the model current value is given by the following relation:

$$I_{mod} = I_{exp} + k_{fb}e_v = I_{exp} + k_{fb}\left(V_{exp} - V_{est}\right) \tag{5}$$

Signal e_i tends to correct the errors on V_{est}, keeping well-aligned V_{est} with V_{exp}. If this happens, the corresponding estimation of SOC_{est} well approximates the cell's real SOC.

To have good performance, it is firstly important to have an accurate set of electrical circuit parameters. Then, also to fine-tune the feedback constant k_{fb}, higher values give higher voltage corrections.

Since the battery model has a limited precision, choosing a too high value of k_{fb} would imply changes on the SOC estimated due to SOC differences being not real between battery and model internal states and the model's limited accuracy. A too small value, instead, would recover SOC differences in a longer time. A value expressed in A/V numerically equal to the battery nominal current is in general a good first orientation.

4 Results

The quality of the SOC estimation for the NMC lithium cell considered in this paper was checked on two different tests. The first one is represented by the already seen MST current profile, shown in Figs. 2 and 3, which was applied to the real NMC cell and to the model having one single R-C block, whose parameters have been shown in Table 2. Then, SOC estimation has been performed accordingly to the Luenberger-style SOC estimation technique, depicted in Fig. 4.

As starting point, the feedback constant k_{fb} was not set, leaving it to zero. This corresponds to the situation shown in Fig. 3, with the initial SOC correctly set from experimental measurement of the OCV (in the shown case, exactly at the OCV corresponding to the cell fully charged). Result is shown also in Fig. 5, in the case of k_{fb} null.

When the feedback constant k_{fb} is progressively increased up to 1000 A/V, I_{mod} (e.g. the input of the cell model) is then corrected proportionally to the voltage error (see Fig. 4). As visible from results, voltage profiles tend to be much more flattened, nearly coincident, to the voltage profile, thus introducing a corresponding error on SOC estimation, as visible in the range between 500 and 1000 s, in the bottom part of Fig. 5. This means that in case of high values of k_{fb}, the correction is very sensitive during rapid transients, when the cell model cannot perfectly reproduce the cell behaviour. However, the actual SOC is immediately recovered at the end of the

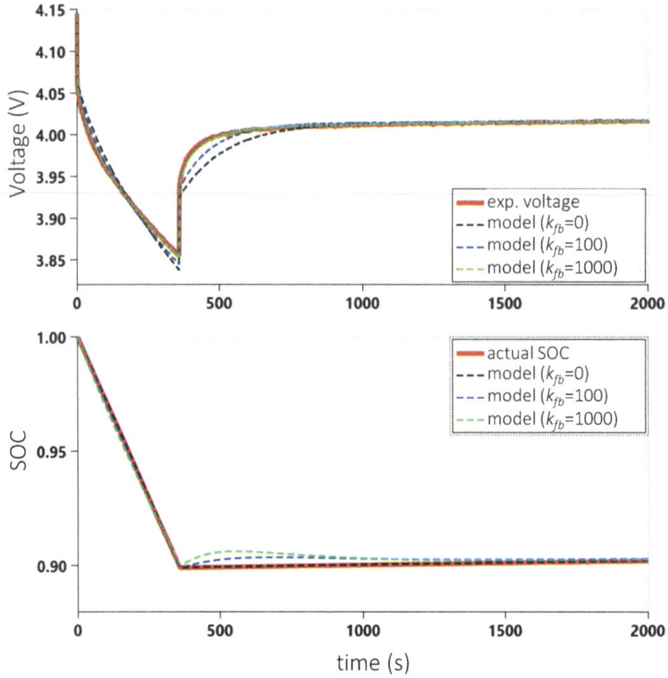

Fig. 5 Simulated model and experimental lithium cell voltage with exact initial SOC and different k_{fb} values, MST cycle

subsequent long-rest phase, as visible in the bottom part of Fig. 5, when the green dashed curve overlaps the red one.

The other considered case is in the presence of an offset error on the initial SOC, as it can result from self-discharge phenomena. If this error is set to 10%, the situation changes as shown in Fig. 6, always depending on the feedback constant k_{fb}. As visible, the error is progressively recovered in increasingly shorter time durations, by acting on its value up to 1000 A/V.

The test was then repeated by changing the initial SOC error offset, in the range of 5–20%. Results of time durations required to reduce the considered offset error within 1%, for different feedback constants k_{fb}, are shown in Table 3. In some of the considered cases, no rapid convergence can exist, i.e. the error offset can be recovered only for a long-time duration, e.g. over 1 day when k_{fb} is posed equal to 1 A/V.

As visible from results, an intermediate k_{fb} value of 100 A/V can be selected. In this way, the time to recover the initial error is lower than 1 h in the most common situations. As said, we must also consider that this offset error is in real applications typically given by the self-discharge of the NMC cell, typically within a few percentages, i.e. about 0.1% per day. According to that, k_{fb} could be further

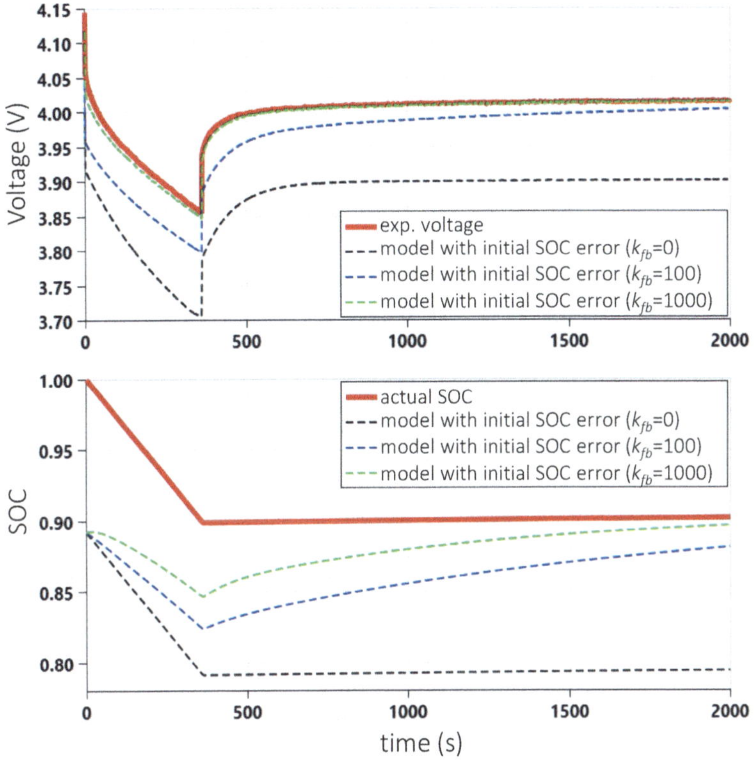

Fig. 6 Simulated model and experimental lithium cell voltage with initial SOC error of 10% and different k_{fb} values, portion of MST cycle

Table 3 Time durations (in seconds) to reduce the SOC offset error within 1%, for different feedback constants k_{fb}, MST cycle

k_{fb} (A/V)	Initial SOC error (%)		
	5	10	20
1	–	–	–
10	15,000	32,000	–
100	1800	2700	3700
1000	600	1500	2000

reduced, thus guaranteeing to recover the small initial offset error in a small time, being not influenced from the model inaccuracy during rapid transients.

The second reference test wants to better represent the realistic use of a lithium battery. It was considered a battery installed on-board a hybrid vehicle performing the NEDC standardised driving cycle [17]. Following a common energy strategy, the battery delivers or adsorbs only the higher variable ripple load power demand, whilst the average value is delivered by the vehicle's internal combustion engine [18]. Figure 7 shows the corresponding current and voltage profiles.

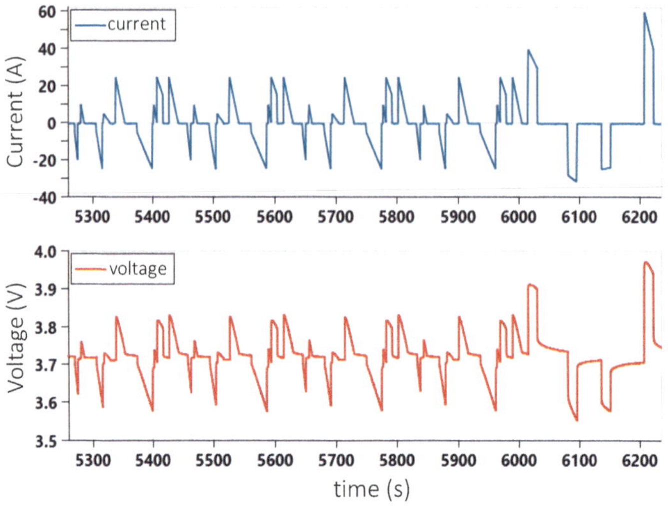

Fig. 7 Reference cycle for the NMC lithium cell under test

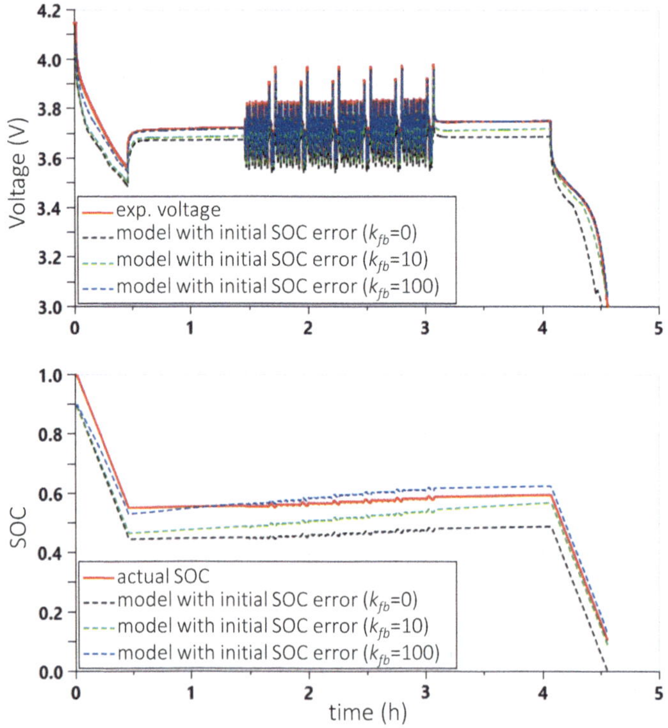

Fig. 8 Simulated model and experimental lithium cell voltage with initial SOC error (about 10%) and different k_{fb} values, NEDC-based test

Table 4 Time durations (in seconds) to reduce the offset error within 1%, for different feedback constants k_{fb}, NEDC-based test

k_{fb} (A/V)	Initial SOC error (%)		
	5	10	20
1	–	–	–
10	11,600	–	–
100	1300	3100	4800
1000	700	1100	1600

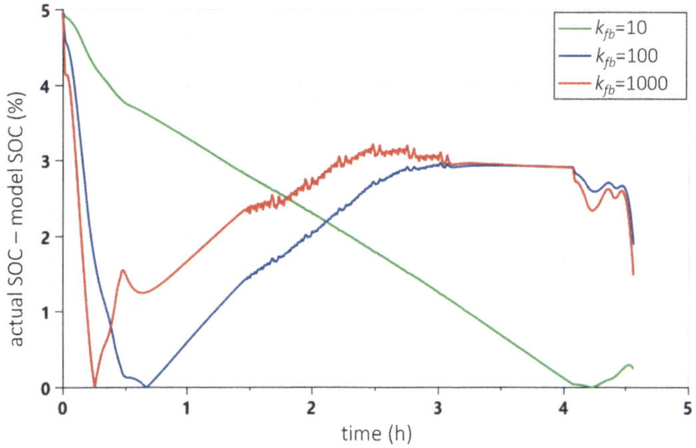

Fig. 9 Difference between simulated and actual SOC starting from an initial error of 5%, with different values of k_{fb} constant

In order to verify the accuracy of the proposed algorithm, the test started from the fully charged cell (e.g. 4.2 V). The test was subdivided into the following main phases: discharge until SOC 0.55, rest phase to measure OCV, series of NEDC cycles for a total duration of 2 h, rest phase to measure OCV and discharge up to the minimum admitted cell voltage (e.g. 3 V).

As previously, the presence of an initial SOC offset error of 10% was considered, as in the top part of Fig. 8. Also, in this case, the error can be rapidly recovered by increasing the feedback constant k_{fb}: with the already considered value of 100 A/V, the error can be recovered within 1 h.

The whole set of results always obtained by changing the initial error and acting on different values of k_{fb} is shown in Table 4.

The situation generally confirms what is already observed in the case of the MST test. In fact, higher values of k_{fb} can recover the initial SOC error faster. However, when high values are used, the high sensitivity tends to re-increase the distance between the actual and the estimated SOC during rapid transients, due to some unavoidable inaccuracy of the model, reproducing a voltage not perfectly aligned to the actual cell one.

Therefore, considering the presence of small errors in practical operations, given by the self-discharge of the NMC cell, it is preferable to use a small k_{fb} feedback constant, in the range of 10–100 A/V. As additional clarification, Fig. 9 shows that the initial offset error of 5% can be recovered by acting on the k_{fb} feedback constant itself: the reduced value of 10 A/V (green curve) tends to nullify the error on SOC with a monotonous trend, although in larger time durations. On the other hand, higher values (blue and red curves) are able to recover rapidly the initial error, which however tends to grow back in the presence of rapid transients, due to the model's inaccuracy and the highly sensitive feedback correction from the proposed Luenberger-style technique.

5 Conclusions

This paper has shown how to correctly apply a technique capable of correctly estimating the SOC of lithium cells, whose state-of-charge cannot be adequately evaluated only by ampere-hour counting method and OCV-SOC correlation. The Luenberger approach has proved to be adequate also with highly variable current profiles. A suitable mathematical model is required, and numerical parameters need to be calibrated on experimental results. The Luenberger-style technique can easily avoid SOC estimation drifts, allowing also to quickly recover initial offset errors. Although the results presented here refer to a particular NMC lithium cell, the approach is general and can be applied with no or very small modifications to other kinds of lithium or non-lithium cells.

References

1. G.L. Plett, Extended Kalman filtering for battery management systems of LiPB-based HEV battery packs; Part 1. Background. J. Power Sources **134**, 252–261 (2014)
2. G.L. Plett, Extended Kalman filtering for battery management systems of LiPB-based HEV battery packs; Part 2. Modeling and identification. J. Power Sources **134**, 262–276 (2014)
3. G.L. Plett, Extended Kalman filtering for battery management systems of LiPB-based HEV battery packs: Part 3. State and parameter estimation. J. Power Sources **134**, 277–292 (2014)
4. H.S. Ramadan, M. Becherif, F. Claude, Extended Kalman filter for accurate state of charge estimation of lithium-based batteries: a comparative analysis. Int. J. Hydrogen Energy **42**, 29033–29046 (2017)
5. G. Dong, J. Wei, Z. Chen et alii, Remaining dischargeable time prediction for lithium-ion batteries using unscented Kalman filter. J. Power Sources **364**, 316–327 (2017)
6. X. Hu et al., Estimation of state of charge of a lithium-ion battery pack for electric vehicles using an adaptive Luenberger observer. Energies **3**, 1586–1603 (2010)
7. L. Zheng, J. Jiang, et al., Embedded implementation of SOC estimation based on the Luenberger observer technique, in *IEEE Conference and Expo Transportation Electrification Asia-Pacific, 31 Aug–03 Sept 2014, Beijing*, (IEEE, Piscataway, NJ, 2014)
8. T. He, D. Li, et al., A modified Luenberger observer for SOC estimation of lithium-ion battery, in *36th Chinese Control Conference (CCC), 26-28 July 2017, Dalian*, (IEEE, Piscataway, NJ, 2017)
9. M. Lagraoui, S. Dobabi, et al., SOC estimation of Lithium-ion battery using Kalman filter and Luenberger observer: a comparative study, in *International Renewable and Sustainable Energy Conference (IRSEC), 17–19 Oct. 2014, Ouarzatate*, (IEEE, Piscataway, NJ, 2014)

10. M. Ceraolo, G. Lutzemberger, D. Poli, State-of-charge evaluation of supercapacitors. J. Energy Storage **11**, 211–218 (2017)
11. T. Huria, G. Ludovici, G. Lutzemberger, State of charge estimation of high power lithium iron phosphate cells. J. Power Sources **249**, 92–102 (2014)
12. M. Ceraolo, G. Lutzemberger, T. Huria, *Experimentally-Determined Models for High-Power Lithium Batteries. SAE Technical Paper, 2011-01-1365* (SAE, Warrendale, PA, 2011)
13. M. Ceraolo, New dynamical models of lead-acid batteries. IEEE Trans. Power Syst. **15**(4), 1184–1190 (2000). 0885-8950
14. S. Barsali, M. Ceraolo, Dynamical models of lead-acid batteries: implementation issues. IEEE Trans. Energy Convers. **17**(1), 16–23 (2002)
15. EIG cell official site. Available at: https://www.eigbattery.com/
16. M. Ceraolo, G. Lutzemberger, C. Scarpelli et alii, Model parameters evaluation for NMC cells, in *19th International Conference on Environment and Electrical Engineering (EEEIC), 2019 IEEE, 11-14 June 2019, Genoa*, (IEEE, Piscataway, NJ, 2019)
17. http://www.dieselnet.com/standards/cycles/ece_eudc.html
18. M. Ceraolo, G. Lutzemberger, N. Doveri, Experiences of realisation and test of a fuel-cell based vehicle, in *SPEEDAM 2010, International Symposium on Power Electronics, Electrical Drives, Automation and Motion, 14-16 June 2010, Pisa*, (IEEE, Piscataway, NJ, 2010)

Coordinated Control of Supercapacitor-Battery Tandem by Smart Converters in Microgrid Scenario

Giovanna Adinolfi, Roberto Ciavarella, Giorgio Graditi, Angelo Merola, and Maria Valenti

Abstract Microgrids represent a promising energetic scenario applicable in different contexts, especially in residential clusters. In this paper, authors propose a novel control logic to implement a coordinated management of generators, loads, and hybrid energy storage systems (HESS) in a microgrid by means of a hierarchical smart converter architecture. The innovative algorithm is embedded in a master converter. It allows the online management of energetic fluxes in cooperation with slave converters distributed among the microgrid resources. They carry out a smart coordination of microgrid generation, absorption, and battery-supercapacitor storage systems with the aim to improve the availability of the storage systems for providing ancillary services to the power grid. The effectiveness of the control is tested applying the smart converter master-slave architecture, including the combined management BESS-supercapacitor algorithm, to a grid-connected residential microgrid.

1 Introduction

The threat of climate change demands a rapid global shift to a sustainable energy system. Increasing renewable energy is necessary to progress on decarbonization, as well as the need for more efficient use of the grids, in order to sustain very high penetration of variable renewable electricity. Thus, grid flexibility and operational support for the electricity grids are expected to increase. In such a scenario, battery energy storage systems (BESS) could play a critical role in providing ancillary services. Supporting power grids, enabling microgrid operation, and participating in restoring unintentional islanding are just some examples of the new possibilities for BESS application. Major manufacturers are already developing BESS solutions

G. Adinolfi · R. Ciavarella · G. Graditi (✉) · A. Merola · M. Valenti
Energy Technologies Department, ENEA, Portici, NA, Italy
e-mail: giovanna.adinolfi@enea.it; roberto.ciavarella@enea.it; giorgio.graditi@enea.it; angelo.merola@enea.it; maria.valenti@enea.it

W. Zamboni, G. Petrone (eds.), *ELECTRIMACS 2019*, Lecture Notes in Electrical Engineering 615, https://doi.org/10.1007/978-3-030-37161-6_50

for the support and the safety of the grid [1]. In literature, many works study the integration of BESS in power systems, including technical and market issues [2–4]. In general, most researchers agree that the effective integration and use of BESS depends on efficient control capabilities of the storage systems. Thus, many works concentrate on developing intelligent control systems for providing ancillary services to the grids by means of distributed BESS [5] or to control grid connected hybrid power system including BESS [6] also in microgrid configuration [7]. In this paper, authors propose a novel control logic to implement a coordinated management of generators, loads, and hybrid energy storage systems (HESS) in a microgrid by a hierarchical smart converter architecture aiming to improve the availability of microgrid storage systems to provide ancillary services to the power grid. To this aim, the developed control logic integrates an innovative algorithm to evaluate the capability for the BESS to participate in providing ancillary services to the power grid and manage the charge/discharge of the couple BESS-supercapacitor in coordinated manner in order to ensure BESS ability to keep the needed level of energy. From a conceptual point of view, the work has been developed by means of three steps. First, a master-slave smart converter architecture was configured and developed by using the tool PiCon-RET, implemented in [8]. In particular, PiCon-RET is a software tool assisting the designers in configuring optimal smart converters for different energetic scenarios; it is based on a multi-objective optimization obtained by combining a SPEA-II algorithm and an electro-thermal design procedure. Then, a control algorithm for the coordinated management of microgrid resources and smart control of the couple BESS-supercapacitor was programmed by the software environment. Finally, as proof of concept for the proposed control, MC effectiveness is tested on a case study representative of a grid-tied residential microgrid.

The paper is structured in four sections, as follows. Section 2 introduces the proposed control logic. Section 3 summarizes a case study to highlight effectiveness of the proposed controller, whereas Sect. 4 recaps main conclusions and possible development for further works.

2 The Control Logic

2.1 Master-Slave Architecture Design

The proposed master-slave architecture is dedicated to the coordinated control of microgrid systems and devices. The considered energetic context is schematically represented in Fig. 1. All the resources in the microgrid (BESS, supercapacitor, AC and DC loads, and distributed generation units—DER) are controlled by means of a master converter (MC), which manages a cluster of smart converter slaves (SCSs), distributed among the resources. The novel HESS algorithm, programmed by the

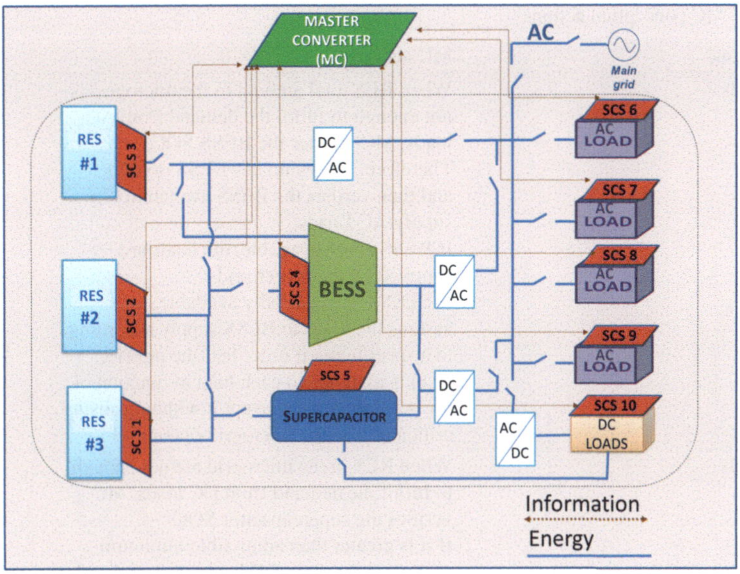

Fig. 1 Microgrid master-slave functional scheme

software environment, is integrated in the more general MC control flow and allows BESS and supercapacitor to operate in a coordinated manner.

More in detail, MC continuously receives RES data from SCS1, SCS2, and SCS3 (Fig. 1) and loads data from SCS6, SCS7, SCS8, SCS9, and SCS10 (Fig. 1). In addition, it verifies the BESS SOC instantaneous value (SCS4) and the supercapacitor status (SCS5). Starting from this information, MC is able to carry out a continuous "online match" (communication time = 5″) among energetic fluxes and to apply the proposed control logic. In order to maximize the use of RES, reduce electricity withdrawn from the grid, and apply the BESS-supercapacitor coordinated control for improving the capacity to provide ancillary services, MC implements the conceptual steps synthetized below (Table 1).

2.2 BESS-Supercapacitor Coordinated Management

In the current energetic scenario, storage systems are going to be more and more applied for supporting reliable operation of power networks. Since the last 10 years, BMS has included algorithms to verify the SOC's instantaneous value falls within an admissible interval (interval between the minimum value of discharge (minSOC) and the maximum value of charge (maxSOC)) to guarantee the battery safeguard. Many works in literature have also proposed BESS control algorithms to improve BESS participation in ancillary services [5–7]. In [9], authors propose a control logic to ensure the BESS SOC falls within an interval (energy restoration band)

Table 1 MC conceptual actions

Resources	MC actions
AC loads	When RES total amount in the microgrid is not enough to fulfill the demand from AC loads, MC verifies the BESS SOC. Therefore, MC evaluates BESS mxrSOC and then verifies the BESS availability to supply AC Loads If BESS is not available, the loads are supplied by the power grid If BESS is just partially available, MC assigns the loads to BESS supply according to a specific merit order list (the priority level is assigned to each load in function of the possibility to interrupt it without causing criticalities in the microgrid operation)
DC loads	When RES in the microgrid are not enough to fulfill the demand from DC loads, MC verifies the supercapacitor SOC If it is greater than admissible minimum state of charge, the DC loads are supplied by the supercapacitor; otherwise, the loads are fulfilled by the BESS, if it is available, or conversely from the power grid if also BESS is not available If BESS is just partially available, MC assigns the loads to BESS supply according to a specific merit order list (the priority level is assigned to each load in function of the possibility to interrupt it without causing criticalities in the microgrid operation). Distinguishing the use of supercapacitor for DC loads from the use of BESS for AC loads allows improving the performance of the overall system so reducing the need of DC/AC conversions
BESS	When BESS SOC does not fall within energy restoration band, the supercapacitor charges/discharges BESS up to BESS SOC coming back within the restoration band
RES	MC controls RES SCSs in order to improve generator performance. More in detail, maximum power point tracking (MPPT) algorithms have been embedded in the SCS control logic to maximize power extraction under all conditions with wind turbines and photovoltaic solar systems

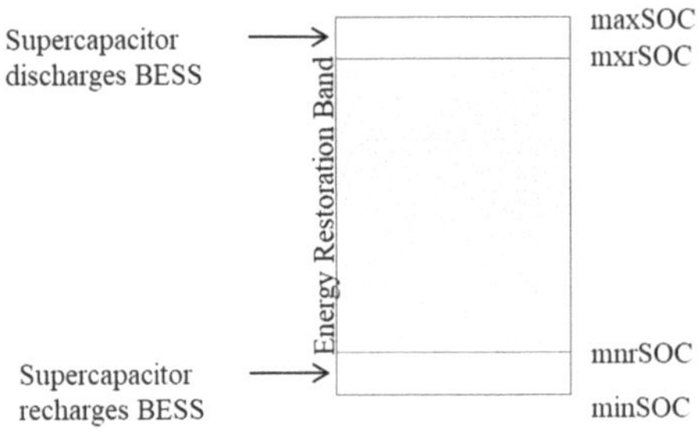

Fig. 2 BESS SOC virtual bands

where the BESS has a sufficient amount of energy reserves to participate, at each time, in ancillary services during the microgrid operation. To this aim, whenever the BESS has not a sufficient level of energy reserves, the control algorithm requires the needed energy from the power grid. In this paper, an additional control is integrated to the previous one, at a higher hierarchical level, to extend the logic described in [9] to a HESS. More in detail, an innovative control algorithm is proposed to coordinate BESS and supercapacitor operation, allowing faster BESS energy reserve restoration from supercapacitor (Fig. 2), getting more efficient the process for providing ancillary services to the grid. The algorithm allows keeping the BESS SOC within the "energy restoration band" (gray band with interval limits: mnrSOC and mxrSOC in Fig. 2), which is the SOC interval where the BESS is certainly able to participate to the frequency restoration process. The latter allows ensuring the needed amount of energy restoration reserves during the microgrid operation for next frequency deviation events.

3 The Proof of Concept

In this section, the microgrid, managed by the MC developed in this work, has been tested on a case study described as follows. It is worth noting that the main aim of the test consists in showing the effectiveness of the MC control applied to the proposed smart converter master-slave architecture, including the combined management BESS-supercapacitor. The improved capability of the BESS to participate in grid balancing, after the introduction of the control algorithm, was already demonstrated in [9]; therefore, this point is not the main focus of this proof of concept.

The simulation of the control logic developed and embedded in the microgrid master-slave converter architecture has been carried out applying it to a grid-tied residential microgrid covering one building composed of eight apartments. A photovoltaic generation system, a battery storage system, and a supercapacitor operate to fulfill energetic requirements of AC and DC residential loads.

The BESS included in the microgrid is a multi-stack Li-ion battery, designed to store 25% of the energy produced by the photovoltaic system during July at Napoli City (latitude coordinate, 40.853294°; longitude coordinate, 14.305573°—optimal tilt). Each stack is conceived to cover the AC absorption for one apartment. The

Table 2 Simulation parameters

Resource	
Connection to the external grid	YES
Photovoltaic	25 kWpSc = 157 $m^2 \eta = 0.15$
BESS	Capacity: 13 kWhLi-ion—multi-stack
Supercapacitor	XLR supercapacitors

BESS, connected to the electric grid by a bidirectional DC/AC converter, covers AC loads but it cannot cover the DC loads, since they are connected both to the supercapacitor and the power grid. More in detail, the supercapacitor, connected to the electric grid by a bidirectional DC/AC, fulfills DC loads but also acts on BESS charging/discharging to bring the BESS SOC in the energy restoration band, as shown in Sect. 3.2.

The MC manages the SCs cluster, distributed among the resources of the microgrid, including BESS, supercapacitor, AC and DC loads, and PV system, according to the functional schema reported in Fig. 1 and logical steps described in Table 1.

Main features and significant microgrid parameters are reported in Table 2.

The load profile of each apartment has been constructed considering the average trends in load demands based on a customer consumption behavior of an Italian typical four-person family [10–12], also including average power consumption for AC home appliances such as one oven, one watching machine, four air-conditioners, and one dishwasher and DC home appliances such as four smartphones, one tablet, and one laptop.

To verify the effectiveness of the proposed model, a validation process based on two simulations has been conducted. The first simulation evaluates the μg operation during the normal operating conditions; the second one simulates, at a certain time, the simultaneous disconnection of six apartments (a significant load) to verify the BESS ability to participate in the restoration process when an imbalance occurs and, then, the BESS capability to come back at the energy restoration band. The energy restoration band is set between 40% (mnrSOC) and 60% (mxrSOC), whereas the BESS security SOC limits are 20% (minSOC) and 90% (maxSOC).

3.1 *First Simulation*

The study lasts 24 h and starts at midnight with the SOC BESS at 50% (Fig. 3). The nighttime demand is contained and it is fulfilled by the BESS. During the time period between 6:00 a.m. and the 7:00 p.m., the supercapacitor starts assuring the DC loads charge, so fulfilling citizens' requirements. Then the BESS satisfies the AC energetic needs. During the day, the PV generation firstly charges the battery until the maximum SOC is reached and then it charges the supercapacitor. This

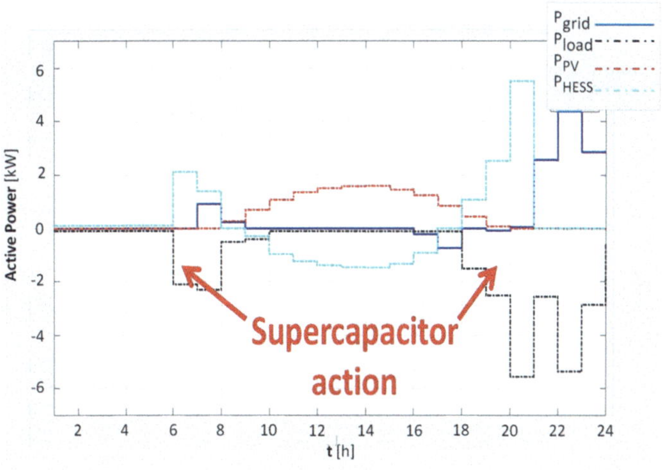

Fig. 3 Adopted control applied to the proposed case study

operation contributes to reduce the μg losses since a further AC/DC conversion is avoided. Once the family returns home at 6:00 p.m., energetic demand increases reaching peak consumption at 8:00 p.m. and 10:00 p.m. In general, citizens who returned home require mobile and portable devices to be recharged. As shown in Fig. 3, from 7:00 p.m. to 8:00 p.m., the MC demands the supercapacitor action. Then the BESS fulfills the energetic absorption until 9:00 p.m. and the grid intervention is necessary until midnight.

In Fig. 4 details about the BESS SOC graphs are reported. The adopted control strategy assures satisfaction of BESS SOC limits during normal operating conditions (step 1).

3.2 Second Simulation

The second simulation supposes an imbalance occurs in the μg at 12.57 a.m. (Fig. 4). To simulate the imbalance, a sudden load decrease is applied (load curtailment: −75%). After the grid balance-restoring process, when the BESS SOC is equal to 61%, this simulation evaluates the capability of the proposed control strategy to bring the BESS SOC in energy restoration band in order to ensure its potential participation in other grid-balancing processes.

In detail, to verify the benefit in applying both the proposed controller and the supercapacitor for the fast BESS recharging, three simulations have been conducted (Fig. 5). In the first case, the controller is not active (controller OFF) and the supercapacitor just works to fulfill DC loads (supercapacitor is disconnected from BESS); in the second case, the controller is active (controller ON) and

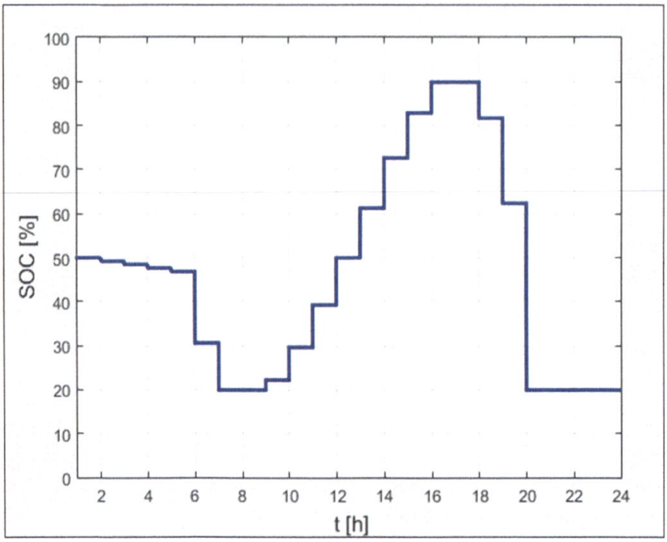

Fig. 4 BESS SOC graph during normal operating conditions

the supercapacitor just works to fulfill DC loads (supercapacitor is disconnected from BESS); in the last case, the controller is active (controller ON) and the supercapacitor is connected with BESS. As shown in Fig. 5, in the first case, the BESS, after participating in balance restoring, moves away from mxrSOC (the loads are disconnected and the controller is OFF). In the second case, MC acts to bring BESS SOC back into the energy restoration band; the BESS provides the surplus of energy to the grid and the BESS SOC comes back to the energy restoration band in about 2–5 min. In the last case, MC again acts to bring BESS SOC back into the energy restoration band, but the BESS surplus of energy is taken by the supercapacitor; thus the process is much faster than in the second case; the BESS SOC comes back to the energy restoration band in less than a minute. Finally, it is clear that the combined use of the proposed control algorithms embedded in the MC and HESS is really useful to ensure BESS capability to provide ancillary services to the grid.

4 Conclusions

In this paper an innovative control logic is introduced to operate the smart management of generators, loads, and hybrid storage devices in a microgrid by means of distributed smart converters in a hierarchical master-slave architecture. The proposed algorithm, operating at the higher control level, evaluates the more suitable match between the hybrid storage system and all resources (generation/demand)

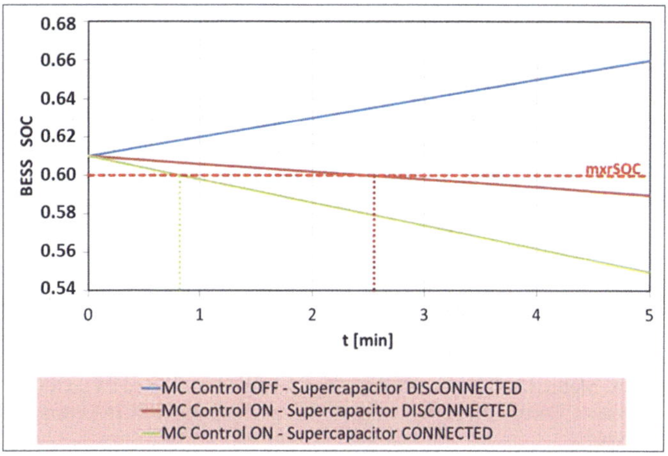

Fig. 5 BESS SOC graph after balance-restoring process

to obtain both the user energetic satisfaction and the availability of a sufficient energetic level to guarantee the BESS participation in grid balance-restoring processes. In order to show the effectiveness of the proposed approach, the control logic is applied to a grid-connected residential microgrid. Results show that, after a frequency instability event, in absence of the proposed algorithm, the BESS energy reserves are not restored during the simulation period (5 min), whereas they are restored in a few minutes when the algorithm works. Furthermore, the simulations show that the restoration time is about 2–5 min when the controller operates only on BESS, and it is reduced to less than 1 min when the algorithm is applied to the couple BESS-supercapacitor. Finally, these results show that the combined use of the proposed control algorithms embedded in the MC and HESS is really useful to improve microgrid storage system capability to provide both user services and ancillary services to the grid.

Acknowledgments This work was supported by MIUR (Ministero dell' Istruzione dell' Università e della Ricerca) in the frame of Progetti di Ricerca Industriale e Sviluppo Sperimentalenelle 12 Aree di Specializzazione Individuate dal PNR 2015–2020 with MIUR notification 1735 on 13.07.2017: ComESto (Community Energy Storage: Aggregated Management of Energy Storage Systems in Power Cloud).

References

1. ABB Energy Storage Solutions Brochure, *EssPro™ energy storage Power Conversion System (PCS) - the power to control energy*. Available at: https://library.e.abb.com/public/a9b8ede76a664383b8ecffaad161730b/EssPro-PCS_Brochure.pdf

2. A.B. Attya, J.L. Domínguez-García, F.D. Bianchi, O. Anaya-Lara, Enhancing frequency stability by integrating non-conventional power sources through multi-terminal HVDC grid. Int. J. Elect. Power Energy Syst. **95**, 128–136 (2018). https://doi.org/10.1016/j.ijepes.2017.08.032. ISSN 0142-0615
3. M. Brenna, F. Foiadelli, M. Longo, D. Zaninelli, Ancillary services provided by BESS in a scenario characterized by an increasing penetration of unpredictable renewables, in *2017 IEEE PES Innovative Smart Grid Technologies Conference Europe (ISGT-Europe), Torino, 2017*, (IEEE, Piscataway, NJ, 2017), pp. 1–6. https://doi.org/10.1109/ISGTEurope.2017.8260215.
4. C. Brivio, S. Mandelli, M. Merlo, Battery energy storage system for primary control reserve and energy arbitrage. Sustain. Energy Grids Netw. **6**, 152–165 (2016). https://doi.org/10.1016/j.segan.2016.03.004. ISSN 2352-4677
5. G. Foggia, A. Neto, A. Michiorri, A. Bocquet, Coordinated control of dispersed battery energy storage systems for services to network operators, in *23rd International Conference on Electricity Distribution - CIRED 2015, 2015, Lyon, France*, (CIRED, Liège, 2015)
6. N. Chettibi, A. Mellit, Intelligent control strategy for a grid connected PV/SOFC/BESS energy generation system. Energy **147**, 239–262 (2018). https://doi.org/10.1016/j.energy.2018.01.030. ISSN 0360-5442
7. C.-H. Yoo, I.-Y. Chung, H.-J. Lee, S.-S. Hong, Intelligent control of battery energy storage for multi-agent based microgrid energy management. Energies **6**, 4956–4979 (2013). https://doi.org/10.3390/en6104956
8. G. Graditi, G. Adinolfi, R. Ciavarella, V. Palladino, Design support tool for multi-DER residential microgrids, in *ICRERA 2018 - 7th International Conference on Renewable Energy Research and Applications*, (IEEE, Piscataway, NJ, 2018), pp. 667–672. https://doi.org/10.1109/ICRERA.2018.8566729
9. G. Graditi, R. Ciavarella, M. Valenti, An innovative BESS management for dynamic frequency restoration, in *17th IEEE International Conference on Environment and Electrical Engineering and 2017 1st IEEE Industrial and Commercial Power Systems Europe, EEEIC/I and CPS Europe 2017*, (IEEE, Piscataway, NJ, 2017). https://doi.org/10.1109/EEEIC.2017.7977864. Article number 7977864
10. Istat, *Data Warehouse, Censimento Popolazione Abitazioni*. Available at: http://dati-censimentopopolazione.istat.it/index.aspx
11. C. Ricerche, *Determinazione dei fabbisogni e dei consumi energetici dei sistemi edificio-impianto. Caratterizzazione del parco immobiliare ad uso residenziale - Ricerca Di Sistema Elettrico, Report RdS/2012/109* (ENEA, Rome, 2012)
12. S. Maggiore, *Studi e valutazioni sull'uso razionale dell'Energia Elettrica - Impatto su comportamenti e consumi delle famiglie di un sistema di prezzi biorari dell'energia elettrica: RSE reference number 11000398* (RSE, Milan, 2011)

Part VI
Advances in Batteries and Supercapacitor Technologies, Graphene-Based Devices, and Safety Issues in Electrical Engineering

Comparison and Modeling of Commercial Supercapacitors via Standardized Potentiostatic Electrochemical Impedance Spectroscopy

Filippo Gherdovich, Giuseppe Taddia, Sandro Maria Tenconi, Miguel Pretelli, Alessandro Lampasi, Francesca Soavi, and Maria Luisa Di Vona

Abstract The main scope of the study is the characterization of the capacitive and resistive behavior of two supercapacitor cells and one hybrid supercapacitor available on the market, through potentiostatic electrochemical impedance spectroscopy (PEIS). The PEIS tests were performed by applying to all cells the same voltage perturbation in the same frequency range. In a first phase, the instrumentation used for the acquisitions was optimized, with particular care to the connections between the potentiostat and the supercapacitor cell. The Nyquist diagrams obtained for each sample are compared and capacitance/frequency graphs are deduced. The technological differences between various devices are then discussed in relation to the results. The characterization of the sample cells and the collected data are used to propose the corresponding models conceived for circuit simulation. These models are based on simple electronic components available in the standard circuit simulation software tools.

F. Gherdovich (✉) · G. Taddia · S. M. Tenconi · M. Pretelli
Energy Technology Srl, Bologna, Italy
e-mail: filippo.gherdovich@ocem.com; giuseppe.taddia@ocem.com; sandro.tenconi@ocem.com; miguel.pretelli@ocem.com

A. Lampasi
ENEA - Agenzia nazionale per le nuove tecnologie, l'energia e lo sviluppo economico sostenibile, Rome, Italy
e-mail: alessandro.lampasi@enea.it

F. Soavi
University of Bologna, Bologna, Italy
e-mail: francesca.soavi@unibo.it

M. L. Di Vona
University of Roma "Tor Vergata", Rome, Italy
e-mail: divona@uniroma2.it

W. Zamboni, G. Petrone (eds.), *ELECTRIMACS 2019*, Lecture Notes in Electrical Engineering 615, https://doi.org/10.1007/978-3-030-37161-6_51

1 Introduction

Potentiostatic electrochemical impedance spectroscopy (PEIS) is a powerful method of characterizing many electrical and electrochemical properties of materials and interfaces.

PEIS is useful to investigate the dynamics and kinetics of electric charges in the bulk or interface regions of any type of solid or liquid material (ionic, semiconductor, ion-electronic mixed, and even insulator). This non-destructive method was chosen to swiftly characterize the sample devices with defined low-voltage signals, in order to obtain electrical models from experimental results.

The general approach is to apply a sinus (perturbation voltage) around a DC voltage (Fig. 1) to the electrodes, ideally assuming that the properties of the electrode material do not change over time:

$$V(t) = V_{\mathrm{DC}} + V_{\mathrm{p}} \sin(\omega t). \tag{1}$$

ω is the angular frequency. Induced current measured is:

$$I(t) = I_{\mathrm{DC}} + I_{\mathrm{p}} \sin(\omega t + \varphi). \tag{2}$$

φ is the phase difference between $V(t)$ and $I(t)$. The expression of complex impedance is:

$$Z(i\omega) = \frac{V_{\mathrm{p}} \sin(\omega t)}{I_{\mathrm{p}} \sin(\omega t + \varphi)} = \mathrm{ESR} + \frac{1}{i\omega C}. \tag{3}$$

ESR is the equivalent series resistance of sampled device and C is its capacitance [1].

Fig. 1 Voltage over time during PEIS

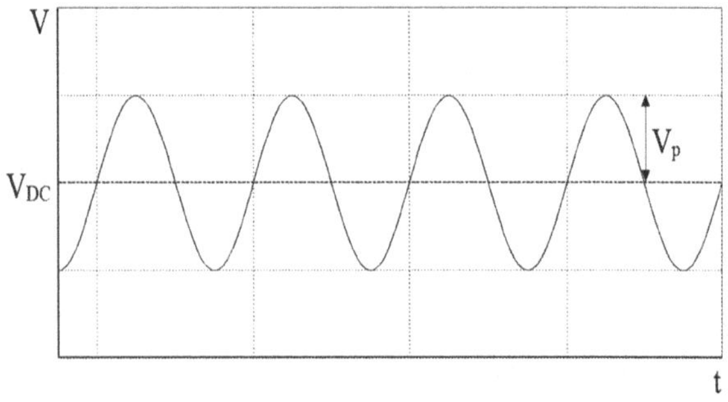

2 Sample Cells Used for Tests

Three different commercial cells were tested [2]: two supercapacitors of about 3000 F (mentioned as "A" and "B") and a 3300 F hybrid supercapacitor, called "C". Table 1 reports the nominal characteristics of the three sample devices.

3 Preparation for the Test Optimization of Input/Output Cables

3.1 Test Instrumentation and Procedures

The measurement equipment includes:

- Five-channel PAR VSP galvanostat/potentiostat
- "EC-Lab V1023" software
- Four-terminal sensing cable
- Room temperature, 20 ± 1 °C
- Faraday cage

Table 1 Rated characteristics of the cells

Product specifications	Value
"A" cell	
Capacitance	3000 F
Voltage	2.7 V
ESR (DC)	0.29 mΩ
Specific energy	6 Wh/kg
Storable energy	3.04 Wh
Temperature range	[−40; 65] °C
"B" cell	
Capacitance	3200 F
Voltage	2.85 V
ESR (DC)	0.095 mΩ
Specific energy	6.8 Wh/kg
Storable energy	3.61 Wh
Temperature range	[−40; 70] °C
"C" cell	
Capacitance	3300 F
Voltage	3.8 V
ESR (DC)	1 mΩ
Specific energy	13 Wh/kg
Storable energy	6.62 Wh
Temperature range	[−30; 70] °C

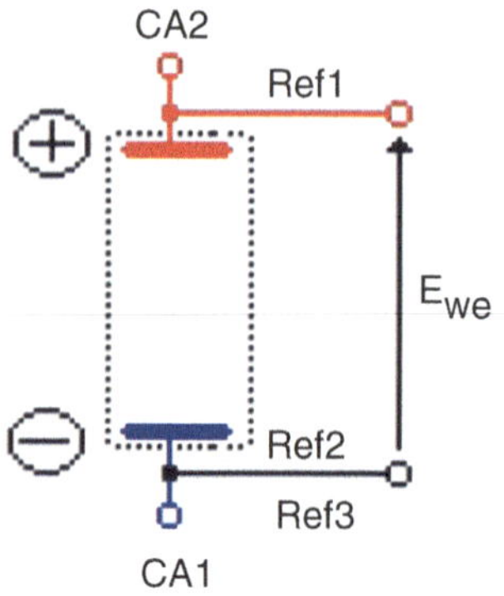

Fig. 2 Cell connection set-up (from EC-Lab V1023 software manual)

Test frequency range was set in the frame of 100 Hz–10 mHz, to focus interest on resistive-capacitive behavior of the sample, by applying a sinusoidal stimulus around the DC potential of 1 mV. DC is set to the cell open circuit voltage in the discharge state. The cell was placed in Faraday cage and connected to the potentiostat/galvanostat by five probes described in Fig. 2. The terminals CA1 and CA2 measured the current. The terminals REF1, REF2, and REF3 were used to measure the voltage. REF1 + CA2 were connected to the cell positive electrode. REF2, REF3, and CA1 were connected to the negative electrode. The PEIS voltage signal is controlled between REF1 and REF2 (REF2 and REF3, depicted in Fig. 2, are connected on the same electrode). The current is measured from CA2 to CA1. This cell connection was used for all the experiments reported below.

3.2 First Test with Hose Clamps

At first, hose clamps were used for the cell. The preliminary tests were performed on the supercapacitor "A".

Qualitatively, the tests give reliable results; however, quantitatively the cable connection included a stand-out resistance of around 22 mΩ, one order of magnitude higher than that reported by the manufacturer data sheet of cell "A", as shown in Table 1.

The obvious conclusion was to improve, first, the connections, by replacing the hose clamps on cell terminals with lugs. It was necessary to remove the plastic sheath, to which the clips, joined to hose clamps, were connected.

3.3 Test with Lugs

During this test, the above-described instrumentation and settings were unchanged, except for the connection to the supercapacitor terminals. Lugs were interposed between clips and cell terminals.

The resistance value obtained by this connection improved and decreased to 4.05 mΩ, 5 times less than the previous value obtained, but still 14 times the value declared by "A" manufacturer. This result showed that measurement connections are critical. Therefore, a further improvement of contact resistance was pursued.

3.4 Optimized Cable Design

Measurement wires were soldered directly to ring lugs. Keeping the test setting unchanged (except for PEIS AC that was set to 0.1 mV), the measured ESR reduced to 0.664 mΩ, 6 times less than the value obtained by lugs, but still about 2 times the value declared by "A" cell manufacturer.

Figure 3 reports the Nyquist plots obtained by PEIS run on cell "A" with the three different cable arrangements. The ESR corresponds to the high-frequency intercept on the real axis of the Nyquist plot.

4 Results and Interpretation of PEIS Tests

The following sections report the results of PEIS tests on the three commercial cells "A", "B", and "C" carried out by using optimized cables.

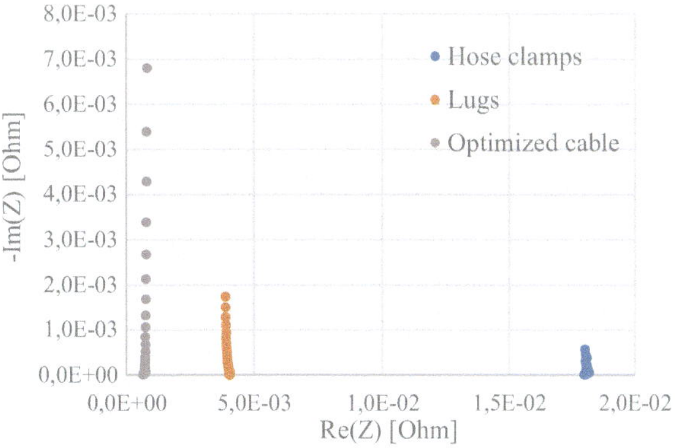

Fig. 3 Nyquist plots obtained from PEIS measurement of cell "A" with different connections

4.1 "A" Supercapacitor Cell

The impedance spectroscopy tests were conducted respecting the following test parameters:

- Initial frequency: 100 Hz
- Final frequency: 10 mHz
- AC voltage: 0.1 mV

The Nyquist diagram obtained from the PEIS test is shown in Fig. 4.

Each point in Fig. 4 corresponds to a specific frequency [3]. At the lower sampled frequency (0.01 Hz), the measured capacitance is equal to [4]:

$$C = \frac{1}{\omega \cdot \left[-\operatorname{Im}(Z)_{0.01\ \mathrm{Hz}}\right]} = 2340.5\ \mathrm{F}. \tag{4}$$

Cell "A" impedance response deviates from that of an ideal capacitor. Indeed, for an ideal capacitor, the Nyquist plot is a straight line parallel to the imaginary axis. In Fig. 4, the straight line has a slope (phase) <90°. This suggests a frequency dependence of capacitance that is highlighted by Fig. 5. The figure reports the capacitance vs. frequency Bode plot of cell "A". In Fig. 5, the capacitance decreases with the increase of the frequency. This has to be related to the diffusion rate of electrolyte ions through the porous architecture of the carbonaceous electrodes. At the lowest frequencies and lowest times, more charges can reach the porous interface, and then a higher capacitance is achieved. Consequently, in these conditions, the device stores more energy.

The curve in Fig. 5 can be fitted by a logarithmic function:

$$y = -308.1\ln(x) + 1454.6 \tag{5}$$

For frequencies above 100 Hz, the behavior of the sample cell is inductive (see Fig. 10).

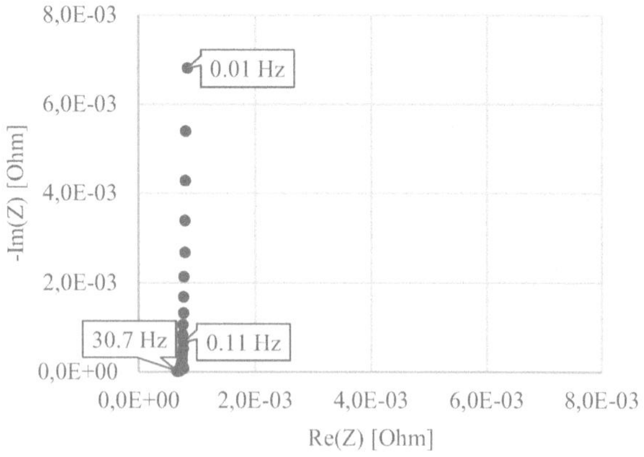

Fig. 4 Nyquist plot of cell "A." ESR is evaluated by the high-frequency intercept on the real impedance axis. The cell capacitance is obtained by the lowest-frequency imaginary impedance (see Eq. 4)

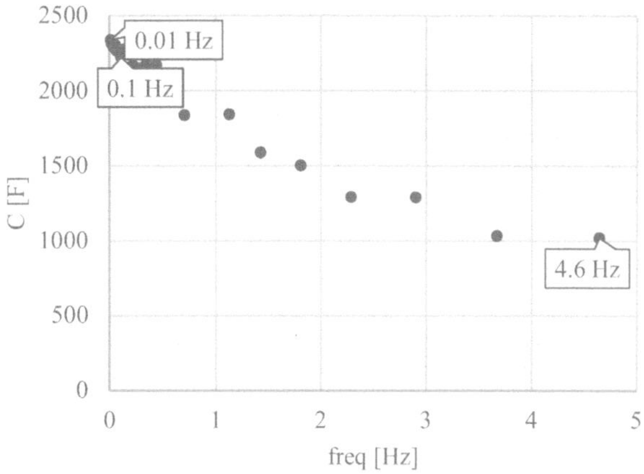

Fig. 5 Capacitance vs. frequency plot of cell "A"

4.2 "B" Supercapacitor Cell

The impedance spectroscopy tests were conducted by setting the following test parameters:

- Initial frequency: 100 Hz
- Final frequency: 10 mHz
- AC voltage: 0.25 mV

The connection to the terminals was harder because there is no thread for bolts on the terminals. As a rule, for very capacitive or low-impedance electrochemical systems, the potential amplitude can lead a current overflow that can stop the experiment in order to protect the unit from overheating. Using galvanostatic electrochemical impedance spectroscopy (GEIS), instead of PEIS, can avoid this inconvenient situation. The setback was solved reducing the perturbation voltage compared to the previous test. This was required for the lower internal resistance of "B" supercapacitor with respect to cell "A". Fig. 6 Nyquist plot of cell "B". ESR is evaluated by the high-frequency intercept on the real impedance axis. The cell capacitance is obtained by the lowest-frequency imaginary impedance (see Eq. 4). Figure 6 illustrates Nyquist diagram of "B" cell.

Basically, cell "B" Nyquist plot is similar to that of cell "A". The former mainly differs from that of the latter for the lower ESR. The value measured between the two frequency points 17–11 Hz is equal to 0.612 mΩ. The capacitive behavior starts to be appreciable at frequencies lower than 5 Hz. At the lower sampled frequency (0.01 Hz), the measured capacitance is equal to 2514 F, as shown in Fig. 6. Figure 7 reports the capacitance vs. frequency plot of cell "B".

Also for "B" cell, best fitting curve is gained with logarithmic equation:

$$y = -45.88 \ln(x) + 2337.1 \tag{6}$$

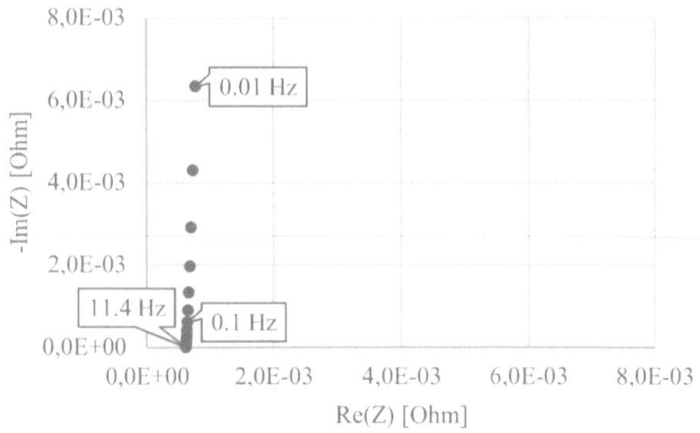

Fig. 6 Nyquist plot of cell "B." ESR is evaluated by the high-frequency intercept on the real impedance axis. The cell capacitance is obtained by the lowest-frequency imaginary impedance (see Eq. 4)

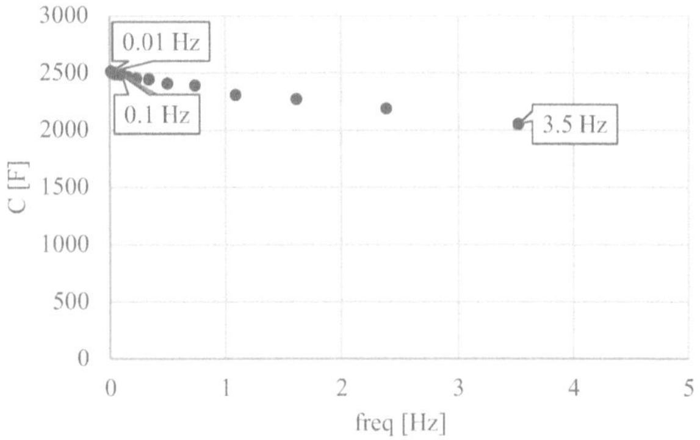

Fig. 7 Capacitance vs. frequency plot of cell "B"

4.3 "C" Hybrid Supercapacitor Cell

Unlike the two previous devices, the "C" cell is a lithium-ion capacitor (LIC). It is a hybrid supercapacitor composed of a pre-lithiated carbon anode, similar to a lithium-ion battery negative electrode. The positive electrode is an activated carbon, like for conventional supercapacitors [5]. The impedance spectroscopy tests were conducted setting the following test parameters:

- Initial frequency: 100 Hz
- Final frequency: 50 mHz
- AC voltage: 1 mV

The choice of a higher AC voltage is due to the fact that the cell has greater internal resistance than the "A" and "B" devices (Fig. 8). This gives no risk of over-current protections on the potentiostat.

The value of the measured series resistance is equal to 0.85 mΩ, similar to the nominal value (6 times as much).

In the "C" cell, due to its hybrid nature, the effect of electrolyte ion diffusion is more evident. Indeed, the low-frequency lines of the Nyquist plots significantly deviate from the ideal capacitor case discussed above. This is an important difference that acts as a watershed between the pure and the hybrid supercapacitor.

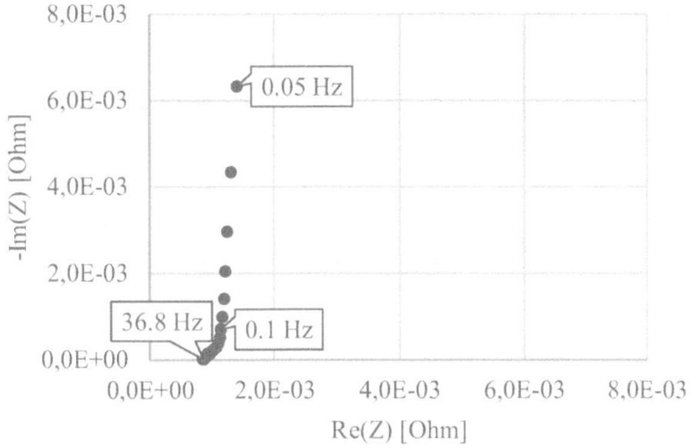

Fig. 8 Nyquist plot of cell "C." ESR is evaluated by the high-frequency intercept on the real impedance axis. The cell capacitance is obtained by the lowest-frequency imaginary impedance (see Eq. 4)

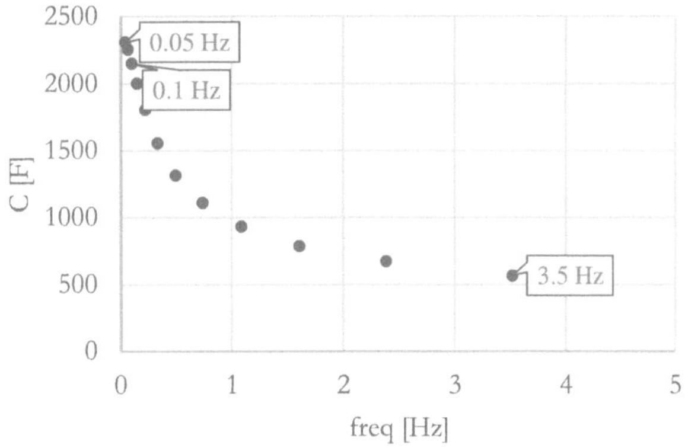

Fig. 9 Capacitance vs. frequency graph of the "C" sample cell

The ESR was evaluated at 54 Hz, while the capacitive behavior begun to be appreciated below 0.15 Hz. At the lowest sampled frequency (0.01 Hz), the measured capacitance was equal roughly to 2450 F (Fig. 9).

Behavior of hybrid cell acts as a power function:

$$y = 953.64x^{-0.343} \tag{7}$$

5 Test of Standardized Impedance Spectroscopy on All Three Devices

In order to be able to compare the "A", "B", and "C" sample cell behaviors, the following parameters have been "standardized":

- Initial frequency: 200 kHz
- Final frequency: 10 mHz
- AC voltage: 0.1 mV

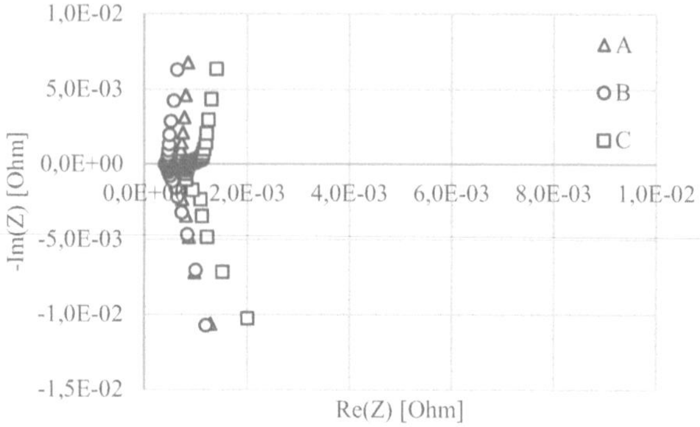

Fig. 10 Nyquist plots of cells "A," "B," and "C" obtained by standardized EIS test

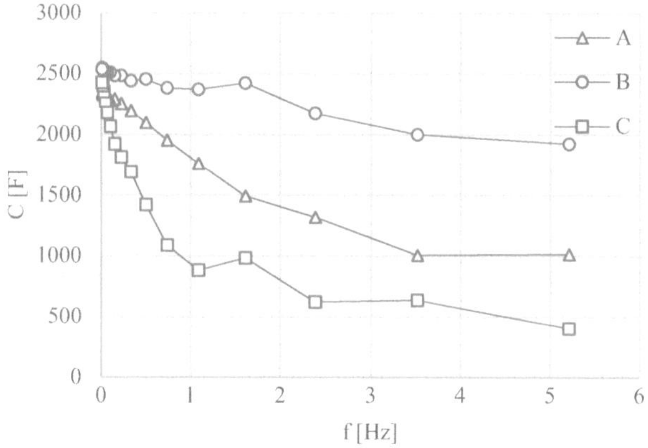

Fig. 11 Comparison of the capacitance vs. frequency trends of cells "A," "B," and "C" obtained by standardized EIS test

Figure 10 reports the Nyquist plots of the three cells and highlights that the three devices behave in approximately the same way. The main difference lies in the resistance of the device: the greater resistance of the "C" cell, where ion diffusion contribution is no longer negligible, stands out.

At the same time, the three cells' capacitance behaviors with frequency give different responses. This is evinced in Fig. 11 that compares the capacitance vs. frequency plots for the supercapacitor cells.

The cell "B" is the one with the highest performance, i.e., with the highest capacitance retention over the investigated frequency range. The corresponding plot has the lowest slope dC/df. In turn, the cell "C" (LIC, hybrid cell) features the lowest capacitance retention and the highest slope. The higher resistance value of cell "C" affects its power performance and its high-frequency energy response. In fact, cell "C" faradaic charge/discharge involves lithium-ion diffusion in the negative electrode, which is a slow process compared to the electrostatic charge/discharge of the conventional "A" and "B" cells.

6 Modeling Capacitive/Resistive Behavior Starting from Impedance Spectroscopy Tests

PEIS permits to convert experimental results in circuital model of energy storage device behavior. This is of paramount importance to address power electronics industrial needs. However, the non-linear behavior of supercapacitors makes modeling over a sufficiently wide frequency range difficult [6]. The application of supercapacitor models into software for electrical circuit simulation and design requires a realistic behavior of the component inside the simulated system. Implementation of the supercapacitor circuit model must take into account that many simulation software do not include elements like Warburg element, constant phase element (CPE), Gerischer element, etc. Given that supercapacitor capacitance is frequency dependent, it is suggested, in software developed considering CPE, to implement the Nyquist plot analysis substituting capacitance with these elements. CPE indeed takes into account how frequencies affect the imaginary component of impedance. Probably, the best way to connect experimental results with models applicable to electrical circuit simulator is to use the simplest electrical elements, remembering that the best circuit model is the simplest possible one, in order to achieve immediate results [7]. In case of supercapacitors, this means resistances, capacitors, and inductances. The kind of application determines the complexity of the model, and every element must have a physical meaning. Generally, supercapacitor could be represented by not more than five RC meshes, each one with different time constants [8]. Figure 12 depicts a solution for the resistive/capacitive behavior of analyzed supercapacitors samples.

A mono-branch model, $R_1 + R_2/C_1$, where R_1 is the device ESR while R_2 is self-discharge resistance, could represent "A" sample. C_1 is the capacitance between 0.01 and 100 Hz. In power electronic simulation, this model performs in a reliable manner not only supercapacitor cell dynamics but also modules. The self-discharge resistance can be inferred by the leakage current value that is often declared in the cell datasheet.

The best fit for "B" cell is achievable with a cogent structure model (Fig. 13).

Two RC meshes in series, $R_1/C_1 + R_2/C_2$, act as "B" supercapacitor cell. The first mesh has time constant $\tau_1 = 0.24$ ms, and the second mesh has $\tau_2 = 498$ s.

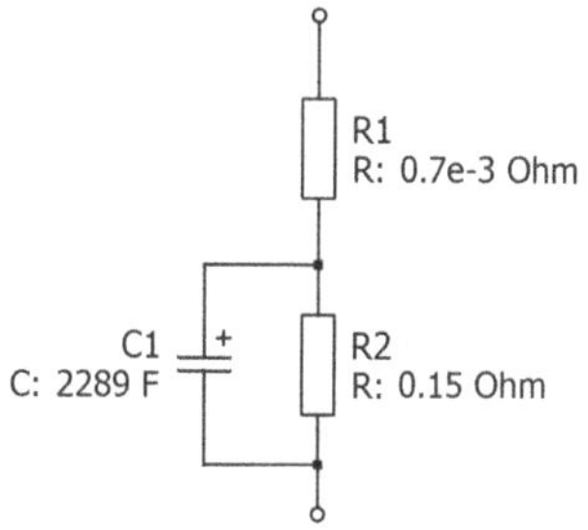

Fig. 12 Cell "A" equivalent circuit model

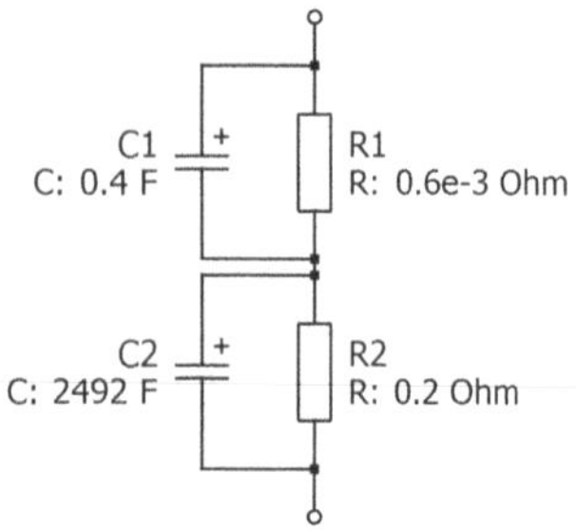

Fig. 13 Cell "B" equivalent circuit model

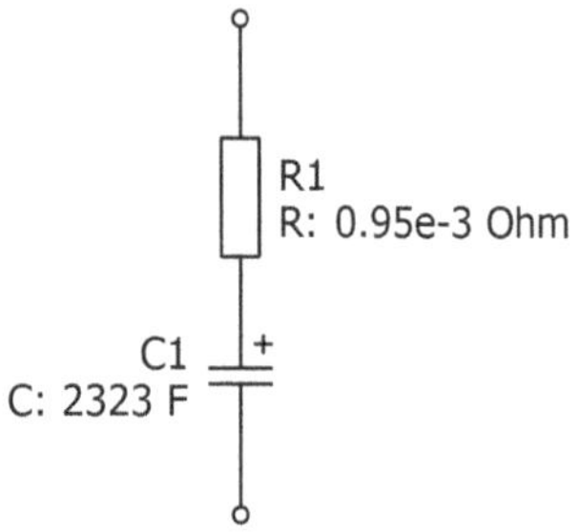

Fig. 14 Cell "C" equivalent circuit model

An overall behavior for "C" cell could be described by the equivalent circuit reported in Fig. 14.

"C" sample model described in Fig. 14 is not sufficiently representative of the hybrid cell. There is a way to improve fitting, but this implies the use of theoretical components such as Warburg element and so on. Indeed, an R(RC)W Randles circuit that takes into account the faradaic process occurring at the negative electrode should be used. Hybrid supercapacitors request further studies to be well modeled.

7 Conclusions

There is a difference of about 20% between the capacitance measured by PEIS and the results reported by the manufacturers of the tested supercapacitors. This difference may be related to the fact that the capacitance measurements were carried out without cell polarization (i.e., around 0 V).

It is generally recognized that the capacitance of the cells is partly voltage dependent: $C(V) = C_0 + kV$ [9]. Rated capacitances are generally meant with the charge at full voltage. This might explain the difference. Also, datasheet capacitance values are obtained by different techniques. Typically they result from galvanostatic tests.

Concerning the ESR, the values that we obtained by PEIS are 5–6 times higher than the values declared by the manufacturers. This is probably due to the extremely low values of the resistance to be measured, at the lower limit of the available instruments.

As said in the introduction, many other tests, such as charge-discharge test, must endorse models obtained by PEIS. To enlarge supercapacitor knowledge, it is useful to apply PEIS test also for supercapacitor modules, identifying right instrumentation.

Acknowledgments Thanks go to professors John R. Miller (Case Western Reserve University) and Philippe Knauth (Aix-Marseille Universitè) for their suggestions and willingness.

References

1. M. Grossi, B. Riccò, Electrical impedance spectroscopy (EIS) for biological analysis and food characterization: a review. J. Sens. Sens. Syst. **6**, 303–325 (2017)
2. F. Rafik, H. Gualous, R. Gallay, A. Crausaz, A. Berthon, Frequency, thermal and voltage supercapacitor characterization and modeling. J. Power Sources **165**, 928–934 (2007)
3. W. Lajnef, J.-M. Vinassa, O. Briat, S. Azzopardi, E. Woirgard, Characterization methods and modelling of ultracapacitors for use as peak power sources. J. Power Sources **168**, 553–560 (2007)
4. P. Navalpotro, M. Anderson, R. Marcilla, J. Palma, Insights into the energy storage mechanism of hybrid supercapacitors with redox electrolytes by Electrochemical Impedance Spectroscopy. Electrochim. Acta **263**, 110–117 (2018)
5. J. Zhang, J. Wang, Z. Shi, Z. Xu, Electrochemical behavior of lithium ion capacitor under low temperature. J. Electroanal. Chem. **817**, 195–201 (2018)
6. S.V. Rajania, V.J. Pandya, V.A. Shah, Experimental validation of the ultracapacitor parameters using the method of averaging for photovoltaic applications. J. Energy Storage **5**, 120–126 (2016)
7. N. Mohan, T.M. Undeland, W.P. Robbins, *Power Electronics: Converters, Applications and Design* (Wiley, New Delhi, 2017)
8. L. Zhang, X. Hu, Z. Wang, F. Sun, D.G. Dorrell, A review of supercapacitor modeling, estimation, and applications: a control/management perspective. Renew. Sustain. Energy Rev. **81**, 1868–1878 (2018)
9. L.E. Zubieta Bernal, *Characterization of Double-Layer Capacitors for Power Electronics Applications*, Master of applied science thesis, University of Toronto, 1997

Economic Analysis of a Hybrid Storage System Associated to PV Sources and Supervised by Fuzzy Logic Power Management

Xingyu Yan, Dhaker Abbes, Antoine Labrunie, Youssef Krim, and Benoit Robyns

Abstract This study concerns a grid-tied photovoltaic (PV) generator related to a hybrid storage system composed of lithium NCA battery (energy source) and Maxwell supercapacitor (power source). Two supervision algorithms have been proposed for energy management system (EMS): a Boolean and a fuzzy logic EMS. Moreover, a comparative study between both supervision algorithms based on levelized cost of energy (LCOE) and the lifespan of the storage system has been suggested. The economic analysis is done with two different planned PV power production profiles: one with a "clear sky" bell curve and a second with an ideal forecast. The supervisor based on Boolean method is simple and easy to understand, while the fuzzy logic method offers more flexibility in supervision. It improves the battery lifespan and system performance a little and reduces significantly the system penalties. The simulation results show for all scenarios the achievement of the planned aims in terms of respect of the production program taking into account the constraints of the electrical network manager with an LCOE below 130 €/MWh.

1 Introduction

Worldwide, renewable energy sources (RES) such as photovoltaic (PV) and wind generators are increasingly becoming significant sources for power and will consequently minimize the use of fossil fuels as energy production means. However, being highly meteorologically sensitive, RES face the challenges of large-scale using and

X. Yan · D. Abbes (✉) · Y. Krim · B. Robyns
Laboratory of Electrical Engineering and Power Electronics of Lille (L2EP), EA 2697, Univ. Lille, Centrale Lille, Arts et Metiers ParisTech, HEI, Ecole des Hautes Etudes d'Ingénieur (HEI), Lille, France
e-mail: dhaker.abbes@yncrea.fr; youssef.kraiem@yncrea.fr; benoit.robyns@yncrea.fr

A. Labrunie
GB SOLAR, Paris Cedex, France
e-mail: antoine.labrunie@gb-solar.com

W. Zamboni, G. Petrone (eds.), *ELECTRIMACS 2019*, Lecture Notes in Electrical Engineering 615, https://doi.org/10.1007/978-3-030-37161-6_52

are withdrawn by their intermittent form with uncertain availability. To maintain secured operation and system reliability with a high penetration level of RES, energy storage system (ESS) is required to mitigate uncertainties of RES power production in more and more grid codes to deal with their intermittent nature [1]. If suitably deployed, ESS can help operators to manage the fluctuation in generation and load.

However, batteries' energy storage faces limitations, such as high cost, lifespan limited by the number of charge/discharge cycles, and inverse proportionality between specific energy and specific power. Instead, several storage devices with different capacities will be required [2]. The combination of two storage technologies and development of an intelligent power management overcome the intermittences of the renewable PV generator, facilitate the integration of the RES into the grid, and increase the longevity of storage elements.

Despite the advanced installation of several large-capacity storage systems in various industrial applications, there is not yet a storage device characterized by both a high density of power and energy. These storage technologies are actually in exploration. Therefore, several research projects are actually established by the European Union and suggest ancillary services using hybrid storage systems, such as the GROW-DERS project [3], which combines Li-ion battery and flywheel energy storage. In [4], by inserting a supercapacitor power storage device with battery, ESS achieves the improvement of the battery lifespan and a good frequency adjustment [5, 6].

Management of electrical ESS is a significant research field, more precisely in association with renewable distributed generators, such as photovoltaic (PV) power systems. Indeed, besides energy management of power flow, many purposes (grid services, state of charge of storage systems, lifespan and aging, etc.) are considered altogether. In addition, integration of two storage systems with complementarity of energy and power characteristics can support system operators to reach energy management goals much easier while ensuring an improvement of lifespan to some sensitive components (such as supercapacitors and batteries, flywheel and batteries, etc.).

In this context, a supervision algorithm based on fuzzy logic is suggested to manage the power flow between different energy sources. The fuzzy logic supervision (FLS) methodology for energy management system (EMS) is now in development in many applications like EMS for polygeneration microgrids [7] and for islanded hybrid renewable generator [8] and EMS for DC microgrid systems [9]. Furthermore [10], depict an EMS where the FLS considers the variation of the energy demand, total cost in 1 day, and the production in order to ensure an accessible grid. In this concern, it can be marked that the respect of production schedule and participation in frequency control are not explored in these aforementioned studies as principal objectives of the EMS.

This study presents a PV generator that is associated with two storage systems (lithium NCA battery for energy needs and lithium NCA battery and Maxwell supercapacitor for power demands) to establish a hybrid storage system (HSS). To manage the powers in this hybrid PV and storage system, two supervision algorithms are developed. The first supervisor is with Boolean logic. The second is an FL EMS that manages the HSS and includes power demand forecasting and frequency regulation.

In this framework, a comparative analysis based on economic criteria is put forward. It is made between both supervision methods using LCOE and battery lifespan as comparison criteria. Generally, RES generation forecasting is needed for scheduling and system operation [11]. Therefore, two different PV production profiles are proposed. They are based on two extreme scenarios: the first one is planned PV power production with "clear sky" bell curve profile, while the other is with ideal forecast.

2 Hybrid Power System Description

Figure 1 depicts the architecture of the PV power generator combined with energy storages in a DC coupled structure. The HSS station is composed of two complementary storage technologies: SC for short-term storage power and battery for long-term storage energy.

Figure 2 illustrates the structure of the energy management supervisor to manage the power flow. It ensures many purposes at once (lifespan of storage system, auxiliary services, availability and levels of storage). In this study, the EMS aims to satisfy 1-day-ahead production planning and control of the state of charge (SoC) of each storage system and participate in frequency regulation.

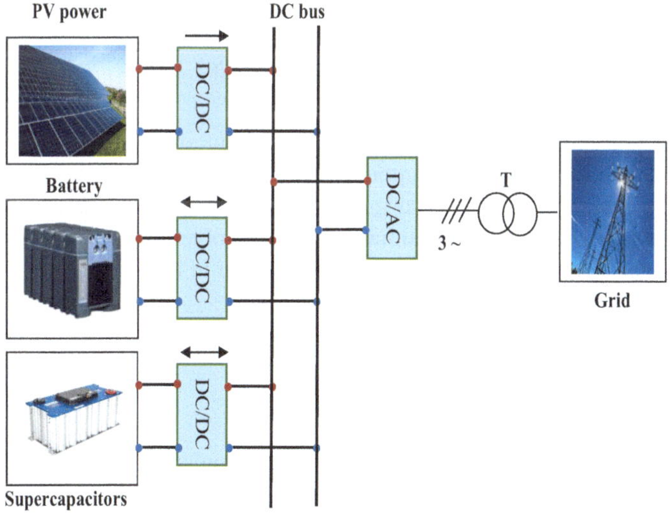

Fig. 1 Architecture of PV system with storages

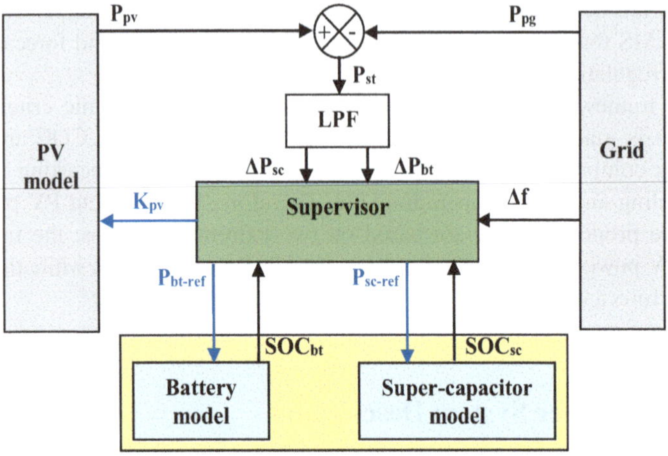

Fig. 2 Management structure of the studied system

The daily operation of the PV power plant should follow the scheduled production to maintain system reliability. It has to satisfy operational constraints and to respond to dynamic disturbances, such as frequency control with primary and secondary reserves [12]. Thus, as an innovative part of our application, the storage system must be able to provide active support power based on frequency deviation.

3 Hybrid Storage System Supervisor

3.1 Supervisor Methodology and Objectives

The supervisor is designed for energy management of the studied PV system associated with battery/supercapacitor hybrid storage system. The purpose of the EMS is to generate a reference power minimizing system cost and fulfilling the commitment to the grid. Table 1 outlines supervisor objectives, constraints, as well as means of actions for the supervision system shown in Fig. 2.

3.2 Supervisor with Boolean Method

Figure 3 depicts the supervisor algorithm with Boolean method for validating the models. The power difference (ΔP) between real-time produced power (P_{pv}) and day-ahead programmed power (P_{pg}) is covered by energy storages separated through a first-order low-pass filter: the high variation power in short time can be absorbed by supercapacitor (P_{sc}), while the other is captivated by battery.

Table 1 EMS operating specifications

Objectives	• Meeting a production program and smoothing the power injected to the grid • Optimizing storage energy management and improving storage lifespan
Constraints	• PV production intermittency • Storage lower/upper capacity and charge/discharge power limitations • Penalties will be applied if the programmed power is not respected
Means of actions	• Injecting to the electrical network the planned PV power with the help of the two storage systems • Degradation of the PV production • Power storage

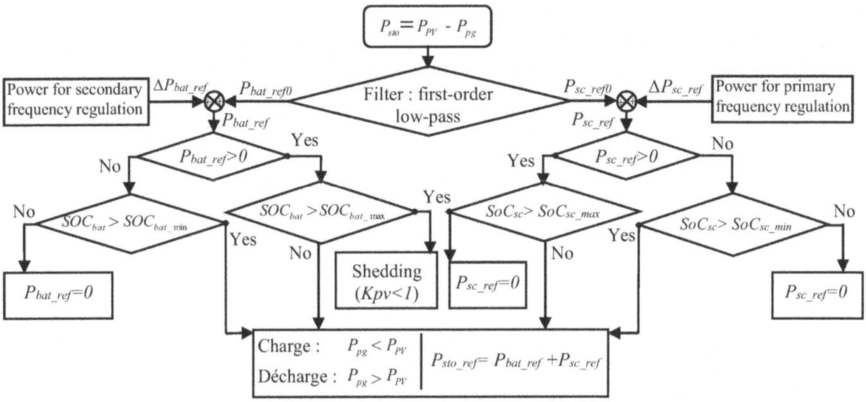

Fig. 3 Hybrid storage system supervisor with Boolean method

Supercapacitor power reference $P_{\mathrm{sc\_ref}}$ is the sum between the initial reference $P_{\mathrm{sc\_ref0}}$ and the power $\Delta P_{\mathrm{sc\_ref}}$ dedicated for primary frequency regulation.

Battery power reference $P_{\mathrm{bat\_ref}}$ includes initial reference $P_{\mathrm{bat\_ref0}}$ and frequency support power $\Delta P_{\mathrm{bat\_ref}}$. In case of storage overcharge, a security function rapidly degrades photovoltaic production to protect the storage system.

The Boolean logic has limited performances, such as significant penalty. Thus, the purpose of the next part is to develop an intelligent approach based on fuzzy logic.

3.3 Supervisor with Fuzzy Logic

Instead of using the Boolean method, in this part, a supervisor with a fuzzy logic methodology is developed. Figure 4 presents the structure of the developed EMS.

Fuzzy logic approach is chosen since linguistic rules can simplify the control and management of such HSS, with many objectives and constraints.

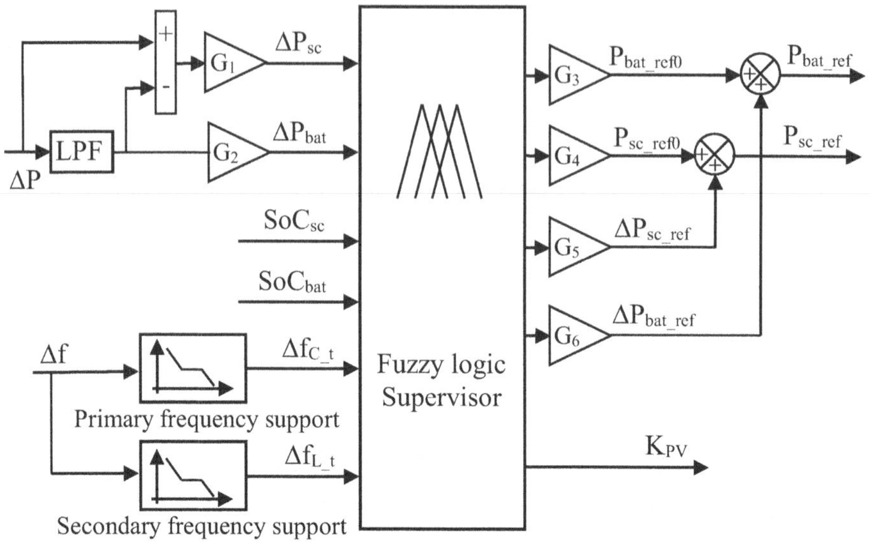

Fig. 4 Block diagram of the fuzzy logic supervisor

Normalization gains ($G_{i=1:6}$) are defined according to input and output membership functions as follows: $G_1 = 1/P_{\text{sc-max}}$, $G_2 = 1/P_{\text{bat-max}}$, $G_3 = P_{\text{sc-max}}$, $G_4 = P_{\text{bat-max}}$, $G_5 = 0.08 \times P_{\text{sc-max}}$, and $G_6 = 0.02 \times P_{\text{bat-max}}$. $P_{\text{sc-max}}$ is the maximum power of supercapacitor and $P_{\text{bat-max}}$ is the maximum power of battery.

Inputs are battery (ΔP_{bat}) and supercapacitor (ΔP_{sc}) powers from demand and production difference and battery SOC (SOC_{bat}) and supercapacitor SOC (SOC_{sc}) powers for first frequency regulation (Δf_{C_t}) and for secondary frequency regulation (Δf_{L_t}). Outputs are the references sent to the means of actions: battery power reference ($P_{\text{bat\_ref}}$), supercapacitor power reference ($P_{\text{bat\_ref}}$), and degradation factor K_{pv} used for safe operation of the battery.

According of the SOC of storage elements, three operation modes are considered [13].

The variation margin of battery SOC is chosen between 20% and 90% of its capacity. These limits offer a safety margin against possible overcharge and excess of discharge.

4 Financial Parameters of Production and Remuneration

4.1 Remuneration and Penalties

The economic indicators suggested in this paper are remunerations and penalties. Indeed, the remuneration of produced power is calculated every minute. In addition, if the production, for each minute, does not respect this announcement (commitment production in day 1) with a tolerance of ±5% of installed capacity, penalties will be applied. The calculation method of the penalties is detailed in [14] and the Annex 9 of the document [15].

4.2 Notion of LCOE

In this sub-section, we calculate the total cost LCOE. Parameters taken into account are the following:

Lifespan of the PV plant with 30 kW rated power: nb_years = 25 years.

Price of building the PV plant (including PV panels and the costs of installation): $C_{PV} = 1$ €/Wc.

Cost of lithium ion batteries (including the costs of transformer station, converter, installation, and transportation): $C_{bat} = 550$ €/kWh, with a replacement cost of 200 €/kWh after 12–15 years.

Price of Maxwell supercapacitor (including the costs of converter, transportation, installation, and transformer stations): $C_{sc} = 500$ €/kW (estimated lifespan is 14 years), with a replacement cost of 200 €/kW. In our case, the nominal power of the supercapacitor is 5 kW and its capacity is 52.8 Wh.

Costs of maintenance and operation (M&O) of PV system each year: $O\&M_{PV} = 20$ k€/MWc (taking into account a discount rate of $r = 2\%$).

Estimated costs of M&O of storage system each year: $M\&O_{sto} = 20$ k€/MWc.

The LCOE is calculated as total cost divided by total energy. $LCOE_p$ represents the cost with penalties, while LCOE represents the cost without penalties.

$$\mathrm{LCOE} = \mathrm{Cost} / \sum E_{\mathrm{Prod}} \tag{1}$$

with: $\mathrm{Cost} = (C_{PV} + C_{bat} + C_{sc} + M\&O_{PV\_total} + M\&O_{sto\_total})$

$$\mathrm{LCOE_p} = \mathrm{Cost_p} / \sum E_{\mathrm{Prod}} \tag{2}$$

with: $\mathrm{Cost_p} = (C_{PV} + C_{bat} + C_{sc} + M\&O_{PV\_total} + M\&O_{sto\_total} + \mathrm{Penalty})$

The total M&O cost of storages during 25 years (nb_years) is calculated as $M\&O_{sto\_total} = M\&O_{sto} \times (1 + r)^{nb\_years}$, and so as to the total costs of PV M&O.

The total energy production is calculated as $\Sigma E_{prod} = E_{prod} \times nb_{years} \times (d)^{nb\_years}$, with $d = 0.5\%$/year being the degradation rate of PV production per year.

5 Case Studies and Simulation Results

In this section, three detailed comparative studies based on economic criteria are presented. They aim to analyze:

- The influence of the filter constant on battery lifespan, penalties, and $LCOE_p$. We change the filter parameter.
- The influence of production schedule profiles, profile without forecast and profile with forecast, on battery lifespan.
- The performances of fuzzy logic method compared with Boolean method.

5.1 PV Production Profiles

A system with 30 kW rated PV power and an energy storage system are taken into consideration. For production programming, we studied two extreme configurations as shown in Fig. 5.

Real PV power production P_{PV} is on blue color, planned PV power P_{pg} without forecasting is on yellow, and broken black line is the ideal forecasted PV power production.

Absence of production forecast: the forecast corresponding to a reference irradiation ("clear sky," bell curve) multiplied by a gain equal to the ratio between the annual production and the annual reference irradiation.

Ideal production forecast: the committed production is produced on the assumption that there is a very good forecast. This planned profile is obtained based on the real production with a filter which follows the following rule: the difference between two successive values (in 1 min) must be below ±5% of installed power, which is 30 kW in this case, as described in Annex 10 of the document [15].

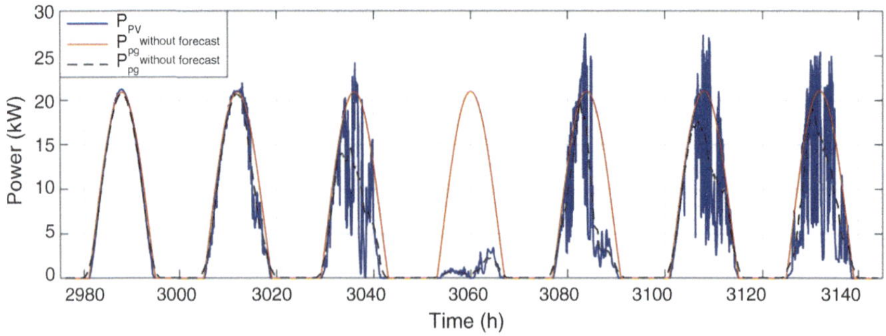

Fig. 5 PV power actual production (in blue), production without forecasting (in red), and ideal production forecast (in orange), example over 7 days

5.2 Filter Choosing

According to the two proposed methodologies for supervisor, there will be four scenarios. The acronyms for different scenarios are defined as follows:

FBC: Supervisor with fuzzy logic, planned PV with bell curve (without prediction)
BBC: Supervisor with Boolean, planned PV with bell curve (without prediction)
FIP: Supervisor with fuzzy logic, planned PV with ideal prediction
BIP: Supervisor with Boolean, planned PV with ideal prediction

In this case, different filter durations (from 60 to 180 s) are studied with the Boolean methodology. Table 2 shows the comparison results of two scenarios: without prevision (BBC) and with ideal prevision (BIP).

Compared with the BBC scenario, the total remuneration in 25 years of scenario BIP is larger because of the precise prediction of the PV power production. For the same reason, the penalty in 25 years is much less.

For lifespan of battery lithium NCA (18 kW), scenario BIP (14.89 years for filter time of 120 s) has a much better performance than scenario BBC (12.34 years for filter time of 120 s).

LCOE is the same for both scenario BIP and BBC, as 96.8 €/MWh. Nevertheless, since the penalty is bigger, the $LCOE_p$ is greater for scenario BBC (105.2 €/MWh for filter time 120 s) than the scenario BIP (98.3 €/MWh for the same filter) and batteries have more extreme situations and thus a shorter lifespan. However, to relativize, we have to include the extra cost for forecasting. In addition, the actual forecast will have a much worse performance than this ideal forecast scenario.

5.3 Comparison of Different Storage Configurations

In this part, the battery sizing effect is studied with Boolean method according to different battery power: from 9 to 31.5 kW. The filter time is chosen as 120 s. The capacity of the supercapacitor is 52.8 Wh and its nominal power is 5 kW.

The comparative framework is between ideal and bell curve inputs for solar power production: it is a comparison based on economic criteria, which allows us to choose between paying penalties and buying good forecasts. For the scenario BBC, as the battery power is increasing, the remuneration increases slightly, while penalty has a significant decrease (Fig. 6). Without forecast, the errors between the planned PV production and real PV production could be much greater than the maximum energy storage (mainly battery) size and batteries will saturate more frequently. Therefore, this scenario could cause bigger penalties and shorter battery lifespan.

LCOE is the same for both scenarios, while $LCOE_p$ for the scenario BBC is bigger than the scenario BIP. It is not a surprise because the penalty in scenario BBC is bigger than in scenario BIP, as shown in Fig. 7.

Table 2 Studied results of different filter times with Boolean method

	BBC					BIP				
Filter (s)	Rem. in 25 years (k€)	Penalty in 25 years (k€)	Lifespan (year)	$LCOE_p$ (€/MWh)	LCOE (€/MWh)	Rem. in 25 years (k€)	Penalty in 25 years (k€)	Lifespan (year)	$LCOE_p$(€/MWh)	LCOE (€/MWh)
60	158	8.0	12.32	106.8	96.8	172	11	14.83	98.1	96.8
90	157	7.0	12.33	105.5	96.8	172	11	14.87	98.2	96.8
120	157	6.8	12.34	105.2	96.8	172	1.2	14.89	98.3	96.8
150	157	6.7	12.34	105.1	96.8	172	1.2	14.90	98.3	96.8
180	157	6.7	12.34	105.1	96.8	172	1.2	14.91	98.3	96.8

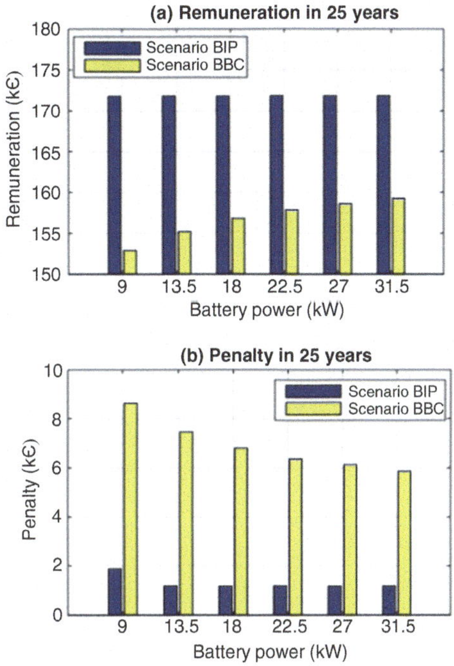

Fig. 6 (**a**, **b**) Remuneration and accumulated penalties over 25 years depending on battery sizing and production profile type

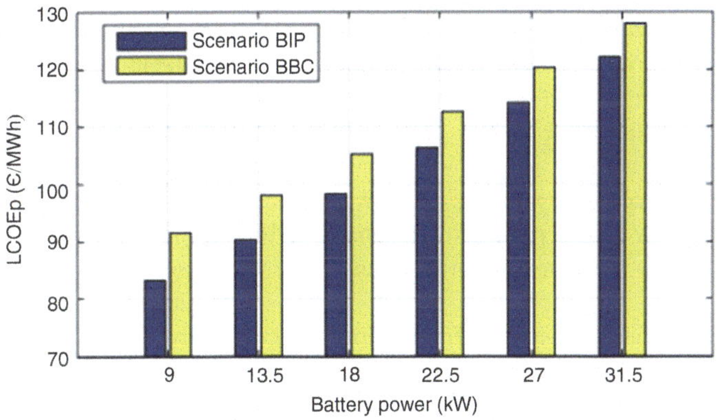

Fig. 7 LCOEp depending on the different battery sizing

5.4 Comparison of the Boolean method with the Fuzzy Logic Method

For the remuneration during 25 years, scenario BBC has a slightly better performance than scenario FBC, as shown in Fig. 8a. At the same time, the penalty during 25 years is less for scenario FBC than scenario BBC (Fig. 8b). As battery size increases, remuneration increases, while penalty decreases due to less extreme

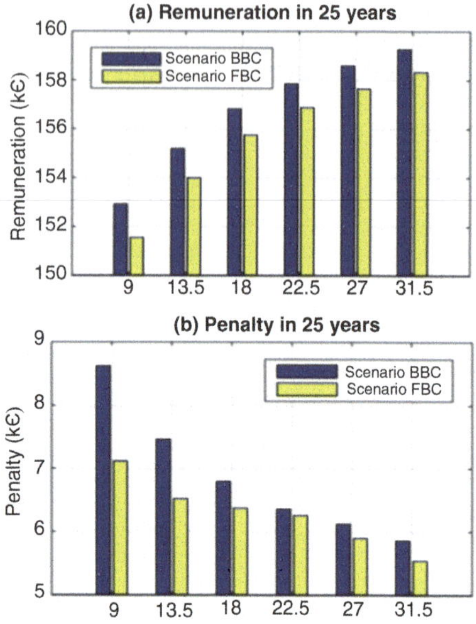

Fig. 8 (**a**, **b**) Remuneration and accumulated penalties over 25 years depending on the battery sizing

situations for battery charge and discharge. The remuneration gap between scenario BBC and scenario FBC is almost constant when the battery sizing increases, while penalty gap decreases for battery power from 9 to 22.5 kW and then increases a little at battery power of 31.5 kW. Therefore, we can conclude that increasing battery size decreases system penalty.

Figure 9 illustrates the battery lifespan in different scenarios. The lithium NCA battery lifespan with an ideal PV power forecasting (scenarios BIP and FIP) is longer than in scenarios without PV power forecasting (BBC and FBC). For the same planned PV power production (scenarios BIP and FIP or scenarios BBC and FBC), the battery lifespan is slightly longer for the fuzzy logic method than Boolean method.

Since LCOE has almost no difference for all the scenarios (from 81 €/MW h with a battery size of 9 kW to 121 €/MW h with a battery size of 31.5 kW), only $LCOE_p$ has been studied in Fig. 10. The same as the battery lifespan, scenarios with ideal prediction (BIP and FIP) have smaller $LCOE_p$ than scenarios without PV power forecasting (BBC and FBC). Moreover, even though $LCOE_p$ for scenarios BIP and FIP has no big difference, it is smaller for scenario FBC than scenario BBC.

Consequently, we could conclude that the supervisor based on the fuzzy logic method performs better than that based on Boolean method even though it is not much. In the future, the performance could be improved by applying optimizing fuzzy membership functions and thus rules.

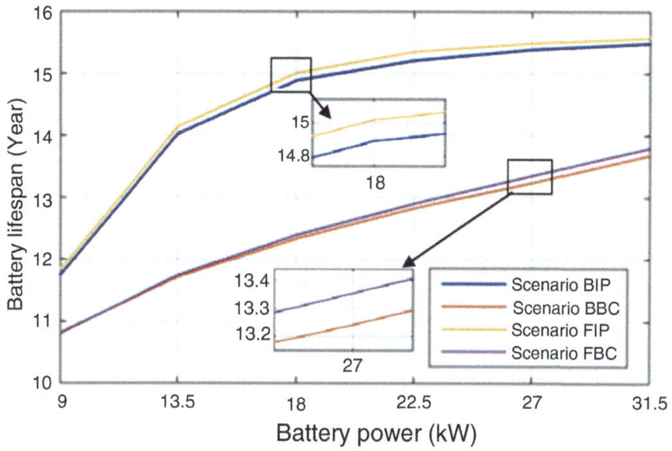

Fig. 9 Battery lifespan for different scenarios

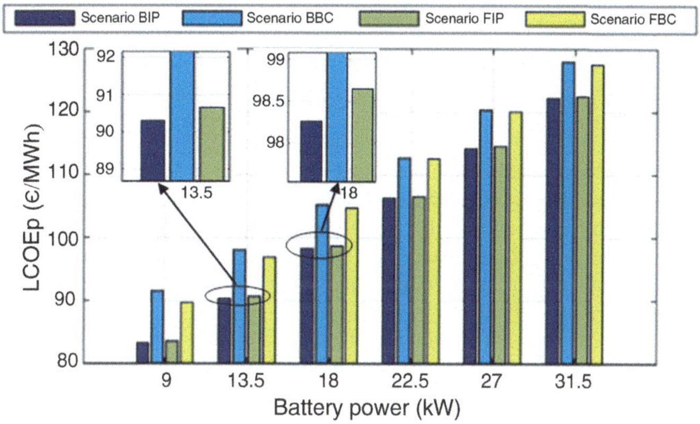

Fig. 10 $LCOE_p$ for different scenarios depending on battery sizing

6 Conclusions

This paper has proposed a detailed comparative study for a hybrid storage system associated to photovoltaic power with:

- Two complementary storage systems (lithium NCA battery for energy needs and lithium NCA battery and Maxwell supercapacitor for power needs).
- Two different PV production profiles proposed: the first one is planned PV power production with "clear sky" bell curve profile, while the other is with ideal forecast.
- Two supervision algorithms developed: one with Boolean logic and another with fuzzy logic.

- New criteria of comparison: battery lifespan, penalties, and LCOE.

The supervisor based on Boolean method is simple and easy to understand, while the fuzzy logic method offers more flexibility in supervision and improves battery lifespan and system performance a little. It also reduces significantly system penalties.

In BBC scenario, the errors between the planned PV production and real PV production are more important than the battery capacity, which causes a reduction of the battery lifespan and bigger penalties. However, to relativize, we have to include the extra cost for forecasting. In addition, the actual forecast will have a much worse performance than the ideal forecast used in this paper.

Finally, a perspective of this study is to explore a joint optimization method of the battery sizing and the membership functions of the fuzzy logic EMS to reduce the LCOE.

Acknowledgments This work was founded as a part of the CRE PV call of 2015 (research component) and also of the DEESSE Project founded by ADEME, France.

References

1. A. Ipakchi, F. Albuyeh, Grid of the future. IEEE Power Energy Mag. **7**, 52–62 (2009)
2. D. Stenclik, P. Denholm, B. Chalamala, Maintaining balance: the increasing role of energy storage for renewable integration. IEEE Power Energy Mag. **15**, 31–39 (2017)
3. M. Schreurs, P. De Boer, R. Hooiveld, GROW-DERS; Grid reliability and operability with distributed generation using flexible storage, in *Proceeding of China International Conference on Electricity Distribution*, (2010)
4. Y. Krim, D. Abbes, S. Krim, M.F. Mimouni, Intelligent droop control and power management of active generator for ancillary services under grid instability using fuzzy logic technology. Contr. Eng. Pract. **81**, 215–230 (2018)
5. V. Courtecuisse, J. Sprooten, B. Robyns, M. Petit, B. Francois, J. Deuse, A methodology to design a fuzzy logic based supervision of hybrid renewable energy systems. Math. Comput. Simulat. **81**, 208–224 (2010)
6. B. Robyns, A. Davigny, C. Saudemont, Methodologies for supervision of hybrid energy sources based on storage systems–a survey. Math. Comput. Simulat. **91**, 52–71 (2013)
7. G. Kyriakarakos, A.I. Dounis, K.G. Arvanitis, G. Papadakis, A fuzzy logic energy management system for polygeneration microgrids. Renew. Energy **41**, 315–327 (2012)
8. O. Erdinc, M. Uzunoglu, The importance of detailed data utilization on the performance evaluation of a grid-independent hybrid renewable energy system. Int. J. Hydrogen Energy **36**, 12664–12677 (2011)
9. Y. Chen, Y. Wu, C. Song, Y. Chen, Design and implementation of energy management system with fuzzy control for DC microgrid systems. IEEE Trans. Power Electron. **28**(4), 1563–1570 (2013)
10. D. Arcos-Aviles, J. Pascual, L. Marroyo, P. Sanchis, F. Guinjoan, Fuzzy logic-based energy management system design for residential grid-connected microgrids. IEEE Trans. Smart Grid **9**(2), 530–543 (2018)
11. A. Mills, *Integrating Solar PV in Utility System Operations* (Argonne National Laboratory, Lemont, IL, 2014). LBNL-6525E

12. K. Divya, J. stergaard, Battery energy storage technology for power systems—an overview. Electr. Pow. Syst. Res. **79**, 511–520 (2009)
13. A. Houari, D. Abbes, A. Labrunie, B. Robyns, Hybridization of electrical energy storage for intelligent integration of photovoltaics in electric networks, in *17th European Conference on Power Electronics and Applications (EPE'15 ECCE-Europe)*, (2015), pp. 1–10
14. D. Abbes, A. Labrunie, B. Robyns, Sizing and techno-economic analysis of a grid connected photovoltaic system with hybrid storage, in *51st International Universities PowerEngineering Conference (UPEC), Coimbra, Portugal, 6–9 Sept, 2016*, (2016)
15. CRE, *Cahier des charges de l'appel d'offres portant sur la réalisation et l'exploitation d'installations de production d'électricité à partir de techniques de conversion du rayonnement solaire d'une puissance supérieure à 100 kWc et situées dans les zones non interconnectées* (2015). Pdfdocument

Development of High-Capacity Lithium Sulfur Batteries

Mariasole Di Carli, Margherita Moreno, Gabriele Tarquini, Alfonso Pozio, Annalisa Aurora, Livia Della Seta, and Pier Paolo Prosini

Abstract In this paper we want to show some of the most recent results obtained in our laboratory concerning the fabrication of lithium sulfur batteries. For their construction we used two different binders and two carbons with different surface areas, deposited directly on the separator. Sulfur was introduced mixed with the electrolyte in the form of polysulfide. The particular cell configuration has allowed to obtain stable specific capacities after numerous charge and discharge cycles of more than 800 and 1200 mA h/g and low cell resistances.

1 Introduction

In a rapidly changing electrical system scenario, where the energy mix sees an increasingly massive penetration of energy from renewable sources, technological solutions for the accumulation of electrical energy will progressively acquire a fundamental role. In recent years, innovative storage systems have been developed, in particular lithium-ion batteries, with a double or triple energy density compared to other previous technologies [1, 2]. Nevertheless, further technological progress and greater cost competitiveness are required for the development of storage systems for electro-mobility or stationary energy storage. To increase the energy density of the energy storage systems, it is necessary to use innovative electrochemical materials and systems. Lithium sulfur batteries (Li-S) are promising candidates to replace lithium-ion batteries because cheaper and lighter and, weight for weight, can store almost twice as much energy [3, 4]. The development of this type of battery would not only allow the development of electric vehicles but would also allow

M. Di Carli · M. Moreno · A. Pozio · A. Aurora · L. Della Seta · P. P. Prosini (✉)
ENEA - C.R. Casaccia, Santa Maria di Galeria, RM, Italy
e-mail: mariasole.dicarli@enea.it; margherita.moreno@enea.it; alfonso.pozio@enea.it; annalisa.aurora@enea.it; livia.dellaseta@enea.it; pierpaolo.prosini@enea.it

G. Tarquini
University "La Sapienza", Rome, RM, Italy
e-mail: gabriele.tarquini@uniroma1.it

W. Zamboni, G. Petrone (eds.), *ELECTRIMACS 2019*, Lecture Notes in Electrical Engineering 615, https://doi.org/10.1007/978-3-030-37161-6_53

the construction of large storage facilities to support the electricity grid. This work reports the research activities carried out in ENEA during the last year to develop lithium sulfur batteries with high energy density.

2 Materials and Methods

2.1 Preparation of the Electrode Tape

The electrodes were prepared directly on glass microfiber filters used as battery separators (Whatman® glass microfiber filters, Grade GF/A). The electrodes have been prepared with two different binders: ethylene vinyl acetate (EVA) or a self-cross-linking acrylic dispersion (CRILAT) both from Vinavil SpA. The slurry was prepared by dissolving the Byk 2061 dispersant in water, which was followed by the addition of Ketjenblack (KJB) carbon or Super P carbon. After 5 min of mixing at 2000 rpm, using the Thinky Planetary Mixer (Thinky ARE-310), the binder, the plasticizer, and finally the required amount of water were added. The resulting mixture was deposited on the support. Circular electrodes with a diameter of 10 mm were punched from the tape and used as the cathode for the electrochemical characterization.

2.2 Synthesis of Li_2S_8

Lithium sulfide (Li_2S, 99.9%, Acros Organics) and elemental sulfur (Furuuchi Chem.; 99.9%) were weighed in a 1:7 mole ratio and suspended in a 1:1 mixture of 1,3-dioxolane (DOL, 99.5%, Sigma-Aldrich) and 1,2-dimethoxyethane (DME, 99%, Sigma-Aldrich). The DOL/DME mixture was heated at 80 °C under magnetic stirring for 48 h to obtain a homogeneous dark red solution. The concentration of Li_2S_8 was 0.5 M. A solution of 1.0 M of lithium bis-trifluoromethane-sulfonyl-imide (LiTFSI) and 0.1 M of lithium nitrate ($LiNO_3$) in a 1:1 mixture of DOL/DME was used as the electrolyte. The catholyte solution was prepared by adding 10 ml of the Li_2S_8 solution to 40 ml of electrolyte: the final concentration of Li_2S_8 in the catholyte solution was 0.1 M.

2.3 Chemical-Physical Characterization

The morphology of the working electrode was studied by scanning electron microscopy (SEM). High-magnification microphotographs were obtained using a VEGA3 TESCAN SEM. The specimens were directly mounted onto a conductive double-face carbon tape, which was previously mounted on a slab. The TGA was carried out with the TA Instruments Q600 SDT analyzer in a temperature range of 20–750 °C, at a heating rate of 10 °C/min, in air flow.

2.4 Electrochemical Measurements

Electrochemical characterizations were performed with LIR2016-type coin cell, using lithium foil as a counter-reference electrode and the carbon tape as the working electrode. A Whatman glass microfiber filter (Grade GF/A) was used as separator. The separator was soaked by 30 μL of catholyte solution. The sulfur content in the catholyte was about 1.20 mg/cm^2. Cycling tests were automatically carried out with a battery cycler (Maccor 4000). Electrochemical impedance spectroscopy (EIS) was performed with a model 1260 impedance/gain-phase analyzer (Solartron Analytical) using an electrochemical workstation (Solartron Analytical 1287 electrochemical interface). All materials and cells were handled and assembled in a controlled atmosphere in a dry room (R.H. < 0.1% at 20 °C).

3 Results and Discussion

3.1 Morphological Characterization of the Electrodes

The carbon-based electrodes were morphologically characterized by SEM as shown in Fig. 1. The KJB film extends in a very inhomogeneous way and a very fragmented electrode surface is obtained. The electrode prepared with the Super P appears instead very homogeneous.

3.2 Thermal Characterization of the Electrodes

The purpose of the measurement was to determine the amount of material deposited on the glass fiber. For this reason the measurements were carried out in air, in order to induce total combustion of the carbonaceous material and the above deposited additives. Figure 2 reports the result of the TGA.

The glass fiber without any deposit (green curve) shows a weight loss of less than 2% due to moisture and impurities. The thermogravimetric curve for the KJB/EVA electrode (red curve) shows a weight loss of 0.7765 mg. The initial weight of the sample was 5.820 mg. Assuming a physiological loss of the glass fiber by 1.8%, as shown by the green curve, the contribution of the glass fiber to the total weight loss is about 0.105 mg. The KJB/EVA electrode, therefore, weighs 0.672 mg and

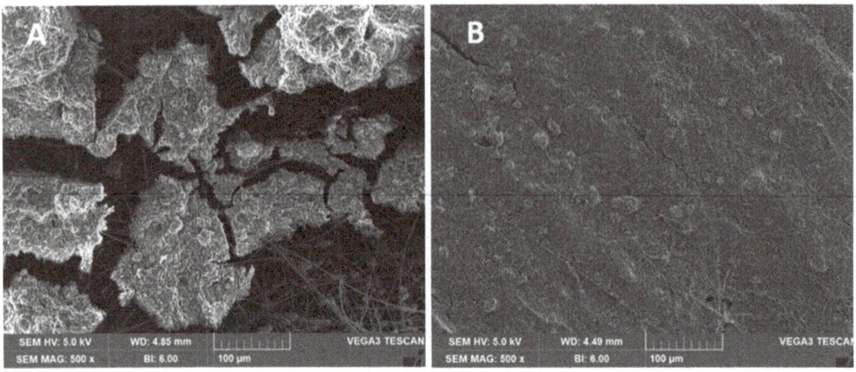

Fig. 1 SEM image of the glass fibrer coated with a layer of KJB/CRILAT (**a**) and Super P/CRILAT (**b**)

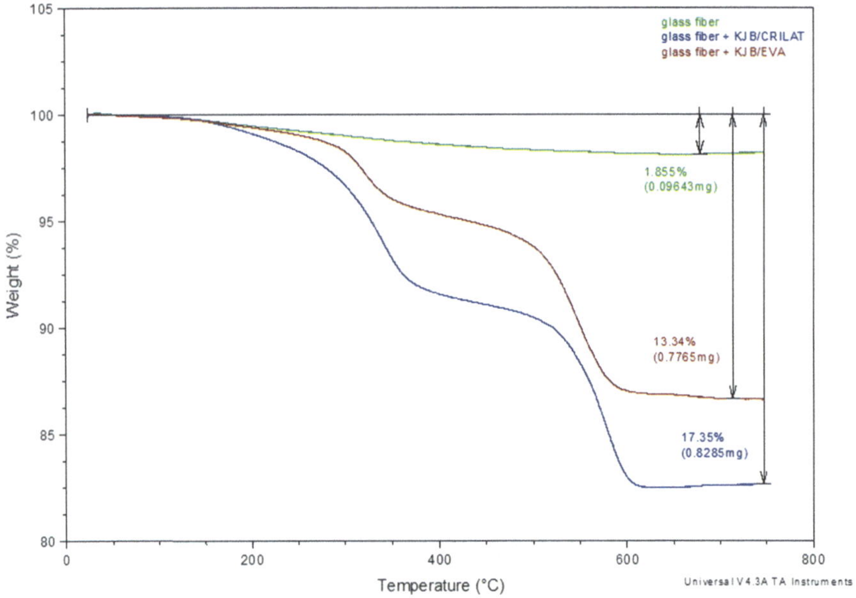

Fig. 2 TGA of the glass fibrer without KJB (green curve) and with KJB. Two different binders (EVA, red curve, and CRILAT, blue curve) have been tested

it corresponds to 11.5% by weight of the entire electrode. A similar argument can also be made for the KJB/CRILAT electrode. The initial weight of the sample is 4.774 mg. The contribution of the glass fiber is calculated to be 0.086 mg that has to be subtracted from the total weight loss of 0.8285 mg. The KJB/CRILAT electrode, therefore, weighs 0.743 mg and corresponds to 15.5% by weight of the entire electrode.

3.3 Electrochemical Characterization of the Electrodes

The Super P/EVA-based electrodes deposited on the glass fiber showed an initial specific capacity of about 1350 mA h/g at the first cycle and then undergo a continuous and gradual decrease up to about 900 mA h/g at the tenth cycle. The capacity loss decreases in subsequent cycles and the specific capacity evaluated at the 60th cycle was around 825 mA h/g (Fig. 3).

The KJB/CRILAT-based electrodes deposited on the glass fiber showed a higher first-cycle specific capacity. Also in this case a capacity reduction upon cycling is observed. After 10 cycles the specific capacity is reduced to about 1200 mA h/g (Fig. 4).

3.4 EIS Analysis

The EIS analysis was performed by using the model developed by Deng et al. [5] and Fig. 5 shows the equivalent circuit used to fit the results.

In this model, Re represents the resistance of the electrolyte. Rint//CPEint is the contact resistance at the interphase and its relative capacity in the body of the electrode. These elements simulate the electronic transfer between the carbon grains. Rct//CPEdl represents the resistance to charge transfer of the electro-active species and its capacity relative to the electrode/electrolyte interface. Finally, CPEdif represents the diffusive impedance referred to the lithium-ion diffusion process. Impedance analysis was performed on the Super P/EVA electrode. The cells assembled with this electrode show that the electrode interface changes drastically after the first charge/discharge cycles.

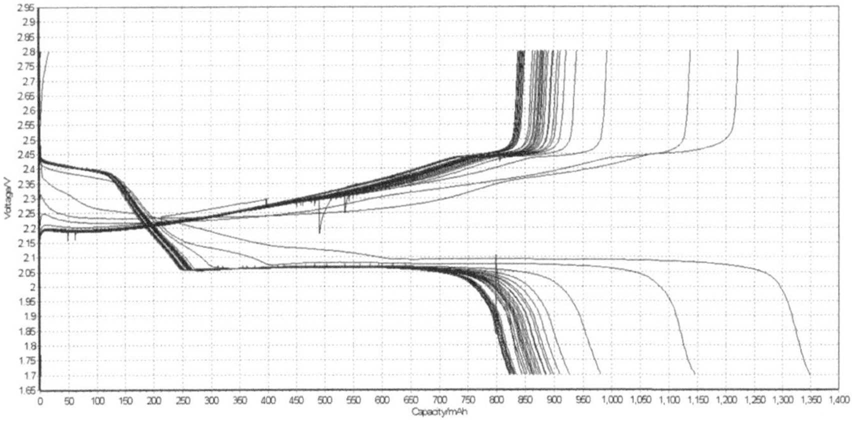

Fig. 3 Voltage profiles for the first 60 cycles of an electrode of Super P/EVA

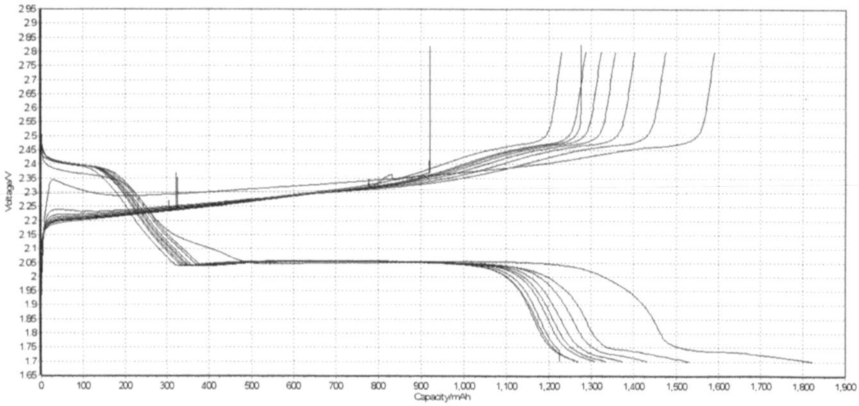

Fig. 4 Voltage profiles for the first 10 cycles of an electrode of KJB/Crilat

Fig. 5 Equivalent circuit for the Li-S cell

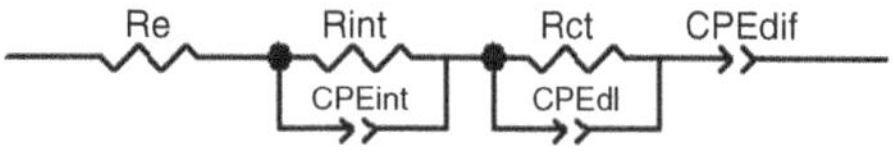

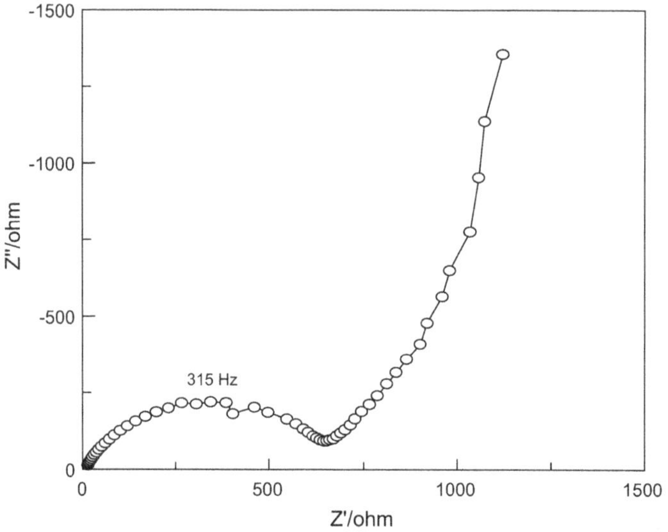

Fig. 6 Initial Nyquist diagram for cell with Super P/EVA electrode

Initially (Fig. 6) the Nyquist diagram shows a single semicircle with a maximum at a frequency of about 315 Hz and a diameter of about 760 Ω followed by a characteristic increase in the typical imaginary impedance of an electrode limited by a diffusion process.

In this situation in which the battery has not yet received the first discharge/charge, both the processes of electronic transfer between the grains of

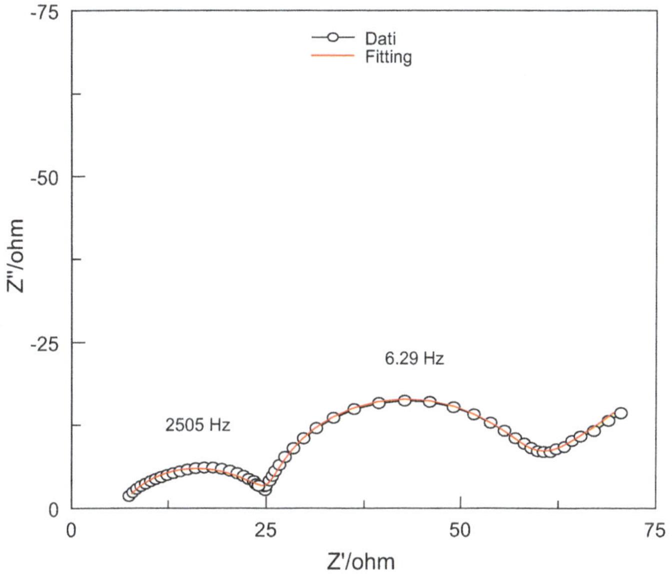

Fig. 7 Nyquist diagram for cell with Super P/EVA electrode after 10 cycles

which the carbon-based electrode is composed and the diffusion of the redox species to the electrode/solution phase, in this case Li_2S_4 presumably, are limited. After 10 cycles the inter-granular contact resistance, highlighted by the semicircle diameter at high frequencies, decreased at about 20 Ω. The resistance to charge transfer, highlighted by the second semicircle, also decreased (30 Ω, Fig. 7). The impedance indicates quite clearly that the charging and discharging process is limited by the charge transfer to the electrode/solution phase of the electro-active species. In any case, an improvement both in the process of charging and discharging (decrease in Rct) and in the process of electronic transfer between the grains of the carbon (decrease of Rint) is observed.

4 Conclusions

The cells assembled with the KJB- and Super P-based electrodes show excellent electrochemical performances in terms of specific capacity and cell resistance. In particular, the KJB-based electrodes, thanks to their characteristics of high specific capacity and stability, represent potential promising candidates for the construction of high-energy-density Li-S cells.

Acknowledgments Part of this work was carried out within the research program "Ricerca di Sistema Elettrico" inside the "Accordo di Programma MiSE - ENEA."

References

1. P.G. Bruce, B. Scrosati, J.M. Tarascon, Angew. Chem. Int. Ed. **47**, 2930–2946 (2008)
2. R. Van Noorden, Nature **507**, 26–28 (2014)
3. N. Moreno, A. Caballero, J. Morales, E. Rodriguez-Castellon, J. Power Sources **313**, 21–29 (2016)
4. M. Wei, P. Yuan, W. Chen, J. Hu, J. Mao, G. Shao, Electrochim. Acta **178**, 564–570 (2015)
5. Z. Deng, Z. Zhang, Y. Lai, J. Liu, J. Li, X. Liu, J. Electrochem. Soc. **160**, A553–A558 (2013)

An Impedance-Based Life-Monitoring Technique for a Graphene Water Filter

Stefano Bellucci, Antonino Cataldo, Luigi Ferrigno, Samuele Giovannetti, and Antonio Maffucci

Abstract The paper deals with the design and verification of a monitoring system for the analysis of the useful life of a water filter. The filter is made of pressed graphene nanoplatelets, obtained from commercial graphite with a low-cost fabrication procedure. The state of the filter is monitored by measuring the electrical impedance at the port of a suitable circuit embedding the graphene filter. It has demonstrated a good sensitivity of the impedance with respect to the saturation level of the pollutants into the filter. The technique is also shown to provide a high level of reproducibility and stability with environmental conditions.

1 Introduction

Due to its outstanding properties, the graphene has been recently proposed for environmental applications, such as water remediation [1] and water desalinization [2]. Delocalized π-electrons, surface area, and functionalization are the physico-chemical peculiarities of graphene that are most involved in the process of metal ion and organic compound pollutant adsorption [3]. In fact, one of the available mechanisms of adsorption of cations is the interaction with the π-electron system to form a complex. Moreover, the same interaction is involved between the π-electrons and organic aromatic compounds or organic compounds having high electronic density (like double or triple bonds).

The use of graphene filters for water and other liquids has been successfully demonstrated in the last years [1, 4, 5], with excellent results: the graphene in [1] has been demonstrated to filter water and other liquids about 10 times faster than the current commercial filters. Given these results on prototypes, the present stage

S. Bellucci · A. Cataldo
INFN-LNF, Italian Institute of Nuclear Physics, Frascati, RM, Italy
e-mail: stefano.bellucci@lnf.infn.it; antonino.cataldo@lnf.infn.it

L. Ferrigno · S. Giovannetti · A. Maffucci (✉)
DIEI, University of Cassino and Southern Lazio, Cassino, FR, Italy
e-mail: ferrigno@unicas.it; maffucci@unicas.it

W. Zamboni, G. Petrone (eds.), *ELECTRIMACS 2019*, Lecture Notes in Electrical Engineering 615, https://doi.org/10.1007/978-3-030-37161-6_54

of research is aimed at solving some crucial challenges in order to move to industrial applications, such as the scaling-up of membranes to large area. Recently, a solution to such a problem has been proposed, based on an industrially adaptable method of blade coating [6].

In view of the industrial use of graphene as water filter, it is important to assess suitable procedures to measure its state, eventually addressing the paradigms of the predictive maintenance. In this paper, we propose a monitoring technique based on the measurement of the electrical impedance of the graphene filter. Section 2 presents the fabrication of a low-cost graphene filter, whereas Sect. 3 is devoted to the proposed experimental setup. The results of the analysis are discussed in Sect. 4: specifically, we monitored the adsorption of acetonitrile, a water-soluble organic compound having a triple bond, widely adopted in industrial processes.

2 The Graphene Filter

The water filters analyzed in this paper have been realized by means of low-cost graphene produced in the form of graphene nanoplatelets (GNPs), by using the simple and industrially scalable procedure developed at the INFN nanotechnology laboratory in Frascati [7–9]. The starting material is a commercial expandable graphite, i.e., a type of graphite intercalated with chemical substances, sulfates, and nitrates, inserted between the graphene planes of the graphite. The graphite is irradiated in a microwave oven for a few seconds and is expanded by heat shock, due to the sublimation of the intercalating molecules. The result is a worm-like graphene that is then placed in isopropyl alcohol and is subjected to sonication, which disperses nanoparticles in a solution consisting of a few layers. Next, a vacuum filtration is carried out, with a filter of adequate porosity, so that most of the isopropyl alcohol is aspirated and a compact and rather thick layer of GNPs is created on the filter. The obtained compound is placed in the oven for some hours, in order to guarantee the complete evaporation of alcohol. Finally, the compound is pressed to obtain the final filter, showed in Fig. 1, with a diameter of about 3.5 cm and thickness of 0.5 mm. A SEM image is also shown in Fig. 2 to put into evidence the surface detail of the filter.

3 The Monitoring System: Principles, Design, and Realization

The monitoring system proposed here is intended to check the status of the filter during its normal operation, namely, without interfering with the filtering activity. The basic idea is that of exploiting the high sensitivity of the electrical properties of the graphene to the presence of external elements adsorbed in its lattice. Indeed,

Fig. 1 The final GNP filter, obtained after the steps of sonication, alcohol filtering, and pressing

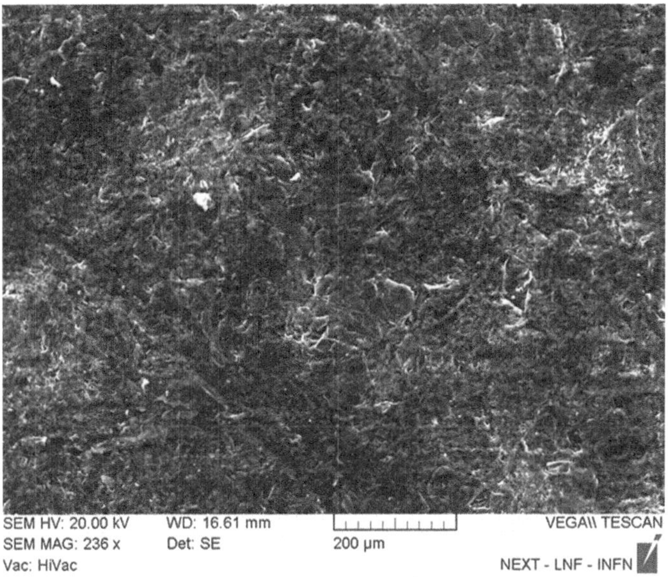

Fig. 2 SEM detail of the filter surface

intentional doping is a widely diffused technique aimed at tailoring the electrical properties of graphene, such as the carrier mobility [10], the electrical conductivity [11], and the electrical permittivity and magnetic permeability [12]. Therefore, the adsorption of pollutant molecules into the graphene filter must change its electrical impedance, so that its variation can be assumed as an index of the quantity of pollutant adsorbed by the filter.

The design of the proposed system is reported in Fig. 3: the solution to be depurated flows in a hollow plastic cylinder, passing through the graphene filter. Two metal rings are put directly in contact with the filter, anchored at the ends of the two tubes. The overall system is geometrically fixed and sealed to prevent any loss of load: the realized setup is shown in Fig. 4.

The electrical impedance is measured between the two rings shown in Fig. 3. The two wires for the impedance measurement are connected to each metal ring (these wires are visible in Fig. 4). The impedance measurements were performed by using a commercial chip (SENSIPLUS® chip from Sensichips [13]) and a PC for data processing. This chip allows performing the analysis at different frequency values, in the range of 20–75 kHz. Before each measurement, a proper calibration procedure is executed, able to compensate the effects of contacts, wire impedances, and parasitic circuits. This procedure allows to improve the reproducibility of the hand-made realized setup.

The measurement system has been verified in terms of reproducibility of the results and of stability with environmental parameters. The reproducibility has been verified by measuring the value of the imaginary part of the impedance with five different filters, without pollutants, at the three operating frequencies of the adopted chip. The results reported in Table 1 clearly show that using the higher available frequency (75 kHz) leads to the best performance. Next, the stability of the measurement results against the variation of the temperature has been checked, by measuring the impedance values for one of the filters, in the temperature range of 24–45 °C, at five different frequency values. The results are reported in Fig. 5: once again, the best performance is obtained when choosing a frequency of 75 kHz, which provides the smaller variation of reactance with respect to the temperature variation, compared to other frequencies. Therefore, in the following, such a frequency value is chosen to perform the analysis of the filter status.

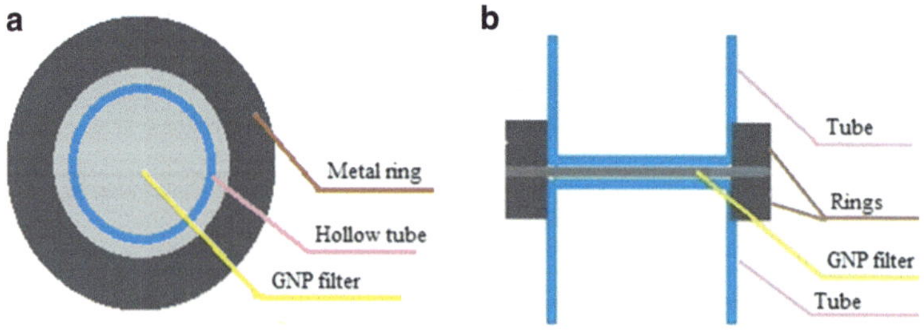

Fig. 3 The proposed setup for the impedance measurement of the graphene filter: (**a**) section view, (**b**) side view

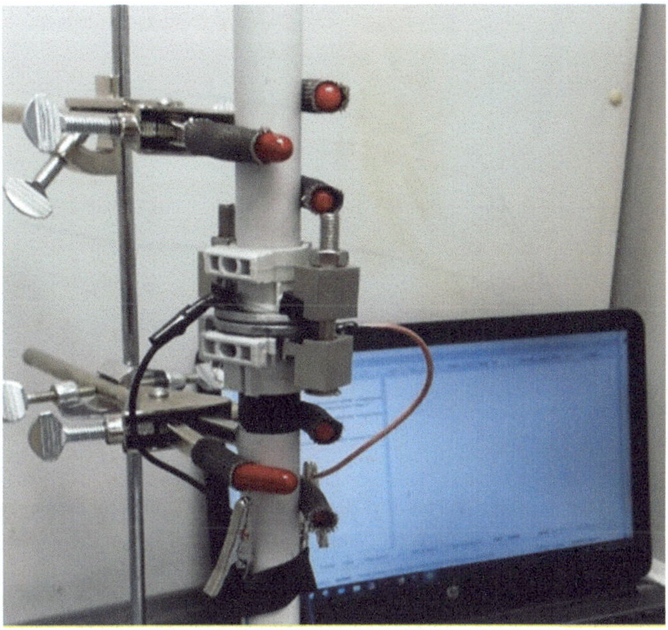

Fig. 4 The realized setup for the impedance measurement of the graphene filter

Table 1 Reproducibility test: the obtained standard deviation values for the imaginary part of the impedance

Frequency (kHz)	Standard deviation
25	6.11
50	3.99
75	1.47

4 Results and Discussion

The system in Sect. 3 has been employed to detect the state of the graphene filter, when it is used to remove acetonitrile, a toxic solvent of organic nature of wide industrial use.

Three different case studies have been analyzed: in the first two cases, the filter is used to adsorb the pure pollutant, whereas in the third case, the filter is used to clean a solution of water and pollutant. In all cases, even if the experimental results have shown appreciable sensitivities both in the real and imaginary part of the impedance, due to the adopted setup, the variation of imaginary part seems to show a better repeatability. Hence, in the following, we will refer to the measured reactance. Table 2 reports the initial values of the measured reactance, X_0, for the considered case studies, confirming the good reproducibility observed in Sect. 3.

The impedance measurement on the graphene filters has been then carried out, by varying the pollutant concentration until the saturation of the filter is reached.

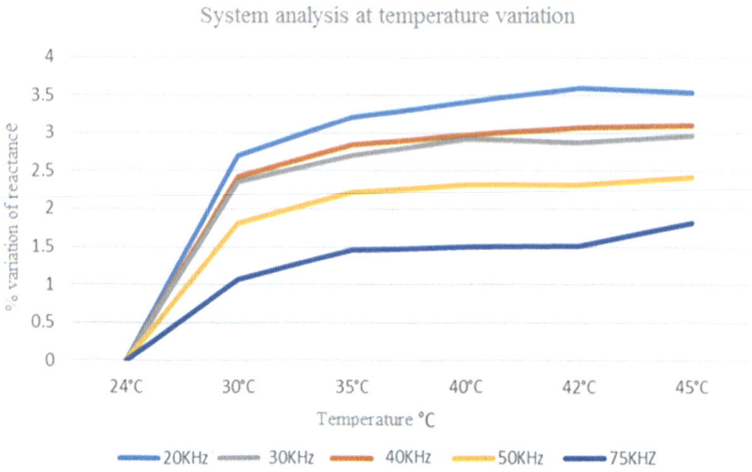

Fig. 5 Variation of the measured reactance vs. temperature, for five different frequency values

Table 2 Measured initial value of the reactance (at 75 kHz) of the filters used for the analyzed case studies (0% pollutant concentration)

Case study	Initial reactance $X_0(\Omega)$
1. Filter with pollutant only	16.36
2. Filter with pollutant only	16.37
3. Filter with polluted water	16.62

Specifically, the saturation is detected when the reactance is found to reach a stabilized value.

The results for the first two case studies are given in Fig. 6, which shows the time evolution of the relative reactance variation, defined as:

$$x_r(t) = \frac{X(t) - X_0}{X_0} \tag{1}$$

where $X(t)$ is the reactance measured at the time instant t. The filters 1 and 2 exhibit a good sensitivity to the presence of pollutant, since the relative variations of its reactance are of the order of some percent. In addition, a maximum relative variation of about 4.6% is reached when the two filters saturate. By weighing the filters before and after the experiments, it has been evaluated that they saturate after adsorbing about 0.18 ml of pollutant.

The third case study refers to the use of the same kind of filter to clean a solution of water and pollutant. The time evolution of the measured relative reactance value is given in Fig. 7. The relative variation exhibits the same qualitative behavior as for Case Studies 2 and 3, with a slightly higher sensitivity level, since now its value at saturation is about 6.5%.

These results confirm that the proposed monitoring system retains enough sensitivity to sense the presence of the pollutant in the filter and is also able to detect its saturation. In addition, the qualitative and quantitative behavior is very

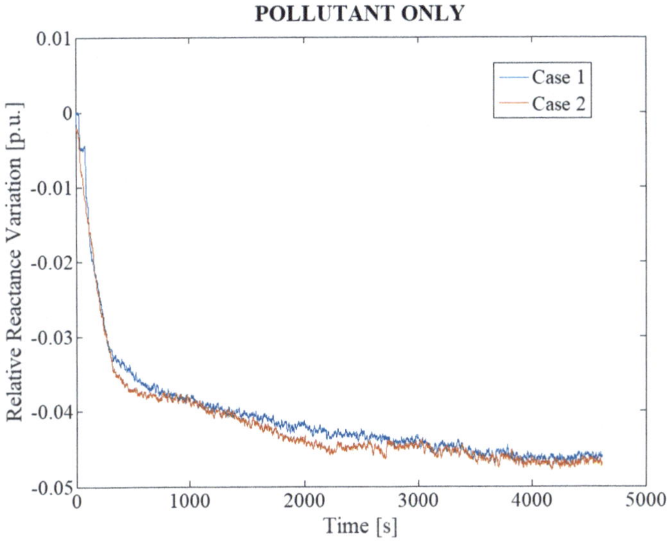

Fig. 6 Time evolution of the measured relative reactance variation (Eq. 1) for Case Studies 1 and 2 (filters with pollutant only). The reference values are given in Table 2

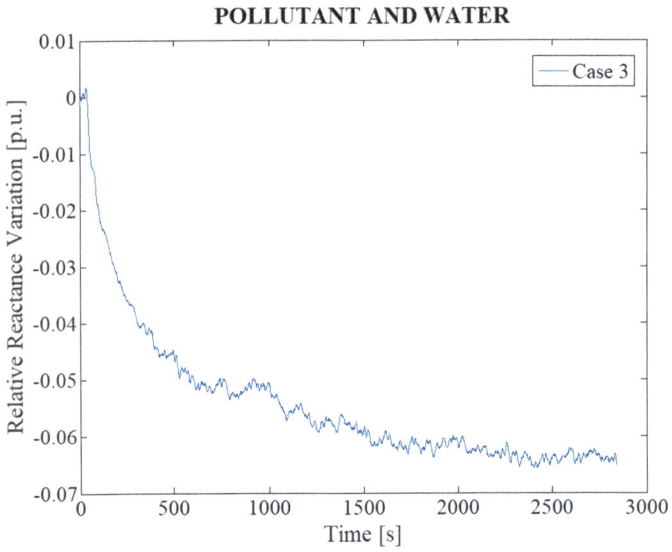

Fig. 7 Time evolution of the measured relative reactance variation (Eq. 1) for Case Study 3 (filter with polluted water). The reference value is given in Table 2

stable from sample to sample, so that the good reproducibility shown in Sect. 3 is here confirmed.

5 Conclusions

An experimental setup has been proposed and validated, to monitor the state of a graphene water filter, based on the measurement of the electrical impedance as a function of the pollutant concentration.

The graphene filter has been fabricated from a low-cost material (intercalated graphite), by means of a simple and industrially scalable technique based on heating, sonication, and pressing.

The proposed monitoring system has been shown to be sensitive to the pollutant presence, with a relative variation of the reactance of the order of some percent (up to about 5%). In addition, the system can detect the state of saturation. Experiments carried out on different filters have also demonstrated a good degree of reproducibility of the results as well as their stability with environmental conditions.

Acknowledgments The results here presented are developed in the framework of the project "Smart Distributed Systems," funded by the Italian Ministry of Education, University and Research (MIUR) program "Dipartimenti di Eccellenza 2018-2022."

References

1. H. Huang, Z. Song, N. Wei, Ultrafast viscous water flow through nanostrand-channelled graphene oxide membranes. Nat. Commun. **4**, 2979 (2013)
2. D. Cohen-Tanugi, J.C. Grossman, Water desalination across nanoporous graphene. Nano Lett. **12**, 3602–3608 (2012)
3. F. Perreault, A.F. De Faria, M. Elimelech, Environmental applications of graphene-based nanomaterials. Chem. Soc. Rev. **44**, 5861–5896 (2015)
4. R.R. Nair, H.A. Wu, P.N. Jayaram, Unimpeded permeation of water through helium-leak-tight graphene-based membranes. Science **335**, 442–444 (2012)
5. R.K. Joshi, P. Carbone, F.C. Wang, Precise and ultrafast molecular sieving through graphene oxide membranes. Science **343**, 752–754 (2014)
6. A. Akbari et al., Large-area graphene-based nanofiltration membranes by shear alignment of discoticnematic liquid crystals of graphene oxide. Nat. Commun. **7**, 10891 (2016)
7. A. Dabrowska, S. Bellucci, A. Cataldo, F. Micciulla, A. Huczko, Nanocomposites of epoxy resin with graphene nanoplates and exfoliated graphite: synthesis and electrical properties. Phys. Stat. Sol. B. **251**, 2599–2602 (2014)
8. A. Maffucci, F. Micciulla, A. Cataldo, G. Miano, S. Bellucci, Bottom-up realization and electrical characterization of a graphene-based device. Nanotechnology **27**, 095204–0951-9 (2016)
9. M. Potenza, A. Cataldo, G. Bovesecchi, S. Corasaniti, P. Coppa, S. Bellucci, Graphene nanoplatelets: thermal diffusivity and thermal conductivity by the flash method. AIP Adv. **7**, 075214 (2017)
10. D. Wei, Y. Liu, Y. Wang, H. Zhang, L. Huang, G. Yu, Synthesis of N-doped graphene by chemical vapor deposition and its electrical properties. Nano Lett. **9**, 1752–1758 (2009)
11. K.K. Kim et al., Enhancing the conductivity of transparent graphene films via doping. Nanotechnology **21**, 285205 (2010)
12. F. Shahzad et al., Sulfur doped graphene/polystyrene nanocomposites for electromagnetic interference shielding. Compos. Struct. **133**, 1267–1275 (2015)
13. https://www.sensichips.com/

Multicast Routing Protocol for LoRa Mesh Networks in Safety Critical Communications

Roberto Di Stefano and Fabrizio Marignetti

Abstract The risk reduction, in case of catastrophic events, is strongly conditioned by the possibility of performing mechanical actuation in extreme conditions and without electric energy. Many devices, which implement security features and fulfill these specifications, are on the market, however the capability to transfer data and information through reliable technologies is also required in order to monitor and coordinate the safety systems. A possible candidate that could provide a communication link, even in the absence of electricity, is LoRa technology, even if it has some problems and limitations mainly due to the topological configuration of the LoRaWAN communication network and the increasing number of devices and users. This paper reports a proposal to improve the reliability of LoRa data transfer in very severe environmental conditions and on large areas. Some accurate simulations conducted on a LoRa network, make it possible to verify the effectiveness of the proposed communication strategy.

1 Introduction

When an extraordinary event occurs in a high-risk environment, it is often necessary to promptly carry out maneuvers aimed at mitigating possible damage and increasing the overall level of safety with regard to people, things, and environment in general. This is especially the case of industrial plants, in which it is necessary to implement fire protection systems, actuate valves to close the flow of dangerous substances, or isolate compartmentalized environments.

Different types of devices able to perform the actuation even in the absence of electricity are present on the market. Their operation is based on the accumulation

R. Di Stefano (✉) · F. Marignetti
DIEI - University of Cassino, Cassino (FR), Italy
e-mail: distefano@unicas.it; marignetti@unicas.it

W. Zamboni, G. Petrone (eds.), *ELECTRIMACS 2019*, Lecture Notes in Electrical Engineering 615, https://doi.org/10.1007/978-3-030-37161-6_55

of mechanical energy in the form of compressed air, or in an elastic form. These actuators are subjected to very strict standards (i.e., ISO 26262, EN 50129, EN ISO 13849-1, IEC 62061, IEC 60601) that regulate the component functional safety on which the system safety as a whole depends. Manufacturers must therefore take all the necessary precautions to minimize the risk of malfunction even when extreme conditions are present, such as high temperatures caused by a fire.

In order to increase the reliability and robustness, during the design and development of the product, it is necessary to implement redundant functions and perform a preventive analysis of the possible causes of failure caused by random or systematic events or errors. As is known, the risk assessment process is essential to define the functional safety requirements. One of the intermediate results of the evaluation process is a list of the functions that allow to mitigate the consequences of failures, each of them characterized by its degree of reliability and the degree of criticality or associated risk [1–3]. The risk assessment process can be completed during the design and testing phase before the final installation. In other cases, it can also be extended during operation period, analyzing and studying the statistics of operation, intervention, maintenance, and failure [4, 5]. This further analysis complements the previous one, constituting in fact a feedback of considerable importance, which allows to detect causes of malfunctioning, which had not been taken into consideration and therefore to carry out, with greater precision, the evaluation of the risk. The result is the possibility of significantly improving the safety integrity level (SIL), increasing the risk reduction factor (RRF), both in a further design phase, or with extraordinary maintenance upgrading the product [6, 7]. It is therefore clear the need to equip these elements with an apparatus capable of detecting in detail the overall state and of carrying out an equally reliable and robust communication [8]. Then the following points worthy of study were identified:

- to study and design a system for detecting the component status;
- to identify an appropriate wireless data transmission and methodology;
- to study and implement a reliable and non-redundant communication protocol that can cover areas of considerable size.

2 LoRa Technology as Candidate in Critical Applications

2.1 LoRa Technology at a Glance

LoRa stands for long-range communication and is a proprietary modulation technique by Semtech, it allows the radio connection across long distance (some kilometers) because of its high sensitivity (10 dB better than GFSK) [9] with low energy consumption. It is derived by the chirp spread spectrum (CCS) [10] with an embedded forward error correction and allows to encode multiple bits (*SF*) per symbol. The transmission frequencies that are usually used are allocated just below 869 or 928 MHz, other frequencies being used in Asia. A LoRa device requires the setting of many parameters: spreading factor (*SF*), transmission power (*TP*), bandwidth (*BW*), coding rate (*CR*), carrier frequency (*CF*). This makes this technology very flexible as it allows to identify several possible options that have influence on the transmission distance, transmission time, integrity of data, and energy consumption. For this reason, researchers [11–16] have tried to identify general criteria for obtaining the best performances by a suitable combination of parameters setting, also in relation to the environment and the number of devices involved.

2.2 *Merits and Defects of LoRaWAN and Others Protocols*

LoRaWAN is the link layer protocol designed for LoRa. Its topology is deployed as star of stars, this implies the presence of gateways (nodes) that concentrate the uplink and downlink packets, respectively, from and to end-devices, routing them towards network servers. Since the gateways are connected to a physical network (internet) this implies the functional mix of the two networks: the wireless network is limited in its geographical extension by the possibility of installing gateways; even the functions and reliability of LoRaWAN depend heavily on gateways. In order to better meet the needs of different applications, three different classes of operation have been defined for LoRa devices which are part of a LoRaWAN: Class *A*, which provides for the management of two listening downlink time slots after the transmission of an uplink packet to the gateway. The class *B* which, in addition to slots considered in the operation of class *A*, defines additional listening time slots programmed on the basis of a time synchronization received from the gateway. The class *C* in which the device is continuously listening. Of course, these classes of operation involve a different energy consumption (Class *C* is usually powered by mains) and a different chance that data exchange can be successful. Because each gateway acts as a bridge among the end-devices and the backhaul access, the concentration of the data in a single node has strong implications on the transmission capacity and reliability. There are numerous recent studies that clear up the real limits of LoRaWAN technology in different environmental conditions, also defining analytical models that allow to calculate the real performance of network [16–19]. The probability of packet collision, which could be measured by using packet reception rate(PRR), derives by the difficulty of coordinating uplink transmissions, this dramatically compromises the scalability of network. A further factor that can limit the network size concerns the relation between spreading factor and bit rate is

$$BR = SF \cdot \frac{4}{(4 + CR)\left(\frac{2^{SF}}{BW}\right)}. \tag{1}$$

In fact, in order to limit the time required to transmit the payload, it is necessary to consider lower values of the spreading factor. In Europe, other limitations derive from the compliance with the requirements imposed by the European Telecommunications Standards Institute (ETSI) which requires a duty-cycle for each device not exceeding 1% and a maximum output power of +14 dBm. HART communication protocol born as an industrial automation control protocol on wired infrastructure by means of FSK modulation, superimposed on analog 4–20 mA communication lines. The line has the possibility to transfer both analog and digital information. The protocol has been ported in a mesh wireless 2.4 GHz infrastructure (WirelessHART), of course missing the analog transport, becoming very popular in industrial environment. The network is able to reconfigure itself and has healing capabilities, within certain limits, since it guarantees transmission even in the presence of any faults.

2.3 *Other Solution*

A mesh-like network is based on a different paradigm, each device also acts as a gateway, propagating the packet received towards the other devices, this allows to overcome the limitations in covering large areas, or characterized by environments that shield the radio waves, such as dull industrial surroundings, or tunnels, or subways. The mesh-type network can be organized, from the logical point of view, into flexible way in order to avoid the concentration of transmissions in a single node, moreover it can dynamically reconfigure itself facing unexpected situations, such as failures, node movements, loss of connection, etc. This property makes transmission of packets much more reliable, reducing the risk of collision, but has the negative effect of increasing latency times, energy consumption, bandwidth limitation. There are three types of multicast routing protocols: the first, called proactive, maintains the routes of group members and those of single nodes that are not part of groups; the second, called reactive, constructs the route when it is required for the transmission of a packet; the third is known as hybrid because it implements a combination of first two. Of course a complete survey of these technologies cannot be reported here, many papers are present in scientific literature, some of them are reported in [20–22]. In [23], Lundell et al. propose an hybrid communication protocol suitable for LoRa networks, however the specifications referred to in these proposals are mainly oriented towards energy saving and rapid dynamic reconfiguration of routing tables.

3 A Different Communication Strategy

3.1 Technical Specifications

A network of sensors for industrial and environmental monitoring is usually not powered by the electricity grid, except in emergency conditions where electricity can be missing. The network topology is substantially static under normal conditions, while it may require a reconfiguration of its paths in case one of the nodes is not available or intentionally excluded. This unavailability can involve even more contiguous nodes, so the ability of the logical network to reconfigure itself must resort to LoRa's technological features in an attempt to establish transmission paths even at considerable distances. The data traffic may not be symmetrical, that is, it may be unbalanced in uplink rather than in downlink, in fact the latter may be related to control remote security systems, while the uplink mainly conveys the communication of the status of devices and environmental conditions. From this point of view, the protocol that seems to be most suitable for this type of network is proactive, such as optimized link state routing protocol (OLSR) that fixes a number of multi-point relays, each of which connects a single pair of nodes. The routing tables are updated only if necessary in order to minimize communication overhead. In the specific case, under normal conditions, the network does not present problems of high traffic density, and does not require special measures to limit the energy required for communications, so it is possible to define a configuration methodology that engages the network especially when the sensors and the actuators are powered regularly and instead have the ability to quickly reconfigure communication paths without performing complex operations when the state of emergency is in place.

3.2 Sensor and Actuators Network

The network topology, which is taken into account in this work, is basically composed of static nodes. They can be connected to sensors and actuators located in an area and installed for different purposes. As an example, a large industrial area may be considered, in which it is necessary to control emergency actuators and monitor some environmental variables, or a metropolitan area where it is necessary to measure the pollution, traffic or, in general, other conditions or variables of interest. The nodes could also be installed in closed environments, such as tunnels, mines, foundries, etc. So it is a network that, from the geometric point of view, is not affected by important changes. The network can be strongly influenced by extraordinary events that can make some nodes inactive, for example, fires, explosions, even simply uncontrolled temperature increases [24]. The proposed communication protocol was designed to allow, as far as technically possible, the transfer of information from one node to another, identifying the best possible route and minimizing the network when a critical situation occurs, exploiting at best the

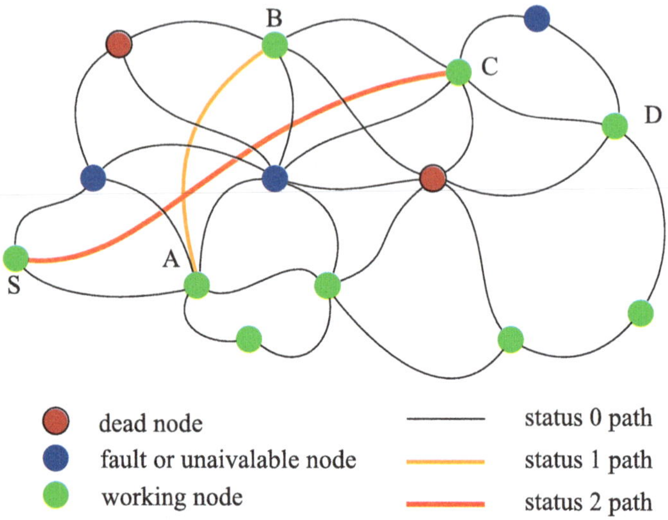

Fig. 1 Possible LoRa network form

characteristics of the LoRa technology. In Fig. 1 a sketch of a possible network is reported. By the way, none of the nodes has backhaul access to internet, even if it could be connected to a local server.

3.3 General Communication Strategy

The proposed communication strategy considers two different operating states, the normal one, in which the nodes are powered by the electric grid or in any case do not have particular energy saving requirements, and the emergency one, during which the network may be affected by losses of connectivity, lack of power, and inactive nodes. When the network operates in normal condition, the communication strategy aims to identify the optimal routes, but it also identifies possible alternatives to be used when alterations of topology occur. In this state, it is preferable that the nodes make periodic reconfiguration attempts to establish a certain number of alternative routes to be stored systematically in hierarchical tables that will be used if the default paths are unavailable. When the network is in emergency state the communication path is chosen among the possible paths stored in the hierarchical tables, it will be discarded if no acknowledge will return back by the destination end-node. In this case another path will be chosen in place. The route switching process will be described in more detail afterwards. This strategy relieves the network from self-commitment and reconfiguring itself when the emergency state is in place, it offers an immediate alternative to communications reducing energy needs just when energy should not be wasted.

4 Communication Handling and Protocol Design

The design of the protocol must consider some aspects that concern both LoRa technology and the limitations imposed by regional regulations. In Europe, there are significant limitations on the duty-cycle, which cannot exceed 1% for each device. This limits the amount of information that can be transmitted for the purpose of determining communication paths and packet size. Considering the node numerosity in the network, the identification of the nodes can be done not using a mac-address formed by 8 bytes, but by an ID formed by a lower number of bytes (i.e., only two) that are set unambiguously during installation. This allows to save on the payload size favoring a higher transmission frequency.

The state of the network can assume different degrees of emergency. The lowest is that in which the status of the nodes is normal operation or in a programmed suspension condition, this state corresponds to the use of the default set of routing tables. The following states correspond to a situation in which one or more nodes cannot guarantee the transfer of information due to unforeseen events. In this case, all nodes use the second set of routing tables. If there is an additional set of routing tables, this will be adopted by all nodes if the second set does not guarantee the transfer of information. The switching mechanism among the statuses must involve all the nodes, in fact it is necessary to maintain consistency between the routing tables adopted by the nodes, because the information stored in the routing tables regards also TX/RX parameters which should have to correspond for all the nodes of the network.

This circumstance is guaranteed only if all nodes use the same tables set. Each node must decide autonomously in what condition to operate on the basis of information received from other nodes or on the basis of information that has not been received.

When the network is in normal operating conditions, it is likely that the maximum distance between a pair of nodes is enough to guarantee efficient communication with a low spreading factor. This allows the frequency of the transmitted packets to be increased, respecting the duty-cycle limit, in order to identify all the possible routes and to create the first routing tables, i.e., those with the highest priority.

When the short communication channels are stable, limited time windows will be set during which a pair of nodes at a greater distance will try to establish a new hop with different TX/RX parameters, for example, with a greater SF. This is achieved by transmitting a packet of Route REQuest (RREQ) to the node with which to connect.

The RREQ defines the TX/RX parameters for test communication between the two nodes and also defines the time parameters within which this transmission will take place, these data are written into payload of the RREQ packet. Therefore, after this exchange of information, there will be a finite time window during which two nodes can establish a test transmission useful to verify the effectiveness of the connection. If the test is positive, the respective routing tables will be updated. The test can fail either because of the weakness of the received signal, or because of the difficulty in receiving a message without errors. Of course also the receiver time-out produces the same result and therefore the information dumping. Each node then holds two or more sets of routing tables and periodically performs operations that are intended to create and update tables.

The quality of the transmission was performed by considering the values of the RSSI (signal strength indicator received), and the number of errors corrected during transmission of the test messages.

Figure 2 reports the general flowchart of the procedure which is scheduled in the logic of each node.

In this experimental work, the number of routing tables sets is only two, but it could be increased, although it does not seem that more than three sets could further improve the probability that a packet can be successfully transferred. The packet format used for this purpose is shown in Fig. 3.

The packet contains the same fields of standard packets: the identifier of source (S-ID, two bytes), the identifier of destination (D-ID, two bytes), the transmitter identifier (T-ID, two bytes), the receiver identifier (R-ID, two bytes), the type of message (T, one byte), the priority level of the routing table that is attempted to create (Lev, one, byte), the new transmission and reception parameters (TR-Par, six bytes), the temporal characteristics of the window to be used (Window, four bytes).

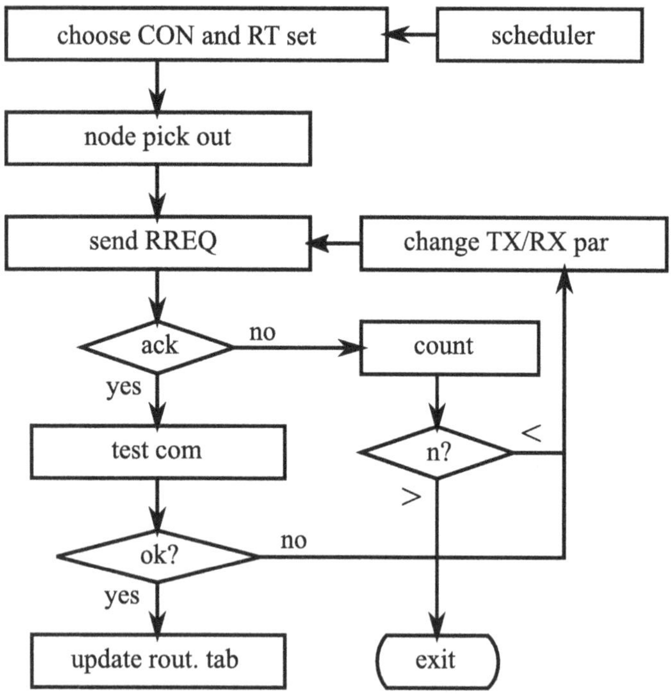

Fig. 2 Diagram of the scheduled procedure to update the routing tables

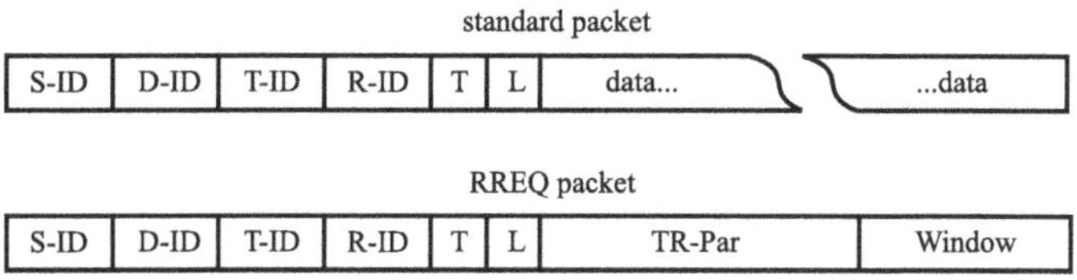

Fig. 3 Route Request (RREQ) packet and acknowledge

Table 1 Routing table

Status	Type T	Dest D-ID	Next relay T-ID	TX/RX parameters TX/RX
0	C	12	8	short
0	D	16	3	short
0	D	16	10	short
1	D	16	16	long
0	S	12	10	short
0	D	12	10	short

The packet has the same format of a standard packet, except for the contents of the identifiers, in fact S-ID and T-ID are the same, as are also D-ID and R-ID. It will be transmitted to a non-contiguous node to ask to establish a new path. The receiving node responds to the request with an acknowledge message sending back the same packet, of course with swapped sending and receive IDs.

4.1 Structure of the Routing Tables

Table 1 shows a possible routing table related to a certain node. The first column shows the status of the node, it is possible to notice how there are two possible paths to the destination 16. The first, relative to the state 0, passes through the relay 10, the second, relative to the state 1, comes directly to the destination node 16 with suitable transmission parameters.

4.2 Management of Communications in States Other than 0

In case a node is unable to communicate with an adjacent one, listed in its RT and marked in 0 state, a verification procedure is activated to assess that such malfunction is due to a permanent failure rather than to a temporary irregularity. At the end of this phase, the node can enter the next state (in this example state 1) and use a different path for routing. State 1 implies different parameters of transmission and reception for some nodes too, so two time windows will be established, even of different sizes, during which the node will switch its TX/RX mode. Naturally, the information necessary to communicate in these time windows, with the correct transmission and reception parameters, must be transmitted to the adjacent still functioning nodes.

5 Performance Evaluation

The communication strategy and relative protocol have been tested simulating a set of nodes each composed by an embedded LoRa SX1276 module mounted on a nucleo STM32L476RG development board which features an ultra-low-power micro-controller. Some of this modules are also equipped with an IKS01A1 sensor board, which provides numerous environmental data (see Fig. 4). In order to perform a complete and accurate simulation, OMNeT++ has been used. This tool contains a complete set of c++ libraries for simulate many kind of networks topologies also building custom nodes, sensors, and protocols. It allows to evaluate the performance of implemented systems. The most important feature is represented by its modularity, in fact it could be integrated by other frameworks that give facility to simulate off standard networks. The proposed network uses the libraries for LoRa devices (FLoRa) developed at Aalto University School of Science (Finland). The simulation allows to check the statistics of packets lost due to possible collisions, and to measure the dynamics of creation and update of the routing tables. The use of an accurate virtual system is particularly useful for verifying network performance in situations that can hardly be verified experimentally.

Fig. 4 LoRa modules used for experimental tests

6 Conclusions

A methodology for handling a wireless network based on LoRa technology has been designed and implemented. The aim of the proposed technique is to maximize the probability that a given message reach its destination in the same network, even in emergency conditions, i.e., when some nodes in the network are faulty or inefficient.

The communication methodology presents some implementation difficulties that can be overcome. Improvements may be implemented in order to verify the performance and reliability of communications even on networks composed of LoRa nodes with less homogeneous characteristics. Further improvements could include: the management of information security, providing the network with encryption capabilities, and the management of unexpected events such as failures or temporary inefficiency of transmission (for example, due to poor weather conditions).

Acknowledgements The authors acknowledge the financial support of the Regione Lazio thanks to the grant "Progetti di Gruppi di Ricerca finanziati ai sensi della L.R.Lazio13/08."

References

1. Z.E. Bhatti, P.S. Roop, R. Sinha, Unified functional safety assessment of industrial automation systems. IEEE Trans. on Ind. Inf. **13**(1), 17–26 (2017)
2. P. Stirgwolt, Effective management of functional safety for ISO 26262 standard, in *2013 Proceedings Annual Reliability and Maintainability Symposium (RAMS)*, pp. 1–6, Jan 2013
3. T. I. of Electrical and E. E. USA, 2017 National Electrical Safety Code(R) (NESC(R)), in *2017 National Electrical Safety Code(R) (NESC(R))*, pp. 1–405, Aug 2016
4. D. Etz, T. Frühwirth, A. Ismail, W. Kastner, Simplifying functional safety communication in modular, heterogeneous production lines, in *2018 14th IEEE International Workshop on Factory Communication Systems (WFCS)*, pp. 1–4, June 2018
5. J.V. Bukowski, Modeling and analyzing the effects of periodic inspection on the performance of safety-critical systems. IEEE Trans. Reliab. **50**(3), 321–329 (2001)
6. J.V. Bukowski, A unified model for evaluating the safety integrity level of safety instrumented systems, in *2008 Annual Reliability and Maintainability Symposium*, pp. 137–142, Jan 2008
7. J. Blanquart, P. Baufreton, J. Boulanger, J. Camus, C. Comar, H. Delseny, J. Gassino, E. Ledinot, P. Quéré, B. Ricque, Software safety assessment and probabilities, in *2016 46th Annual IEEE/IFIP International Conference on Dependable Systems and Networks Workshop (DSN-W)*, pp. 213–214, June 2016
8. E. Babeshko, V. Kharchenko, K. Leontiiev, E. Ruchkov, V. Sklyar, Reliability assessment of safety critical system considering different communication architectures, in *2018 IEEE 9th International Conference on Dependable Systems, Services and Technologies (DESSERT)*, pp. 17–20, May 2018
9. Semtech, LoRa (TM) modulation basics application note an1200.22. Semtech Corporation - Wireless Sensing and Timing Products Division, Tech. Rep., (2015), https://www.semtech.com/uploads/documents/an1200.22.pdf
10. B. Reynders, S. Pollin, Chirp spread spectrum as a modulation technique for long range communication, in *2016 Symposium on Communications and Vehicular Technologies (SCVT)*, pp. 1–5, Nov 2016

11. M. Bor, U. Roedig, LoRa transmission parameter selection, in *2017 13th International Conference on Distributed Computing in Sensor Systems (DCOSS)*, pp. 27–34, June 2017
12. J. Lim, Y. Han, Spreading factor allocation for massive connectivity in LoRa systems. IEEE Commun. Lett. **22**(4), 800–803 (2018)
13. Z. Qin, J.A. McCann, Resource efficiency in low-power wide-area networks for IOT applications, in *GLOBECOM 2017 - 2017 IEEE Global Communications Conference*, pp. 1–7, Dec 2017
14. F. Cuomo, M. Campo, A. Caponi, G. Bianchi, G. Rossini, P. Pisani, EXPLoRa: Extending the performance of LoRa by suitable spreading factor allocations, in *2017 IEEE 13th International Conference on Wireless and Mobile Computing, Networking and Communications (WiMob)*, pp. 1–8, Oct 2017
15. F. Yu, Z. Zhu, Z. Fan, Study on the feasibility of LoRaWAN for smart city applications, in *2017 IEEE 13th International Conference on Wireless and Mobile Computing, Networking and Communications (WiMob)*, pp. 334–340, Oct 2017
16. O. Georgiou, U. Raza, Low power wide area network analysis: Can LoRa scale? IEEE Wireless Commun. Lett. **6**(2), 162–165 (2017)
17. A. Pop, U. Raza, P. Kulkarni, M. Sooriyabandara, Does bidirectional traffic do more harm than good in LoRaWAN based LPWA networks? in *GLOBECOM 2017 - 2017 IEEE Global Communications Conference*, Dec 2017, pp. 1–6.
18. D. Bankov, E. Khorov, A. Lyakhov, Mathematical model of LoRaWAN channel access, in *2017 IEEE 18th International Symposium on a World of Wireless, Mobile and Multimedia Networks (WoWMoM)*, pp. 1–3, June 2017
19. K.A. Rahman, K.E. Tepe, Towards a cross-layer based MAC for smooth V2V and V2I communications for safety applications in DSRC/WAVE based systems, in *2014 IEEE Intelligent Vehicles Symposium Proceedings*, pp. 969–973, June 2014
20. P. Garnepudi, T. Damarla, J. Gaddipati, D. Veeraiah, Proactive, reactive and hybrid multicast routing protocols for wireless mesh networks, in *2013 IEEE International Conference on Computational Intelligence and Computing Research*, pp. 1–7, Dec 2013
21. K. Pavai, A. Sivagami, D. Sridharan, Study of routing protocols in wireless sensor networks, in *2009 International Conference on Advances in Computing, Control, and Telecommunication Technologies*, pp. 522–525, Dec 2009
22. T.N. Nagabhushan, S.P.S. Prakash, K. Krinkin, Power-saving routing algorithms in wireless mesh networks: a survey, in *2012 11th Conference of Open Innovations Association (FRUCT)*, pp. 107–115, April 2012
23. D. Lundell, A. Hedberg, C. Nyberg, E. Fitzgerald, A routing protocol for LoRa mesh networks, in *2018 IEEE 19th International Symposium on "A World of Wireless, Mobile and Multimedia Networks" (WoWMoM)*, pp. 14–19, June 2018
24. C.A. Boano, M. Cattani, K. Römer, An experimental evaluation of the reliability of LoRa long-range low-power wireless communication. J. Sensors Actuator Netw. (2017). https://doi.org/10.3390/jsan6020007

Investigation on the Fire Hazards of Li-Ion Cells

Paola Russo, Maria Luisa Mele, Giovanni Longobardo, Michele Mazzaro, and Cinzia Di Bari

Abstract Li-ion batteries had a great development in recent years, and their use has grown massively because of their higher energy and power density with respect to traditional ones.

However, their high energy density implies great danger in the event of malfunctions or failures, due to the emission of toxic and highly flammable substances. In the worst case, thermal runaway can occur. It is a chain reaction where unwanted reactions take place that leads to an uncontrolled and unstoppable increase in temperature. It can cause uncontrolled combustion and then explosion with great danger.

In order to identify the conditions that lead to the thermal runaway and to limit its occurrence, thermal stability of Li-ion batteries is here investigated. Thermal abuse tests are performed on lithium nickel manganese cobalt oxide cells from Panasonic in an ISO 5660 cone calorimeter.

Heat release rate is measured by changing the state of charge (SoC) of the cells and the radiant power of the cone calorimeter. The relationship between the SoC and the onset of the heat release is clearly revealed.

P. Russo (✉) · M. L. Mele
DICMA, Sapienza University of Rome, Rome, Italy
e-mail: paola.russo@uniroma1.it; marialuisa.mele@uniroma1.it

G. Longobardo · M. Mazzaro
DCPST, Italian National Fire Rescue and Service, Rome, Italy
e-mail: michele.mazzaro@vigilfuoco.it

C. Di Bari
ENEA DTE-PCU-STMA, CR Casaccia, Rome, Italy
e-mail: cinzia.dibari@enea.it

W. Zamboni, G. Petrone (eds.), *ELECTRIMACS 2019*, Lecture Notes in Electrical Engineering 615, https://doi.org/10.1007/978-3-030-37161-6_56

1 Introduction

Lithium-ion batteries are the most promising solution for feeding portable electronics, for battery electric vehicles, and for aerospace applications: they charge faster, last longer, and have a higher energy density than traditional batteries.

Their success is due to their main features: high energy and power density [1].

Despite its overall advantages, the greatest limitation to lithium-ion technology development is due to safety problems. In normal use conditions, lithium ions move from the anode to the cathode passing through the electrolyte, in a quasi-reversible process. If accidentally a Li-ion cell failure occurs, the power generated can become greater than the heat that the cell is able to dissipate. Consequently, the temperature rapidly increases, and a mechanism called thermal runaway can start [2, 3]. With the increase of internal temperature, the organic electrolyte begins to react with the other parts of the cell. The further reactions that take place contribute to the production of additional heat. Therefore, this mechanism could lead to uncontrolled combustion and then to explosion, with consequent projection of fragments and toxic gases and vapor release, mostly composed of carbon, hydrogen, and oxygen [2–5]. This can lead to disastrous consequences [6, 7].

The aim of this study is to examine the response to thermal abuse of NMC cells from Panasonic at different state of charge (SoC). To this aim heat release rate (HRR) in an ISO 5660 cone calorimeter is measured. The measurement of heat release rate is the key test used to evaluate the risk of fire of materials and products and to quantify the size of a fire, the rate of fire growth, and consequently the release of smoke and toxic gases [8, 9].

2 Experimental

2.1 Cell Composition

The tests are performed on Li-ion cells from Panasonic: NCM 18650 BM. The positive electrode is composed of lithium nickel aluminium cobalt oxide ($LiNi0.80Co0.15Al0.05O_2$, LNCA) while the negative electrode is made of graphite. The electrolyte is composed of ethylene carbonate (EC) and diethyl carbonate (DEC) as solvents and lithium hexafluorophosphate ($LiPF_6$) as salt [10].

The cells have the following characteristics: 2700 mA h rated capacity, 3.6 V nominal voltage, 577 W h/L volumetric energy density, 214 W h/kg gravimetric energy density, 65 mm length, and 18 mm width.

2.2 *Experimental Methods*

Heat release rate is measured by means of an ISO 5660 cone calorimeter. The calculation method is based on the measurement of the amount of oxygen consumed during the test. It has been observed that the thermal power released during the combustion is directly proportional to the consumed oxygen flow rate, as illustrated in Fig. 1. This calculation method is proposed by ISO 5660-1:2015 [11].

$$\dot{q} = E\left(\dot{m}_a Y^{a}_{O_2} - \dot{m}_e Y^{e}_{O_2}\right) \tag{1}$$

where

E = heat release per mass unit of oxygen consumed (≈13.1 kJ/g)

$Y^{a}_{O_2}$ = mass fraction of oxygen in the combustion air (0.232 g/g in dry air)

$Y^{e}_{O_2}$ = mass fraction of oxygen in the combustion products (g/g)

The main advantage of this methodology is that the composition of the tested material is not necessary to the evaluation of HRR. Measurements can be performed also on non-homogeneous materials as Li-ion batteries.

2.2.1 Cone Calorimeter

A cone calorimeter is a fire testing tool. A schematic view is shown in Fig. 2.

The sample to be tested is placed on a load cell below a radiant source with a cone shape and the radiant flow can be varied in the range of 10–100 kW/m^2. The ignition

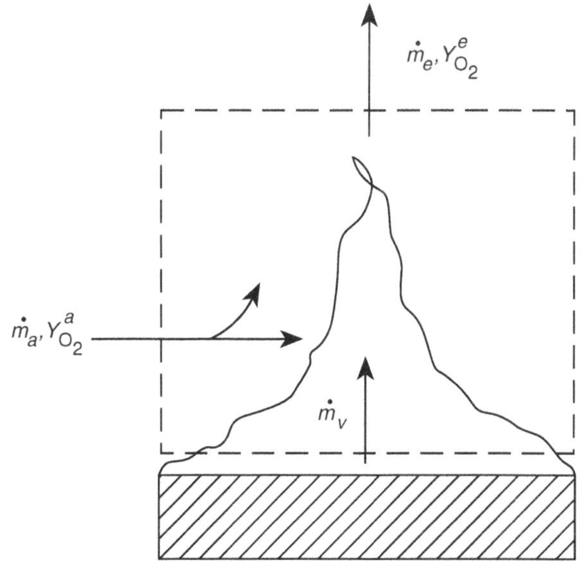

Fig. 1 Balance of oxygen during the combustion of a material

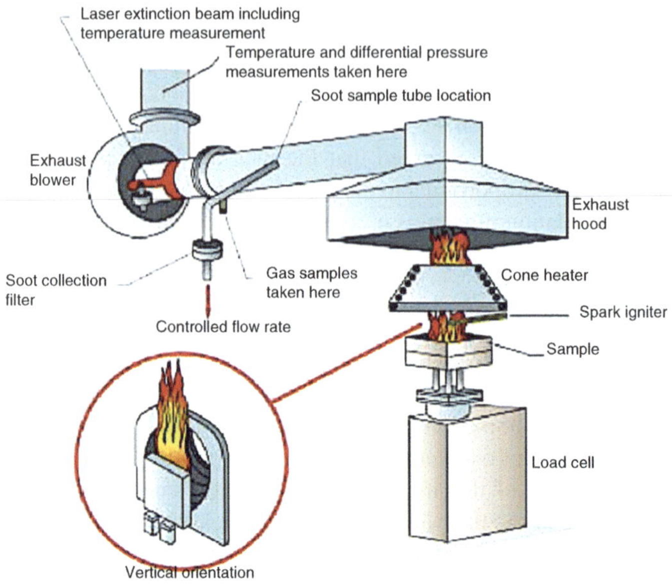

Fig. 2 Cone calorimeter components

occurs by means of a spark igniter, and during the fire, the smoke and the exhausted gases (i.e., O_2, CO, CO_2) are collected by an instrumented tube and measured by specific analyzers [11].

A photoionization detector (PID) is also used for the VOC measurement. Moreover, all the tests are registered through a video camera.

2.2.2 PID: Photoionization Detector

In Fig. 3 the photoionization detector (PID) (RAE Systems) used in the experiments for the VOC measurement is shown. The PID is a very sensitive and precise wide-spectrum sensor, with a sensitivity range of 1 ppb (parts per billion) to 10,000 ppm (parts per million). A PID uses an ultraviolet (UV) light source to break down chemicals into positive and negative ions (ionization) that are easily measurable with a detector. If the ionization potential (IP) of the gas is lower than the emitted power (EV) of the lamp, the gas is ionized and the charged ions flow into the sensor that detects the charge and converts the signal into current. The current is then amplified, measured, and displayed on the counter as a concentration (ppb or ppm). On the other hand, if the gas has an IP above the EV power of the lamp, it is not detected.

Fig. 3 Photoionization detector during a test

2.2.3 Test Procedure

Cells at different state of charge (i.e., SoC of 0%, 50%, and 100%) are tested. In each test the sample is exposed to a given radiant heat flux, and the quantity of oxygen consumed is measured together with the surface temperature and the VOC emitted. Each cell is measured at the start and the end of each test to evaluate the weight loss. Tests are performed at various radiant heat fluxes in the range of 15–50 kW/m^2. In Fig. 4 the cone calorimeter is shown at the beginning of a test.

3 Results

In Fig. 5e and in Fig. 6 the results of the tests in terms of venting time and explosion time as a function of the radiant power are shown. The venting time is defined as the time at which the release of gas and vapor is observed, while the explosion time is the time at which the failure of the cell occurs.

For both SoC values, a reduction of the venting time is observed with increasing radiant power, up to a value of 35 kW/m^2, above which the value remains almost constant.

A similar trend is observed for explosion time.

In Fig. 7 the results in terms of HRR as a function of the radiant power are shown. For cells at 50% SoC the HRR value increases with the increase of external radiant flux, but it remains almost constant for a radiant power greater than 35 kW/m^2. For the fully charged cells, instead, the HRR increases as the radiant power grows.

Moreover, for the fully charged cells, HRR values of about 1 kW/cell are measured at lower radiant power than the cells with SoC equal to 50%.

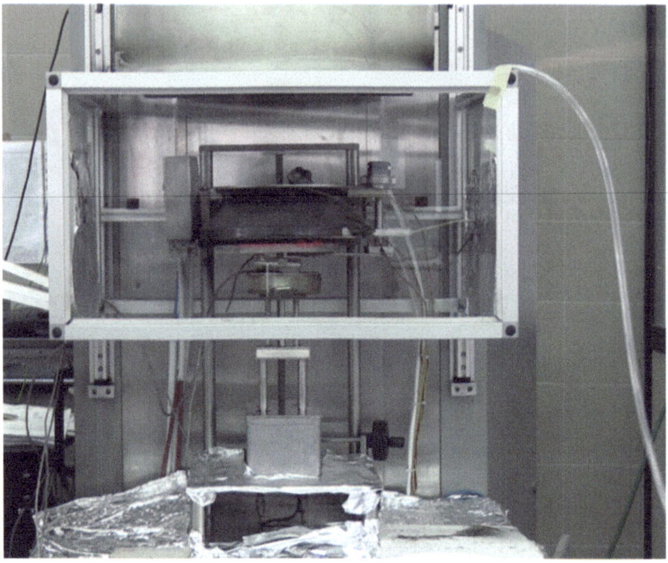

Fig. 4 The cone calorimeter during a test

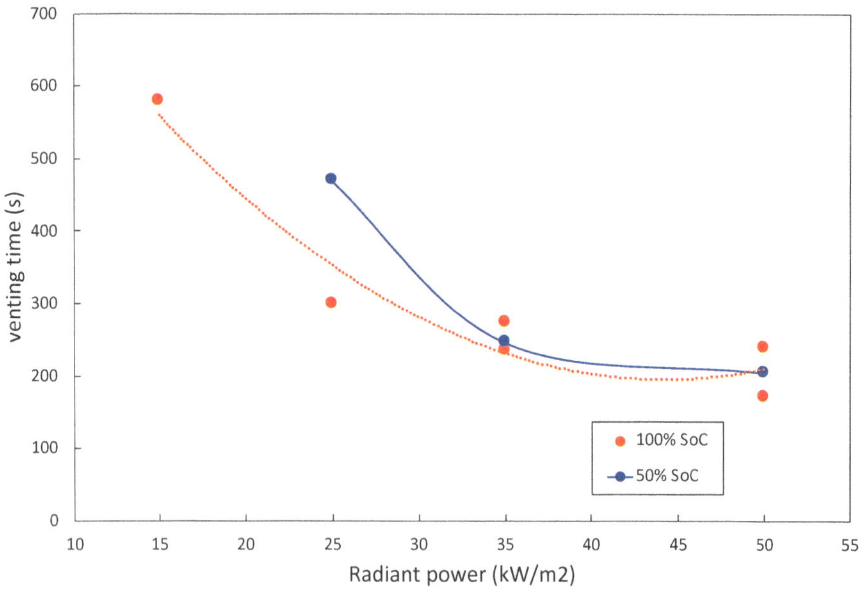

Fig. 5 Venting time as a function of the radiant power for cells at different SoC

The dependence of the onset of the heat release by the SoC is also clearly revealed. From the tests a rapid increase of HRR for 100% charged cells in a shorter time than for 0% and 50% charged cells is observed.

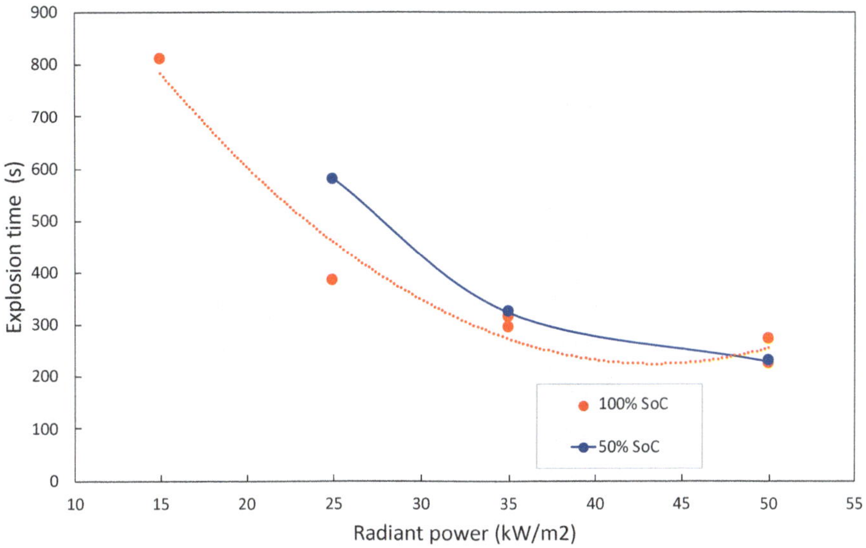

Fig. 6 Explosion time as a function of the radiant power for cells at different SoC

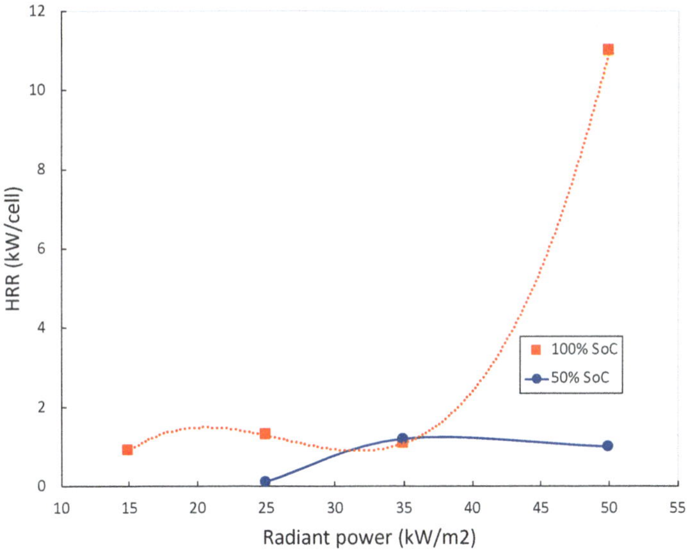

Fig. 7 HRR as a function of the radiant power for cells at different SoC

Specifically, for a fully charged cell exposed to 50 kW/m^2, the results show first the occurrence of very rapid HRR at short time and then a secondary peak when catastrophic cell rupture occurs. The calculated HRR value corresponding to this test is very high (11 kW/cell—Fig. 7). On the contrary a considerably slower HRR

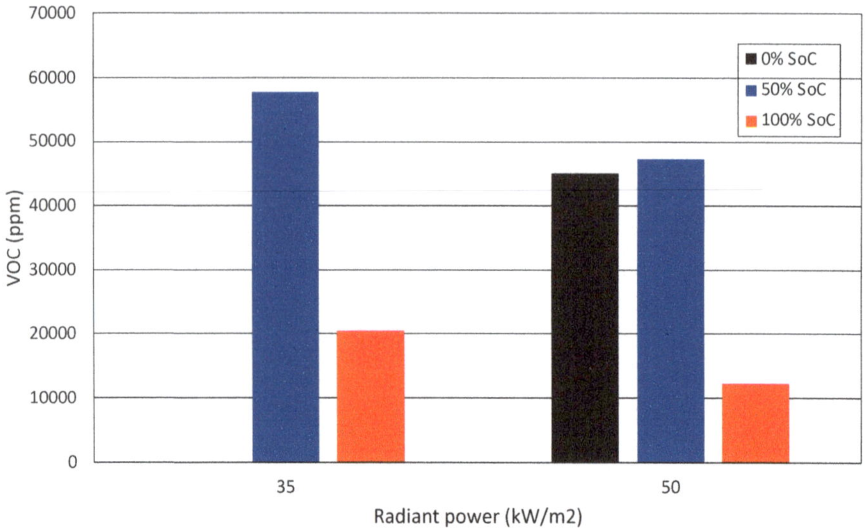

Fig. 8 VOC concentration vs. radiant power at different state of charge

and a long period of heat evolution before the cell explosion for cell at 0% and 50% of SoC are observed.

This behavior can be explained assuming that at the first peak flammable gases are produced by the decomposition of electrolytes and/or their reaction with other components of the cells, while at the second peak VOCs are released, indicating an incomplete oxidation (Fig. 8).

Further studies must be conducted to ascertain the nature and repeatability of this trend.

In Fig. 8 the results of VOC peak concentration during the tests are shown. For cells at a given SoC, a lower quantity of VOCs is released in the tests at higher radiant power, because in these conditions the combustion is almost complete. Similar considerations can be done comparing cells differently charged at a given radiant power, because the higher the SoC, the higher is the combustion conversion.

In Figs. 9, 10, and 11 the key moments identified during a thermal abuse test performed on a fully charged cell (SoC = 100%) subjected to a radiant power of 50 kW/m^2 are shown. Few seconds after the venting of flammable gases and/or vapors (171 s), they are ignited and flames appear. About a minute later, the explosion of the cell occurs with the ejection of flames and fragments.

Fig. 9 Test on fully charged cell at radiant power of 50 kW/m^2: venting, 171 s

Fig. 10 Test on fully charged cell at radiant power of 50 kW/m^2: flame, 173 s

4 Conclusions

In this study thermal abuse tests are carried out on NCR 18650 cells from Panasonic in an ISO 5660 cone calorimeter. The calorimetric tests allowed to determine the behavior of cells in critical conditions by exposure to radiant power of up to 50 kW/m^2.

For fully charged cells, it is observed that the venting of the gases and the subsequent explosion of the cell occur at shorter times and with higher HRR than

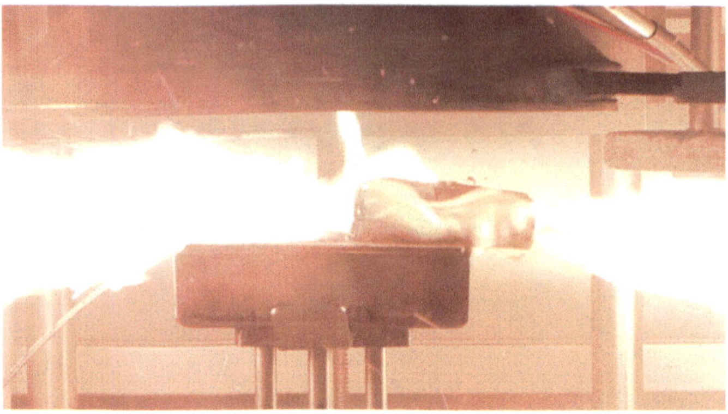

Fig. 11 Test on fully charged cell at radiant power of 50 kW/m^2: explosion, 226 s

semi-charged or completely discharged cells. Moreover, for fully charged cells, the combustion is almost complete.

The results indicate that the initial energy content of the cell provides the activation energy necessary to start the thermal runaway.

Acknowledgments This work has been financed by the Italian Ministry of Economic Development, RSE PAR 2017, in the frame of a cooperation agreement between ENEA and Sapienza University of Rome.

References

1. J.M. Tarascon, M. Armand, Issues and challenges facing rechargeable lithium batteries. Nature **414**, 359–367 (2001)
2. G. A. Nazri, G. Pistoia (eds.), *Lithium Batteries: Science and Technology* (Springer, New York, NY, 2003)
3. Q. Wang, P. Ping, X. Zhao, G. Chu, J. Sun, C. Chen, Thermal runaway caused fire and explosion of lithium ion battery. J. Power Sources **208**, 210–224 (2012)
4. T. Maloney, *Lithium Battery Thermal Runaway Vent Gas Analysis*. Report No. DOT/FAA/TC-15/59, U.S. Department of Transportation William J. Hughes Technical Center Aviation Research Division (2016)
5. W. Zhang, X. Chen, Q. Chen, C. Ding, J. Liu, M. Chen, J. Wang, Combustion calorimetry of carbonate electrolytes used in lithium ion batteries. J. Fire Sci. **33**(I), 22–36 (2015)
6. P. Russo, P. Papillo, *La ricarica di veicoli elettrici: studio preliminare alla modellazione delle distanze di sicurezza*. Report RdS/PAR2016/241 (ENEA, 2017)
7. Federal Aviation Administration, *Lithium Batteries & Lithium Battery-Powered Devices*. Available at: www.faa.gov/about/office_org/headquarters_offices/ash/ash_programs/hazmat/aircarrier_info/media/battery_incident_chart.pdf
8. H. Biteau, T. Steinhaus, C. Schemel, A. Simeoni, G. Marlair, N. Bal, J.L. Torero, Calculation methods for the heat release rate of materials of unknown composition. Fire Saf. Sci. **9**, 1165–1176 (2008)

9. S. Bakhtiyari, L. Taghi-Akbari, M.J. Ashtiani, Evaluation of thermal fire hazard of 10 polymeric building materials and proposing a classification method based on cone calorimeter results. Fire Mater. **39**, 1–13 (2015)
10. Panasonic Batteries, *Product Information Sheet*. Available at: https://na.industrial.panasonic.com/sites/default/pidsa/files/downloads/files/panasonic_liion_battery_product_information_sheet.pdf
11. ISO, *ISO 5660-1:2015 Reaction to Fire Tests - Heat Release, Smoke Production and Mass Loss Rate - Part 1: Heat Release Rate (Cone Calorimeter Method) and Smoke Production Rate (Dynamic Measurement)* (2019)

Correction to: Hardware-in-the-Loop Simulation of High-Power Modular Converters and Drives

Christian Dufour, Karthik Palaniappan, and Brian J. Seibel

Correction to:
Chapter 2 in: W. Zamboni, G. Petrone (eds.), *ELECTRIMACS 2019*, Lecture Notes in Electrical Engineering 604, https://doi.org/10.1007/978-3-030-37161-6_2

The original version of this chapter was revised due to the term "20–50 μs" being erroneously changed to "20–50 μm". The occurrence appears on the first page, line 17, of the chapter.

The updated online version of this chapter can be found at
https://doi.org/10.1007/978-3-030-37161-6_2

W. Zamboni, G. Petrone (eds.), *ELECTRIMACS 2019*, Lecture Notes in Electrical Engineering 615, https://doi.org/10.1007/978-3-030-37161-6_57

The manufacturer's authorised representative in the EU is Springer Nature Customer Service Centre GmbH, Europaplatz 3, 69115 Heidelberg, Germany. If you have any concerns regarding our products, please contact ProductSafety@springernature.com

Printed and bound by CPI Group (UK) Ltd, Croydon, CR0 4YY
30/09/2026
02225648-0019